P9-CBI-292

PERGAMON INTERNATIONAL LIBRARY
of Science, Technology, Engineering and Social Studies
The 1000-volume original paperback library in aid of education, industrial training and the enjoyment of leisure
Publisher: Robert Maxwell, M.C.

SYSTEMS: DECOMPOSITION, OPTIMISATION AND CONTROL

OTHER PERGAMON TITLES OF INTEREST

ANAND	Introduction to Control Systems
ATHERTON	Multivariable Technological Systems
BANKS & PRITCHARD	Control of Distributed Parameter Systems
CICHOCKI & STRASZAK	Systems Analysis Applications to Complex Programs
SAWARAGI & AKASHI	Environmental Systems, Planning, Design and Control
SINGH & TITLI	Control and Management of Integrated Industrial Complexes
SMEDEMA	Real Time Programming 1977
WEXLER	Large Engineering Systems

PERGAMON RELATED JOURNALS

Automatica
Computers and Electrical Engineering
Computers and Industrial Engineering
Operations Research Quarterly

SYSTEMS: DECOMPOSITION, OPTIMISATION AND CONTROL

Edited by

M. G. SINGH

Maître de Conférences Associé
Université Paul Sabatier
Toulouse

A. TITLI

Professeur sans Chaire
Institut National des Sciences
Appliquées - Toulouse

and

LABORATOIRE D'AUTOMATIQUE ET D'ANALYSE DES SYSTEMES du C.N.R.S.

PERGAMON PRESS

OXFORD · NEW YORK · TORONTO · SYDNEY · PARIS · FRANKFURT

U.K.	Pergamon Press Ltd., Headington Hill Hall, Oxford OX3 0BW, England
U.S.A.	Pergamon Press Inc., Maxwell House, Fairview Park, Elmsford, New York 10523, U.S.A.
CANADA	Pergamon of Canada Ltd., 75 The East Mall, Toronto, Ontario, Canada
AUSTRALIA	Pergamon Press (Aust.) Pty. Ltd., 19a Boundary Street, Rushcutters Bay, N.S.W. 2011, Australia
FRANCE	Pergamon Press SARL, 24 rue des Ecoles, 75240 Paris, Cedex 05, France
FEDERAL REPUBLIC OF GERMANY	Pergamon Press GmbH, 6242 Kronberg-Taunus, Pferdstrasse 1, Federal Republic of Germany

British Library Cataloguing in Publication Data

Singh, Madan G
Systems - decomposition, optimisation and control.
- (Pergamon international library).
1. Control theory 2. Mathematical optimization
I. Title II. Titli, Andre
003 QA402.3 78-40604

ISBN 0-08-022150-5 Hardcover
ISBN 0-08-023238-8 Flexicover

In order to make this volume available as economically and as rapidly as possible the authors' typescripts have been reproduced in their original forms. This method unfortunately has its typographical limitations but it is hoped that they in no way distract the reader.

Printed in Great Britain by William Clowes & Sons Limited London, Beccles and Colchester

to Pushpa and
Gurbachan

(M.G.SINGH)

to Evelyne,
to Karine

(A.TITLI)

Contents

PART 3 : STOCHASTIC PROBLEMS

PART 4 : ROBUST DECENTRALISED CONTROL

Acknowledgements

This book represents a synthesis of both our teaching and our research interests. We are grateful to the C.N.R.S., and the L.A.A.S., for having provided a climate favourable for preparing the manuscript and our students at the Institut National des Sciences Appliquées of Toulouse, at the Ecole Nationale Supérieure de l'Aéronautique et de l'Espace, the University of Toulouse, and our students and colleagues at the Ecole Polytechnique of Montreal (Canada), the Federal University of Rio de Janeiro (Brazil), the Catholic University of Rio, the Federal University of Campinas, for their comments and criticisms which were invaluable when we tested the material of the book on them. We are also grateful to the researchers of the hierarchical control group at the L.A.A.S. some of whose work is treated in the book. We should like to thank in particular Dr. M.F. HASSAN with whom we have collaborated for many years, not only for the invaluable discussions that we had with him but also for his help in proof reading the final manuscript.

We are grateful too to Prof. G. ALENGRIN for having permitted us to draw on his excellent course notes when we were writing chapters 9 and 10.

We are grateful to Miss Wendy ROTHE who helped in translating a chapter of the book into English, for preparing the index and for proofreading the book. We are also grateful to Mrs GRIMA for her excellent typing of the final draft and to Mr ROUSSEL for drawing the figures. The errors that remain are our own.

M. SINGH - A. TITLI

Toulouse - March 1978

Preface

Although courses are now beginning to be taught on Large systems theory and practice, there is still a gulf between standard optimisation and control theory and Large Systems theory. This book attempts to bridge this gulf by treating in one volume both the standard theory and Large Systems theory. Again, it is generally agreed that control theorists should be taught some basic aspects of operational research and that operational research students should learn a bit of control theory. This book is aimed at both control theory students as well as operational research students.

The book introduces optimisation and control of systems at a level appropriate for a first or second year graduate course, an undergraduate honours course or for self study. The decomposition-coordination aspect should be of interest also to researchers working in the field of Large Scale Systems as well as to Decision makers who currently manage such systems. In the interest of flexibility, the book is divided into four main parts : Static optimisation and control (chapters 1 to 4), Dynamic optimisation and control (chapters 5 to 8), Identification, estimation and control (chapters 9 to 11), and robust control (chapter 12). In each part, we deal side by side with the standard approaches and how such approaches can be extended using decomposition to tackle large systems problems. Thus, in chapter 1, we introduce the mathematical programming problem and in chapter 2 we treat linear programming for both low order problems as well as for high order problems using the Dantzig-Wolfe decomposition. In chapter 3, we tackle the non-linear programming problem whilst in chapter 4 we introduce decomposition-coordination techniques for tackling such problems. In our experience, these four chapters could be taught at a modest pace in a 30 contact hours course.

In part 2, we start in chapter 5 by introducing the dynamic optimisation problem and we formulate the related two point boundary value problem. As a special case, we consider the linear regulator and servomechanism problems. In chapter 6 we apply decomposition-coordination to linear-quadratic problems. In chapter 7, we study non-linear two point boundary value problems and provide iterative techniques for solving them. In chapter 8, we see how non-linear dynamical systems problems can be tackled by decomposition-coordination. These four chapters could also be taught in a 30 contact hours course.

In part 3, we begin in chapter 9 by introducing probability theory and stochastic processes. In chapter 10, we introduce parameter estimation techniques, state estimation techniques and stochastic control whilst in chapter 11 we apply decomposition-coordination to such problems. These three chapters could also be covered in a 30 contact hours course.

In part 4, we treat the practical robust control problem. This part could be tagged on to part 2 and/or 3.

Part 1, 2 are reasonably self contained and may be read in any order without seriously affecting the treatment in the other. It is recommended that part 3 be read after part 2.

PART 1

Static Optimisation

CHAPTER 1

Introduction

As the non-renewable resources of the world begin to diminish, there is increasing interest in ensuring that the best possible performance is achieved. The performance may be that of a motor car, a business, a chemical process or an economy, just to take a few examples. In each case, the performance depends upon a large number of decisions and the problem that we study in this book is how to choose the "best" set of decisions to achieve a particular objective.

In order to know what the "best" decision is, we require :

(a) a cost function which enables us to quantify the effects of any decision

(b) a model which enables us to predict the effect of any decision

(c) a knowledge of all environmental factors influencing the past, present and future

So the essential optimisation problem is to find the set of decisions which minimise the cost function in (a) above by using the model in (b) and the knowledge of the environment in (c). The model in (b) is necessary because the number of possible decisions for even the simplest processes is relatively large and it is often extremely expensive to experiment with the system itself to find out the effect of a particular decision and in certain cases it might be dangerous (e.g. in the case of socio-economic systems) to do so.

As we will see in this book, both the model building and the optimisation phases require large amounts of calculation and these calculations increase dramatically as the order (or dimension) of the problem increases, and for many problems of practical interest, the computations required surpass the capabilities of a single computer. To get around this difficulty, we treat in this book not only the standard methods of optimisation and control but also ways of dealing with large scale problems through decomposition.

Thus, we examine side by side the methods for the low order decomposed subproblems and of the larger problems. This is done by following each chapter on low order systems by a corresponding one on large scale systems. In this chapter, we give an outline of the general problem and in subsequent chapters we see how we can tackle it for both low order and high order systems.

Let us begin by giving some general notions about the control problem.

1.1 GENERAL NOTIONS OF CONTROL OF PROCESSES [1, 2]

1.1.1 The control problem

Some definitions

Amongst the measurable quantities, we distinguish between the outputs and the controls of the process.

The outputs are the quantities whose evolution is of interest to us a priori and with these we could associate a task. This task which is usually called the set point often represents something that we would like the outputs to do.

Example : For the regulator problem, the outputs should arrive at and subsequently follow the fixed set points.

In the servomechanism or tracking problem, the outputs follow the set points which vary in time.

The controls are the variables which enable us to perform the required task by modifying the behaviour of the outputs.

The evolution of the controls in time are called the policies or control trajectories and the time period of interest is called the horizon.

The horizon could be bounded or it could be infinite. It could be chosen a priori or it may be linked with the evolution of the process. If the horizon is not fixed a priori, the horizon is said to be free. (Example : case of a horizon which is determined by the satisfaction of the set points).

Often, the controls and the outputs are not permitted to evolve arbitrarily but are required to satisfy certain conditions called constraints. The controls and outputs which satisfy the constraints are termed admissible.

The constraints could be equalities ($\underline{h}(\underline{x}, \underline{y}) = \underline{0}$) although the most common ones are of the inequality type ($\underline{h}(\underline{x}, \underline{y}) \leqslant \underline{0}$). These constraints may be satisfied instantaneously for certain cases whilst in other cases integrals of such constraints need to be satisfied.

The instantaneous inequality constraints usually correspond to physical limitations.

Example : $|u_1| \leqslant M, \qquad |u_2| \leqslant N$

If u_1, u_2 are the controls, then these constraints define a closed region in the control space called the region of admissible constraints $\mathcal{D}$.

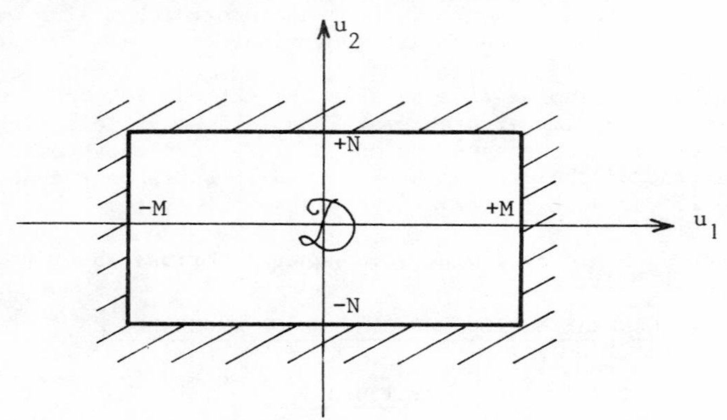

Fig. 1.1

Integral inequality constraints ($\int_0^T f(t)\,dt < k$) often correspond to a limitation of some variable which depends on the control trajectories. Ex : limited fuel for a rocket.

We note that constraints on the output could be viewed in a similar way to set points.

Notion of optimisation

Upto now, the task to be fulfilled by the process was uniquely characterised by the set point. However, in general, there will exist a large number of control trajectories which will enable the set point to be achieved whilst at the same time satisfying the constraints. We need, therefore, to choose amongst all these control trajectories and this leads to the notion of optimisation.

To enable us to make a rational choice amongst the various possible admissible controls, it is necessary to define a cost function (or a criterion function). The choice of the cost function is an important and difficult problem for which one often has to use one's intuition and/or experience.

In static systems, the criterion function will have an instantaneous character. On the other hand, for dynamic problems, the cost function will depend upon the control trajectories and the evolution of the system in an integral fashion.

Example : - cost function $J = \int_{t_o}^{t_f} r\,dt$ (an integral cost function), r : elementary pay off.

- cost function $J = g(t_f)$ terminal criterion which involves only the values of the variables at the final instant.

A control trajectory which achieves the set point, optimises the pay off and satisfies the constraints is called the optimal control trajectory.

The cost function is generally a scalar. However, if it is a vector, it may still be possible to use optimisation techniques but the theory and practice of optimisation for vector valued cost functions is beyond the scope of the present book.

The different mathematical forms of the criterion functions which are most commonly used are given in Table 1.1.

Control system classification

Control systems can be conveniently classified as follows :

a) Open loop control systems

This is a predictive control where there is no regulation of the result of application of a control as shown in Fig. 1.2.

Static processes	
$J_1 = \sum_i C_i x_i$	J_1 is a linear function of the x_i variables C_i are constants.
$J_2 = \sum_i C_i x_i^2$	A quadratic cost function
$J_3 = \phi_1(x_1) + \ldots + \phi_n(x_n)$	Separable cost function (i.e. each of the $\phi_i(x_i)$ functions which may or may not be linear, are functions of a single variable x_j)
$J_4 = \phi(\underline{x})$	This is the most general form. ϕ here is a non-linear function of $\underline{x}^T = (x_1, x_2, \ldots, x_n)$.
Dynamical processes	
$J_5 = \int_0^T dt = T$	Minimal time problem. The horizon T here is free and it is to be determined using an optimisation procedure.
$J_6 = \int_0^T (\underline{x}-\overline{\underline{x}})^T Q(\underline{x}-\overline{\underline{x}})dt$	The horizon here is fixed. This is a problem of minimising the mean square error. Here $\overline{\underline{x}}$ is a reference state vector.
$J_7 = \int_0^T dt + \mu \int_0^T (\underline{x}-\overline{\underline{x}})^T Q(\underline{x}-\overline{\underline{x}})dt$	T is free. μ is a weighting factor. This is a combination of the criteria J_5 and J_6.
$J_8 = \int_0^T L(\underline{x}, \underline{u}, t)dt$	$\underline{x}$: state, $\underline{u}$: control, L : a non linear functional (in general)
$J_9 = G[\underline{x},\underline{u},t]_{t_o}^{t_f} + \int_{t_o}^{t_f} L(\underline{x},\underline{u},t)dt$	The general form which is often used in dynamical optimisation

Table 1.1

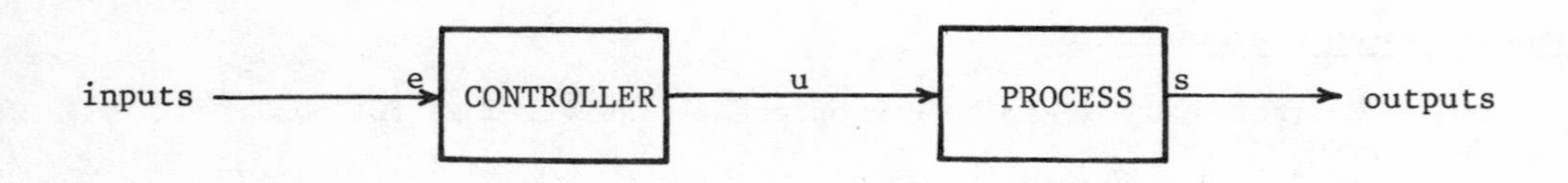

Fig. 1.2

b) Closed loop control systems

Such systems comprise a direct link and a feedback link such that the control is a function of the state of the controlled system. Fig. 1.3 shows a closed loop control system.

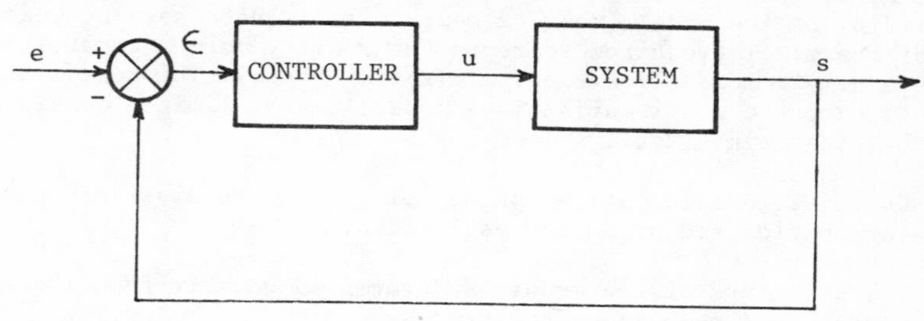

Fig. 1.3

c) Self adaptive control systems

In such systems, the control adapts itself as the model of the system changes.

d) Self organising control systems

Such systems choose their structure as a function of the changing environment.

e) Hierarchical or multi-level systems

The control action in this case is distributed over many levels.

Remarks

1. Whilst an open loop control problem could be a static or a dynamic one, for the case of closed loop control, since the control is a function of the state, the problem is in general one of dynamical optimisation.

2. The control problems which include random factors are called stochastic problems and these are considered in chapters 9, 10 and 11. The control problems where no stochastic inputs exist are termed deterministic and the first eight chapters of this book are devoted to the solution of the deterministic problem.

Next we will discuss modelling.

1.1.2. Mathematical models and identification

Optimisation techniques generally consist of a mathematical procedure and for this procedure it is necessary to have a mathematical representation of the system under consideration. In other words it is necessary to represent the behaviour of the real process by a set of mathematical relations which constitute the mathematical model of the process.

The building of the mathematical model for a process is called identification. The accuracy of identification could be measured by the difference between the output of the real system and that of the model.

Identification is the first important step before optimisation can be performed since the results of the optimisation will be critically dependent upon the validity of the model. Here, a compromise is necessary between the complexity of the model (required to represent reality more accurately) and the need to have a simple model (for the optimisation phase). The models are simplified using various factors e.g. linearisations possible, the limited domain of interest defined by the constraints, etc.

Usually, identification can be split into two distinct phases : structure determination and parameter estimation.

The structure of the model is determined usually from the a priori physical knowledge that we have on the process.

Parameter estimation enables us to put precise values on the parameters such that the model describes the real process and not a class of such processes. Parameter estimation usually requires experimental data on the inputs and outputs of the system.

In the adaptive modelling approach, a combination of the structure determination and the parameter estimation is performed simultaneously [3] as shown in Fig. 1.4.

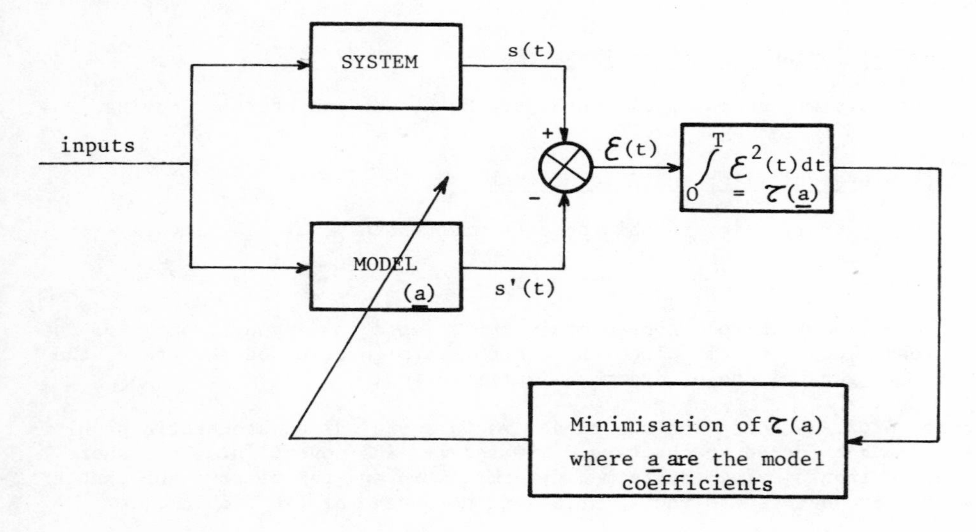

Fig. 1.4

Mathematical models could consist of :

a - algebraic equations (for static processes)
b - integro-differential equations (for dynamical systems)
c - partial differential equations (for distributed parameter systems) and

d - difference equations (for discrete time systems).

Next we give a table (Table 1.2) which summarises the principal types of models that one is likely to run into in this book.

Static linear systems : $\underline{y} = A\,\underline{x}$	input $\underline{x} \in R^n$, output $\underline{y} \in R^m$, A : m x n matrix.
Static non-linear systems : $\underline{y} = \underline{g}(\underline{x})$	$\underline{g}$: non-linear vector valued function
Autonomous linear dynamical system : $\dot{\underline{x}} = A\,\underline{x}(t)$	$\underline{x} \in R^n$, A : n x n constant matrix
Linear dynamical system : $\dot{\underline{x}}(t) = A\,\underline{x}(t) + B\,\underline{u}(t)$	Control $\underline{u} \in R^m$, B : n x m matrix
Time varying linear system : $\dot{\underline{x}}(t) = A(t)\,\underline{x}(t) + B(t)\,u(t)$	A, B are no longer constant
Linear closed loop control system : $\dot{\underline{x}}(t) = A(t)\,\underline{x}(t) + B(t)\,\underline{u}\left[\underline{x}(t)\right]$	
Linearised form for a non-linear system : $\dot{\underline{x}}(t) = A(t)\,\underline{x}(t) + B(t)\,\underline{u}(t) + \underline{g}(\underline{x},t)$	$\underline{g}(\underline{x},\underline{t})$ is a non linear vector valued function
General form of a non-linear dynamical system : $\dot{\underline{x}} = \underline{g}(\underline{x}, \underline{u}, t)$	
Discrete time system : $\underline{x}_{n+1} = \underline{f}_n(\underline{x}_n, \underline{u}_n, n)$	

Table 1.2 : Table of principal types of models

Next we consider optimisation methods.

1.1.3 Optimisation methods

A number of relatively recently developed methods are available for solving the optimisation problem i.e. of determining the controls which satisfy the set points whilst optimising the criterion and satisfying the constraints.

When a mathematical formulation of the problem is available (mathematical model, criterion function, constraints), the techniques used come from classical theories and could be classified in the following three categories :

a - mathematical programming
b - dynamic programming
c - variational methods : - maximum principle
- classical variational calculus.

On the other hand, when a mathematical model is not available, we can't use a mathematical approach and we are obliged to use direct optimisation methods (extremum seeking methods). Such methods are beyond the scope of this book.

Next we give the relationship between some problem formulations and the methods that we can use to solve them (Table 1.3).

Problem formulation	Method
1. Static processes	
1.1 Unconstrained problems max $J = J(\underline{x})$ (criterion), $\underline{x}$ vector	Ordinary theory of maxima and minima
1.2 Problems with equality constraints max $J = J(\underline{x})$ (criterion) subject to $\underline{g}(\underline{x}) = \underline{0}$ (equality constraints)	Lagrange multiplier method
1.3 Problems with inequality constraints $J = J(\underline{x})$ (criterion) $\underline{h}(\underline{x}) \leqslant 0$ (inequality constraints)	Kuhn-Tucker multiplier method
1.4 Linear problems $J = \sum_i C_i x_i \quad i = 1,2,\ldots,n$ subject to $\sum_i a_{ij} x_i = b_j \quad j = 1,2,\ldots m$ $x_i \geqslant 0 \quad i = 1,2,\ldots n$	Simplex method
1.5 General non-linear problems $J = J(\underline{x})$ subject to $g_j(\underline{x}) \ (\leqslant = \geqslant)\ b_j \quad j = 1,2,\ldots m$	Non-linear programming. Kuhn-Tucker theory in particular
2. Dynamical systems	
2.1 Linear problems $J = J(x, u, t)$ subject to $\underline{\phi} = \dot{\underline{x}} - A(t)\,\underline{x} - B(t)\,\underline{u} = \underline{0}$	- variational methods - dynamic programming - gradient method
2.2 Non-linear problems $J = J(\underline{x}, \underline{u}, t)$ subject to $\underline{\phi} = \dot{\underline{x}} - \underline{g}(\underline{x}, \underline{u}, t) = \underline{0}$ J, $\underline{g}$ non-linear	" + quasilinearisation
3. Complex static or dynamical systems	- decomposition-coordination methods in mathematical programming and in variational calculus - hierarchical control and optimisation.

Table 1.3

In the choice of the method to be used we will thus take into consideration factors such as :

a) The static or dynamical nature of the problem.
b) The linear or non-linear nature of the problem
c) Whether it is monovariable or multivariable
d) Existence of constraints.

The approach that we will use to solve optimisation problems will require us to put two different kinds of questions, i.e.

a) On the mathematical level :

1. existence of the solution
2. uniqueness of the solution
3. necessary conditions for optimality
4. sufficient conditions for optimality
5. nature of the extremum

As far as the above five points are concerned, we will see in subsequent chapters that certain properties of the criterion and of the constraints play an important role.

b) On the computational level :

1. existence of a numerical solution approach
2. type and size of the calculator or computer necessary for the job
3. if an iterative procedure is used, its convergence
4. calculation time.

In optimisation practice, all these computational factors are important.

To summarise this discussion, we next give a flow chart which describes the different stages in the formulation and solution of optimisation problems (Fig. 1.5).

Next, we consider the problem of synthesising the control system.

1.1.4. Synthesis of the control system

The final stage in the control of a process consists of implementation of the solution that has been found, i.e. the design of the physical system which will provide at its output the required control for the process under study.

If the mathematical model was exact, an open loop controller could be used.

However, exact solutions are often unrealisable for various reasons (approximations in the modelling, disturbances, approximations in the solution, etc.). It is thus usually necessary to utilise a closed loop solution in order to make the real process insensitive to all these imperfections.

With a closed loop control, one can in fact often use a controller which is relatively simple (and thus cheap) in order to provide near optimal performance.

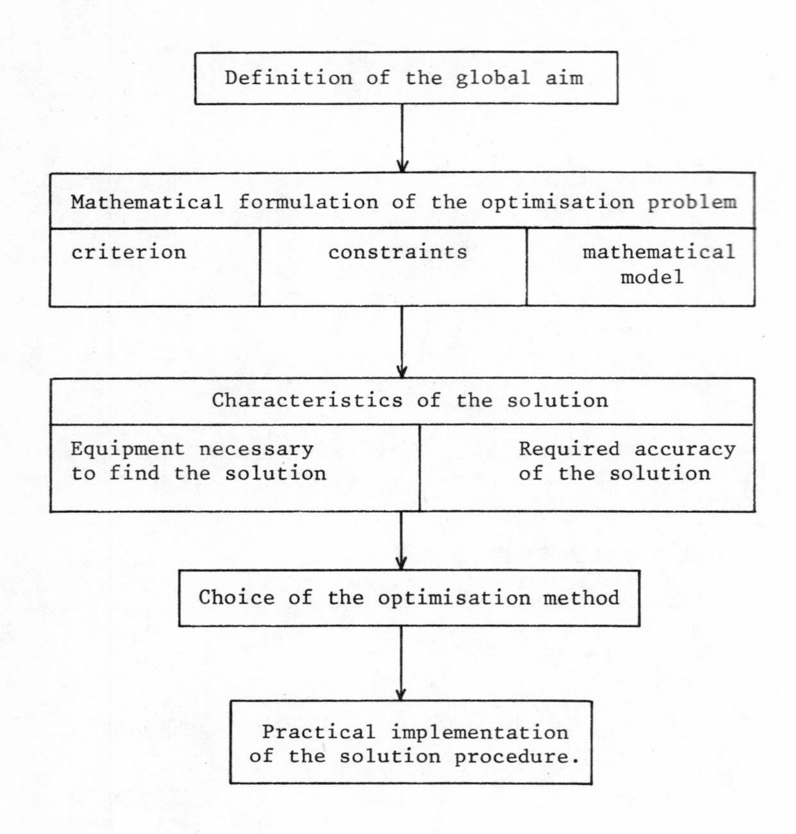

Fig. 1.5

Having discussed in a general way the problem of optimisation and control, next we go on to study how in recent years the field of Control and Systems Engineering has evolved for tackling complex system problems.

1.2 EVOLUTION OF SYSTEMS ENGINEERING TECHNIQUES TO SOLVE COMPLEX SYSTEM PROBLEMS

1.2.1 Introduction

Complex system (i.e. high order systems comprising interconnected subsystems with possibly conflicting objectives) problems raise significant difficulties on the level of analysis, decomposition, aggregation and control. Such systems are found not only in the industrial world (large scale production systems of various kinds), but also in the socio-economic field (transportation and distribution, energy systems, etc.).

Whilst it may be possible to attack directly the analysis phase for the case of low order systems (analysis implying here the definition of the inputs and outputs, controls, construction of the model, estimation of the parameters, definition of the criterion, etc.) as well as the control phase (i.e. synthesis and implementation of the control algorithm) this is not usually possible for the case of high order systems. The difficulty could be purely theoretic (bad convergence or even divergence of the algorithms for large scale systems) or it may arise from economic considerations (investment in new computers may be disproportionate to the expected profits).

In addition, treating complex system problems requires :

a) New analytical methods, given that the dimensionality is now very high.

b) In addition, it is necessary to have an intermediate stage before the control, which could consist of :

   * either the reduction of the dimension of the problem by an aggregation procedure and then the application of standard techniques like mathematical programming, dynamic programming, etc. to the reduced order problem

   or * the decomposition of the system by defining suitable subsystems in order to use decomposition-coordination methods and a multi-level control structure.

c) The synthesis of the control algorithms using the principles of decomposition-coordination. We can synthesise the multilevel control structure which could enable us to solve the problem efficiently.

The relationship between these different phases is given in Table 1.4.

The tools which could be used for each of these steps are outlined in Table 1.4 and we next examine them briefly.

1.2.2. Analysis [4, 5]

In a high order system in general, all the variables are not related to each other and even if they are, there exists a hierarchy in the strength of the coupling between the variables. The aim of the analysis phase is to start from this hierarchy assumption and to study the complex system which is basically characterised by the data in order to define the structure of the system by analysing its data base.

The aim of analysis is to extract a coupling matrix whose elements (i, j) give a measure of the coupling (statistical dependence) between the variables X_i and X_j and using this information, we will be able to partition or decompose the system.

The coupling measure could be 0 or 1 which corresponds to the qualitative approach of non-valued graphs or it could be between 0 and 1 which corresponds to the quantitative approach of complex system representation using a valued graph.

For continuous systems, the coupling is expressed, in the static case, by a correlation coefficient and in the dynamic case by the correlation functions. Here, all the tools of statistical data analysis (normed principle components analysis, etc.) could be used.

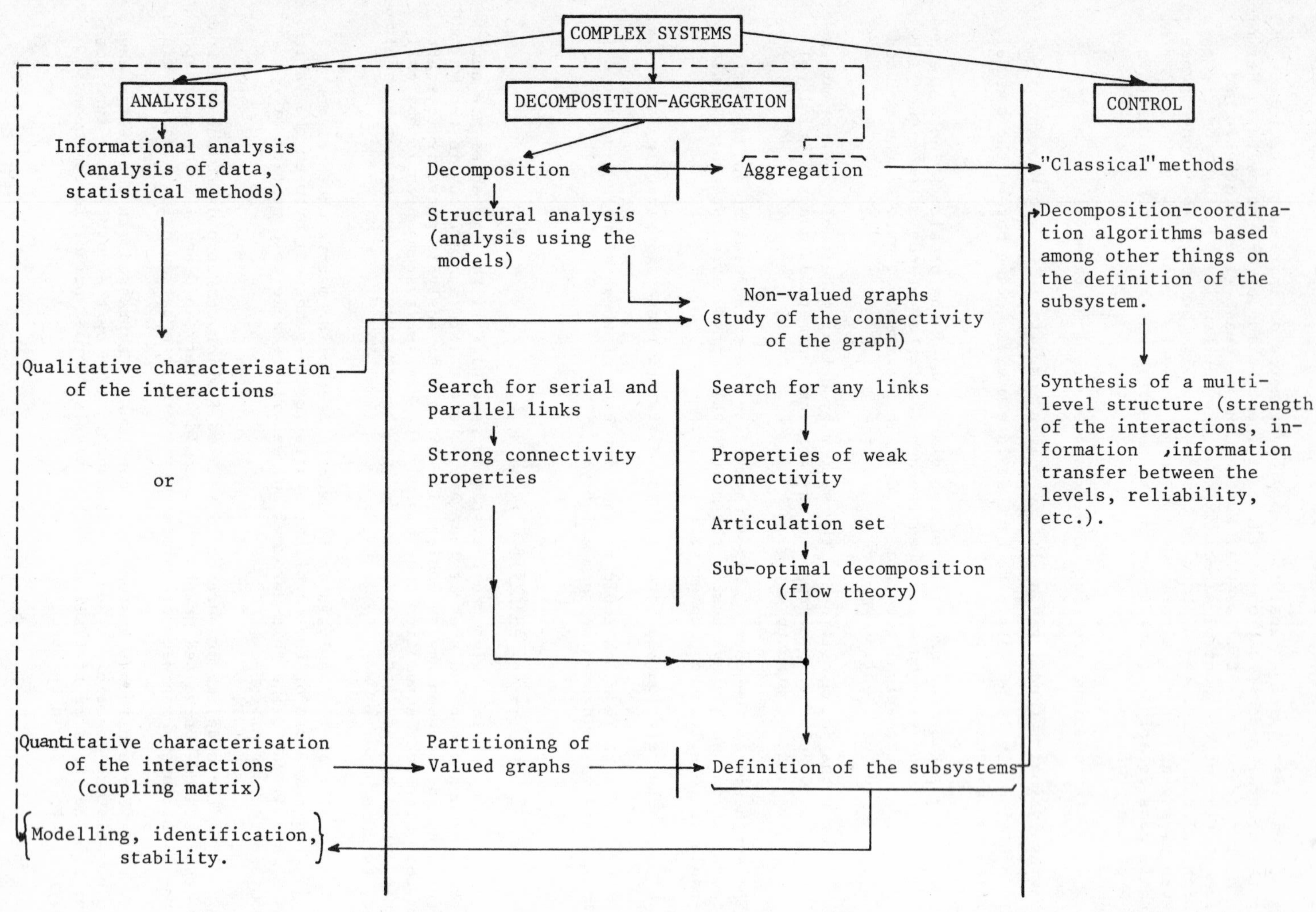
COMPLEX SYSTEMS
ANALYSIS
DECOMPOSITION-AGGREGATION
CONTROL
Informational analysis
(analysis of data,
statistical methods)
Decomposition
Aggregation
"Classical" methods
Structural analysis
(analysis using the
models)
Decomposition-coordination algorithms based among other things on the definition of the subsystem.
Non-valued graphs
(study of the connectivity
of the graph)
Qualitative characterisation
of the interactions
Search for serial and
parallel links
Search for any links
Synthesis of a multi-level structure (strength of the interactions, information ,information transfer between the levels, reliability, etc.).
Strong connectivity
properties
Properties of weak
connectivity
or
Articulation set
Sub-optimal decomposition
(flow theory)
Quantitative characterisation
of the interactions
(coupling matrix)
Partitioning of
Valued graphs
Definition of the subsystems
Modelling, identification,
stability.

Fig. 1.4

For discrete systems, we could use probabilistic concepts which come from the theory of information [6].

1.2.3 Aggregation [7]

This is an approach which complements the multi-level approach and which could be used for suboptimal control, state estimation, parameter estimation, etc.

To give a brief idea of the concept of aggregation, let us consider two dynamical systems S_1 and S_2 where the dimension of $S_1 = n >$ dimension of $S_2 = \ell$. Here S_1 could be the process to be controlled and S_2 the model to be used by the controller or S_1, S_2 could be two models of different complexity. Let $\underline{x}$ be the state of S_1 and $\underline{z}$ the state of S_2. These states satisfy, in the case of linear aggregation, the relation :

$$\underline{z} = C\,\underline{x} \qquad (C : \ell \times n \text{ constant matrix})$$

Thus aggregation involves the choice of C and this choice determines the suboptimality of the control structure and the stability of S_1 which is controlled using S_2.

Next let us consider decomposition.

1.2.4 Decomposition or partitioning of complex systems [4, 5, 8]

It is convenient to represent complex systems by graphs using the following convention :

- the vertices of the graph are the variables
- the lines in the graph represent elementary systems (non-valued).

The problem of partitioning of complex systems into interconnected subsystems reduces thus to the utilisation of simple connectivity properties of the graph thus obtained.

To the coupling variables corresponds the articulation set and to the sub-systems, the simply connected components of the subgraphs obtained by suppressing the articulation set.

For high order graphs, the number of the articulation sets (and thus the number of decompositions) possible could be large and a choice must be made using some criterion (minimal interactions for example).

On the other hand, the use of strong connectivity properties of the graph enables us to delimit series-parallel structures which exist, the order of resolution of the subsystems having been fixed by the rank of each strongly connected component which represents the subsystem.

Finally, if it is possible to construct a valued graph, which is associated with our complex system, the decomposition problem in this case becomes one of optimal partitioning of the graph and algorithms exist for this purpose.

Next we give a brief overview of hierarchical control techniques.

1.2.5 General principles of hierarchical control

1.2.5.1 The multi-level - multi-objective structure

The global control problem could be represented as shown in Fig. 1.6.

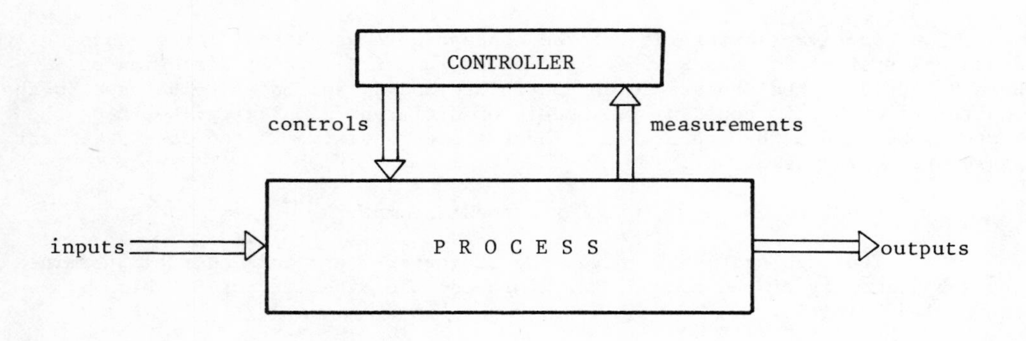

Fig. 1.6

We have already seen that the implementation in this form is difficult for complex systems. Such a controller is called a "simple level - simple objective" controller. If we consider the global system to consist of independent subsystems, we could define a control structure called "simple level-multiple objective" as shown in Fig. 1.7.

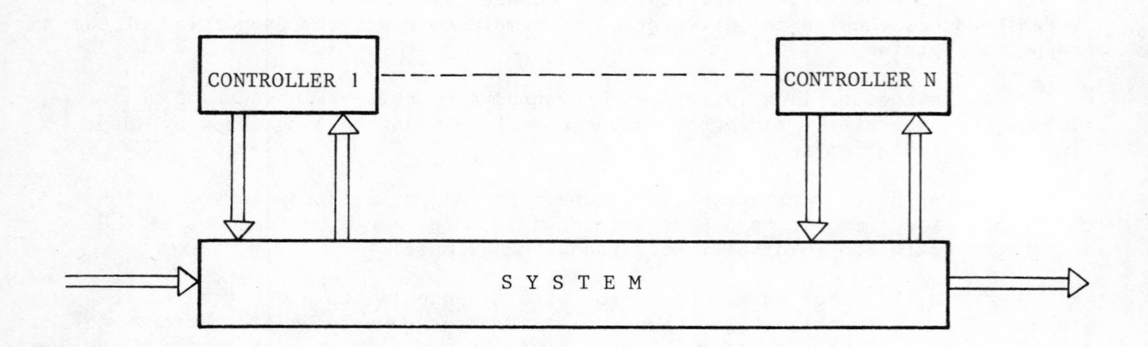

Fig. 1.7

However, since the independence assumption is clearly unrealistic, the solution obtained by the global controller is sub-optimal, whose degree of suboptimality often depends on the strength of the interactions between the subsystems.

When the interactions between the subsystems are strong (and this is often the case for large scale systems) and a criterion function is associated with each controller, conflicts could arise between the controllers if none of them have priority of action. To resolve these conflicts, we envisage a second level of control which takes into account the interactions and modifies if necessary the objectives of any controller. We thus have a control structure having multiple levels and multiple objectives.

Basically, we consider the control to consist of controllers which are distributed in a hierarchy having a pyramid structure such that on the first level there is a controller to control each one of the interconnected subsystems. Fig. 1.8 shows the "multi-level multi-objective" control structure.

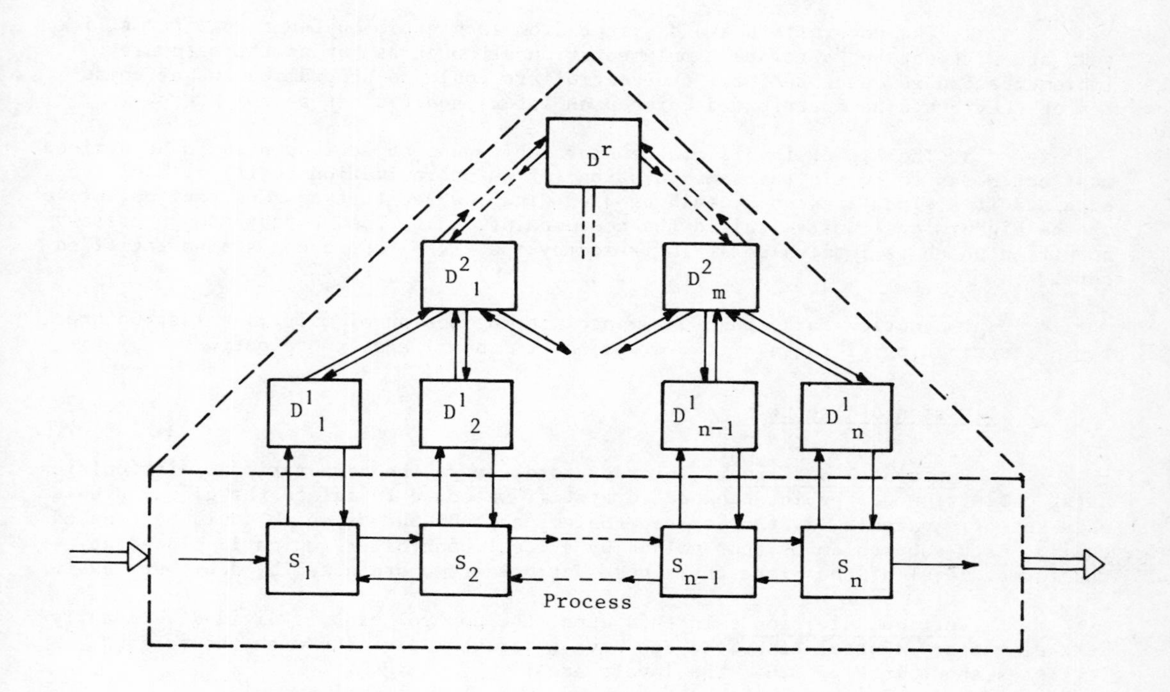

Fig. 1.8

In this structure the controls are distributed amongst two or more levels and some of the controllers only have an indirect access to the process to be controlled. Such controllers receive information from controllers which are superior to them in the hierarchy and use this to control the controllers (or subsystems) inferior to them.

The system is called "multi-objective" since the controllers have different objectives which may even be partially in conflict.

However, here we have controllers which can play the role of the coordinator with respect to the lower level units and this could enable us to resolve the conflicts in order to reach the solution.

It is useful to emphasise some of the characteristic aspects of this control structure :

1. The objectives of the controllers at lower levels are independent of each other and are in general unknown to other units.

2. The controllers are distinct from each other in functional terms, i.e. they are distinct in "software" implementation although as far as the hardware implementation is concerned, all the controllers could be programmed on one computer or they could be distributed between an interconnected set of computers.

3. The higher levels of the hierarchy whose objectives need to be defined must act so as to obtain the same solution (or an approximation to it) as could be obtained if a global approach could be used. Thus we see that an important objective of the higher level units (called the coordinator) is to ensure that some necessary condition which is intrinsically satisfied by the global approach is also satisfied here.

Two notions are thus fundamental in the design of multi-level structures, i.e. "distribution of tasks" (i.e. division of labour) and "coordination".

1.2.5.2 Division of labour

Horizontal division : In order to decrease the computational difficulties after definition of the interconnected system, we could formulate the global problem as a set of separable lower order subproblems each of one of which could be treated easily. Each subproblem is thus solved by a local controller (which is placed at the lowest level of the hierarchy) whose actions are coordinated by a higher level.

Vertical division : In this case, the control task is divided vertically into elementary control tasks. We can have a vertical division with the various levels as shown in Fig. 1.9. The levels are :

a) regulation or direct control

b) optimisation (determination of the set points of the regulators)

c) adaptation (self adaptation of the model and the control law directly)

d) self organisation (choice of structure of the model and the control as a function of the changing environment).

Time division or functional division : A complex process could be in many different situations and it may be useful to explicitly recognise these situations and to "coordinate" between the different possibilities. For example, we could have one strategy for the transient and another for the steady state in the case of a stable dynamical system.

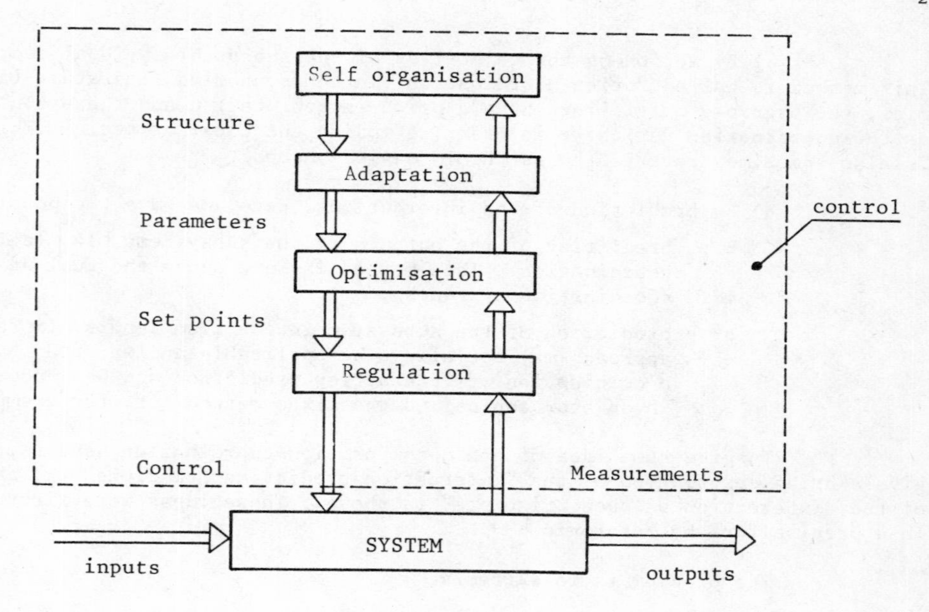

Fig. 1.9

Remark

These different divisions of work could be used separately or together thus leading to general multi-level multi-objective structures.

1.2.5.3 Coordination

We have seen that the principle behind hierarchical control is to decompose the global problem P, whose objective is to extremise a global criterion associated with a complex system into a certain number of subproblems P_i. The solutions of P_i are found locally such that

Solution $\left[P_1, P_2, \ldots P_n \right]$ implies the solution of the global problem P
1.2.1

In fact, such a relation as 1.2.1 can not, in general, be satisfied because of the existence of the interactions between the P_i's and the resulting conflicts. For this reason it is necessary to introduce an "intervention vector" or "coordination parameter" and to replace P_i by $P_i(\underline{\lambda})$ such that [3] the relation 1.2.2 is satisfied, i.e.

$$\text{Solution}\left[P_1(\underline{\lambda}), P_2(\underline{\lambda}), \ldots P_n(\underline{\lambda})\right]_{\underline{\lambda}=\underline{\lambda}^*} \Rightarrow \text{solution } P \qquad 1.2.2$$

Hierarchical optimisation involves choosing $\underline{\lambda}$ from some initial values $\underline{\lambda}^o$ and iterating to the final values $\underline{\lambda}^*$ which solve the coordination problem in hierarchical control.

Many coordination methods have been suggested. Here, to simplify the presentation, we will describe two approaches, i.e.

a) By action on the objectives of the sub-problems (Goal coordination). This method is called "price coordination" in the economics literature (in economics, the Lagrange multipliers have a price interpretation and these could be used as the coordination variables in order to modify the local criteria). This method is also sometimes referred to as the infeasible method.

b) By prediction of the interactions. Here, we have two possibilities :

b_1 - prediction of the outputs of the subsystems [14] (called model coordination). This method is in a sense the dual of the Goal coordination approach ;

b_2 - prediction of the subsystem interaction inputs. In fact, this approach may not always be applicable and we often have to use a combination of interaction prediction and Goal coordination in an approach called the mixed method (cf. for example [12]).

To give some idea of the decomposition-coordination aspects, we will give a brief description of the "interaction prediction principle" (b_2 above) and of the "interaction balance principle" (a above). These ideas were first formulated in a precise form by Mesarovic [9].

The problem is to extremise

$$v = F(\underline{m}, \underline{y}) \qquad \text{(global criterion)}$$

subject to

$$\underline{y} = P(\underline{m})$$

Interaction prediction principle

Here each subsystem is defined by

$$\underline{y}_i = \underline{P}_i\ (\underline{m}_i, \underline{u}_i) \qquad \text{(the local model)} \qquad 1.2.3$$

and

$$v_i = F_i\ (\underline{m}_i, \underline{y}_i, \underline{u}_i) \qquad \text{(the local objective)} \qquad 1.2.4$$

Here $\underline{u}_i$ represents the interactions between the subsystems. The lower level problem Δ_i is thus defined by

$$\Delta_i \begin{cases} \underline{y}_i = \underline{P}_i(\underline{m}_i, \underline{\alpha}_i) & 1.2.5 \\ v_i = F_i(\underline{m}_i, \underline{y}_i, \underline{\alpha}_i) & 1.2.6 \end{cases}$$

Here, $\underline{\alpha}_i$ is the predicted interaction and this is considered known at the lower level and can be used to determine the local control $\hat{\underline{m}}_i(\underline{\alpha}_i)$. These controls $\underline{m}_i$ are sent to the coordinator which checks whether $\underline{\alpha}_i$ was correct or not and based on this, improves $\underline{\alpha}_i$ for the next iteration.

This coordination mode which is usually called one with "direct intervention" since the higher control levels act directly on the model, has given rise to the following coordination principle [9] :

"Let $\underline{\alpha}_1, \ldots \underline{\alpha}_n$ be the interactions predicted by the coordinator and $\underline{u}_1(\hat{\underline{m}}(\underline{\alpha})) \ldots \underline{u}_n(\hat{\underline{m}}(\underline{\alpha}))$ the real interactions which are observed when the optimal control (based on the predicted interactions) is applied. Then the global optimum is obtained when

$$\underline{\alpha}_i = \underline{u}_i(\hat{\underline{m}}(\underline{\alpha})) \qquad \forall_i \qquad 1.2.7$$

i.e. when the real and predicted interactions are identical".

The application of this principle gives us directly the higher level problem which is to find $\underline{\alpha}_i$ such that

$$\underline{\epsilon}_i = \underline{\alpha}_i - \underline{u}_i\ (\underline{\hat{m}}(\alpha)) = \underline{0} \quad \forall_i \qquad 1.2.8$$

This is illustrated in Fig. 1.10.

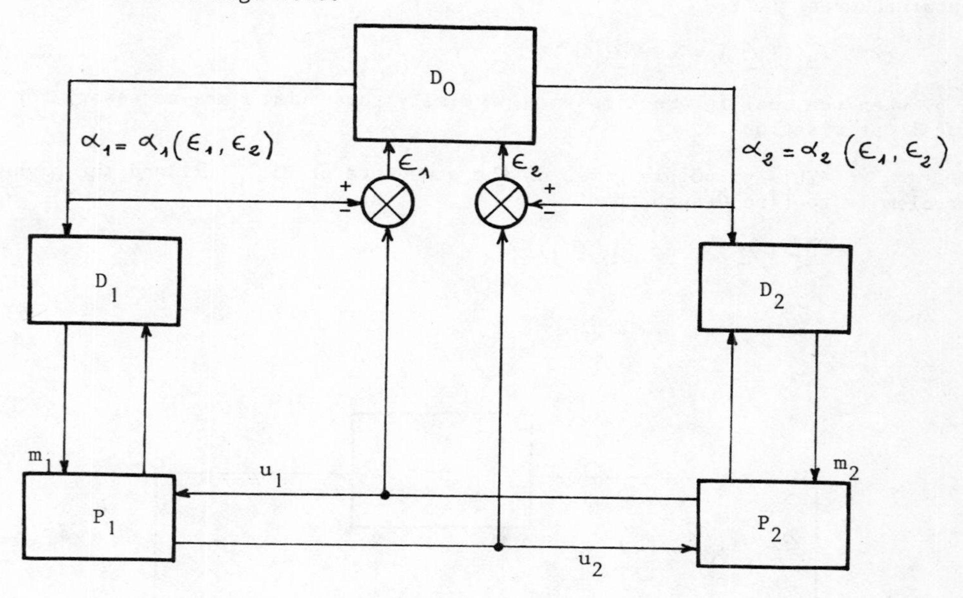

Fig. 1.10 : Interaction Prediction

We note that in such a procedure the higher level does not need to know the subsystems in detail since it merely needs to calculate the interactions $\underline{u}_i(\underline{\hat{m}}(\underline{\alpha}))$ resulting from the application of the local controls $\underline{\hat{m}}(\underline{\alpha})$. In addition, the information exchange between the higher and the lower level is quite limited.

Interaction balance principle

In this case the lower level problem Δ_i is given by

$$\Delta i \begin{cases} \underline{y}_i = \underline{P}_i\ (\underline{m}_i,\ \underline{u}_i) & 1.2.9 \\ \underline{v}_i = F_i\ (\underline{m}_i,\ \underline{y}_i, \underline{\beta}_i,\ \underline{u}_i) & 1.2.10 \end{cases}$$

$\underline{\beta}_i$ is again called the intervention or coordination vector. Coordination is achieved by operating on the cost function and it is for this reason that the method is also often called the Goal coordination method. Obviously, the modification of the global criterion must be zero at the end of the coordination. Then our final hierarchical solution corresponds to the global one.

The model $\underline{y}_i$ involves in this case two independent vectors $\underline{m}_i$ and $\underline{u}_i$ so that the solution of the local problem will be of the form :

$$\underline{\tilde{u}}_i(\underline{\beta}_i), \quad \underline{\tilde{m}}_i(\underline{\beta}_i)$$

The corresponding coordination principle is [9]:

"Let $\underline{\tilde{u}}_1(\underline{\beta}) \ldots \underline{\tilde{u}}_n(\underline{\beta})$ be the necessary interactions which enable the lower level to obtain the local optimum and $\underline{u}_1(\underline{\tilde{m}}(\underline{\beta})) \ldots \underline{u}_n(\underline{\tilde{m}}(\underline{\beta}))$ the interactions that result when $\underline{\tilde{m}}(\underline{\beta})$ is actually applied. Then the global optimum is obtained when

$$\underline{\tilde{u}}_i(\underline{\beta}) = \underline{u}_i(\underline{\tilde{m}}(\underline{\beta})) \quad \forall_i \tag{1.2.11}$$

i.e. when the real interactions are precisely those which are necessary for the local optimisation".

This principle gives us the structure of Fig. 1.11 and the higher level problem is to find $\underline{\beta}$ such that

$$\underline{\tilde{u}}_i(\underline{\beta}) = \underline{u}_i(\underline{\tilde{m}}(\underline{\beta})) \tag{1.2.12}$$

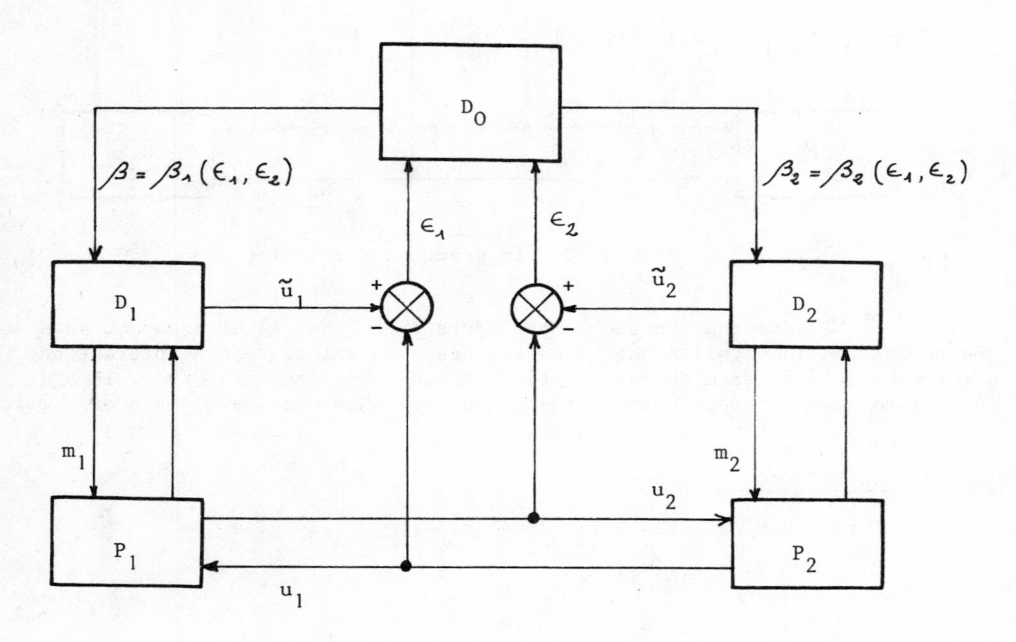

Fig. 1.11

Here again, for a given $\underline{\beta}$, each local controller solves its problem and this gives $\underline{\tilde{u}}_i(\underline{\beta})$ and $\underline{\tilde{m}}_i(\underline{\beta})$ and these results are sent to the second level. The higher level computes the interaction which would result if the control $\underline{\tilde{m}}(\underline{\beta})$ was applied and compares this result with $\underline{\tilde{u}}_i(\underline{\beta})$. The difference is used to improve $\underline{\beta}$ for the next iteration.

In this chapter, we have so far outlined the different problems which need to be tackled in order to optimise and control low order and complex systems and we now go on to finally give some generalities on mathematical programming. This would enable us to develop concrete algorithms in the next chapter.

1.3 MATHEMATICAL GENERALITIES FOR MATHEMATICAL PROGRAMMING

1.3.1 Aim of mathematical programming

Consider a scalar multivariable function with components x_1, x_2,... x_m denoted by $f(\underline{x})$. This function is the cost function or the objective function.

$\underline{x}$ belongs to a given domain $\mathcal{D} \subset R^m$ which is defined by the constraint relations :

(a) of the equality type : $g_i(\underline{x}) = 0$ $\quad (i = 1, 2, \ldots p)$

and/or

(b) of the inequality type : $h_j(\underline{x}) \leqslant 0$ $\quad (j = 1, 2, \ldots q)$

Note that if $h_j(\underline{x}) = 0$, the corresponding constraint is said to be saturated.

If $\underline{x} \in \mathcal{D}$, $\underline{x}$ is said to be admissible.

The object of mathematical programming is to find, amongst the $\underline{x} \in \mathcal{D}$ those $\underline{x}$ which are called :

(a) the absolute or global minima and which satisfy

$$f(\hat{\underline{x}}) \leqslant f(\underline{x}) \quad \forall \quad \underline{x} \in \mathcal{D}$$

and

(b) the local or relative minima $\tilde{\underline{x}}$ which satisfy

$$f(\tilde{\underline{x}}) \leqslant f(\underline{x}) \; \forall \, x \in \mathcal{V}(\tilde{\underline{x}}) \subset \mathcal{D}$$

where $\mathcal{V}$ denotes a small neighbourhood of $\underline{x}$.

The minimum is said to be strict or strong if the inequality is strictly satisfied $\forall$ $\underline{x} \neq \hat{\underline{x}}$ or $\tilde{\underline{x}}$. Otherwise it is called a weak minimum.

For a single variable x, the different minima which could arise are shown in Fig. 1.12.

In Fig. 1.12,

- x_1 is a strong relative minimum
- x_2, x_5 are weak global minima
- $x \in [x_3, x_4]$ represents weak local minima
- x_6 is a weak global minimum.

We note that the absolute minimum is not necessarily unique.

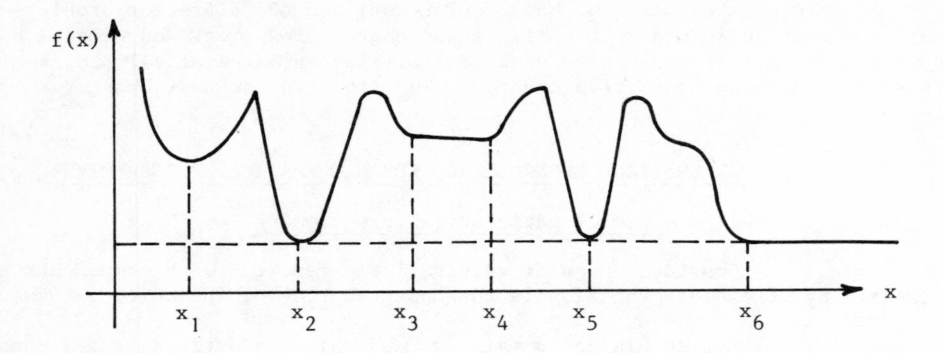

Fig. 1.12

If $f(\underline{x})$ possesses many extremal points, $f(\underline{x})$ is said to be multimodal. Otherwise, $f(\underline{x})$ is said to be unimodal.

In general, it is difficult to find the absolute minimum of a multimodal function [13]. Faced with this problem, basically one isolates all the different local minima $\underline{x}_i$ and picks out the smallest one, i.e.

$$f(\hat{\underline{x}}) = \min_i \left[f(\tilde{\underline{x}}_i) \right]$$

One possibility, which is however only applicable to very special cases is shown in Fig. 1.13

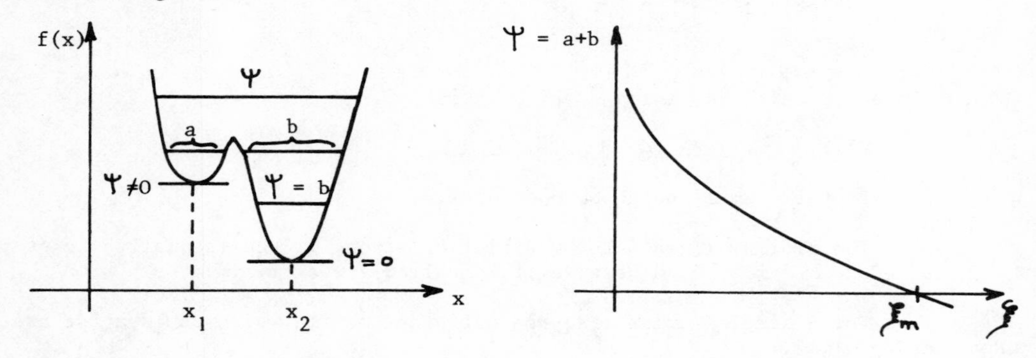

Fig. 1.13

In this method we replace the search for the absolute minimum $\hat{x}_2$ of the function $f(x)$ by a search for the zero of the function $\Psi(\xi)$ which become zero only at the absolute minimum of $f(x)$. Indeed, to obtain this, it is necessary in this simple case to take $\Psi = a + b$.

However, evidently the function Ψ is difficult to obtain in the multivariable case and this limits the utility of such an approach.

Fortunately, unimodal functions (i.e. those with a single extremum) are quite wide spread particularly since we often limit the domain of search for physical reasons.

More rigorously, a function $f(\underline{x})$ is said to be unimodal on an interval $[a, b]$ if

1. $\exists\ \hat{x}$ such that $f(\hat{x}) \leqslant f(x) \quad \forall\ x \in [a, b]$

2. Let $y_1 = f(x_1), y_2 = f(x_2)$ and $x_1 < x_2$
   Then $x_2 < \hat{x}$ implies that $y_1 > y_2$ 1.3.1
   $x_1 > \hat{x}$ implies that $y_1 < y_2$

This definition enables us to consider unimodal functions which sometimes have jumps, discontinuities and which may even be undefined over certain intervals.

If the function is continuous, the b in relation 1.3.1 reduces to (Fig. 1.14) :

- $f(x)$ decreasing on $[a, \hat{x}]$
- $f(x)$ increasing on $[\hat{x}, b]$

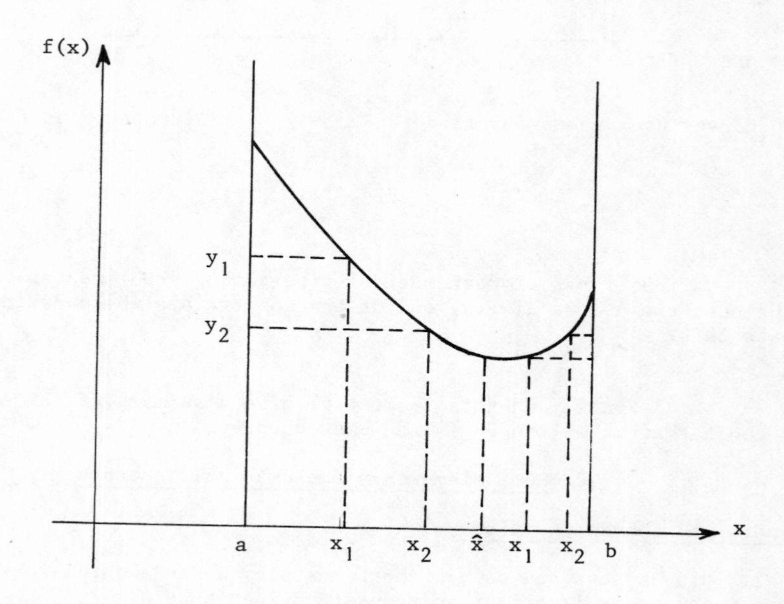

Fig. 1.14

In Fig. 1.15, we give some examples of unimodal functions.

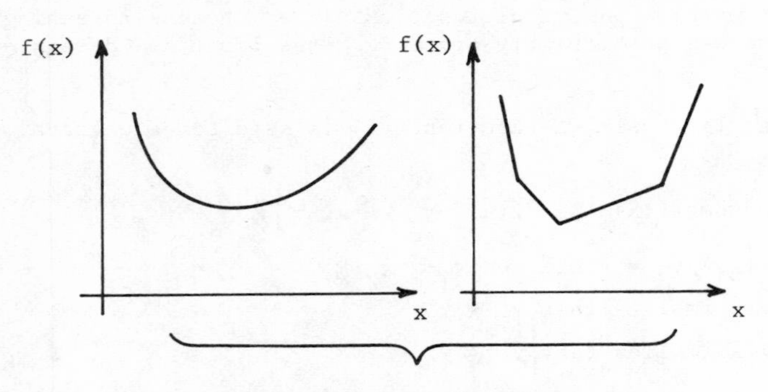

convex continuous functions

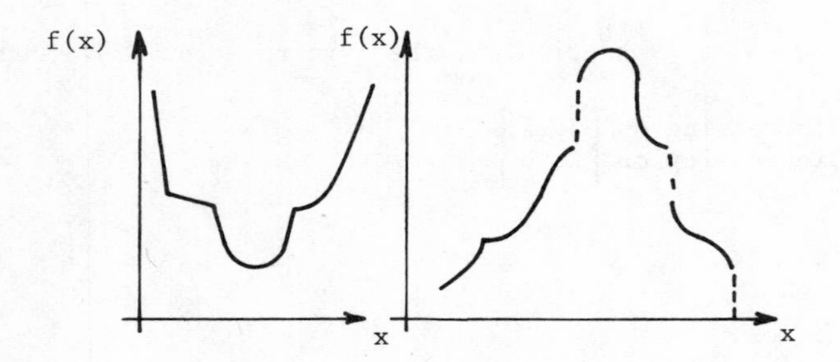

non convex continuous function   discontinuous function

Fig. 1.15

Various types of mathematical programming problems can be defined depending upon the nature of f(x) and of the constraints which define $\mathfrak{D}$. Table 1.5 summarises this.

Finally in this chapter we give a summary of the elementary mathematical tools that we will need in the next chapter.

1.3.2 Some elementary mathematical concepts

Convex sets and convex functions

Convex sets and convex functions play a fundamental role in optimisation theory and it is important to understand what they are.

Definition

A set $\mathcal{C}$ is convex if :

f(x)	constraints	x	type of programming
linear	linear	real	linear programming
linear	linear	integers	integer linear programming
quadratic	linear	real	quadratic programming
convex	convex	real	convex programming
non linear	non linear	real	non-linear programming

Table 1.5

$$\forall\ \underline{x}_1,\ \underline{x}_2 \in C\ ,\ \lambda \in [0,\ 1] \text{ we have}$$
$$\lambda \underline{x}_1 + (1 - \lambda)\ \underline{x}_2 \in C \qquad 1.3.2$$

Geometrically a convex set has the property that the line segment joining any two points in it belongs to C . Fig. 1.16 illustrates this.

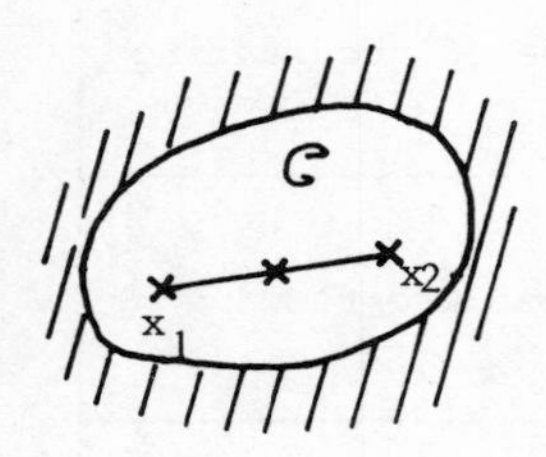

Convex set

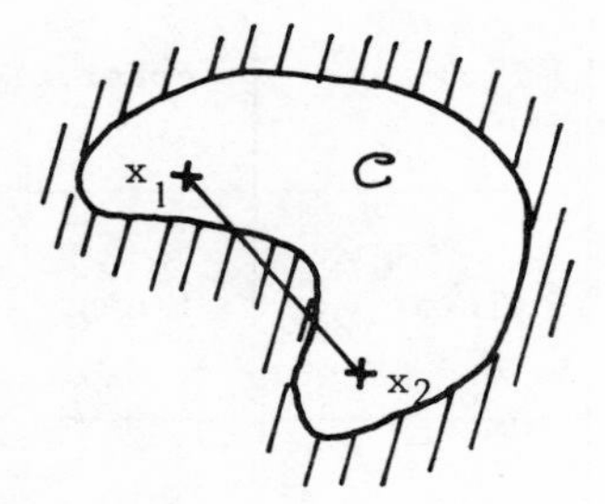

Non convex set

Fig. 1.16

Convex sets have the following two important properties, i.e.

a) All intersections of convex sets are convex.

b) The convexity property is preserved after all linear transformations.

Convex functions

A function $f(\underline{x})$ is convex in a convex set C if :

$$\left.\begin{array}{c}\forall\ \underline{x}_1,\ \underline{x}_2 \in C\\ 0 \leqslant \lambda \leqslant 1\end{array}\right\} f\left[\lambda \underline{x}_1 + (1-\lambda)\ \underline{x}_2\right] \leqslant \lambda f(\underline{x}_1) + (1-\lambda)\ f(\underline{x}_2) \quad 1.3.3$$

This property is illustrated in Fig. 1.17.

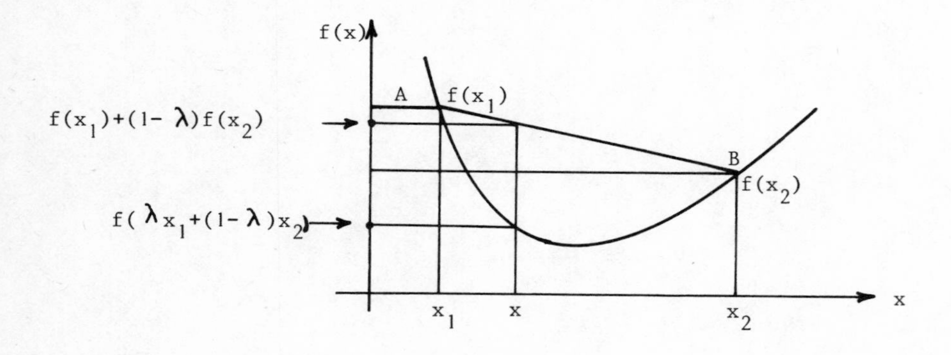

Fig. 1.17

Geometrically, all points on the curve between A and B have an ordinate which is less than or equal to the point with the same abscissa on the segment AB.

If $-f$ is convex, f is said to be concave and is defined by

$$\left.\begin{array}{l} \forall\, \underline{x}_1, \underline{x}_2 \in \mathcal{C} \\ 0 \leqslant \lambda \leqslant 1 \end{array}\right\} f(\lambda \underline{x}_1 + (1-\lambda)\underline{x}_2) \geqslant \lambda f(\underline{x}_1) + (1-\lambda) f(\underline{x}_2) \qquad 1.3.4$$

We note that a linear function is both convex and concave at the same time.

If the inequalities in 1.3.3 and 1.3.4 are strictly satisfied, the functions are said to be strictly convex (or concave).

Convex functions have the following important properties :

a - A function formed by a linear combination with coefficients > 0 of convex functions is a convex function.

b - If $f(\underline{x})$ is convex and $\underline{x}_1, \underline{x}_2 \in \mathcal{C}$, then $f(\lambda \underline{x}_1 + (1-\lambda) \underline{x}_2)$ is a convex function of λ on $[0, 1]$

c - If $f(\underline{x})$ is convex, then the set $\mathcal{P}$ of points which satisfy $f(\underline{x}) \leqslant 0$ is also convex.

This property c- is important in mathematical programming since it enables us to deduce that if the constraint relations $h_i(x)$ are all convex, then the admissible set $\mathcal{D}$ which is defined by $\underline{h}_i(\underline{x}) \leqslant \underline{0}$ is also convex.

Next we give the fundamental property of convex systems which provides the reason for their importance in mathematical programming.

Fundamental property

Let $f(\underline{x})$ be a convex function on a closed convex set $\mathcal{C} \subset R^m$ (closed i.e. the boundary points form a part of the set). Then all relative minima are also absolute minima on $\mathcal{C}$.

Proof

Let $\underline{x}^*$ be a relative minimum and $\hat{\underline{x}}$ the absolute minimum.

Then $\qquad f(\hat{\underline{x}}) < f(\underline{x}^*)$

Using the definition of convex functions, we have :

$$f(\underline{x}) = f(\lambda \underline{x}^* + (1-\lambda)\, \hat{\underline{x}}) \leqslant f(\underline{x}^*) + (1-\lambda)\, f(\hat{\underline{x}}) < \lambda f(\underline{x}^*) + (1-\lambda)\, f(\underline{x}^*) = f(x^*)$$

Then $\qquad \forall\, \underline{x} \in [\underline{x}^*, \hat{\underline{x}}]$

$$f(\underline{x}) < f(\underline{x}^*)$$

which is a contradiction. Hence the result.

Next we study another important property of such functions.

Let $f(\underline{x})$ be a doubly differentiable continuous function. We can then expand $f(\underline{x})$ in a Taylor series upto second order around $\underline{x}$, i.e. :

$$f(\underline{x}+\underline{\delta}) = f(\underline{x}) + \sum_i \frac{\partial f}{\partial x_i} \delta_i + \frac{1}{2} \sum_i \sum_j \frac{\partial^2 f}{\partial x_i x_j} \delta_i \delta_j + 0(\| \underline{\delta} \|^3) \qquad 1.3.5$$

This equation can be rewritten in a vector form as :

$$f(\underline{x} + \underline{\delta}) = f(\underline{x}) + \underline{f}_x{}^T \underline{\delta} + \frac{1}{2} \underline{\delta}^T F_{xx} \underline{\delta} + 0(\| \underline{\delta} \|)^3 \qquad 1.3.6$$

Here, $\underline{f}_x$ is the gradient of $f(\underline{x})$ w.r.t. $\underline{x}$ and F_{xx} is the matrix of second derivatives of $f(\underline{x})$ w.r.t. x.F_{xx} is commonly called the Hessian.

Let us next assume that $f(\underline{x})$ is convex in a region $\mathfrak{D}$ and we want to see what effect this has.

Let $\underline{x}_1$, $\underline{x}_2$ be such that $\underline{x}_2 = \underline{x}_1 + \underline{\delta}$, where $\|\underline{\delta}\|$ is small and $\underline{x}_1, \underline{x}_2 \in \mathfrak{D}$

Let $\underline{x} = \lambda \underline{x}_1 + (1-\lambda)\underline{x}_2 = \lambda \underline{x}_1 + (1-\lambda)(\underline{x}_1 + \underline{\delta}) = \underline{x}_1 + (1-\lambda)\underline{\delta}$ with $0 \leqslant \lambda \leqslant 1$.

Since $f(\underline{x})$ is convex :

$$f(\underline{x}) \leqslant \lambda f(\underline{x}_1) + (1-\lambda) f(\underline{x}_2)$$

But $f(\underline{x}) = f(\underline{x}_1 + (1-\lambda)\underline{\delta}) = f(\underline{x}_1) + (1-\lambda) \underline{f}_x{}^T (\underline{x}_1) \underline{\delta}$

$$+ \frac{(1-\lambda)^2}{2} \underline{\delta}^T F_{xx}(\underline{x}_1) \underline{\delta} + (1-\lambda)^3 \; 0(\|\underline{\delta}\|^3)$$

and

$$f(\underline{x}) \leqslant \lambda f(\underline{x}_1) + (1-\lambda) f(\underline{x}_1 + \underline{\delta}) = \lambda f(\underline{x}_1) + (1-\lambda) \; f(\underline{x}_1)$$

$$+ [\underline{f}_x{}^T(x_1) \underline{\delta} + \frac{1}{2} \underline{\delta}^T F_{xx}(\underline{x}_1) \underline{\delta} + 0(\|\underline{\delta}\|^3)]$$

from which we have

$$\frac{1}{2} \underline{\delta}^T F_{xx}(x_1) \underline{\delta} \, [(1-\lambda) - (1-\lambda)^2] + 0(\|\underline{\delta}\|^3) [(1-\lambda) - (1-\lambda)^3] \geqslant 0$$

Neglecting the third and higher order terms and noting that the term $[(1-\lambda)-(1-\lambda)^2]$ is $\geqslant 0$ since $\lambda \in [0,1]$, we obtain :

$$\frac{1}{2} \underline{\delta}^T F_{xx}(x_1) \underline{\delta} \geqslant 0 \quad \forall \underline{\delta} \text{ such that } \underline{x}_1, \underline{x}_2 \in \mathfrak{D} \qquad 1.3.7$$

i.e. F_{xx} is non negative definite on the restriction of $\mathfrak{D}$.

Next we recall some results on positive definite matrices.

Positive definite matrices

Definitions

The m x m square symmetric matrix M is said to be

- non negative definite if $\underline{v}^T M \underline{v} \geqslant 0 \quad \forall \underline{v} \in R^m$
- positive definite if $\underline{v}^T M \underline{v} > 0 \quad \forall \underline{v} \neq \underline{0}$ and $\underline{v} \in R^m$

Consider the subspace $\mathcal{E} \subset R^m$. If :

$$\left.\begin{array}{ll} * \ \underline{v}^T M \underline{v} \geq 0 & \forall \ \underline{v} \in \mathcal{E} \subset R^m \\ * \ \underline{v}^T M \underline{v} > 0 & \forall \underline{v} \neq \underline{0}, \ \underline{v} \in \mathcal{E} \subset R^m \end{array}\right\} \text{the restriction of } M \text{ on } \mathcal{E} \text{ is } \left\{\begin{array}{l} \text{non negative definite} \\ \text{positive definite} \end{array}\right.$$

If we change the direction of these inequalities, we obtain the definitions of non positive definite and negative definite matrices.

We will denote these symbolically by $M > 0$, $M \geq 0$, $M < 0$, $M \leq 0$. We see directly from the above inequalities that if $M > 0$ or $M \geq 0$, then $(-M) < 0$ or $(-M) \leq 0$.

Criteria for definition without restriction

These criteria are very simple and are in relation to the eigenvalues or the principal determinants taken successively.

If :	Then	$M > 0$	$M \geq 0$	$M < 0$	$M \leq 0$
Eigenvalues ω_i =1 to m		$\omega_i > 0$	$\omega_i \geq 0$	$\omega_i < 0$	$\omega_i \leq 0$
or : Principal determinants D_i		$D_i > 0$ taken successively	$D_i \geq 0$	$(-1)^i D_i > 0$ taken successively	$(-1)^i D_i \geq 0$

Table 1.6

How do we use this table ? For example, if all the eigenvalues ω_i, i = 1 to m, of M are > 0, or if all the principal determinants taken successively of D_i are > 0, then $M > 0$.

We recall that : the eigenvalues ω_i are the roots of the equation $|M - \omega \mathbb{1}| = 0$

Consider the matrix $M = \begin{bmatrix} m_{11} & \dots & m_{1m} \\ m_{21} & \dots & m_{2m} \\ m_{m1} & \dots & m_{mm} \end{bmatrix}$. The principal determinants taken successively can be written as :

$$D_1 = m_{11},\ D_2 = \begin{vmatrix} m_{11} & m_{12} \\ m_{21} & m_{22} \end{vmatrix},\ D_3 = \begin{vmatrix} m_{11} & m_{12} & m_{13} \\ m_{21} & m_{22} & m_{23} \\ m_{31} & m_{32} & m_{33} \end{vmatrix},\ \text{etc.}\ D_m = |M|$$

and the principal determinants are those minors which have as their diagonal, the diagonal of M.

Example : $M = \begin{vmatrix} 1 & 2 \\ 2 & 5 \end{vmatrix}$

$$|M - \omega \mathbb{1}| = \begin{vmatrix} 1-\omega & 2 \\ 2 & 5-\omega \end{vmatrix} = (1-\omega)(5-\omega) - 4 = 0 = \omega^2 - 6\omega + 1 = 0$$

or $\omega = 3 \pm \sqrt{8} > 0$

$\omega_1 > 0$, $\omega_2 > 0$, the matrix is thus positive definite. We find the same result using the principal determinants i.e.

$$D_1 = |1| > 0, \quad D_2 = \begin{vmatrix} 1 & 2 \\ 2 & 5 \end{vmatrix} = 1 > 0.$$

Criterion with restriction

We will give only one method here. Consider the subspace defined by :

$$\mathcal{E} = \left\{ \underline{x} \;/\; A\underline{x} = \underline{0} \right\}$$

Here A is a p x m matrix of rank p.

The restriction of M on $\mathcal{E}$ is positive definite if the roots of the polynomial in ω of degree m-p given by :

$$\det \begin{vmatrix} \omega \mathbb{1} - M & A^T \\ A & 0 \end{vmatrix} = 0 \qquad 1.3.8$$

are all real and positive.

If $\omega_i < 0$, i = 1 to m-p, M is negative definite on the restriction $\mathcal{E}$, etc.

Example :

$$M = \begin{bmatrix} 1 & 2 \\ 2 & 1 \end{bmatrix} \quad A = \begin{bmatrix} 1, & 2 \end{bmatrix} \qquad m = 2, \; p = 1$$

The equation of degree m-p = 1, in ω, is

$$\det \begin{vmatrix} \omega-1 & -2 & 1 \\ -2 & \omega-1 & 2 \\ 1 & 2 & 0 \end{vmatrix} = 0$$

Thus, expanding for example by using the terms of the last row, we have :

$$[-4 - (\omega-1)] - 2\,[2(\omega-1) + 2] = 0 \text{ or } 5\omega+3=0 \text{ or } \omega = -3/5 < 0$$

Thus M is negative definite in $\mathcal{E}$.

1.4 CONCLUSIONS

In this chapter we have recapitulated the different notions of optimisation and control as well as the general principles of control of complex systems. We thus have the basis now for developing the optimisation aspect of both static (high order or low order) and dynamic systems using global techniques or decomposition-coordination techniques.

REFERENCES FOR CHAPTER 1

[1] PUN, L., 1972, "Introduction à la pratique de l'optimisation". Dunod, Paris.

[2] BOUDAREL, R., DELMAS, J., GUICHET, P. "Commande optimale des processus".
a) T.1 Concepts fondamentaux de l'Automatique, Dunod, 1967
b) T.2 Programmation non linéaire et ses applications, Dunod, 1968

[3] RICHALET, J., RAULT, A., POULIQUEN, R., 1972, "Identification des processus par la méthode du modèle". Gordon and Breach, Théorie des Systèmes, vol. 4.

[4] HIMMELBLAU, D.M. (editor), 1973, "Decomposition of large scale problem". North Holland Publishing co.

[5] RICHETIN, M. : "Analyse structurale des systèmes complexes en vue d'une commande hiérarchisée". Thèse de Doctorat d'Etat, Université de Toulouse, 1975.

[6] CONANT, R.C. : "Information transfer in complex systems with application to regulation". Ph.D.Univ. of Illinois, USA, 1968.

[7] AOKI, M. : "Control of large scale dynamic systems by aggregation". IEEE Trans. on A.C. vol. AC 13. p. 246-293, june 1968

[8] MILGRAM, M. "Méthode de décomposition de systèmes". Thèse de Spécialité, U. de Technologie de Compiègne, 1975.

[9] MESAROVIC, M.D., MACKO, D., TAKAHARA, Y., 1970, "Theory of hierarchical multilevel systems". Academic Press, New-York.

[10] FOSSARD, A.J., CLIQUE, M., IMBERT, N. : "Aperçus sur la commande hiérarchisée". RAIRO Automatique, n° 3, 1972.

[11] TITLI, A., 1975, "Commande hiérarchisée et optimisation des systèmes complexes". Dunod Automatique, Paris.

[12] BERNHARD, P., 1976, "Commande optimale. Decentralisation et jeux dynamiques". Dunod, Paris.

[13] DIXON, L.C., SZEGO, G.P. (Editor), 1975, "Towards global optimization". North Holland P. Company.

[14] GANTMACHER, F.R., 1966, "Th. des matrices". Tome 1, Dunod.

CHAPTER 2

Linear Programming

2.1 BASIC CONCEPTS : INTRODUCTORY EXAMPLES

Linear programming is a powerful approach to tackling optimisation problems with linear cost functions and linear constraints. Such problems arise often in economics, transportation, etc. Linear programming problems, even of high dimension, are easy to solve using modern digital computers. To understand how such problems arise, we begin by considering some elementary examples.

2.1.1 Example 1

A company produces 2 products A and B using 3 different raw materials M_1, M_2, and M_3.

At the beginning of the year, the company has the following stocks of raw materials : 300 tons of M_1, 400 tons of M_2 and 250 tons of M_3.

In order to produce 1 ton of A, it is necessary to use 1 ton of M_1 + 2 tons of M_2. To produce 1 ton of B, it is necessary to have 1 ton of M_1 + 1 ton of M_2 + 1 ton of M_3.

The sale of 1 ton of A provides a profit of \$ 50 and that of B \$ 100.

The problem is to find how much of A and B the company should produce to maximise its profits.

Mathematical formulation of the problem

Let x and y be the unknown variables, i.e. x is the quantity of A produced and y that of B. Then the problem could be written as

$$\max_{x,y} P = 50\,x + 100\,y$$

$$\text{subject to} \quad \left[\begin{array}{l} x+y \leqslant 30 \\ 2x+y \leqslant 400 \\ y \leqslant 250 \\ x \geqslant 0,\ y \geqslant 0 \end{array}\right\} \quad \text{limitations on the stocks}$$

Graphical solution

This simple 2 variables problem can be solved using a graph. Each manufacturing option could be represented by a point with (x, y) coordinates. The stock constraints could be represented by straignt lines and these lines delimit the feasible region in which the solution must lie.

Fig. 2.1 shows the constraint set graphically.

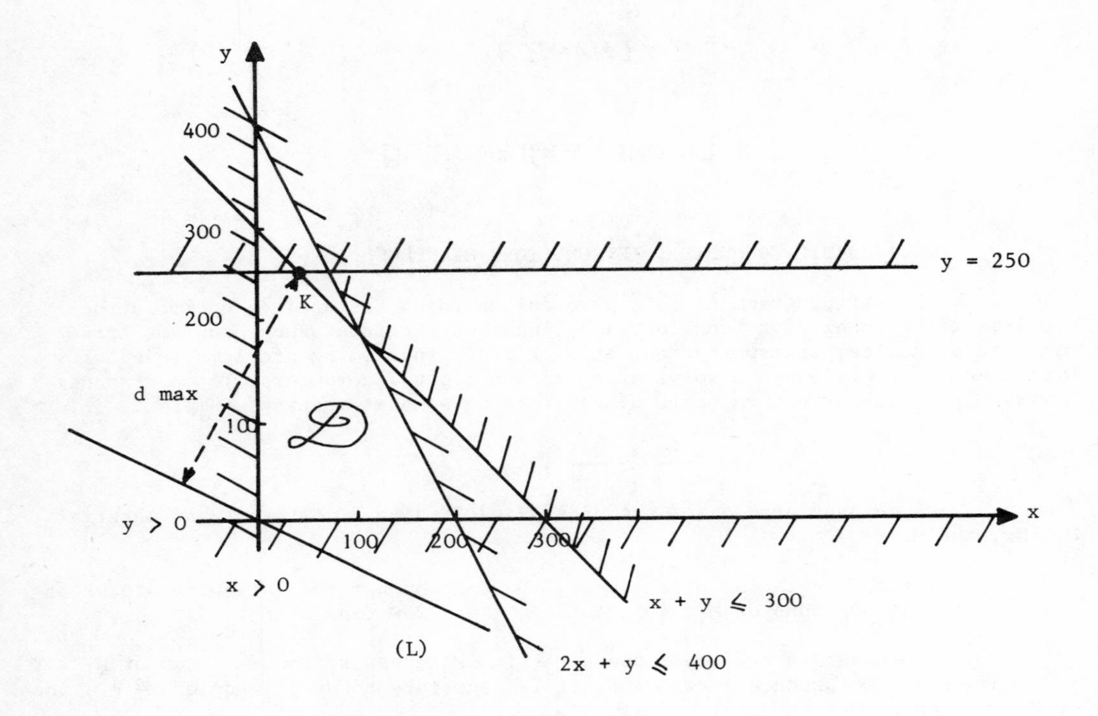

Fig. 2.1

The $50x + 100y = P$ are lines parallel to (L) ($x + 2y = 0$) and these represent more profit the further they are from the origin.

The problem is thus to find the point in the non-hatched region $\mathcal{D}$ in Fig. 2.1 through which a line which is parallel to (L) and which is furthest from (L) could pass. This gives, as the optimal solution, the point K.

Although this reasoning could be extended to the multivariable case, it is not possible to use simple geometry as was done in this two dimensional example. We will have to use some analytical method like the simplex method that we will develop later on in the chapter.

As we will see, the simplex method consists of finding an admissible point and starting from this admissible point, finding another one which improves the criterion function. In this way, the optimum is achieved in a finite number of steps.

We note that linear programming problems are quite common in the area of economics and management. Typical examples of such problems are :

- Definition of a manufacturing program to maximise the profit by taking into account the prices, sales, manufacturing possibilities, commercial constraints, etc.

- Definition of the optimal composition of a mixture (for example petrol, animal feedstuffs, etc.) such that its price is minimal whilst satisfying certain constraints (e.g. octane index, vitamin doses, protein levels, etc.).

- Organisation of transportation networks which minimise distribution costs.

- To fix the schedules for different machines operating on different lots with different costs.

- Optimal control of discrete linear systems with linear cost function.

Next let us consider another introductory example.

2.1.2 Example 2

The problem is to maximise

$$Z = x_1 + 3x_2$$

subject to
$$\begin{cases} -x_1 + x_2 \leqslant 1 \\ x_1 + x_2 \leqslant 2 \\ x_1 \geqslant 0,\ x_2 \geqslant 0 \end{cases}$$

Fig. 2.2 shows the feasible region for this problem :

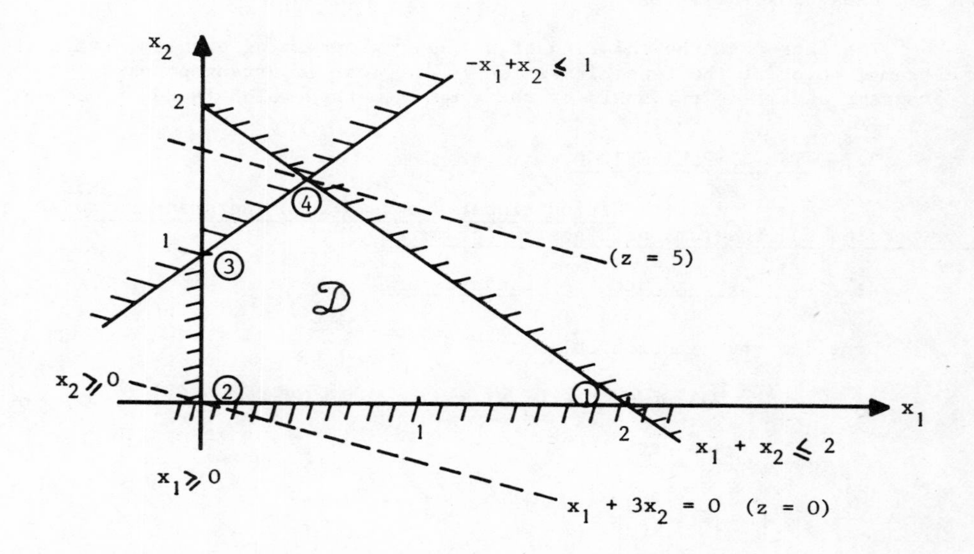

Fig. 2.2

The points (1) to (4) in Fig. 2.2 are called the extremal points of the constraint set. For linear constraints (which is the case considered here), the number of such points is finite.

The equal value contours here are the lines $x_1 + 3x_2 = Z =$ constant, parallel to the line Z = 0. The maximum of Z then corresponds to the largest value of Z for which this line has at least one point in the feasible region and this gives as the solution the point (4).

We note that the maximum is an extremal point of the feasible region. If the problem had been to minimise Z, the solution would have been the point (2) which is also an extremal point.

If the problem had been to minimise $Z = 2x_1 + 2x_2$, the solution would have been (1) and (4), i.e. the two extremal points or rather all the points of the segment joining these 2 extremal points.

If we suppress the constraint $x_1 + x_2 \leqslant 2$, the admissible region becomes infinite and the solution would be $Z \longrightarrow \infty$.

If we replace the constraint $x_1 + x_2 \leqslant 2$ by $x_1 + x_2 \leqslant -1$, the feasible set would be empty and there would be no solution.

From the above discussion we see that a linear programming problem :

1. May not have a solution
2. May have a finite solution
3. May have an infinite solution (unbounded)

or 4. May have an infinite number of solutions.

Any method for solving linear programming problems should take into account all these possibilities.

The fact that the solution of a linear programming problem always lies at an extremal point of the feasible region is the most important property of linear programs and it is the basis of the simplex method which is described next.

2.2. THE SIMPLEX METHOD*

2.2.1 Putting linear programs into the standard form. Basic properties and theorems of Linear Programming

2.2.1.1 Putting into the standard form

Consider the linear problem :

$$\max_{x_j} Z = \sum_{j=1}^{n} C_j x_j \qquad 2.2.1$$

subject to

$$\sum_{j=1}^{n} a_{ij} x_j = b_i \qquad i = 1, 2, \ldots m \qquad 2.2.2$$

$$x_j \geqslant 0 \qquad j = 1, \ldots n \qquad 2.2.3$$

* Here we rely heavily on the description of L.S. LASDON [1] which appears to us to be a pedagogically sound way of explaining the method.

This problem can be put into the matrix form

Maximise $Z = \underline{C}^T \underline{x}$

subject to $\left[\begin{array}{l} A\underline{x} = \underline{b} \\ \underline{x} \geqslant \underline{0} \end{array}\right.$ A is an m x n constant matrix

If any of the equations 2.2.2 are redundant (for example if there are equations which are linear combinations of other equations), we could eliminate these without changing the solution of the problem.

If there are no solutions or only a single solution to the equations 2.2.2, there is no optimisation problem. Thus, for non-trivial linear programming problems we have :

$$\boxed{\begin{array}{c} \text{rank } A = m \\ n > m \end{array}}$$

Next we assume that the constraints are of the form $\underline{x} \geqslant \underline{0}$ and $A\underline{x} = \underline{b}$ since this is the easiest case to treat. We can convert any constraints which are not originally in this form to this form as follows :

Transformation of inequality constraints

This is done by the addition of slack or surplus variables. Thus

$$\sum_{j=1}^{n} a_{ij}\, x_j \leqslant b_i \qquad \text{becomes} \qquad \sum_{j=1}^{n} a_{ij}\, x_j + x_{n+i} = b_i$$

x_{n+i} being a slack variable. $\qquad x_{n+i} \geqslant 0$

For the inequality in the opposite direction, we have :

$$\sum_{j=1}^{n} a_{ij}\, x_j \geqslant b_i \qquad \text{becomes} \qquad \sum_{j=1}^{n} a_{ij}\, x_j - x_{n+i} = b_i$$

$$x_{n+i} \geqslant 0$$

where x_{n+i} here is a surplus variable.

Unconstrained variables

If any of the variables are unconstrained (i.e. not non-negative), we can put them in the standard non-negative constraint form by defining additional variables. Thus x_k which has an arbitrary sign could be converted to

$$x_k = x'_k - x''_k \qquad \text{with } x'_k \geqslant 0,\ x''_k \geqslant 0$$

This obviously increases the number of variables. However, this is usually less of a problem than the difficulties caused by having variables with an arbitrary sign.

Example : Put into standard form the following problem :

$$\text{Min } Z = x_1 + x_2$$

$$\text{subject to} \quad \begin{cases} 2x_1 + 3x_2 \leqslant 6 \\ x_1 + 7x_2 \geqslant 4 \\ x_1 + x_2 = 3 \\ x_1 \geqslant 0, \; x_2 \text{ arbitrary} \end{cases}$$

Solution : $x_2 = x'_2 - x''_2$, $x'_2 \geqslant 0$, $x''_2 \geqslant 0$

$$\text{Min } Z = x_1 + x'_2 - x''_2$$

$$\text{subject to} \quad \begin{cases} 2x_1 + 3x'_2 - 3x''_2 + x_3 = 6 \\ x_1 + 7x'_2 - 7x''_2 - x_4 = 4 \\ x_1 + 3x'_2 - 3x''_2 = 3 \\ x_1 \geqslant 0, \; x'_2 \geqslant 0, \; x''_2 \geqslant 0, \; x_3 \geqslant 0, \; x_4 \geqslant 0 \end{cases}$$

2.2.1.2 Basic theorems of linear programming

Definitions

1. A feasible solution is one which satisfies all the equations 2.2.2 and the inequalities 2.2.3.

2. A basis matrix is a non-singular m x m matrix which is formed with m columns of the A matrix (since A has rank m, it comprises at least one basis matrix).

3. A basic solution is the unique vector determined by choosing a basis matrix, cancelling the (n-m) variables of the A matrix which do not form a part of the basis matrix and solving the non-singular system of equations for the m variables that remain.

4. A feasible basic solution is a basic solution with non-negative variables.

5. A feasible non-degenerate solution is a basic solution with exactly m variables $x_i > 0$.

6. An optimal solution is a feasible solution which minimises Z.

Example

$$\text{Consider :} \quad \begin{cases} -x_1 + x_2 + x_3 = 1 \\ x_1 + x_2 + x_4 = 2 \\ x_i \geqslant 0, \qquad i = 1, \ldots, 4 \end{cases}$$

In the matrix form, this becomes

$$\underline{Ax} = \underline{b}, \; \underline{x} \geqslant 0 \text{ with}$$

$$A = \begin{bmatrix} -1 & 1 & 1 & 0 \\ 1 & 1 & 0 & 1 \end{bmatrix}$$

Here we could extract $B = \begin{bmatrix} 1 & 0 \\ 0 & 1 \end{bmatrix}$ which is non-singular and which could thus be used as a basis matrix.

The basic solution corresponding to this is

$$\left.\begin{array}{l} x_3 + 0 = 1 \\ 0 + x_4 = 2 \end{array}\right] \quad \text{or} \quad \begin{array}{ll} x_3 = 1 \;; & x_4 = 2 \\ x_4 = 0 \;; & x_2 = 0 \end{array}$$

This is a non-degenerate basic solution since it has two strictly positive components (x_3 and x_4).

The matrix $B' = \begin{bmatrix} -1 & 0 \\ 1 & 1 \end{bmatrix}$ is also a basis matrix. The corresponding solution is :

$$\left.\begin{array}{l} -x_1 + 0 = 1 \\ x_1 + x_4 = 2 \end{array}\right] \quad \text{or} \quad \begin{array}{ll} x_1 = -1 \;; & x_4 = 3 \\ x_2 = 0 \;; & x_3 = 0 \end{array}$$

This solution is infeasible ($x_1 < 0$).

The importance of these definitions becomes evident when we consider the following two theorems which are given here without proof.

Theorem 2.1

The objective function Z attains its minimum at an extremal point of the feasible region. If this minimum lies on more than one extremal point, Z has the same value Z_m at all the points of the segment joining these two extremal points.

This is a generalisation of what we saw on our two introductory examples. From theorem 2.1, we see that we need only examine extremal points.

Theorem 2.2

A vector $\underline{x} = (x_1, \dots x_n)^T$ is an extremal point of the feasible region of a linear programming problem if and only if $\underline{x}$ is a feasible basic solution.

This theorem can be verified on example 2 : $x_1 = 0$, $x_2 = 0$ corresponds to an extremal point with $x_3 = -1$, $x_4 = 2$; this is also a feasible basic solution.

We can also verify that the optimum ($x_1 = \frac{1}{2}$, $x_2 = \frac{3}{2}$, $x_3 = x_4 = 0$) corresponds to a feasible basic solution with the base matrix

$$B = \begin{bmatrix} -1 & 1 \\ 1 & 1 \end{bmatrix} \quad \text{which gives}$$

$$\left.\begin{array}{l} -x_1 + x_2 = 1 \\ x_1 + x_2 = 2 \end{array}\right] \Longrightarrow \begin{array}{l} x_1 = \frac{1}{2},\; x_2 = \frac{3}{2} \\ x_3 = x_4 = 0 \end{array}$$

These two theorems thus show that to find the optimal solution, it is only necessary to consider extremal points or feasible basic solutions.

However, for n, m large, the number of extremal points becomes very large and it is not practical to examine them all. It is therefore necessary to do an ordered search which enables us to pass from one point to the next whilst improving Z. This is the aim of the simplex method which converges in a finite number of iterations (between m and 2m) and which also enables us to distinguish between the cases where there is no solution or the solution is unbounded.

The next step is to perform simple manipulations on linear equations.

2.2.1.3 Linear systems of equations and equivalent systems

Consider the m linear equations in n unknowns :

$$\left[\begin{array}{l} a_{11} x_1 + a_{12} x_2 + \quad \dots \quad + a_{1n} x_n = b_1 \\ \vdots \\ a_{m1} x_1 + a_{m2} x_2 + \quad \dots \quad + a_{mn} x_n = b_n \end{array}\right. \qquad 2.2.4$$

If these equations have a solution set which is empty, the equations are said to be inconsistent.

Two systems of equations are said to be equivalent if they have the same solution set. The following operations transform a system of linear equations into an equivalent system :

1. Multiplication of an equation E_i by a constant $k \neq 0$

2. Replacement of an equation E_t by an equation $E_t + kE_i$ where E_i is any other equation of the system.

These operations are called elementary row operations.

Example :

$$\left[\begin{array}{ll} -x_1 + x_2 + x_3 = 1 & (a) \\ x_1 + x_2 + x_4 = 2 & (b) \end{array}\right. \Longrightarrow \left[\begin{array}{ll} -x_1 + x_2 + x_3 = 1 & (a) \\ 2x_1 - x_3 + x_4 = 1 & (a)+(b) \end{array}\right.$$

A solution of the two systems is :

$$x_1 = 0,\ x_3 = 0,\ x_2 = 1,\ x_4 = 1$$

Indeed, all solutions of one system are solutions of the other.

The above transformations are used in the Pivot method which consists of replacing a linear system in m elementary operations by an equivalent linear system in which a specific variable has unity coefficient in an equation and zero in all others.

The steps in the pivot method are :

Step 1 : choose a term $a_{rs} x_s$ of the equation r column s with $a_{rs} \neq 0$ (called the pivot term).

Stem 2 : replace each E_i by $E_i - \dfrac{a_{is}}{a_{rs}} E_r$ for

$i = 1, 2, \dots m$ for $i \neq r$.

<u>Example</u>

$$\begin{cases} 2x_1 + 3x_2 - 4x_3 + x_4 = 1 & E_1 \\ x_1 - x_2 \qquad + 5x_4 = 6 & E_2 \\ 3x_1 + x_2 + x_3 \qquad = 2 & E_3 \end{cases}$$

Choosing the term $2x_1$ in E_1 as the pivot, we have :

$$\text{I} \begin{cases} x_1 + \frac{3}{2}x_2 - 2x_3 + \frac{1}{2}x_4 = \frac{1}{2} & \\ 0 - \frac{5}{2}x_2 + 2x_3 + \frac{1}{2}x_4 = \frac{11}{2} & \\ 3x_1 + x_2 + 2x_3 \qquad = 2 & E_2 - E_1 \end{cases}$$

$$\text{II} \begin{cases} x_1 + \frac{3}{2}x_2 - 2x_3 + \frac{1}{2}x_4 = \frac{1}{2} & \\ 0 - \frac{5}{2}x_2 + 2x_3 + \frac{1}{2}x_4 = \frac{11}{2} & \\ 0 - \frac{7}{2}x_2 + 7x_3 - \frac{3}{2}x_4 = \frac{1}{2} & E_3 - 3E_1 \end{cases}$$

Next let us consider canonical systems.

<u>Canonical systems</u>

Let us assume that the first m columns of the linear system 2.2.4 form a base matrix B :

$$A\,\underline{x} = \underline{b} \implies [B \;\vdots\; M]\,\underline{x} = \underline{b}$$

On multiplying 2.2.4 by B^{-1}, we obtain an equivalent transformed system in which the coefficients of the variables $x_1, \dots x_m$ are unity. Such a system is called a canonical system.

$$[B^{-1}]\,[B \;\vdots\; M]\,\underline{x} = [B^{-1}]\,\underline{b} \implies [I \;\vdots\; B^{-1}M]\,\underline{x} = [B^{-1}\,\underline{b}]$$

or in a detailed form :

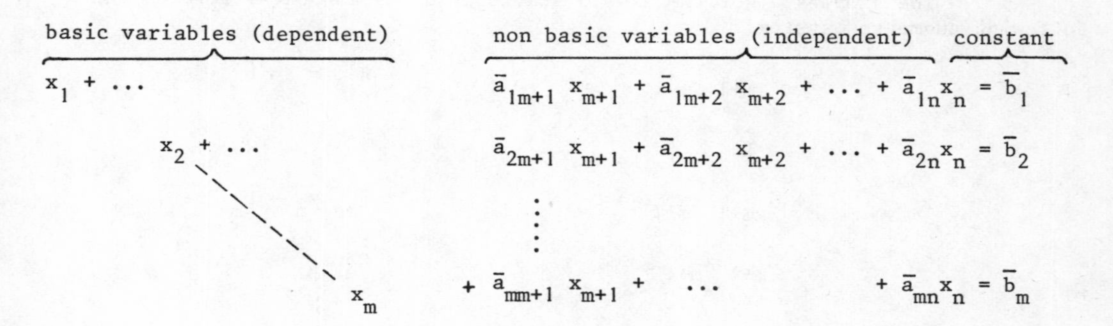

$$\begin{array}{llll} x_1 + \dots & \bar{a}_{1m+1}x_{m+1} + \bar{a}_{1m+2}x_{m+2} + \dots + \bar{a}_{1n}x_n = \bar{b}_1 \\ \quad x_2 + \dots & \bar{a}_{2m+1}x_{m+1} + \bar{a}_{2m+2}x_{m+2} + \dots + \bar{a}_{2n}x_n = \bar{b}_2 \\ \qquad \ddots & \vdots \\ \qquad\quad x_m & + \bar{a}_{mm+1}x_{m+1} + \dots + \bar{a}_{mn}x_n = \bar{b}_m \end{array}$$

The variables $x_1 \dots x_m$ associated with the columns of B are called the <u>basic variables</u> ; they are also sometimes called dependent variables since once the non-basic variables are fixed, the basic variables can be determined directly. For example, if :

$$x_{m+1} = \dots = x_n = 0$$

$$x_i = b_i \qquad i = 1 \text{ to } m$$

If $\bar{b}_i \geqslant 0$, $i = 1$ to m, the basic solution is feasible. If one or more $\bar{b}_i$ are zero, the feasible basic solution is degenerate.

Instead of calculating B^{-1} every time in order to multiply the system 2.2.1, we prefer to put the system 2.2.4 in the canonical form using the pivot method. To do this, we choose a pivot $a_{ij}x_j$ such that $a_{ij} \neq 0$. Since B is non singular, we can always find a non zero a_{ij} and we can use this as our element (1, 1) by rearranging the columns. We perform a pivot operation with this term and we then restart with $a_{22}x_2$ until a_{mm} x_m. This provides us with the canonical form which will be used in the simplex method.

Having described some of the elementary operations, we are now in a position to describe the simplex algorithm.

2.2.2 The simplex algorithm

The simplex method is essentially a two phase procedure.

In the first phase, we find a feasible basic solution, if it exists, or provide the information that it does not exist (for the case of inconsistant constraints), so that the problem has no solution.

In the second phase, we use this solution as the starting point and either

- find the optimal solution

or - indicate that the optimum is unbounded.

The beauty of the method lies in the fact that both these phases use the simplex algorithm which we will describe next.

In the simplex method, we begin by treating the objective function $Z = C_1 x_1 + \dots + C_n x_n$ as an additional equation $-Z + C_1 x_1 + \dots + C_n x_n = 0$ which is added on to the set of constraint equations.

The simplex algorithm always starts with the canonical form of the following augmented system :

$$\begin{array}{llllllll}
x_1 & & & + \bar{a}_{1m+1} x_{m+1} & + & \dots & + \bar{a}_{1n} x_n & = \bar{b}_1 \\
& x_2 & & \vdots & & & & \\
& & \ddots\, x_m & + \bar{a}_{mm+1} x_{m+1} & + & \dots & + \bar{a}_{mn} x_n & = \bar{b}_m \\
& & -Z & + \bar{C}_{m+1} x_{m+1} & + & \dots & + \bar{C}_n x_n & = -\bar{Z}
\end{array}$$

With this canonical form, the basic solution is :

$$Z = \bar{Z},\ x_1 = \bar{b}_1,\ \dots\ x_m = \bar{b}_m,\ x_{m+1} = x_{m+2} = \dots\ x_n = 0 \qquad 2.2.5$$

Let us assume that this basic solution is feasible, i.e.

$$\bar{b}_1 \geqslant 0 \quad \ldots \quad \bar{b}_m \geqslant 0 \qquad 2.2.6$$

In that case, the above canonical form is said to be feasible.

Next let us define the optimality test which enables us to finish the computation.

2.2.2.1. Optimality test

The optimality test is given by theorem 2.3.

Theorem 2.3

A feasible basic solution is the optimal (minimal solution with cost Z, if all the constants $\bar{C}_{m+1}, \ldots \bar{C}_n$ (the relative cost factors) are non-negative

$$\bar{C}_j \geqslant 0 \qquad j = m + 1 \text{ to } n$$

We assume in the statement of the above theorem that the problem is one of minimisation. In the case where the problem is max Z, we change the sign of the $\bar{C}_j$ or we could write :

$$\max Z = \min (-Z)$$

Proof : Let us write the last equation of the canonical system, i.e.

$$Z = \bar{Z} + \bar{C}_{m+1} x_{m+1} + \ldots + \bar{C}_n x_n$$

Since the variables x_j, j = m+1 to n are zero here and since they must satisfy $x_j \geqslant 0$, they can not change except to become positive and this won't decrease the cost function Z since $\bar{C}_j \geqslant 0$ ($\bar{C}_j x_j \geqslant 0$).

Since no change is possible in the non-basic solution to improve Z, the solution for which $\bar{C}_j \geqslant 0$ must be optimal.

We can also conclude if the optimum is multiple or not. Consider $\bar{C}_j \geqslant 0$ except for $\bar{C}_k$ which is zero. Then if the constraints allow x_k to be positive, there will be no change in Z and we have a multiple optimum.

However, it is possible that the constraints do not allow us to have $x_k \geqslant 0$. This is the degenerate case which will be examined later.

Corolary : A feasible basic solution is the unique optimal solution if $\bar{C}_j > 0$ for all the non-basic variables.

If $\bar{C}_j < 0$, the solution is not optimal.

Next, we see how we can go from one basic solution to another improved basic solution. To fix our ideas, we consider a simple example.

2.2.2.2. Improvement of a non-optimal basic solution : an example

Consider the augmented system :

$$\left[\begin{array}{l} 5x_1 - 4x_2 + 13x_3 - 2x_4 + x_5 = 20 \\ x_1 - x_2 + 5x_3 - x_4 + x_5 = 8 \\ x_1 + 6x_2 - 7x_3 + x_4 + 5x_5 - Z = 0 \end{array}\right\} \quad 2.2.7$$

$$x_j \geqslant 0 \qquad j = 1, 2, \ldots, 5 \qquad 2.2.8$$

We assume that we know that we can use x_5, x_1, $(-Z)$ as the pivot (i.e. as the feasible basic solution) in order to find the feasible canonical form (using x_5 in the first equation and x_1 in the second) :

$$\left.\begin{array}{l} x_5 \quad - \frac{1}{4} x_2 + 3x_3 - \frac{3}{4} x_4 = 5 \\ x_1 \quad - \frac{3}{4} x_2 + 2x_3 - \frac{1}{4} x_4 = 3 \\ -Z + 8\, x_2 - 24x_3 + 5\, x_4 = -28 \end{array}\right\} \quad 2.2.9$$

The corresponding basic solution is :

$$x_5 = 5,\ x_1 = 3,\ x_2 = x_3 = x_4 = 0,\ Z = 28 \qquad 2.2.10$$

In the case of this solution 2.2.10, one relative cost factor is negative ($\bar{C}_3 = -24 < 0$). Thus this solution is non optimal although it is feasible (theorem 2.3 not satisfied).

Indeed, if x_3 is increased from its present value of zero (the other basic variables remaining zero), Z could decrease since by 2.2.9 :

$$Z = 28 - 24x_3 \qquad 2.2.11$$

In view of 2.2.11, it would be desirable to increase x_3 as much as possible in order to obtain the smallest possible value of Z. However, the non-negativity constraints fix a limit for x_3.

If $x_2 = x_4 = 0$, using 2.2.9, x_1, x_5, x_3 are related to each other by :

$$\left.\begin{array}{l} x_5 = 5 - 3x_3 \\ x_1 = 3 - 2x_3 \end{array}\right\} \quad 2.2.12$$

If $x_3 = \frac{5}{3}$, $x_5 = 0$ but $x_1 < 0$. If $x_3 = \frac{3}{2}$, $x_1 = 0$ with $x_5 > 0$. Thus the biggest possible value for x_3 is 3/2. Putting this value of $x_3 = 3/2$ in 2.2.11 and 2.2.12, we obtain a new basic solution and a new cost, i.e.

$$x_5 = \frac{1}{2},\ x_3 = \frac{3}{2},\ x_1 = x_2 = x_4 = 0,\ Z = -8$$

and we see that the cost has been reduced from 28 to -8.

Is this the optimal solution ? To find this out, it is necessary to write the system equations in a canonical form with x_5, x_3, $-Z$ as the basic variables.

It is thus necessary to replace x_1 by x_3 in the basis using the pivot method. For this purpose, using 2.2.9, we choose $2x_3$ as the pivot in the row of x_1 and in the column corresponding to the negative cost term. We thus have :

$$\left.\begin{array}{l} x_5 \quad - \frac{3}{2} x_1 + \frac{7}{8} x_2 - \frac{3}{8} x_4 = \frac{1}{2} \\ x_3 \quad + \frac{1}{2} x_1 - \frac{3}{8} x_2 - \frac{1}{8} x_4 = \frac{3}{2} \\ -Z + 12x_1 - x_2 + 2 x_4 = 8 \end{array}\right\} \qquad 2.2.14$$

and this gives the basic solution of 2.2.13 as stated.

2.2.14 shows that although this solution is much better than the previous one, it is not optimal since $\bar{C}_2 = -1 < 0$. The solution can thus be still improved by increasing x_2 and keeping $x_1 = x_4 = 0$. By 2.2.14, we have :

$$x_5 = \frac{1}{2} - \frac{7}{8} x_2 \qquad 2.2.15$$

$$x_3 = \frac{3}{2} + \frac{3}{8} x_2 \qquad 2.2.16$$

$$Z = -8 - x_2 \qquad 2.2.17$$

2.2.16 does not limit the variation of x_2 but 2.2.15 does since x_5 must be $\geqslant 0$:

$$x_5 = 0 \Longrightarrow x_2 = \frac{4}{7}$$

As before on using the term $\frac{7}{8} x_2$ of 2.2.14 as the pivot, we obtain the new canonical form :

$$\left.\begin{array}{l} x_2 \quad - \frac{12}{7} x_1 - \frac{3}{7} x_4 + \frac{8}{7} x_5 = \frac{4}{7} \\ x_3 \quad - \frac{1}{7} x_1 - \frac{2}{7} x_4 + \frac{3}{7} x_5 = \frac{12}{7} \\ -Z + \frac{72}{7} x_1 + \frac{11}{7} x_4 + \frac{8}{7} x_5 = \frac{60}{7} \end{array}\right\} \qquad 2.2.18$$

and the basic solution $x_2 = \frac{4}{7}$, $x_3 = \frac{12}{7}$, $Z = -\frac{60}{7}$, $x_1 = x_4 = x_5 = 0$ is optimal ($\bar{C}_j > 0$). It was obtained here in only 3 iterations.

Degeneration : Let us assume that the 2nd equation of 2.2.9 was

$$x_1 - \frac{3}{4} x_2 + 2x_3 - \frac{1}{4} x_4 = 0$$

The second equation 2.2.12 thus becomes $x_1 = -2x_3$ and any positive change in x_3 makes x_1 negative. Thus x_3 remains zero and Z can not decrease even though it is non-optimal. This is a degenerate case which requires a new pivot transformation.

Unbounded solution : If the equations 2.2.15 to 2.2.17 had been

$$x_5 = \frac{1}{2} + \frac{7}{8} x_2$$

$$x_3 = \frac{3}{2} + \frac{3}{8} x_2$$

$$Z = -8 - x_2$$

we could have chosen x_2 arbitrarily > 0 without having $x_3 < 0$, $x_5 < 0$, and Z could have taken an arbitrary negative value. This is a case of an unbounded solution.

We note that this happens whenever all the coefficients of a column of the canonical system which correspond to $\bar{C}_j < 0$ are also < 0 or $= 0$.

Indeed, 2.2.14 becomes

$$\begin{aligned} x_5 \quad & - \frac{3}{2} x_1 - \frac{7}{8} x_2 - \frac{3}{8} x_4 = \frac{1}{2} \\ x_3 \quad & + \frac{1}{2} x_1 - \frac{3}{8} x_2 - \frac{1}{8} x_4 = \frac{3}{2} \\ Z \quad & + 12 x_1 - x_2 + 2 x_4 = 8 \end{aligned}$$

2.2.2.3. Improvement of a non optimal but feasible basic solution. General case

If there exists a $\bar{C}_j < 0$ and if we neglect the degenerate case (all the $\bar{b}_i > 0$), then, it is always possible to use a pivot operation to go from one basic feasible solution to another one with a lower cost.

If many $\bar{C}_j < 0$, the variable x_s which needs to be increased to make it positive is the one for which $\bar{C}_j$ is most negative, i.e.

$$\bar{C}_s = \min_j \bar{C}_j < 0 \qquad 2.2.19$$

Having retained x_s, we increase it from zero whilst having the other non-basic variables at zero and we observe the basic variables which are related to x_s by :

$$\left.\begin{aligned} x_1 &= \bar{b}_1 - \bar{a}_{1s} x_s \\ x_2 &= \bar{b}_2 - \bar{a}_{2s} x_s \\ &\vdots \\ x_m &= \bar{b}_m - \bar{a}_{ms} x_s \\ Z &= \bar{Z} + \bar{C}_s x_s \qquad \bar{C}_s < 0 \end{aligned}\right\} \qquad 2.2.20$$

If $x_s \nearrow$, $Z \searrow$ and the only limitation is due to the constraints $x_i \geqslant 0$, i = 1 to m. However, if $\bar{a}_{is} \leqslant 0$, i = 1 to m, x_s can $\rightarrow$ indefinitely. Hence the theorem

Theorem 2.4

(Unbounded solution). If in the canonical system, for $\bar{C}_s < 0$, all the $\bar{a}_{is} \leqslant 0$, then a class of feasible solutions can be constructed for which the set of values of Z has no lower bound.

This class of solutions is :

$$x_i = \bar{b}_i - \bar{a}_{is} x_s \qquad i = 1, \ldots m$$

$$x_s \text{ arbitrary and } x_j = 0 \qquad j = m+1 \text{ à } n, \; j \neq s.$$

However, if at least one $\bar{a}_{is} > 0$, x_s is limited by $x_i \geqslant 0$ and its limiting value is :

$$x_s = \frac{\bar{b}_i}{\bar{a}_{is}} \qquad \bar{a}_{is} > 0 \tag{2.2.21}$$

The greatest permitted value of x_s will thus be

$$\boxed{x_s^* = \frac{\bar{b}_r}{\bar{a}_{rs}} = \min_{\bar{a}_{is} > 0} \frac{\bar{b}_i}{\bar{a}_{is}}} \tag{2.2.22}$$

The variable x_r should no longer be in the basis and it is to be replaced by x_s. This could be done using the term $\bar{a}_{rs}\, x_s$ as the pivot.

The previous operations thus enable us to locate the pivot, i.e.

- from equation 2.2.19 we know that the pivot is in the column s

and - from 2.2.22, we know that it is in row r.

Degeneration

If x_s given by 2.2.22 is zero which implies that $\bar{b}_i = 0$, then although we could still use the pivot equation, Z is unchanged. This could cause cycling. There are procedures which enable us to avoid this case. However, since this case is so rare in practice, we will not treat it here. Suffice it to say that the efficiency of the simplex method may be slightly reduced because of this since it may be necessary to perform a few additional iterations.

The iterative procedure

We have seen that the above procedure enables us to go from one feasible solution to another in which the cost Z is lesser (or equal in the degenerate case). This procedure can be repeated until the optimality condition is satisfied or until we obtain an unbounded solution. The number of steps in the iterative procedure is finite as we show next.

Theorem 2.5

If we assume that degenerate solutions do not arise at any iteration, then the simplex algorithm will converge to the optimum in a finite number of iterations.

Proof : Since the number of feasible basic solutions (extremal points of $\mathcal{D}$) is finite, the algorithm can only fail if the value of Z is repeated and this case will not arise according to the non-degeneration assumption.

2.2.3 The two phases of the simplex method

In order to start the simplex procedure, it is necessary to have a feasible basic solution. Such solutions are not always easy to find and in certain cases they may not even exist (e.g. when the constraints are inconsistent).

In the first phase of the simplex method, we find a basic feasible solution or supply the information that such a solution does not exist.

In the second phase, we begin with this information and seek the optimal solution or show that the optimal solution is unbounded.

Both the phases use the simplex method described above. Let us next see how phase 1 works.

Phase 1 : This begins with the augmented system (constraints + cost function equation) and we introduce the artificial variables $x_{n+1}, \ldots x_{n+m}$ as follows :

$$\begin{cases} a_{11} x_1 + a_{12} x_2 + \ldots a_{1n} x_n + x_{n+1} = b_1 \\ a_{22} x_1 + a_{22} x_2 + \ldots a_{2n} x_n + x_{n+2} = b_2 \\ \vdots \\ a_{m1} x_1 + a_{m2} x_2 + \ldots a_{mn} x_n + x_{n+m} = b_m \\ C_1 x_1 + C_2 x_2 + \ldots C_n x_n - Z = 0 \end{cases} \quad 2.2.23$$

All the b_i have been made ≥ 0, if necessary by multiplying the equations by (-1).

The system 2.2.23 has the obvious (and feasible : $b_i \geq 0$) basic solution

$$Z = x_1 = x_2 = \ldots x_n = 0, \; x_{n+1} = b_1, \; x_{n+2} = b_2 \ldots x_{n+m} = b_m \quad 2.2.24$$

All solutions of 2.2.23 with $x_{n+i} = 0$, $i = 1$ to m, constitute with x_j, $j = 1$ to n, a solution of the initial system.

One way of finding a feasible basic solution of the initial problem is to start with 2.2.24 and to use the simplex method so as to force the artificial variables to go to zero by for example doing

$$\min W = x_{n+1} + x_{n+2} + \ldots x_{n+m}$$

If $\min W = 0$, then $x_{n+i} = 0$ and vice versa, since $x_{n+i} \geq 0$.

If $\min W > 0$, the original problem has no solution since all the artificial variables can not be made zero.

It is not always necessary to introduce m artificial variables : we could use any of the obvious basic variables of the initial problem.

Ex. : $2x_1 + 3x_2 - x_3 + x_4 = 2$

$-x_1 + 2x_2 + 3x_3 = 1$

x_4 can be used as a basic variable so it is only necessary to introduce x_5.

$$2x_1 + 3x_2 - x_3 + x_4 = 2$$

$$-x_1 + 2x_2 + 3x_3 + x_5 = 1$$

2.2.4 Examples of linear programming

Example 1 : Let us reconsider the problem :

$$\max Z' = x_1 + 3x_2 \longrightarrow \min Z = -x_1 - 3x_2$$

subject to $\begin{cases} -x_1 + x_2 + x_3 = 1 \\ x_1 + x_2 + x_4 = 2 \end{cases}$

$$x_i \geqslant 0 \qquad i = 1 \text{ to } 4$$

The augmented system is

$$\begin{cases} -x_1 + \textcircled{x_2} + x_3 = 1 \\ x_1 + x_2 + x_4 = 2 \\ -x_1 - 3x_2 - Z = 0 \end{cases}$$

Here, phase 1 is unnecessary since we have an obvious basic feasible solution, i.e.

$$x_3 = 1,\ x_4 = 2,\ Z = 0,\ x_1 = x_2 = 0$$

Iteration 1 (phase 2) : $\exists\, C_j < 0 \qquad C_1 = \min C_j < 0 = -3$, x_2 becomes thus a basic variable.

$$x_2^* = \min \left[\frac{b_1}{a_{12}} = 1,\ \frac{b_2}{a_{22}} = 2 \right] = 1$$

x_2 will replace x_3 in the first equation and thus become the pivot term. On using this pivot we obtain :

$$\begin{aligned} -x_1 + x_2 + x_3 &= 1 \\ \textcircled{2x_1} - x_3 + x_4 &= 1 \\ -4x_1 + 3x_3 - Z &= 3 \end{aligned}$$

Iteration 2 : $\overline{C}_1 = -4$, x_1 becomes a basic variable. The only relationship $\frac{b_i}{a_{i1}}$ with $a_{ij} > 0$ is for $i = 2$, hence the encircled point ; x_4 thus leaves the basis and using a new pivot transformation, we obtain :

$$\begin{aligned} x_2 + \frac{1}{2} x_3 + \frac{1}{2} x_4 &= \frac{3}{2} \\ x_1 - \frac{1}{2} x_3 + \frac{1}{2} x_4 &= \frac{1}{2} \\ Z + x_3 + \frac{1}{2} x_4 &= 5 \end{aligned}$$

hence we obtain the optimal solution : $x_1 = \frac{1}{2}$, $x_2 = \frac{3}{2}$, $x_3 = x_4 = 0$, $Z = -5$.

<u>Example 2</u> : (with phase 1 to find a basic feasible solution)

Consider the augmented system :

$$\begin{aligned} 5x_1 - 4x_2 + 13\,x_3 - 2x_4 + x_5 &= 20 \\ x_1 - x_2 + 5\,x_3 - x_4 + x_5 &= 8 \\ x_1 + 6x_2 - 7\,x_3 + x_4 + 5x_5 - Z &= 0 \end{aligned}$$

$$x_i \geqslant 0$$

Here, a basic feasible solution is not easy to find and we have to use the 2 phases of the simplex method. The terms b_i are already $\geqslant 0$ and we could form the augmented system with the criterion W :

$$
\begin{array}{l}
5x_1 - 4x_2 + 13\,x_3 - 2x_4 + x_5 + x_6 = 20 \\
x_1 - x_2 + 5\,x_3 - x_4 + x_5 + x_7 = 8 \\
x_1 + 6x_2 - 7\,x_3 + x_4 + 5x_5 - Z = 0 \\
x_6 + x_7 - W = 0
\end{array}
$$

In the calculations which follow, the pivots are encircled : a little circle (full : ● or empty : o) under a column designates a basis variable. The variable which enters the basis is marked ★ , and the one which leaves the basis is marked o.

The steps in the optimisation of W are identical to those of Z. In the 1st iteration, W went from 28 to $\frac{4}{3}$. In the 2nd iteration, W evolved from $\frac{4}{13}$ to zero thus yielding the feasible basic solution ($x_3 = 3/2$, $x_5 = 1/2$, $Z = -8$), and enabling us to eliminate x_6, x_7 which along with W are now unnecessary. The minimisation of Z begins now and is done in one iteration.

<u>Iteration 1</u> (phase 1)

$$
\begin{array}{l}
5x_1 - 4x_2 + 13x_3 - 2x_4 + x_5 + x_6 = 20 \\
x_1 - x_2 + 5x_3 - x_4 + x_5 + x_7 = 8 \\
x_1 + 6x_2 - 7x_3 + x_4 + 5x_5 - Z = 0 \\
-6x_1 + 5x_2 - 18x_3 + 3x_4 - 2x_5 - W = -28
\end{array}
$$

★ (under x_3)   o (under x_5)   ● ● ●

<u>Iteration 2</u> (phase 1)

$$
\begin{array}{l}
\frac{5}{13}x_1 - \frac{4}{13}x_2 + x_3 - \frac{2}{13}x_4 + \boxed{\frac{1}{13}x_5} + \frac{1}{13}x_6 = \frac{20}{13} \\
\frac{12}{13}x_1 + \frac{7}{13}x_2 - \frac{3}{13}x_4 + \frac{8}{13}x_5 - \frac{5}{13}x_6 + x_7 = \frac{4}{13} \\
\frac{48}{13}x_1 + \frac{50}{13}x_2 - \frac{1}{13}x_4 + \frac{72}{13}x_5 + \frac{7}{13}x_6 - Z = \frac{140}{13} \\
\frac{12}{13}x_1 - \frac{7}{13}x_2 + \;+ \frac{3}{13}x_4 - \frac{8}{13}x_5 + \frac{18}{13}x_6 - W = -\frac{4}{13}
\end{array}
$$

● (under x_3)   ★ (under x_5)   o ● ●

<u>Iteration 3</u> (phase 1 and 2)

$$
\begin{array}{l}
\frac{1}{2}x_1 - \frac{3}{8}x_2 + x_3 - \frac{1}{8}x_4 + \frac{1}{8}x_6 - \frac{1}{8}x_7 = \frac{3}{2} \\
-\frac{12}{8}x_1 + \boxed{\frac{7}{8}x_2} - \frac{3}{8}x_4 + x_5 - \frac{5}{8}x_6 + \frac{13}{8}x_7 = \frac{4}{8} \\
12\,x_1 - x_2 + 2\,x_4 + 4\,x_6 - 9\,x_7 - Z = 8 \\
x_6 + x_7 - W = 0
\end{array}
$$

Iteration 4 (phase 2 optimum)

$$-\frac{1}{7}x_1 \qquad + x_3 - \frac{2}{7}x_4 + \frac{3}{7}x_5 \qquad = \frac{12}{7}$$

$$-\frac{12}{7}x_1 + x_2 - \frac{3}{7}x_4 + \frac{8}{7}x_5 \qquad = \frac{4}{7}$$

$$\frac{72}{7}x_1 \qquad + \frac{11}{7}x_4 + \frac{8}{7}x_5 - Z = \frac{60}{7}$$

Optimal solution : $x_3 = \frac{12}{7}$, $x_2 = \frac{4}{7}$, $x_1 = x_4 = x_5 = 0$, $Z = -\frac{60}{7}$

Next let us consider the revised simplex method.

2.3 THE REVISED SIMPLEX METHOD

In the use of the simplex method in any one iteration, a lot of information about the original problem which is contained in A is not used.

Indeed, only the following information is necessary :

1. Relative cost factors $\bar{C}_j$: on using these we find :

$$\bar{C}_s = \min_{\bar{C}_j < 0} \bar{C}_j$$

2. If there is a $\bar{C}_s < 0$, we find the new column :

$$\underline{\bar{P}}_s = (\bar{a}_{1s} \ \dots \ \bar{a}_{ms})^T$$

and the values of the basic variables i.e.

$$\underline{x}_B = (\bar{b}_1 \ \dots \ \bar{b}_m)^T$$

This enables us to compute :

$$\frac{\bar{b}_r}{\bar{a}_{rs}} = \min_{\bar{a}_{is} > 0} \frac{\bar{b}_i}{\bar{a}_{is}}$$

and this defines the new pivot $\bar{a}_{rs}$ x_s.

We remark that doing the above, only one non basic column of the canonical system $\underline{\bar{P}}_s$ is used.

Since it is common in Linear Programming problems to have more columns than rows, a lot of calculation time and memory are badly used if we keep $\underline{\bar{P}}_j$, $j \neq s$.

A more efficient approach may be to generate at each iteration, using the original problem and its information, the $\bar{C}_j$ and $\underline{\bar{P}}_s$.

The revised simplex method does precisely this and it uses the inverse of the basis matrix to generate the desired quantities.

Consider a linear program written in column form i.e.

$$\text{Min } Z = C_1 x_1 + C_2 x_2 + \dots C_n x_n \tag{2.3.1}$$

$$\text{subject to } \underline{P}_1 x_1 + \underline{P}_2 x_2 \dots + \underline{P}_n x_n = \underline{b} \tag{2.3.2}$$

$$x_i \geqslant 0 \qquad i = 1 \text{ to } n \tag{2.3.3}$$

$$\text{where } \underline{P}_j = (a_{1j} \dots a_{mj})^T \tag{2.3.4}$$

$\underline{P}_j$ is the jth column of the matrix of coefficients A. We assume as before that the rank of A = m.

$$\text{Let } B = \left[P_{j_1} \dots P_{j_m} \right] \tag{2.3.5}$$

be a basis matrix (we take m columns to form B, denoted by j_1 to j_m) and let

$$\underline{x}_B = (x_{j_1} \dots x_{j_m})^T \tag{2.3.6}$$

$$\underline{C}_B = (C_{j_1} \dots C_{j_m}) \qquad \text{(row vector)} \tag{2.3.7}$$

be the vectors corresponding to the basic variables and the corresponding relative cost factors.

$\underline{x}_B$, which we assume to be feasible ($\underline{x}_B \geqslant \underline{0}$), is given by :

$$\underline{x}_B = B^{-1} \underline{b} = \overline{\underline{b}} \tag{2.3.8}$$

As we have seen, it is useful to consider the criterion as the $(m+1)^{th}$ equation with -Z as a permanent basis variable. Thus the augmented system yields the following columns :

$$\hat{\underline{P}}_j = (a_{1j}, \dots a_{mj}, C_j)^T \qquad j = 1, \dots n \tag{2.3.9}$$

$$\hat{\underline{P}}_{n+1} = (0, \dots 0, 1)^T \tag{2.3.10}$$

$$\hat{\underline{b}} = (b_1 \dots b_m, 0)^T \tag{2.3.11}$$

and could be written as

$$\sum_{j=1}^{n} \hat{\underline{P}}_j x_j + \hat{\underline{P}}_{n+1} (-Z) = \hat{\underline{b}} \tag{2.3.12}$$

Since B is a basis matrix, the (m+1)·(m+1) matrix :

$$\hat{B} = \left[\hat{P}_{j_1} \dots \hat{P}_{j_m} \ \hat{P}_{n+1} \right] = \left[\begin{array}{c|c} B & 0 \\ \hline \underline{C}_B & 1 \end{array} \right] \tag{2.3.13}$$

is also a basis matrix for the augmented system 2.3.12. Indeed, we can easily show that

$$\hat{B}^{-1} = \left[\begin{array}{c|c} B^{-1} & 0 \\ \hline -\underline{C}_B B^{-1} & 1 \end{array} \right] \tag{2.3.14}$$

exists if B^{-1} exists.

<u>Definition</u> : The row vector $\underline{\pi} = (\pi_1 \ldots \pi_m) = \underline{C}_B . B^{-1}$ 2.3.15

is called the vector of simplex multipliers associated with the basis B.

<u>Interpretation of $\underline{\pi}$</u> : Let us multiply each equation "i" of 2.3.2 by π_i and the criterion equation

$$\sum_{i=1}^{n} C_i x_i - Z = 0 \quad \text{by } -1.$$

Forming the sum of these equations, we find that the coefficient of x_j is

$$\underline{\pi}\,\underline{P}_j - C_j \qquad 2.3.16$$

Putting $\underline{\pi}\,\underline{P}_j - C_j = 0 \;\forall\; x_j$ which are the basic variables, this yields the equation

$$\underline{\pi}\left[P_{j_1} \ldots P_{j_m}\right] - \left[C_{j_1} \ldots C_{j_m}\right] = \underline{0} \qquad 2.3.17$$

or

$$\underline{\pi}\,B - \underline{C}_B = \underline{0} \qquad 2.3.18$$

Here the solution is $\underline{\pi} = \underline{C}_B\, B^{-1}$ which is the simplex multiplier vector. Thus $\underline{\pi}$ could be defined as the row vector which enables us to eliminate the coefficient of the basic variables by multiplying the columns of the original system, and summing these new equations and subtracting the Z equation.

Using 2.3.14, we could write :

$$\hat{B}^{-1} = \left[\begin{array}{c|c} B^{-1} & 0 \\ \hline -\underline{\pi} & 1 \end{array}\right] \qquad 2.3.19$$

On rearranging the coefficients a_{ij} in order to make the B matrix appear in the first m columns, the system could be written in the matrix form as :

$$\left[\begin{array}{c|c|c} B & \{\underline{P}_j\} & 0 \\ \hline \underline{C}_B & \{C_j\} & 1 \end{array}\right] \left[\begin{array}{c} \underline{X} \\ \hline -Z \end{array}\right] = \left[\begin{array}{c} \underline{b} \\ \hline 0 \end{array}\right] \qquad 2.3.20$$

B, $\underline{C}_B$ correspond thus to the basis variables.

Multiplying 2.3.20 by 2.3.19, we have :

$$\left[\begin{array}{c|c} B^{-1} & 0 \\ \hline -\underline{\pi} & 1 \end{array}\right] \left[\begin{array}{c|c|c} B & \{\underline{P}_j\} & 0 \\ \hline \underline{C}_B & \{C_j\} & 1 \end{array}\right] \left[\begin{array}{c} \underline{X} \\ \hline -Z \end{array}\right] = \left[\begin{array}{c|c} B^{-1} & 0 \\ \hline -\underline{\pi} & 1 \end{array}\right] \left[\begin{array}{c} \underline{b} \\ \hline 0 \end{array}\right] \qquad 2.3.21$$

Expanding each term :

$$\left[\begin{array}{c|c|c} B^{-1} + 0 \,.\, \underline{C}_B & B^{-1}\{\underline{P}_j\} + 0 \,.\, \{C_j\} & B^{-1}\, 0 + 0.\, 1 \\ \hline -\underline{\pi}\, B + \underline{C}_B & -\underline{\pi}\{\underline{P}_j\} + 1 \,.\, \{C_j\} & -\underline{\pi} \,.\, 0 + 1 \end{array}\right] \left[\begin{array}{c} \underline{X} \\ \hline -Z \end{array}\right] = \left[\begin{array}{c} B^{-1}\, \underline{b} + 0.\, 0 \\ \hline -\underline{\pi}\, \underline{b} + 1.\; 0 \end{array}\right] \qquad 2.3.22$$

$$\left[\begin{array}{c|c|c} \mathbb{1} & B^{-1}\{P_j\} & 0 \\ \hline -\underline{\pi}B + \underline{C}_B & -\underline{\pi}\{P_j\} + C_j & 1 \end{array}\right] \left[\begin{array}{c} \underline{X} \\ \hline -Z \end{array}\right] = \left[\begin{array}{c} B^{-1}\, \underline{b} \\ \hline -\underline{\pi}\, \underline{b} \end{array}\right] \qquad 2.3.23$$

But $-\underline{\pi} B + \underline{C}_B = \underline{0}$ for the basic variables. Hence we arrive at the canonical system

$$\begin{matrix} x_{j_1} & & \\ & \ddots & \\ & & x_{j_m} \end{matrix} + \left[\sum_{\substack{j \text{ non} \\ \text{basic}}} \underline{\bar{P}}_j x_j \right] = \begin{bmatrix} \bar{b}_1 \\ \vdots \\ \bar{b}_m \end{bmatrix} \qquad 2.3.24$$

$$-Z + \sum_{\substack{j \text{ non} \\ \text{basic}}} \bar{C}_j x_j = -Z_o$$

We have thus by identification :

$$\left.\begin{aligned} \underline{\bar{b}} &= B^{-1} \underline{b} \\ Z_o &= \underline{\pi}\, \underline{b} \end{aligned}\right\} \qquad 2.3.25$$

$$\underline{\bar{P}}_j = B^{-1} \underline{P}_j \qquad 2.3.26$$

$$\bar{C}_j = C_j - \underline{\pi}\, \underline{P}_j \qquad 2.3.27$$

These last two formulae are fundamental. They show how, given $\hat{B}^{-1}$ (or equivalently B^{-1} and $\underline{\pi}$), the quantities necessary for one iteration of simplex ($\bar{C}_j$ and $\underline{\bar{P}}_j$) could be calculated using the initial information C_j and $\underline{P}_j$.

Indeed, assume that $\bar{C}_j$ is calculated using 2.3.27 and let $\bar{C}_s = \min_{\bar{C}_j < 0} \bar{C}_j$

Equation 2.3.26 enables us to compute $\underline{\bar{P}}_s$ and 2.3.25 gives $\underline{\bar{b}}$. These quantities are used in :

$$\frac{\bar{b}_r}{\bar{a}_{rs}} = \min_{\bar{a}_{is} > 0} \frac{\bar{b}_i}{\bar{a}_{is}}$$

It remains thus to introduce $\underline{\bar{P}}_s$ in the basis and to take out $\bar{P}_{j_r}$, i.e. to generate the new basis matrix.

For this purpose, let us consider the following partitioned matrix

$$\left[\begin{array}{ccc|c|c} \hat{P}_{j_1} \; \dots \; \hat{P}_{j_m} & & \hat{P}_{n+1} & \underline{0} & \begin{matrix} a_{1s} \\ a_{2s} \\ \vdots \\ a_{m_s} \\ C_s \end{matrix} \end{array} \right] \qquad 2.3.28$$

Multiplying 2.3.28 by $\hat{B}^{-1}$ we obtain :

$$\left[\begin{array}{c|c|c} \Lambda & \hat{B}^{-1} & \begin{array}{c} \bar{a}_{1s} \\ \bar{a}_{2s} \\ \vdots \\ \bar{a}_{ms} \\ \bar{C}_s \end{array} \end{array}\right] \qquad 2.3.29$$

Doing a pivot operation on $\bar{a}_{rs}$ for the matrix 2.3.29, we obtain :

$$\left[\underline{u}_i \cdots \underline{u}_{r-1} \underline{\alpha}\ \underline{u}_{r+1} \cdots \underline{u}_{m+1} \ \middle|\ Q \ \middle|\ \underline{u}_r \right] \qquad 2.3.30$$

Here, $\underline{u}_i$ is a zero vector with one in position i only.

Next we show that $Q = \hat{B}^{-1}_{new}$. Indeed, the pivot operation of 2.3.29 is equivalent to multiplying 2.3.28 by $\hat{B}^{-1}_{new}$ directly which gives :

$$\left[\hat{B}^{-1}_{new} \quad \hat{B} \ \middle|\ \hat{B}^{-1}_{new} \ \middle|\ \underline{u}_r \right]$$

$$\text{Then } Q = \hat{B}^{-1}_{new}$$

$\hat{B}^{-1}_{new}$ is thus obtained by a pivot operation on $\bar{a}_{rs}$, for the matrix $\left[\begin{array}{c|c} \hat{B}^{-1} & \begin{array}{c} \bar{a}_{1s} \\ \vdots \\ \bar{a}_{rs} \\ \vdots \\ \bar{a}_{ms} \\ \bar{C}_s \end{array} \end{array}\right]$ and on keeping the first (m+1) columns.

<u>Summary of the Revised simplex method</u> : Let A, $\underline{b}$, $\underline{c}$, $\hat{B}^{-1}$ be the initial information

1. The row (m+1) of $\hat{B}^{-1} = (-\pi_1 \ldots -\pi_m, 1)$ gives the vector of simplex multipliers. For each non-basic variables, we could thus form the relative cost factors :

$$\bar{C}_j = C_j - \underline{\pi}\,\underline{P}_j = C_j - \sum_{i=1}^{m} \pi_i\, a_{ij}$$

2. Using the rule on column selection, we find that

$$\bar{C}_s = \min_{\bar{C}_j < 0} \bar{C}_j$$

3. If $\bar{C}_s > 0$, stop : the solution is optimal with the optimum being given by 2.3.25.

4. If $\bar{C}_s < 0$, calculate the transformed column :

$$\begin{bmatrix} \underline{\overline{P}}_s \\ \\ \overline{C}_s \\ \end{bmatrix} = \hat{B}^{-1} \begin{bmatrix} \underline{P}_s \\ \\ C_s \\ \end{bmatrix} = \begin{bmatrix} \overline{a}_{1s} \\ \vdots \\ \overline{a}_{ms} \\ \overline{C}_s \end{bmatrix}$$

5. If all the $\overline{a}_{is} \leqslant 0$: stop ; the optimum is unbounded.

6. If $\exists\ \overline{a}_{is} > 0$, calculate $\underline{\overline{b}}$ using 2.3.25 and use the rule

$$\frac{\overline{b}_r}{\overline{a}_{rs}} = \min_{\overline{a}_{is} > 0} \frac{\overline{b}_i}{\overline{a}_{is}} = \theta$$

7. Construct the matrix $\begin{bmatrix} \hat{B}^{-1} & \vline & \underline{\overline{P}}_s \\ & \vline & \overline{C}_s \end{bmatrix}$, transform it by a pivot operation on $\overline{a}_{rs}$.

The first (m+1) columns give the inverse of the new basic matrix.

8. Write the basis solution : $(x_B)_i = (x_B) - \theta\,\overline{a}_{is} \qquad i \neq r$

$(x_B)_r = \theta$ and go back to step 1.

Example

Let us reconsider the example : $\min Z = -x_1 - 3x_2$

subject to
$$\begin{cases} -x_1 + x_2 + x_3 = 1 \\ x_1 + x_2 + x_4 = 2 \\ x_i \geqslant 0 \quad i = 1 \text{ to } 4 \end{cases}$$

The initial information is :

$$A = \begin{bmatrix} -1 & 1 & 1 & 0 \\ 1 & 1 & 0 & 1 \end{bmatrix} \qquad C = \begin{bmatrix} -1 & -3 \end{bmatrix} \qquad b = \begin{bmatrix} 1 \\ 2 \end{bmatrix}$$

An initial basic solution is : $x_3 = 1,\ x_4 = 2$ (with $B = \begin{bmatrix} 1 & 0 \\ 0 & 1 \end{bmatrix}$)

For this solution, $C_i < 0$, $i = 1, 2$, thus it is not optimal.

Iteration 1 : We calculate

$$\theta = \min_{\overline{a}_{is} < 0} \frac{\overline{b}_i}{\overline{a}_{is}} = \min\left[\ 1/1,\ 2/1\right] = 1 \qquad \text{for } \overline{C}_s = \overline{C}_2 = -3$$

We form : $\begin{bmatrix} \hat{B}^{-1} & \vline & \underline{\overline{P}}_s \\ & \vline & \overline{C}_s \end{bmatrix} = \left[\begin{array}{cc|c|c} 1 & 0 & 0 & \textcircled{1} \\ 0 & 1 & 0 & 1 \\ \hline 0 & 0 & 1 & -3 \end{array}\right]$ on which we perform a pivot transformation for $\bar{a}_{rs} = \bar{a}_{12}$.

This gives : $\begin{bmatrix} 1 & 0 & 0 & 1 \\ -1 & 1 & 0 & 0 \\ 3 & 0 & 1 & 0 \end{bmatrix}$ and the inverse of the new augmented basic matrix :

$$\hat{B}^{-1} = \left[\begin{array}{cc|c} 1 & 0 & 0 \\ -1 & 1 & 0 \\ \hline 3 & 0 & 1 \end{array}\right]$$

Thus $\underline{\pi} = \begin{bmatrix} -3 & 0 \end{bmatrix}$

The new basic solution is : $\begin{cases} x_2 = 1 \\ x_4 = 2 - \theta = 1. \end{cases}$ $\left(\begin{array}{l} (x_B)_r = \theta \\ (x_B)_i = (xB) - \theta\, \bar{a}_{is} \end{array} \right)$ $i \neq r$

We calculate $\bar{C}_j = C_j - \underline{\pi}\, \underline{P}_j$

Here $\bar{C}_1 = C_1 - \underline{\pi}\, \underline{P}_1 = -1 - \begin{bmatrix} -3 & 0 \end{bmatrix} \begin{bmatrix} -1 \\ 1 \end{bmatrix} = -4$

$$\bar{C}_3 = C_3 - \underline{\pi}\, \underline{P}_3 = - \quad \begin{bmatrix} -3 & 0 \end{bmatrix} \begin{bmatrix} 1 \\ 0 \end{bmatrix} = +3$$

Since $\bar{C}_1 = -4 < 0$, we have still not arrived at the optimum.

<u>Iteration 2</u> :

Calculate : $\underline{\bar{P}}_1 = B^{-1}\, \underline{P}_1 = \begin{bmatrix} 1 & 0 \\ -1 & 1 \end{bmatrix} \begin{bmatrix} -1 \\ 1 \end{bmatrix} = \begin{bmatrix} -1 \\ 2 \end{bmatrix}$, $\underline{\bar{b}} = B^{-1} \underline{b} = \begin{bmatrix} 1 & 0 \\ -1 & 1 \end{bmatrix} \begin{bmatrix} 1 \\ 2 \end{bmatrix} = \begin{bmatrix} 1 \\ 1 \end{bmatrix}$

Thus $\theta = 1/2$ and $\bar{a}_{rs} = \bar{a}_{21}$

We form :

$$\left[\begin{array}{c|c} \hat{B}^{-1} & \underline{\bar{P}}_s \\ & \bar{C}_s \end{array}\right] = \left[\begin{array}{ccc|c} 1 & 0 & 0 & -1 \\ -1 & 1 & 0 & \textcircled{2} \\ 3 & 0 & 1 & -4 \end{array}\right]$$

and by a pivot transformation on $\bar{a}_{21}$, we obtain : $\left[\begin{array}{ccc|c} 1/2 & 1/2 & 0 & 0 \\ -1/2 & 1/2 & 0 & 1 \\ 1 & 2 & 1 & 0 \end{array}\right]$ Thus $\hat{B}_i^{-1} = \left[\begin{array}{cc|c} 1/2 & 1/2 & 0 \\ -1/2 & 1/2 & 0 \\ \hline 1 & 2 & 1 \end{array}\right]$

The new basis variables are : $x_1 = 1/2 \qquad = 1/2$

$x_2 = 1 + 1/2 = 3/2$

The new relative cost factors for this solution are :

$$\bar{C}_3 = C_3 - \underline{\pi} \cdot \underline{P}_3 = 0 - \begin{bmatrix} -1 & -2 \end{bmatrix} \begin{bmatrix} 1 \\ 0 \end{bmatrix} = 1 > 0$$

$$\bar{C}_4 = C_4 - \underline{\pi} \cdot \underline{P}_4 = 0 - \begin{bmatrix} -1 & -2 \end{bmatrix} \begin{bmatrix} 0 \\ 1 \end{bmatrix} = 2 > 0$$

Thus the solution is optimal and the value of the cost function is :

$$z^{*} = \underline{\pi}\, \underline{b} = \begin{bmatrix} -1 & -2 \end{bmatrix} \begin{bmatrix} 1 \\ 2 \end{bmatrix} = -5$$

2.4 DUALITY IN LINEAR PROGRAMMING

To all Linear programming problems called the "primal" problem, there corresponds another Linear programming problem called the "dual". These two problems are related to each other. The relationship between these 2 problems is given by the following table :

Primal	Dual
$\min Z = \underline{C}^T \underline{x}$	$\max v = \underline{\pi}\, \underline{b}$
subject to $\left[\begin{array}{l} A\underline{x} \left\{\begin{array}{c} = \\ \geqslant \end{array}\right\} \underline{b} \\ x_i \geqslant 0 \end{array}\right.$	subject to $\left[\begin{array}{l} A^T \underline{\pi}^T \left\{\begin{array}{c} = \\ \leqslant \end{array}\right\} \underline{C} \\ \pi_i \geqslant 0 \end{array}\right.$
	($\underline{\pi}$: row vector)

We have thus :

Cost function $\underline{C}^T \underline{x} \rightarrow$ min $\Longrightarrow$	cost function $\underline{\pi}\,\underline{b} \longrightarrow$ max
variables $x_i \geqslant 0$	constraints $\underline{\pi}\,\underline{P}_j \leqslant C_j$ (inequality)
$A_i\underline{x} = b_i$ (equality)	π_i unconstrained
$A_i x \geqslant b_i$ (inequality)	$\pi_i \geqslant 0$
matrix A	matrix A^T
$\underline{b}$	$\underline{C}$
$\underline{C}$	$\underline{b}$

We thus see that the dual of the dual is the primal.

Theorem 2.6 (without proof)

If $\bar{\underline{x}}$ and $\bar{\underline{\pi}}$ are feasible solutions of the primal and dual problems, then :

$$\bar{Z} = \underline{C}^T \bar{\underline{x}} \geqslant \bar{\underline{\pi}}\,\underline{b} = \bar{v}$$

Theorem 2.7 (without proof)

At the optimum min Z = max v.

We can give the following geometric interpretation of these two theorems (Fig. 2.3)

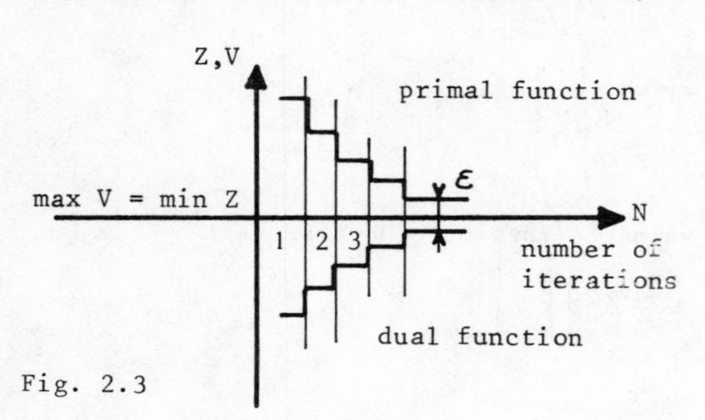

Fig. 2.3

The cost function of the primal and dual approach the same value from opposite directions.

Thus the number Z - v could be used as a stopping criterion in an optimisation procedure.

We can also show that the dual constraints are merely the primal optimality conditions and vice versa.

Also, solving the primal yields the $\underline{\pi}$ vector and thus the solution of the dual (and vice versa).

Thus, given a linear programming problem, we can solve it either via the primal or via the dual.

Example :

$$\min(C_1 x_1 + C_2 x_2)$$

subject to

$$\left| \begin{array}{l} a_{11} x_1 + a_{12} x_2 \geqslant b_1 \\ \vdots \\ a_{201} x_1 + a_{202} x_2 \geqslant b_{20} \end{array} \right.$$

Here 20 slack variables are necessary and this gives us a problem with 22 variables and 20 equations, and requires us to handle a 20 x 20 basis matrix.

On the other hand, the dual can be written as :

$$\max(\pi_1 b_1 + \ldots + \pi_{20} b_{20})$$

subject to

$$\left[\begin{array}{l} a_{11} \pi_1 + \ldots + a_{201} \pi_{20} \leqslant C_1 \\ a_{12} \pi_2 + \ldots + a_{202} \quad {}_{20} \leqslant C_2 \end{array} \right.$$

and the basis matrix will be 2 x 2.

Thus, for an n variable problem with m constraint equations, it would be desirable to solve the dual if $n < m$ and the primal if $m < n$.

Finally in this chapter, we describe the Dantzig-Wolfe decomposition algorithm.

2.5 THE DANTZIG-WOLFE DECOMPOSITION ALGORITHM

Although the simplex algorithm is an efficient one, we run into computational difficulties if the order of the system is high. To avoid these difficulties in certain cases, we could use whatever structural information that there exists about the problem. This is the basis of decomposition and partitioning methods. Here we will only give a brief description of the Dantzig-Wolfe method.

The Dantzig-Wolfe method was the first of the decomposition techniques and it has certainly influenced most of the literature. However, the difference between it and the methods in chapters 1 and 4 is that here we find the global solution by doing a linear combination of local solutions whilst in chapters 1 and 4, the local solutions are coordinated in a different way.

The method applies to linear problems of the form :

$$\left. \begin{array}{l} \min \left[\underline{C}_1 \underline{x}_1 + \ldots \underline{C}_p \underline{x}_p \right] \\ \text{subject to} \left[\begin{array}{l} A_1 \underline{x}_1 + \ldots + A_p \underline{x}_p = \underline{a}_1 \\ B_1 \underline{x}_1 \qquad\qquad\quad = \underline{b}_1 \end{array} \right. \end{array} \right\} \qquad 2.5.1$$

$$\left|\begin{array}{l} B_2\underline{x}_2 \quad\quad = \underline{b}_2 \\ \quad\ddots \quad\quad \vdots \\ \quad\quad B_p\,\underline{x}_p = \underline{b}_p \\ \quad\quad\quad \underline{x} \geqslant \underline{0} \end{array}\right\} \qquad 2.5.1$$

Let L be the Lagrangian associated with problem 2.5.1 (cf. Chapter 3) :

$$L = \underline{C}_1^T\,\underline{x}_1 + \ldots + \underline{C}_p^{\,T}\,\underline{x}_p + \underline{\pi}_1^{\,T}\left[\,\underline{a}_1 - A_1\underline{x}_1 - A_2\underline{x}_2 \ldots A_p\underline{x}_p\,\right]$$
$$+ \underline{\alpha}_1^{\,T}\left[\,\underline{b}_1 - B_1\underline{x}_1\,\right] + \ldots + \underline{\alpha}_p^{\,T}\left[\,\underline{b}_p - B_p\,\underline{x}_p\,\right] \qquad 2.5.2$$

$\underline{\alpha}_i$, i = 1 to P, $\underline{\pi}_1$ being the Lagrange multipliers (cf. Chapter 3 for the definition of Lagrange multipliers).

We note that this Lagrangian becomes additively separable if we fix $\underline{\pi}_1$.

$$L = \sum_i L_i = \sum_i \left\{ \underline{C}_i^{\,T}\underline{x}_i - \underline{\pi}_i^{\,T} A_i\,\underline{x}_i + \underline{\alpha}_i^{\,T}\left[\,\underline{b}_i - B_i\underline{x}_i\,\right]\right\} \qquad 2.5.3$$

For a given $\underline{\pi}_1$, $\underline{\pi}_1^{\,T}\,\underline{a}_1$ is constant and can be dropped.

Each local Lagrangian yields the sub-problem :

$$\text{for given } \underline{\pi}_1 \qquad \left.\begin{array}{l} \min\ (\underline{C}_i^{\,T} - \underline{\pi}_1^{\,T} A_i\,)\,\underline{x}_i \\ \text{subject to} \begin{cases} B_i\,\underline{x}_i = \underline{b}_i \\ \underline{x}_i \geqslant \underline{0} \end{cases} \end{array}\right\} \qquad 2.5.4$$

It remains to find out how we can determine $\underline{\pi}_1$ using local solutions such that we could solve 2.5.1.

To do this, let us start with the following simplified structure (p=1) which could in fact represent all Linear programs.

$$\min\ \underline{C}^T\,\underline{x} \qquad 2.5.5$$

$$\text{subject to} \left[\begin{array}{ll} A_1\,\underline{x} = \underline{b}_1 & m_1 \text{ constraints} \qquad 2.5.6 \\ A_2\,\underline{x} = \underline{b}_2 & m_2 \text{ constraints} \qquad 2.5.7 \end{array}\right.$$
$$\underline{x} \geqslant \underline{0}$$

We can then state the following theorem :

"Let $X = \{\underline{x}\ /\ A\,\underline{x} = \underline{b},\ \underline{x} \geqslant \underline{0}\}$ be a bounded non-empty set, and $\underline{x}^i$ the extremal points of this set (i.e. feasible basic solutions). Then all elements $\underline{x} \in X$ could be written in the form :

$$\underline{x} = \sum_{i=1}^{r} \lambda_i\underline{x}^i \qquad \lambda_i \geqslant 0 \qquad \sum_{i=1}^{r} \lambda_i = 1 \qquad 2.5.8$$

i = 1 to r, where r is the finite number of extremal points".

Let $S_2 = \left\{ \underline{x} \;/\; A_2\, \underline{x} = \underline{b}_2,\ \underline{x} \geqslant \underline{0} \right\}$. Then from the theorem, for all $\underline{x} \in S_2$:

$$\underline{x} = \sum_j \lambda_j\, \underline{x}^j \qquad \sum_j \lambda_j = 1 \quad \lambda_j \geqslant 0 \tag{2.5.9}$$

where $\underline{x}^j$ is a feasible basic solution.

The problem 2.5.5 to 2.5.7 can thus be rewritten as :

Choose amongst the solutions of 2.5.7 those which satisfy 2.5.7 and which also minimise 2.5.5.

To satisfy 2.5.6, taking into account 2.5.9, we must have :

$$A_1 \sum_j \lambda_j\, \underline{x}^j = \underline{b}_1 \implies \sum_j (A_1\, \underline{x}^j).\, \lambda_j = \underline{b}_1 \tag{2.5.10}$$

For min $\underline{C}^T\, \underline{x}$, we must min $\underline{C}^T \sum_j \lambda_j\, \underline{x}^j = \sum_j (C^T\, \underline{x}^j)\, \lambda_j$

Putting $A_1\, \underline{x}^j = \underline{p}^j \implies \sum_j \underline{p}_j\, \lambda_j = \underline{b}_1$

$$\underline{C}^T\, \underline{x}^j = f_j \implies \underline{C}^T\, \underline{x} = \sum_j f_j\, \lambda_j$$

The problem given by equations 2.5.5 to 2.5.7 is equivalent to

$$\left.\begin{aligned} &\min \sum_j f_j\, \lambda_j \\ &\text{subject to} \quad \sum_j \underline{p}_j\, \lambda_j = \underline{b}_1 \\ &\qquad\qquad\quad\ \sum_j \lambda_j = 1 \\ &\qquad\qquad\quad\ \lambda_j \geqslant 0 \end{aligned}\right\} \tag{2.5.11}$$

and this is called the "Master Program".

This program has only (m_1+1) rows instead of (m_1+m_2) for the global problem 2.5.5 to 2.5.7. It has as many columns as there are extremal points in the region S_2, and since this number could be high, we use the revised simplex method to solve the problem.

The cost factors here can be generated using

$$\overline{f}_j = f_j - \underline{\pi} \begin{bmatrix} \underline{p}_j \\ 1 \end{bmatrix} \qquad (\underline{\pi} : \text{simplex multipliers in row form}) \tag{2.5.12}$$

Let us partition $\underline{\pi}$ into $\underline{\pi} = [\, \underline{\pi}_1 \,\vdots\, \pi_0 \,]$

$$\text{Then } \overline{f}_j = f_j - [\, \underline{\pi}_1 \,\vdots\, \pi_0 \,] \begin{bmatrix} \underline{p}_j \\ 1 \end{bmatrix} = f_j - \underline{\pi}_1 \underline{p}_j - \pi_0 \tag{2.5.13}$$

But from the definition of $\overline{f}_j$ and $\overline{p}_j$

$$\overline{f}_j = (C - \underline{\pi}_1\, A_1)\underline{x}^j - \pi_0 \tag{2.5.14}$$

To locate the column of the new pivot we do :

$$\min_j \bar{f}_j = \bar{f}_S = (C - \underline{\pi}_1 A_1)\underline{x}^S - \pi_0 \qquad 2.5.15$$

which will enable λ_S to enter into the basis of the master program.

We thus have to minimise $\bar{f}_j$ with respect to $\underline{x}^j$. Since the solution of a linear program in a bounded region lies necessarily at an extremal point (in order to satisfy the constraints) 2.5.15 is equivalent to

$$\left.\begin{array}{l} \min \left[\underline{C}^T - \underline{\pi}_1 A_1 \right] \underline{x} \\ \text{subject to} \left[\begin{array}{l} A_2 \underline{x} = \underline{b}_2 \\ \underline{x} \geq \underline{0} \end{array} \right. \end{array} \right\} \qquad 2.5.16$$

Let x^S be the solution of this problem (2.5.15 or 2.5.16). The column $\underline{p}_S = \begin{bmatrix} A_1 \underline{x}^S \\ 1 \end{bmatrix}$ will enter into the basis.

This method becomes particularly useful if $p > 1$ (formulation 2.5.1). In this case we find 2.5.4 as the problem 2.5.16.

We have thus the following algorithm (Fig. 2.4) :

We assume that we have a feasible basic solution for the master program, with the associated basis matrix B and the simplex multipliers $(\underline{\pi}_1 \mid \pi_0)$.

1. Using the simplex multipliers, $\underline{\pi} = (\underline{\pi}_1 \mid \pi_0)$, we solve the problem 2.5.4 :

$$\text{for given } \underline{\pi}_1 \left\{ \begin{array}{l} \min (\underline{C}_i^T - \underline{\pi}_1 A_i) \underline{x}_i \\ \text{subject to} \left\{ \begin{array}{l} B_i \underline{x}_i = \underline{b}_i \\ \underline{x}_i \geq \underline{0} \end{array} \right. \end{array} \right\} \qquad 2.5.17$$

This yields the solution :

$$\begin{array}{l} \underline{x}_i^* (\underline{\pi}_1) \\ Z_i^* = (\underline{C}_i^T - \underline{\pi}_1 A_i) \underline{x}_i^* (\underline{\pi}_1) \end{array} \qquad 2.5.18$$

2. Taking into account all the local solutions that have been computed :

$$\min_j \bar{f}_j = \bar{f}_S = \sum_{i=1}^{p} Z_i^* - \pi_0 \qquad 2.5.19$$

If $\bar{f}_S > 0$ we have the solution $\underline{x}^* = \sum \lambda_j \underline{x}^j$

where $\underline{x}^j$ are the extremal points of the set of sub-problems corresponding to the basis variables λ_j of the master program.

3. If $\bar{f}_S < 0$, form the column $\underline{p} = \begin{bmatrix} \sum_i A_i \underline{x}_i (\underline{\pi}_1) \\ 1 \end{bmatrix}$

transform it by multiplication by B^{-1} and continue using the classical revised simplex method at the level of the master program (finding a new pivot, new multipliers and a new basis matrix).

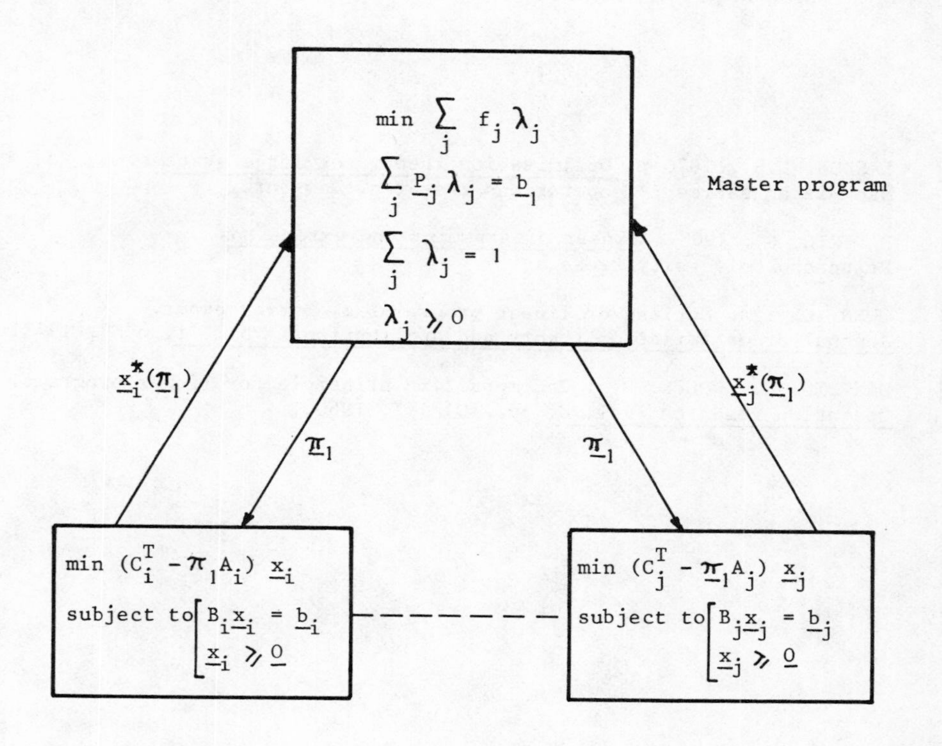

Fig. 2.3

2.6 CONCLUSIONS

Linear programming is certainly the most highly developed area of mathematical programming and it has been so for many years. The reason for this is that this area has had a lot of attention focused on it due to its application potential.

In this chapter, we have described the linear programming algorithms in a hopefully simple and easy to understand way. For a more complete treatment, the reader is referred to the excellent texts listed in the references.

REFERENCES FOR CHAPTER 2

[1] LASDON, L.S., 1970 : "Optimisation theory for large systems. Mac Millan series operations research. New-York.

[2] DANTZIG, G., 1963 : Linear programming and extensions. Princeton University Press.

[3] OWEN, G. : An outline of linear programming. Survey paper. Journal of optimisation theory and applications. Vol. 11, n° 8, 1973.

[4] DANTZIG, G., WOLFE, P. : Decomposition principle for linear programs. Operation research, vol. 8, pp. 101-111, 1960.

PROBLEMS FOR CHAPTER 2

1. Solve the following linear programming problem

$$\max Z = x_1 + x_2$$

$$\text{subject to} \quad \begin{cases} 3x_1 + 2x_2 \leq 6 \\ x_1 + 4x_2 \leq 4 \\ x_1 \geq 0,\ x_2 \geq 0 \end{cases}$$

using the simplex method.

Start by using the obvious basic feasible solution. Verify the result obtained graphically.

2. Solve

$$\min Z = x_2 - 2x_3 + 2x_5$$

$$\text{subject to} \quad \begin{cases} x_1 + 3x_2 - x_3 + \qquad = 7 \\ -2x_2 + 4x_3 + x_4 = 12 \\ -4x_2 + 3x_3 + 8x_5 + x_6 = 10 \\ x_j \geq 0 \quad j = 1 \text{ to } 6. \end{cases}$$

3. An enterprise manufactures on a given machine working 45 hours per week, 3 different products P_1, P_2, P_3. The article P_1 yields a profit of \$ 4, P_2 \$ 12 and P_3 \$ 3. The machine is able to produce 50 P_1 per hour, 25 P_2 per hour and 75 P_3 per hour. We know in addition from a market study that it is not possible to sell more than 1000 P_1, 500 P_2 and 500 P_3 per week. Find the optimal distribution of the production capacity between the 3 products so as to maximise the profits of the enterprise.

4. Consider the linear continuous dynamical process

$$\ddot{s}(t) = u(t)$$

and we wish to compute the control u(t) which enables the system to go from the initial state $s(0) = \dot{s}(0) = 0$, to the final state $s(\tau) = a$, $a > 0$, $\dot{s}(\tau) = 0$, by minimising the criterion

$$c = \int_0^{\tau} |u(t)| \, dt$$

This control will be given by a blocking circuit such that :

$$u(t) = u_n = \text{constant if } nT \leq t < (n+1)T$$

T being the sampling period and u_n the control calculated by a computer using $s(nT)$ and $\dot{s}(nT)$.

Under these conditions, the control law should allow the system to go from $s(0) = 0$, $\dot{s}(0) = 0$ to $s(NT) = a$, $\dot{s}(NT) = 0$, by minimising the criterion

$$r = \sum_{n=0}^{N-1} |u_n|$$

1. Write the discrete state equation for the system. To do this, integrate the equation $\ddot{s}(t) = u(t)$ over the interval $T = t_{n+1} - t_n$ which represents a sampling period, for a control $u(t) = u_n$ = constant.

Put $x_n = \begin{bmatrix} \dot{s}(nT) \\ s(nT) \end{bmatrix}$ and show that we obtain the state equation :

$$\underline{x}_{n+1} = A\underline{x}_n + Bu_n$$

in which we need to specify the A and B matrices.

2. Starting from the initial state $\underline{x}_0 = \underline{0}$, calculate the general solution $\underline{x}_n$.

3. Noting that $A^m B = \begin{bmatrix} T \\ (m+1/2)T^2 \end{bmatrix}$ and using the expression for the general solution $\underline{x}_n$ for n = 2, N = 2, write the constraint relations imposed by the terminal conditions.

4. Show that the discrete problem thus obtained could be formulated in terms of the following linear programming problem

$$\min \sum_{i=1}^{6} y_i$$

$$\text{subject to} \quad \begin{bmatrix} 1 & -1 & 1 & -1 & 1 & -1 \\ 5 & -5 & 3 & -3 & 1 & -1 \end{bmatrix} \underline{y} = \begin{bmatrix} 0 \\ 2a/T^2 \end{bmatrix}$$

On putting for the unconstrained variables :

$$\begin{bmatrix} u_0 = y_1 - y_2 \\ u_1 = y_3 - y_4 \\ u_2 = y_5 - y_6 \end{bmatrix} \qquad y_j \geqslant 0 \qquad j = 1 \text{ to } 6$$

Show in particular that
$$\begin{cases} |u_0| = y_1 + y_2 \\ |u_1| = y_3 + y_4 \\ |u_2| = y_5 + y_6 \end{cases}$$

by examining the constraint matrix of the linear problem.

5. Solve this linear problem in order to find the optimal control using the ordinary simplex method. To start, try as the basis variables (y_1, y_3) and then (y_1, y_4).

CHAPTER 3

Non-Linear Programming

Although many non-linear problems can be linearised and solved as Linear Programming problems using the techniques of the previous chapter, this is not always possible or even desirable. In this chapter we study ways of tackling non-linear programming problems directly. We study problems both with and without constraints.

3.1 SEEKING THE MINIMUM FOR UNCONSTRAINED PROBLEMS

3.1.1 First order conditions

Problem : Find $\underline{x} \in R^m$ which is the absolute or relative minimum of $f(\underline{x})$.

Assumption : $f(\underline{x})$ is assumed to be continuous and continuously differentiable upto first order.

For $\underline{x}$ sufficiently small, we could thus do a Taylor series expansion of $f(\underline{x})$ upto first order, i.e. :

$$f(\underline{x} + \delta\underline{x}) \simeq f(\underline{x}) + \underline{f}_x^T(\underline{x})\ \delta\underline{x} + O(\|\delta\underline{x}\|^2) \qquad 3.1.1$$

If $\hat{\underline{x}}$ is a relative minimum of $f(\underline{x})$ then we should have in the neighbourhood of $\hat{\underline{x}}$:

$$f(\hat{\underline{x}} + \delta\underline{x}) \geqslant f(\hat{\underline{x}}) \quad \forall\ \delta\underline{x} \text{ small, } \delta\underline{x} \in R^m \qquad 3.1.2$$

i.e. using 3.1.1

$$\underline{f}_x^T\ \delta\underline{x} \geqslant 0 \quad \forall\ \delta\underline{x} \in R^m$$

(if we neglect the term $O(\|\delta\|^2)$).

This relation must be satisfied $\forall$ small $\delta\underline{x}$ and in particular when we change $\delta\underline{x}$ to $-\delta\underline{x}$. This implies that :

$$\underline{f}_x\ (\hat{\underline{x}}) = \underline{0} \qquad 3.1.3$$

we have then to 1st order in the neighbourhood of $\hat{\underline{x}}$:

$$f(\hat{\underline{x}} + \delta\underline{x}) \simeq f(\hat{\underline{x}}) = \text{constant}$$

and we can thus state :

> The necessary condition for a local extremum of the continuous and continuously differentiable (to 1st order) function $f(\underline{x})$ is that it be stationary, i.e.
>
> $$\underline{f}_x(\hat{\underline{x}}) = \underline{0}$$

3.1.2 Second order conditions

For $\delta \underline{x}$ small, $f(\underline{x})$ continuous and continuously differentiable to 2nd order, we saw (chapter 1) that we could write :

$$f(\underline{x}+\delta\underline{x}) = f(\underline{x}) + \underline{f}_x^T \, \delta\underline{x} + \frac{1}{2}\, \delta\underline{x}^T F_{XX}(\underline{x})\, \delta\underline{x} + O(\|\delta\|^3) \qquad 3.1.4$$

(notation : $O(\|\delta\|^3) = O^3$)

F_{XX} : the square symmetrical second derivatives matrix with the ijth element $\frac{\partial^2 f}{\partial x_i \partial x_j}$ (often called the Hessian).

If $\hat{\underline{x}}$ is a relative minimum, then we have in the neighbourhood of $\hat{\underline{x}}$:

$$f(\hat{\underline{x}} + \delta\underline{x}) \geqslant f(\hat{\underline{x}}) \qquad 3.1.5$$

Taking into account the first order (necessary) condition ($\underline{f}_x = \underline{0}$), we have from 3.1.5, for $\delta\underline{x}$ small :

$$\delta\underline{x}^T F_{XX}(\hat{\underline{x}})\, \delta\underline{x} \geqslant 0 \qquad \forall\, \delta\underline{x} \text{ small} \in R^m \qquad 3.1.6$$

(here we have neglected the 3rd order term : O^3).

This shows that $F_{XX}(\hat{\underline{x}})$ is non-negative definite.

A necessary condition for a relative minimum is thus
$\underline{f}_x(\hat{\underline{x}}) = \underline{0}$, $F_{XX}(\hat{\underline{x}})$ non-negative definite

However, $F_{XX}(\hat{\underline{x}})$ being merely non-negative definite, there could exist $\delta\underline{x} \neq 0$ for which

$$\delta\underline{x}^T F_{XX}(\hat{\underline{x}})\, \delta\underline{x} = 0 \implies f(\hat{\underline{x}} + \delta\underline{x}) = f(\underline{x}) + O^3$$

and we need the 3rd order term to conclude. The above conditions are therefore not sufficient.

However, if we assume that $F_{XX}(\hat{\underline{x}})$ is positive definite and $\underline{f}_x(\hat{\underline{x}}) = \underline{0}$, then $\delta\underline{x}^T F_{XX}(\hat{\underline{x}})\, \delta\underline{x} > 0$ for $\delta\underline{x} \neq 0$, the term of 3rd order being negligible for small $\delta\underline{x}$, $\hat{\underline{x}}$ is a strict local minimum.

$$f(\hat{\underline{x}} + \delta\underline{x}) > f(\hat{\underline{x}})$$

A sufficient condition for a strict relative minimum is thus :
$\underline{f}_x(\hat{\underline{x}}) = \underline{0}$; $F_{XX}(\hat{\underline{x}})$ positive definite.

The special case of convex functions

We have seen that convex functions always satisfy the $F_{XX}(\hat{\underline{x}})$ non-negative definite condition. For such functions, the necessary conditions for a relative minimum reduce to $\underline{f}_x(\hat{\underline{x}}) = \underline{0}$ and we can show there can't be a maximum.

In addition, all local (or relative) minima are necessarily global minima. Hence we have the property :

If $f(\underline{x})$ is a continuous differentiable function for which $\hat{\underline{x}}$ satisfies $\underline{f}_x(\hat{\underline{x}}) = \underline{0}$, $\hat{\underline{x}}$ is the global minimum.

In addition, if $\hat{\underline{x}}$ corresponds to strict minimum, it is the unique absolute minimum.

Application to quadratic functions

Consider the continuous and continuously differentiable (to 2nd order) function in $\underline{x}$:

$$f(\underline{x}) = a + \underline{b}^T \underline{x} + \underline{x}^T C \underline{x}$$

$$\underline{f}_x = \underline{b} + 2 C \underline{x} \qquad C : \text{symmetrical matrix}$$

If $|C| \neq 0 \Longrightarrow \hat{\underline{x}} = -\frac{1}{2} C^{-1} \underline{b}$

If $|C| = 0$ it is necessary to use the pseudo-inverse of C and the solution is not unique.

The second order conditions give :

$$F_{xx} = 2 C$$

If C is non negative definite, $\hat{\underline{x}}$ can be a minimum
If C is positive definite, $\hat{\underline{x}}$ is an absolute strict minimum (since $\hat{\underline{x}}$ is a relative minimum and it is unique).

However, we note that a non-negative definite C requires that $f(\underline{x})$ be convex and we can thus conclude that $\hat{\underline{x}}$ is the absolute minimum.

We conclude thus that for continuous differentiable functions, the convexity assumption allows us to limit ourselves only to examining first order conditions.

3.1.3 Iterative search methods

The previous conditions do not enable us to compute the value of the minimum directly since they are in general non-linear and unsolvable analytically. We thus have to use iterative techniques which starting from an initial value allow us to calculate the sequence $\underline{x}_1, \underline{x}_2, \ldots, \underline{x}_i \ldots \underline{x}_n$ which converges to $\hat{\underline{x}}$.

Two main types of methods are used :

* The direct methods such as the gradient method where, from $\underline{x}_i$, we determine $\underline{x}_{i+1}$ such that $f(\underline{x}_{i+1}) < f(\underline{x}_i)$ and this leads to

$$f(\underline{x}_o) > f(\underline{x}_1) \ldots > f(\underline{x}_i), \ldots > f(\underline{x}_n)$$

This sequence being bounded below by the solution $f(\hat{\underline{x}})$, indeed converges.

* The indirect methods such as Newton's method in which we solve iteratively the implicit equations obtained from the optimality conditions.

3.1.3.1. Gradient method

Let $\underline{x}_i$ be an arbitrary point

$$f(\underline{x}_i + \delta \underline{x}_i) \simeq f(\underline{x}_i) + \underline{f}_x^T(\underline{x}_i) \; \delta \underline{x}_i \quad \text{for } \|\delta \underline{x}\| < \varepsilon$$

It is obvious that for $\delta \underline{x}_i$ of a given length, the minimum of :

$$f(\underline{x}_i + \delta \underline{x}_i) - f(\underline{x}_i) = \Delta f < 0$$

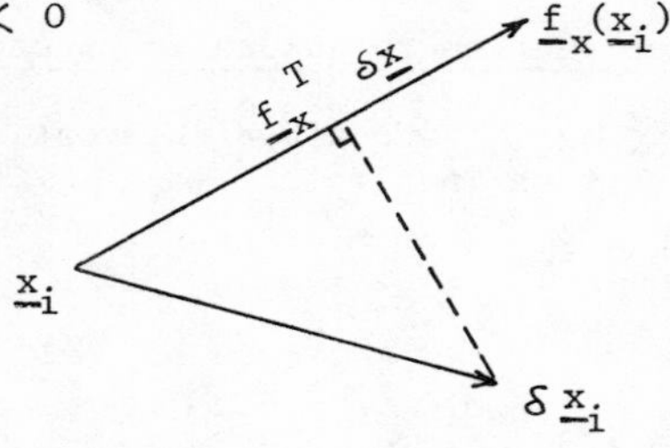

will be obtained for $\delta \underline{x}_i$ colinear with $\underline{f}_x(\underline{x}_i)$

i.e. $\quad \delta \underline{x}_i = \alpha k \underline{f}_x(\underline{x}_i) \qquad \alpha = \pm 1$

Then $\quad \Delta f = \underline{f}_x^T \delta \underline{x}_i = \alpha k \underline{f}_x^T \underline{f}_x < 0$ if

$\alpha = -1$ and $k > 0$

Hence the iterative algorithm ($\underline{x}_{i+1} = \underline{x}_i + \delta \underline{x}_i$)

$$\underline{x}_{i+1} = \underline{x}_i - k \underline{f}_x(\underline{x}_i) \qquad 3.1.7$$

Here k is an iteration constant which is chosen so as to remain in the valid region of the expansion $f(\underline{x}_i + \delta \underline{x}_i)$.

The method is called the Gradient or Steepest descent method since it consists of moving in the gradient or steepest descent direction on the surface $f(\underline{x})$.

3.1.3.2 Newton's method

In this algorithm we apply the method of Newton Raphson in order to solve the stationarity conditions $\underline{f}_x(\underline{x}) = \underline{0}$. We note that the gradient method gives the minimum after convergence, whilst Newton's method merely gives the stationary point. It is then necessary to test the nature of this stationary point.

Recapitulation of the Newton-Raphson method

This algorithm solves the problem :

find $\underline{x}$ such that $\underline{g}(\underline{x}) = \underline{0}$

The algorithm used is :

$$\underline{x}_{i+1} = \underline{x}_i - k G_x^{-1} \underline{g}(\underline{x}_i)$$

where $G_x = \left[\dfrac{\partial g_i}{\partial x_j} \right]$

k can be taken as k = 1 in general. This arises from the convergence properties of the method which we will study later.

Application of the Newton-Raphson method to the solution of $\underline{f}_x(\underline{x}_i) = \underline{0}$

Applying the formula, we obtain directly the recursive relationship :

$$\underline{x}_{i+1} = \underline{x}_i - k F_{XX}^{-1}(\underline{x}_i) \underline{f}_x(\underline{x}_i) \qquad 3.1.8$$

This relation could also have been obtained directly from the Taylor series expansion to second order of $f(\underline{x}_i + \delta \underline{x}_i)$, i.e.

$$f(\underline{x}_i + \delta \underline{x}_i) \simeq f(\underline{x}_i) + \underline{f}_x^T(\underline{x}_i) \delta \underline{x}_i + \frac{1}{2} \delta \underline{x}_i^T F_{XX}(\underline{x}_i) \delta \underline{x}_i$$

The minimum of this quadratic expression in $\underline{x}_i$ is given by differentiating w.r.t. $\underline{x}_i$ and setting to zero, i.e.

$$\underline{f}_x(\underline{x}_i) + F_{XX}(\underline{x}_i)\ \delta\underline{x}_i = \underline{0}$$

so that if $|F_{XX}(\underline{x}_i)| \neq 0$ for

$$\delta\hat{\underline{x}}_i = -F_{XX}^{-1}(\underline{x}_i)\ \underline{f}_x(\underline{x}_i)$$

the relationship

$$\underline{x}_{i+1} = \underline{x}_i + k\ \delta\hat{\underline{x}}_i$$

does indeed give us the formula 3.1.8.

In this case we are not going in the direction of the gradient but rather in the direction of the peak S of the quadratic criterion tangential to $f(\underline{x}_i)$ in $\underline{x}_i$ (we will go to this peak directly if k = 1).

This method will not be efficient unless the quadratic in $\underline{x}_i$ differs only slightly from that in $\hat{\underline{x}}$. Hence the region of convergence of this procedure is quite limited.

To obtain a minimum at each iteration, it is necessary that $F_{XX}(\underline{x}_i)$ be non-negative definite ($\forall\ \underline{x}_i \Longrightarrow f(\underline{x})$ convex) and $F_{XX}(\underline{x}_i)$ be regular and this is achieved if $F_{XX}(\underline{x}_i)$ is positive definite.

Since this method has a limited region of convergence and since it requires a lot of calculation (arising from the matrix inversion) it could be used as the final stage after having used the gradient method since as we will see, the present method converges more quickly than the gradient method.

Next we study the convergence properties of the above two methods.

3.1.3.3. Convergence of the gradient method and Newton's method

General theory

Both the methods yield the following general iterative form :

$$\underline{x}_{i+1} = \underline{x}_i - k\ \underline{\phi}(\underline{x}_i)$$

If they converge to $\hat{\underline{x}}$, $\hat{\underline{x}}$ should satisfy

$$\hat{\underline{x}} = \hat{\underline{x}} - k\ \underline{\phi}(\hat{\underline{x}}) \Longrightarrow \underline{\phi}(\hat{\underline{x}}) = \underline{0}$$

which in the case of both the methods does indeed correspond to the satisfaction of the necessary conditions for a local extremum, i.e.

$$\underline{f}_x(\hat{\underline{x}}) = \underline{0}$$

To prove the convergence of this iterative procedure towards a unique $\hat{\underline{x}}$ (i.e. $\forall\ \underline{x}_o$, $\underline{x}_i \underset{i \Longrightarrow \infty}{\Longrightarrow} \hat{\underline{x}}$) we will use the following contraction mapping theorem :

Theorem : If the transformation $\mathcal{F}(\underline{x})$ is a contraction mapping on $\mathcal{E}$, i.e. there exists an $\alpha < 1$ such that

$$\|\mathcal{F}(\underline{x}_1) - \mathcal{F}(\underline{x}_2)\| < \alpha\ \|\underline{x}_1 - \underline{x}_2\| \quad \forall\ \underline{x}_1, \underline{x}_2 \in \mathcal{E}$$

then the iterative procedure $\underline{x}_{i+1} = \mathcal{F}(\underline{x}_i)$ converges uniformly in $\mathcal{E}$ to the unique solution of $\underline{x} = \mathcal{F}(\underline{x})$.

It is thus necessary in our case to show that $\underline{x} - k\,\underline{\phi}(\underline{x})$ is a contraction mapping but this is virtually impossible to do since $f(\underline{x})$ is often unknown a priori and it could, in that case, only be evaluated for the values obtained after a sequence of tests.

Far from $\hat{\underline{x}}$, the nature of the convergence depends upon the topology of the objective function and we can not do a theoretical study except for the $\underline{x}_i$ which are in the neighbourhood of $\hat{\underline{x}}$; in this neighbourhood, the cost function could be represented by :

$$f(\hat{\underline{x}} + \delta\underline{x}) \simeq f(\hat{\underline{x}}) + \frac{1}{2}\delta\underline{x}^T F_{XX}(\hat{\underline{x}})\,\delta\underline{x} \qquad (\text{since } \underline{f}_x(\hat{\underline{x}}) = \underline{0})$$

and if we put $\underline{x}_i = \hat{\underline{x}} + \delta\underline{x}_i$ we will retain for the convergence study :

$$\left.\begin{aligned} \underline{f}_x(\underline{x}_i) &\simeq F_{XX}(\hat{\underline{x}})\,\delta\underline{x}_i \\ F_{XX}(\underline{x}_i) &\simeq F_{XX}(\hat{\underline{x}}) \end{aligned}\right\} \qquad 3.1.9$$

However, since $\hat{\underline{x}}$ is unknown a priori (being the aim of the search), the convergence study that will be presented aims merely to give some general indications which could guide us in choosing k and we will study the influence of this k on the convergence.

<u>Gradient method</u>

$$\underline{x}_{i+1} = \underline{x}_i - k\,\underline{f}_x(\underline{x}_i) \qquad k > 0$$

Taking into account 3.1.9, we could write :

$$\underline{x}_{i+1} = \hat{\underline{x}} + \delta\underline{x}_{i+1} = \hat{\underline{x}} + \delta\underline{x}_i - k\left[F_{XX}(\hat{\underline{x}})\right]\delta\underline{x}_i$$

or

$$\delta\underline{x}_{i+1} = \left[\mathbb{1} - k\,F_{XX}(\hat{\underline{x}})\right]\delta\underline{x}_i \qquad 3.1.10$$

Here, $\delta\underline{x}_{i+1}$ depends linearly on $\delta\underline{x}_i$. The convergence is said to be linear. There will not be convergence to $\delta\underline{x} = \underline{0}$ unless $\|\delta\underline{x}_{i+1}\| < \|\delta\underline{x}_i\|$ and this requires that the matrix $M = \left[\mathbb{1} - k\,F_{XX}(\hat{x})\right]$ has eigenvalues λ_i which have moduli less than 1.

Let $\lambda_1 < \lambda_2 \ldots < \lambda_m$ be the eigenvalues of $F_{XX}(\hat{\underline{x}})$; we thus have :

$$\Upsilon_i = 1 - k\,\lambda_i$$

F_{XX} being a square symmetrical matrix, it has real eigenvalues. If we assume in addition that $F_{XX}(\hat{\underline{x}})$ is positive definite ($\hat{\underline{x}}$: minimum), then the eigenvalues are also positive.

For convergence we thus have :

$$-1 \underset{(a)}{<} 1 - k\lambda_i \underset{(b)}{<} 1 \qquad 3.1.11$$

The inequality 3.1.11.b gives $k\lambda i > 0 \;\forall\, i = 1$ to m. Hence $k > 0$ which is something that we know already.

The inequality 3.1.11.a gives $2 > k\lambda_i$ or $k < \frac{2}{\lambda i}$ for i = 1 to m. Hence a maximal value for k is

$$k < k_{max} = \frac{2}{\lambda_m} = \frac{2}{\lambda_{max}} \qquad 3.1.12$$

on the other hand from 3.1.10

$$\delta \underline{x}_{i+\ell} = M^{\ell} \delta \underline{x}_i$$

Then if ℓ is large :

$$\| \delta \underline{x}_{i+\ell} \| \simeq |\rho|^{\ell} \quad \| \delta \underline{x}_i \|$$

where $|\rho|$ is the maximum modulus of eigenvalues of M, i.e.

$$|\rho| = \max_i \quad |1 - k\lambda_i|$$

To have fast convergence, we should choose k so as to have as small a $|\rho|$ as possible. It is then necessary to minimise w.r.t. k the largest possible value (in modulus), i.e. :

$$|\rho| = \min_k |\mu_{max}| = \min_k \max_i \quad |1 - k\lambda_i|$$

Consider the limit of $|\mu_i|$: $|\mu_1| = |1 - k\lambda_1|$, $|\mu_m| = |1 - k\lambda_m|$; $\forall$k and i, $|\mu_i|$ will necessarily evolve in the hatched zone (see Fig. 3.1).

$|\rho| = \max_i |\mu_i|$ $\forall$ k is in the part ABC and $|\hat{\rho}| = \min_k |\rho|$ is in B.

$$\text{For } |\hat{\rho}| : -(1 - k\lambda_m) = 1 - k\lambda_1 \Longrightarrow \hat{k} = k_{opt} = \frac{2}{\lambda_1 + \lambda_m} \qquad 3.1.13$$

$$|\hat{\rho}| = 1 - \frac{2\lambda_1}{\lambda_1 + \lambda_m} = \frac{\lambda_m - \lambda_1}{\lambda_m + \lambda_1} = \frac{C - 1}{C + 1} \text{ with } C = \frac{\lambda_m}{\lambda_1}$$

We could again write $\hat{\rho} = \frac{1 - 1/C}{1 + 1/C}$ and for C large :

$$|\hat{\rho}|^{\ell} \simeq (1 - 1/C)^{\ell} \; (1 - 1/C)^{\ell} \simeq 1 - \frac{2\ell}{C}$$

The convergence is thus limited by the dispersion C of the eigenvalues of F_{XX} and the slowest convergence takes place for the predominant eigenvalue of M (coming together of the trajectories on a limit. Fig. 3.2).

The final convergence of Newton's method

The method uses the algorithm :

$$\underline{x}_{i+1} = \underline{x}_i - k\, F_{XX}^{-1} (\underline{x}_i)\, \underline{f}_x(\underline{x}_i)$$

Replacing $\underline{f}_x(\underline{x}_i)$ by $F_{XX}(\hat{\underline{x}})\, \delta \underline{x}_i$ and $F_{XX}^{-1}(\underline{x}_i)$ by $F_{XX}^{-1}(\hat{\underline{x}})$, we have

$$\underline{x}_{i+1} = \hat{\underline{x}} + \delta \underline{x}_{i+1} = \hat{\underline{x}} + \delta \underline{x}_i - k\, F_{XX}^{-1}(\hat{\underline{x}})\, F_{XX}(\hat{\underline{x}})\, \delta \underline{x}_i$$

or

$$\delta \underline{x}_{i+1} = [1 - k]\, \delta \underline{x}_i \qquad 3.1.14$$

If $k \neq 1$ and $0 < k < 2$, the convergence is isotropic and linear (i.e. it is the same in all directions).

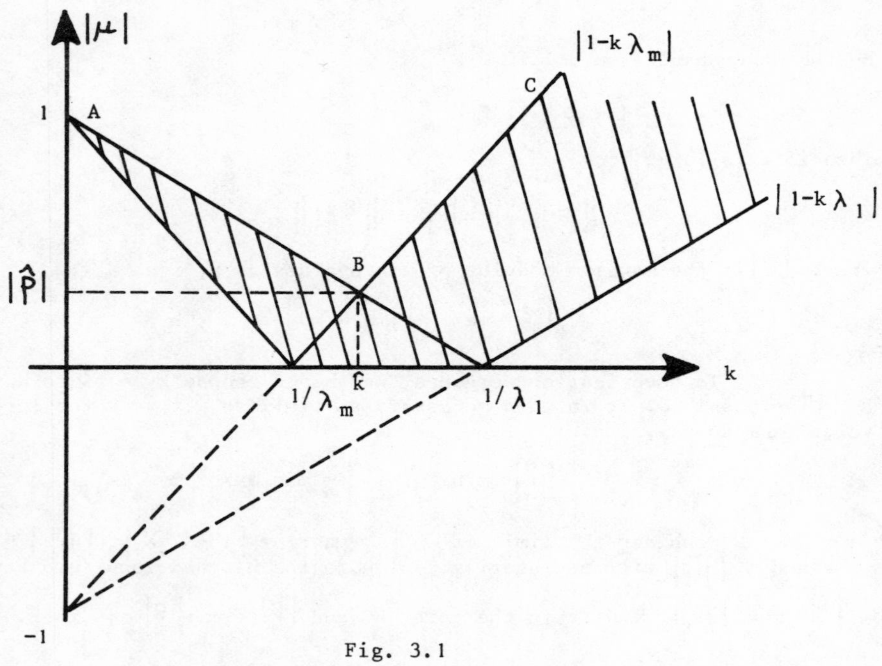

$|\mu|$
$|1-k\lambda_m|$
C
1
A
$|1-k\lambda_1|$
B
$|\hat{\rho}|$
k
$\hat{k}$
$1/\lambda_m$
$1/\lambda_1$
-1

Fig. 3.1

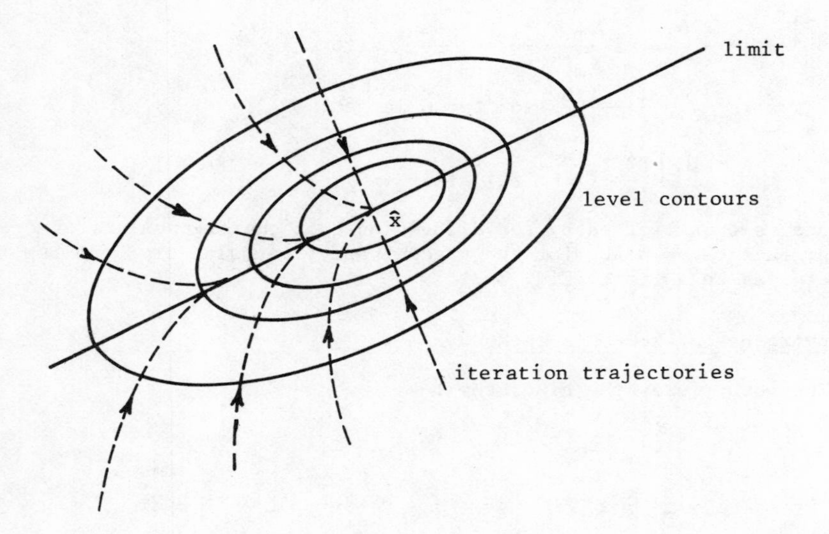

limit
level contours
$\hat{x}$
iteration trajectories

Fig. 3.2

For $k = 1$, the convergence is quadratic, but in order to study it in detail, we need to expand $f(\underline{\hat{x}})$ to third order.

Choice of the iteration constant

We have just seen that k has a big influence on the speed of convergence. However, it can not be determined theoretically since the eigenvalues that we need for it are in general unknown.

Another difficulty lies in the fact that in Newton's method, k is dimensionless and could be compared to 1 whilst in the gradient method, k has a dimension given by

$$[k] = \frac{[\text{dimension of } x]^2}{[\text{criterion dimension}]}$$

If we fix a $\delta \underline{x} : \|\delta \underline{x}\| < \varepsilon$, we could then take $k \leqslant \frac{\varepsilon}{\|f_x\|}$. We could in addition improve the convergence by adapting k as a function of the evolution of Δf (see the flow chart of Fig. 3.3).

If $\Delta f = f(\underline{x}_i) - f(\underline{x}_{i-1}) < 0$ increase k (k = 1.6 k for example)

If $\Delta f = f(\underline{x}_i) - f(\underline{x}_{i-1}) > 0$ decrease k and restart from $\underline{x}_{i-1}$ or from $\frac{\underline{x}_i + \underline{x}_{i-1}}{2}$ (k = 0.6 for example).

Stopping criterion

Since $f(\underline{\hat{x}})$ is unknown, we can not base the stopping test on the comparison of $f(\underline{\hat{x}})$ and $f(\underline{x}_i)$; we can however stop iterating when $f(\underline{x}_{i-N}) - f(\underline{x}_i)$ is smaller than the required accuracy or by using variations around $\underline{x}_i$, i.e. we could use

$$\frac{\|\underline{x}_{i+1} - \underline{x}_i\|}{\|\underline{x}_i\|} < E$$

Another possibility is to use the fact that since the optimisation algorithm must finally satisfy the necessary (1st order) conditions for optimality ($\underline{f}_x = \underline{0}$), we could also use as the stopping criterion

$$\|\underline{f}_x\| < E$$

Fig. 3.3 gives the flow chart of the gradient method.

3.1.4 A few additional algorithms

The algorithms that we will describe next are, partly, derived from the previous ones and they are able, for certain problems, to lead to savings in computation time.

3.1.4.1 Simplified algorithms

The simplified gradient method (Fig. 3.4)

In this method, the gradient is not recalculated until such that any further movement in the direction of the previous gradient causes an increase in $f(\underline{x})$.

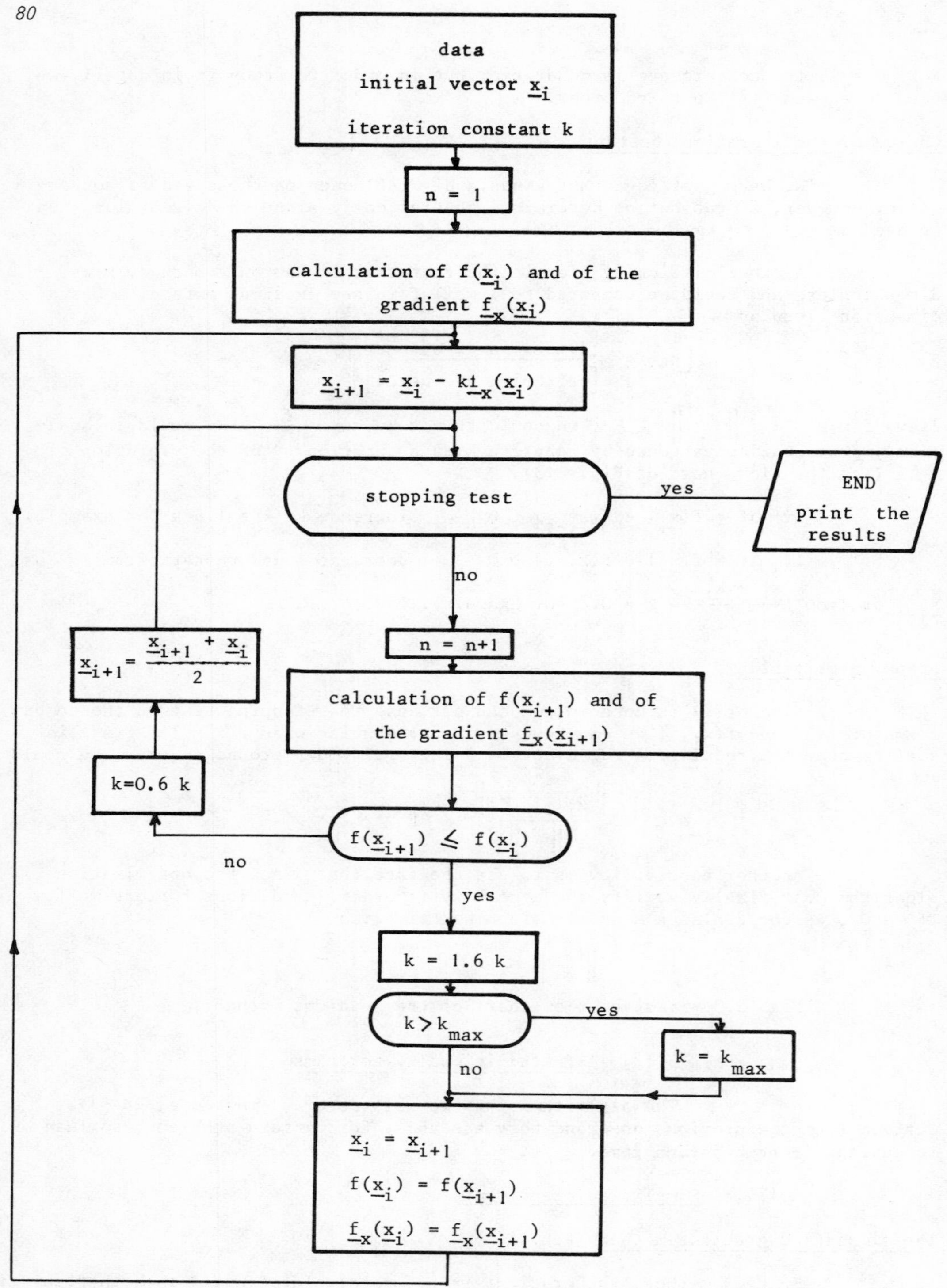

Fig. 3.3 : Flow chart of the gradient method

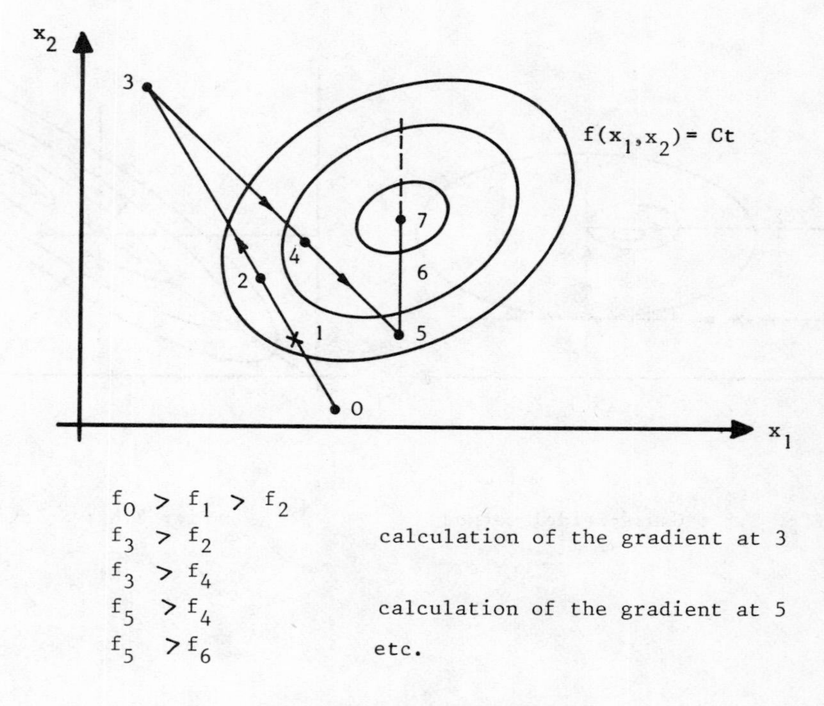

Fig. 3.4 : Simplified gradient method

This principle, which could also be applied to Newton's method, leads to the optimum being searched in a zig-zag way and this is often undesirable.

Sequential or Gauss-Seidel method

This is the ultimate simplification of the gradient method in that here the minimum is searched successively parellel to each axis (Fig. 3.5).

However, this method fails on an edge as shown in Fig. 3.6 where 1 is on an edge and small displacements in the directions (0 x_1) and (0 x_2) lead to 2, 3, 4, 5 and none of these points give a smaller value of the cost than 1 which is the wrong solution. To avoid this difficulty, it is possible to make an automatic coordinate change whenever the search stops at an edge so as to enable us to continue to search in the direction of the true solution $\hat{x}$ as is done in Fig. 3.6 where (1, X_2), (1, X_1) are the new axes (Rosenbrock's method).

3.1.4.2 The PARTAN method

This method uses the properties of conjugate directions of hyper-quadratics. In the two dimensional case these properties are illustrated on Fig. 3.7. Here, let D_1 and D_2 be two parallel directions and A and B be the extrema in these two directions respectively. Then the extremum that we are searching for is on the line AB joining the relative extrema in the 2 parallel directions. Hence the name of the method : PARTAN $\Longrightarrow$ PARallel-TANgent.

If the equivalue contours are really quadratic, we should reach the solution in one iteration. Otherwise, we perform the following cycle shown in Fig. 3.8 for the case of a minimum being searched.

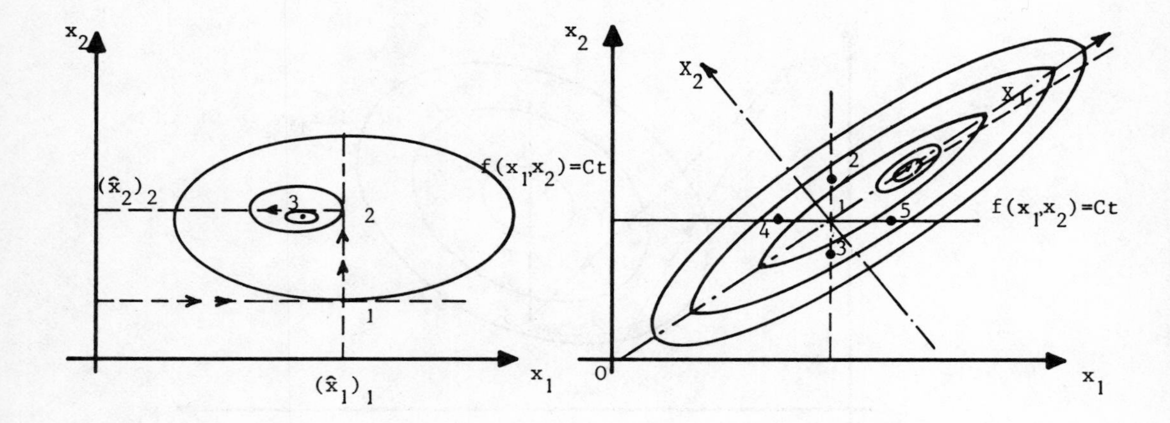

Fig. 3.5 : Gauss-Seidel method

Fig. 3.6 : Solution line

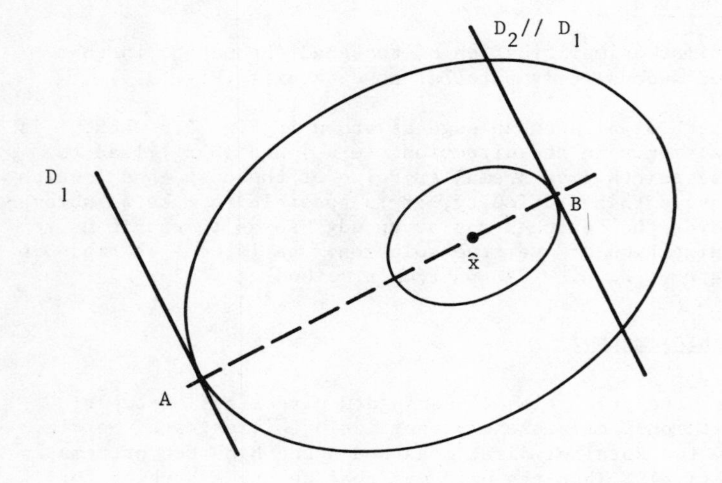

Fig. 3.7

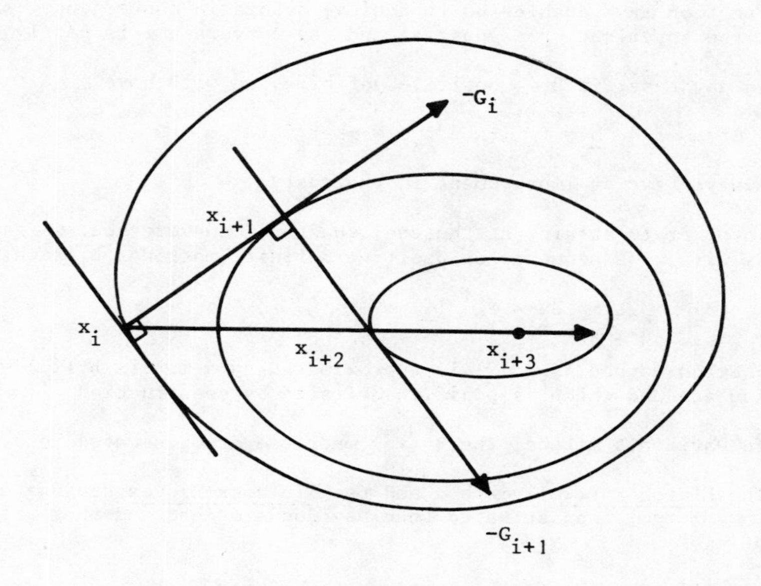

Fig. 3.8 : The PARTAN method

1. Calculate the gradient $\underline{G}_i$ at $\underline{x}_i$
2. Search along $(-\underline{G}_i)$ the minimum $f(\underline{x}_{i+1})$ situated at $\underline{x}_{i+1}$
3. Calculate the gradient $\underline{G}_{i+1}$ at $\underline{x}_{i+1}$
4. Search along $-\underline{G}_{i+1}$ the minimum $f(\underline{x}_{i+2})$ at $\underline{x}_{i+2}$
5. Search along $(\underline{x}_i, \underline{x}_{i+2})$ the minimum $f(\underline{x}_{i+3})$
6. Test if this is the solution. If not, put $\underline{x}_{i+3} = \underline{x}_i$ and return to step 1.

The steps 2, 4, 5 are carried out using single variable search methods (e.g. Fibonacci method, quadratic or cubic interpolation, etc.). Such methods are very efficient.

3.1.4.3 Algorithms having a quadratic final convergence

These methods are an adaptation of Newton's algorithm

$$\underline{x}_{i+1} = \underline{x}_i - k\, F_{XX}^{-1}(\underline{x}_i)\, \underline{f}_X(\underline{x}_i)$$

In this algorithm, it is necessary, in order to converge to a minimum, that $F_{XX}(\underline{x}_i)$ be positive definite and this condition is not always satisfied for all starting points $\underline{x}_i$.

In order to avoid this difficulty let us introduce a modified gradient

$$\underline{g} = B\, \underline{f}_X$$

If $B = F_{XX}^{-1}$, this is equivalent to "spherising" the equi-value contours of the cost function and it enables us to achieve quadratic convergence. If B is close to F_{XX}^{-1}, the spherisation is partial and the convergence is nearly quadratic.

In any case, if B is positive definite, we will have :

$$\Delta f \simeq \underline{f}_x^T \, \underline{\delta} = \underline{f}_x^T(-k\,B\,\underline{f}_x) = -k\,\underline{f}_x^T\,B\,\underline{f}_x < 0$$

and thus we always have an improvement in the cost.

In order to obtain, at the end, quadratic convergence, we generate simultaneously with $\underline{x}_i$ a sequence of positive definite matrices B_i such that

$$\lim_{i\to\infty} B_i = F_{XX}^{-1}\,(\underline{\hat{x}})$$

In the Gauss-Newton method, $F_{XX}^{-1}\,(\underline{\hat{x}})$ is approximated by a matrix $B_i(\underline{x}_i)$ which is easier to calculate and which is positive definite by construction.

In Davidon's method, the $B_i(\underline{x}_i)$ matrices are generated recursively.

The <u>Fletcher-Powell</u> method and the <u>Fletcher-Reeves</u> conjugate gradient method are slightly more sophisticated than Davidon's method and they are certainly very efficient.

3.1.4.4 Choice of the algorithm to be used

So far we have presented a number of algorithms and it is interesting to ask which algorithm should be used in which situation.

The criterion of choice should take into account not only the speed of convergence (number of iterations), but also the calculation time and the length and complexity of the program used to implement the algorithm.

As far as the choice between Newton's method and the Gradient method is concerned, we note that Newton's method requires, in addition to the gradient calculation, the calculation of $F_{XX}(\underline{x}_i)$ and the solution of a linear system of equations (or the inversion of a matrix).

Let us assume that the analytical form of $F(\underline{x})$ is unknown although values of $f(\underline{x})$ could be computed numerically. In most cases of this type, this numerical calculation represents the longest operation. But this calculation is done (m+1) times in the gradient method and m+1 + (m+1) m/2 times in Newton's method. To see this, we note that we can write approximately :

$$\underline{f}_x(\underline{x}_i) \simeq \begin{bmatrix} \dfrac{f(x_i^1 + \Delta x_i^1, x_i^2, \dots x_i^m) - f(x_i^1, x_i^2, \dots x_i^m)}{\Delta x_i^1} \\ \vdots \\ \dfrac{f(x_i^1, \dots x_i^m + \Delta x_i^m) - f(x_i^1, x_i^2, \dots x_i^m)}{\Delta x_i^m} \end{bmatrix}$$

where

$$\underline{x}_i = \begin{bmatrix} x_i^1 \\ x_i^2 \\ \vdots \\ x_i^m \end{bmatrix}$$

We see that this computation requires us to calculate $f(\underline{x})$ (m+1) times.

In the case of Newton's method, we have to compute $F_{XX}(\underline{x}_i)$. In order to obtain the terms of the principal diagonal of $F_{XX}(\underline{x}_i)$, we need to calculate $f(\underline{x})$ m times. To see this, we note for example that :

$$\frac{\partial^2 f(\underline{x}_i)}{\partial x_i^{1\,2}} \simeq \frac{f(x_i^1 + 2\Delta x_i^1, x_i^2, \ldots x_i^m) - 2f(x_i^1 + \Delta x_i^1, x_i^2, \ldots x_i^m) + f(x_i^1, x_i^2, \ldots x_i^m)}{(\Delta x_i^1)^2}$$

For the off diagonal terms in $F_{XX}(\underline{x}_i)$, it is necessary to add on C_m^2 computations of $f(\underline{x})$ (C_m^2 is the number of combinations of m variables taken 2 by 2 , $C_n^p = \frac{n\,!}{(n-p)!p!}$, where we have not taken into account the order of the combinations since the matrix is symmetrical) since

$$\frac{\partial^2 f(x_i)}{\partial x_i^k \partial x_i^j} \simeq \frac{f(x_i^1, \ldots x_i^k + \Delta x_i^k \ldots x_i^j + \Delta x_i^j \ldots x_i^m) - f(x_i^1 \ldots x_i^k + \Delta x_i^k \ldots x_i^j \ldots x_i^m}{\Delta x_i^k \Delta x_i^j}$$

$$\frac{- f(x_i^1 \ldots x_i^k \ldots x_i^j + \Delta x_i^j \ldots x_i^m) + f(x_i^1 \ldots x_i^k \ldots x_i^j \ldots x_i^m)}{\Delta x_i^k \; \Delta x_i^j}$$

Thus the total number of multiplications required for Newton's method is :

$$\underbrace{(m+1)}_{\underline{f}_x(\underline{x})} + \underbrace{m + \frac{(m-1)m}{2}}_{F_{XX}(\underline{x})} = m+1 + (m+1)\frac{m}{2}$$

as stated above.

In addition, to be competetive with the gradient method from the point of view of calculation time only (without taking into account the storage requirements) and neglecting the problem of inverting $F_{XX}(\underline{x}_i)$, the method of Newton must be able to increase the speed of convergence by a ratio greater than

$$1 + \frac{m}{2} = \frac{(m+1) + (m+1)\,m/2}{m+1}$$

We see thus that the choice of the algorithm depends largely on the problem being solved and we next give some qualitative indications which could help in this choice :

* In the beginning of the search, the simplified algorithms usually lead to an edge solution.

* To continue along the edge, we could use Rosenbrock's method or PARTAN.

* To end the convergence in the neighbourhood of the optimum, Newton's method is particularly effective.

* Mixed methods (gradient + Newton), or methods with quadratic final convergence, are often useful since they give quadratic convergence near the end while having at the same time good initial convergence as in the gradient method.

Next we consider optimisation methods where we search for a minimum in the presence of equality constraints.

3.2 MINIMISATION SUBJECT TO EQUALITY CONSTRAINTS

We will assume that the cost function $f(\underline{x})$ is continuous and continuously differentiable upto 1st order (for the 1st order conditions) or to 2nd order (for the second order conditions) in $\mathcal{D}$. The feasible region $\mathcal{D}$ is defined by p constraints, which are continuous and continuously differentiable (to 1st or 2nd order), of the type

$$g_i(\underline{x}) = 0 \qquad i = 1, 2, \ldots p < m \qquad \underline{x} \in R^m$$

$\underline{x}$ is said to be a "regular point" of the constraints if at this point all the constraints g are independent. This implies that the m x p matrix G_x defined below be of rank $\overline{p}$

$$G_x = \left[\frac{\partial g_i}{\partial x_j}\right] = \begin{bmatrix} \frac{\partial g_1}{\partial x_1} & \cdots & \frac{\partial g_1}{\partial x_m} \\ \vdots & & \\ \frac{\partial g_p}{\partial x_1} & \cdots & \frac{\partial g_p}{\partial x_m} \end{bmatrix}$$

3.2.1 First order conditions

As in the unconstrained case, we begin by deriving the first order conditions of optimality. With the above continuity assumptions, we could expand $f(\underline{x} + \delta\underline{x})$ in a Taylor series about $\underline{x}$ upto first order, i.e.

$$f(\underline{x} + \delta\underline{x}) = f(\underline{x}) + \underline{f}_x^T \delta\underline{x} + 0^2$$

The variations around a regular point which is a relative minimum should be compatible with the constraints and this implies, on expanding the constraints to first order

$$g_i(\underline{x} + \delta\underline{x}) = g_i(\underline{x}) + \underline{g}_{ix}^T \delta\underline{x} + 0^2 = 0$$

or in vector form for the p constraints

$$\underline{g}(\underline{x} + \delta\underline{x}) = \underline{g}(\underline{x}) + G_X(\underline{x}) \delta\underline{x} + 0^2 = \underline{0}$$

But if $\underline{x} \in \mathcal{D}$, $\underline{g}(\underline{x}) = \underline{0}$, $\underline{x} + \delta\underline{x} \in \mathcal{D}$, $\underline{g}(\underline{x} + \delta\underline{x}) = \underline{0}$ and we have

$$G_X(\underline{x}) \delta\underline{x} = 0 \qquad 3.2.1$$

3.2.1 defines for $\delta\underline{x}$ a vector space $\mathcal{E} \subset R^m$ of dimension (m-p) since G_X is assumed to be of rank p.

Let $\hat{\underline{x}}$ be a relative minimum. Then :

$$f(\hat{\underline{x}}) \leqslant f(\hat{\underline{x}} + \delta\underline{x}) \simeq f(\underline{x}) + \underline{f}_X(\hat{\underline{x}})\,\delta\underline{x} \quad \forall\ \delta\underline{x} \in \mathcal{E} \text{ and } \|\delta\underline{x}\| \text{ small}$$

Thus $\underline{f}_x^T(\hat{\underline{x}})\ \delta\underline{x} \geqslant 0$ and since this inequality must be satisfied if we change $\delta\underline{x}$ to $-\delta\underline{x}$, we have :

$$\underline{f}_x^T(\hat{\underline{x}})\,\delta\underline{x} = 0 \qquad \forall\ \delta\underline{x} \in \mathcal{E} \tag{3.2.2}$$

Geometrically, $\underline{f}_x(\hat{\underline{x}})$ is thus normal to $\mathcal{E}$ of dimension (m-p). It belongs thus to its complement of dimension p which has as its basis of independent vectors the $g_{i_x}(\hat{\underline{x}})$ (as an example in 2 dimensions cf. Fig. 3.9).

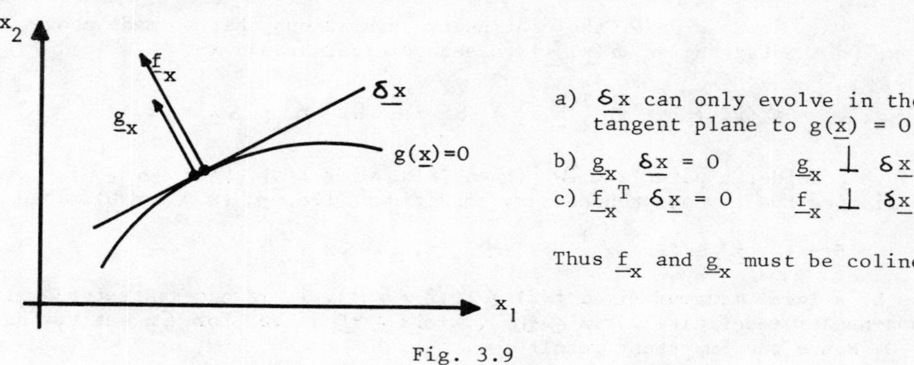

x_2
$\underline{f}_x$
$\underline{g}_x$
$\delta\underline{x}$
$g(\underline{x})=0$
x_1
a) $\delta\underline{x}$ can only evolve in the tangent plane to $g(\underline{x}) = 0$
b) $\underline{g}_x\ \delta\underline{x} = 0$ $\quad \underline{g}_x \perp \delta\underline{x}$
c) $\underline{f}_x^T\ \delta\underline{x} = 0$ $\quad \underline{f}_x \perp \delta\underline{x}$
Thus $\underline{f}_x$ and $\underline{g}_x$ must be colinear.

Fig. 3.9

Following from this, $\underline{f}_x$ is a linear combination of these vectors and this can be written as

$$\underline{f}_x(\hat{\underline{x}}) + \sum_{i=1}^{p} \lambda_i\, \underline{g}_{i_x}(\hat{\underline{x}}) = \underline{0} \tag{3.2.3}$$

This expression is essentially the stationarity condition (1st order) of the function

$$\phi(\underline{x}, \underline{\lambda}) = f(\underline{x}) + \sum_{i=1}^{p} \lambda_i\, g_i(\underline{x})$$

$$= f(\underline{x}) + \underline{\lambda}^T \underline{g}(x)$$

To see this, we note that :

$$\underline{\phi}_x = \underline{f}_x + \sum_{i=1}^{p} \lambda_i\, \underline{g}_{i_x} = \underline{0}$$

and

$$\underline{\phi}_\lambda = \underline{g}(\underline{x}) = \underline{0}$$

From this we have the important result :

If $f(\underline{x})$ and $g_i(\underline{x})$ (i = 1 to p) satisfy the continuity conditions, then for a regular point $\hat{\underline{x}}$ to be a local minimum, it is necessary that there exist multipliers λ_i such that the Lagrangian function

$$\phi(\underline{x}, \lambda) = f(\underline{x}) + \sum_{i=1}^{p} \lambda_i\, g_i(\underline{x})$$

be stationary w.r.t. $\underline{x}$ and $\underline{\lambda}$, i.e.

$$\underline{\phi}_x(\hat{\underline{x}}, \underline{\lambda}) = \underline{0}, \quad \underline{\phi}_\lambda(\hat{\underline{x}}) = \underline{0}$$

The stationarity conditions yield m+p equations in the m+p unknowns ($\underline{x}$ and $\underline{\lambda}$).

The Lagrange multipliers λ_i enable us to treat a constrained optimisation problem with equality constraints in the same way as an unconstrained one. However, the existence of the equality constraints means that the number of variables is increased.

We note that for $\underline{x} \in \mathcal{D}$, $g_i(\underline{x}) = 0$ so that $\phi(\underline{x}, \underline{\lambda}) \equiv f(\underline{x})$

3.2.2 Second order conditions

Under the continuity assumptions that we made above, let us now expand the Lagrangian $\phi(\underline{x}, \underline{\lambda})$ to second order around $\underline{x}$:

$$\phi(\underline{x} + \delta\underline{x}, \underline{\lambda}) = \phi(\underline{x}, \underline{\lambda}) + \underline{\phi}_x^T(\underline{x}, \underline{\lambda})\ \delta\underline{x} + \frac{1}{2}\ \delta\underline{x}^T \phi_{XX}\ \delta\underline{x} + \ldots \qquad 3.2.4$$

Consider a point $\underline{x} \in \mathcal{D}$ (then $\phi(\underline{x}, \underline{\lambda}) = f(\underline{x})$, $\phi(\underline{x} + \delta\underline{x}, \underline{\lambda}) = f(\underline{x}+\delta\underline{x})$) which satisfies the 1st order necessary conditions (i.e. $\underline{\phi}_x(\underline{x}, \underline{\lambda}) = \underline{0}$). Then

$$f(\underline{x} + \delta\underline{x}) \simeq f(\underline{x}) + \frac{1}{2}\ \delta\underline{x}^T\ \phi_{XX}(\underline{x}, \underline{\lambda})\ \delta\underline{x} \qquad 3.2.5$$

For $\hat{\underline{x}}$ to be a local minimum (i.e. $f(\hat{\underline{x}}) \leqslant f(\hat{\underline{x}} + \delta\underline{x})$), it is necessary for $\phi_{XX}(\hat{\underline{x}}, \underline{\lambda})$ to be non-negative definite for $\underline{x} \in \mathcal{D}$, $\underline{x}+\delta\underline{x} \in \mathcal{D}$, i.e. for $\delta\underline{x}$ satisfying $G_x\,\delta\underline{x} = \underline{0}$. Hence the important result :

> The necessary conditions for a relative minimum are :
>
> 1st order : $\underline{\phi}_x(\hat{\underline{x}}, \hat{\underline{\lambda}}) = \underline{0}$ $\qquad \underline{\phi}_\lambda(\hat{\underline{x}}) = \underline{0}$
>
> 2nd order : $\phi_{XX}(\hat{\underline{x}}, \hat{\underline{\lambda}})$ must be non-negative definite on the restriction $\mathcal{E}$ defined by :
>
> $$\mathcal{E} = \left\{ \underline{v},\ G_x\,\underline{v} = \underline{0} \right\}$$

We need thus to examine the roots of the equation in ω (restriction of M on A) :

$$\Delta = \det \begin{vmatrix} \omega I - M & A^T \\ A & 0 \end{vmatrix} = 0 \qquad \text{or in this case}$$

$$\Delta = \det \begin{vmatrix} \omega I - \phi_{XX} & G_X^T \\ G_X & 0 \end{vmatrix} = 0$$

where I is the identity matrix. If the matrix is positive definite, we have a sufficient condition for a local minimum.

Examples

Example 1 : $\min R = \sum_{i=1}^{4} y_1^2$ subject to $\sum_{i=1}^{4} y_i = a > 0$

Solution : $\phi(\underline{y}, \lambda) = \sum_i y_i^2 + \lambda(\sum_i y_i - a)$

$$\phi_{y_i} = 2y_1 + \lambda = 0 \implies y_i = -\frac{\lambda}{2}$$

$$\phi_\lambda = 0 \Longrightarrow y_i = \frac{a}{4}$$

This is the result using the first order conditions. Let us next see what happens when we test the second order conditions :

Here

$$\phi_{YY} = \begin{bmatrix} 2 & 0 & 0 & 0 \\ \vdots & 2 & & \vdots \\ \vdots & & 2 & \vdots \\ 0 & \cdots & \cdots & 2 \end{bmatrix}$$ which is a positive definite matrix.

Hence the solution obtained is at a strict minimum.

We note in addition that the criterion is convex on $\mathcal{D}$ convex (linear equality constraints). Thus, all relative minima are absolute minima and $y_i = \frac{a}{4}$ is thus a unique strict absolute minimum.

Example 2 : Let us next study the nature of the extrema of

$$f(x, y) = 2x + 3y - 1 \quad \text{subject to}$$
$$g(x, y) = x^2 + 1.5\,y^2 - 6 = 0$$

Here $\phi(x, y, \lambda) = 2x + 3y - 1 + \lambda(x^2 + 1.5\,y^2 - 6)$

$$\left.\begin{aligned} \phi_x &= 2 + 2\lambda x = 0 \\ \phi_y &= 3 + 3\lambda y = 0 \end{aligned}\right\} \Longrightarrow x = y = -\frac{1}{\lambda}$$

$$\phi_\lambda = x^2 + 1.5\,y^2 - 6 = 0 \Longrightarrow 2.5 - 6\lambda^2 = 0$$

$$\Longrightarrow \lambda = \pm 0.645$$

Second order conditions : here :

$$\phi_{XY} = \begin{bmatrix} 2\lambda & 0 \\ 0 & 3\lambda \end{bmatrix} \qquad D_1 = 2\lambda \qquad D_2 = 6\lambda^2$$

The nature of the extrema is summarised below :

Value of λ		-0.645	+0.645
Solution	x	1.55	-1.55
	y	1.55	-1.55
Criterion		6.75	-8.75
2nd order conditions	D_1	$-1.29 < 0$	$1.29 > 0$
	D_2	$2.5 > 0$	$2.5 > 0$
Nature of the extremum		max	min

Example 3 : Let us consider the nature of the extrema of

$$f(x_1, x_2) = 4 - x_1^2 - x_2^2 \quad \text{subject to}$$

$$g = 1 - x_1^2 + x_2 = 0$$

Using the Lagrange multipliers method, we have :

$$\phi(x_1, x_2, \lambda) = 4 - x_1^2 - x_2^2 + \lambda(1 - x_1^2 + x_2)$$

$$\phi_{x_1} = x_1(1+\lambda) = 0$$

$$\phi_{x_2} = (-2x_2 + \lambda) = 0$$

$$\phi_\lambda = 1 - x_1^2 + x_2 = 0$$

1st solution : $x_1 = 0$, $x_2 = -1$ (using g) $\lambda = -2$.

2nd solution : $\lambda = 1$, $x_2 = -\frac{1}{2}$, $x_1 = \pm\sqrt{\frac{1}{2}}$ (using g)

We thus have 3 stationary points whose nature must be examined using second order conditions.

Here :

$$\phi_{XX} = \begin{bmatrix} -2-2\lambda & 0 \\ 0 & -2 \end{bmatrix}$$

If we put $\lambda = -1$ or $\lambda = -2$ in ϕ_{XX}, we don't get any special property in R^2 ;

it is therefore necessary to study the definition on the restriction

$$\det \begin{vmatrix} \omega I - \phi_{XX} & G_X^T \\ G_X & 0 \end{vmatrix} = 0 \qquad \det \begin{vmatrix} \omega+2+2\lambda & 0 & -2x_1 \\ 0 & \omega+2 & 1 \\ -2x_1 & 1 & 0 \end{vmatrix} = 0$$

For $x_1 = 0$, $x_2 = -1$, $\lambda = -2$

$$\begin{vmatrix} \omega-2 & 0 & 0 \\ 0 & \omega+2 & 1 \\ 0 & 1 & 0 \end{vmatrix} = 0 \Longrightarrow \omega = 2 > 0$$

and the matrix is positive definite on the restriction so that we have a strict local minimum.

For $\lambda = -1$, $x_2 = -\frac{1}{2}$, $x_1 = \pm \frac{1}{2}$

$$\begin{vmatrix} \omega & 0 & -(\pm\sqrt{2}) \\ 0 & \omega+2 & 1 \\ -(\pm\sqrt{2}) & 1 & 0 \end{vmatrix} = 0 \Longrightarrow \omega = -\frac{4}{3} < 0$$

The matrice is thus negative definite on the restriction and we have a strict local maximum.

Geometric solution

a) $g = 1 - x_1^2 + x_2 = 0 \qquad x_1^2 = 1 + x_2 (x_2 > -1)$

which represents a parabola in the (x_1, x_2) plane (Fig. 3.10) and the extremal points should lie on this curve

$$\text{Min } f = 4 - x_1^2 - x_2^2 \Longrightarrow \min(-(x_1^2 + x_2^2)) = \max \; d^2$$

where d is the distance of any point on the parabola from the origin.

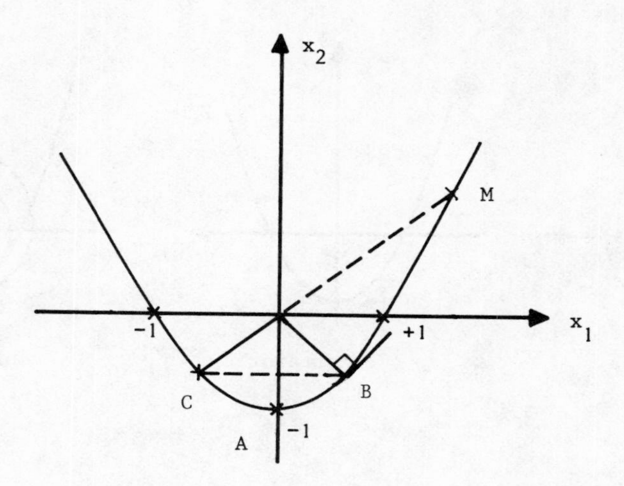

Fig. 3.10

Also, max f $\longrightarrow$ min d^2

Hence the solutions A, B, C where B, C are absolute minima of d and thus absolute maxima of f and A is a local maximum of d and thus a local minimum of f. We thus find once again the results found previously using Lagrange theory.

b) $f = 4 - x_1^2 - x_2^2 = C$

Here the equivalue contours are circles which obey the equation $x_1^2 + x_2^2 = 4 - C$

At the solution :

$$\phi_{x_1} = f_{x_1} + \lambda g_{x_1} = 0 \; ; \quad \phi_{x_2} = f_{x_2} + \lambda g_{x_2} \qquad 3.2.6$$

which enables us to show that at the solution the constraints and the equivalue contours are tangential. To see this, we note that the slope of the tangent to the equivalue contours is :

$$\frac{dx_2}{dx_1} = \frac{\partial f / \partial x_1}{\partial f / \partial x_2}$$

But from 3.2.6

$$\frac{dx_2}{dx_1} = \frac{\partial g / \partial x_1}{\partial g / \partial x_2} = \text{ slope of the tangent to the constraint curve.}$$

This gives us another method for solving optimisation problems, i.e. we seek the points where the constraint curves and the level curves of the criterion are tangential (Fig. 3.11).

Remark : we can, in this last example, extract x_2 from $g(x) = 0$ and thus obtain an unconstrained single variable problem as follows :

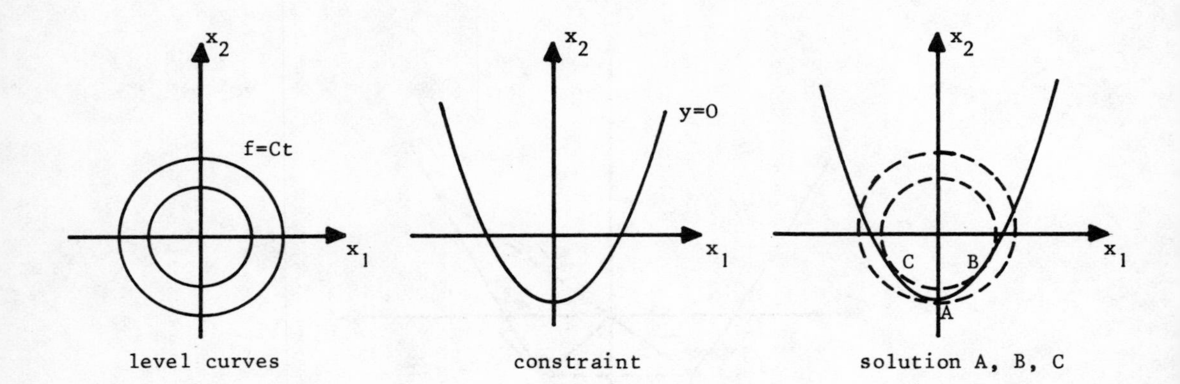

Fig. 3.11

$$x_2 - x_1^2 + 1 = 0 \Longrightarrow f = 4 - x_1^2 - (x_1^2 - 1)^2 = 3 + x_1^2 - x_1^4$$

1st order conditions :

$$f_{x_1} = 0 \quad \text{or} \quad 2x_1 - 4x_1^3 = 0 \Longrightarrow x_1 = 0 \quad \text{or } x_1 = \pm\sqrt{\frac{1}{2}}$$

2nd order conditions :

$$F_{XX} = 2 - 12\, x_1^2 \quad \left| \begin{array}{l} > 0 \text{ for } x_1 = 0 \longrightarrow \text{min} \\ < 0 \text{ for } x_1 = \pm\sqrt{\frac{1}{2}} \longrightarrow \text{max} \end{array} \right.$$

Example 4 : $\min f(x, y) = x^2 + 2y^2 - 4x - 12y + 25$

subject to

$g(y) = y^2 - 7y + 10 = 0$

Here $\phi(x, y, \lambda) = x^2 + y^2(2+\lambda) - 4x - y(12+7\lambda) + 25 + 10\lambda$

$$\frac{\partial \phi}{\partial x} = 2x - 4 = 0 \longrightarrow x^* = 2$$

$$\frac{\partial \phi}{\partial y} = 2y(2+\lambda) - 12 + 7\lambda = 0 \Longrightarrow y^* = \frac{12+7\lambda}{2(2+\lambda)}$$

$\phi_\lambda = y^2 - 7y + 10 = (y-2)(y-5) = 0$ ($y = 2$ and $y = 5$) (Fig. 3.12).
This, on using the expression for y^* gives λ.

2nd order conditions :

$$\phi_{XY} = \begin{vmatrix} 2 & 0 \\ 0 & 2(2+\lambda) \end{vmatrix} \quad ; \quad G_X^T = \begin{vmatrix} 0 \\ 2y-7 \end{vmatrix}$$

$$\Delta = \begin{vmatrix} \omega I - \phi_{XY} & G_X^T \\ G_X & 0 \end{vmatrix} = \det \begin{vmatrix} \omega - 2 & 0 & 0 \\ 0 & \omega - 2\,(2+\lambda) & 2y-7 \\ 0 & 2y-7 & 0 \end{vmatrix} = 0$$

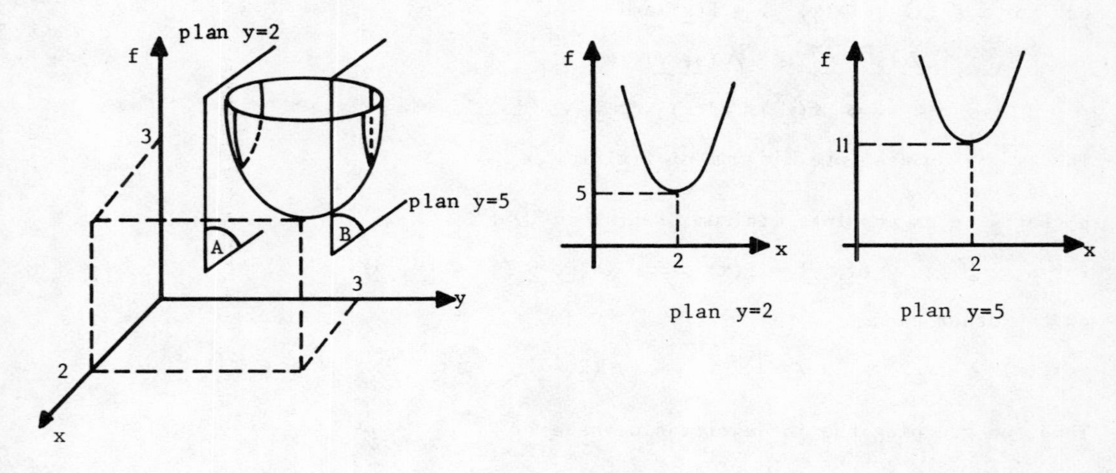

Fig. 3.12

This gives

$$-(\omega-2)(2y-7)^2 = 0 \qquad 2y-7 \neq 0 \text{ (since } g(y) = 0) \quad ; \quad \omega = 2 > 0$$

We have thus 2 strict relative minima. Geometrically, we can easily pick out the absolute minimum.

3.2.3 MINMAX Formula

Let us consider, without making any assumptions on the continuity of $f(\underline{x})$ and $g_i(\underline{x})$, the function

$$\phi(\underline{x}, \underline{y}) = f(\underline{x}) + \sum_i y_i \, g_i(\underline{x}) \qquad \underline{x} \in R^m, \ \underline{y} \in R^n$$

$g_i(\underline{x}) = 0$ defining as before the region $\mathcal{D}$:

a) Let $\underline{x}^*$, $\underline{y}^*$ be the solution of :

$$\phi(\underline{x}^*, \underline{y}^*) = \min_{\underline{x} \in R^m} \ \max_{\underline{y} \in R^n} \ \phi(\underline{x}, \underline{y})$$

i.e.

$$\phi(\underline{x}^*, \underline{y}) \leqslant \phi(\underline{x}^*, \underline{y}^*) \leqslant \phi(\underline{x}, \underline{y}^*) \ \forall \ \underline{x} \in R^m, \ \underline{y} \in R^n$$

Assume that $\underline{x}^* \notin \mathcal{D}$ i.e. $g_i(\underline{x}^*) \neq 0$

Putting

$$y_i = y_i^* + 1 \times \text{sign } (g_i(\underline{x}^*))$$

we have

$$\phi(\underline{x}^*, \underline{y}^*) - \phi(\underline{x}^*, \underline{y}) = \sum_i - \text{sign}\left[g_i(x^*)\right] \ g_i(x^*) \leqslant 0$$

This contradicts

$\phi(\underline{x}^*, \underline{y}^*) \geqslant \phi(\underline{x}^*, \underline{y})$. Thus $\underline{x}^* \in \mathcal{D}$

But for $\underline{x} \in \mathcal{D}$, $\phi(\underline{x}, \underline{y}) = f(\underline{x})$ and

$$\phi(\underline{x}^*, \underline{y}^*) \leqslant \phi(\underline{x}, \underline{y})$$

$$\Longrightarrow f(\underline{x}^*) \leqslant f(\underline{x}) \quad \forall \underline{x} \in \mathcal{D}$$

Thus $\underline{x}^*$ is an absolute minimum of $f(\underline{x})$ in $\mathcal{D}$.

b) Let $\hat{\underline{x}}$ be an absolute minimum of $f(\underline{x})$ in $\mathcal{D}$:

For $\underline{x} \in \mathcal{D}$, $\phi(\underline{x}, \underline{y}) = f(\underline{x}) \Longrightarrow \phi(\hat{\underline{x}}, \hat{\underline{y}}) = \phi(\hat{\underline{x}}, \underline{y})$

on the other hand

$$f(\hat{\underline{x}}) \leqslant f(\underline{x}) \quad \forall \underline{x} \in \mathcal{D} \qquad\qquad \phi(\hat{\underline{x}}, \hat{\underline{y}}) \leqslant \phi(\underline{x}, \hat{\underline{y}}) \quad \forall \underline{x} \in \mathcal{D}$$

Thus, on grouping the inequalities we have :

$$\phi(\hat{\underline{x}}, \underline{y}) = \phi(\hat{\underline{x}}, \hat{\underline{y}}) \leqslant \phi(\underline{x}, \hat{\underline{y}}) \quad \forall \underline{x} \in \mathcal{D}$$

which is included in :

$$\boxed{\phi(\hat{\underline{x}}, \underline{y}) \leqslant \phi(\hat{\underline{x}}, \hat{\underline{y}}) \leqslant \phi(\underline{x}, \hat{\underline{y}}) \quad \forall \underline{x} \in \mathcal{D}}$$

and $\hat{\underline{x}}$ is the solution of $\min\limits_{\underline{x} \in R^m} \max\limits_{\underline{y} \in R^n} \phi(\underline{x}, \underline{y})$

From this, it follows that all solutions of the minimisation problem with equality constraints are solutions of the minimax problem and vice-versa.

3.2.4 The Gradient method

3.2.4.1 The aim of the algorithm

Let $\underline{x}_i$ satisfy the constraints, find $\underline{x}_{i+1}$ which also satisfies the constraints (at least to first order) such that :

$$\underline{x}_{i+1} - \underline{x}_i = \underline{\delta} \quad \text{and } f(\underline{x}_i) - f(\underline{x}_{i+1}) \text{ is maximal}$$

Put $\| \underline{x}_{i+1} - \underline{x}_i \| = d$. $\underline{\delta}$ should then satisfy

$$\underline{\delta}^T \underline{\delta} = d^2$$
$$G_X \underline{\delta} = \underline{0} \quad \text{(motion on the constraint)}$$
$$f(\underline{x}_{i+1}) - f(\underline{x}_i) = \underline{f}_X^T (\underline{x}_i) \underline{\delta} \quad \text{minimum}$$

$\underline{\delta}$ is thus the solution of a minimisation problem having 2 equality constraints. We can solve this problem by introducing Lagrange multipliers. To do this, consider

$$\phi(\underline{\delta}, \underline{\lambda}, \underline{\mu}) = \underline{f}_X^T \underline{\delta} + \underline{\lambda}^T G_X \underline{\delta} + \mu (\underline{\delta}^T \underline{\delta} - d^2)$$

The first order conditions give

$$\underline{\phi}_\delta = \underline{f}_X + G_X^T \underline{\lambda} + 2\mu \underline{\delta} = \underline{0} \Longrightarrow \underline{\delta} = -\frac{1}{2}\mu [\underline{f}_X + G_X^T \underline{\lambda}]$$

Putting this value of $\underline{\delta}$ in $G_X \underline{\delta} = \underline{0}$, we have

$$G_X \left[\underline{f}_x + G_X^T \underline{\lambda} \right] = \underline{0} \Longrightarrow \underline{\lambda} = - \left[G_X G_X^T \right]^{-1} G_X \underline{f}_x$$

We note here that since $\underline{x}_i$ is a regular point, G_X is of full rank (p) and $G_X G_X^T$ is non-singular so that

$$\underline{\delta} = - \frac{1}{2\mu} \left[\underline{f}_x - G_X^T \left[G_X G_X^T \right]^{-1} G_X \underline{f}_x \right]$$

But $\phi_{\delta\delta} = 2 \mu I$. This second order condition requires that μ (function of d) be > 0 and this gives the algorithm :

$$\underline{x}_{i+1} = \underline{x}_i - k \left[I - G_X^T \left[G_X G_X^T \right]^{-1} G_X \quad \underline{f}_X \right] \quad k > 0 \qquad 3.2.7$$

3.2.4.2 Geometrical interpretation

Let us put $Q = G_X^T \left[G_X G_X^T \right]^{-1} G_X$

$\left[1 - Q \right]$ would then act as an orthogonal projection operator on the sub-space defined by $G_X \underline{v} = \underline{0}$ (In the 2 dimensional this corresponds to the projection of the gradient on the tangent plane at the constraint as shown in Fig. 3.13).

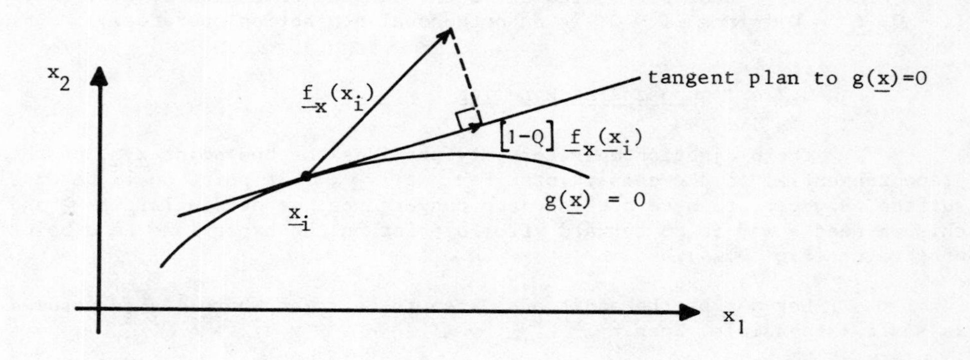

Fig. 3.13

Proof : To see the above geometrical result, let us apply the transformation $\left[1 - Q \right]$ to $\underline{v} \in R^m$; we should obtain a vector $\underline{a} \in \mathcal{E}$ such that $G_X \underline{a} = \underline{0}$. But

$$\underline{a} = \left[1 - Q \right] \underline{v} \qquad G_X \underline{a} = G_X \left[1 - G_X^T \left[G_X G_X^T \right]^{-1} G_X \right] \underline{v} = \left[G_X - G_X G_X^T \left[G_X G_X^T \right]^{-1} G_X \right] \underline{v}$$

$$= \left[G_X - G_X \right] \underline{v} = \underline{0}$$

Next we show that this is an orthogonal transformation. If it was indeed, then $\underline{b} = \underline{v} - \underline{a}$ is perpendicular to all vectors $\underline{w} \in \mathcal{E}$, i.e. to all vectors which satisfy $G_X \underline{w} = \underline{0}$.

Put $\underline{w}^T \underline{b} = 0$. But $\underline{b} = \underline{v} - \underline{a} = Q \underline{v}$

$$\underline{w}^T \underline{b} = \underline{w}^T G_X \left[G_X G_X^T \right]^{-1} G_X \underline{v} = 0 \qquad \text{(since } G_X \underline{w} = 0)$$

Thus $\left[I - Q \right] \underline{f}_x (\underline{x}_i)$ is the projection of the gradient on the hypersurface tangential to the constraints and this corresponds to motion along the steepest slope in the feasible region .

Remark

$$\underline{x}_{i+1} = \underline{x}_i = \hat{\underline{x}} \qquad \text{if a) } k \rightarrow 0 \quad \text{b) } \left[I - Q \right] \underline{f}_x \rightarrow 0$$

The case a) obviously does not correspond to the solution but at the solution, we can indeed show, using the first order necessary conditions, that $\left[I - Q \right] \underline{f}_x = 0$.

To see this in the 2 variable case for example, we have

$$L = f(x, y) + \lambda g(x, y)$$

$$L_x = \frac{\partial f}{\partial x} + \frac{\lambda \partial g}{\partial x} = 0 \quad ; \quad L_y = \frac{\partial f}{\partial y} + \frac{\lambda \partial g}{\partial y} = 0$$

$\dfrac{\partial f/\partial x}{\partial f/\partial y} = \dfrac{\partial g/\partial x}{\partial g/\partial y} \Longrightarrow$ the level curves ($f(x, y)$ = constant) and the constraints ($g(x, y) = 0$) have the same tangent ; the gradient of $f(x, y)$ is thus perpendicular to the tangent to the constraint curve and $(I - Q)\, \underline{f}_x = 0$ since $(I - Q)$ is an orthogonal projection operator.

3.2.4.3 Constraint handling

The projection operator only provides the one point $\underline{x}_{i+1}$ on the hyperplane tangential to the constraints. For small $\underline{\delta}$, this point could be on either surface. However, to have a reasonable convergence, we need a larger $\underline{\delta}$ and for this we need a way to go forward from a point on the hyperplane to a point on the constraint (Fig. 3.14).

Let $\underline{\delta}'_i$ be the additional step to be taken where δ'_i is assumed to be as small as possible. Then :

$$f(\underline{x}_i + \underline{\delta}_i + \underline{\delta}'_i) = 0 = g(\underline{x}_i + \underline{\delta}_i) + \underbrace{G_X(\underline{x}_i + \underline{\delta}_i)}_{\gamma_i} \underline{\delta}'_i + 0^2$$

The problem could be posed as one of optimisation where we

$$\text{minimise } \underline{\delta}'^T_i \underline{\delta}'_i$$

$$\text{subject to } \underline{\gamma}_i + G_X \underline{\delta}'_i = \underline{0}$$

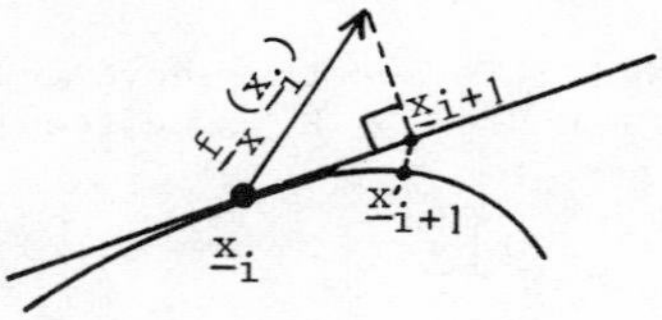

The Lagrangian here is

$$L(\underline{\delta}'_i, \underline{\lambda}) = \underline{\delta}'^T_i \underline{\delta}'_i + \underline{\lambda}^T [\underline{\gamma}_i + G_X \underline{\delta}'_i]$$

Fig. 3.14

The first order conditions (which are in this case also sufficient since we have a convex cost function and linear equality constraints) give

$$\underline{L}_{\delta'_i} = 2 \underline{\delta}'_i + G_X^T \underline{\lambda} = \underline{0} \Longrightarrow \underline{\delta}'_i = -\tfrac{1}{2} G_X^T \underline{\lambda}$$

$$\underline{L}_{\lambda} = \underline{\gamma}_i + G_X \underline{\delta}'_i = \underline{0} \Longrightarrow \underline{\gamma}_i - \tfrac{1}{2} G_X G_X^T \quad = \underline{0}$$

Hence $\underline{\lambda} = 2 \left[G_X \, G_X^T \right]^{-1} \underline{\gamma}_i$

and

$$\boxed{\underline{\delta}'_i = - G_X^T \left[G_X \, G_X^T \right]^{-1} \underline{\gamma}_i}$$

We will thus do at each iteration :

$$\underline{x}_{i+1} = \underline{x}_i - k \left[I - 0 \right] \underline{f}_X$$

$$\underline{x}'_{i+1} = \underline{x}_{i+1} - G_X^T \left[G_X \, G_X^T \right]^{-1} \underline{g}(\underline{x}_{i+1})$$

If $\underline{x}'_{i+1}$ is not sufficiently close to the constraint, we can do another iteration and so on until we obtain the required accuracy of constraint satisfaction.

Remark

In the neighbourhood of $\hat{\underline{x}}$, convergence takes place in the subspace defined by $G_X \, \underline{v} = \underline{0}$; we can show that the convergence is linear and that the rate of convergence is determined by the eigenvalues of the restriction $\left[I - k \, F_{XX} \right]$ on $\mathcal{E}$.

Practical implementation : the flow chart (Fig. 3.15)

In order to implement the method, we require at each step to calculate, in addition to $f(\underline{x})$ and $g_i(\underline{x})$:

- $\underline{f}_X(\underline{x}_i)$ (the gradient of $f(\underline{x})$ at $\underline{x}_i$)
- the gradient of each constraint at $\underline{x}_i$ in order to form the matrix $G_X(\underline{x}_i)$
- the gradient of each constraint at $\underline{x}_{i+1}$ to form the matrix $G_X(\underline{x}_{i+1})$ in order to switch between the constraint and the hyperplane.

In order to accelerate the convergence, here also we can use a self-adaptation of the coefficient k. In addition, in order to simplify the calculations, we don't calculate $G_X(\underline{x}_{i+1})$; we use $G_X(\underline{x}_i)$ in its place.

Next we consider Newton's method.

3.2.5 Newton's method

Here we iteratively solve the following system of equations :

$$\phi_X(\hat{\underline{x}}, \underline{\lambda}) = \underline{0} \qquad \text{(where } \phi = f(\underline{x}) + \underline{\lambda}^T \underline{g}(x))$$

$$\phi_\lambda(\hat{\underline{x}}) = \underline{0}$$

The Newton-Raphson method that is used for this solution gives us the iterative algorithm :

$$\begin{bmatrix} \underline{x}_{i+1} \\ \underline{\lambda}_{i+1} \end{bmatrix} = \begin{bmatrix} \underline{x}_i \\ \underline{\lambda}_i \end{bmatrix} - k \begin{bmatrix} F_{XX}(\underline{x}_i) + \sum_i \lambda_i \, G_{i_{XX}}(\underline{x}_i) & G_X^T(\underline{x}_i) \\ G_X(\underline{x}_i) & 0 \end{bmatrix}^{-1} \begin{bmatrix} \underline{f}_X + \sum_i \lambda_i \underline{g}_{i_X}(\underline{x}) \\ g(\underline{x}) \end{bmatrix}$$

If $(\underline{x}_i, \underline{\lambda}_i)$ are in the neighbourhood of $(\hat{\underline{x}}, \underline{\lambda})$, $\hat{\underline{x}}$ being a strict local minimum, then the above matrix will be positive definite and it will thus be non-singular.

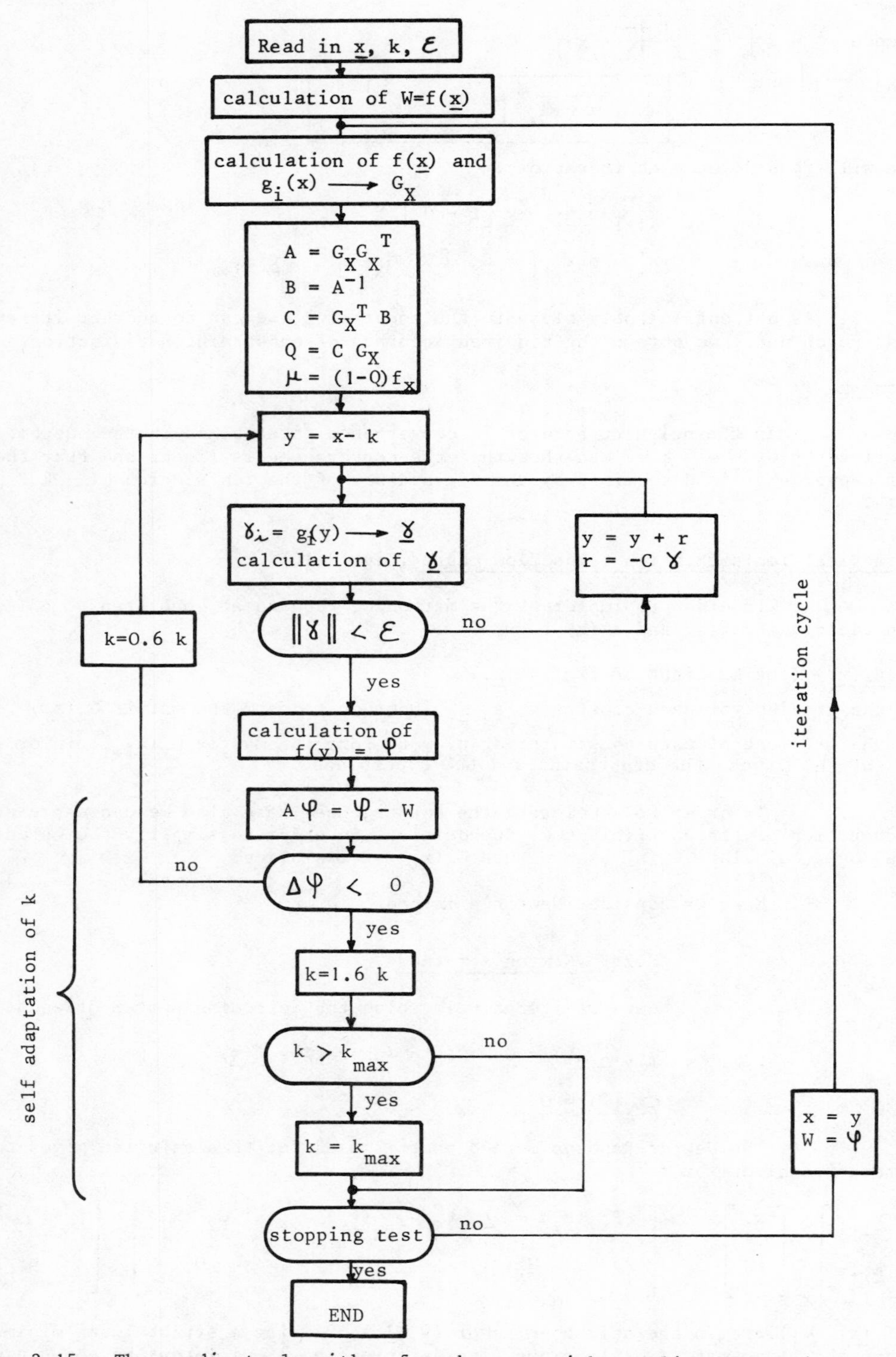

Fig. 3.15 : The gradient algorithm for the case with equality constraints

The calculations here are larger than in the gradient method (since we need to calculate F_{XX}, G_{XX} and solve a system of m+p linear equations), although the convergence is faster. In addition, we could start the iteration with an arbitrary $\underline{x}$ in the region of convergence and we do not necessarily need to satisfy the constraints as in the gradient method.

3.2.6 Penalty function

We have seen above that the search methods for finding a minimum in the presence of equality constraints are relatively complicated. It is therefore interesting to convert the problem somehow to an unconstrained one. This can be done by using a "Penalty function" which penalises constraint violation. The Penalty function could be defined as :

$$P(\underline{x}) = 0 \qquad \text{if } \underline{x} \in \mathcal{D}$$

$$P(\underline{x}) > 0 \qquad \text{if } \underline{x} \notin \mathcal{D}$$

$P(\underline{x})$ is also assumed to be continuous. $P(x) = \sum_i \| g_i(\underline{x}) \|$ for example.

We thus consider the modified cost function :

$$C(\underline{x}) = f(\underline{x}) + K\,P(\underline{x})$$

We can show that when $K \longrightarrow \infty$, the minimum of C goes towards that of $f(\underline{x})$ for $\underline{x} \in \mathcal{D}$. To see this for the case of one constraint only for example, i.e. for $g(\underline{x}) = 0$, we have $P(\underline{x}) = [g(\underline{x})]^2$

The modified cost function is thus :

$$C = f(\underline{x}) + K[g(\underline{x})]^2$$

The stationarity condition yields in this case :

$$\underline{f}_x(x) + 2\,K\,\underline{g}_x(x)\,.\,g(\underline{x}) = \underline{0}$$

But with the Lagrange multipliers, we will have :

$$\underline{f}_x(\underline{x}) + \lambda\,\underline{g}_x(\underline{x}) = \underline{0}$$

Thus the term $2\,K\,g(x)$ plays the role of a Lagrange multiplier : $2\,K\,g = \lambda$. If $K \longrightarrow \infty$, $g \longrightarrow 0$ since λ is non-zero and finite.

If we were to use the gradient method to minimise C, we will have

$$\underline{x}_{i+1} = \underline{x}_i - k[\underline{f}_x(\underline{x}) + K\;\underline{P}_x(\underline{x}_i)]$$

and if $\underline{x}_i$ is far from the constraint, $K\,\underline{P}_x(\underline{x}_i)$ is preponderant and it enables us to switch back to the constraint. This allows us to move along the constraint.

The general flow chart of the method is given in Fig. 3.16.

Next we consider the problem of minimisation subject to inequality constraints.

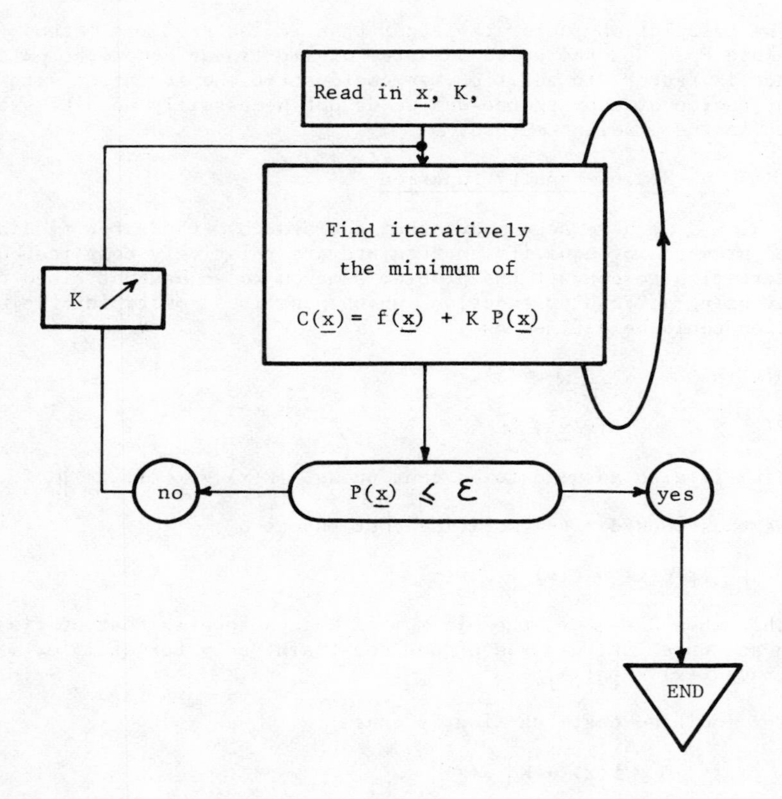

Fig. 3.16 : Use of the Penalty functions

3.3 MINIMISATION SUBJECT TO INEQUALITY CONSTRAINTS

Here the feasible region $\mathcal{D}$ is defined by the inequality relationship

$$h_i(\underline{x}) \leqslant 0 \qquad 1 \leqslant i \leqslant q \qquad \underline{x} \subset R^m$$

If $h_i(\underline{x}) = 0$, the constraint is said to be saturated. q could be $>$ m, although only m constraints could be simultaneously saturated.

We assume that at each point of $\mathcal{D}$, the saturated constraints are independent and that $f(\underline{x})$ and $h_i(\underline{x})$ are continuous and differentiable ; we will denote by $K(\underline{x})$ the set indexed by i of the saturated constraints in $\underline{x} \in \mathcal{D}$. ($K(\underline{x}) = \{$ i such that $h_i(\underline{x}) = 0 \}$).

3.3.1 First order conditions

As before, our analysis begins by the development of the first order conditions. Let us consider the function

$$\varphi(\underline{x}, \underline{y}, \underline{\mu}) = f(\underline{x}) + \sum_{i=1}^{q} \mu_i (h_i(\underline{x}) + y_i^2)$$

The first order conditions for φ can be written as

$$\underline{\varphi}_x = \underline{f}_x + \sum_{i=1}^{q} \mu_i \underline{h}_{i_x} = \underline{0} \qquad 3.3.1$$

$$\left.\begin{aligned} \varphi_{y_i} &= 2\mu_i y_i = 0 \\ \varphi_{\mu_i} &= h_i(x) + y_1^2 = 0 \end{aligned}\right\} i = 1 \text{ to } q \qquad \begin{aligned} 3.3.2 \\ 3.3.3 \end{aligned}$$

Let us also consider the function

$$\phi(\underline{x}, \underline{\mu}) = f(\underline{x}) + \sum_i \mu_i h_i(\underline{x})$$

which is of the same form as the function used in the case of equality constraints. Differentiating,

$$\underline{\phi}_x = \underline{f}_x(\underline{x}) + \sum_{i=1}^{q} \mu_i \underline{h}_{i_x} = 0 \qquad 3.3.4$$

which is equivalent to 3.3.1.

As for 3.3.3, it implies that $\mu_i = 0$ if $y_i \neq 0$, i.e.

$h_i(\underline{x}) < 0$ and $\mu_i \neq o$ if $y_i = 0$, i.e. $h_i(\underline{x}) = 0$

Hence the practical rule :

The first order conditions are identical to those for the case with equality constraints. The multipliers μ_i (called the Kuhn-Tucker multipliers) are zero if the corresponding inequality constraint is satisfied and is unsaturated. If the constraints are saturated, the Kuhn-Tucker multipliers are different from zero.

3.3.2 The second order conditions

Consider the matrix of second derivatives of $\varphi(\underline{x}, \underline{y}, \underline{\mu})$ w.r.t. $\underline{x}$, $\underline{y}$.

$$M = \left[\begin{array}{c|c} F_{XX} + \sum_i \mu_i H_{i_{XX}} & \bigcirc \\ \hline \bigcirc & \begin{matrix} 2\mu_i & \\ & \ddots \\ & & 2\mu_q \end{matrix} \end{array}\right] = \left[\begin{array}{c|c} \phi_{XX} & \bigcirc \\ \hline \bigcirc & \Lambda \end{array}\right]$$

The problem is to find the minimum w.r.t. the $\underline{x}$ and $\underline{y}$ variables and subject to the equality constraints $h_i(\underline{x}) + y_i^2 = 0$. The M matrix should thus be non-negative definite w.r.t. the subspace which corresponds to the linear constraints and which is, for small $\|\delta \underline{x}\|$, the space $\mathcal{E}$ of $\delta\underline{x}$, $\delta\underline{y}$ satisfying

$$\underline{h}_{i_x}(\underline{x})\, \delta\underline{x} + 2 y_i\, \delta y_i = 0 \qquad 3.3.5$$

we must thus have

$$\underline{v}^T M \underline{v} \geqslant 0$$

or for $\underline{v} = \begin{bmatrix} \delta \underline{x} \\ \delta \underline{y} \end{bmatrix}$

$$\delta \underline{x}^T \phi_{XX} \delta \underline{x} + \underline{y}^T \Lambda \delta \underline{y} \geq 0$$

and, given the form of Λ, we have finally :

$$\delta \underline{x}^T \phi_{XX} \delta \underline{x} + \sum_{i=1}^{q} 2\mu_i \, \delta y_i^2 \geq 0 \quad \forall \, \delta \underline{x}, \delta \underline{y} \in \mathcal{E} \qquad 3.3.6$$

We need, thus, to distinguish between 2 cases :

a) $i \in K(\underline{x})$, i.e. constraint number i is saturated. This implies that $y_i = 0$. The definition of $\mathcal{E}$ thus reduces to : $\underline{h}_{i_x}{}^T \delta \underline{x} = 0$ and $\delta \underline{y}$ could be arbitrary.

Since we must have :

$$\delta \underline{x}^T \phi_{XX} \, \delta \underline{x} + \sum_i 2\mu_i \, \delta y_i^2 \geq 0 \qquad \forall \delta \underline{x} \in \mathcal{E} \quad \text{and } \forall \, \delta y_i \text{ arbi-}$$

trary, it is necessary that :

1. $\mu_i \geq 0$ (the case when $\delta y_i \longrightarrow \infty$)
2. ϕ_{XX} is a non-negative definite matrix in $\mathcal{E}$ defined by $\underline{h}_{i_x}{}^T \delta \underline{x} = 0$ (the case when $\delta y_i \longrightarrow 0$).

b) $i \notin K(\underline{x})$ (constraints satisfied in the strict form). Then $\mu_i = 0$ and $\mu_i \, \delta y_i^2 = 0$ in 3.3.1 and the relation 3.3.5 is unrestrictive. To see the latter, we note that $\forall \, \delta \underline{x} \; \exists \; \delta y_i$ such that $\underline{h}_{i_x}{}^T(\underline{x}) \delta \underline{x} + 2y_i \, \delta y_i = 0$ and this equation does not impose a restriction. From this we conclude :

We can eliminate y_i in the statement of the results and the second order conditions are summarised by : $\mu_i \geq 0$, $\phi_{XX}(\underline{x}, \underline{\lambda})$ non-negative definite in the space $\mathcal{E}$ defined by $\underline{h}_{i_x}{}^T(\underline{x}) \, \delta \underline{x} = 0$ for $i \in K(\underline{x})$.

This enables us to state the following necessary condition :

If at a point $\hat{\underline{x}}$, $f(\underline{x})$ has a relative minimum in $\mathcal{D}$ defined by the inequality constraint relations $h_i(\underline{x}) \leq 0$, then, there exist Lagrange multipliers $\mu_i \geq 0$ such that the function :

$$\phi(\underline{x}, \underline{\mu}) = f(\underline{x}) + \sum_i \mu_i \, h_i(\underline{x}) = f(\underline{x}) + \underline{\mu}^T \, \underline{h}(\underline{x})$$

satisfies :

1. $\underline{\phi}_X(\hat{\underline{x}}) = 0$
2. $\mu_i \, h_i \, (\hat{\underline{x}}) = 0$, $h_i(\underline{x}) \leq 0$, $\mu_i \geq 0$ $\quad \left[\begin{array}{l} \mu_i = 0 \text{ if } h_i(\underline{x}) < 0 \\ \mu_i > 0 \text{ if } h_i(\underline{x}) = 0 \end{array} \right.$
3. $\phi_{XX}(\hat{\underline{x}})$ is non-negative definite in the subspace of $\underline{v} \in R^m$ which satisfies $\underline{h}_{i_x}(\hat{\underline{x}})^T \underline{v} = 0$ if $i \in K(x)$ $(h_i(x) = 0)$.

Remark : Let us consider the general problem where we have at the same time both equality constraints and inequality constraints. We introduce a multiplier for each constraint. However, only the inequality constraints are subjected to the sign conditions and they cancel themselves when these constraints are strictly satisfied.

Consider thus the problem :

$\min f(\underline{x})$

$$\text{subject to} \quad \left| \begin{array}{lll} h_i(\underline{x}) \leqslant 0 & & i = 1 \text{ to } q \\ g_j(\underline{x}) = 0 & & j = 1 \text{ to } q \end{array} \right.$$

We have :

$$\phi(\underline{x}, \underline{\lambda}, \underline{\mu}) = f(\underline{x}) + \sum_{j=1}^{p} \lambda_j g_j(x) + \sum_{i=1}^{q} \mu_i h_i(x)$$

The first order conditions are :

$$\underline{\phi}_x = \underline{0}$$

$$\underline{\phi}_\lambda = \underline{g}(\underline{x}) = \underline{0}$$

$$\underline{\mu}^T h(\underline{x}) = 0$$

The second order conditions are :

$\mu_i \geqslant 0$, ϕ_{XX} non-negative definite in the subspace of the constraints defined by

$$G_X \ \delta\underline{x} = \underline{0} \qquad \text{and} \qquad \underline{h}_{iX}^T \ \delta\underline{x} = 0 \quad \text{for } i \in K(x)$$

We note that the sign of the Kuhn-Tucker multipliers μ_i changes with the nature of the problem and the inequality constraints as we see in the table given below :

Nature of the constraints \ Nature of the extremum	Max $f(\underline{x})$	Min $f(\underline{x})$
subject to $h_i(\underline{x}) \leqslant 0$	$\mu_i \leqslant 0$	$\mu_i \geqslant 0$
subject to $h_i(\underline{x}) \geqslant 0$	$\mu_i \geqslant 0$	$\mu_i \leqslant 0$

3.3.3 The case of convex functions

3.3.3.1. Necessary and sufficient conditions for $\mathcal{D}$ to be convex (recapitulation)

We recall that the necessary and sufficient conditions for $\mathcal{D}$ which is defined by the relations $h_i(\underline{x}) \leqslant 0$ to be convex is that all the function $h_i(\underline{x})$ be convex.

One result of this requirement is that if any of the constraints that define $\mathcal{D}$ are equality constraints, they must be linear. To see this, we note that $g_i(\underline{x}) = 0$ implies that $g_i(\underline{x}) \leqslant 0$ and $-g_i(\underline{x}) \leqslant 0$ and $g_i(\underline{x})$, $-g_i(\underline{x})$ can not both be convex unless $g_i(\underline{x})$ is linear.

3.3.3.2 Minimum of a convex function on a convex set

In this case the first order conditions are both necessary and sufficient to define an absolute minimum.

If the objective function and the inequality constraints are convex, continuous and continuously derivable up to first order and the equality constraints are linear, then all points $\hat{\underline{x}}$ which are such that

- all the constraints are satisfied
- the equality constraints and the saturated inequality constraints are independent and there exist multipliers λ_i and μ_i such that

$$\underline{\phi}_x(\hat{\underline{x}}, \underline{\lambda}, \underline{\mu}) = \underline{0} \ ; \quad \mu_i \geqslant 0$$

$$\mu_i h_i(\hat{\underline{x}}) = 0, \quad g_i(\hat{\underline{x}}) = 0$$

Then $\hat{\underline{x}}$ is an absolute minimum on $\mathcal{D}$. To see this, let

$$\phi(\underline{x}, \underline{\lambda}, \underline{\mu}) = F(\underline{x}) + \sum_{i=1}^{p} \lambda_i g_i(\underline{x}) + \sum_{i=1}^{q} \mu_i h_i(\underline{x})$$

$\phi(\underline{x}, \underline{\lambda}, \underline{\mu})$ is a continuous function. It is also convex since $h_i(x)$ is convex and $g_i(x)$ is linear ($\mu_i \geqslant 0$). Since $\underline{\phi}_X(\hat{\underline{x}}, \underline{\lambda}, \underline{\mu}) = \underline{0}$, ϕ has an absolute minimum on R^m at $\hat{\underline{x}}$.

Let $\phi(\hat{\underline{x}}) \leqslant \phi(\underline{x})$

but at $\hat{\underline{x}}$, $\mu_i h_i(\hat{\underline{x}}) = 0 \qquad i = 1$ to q

$g_i(\hat{\underline{x}}) = 0 \qquad i = 1$ to p

This implies that $\phi(\hat{\underline{x}}) = f(\hat{\underline{x}})$

For $\underline{x} \in \mathcal{D}$, $h_i(\underline{x}) \leqslant 0$, $g_i(\underline{x}) = 0$

This implies that $\phi(\underline{x}) \leqslant f(x)$ since $\mu_i \geqslant 0$. Hence :

$$\underline{x} \in \mathcal{D}, \ f(x) \geqslant \phi(x), \quad \phi(\hat{x}) = F(\hat{x})$$

and $f(x)$ has an absolute minimum on $\mathcal{D}$.

3.3.4 The MINMAX Formula

$$\text{Let } \phi(\underline{x}, \underline{y}) = f(\underline{x}) + \sum_{i=1}^{q} y_i h_i(\underline{x}) \qquad \underline{x} \in R^m, \ y_i \geqslant 0$$

$\mathcal{D}$ defined by $h_i(\underline{x}) \leqslant 0$

Let $\underline{x}^*$ and $\underline{y}^*$ be such that

$$\phi(\underline{x}^*, \underline{y}^*) = \min_{\underline{x} \in R^m} \max_{\underline{y} \geq 0} \phi(\underline{x}, \underline{y})$$

we can show that

1. $\underline{x}^* \in \mathcal{D}$

2. $\underline{x}^*$ is the absolute minimum of $f(\underline{x})$ in $\mathcal{D}$.

It follows that for $\hat{\underline{x}}$ to be the absolute minimum of $f(\underline{x})$ in $\mathcal{D}$, it is sufficient for it to equal $\underline{x}^*$ which is the solution of the min-max problem.

However, we can not conclude, without additional assumptions on $f(\underline{x})$ and $\mathcal{D}$ that if $\hat{\underline{x}}$ is the absolute minimum of $f(\underline{x})$ in $\mathcal{D}$, it is the solution of the min-max problem.

On the other hand, we had in the case of the problem with equality constraints that : $\underline{x}^*$ solution of the min-max problem $\underline{x}^* = \hat{\underline{x}}$ the absolute minimum of $f(\underline{x})$ in $\mathcal{D}$.

For the inequality constraints, in the general case, we only have : x^* solution of the min-max problem implies that $\underline{x}^* = \hat{\underline{x}}$ the absolute minimum of $f(\underline{x})$in $\mathcal{D}$.

3.3.5 Examples

I - $\min f = x_1^2 + x_2^2 + x_1x_2 + 2x_1$

subject to
$$\begin{cases} h_1 = x_1^2 + x_2^2 - 1.5 \leq 0 \\ h_2 = x_1 \leq 0 \\ h_3 = -x_2 \leq 0 \end{cases}$$

The admissible region $\mathcal{D}$ is shown in Fig. 3.17

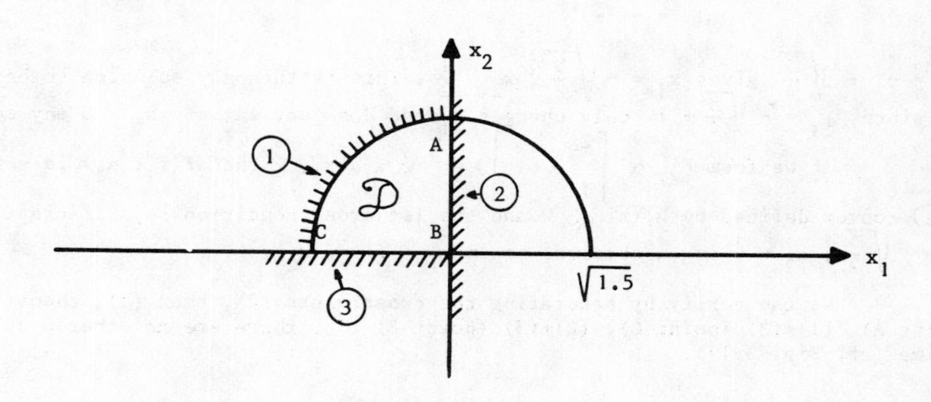

Fig. 3.17

Consider the Lagrangian

$$\varphi(x_1,x_2,\mu_1,\mu_2,\mu_3) = x_1^2 + x_2^2 + x_1x_2 + 2x_1 + \mu_1(x_1+x_2^2 - 1.5)+\mu_2x_1 - \mu_3 x_2$$

a) $\mu_1 = \mu_2 = \mu_3 = 0$ unconstrained problem ($\hat{\underline{x}} \in \mathcal{D}$? is there an extremal point in $\mathcal{D}$?)

$$\varphi = x_1^2 + x_2^2 + x_1x_2 + 2x_1 \Longrightarrow \begin{array}{l} \varphi x_1 = 2x_1 + x_2 + 2 = 0 \\ \varphi x_2 = 2x_2 + x_1 = 0 \end{array} \Longrightarrow \left| \begin{array}{l} x_1 = -\frac{4}{3} \\ x_2 = +\frac{2}{3} \end{array} \right. \notin \mathcal{D}$$

b) constraint 1 saturated, $\mu_1 > 0$, $\mu_2 = \mu_3 = 0$ (constraints 2 and 3 satisfied).

$$\varphi(x_1, x_2,\mu) = x_1^2 + x^2_2 + x_1x_2 + 2x_1 + \mu_1 (x_1^2 + x_2^2 - 1.5)$$

$$\varphi x_1 = 2x_1 + 2\mu_1x_1 + x_2 = 0 \qquad 3.3.7$$

$$\varphi x_2 = 2x_2 + x_1 + 2\mu_1x_2 = 0 \qquad 3.3.8$$

$$x_1^2 + x_2^2 = 1.5 \Longrightarrow \varphi_{\mu_1} = 0 \qquad 3.3.9$$

Solving this set of equations in x_1, x_2, μ_1, 3.3.7 and 3.3.8 gives

$$x_2 = \frac{-2}{1 - 4(1+\mu_1)^2} \qquad x_1 = \frac{4(1+\mu_1)}{1 - 4(1+\mu_1)^2}$$

On putting these values in 3.3.9, we have :

$$1.5(1-\alpha)^2 - 4\alpha - 4 = 0$$

with $\alpha = 4(1+\mu_1)^2$

Let $1.5\alpha^2 - 7\alpha - 2.5 = 0$; a convenient positive solution is $\alpha = 5$ and this gives

$$\mu_1 = -1 \pm \sqrt{\frac{5}{4}}$$

$\mu_1 = -1 + \sqrt{\frac{5}{4}}$ gives $x_1 = -\sqrt{\frac{5}{2}}$, $x_2 = \frac{1}{2}$. This is the only solution to be retained since $\mu_1 = -\sqrt{\frac{5}{4}} + 1$ only changes x_1 and does not satisfy $h_i = 0$ any more.

If we form $F_{XX} = \begin{bmatrix} 2 & 1 \\ 1 & 2 \end{bmatrix} > 0 \quad \forall\, x_1, x_2,$ then $f(x_1, x_2)$ is convex on $\mathcal{D}$ convex defined by $h_i(x) \leqslant 0$ and the 1st order condition is sufficient. Thus $x_1 = -\sqrt{\frac{5}{2}}$, $x_2 = \frac{1}{2}$ is an absolute minimum of $f(x_1, x_2)$.

We can verify by saturating the constraints (2), then (3), then (1)+(2) (point A), (1)+(3) (point C), (2)+(3) (point B) that there are no other absolute minima (cf. Fig. 3.17).

$$\text{II} - \max f = x_1^2 + x_2^2 + x_1x_2 + 2x_1$$

subject to the same constraints

$f(x_1, x_2)$ being convex (and not concave) we can not obtain sufficient conditions and we can not limit ourselves to 1st order conditions ; we must examine all the possibilities for constraint saturation.

There is no stationary point $\in \mathcal{D}$ and $h_1 = 0$ gives a min. We can thus start directly with $h_2 = 0$.

a) $h_2 = 0 \quad \mu_2 < 0, \mu_1 = 0 \quad \mu_3 = 0$

$$\varphi = x_1^2 + x_2^2 + x_1x_2 + 2x_1 + \mu_2 x_1$$

$$\left.\begin{aligned} \varphi x_1 &= 2x_1 + x_2 + 2 + \mu_2 = 0 \\ \varphi x_2 &= 2x_2 + x_1 = 0 \\ \varphi \mu_2 &= x_1 = 0 \end{aligned}\right\} \Longrightarrow \begin{aligned} x_1 &= 0 \\ x_2 &= 0 \\ \mu_2 &= -2 \end{aligned}$$

This point which gives $\mu_2 < 0$ could be solution and we need to examine the 2nd order conditions. However, it corresponds to 2 saturated constraints (point (C)) and we will examine it later.

b) $h_3 = 0 \quad \mu_3 < 0, \quad \mu_1 = \mu_2 = 0$

$$\varphi = x_1^2 + x_2^2 + x_1x_2 + 2x_1 - \mu_3 x_2$$

$$\left.\begin{aligned} \varphi x_1 &= 2x_1 + x_2 + 2 = 0 \\ \varphi x_2 &= 2x_2 + x_1 - \mu_3 = 0 \\ \varphi \mu_3 &= -x_2 = 0 \end{aligned}\right\} \begin{aligned} x_2 &= 0 \\ x_1 &= -1 \\ \mu_3 &= -1 \end{aligned}$$

Here also, the sign condition on μ_3 is satisfied and we must examine the second order conditions :

$$\varphi_{XX} = \begin{bmatrix} 2 & 1 \\ 1 & 2 \end{bmatrix}$$

is positive definite and it is not necessary to examine the restriction which will give

$$\begin{Vmatrix} \omega-2 & -1 & 0 \\ -1 & \omega-2 & -1 \\ 0 & -1 & 0 \end{Vmatrix} = 0 \Longrightarrow \omega = 2 > 0,$$

and the necessary condition for a maximum is not satisfied.

c) $h_1 = 0, h_2 = 0, \mu_1 < 0, \mu_2 < 0, \mu_3 = 0$

$$\varphi = x_1^2 + x_2^2 + x_1x_2 + 2x_1 + \mu_1(x_1^2 + x_2^2 - 1.5) + \mu_2 x_1$$

$$\varphi x_1 = 2x_1 + x_2 + 2 + 2\mu_1 x_1 + \mu_2 = 0 \qquad x_1 = 0$$

$$\varphi x_2 = 2x_2 + x_1 + 2\mu_1 x_2 = 0 \qquad x_2 = + \quad 1.5 \; (-\sqrt{1.5} \notin \mathcal{D})$$

$$\varphi_{\mu 1} = x_1^2 + x_2^2 - 1.5 = 0 \qquad \mu_1 = -1 < 0$$

$$\varphi_{\mu 2} = x_1 = 0 \qquad \mu_2 = -(2+\sqrt{1.5}) < 0$$

Given the sign conditions on μ_1, μ_2, this point could be a maximum and it is indeed since there are 2 saturated constraints for 2 variables and no more variation is possible around this stationary point and the 1st order conditions specifically on the sign of μ_1, μ_2 suffice. The criterion here has a value f = 1.5.

d) $h_1 = 0$, $h_3 = 0$, $\mu_1 < 0$ $\mu_3 < 0$ $\mu_2 = 0$

$$\varphi = x_1^2 + x_2^2 + x_1 x_2 + 2x_1 + \mu_1(x_1^2 + x_2^2 - 1.5) + \mu_3(-x_2)$$

$$\varphi x_1 = 2x_1 + x_2 + 2 + 2\mu_1 x_1 = 0 \qquad x_2 = 0$$

$$\varphi x_2 = 2x_2 + x_1 + 2\mu_1 x_2 - \mu_3 = 0 \qquad x_1 = -\sqrt{1.5} \ (+\sqrt{1.5} \notin \mathcal{D})$$

$$\varphi \mu_1 = x_1^2 + x_2^2 - 1.5 = 0 \qquad \mu_1 = -\frac{(1+\sqrt{1.5})}{\sqrt{1.5}} < 0$$

$$\varphi \mu_3 = -x_2 = 0 \qquad \mu_3 = -\sqrt{1.5} < 0$$

As for the point (C) this point is also a maximum and leads to the cost value :

$$f = 1.5 - 2\sqrt{1.5}.$$

e) $h_2 = 0$, $h_3 = 0$, $\mu_1 = 0$, $\mu_2 < 0$, $\mu_3 < 0$

$$\varphi = x_1^2 + x_2^2 + x_1 x_2 + 2x_1 + \mu_2 x_1 - \mu_3 x_2$$

$$\left.\begin{aligned} \varphi x_1 &= 2x_1 + x_2 + \mu_2 + 2 = 0 \\ \varphi x_2 &= 2x_2 + x_1 - \mu_3 = 0 \\ \varphi \mu_1 &= x_1 = 0 \\ \varphi \mu_3 &= -x_2 = 0 \end{aligned}\right\} \qquad \begin{aligned} x_1 &= 0 \\ x_2 &= 0 \\ \mu_2 &= -2 < 0 \\ \mu_3 &= 0 \end{aligned}$$

$\mu_3 = 0$ does not allow us to conclude directly but allows a variation δx_2 of x_2 which gives $\Delta f = (\delta x_2)^2$ and gives a minimum.

In summary, the point (C) is the absolute maximum and d) is a local maximum.

Next we consider iterative methods for this general case with inequality constraints.

3.3.6 The gradient method

Let $\underline{x}_n$ and $K(\underline{x}_n)$ be such that

$$h_i(\underline{x}_n) = 0 \qquad \forall i \in K_n$$

$$h_i(\underline{x}_n) < 0 \qquad \forall i \notin K_n$$

Let K' be a set indexed by i with $K' \subset K_n$: it defines in the neighbourhood of $\underline{x}_n$ a region $S_{K'}$, whose boundary $\mathcal{D}'$ corresponds to

$$h_i(\underline{x}_n) = 0 \qquad \forall i \in K' \subset K_n$$

If $\| \delta \underline{x}_n \|$ is sufficiently small, the unsaturated constraints are inactive and we have the problem of finding an extremum in the presence of equality constraints and for this case we can apply the gradient method described previously. We define thus

$$Q(K') = M^T(\underline{x}_n) \left[M(\underline{x}_n)\, M^T(\underline{x}_n) \right]^{-1} M(\underline{x}_n)$$

where $M(\underline{x}_n)$ is the matrix with rows $\underline{h}_{i_x}$ for $i \in K'$.

The next step will be $\delta \underline{x}_n = -k \left[1 - Q \right] \underline{f}_x$

But, for this, it is essential that K' be feasible, i.e. the corresponding step does not violate the inequality constraints for $i \in K_n - K'$, or

$$\alpha_i = \underline{h}_{i_x}^T(\underline{x}_n)\, \delta \underline{x}_n \leqslant 0 \qquad \forall_i \in K_n - K'$$

We need thus to solve the following problem :

find $K' \subset K_n$ such that

$$\Delta f = \underline{f}_x^T \left[1 - Q \right] \underline{f}_x \text{ be maximal } (\Delta f = \underline{f}_x^T \, \delta \underline{x}_n) \qquad 3.3.10$$

with $$-\underline{h}_{i_x}^T \left[1 - Q \right] \underline{f}_x \leqslant 0 \qquad \forall_i \in K_n - K' \qquad 3.3.11$$

Let r be the number of elements in K_n $(r \leqslant n)$. To find K', it is necessary to compute 3.3.11 for all ℓ by ℓ different combinations of the r components of K_n from $\ell = 0$ to $\ell = r$. For the different feasible K' that are thus defined, we evaluate Δf by 3.3.10 and we thus find from it $\hat{K}'$ (K' optimal).

Such a search can be long and it is shortened considerably if we note that :

- If $K' \subset K''$ then $\begin{cases} K'' \text{ feasible} \Longrightarrow K' \text{ feasible} \\ K' \text{ non feasible} \Longrightarrow K'' \text{ non feasible} \end{cases}$ 3.3.12

- If $K'' \subset K' \Longrightarrow \quad |\Delta f|_{K'} < |\Delta f|_{K''}$ 3.3.13

(the more constraints we put in, the lower is the value of the optimal solution in Δf).

This leads to the two methods of finding $\hat{K}'$:

- we start with K_n and we consider the sets with decreasing numbers of elements by using 3.3.12.
- we start with the empty set $(S_{K'} = R^m)$ of saturated constraints and we consider the set with increasing number of elements and we stop the search when we find the solution from 3.3.13.

The solution thus found enables us to define the direction $\left[1 - Q(\hat{K}') \right] \underline{f}_x$ in which we take the next step using the gradient projection method with switching and this determines $\underline{x}_{n+1}$.

If $\underline{x}_{n+1}$ satisfies all the constraints for $i \notin \hat{K}'$, it serves for a new iteration with $K_{n+1} = \hat{K}'$

If $\underline{x}_{n+1}$ violates one or more constraints for $i \notin \hat{K}'$, we reduce k up to the saturation of the new constraints and this gives $\underline{x}_{n+1}$ which can be used with $K_{n+1} = \hat{K}'$ + the index of the new saturated constraints

Such an algorithm is illustrated on Fig. 3.18 and the flow chart is given in Fig. 3.19.

To reduce the above calculations, the simplified gradient method avoids switching on the constraint as long as we do not violate a new constraint and the function $f(\underline{x})$ does not increase again. (Fig. 3.20).

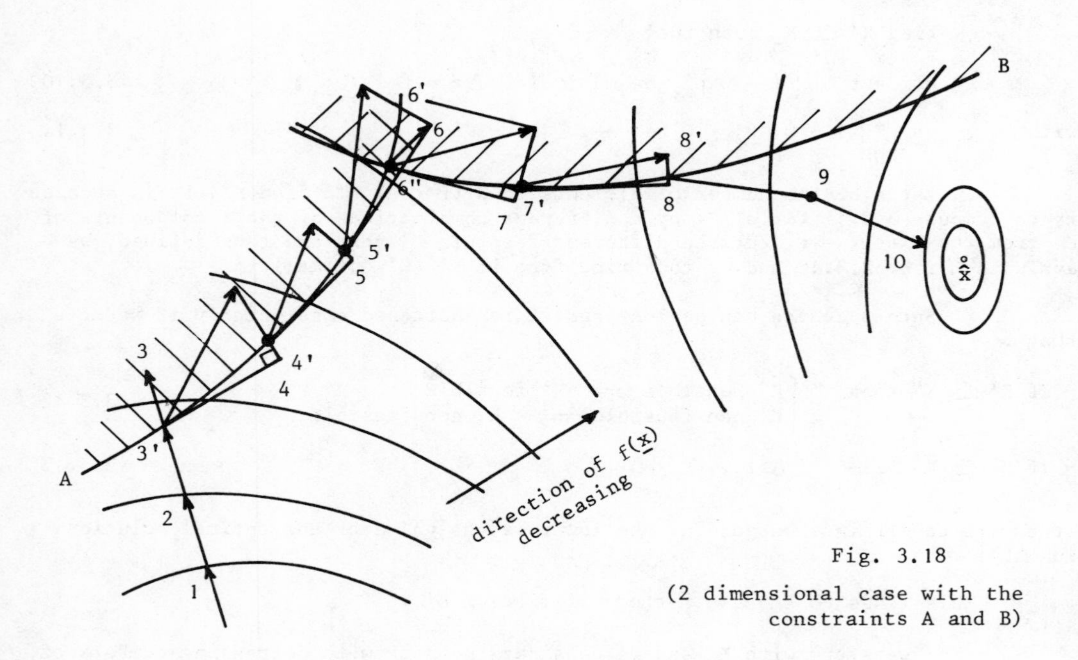

Fig. 3.18

(2 dimensional case with the constraints A and B)

1,2 : unconstrained advance ; 3 : violation of A, reduction of k to satisfy A ⟶ 3' ; 4,5 : advance along A by projection on the tangent plane to the constraint (4,5) and switching (4',5') ; 6 : projection of the gradient on the tangent plane (6), switch (6'), which does not satisfy B, reduction of k to satisfy B (6") ; calculation of the gradient and projection on A (violation of B) and on B which satisfies A. Hence the progression 7, 8.

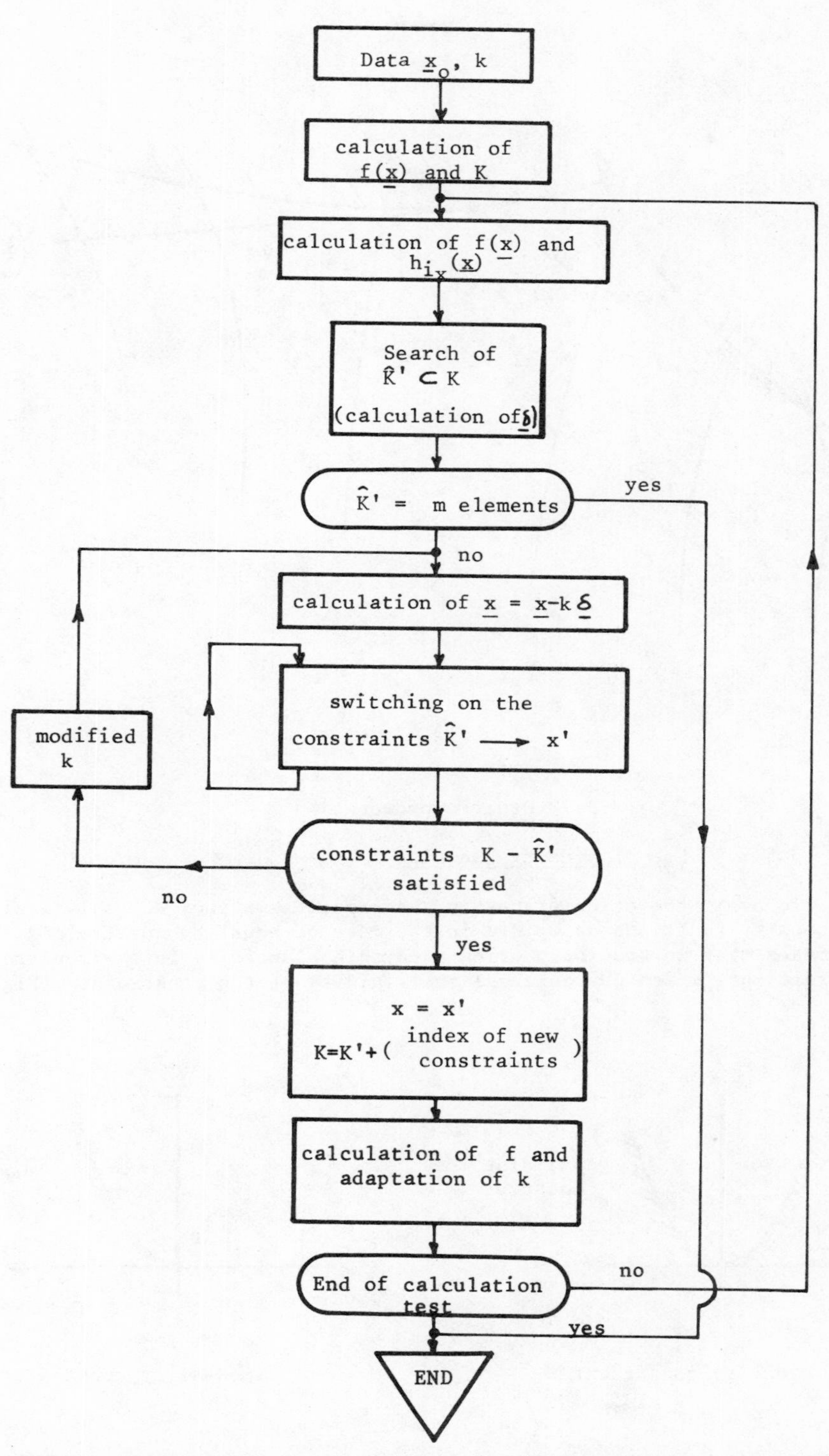

Fig. 3.19 : Flow chart of the gradient method with inequality constraints

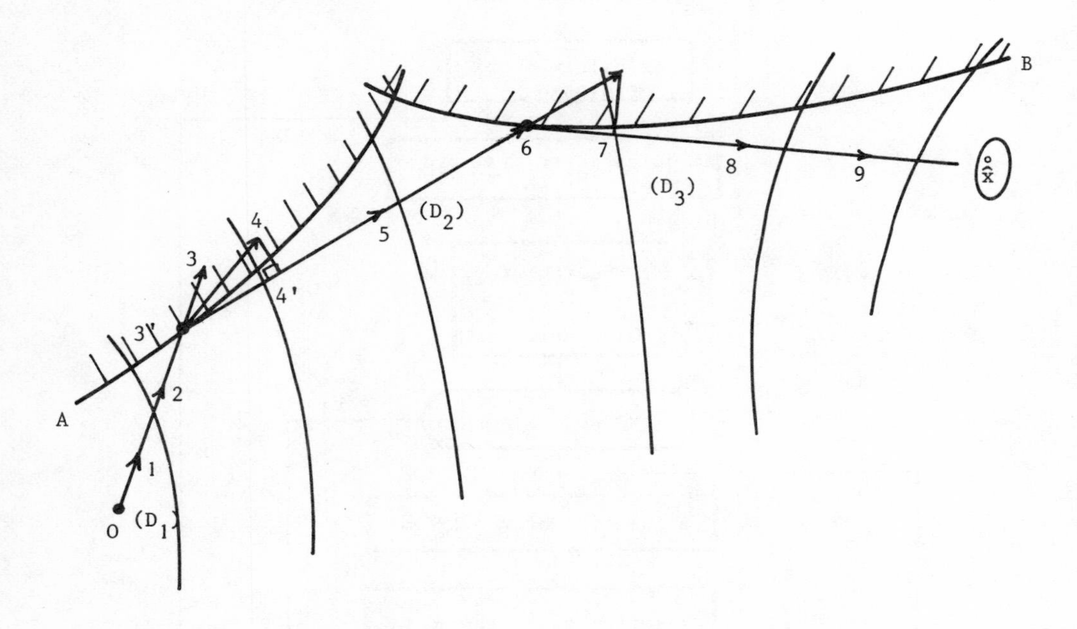

Fig. 3.20

3.3.7 Other approaches

3.3.7.1 Use of penalty functions

To avoid the calculations in the gradient method, we could again consider using penalty functions as we did in the case of equality constraints. This basically means that we are introducing a certain elasticity in the constraints and in the coefficients which characterise the rigidity of the constraints (Fig. 3.21).

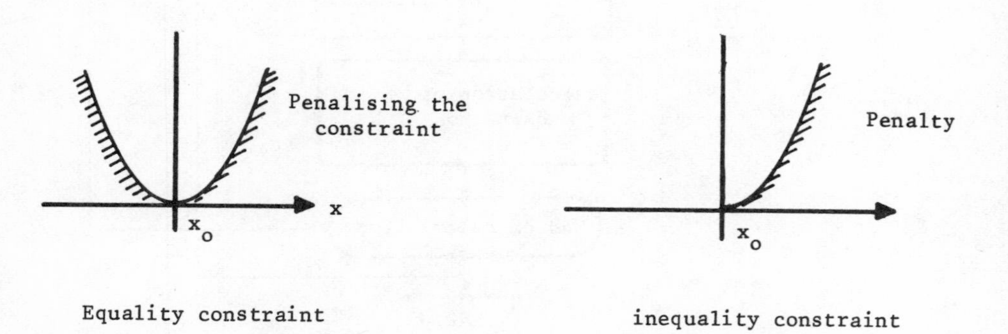

Fig. 3.21

The penalty function approach involves defining a penalty function

$$P_j(\underline{x}) = 0 \quad \text{if} \quad h_j(\underline{x}) \leqslant 0$$

$$P_j(\underline{x}) > 0 \quad \text{if} \quad h_j(\underline{x}) > 0$$

and to extremise the modified criterion

$$\mathcal{C} = f(\underline{x}) + \sum_j K_j P_j(\underline{x})$$

which basically converts it to an unconstrained problem and we have many algorithms to solve such problems as we saw in the beginning of this chapter.

Here also, if $K_j \longrightarrow \infty$, the minimum of $\mathcal{C}$ tends towards the solution of the original problem.

3.3.7.2 <u>The Arrow-Hurwicz method</u>

Consider the problem :

$$\text{Min } f(\underline{x}) \quad \text{subject to } \underline{g}(\underline{x}) = \underline{0},\ \underline{h}(\underline{x}) \leqslant \underline{0}$$

and the associated Lagrangian

$$L = f(\underline{x}) + \underline{\lambda}^T \underline{g}(\underline{x}) + \underline{\mu}^T \underline{h}(\underline{x})$$

The Arrow-Hurwicz method uses the following algorithm to determine $\underline{x}$, $\underline{\lambda}$, $\underline{\mu}$. In equation 3.3.14, we have the continuous time method, and in equation 3.3.15 the discrete time method (t : iteration time) :

$$\left.\begin{aligned} \frac{d\underline{x}}{dt} &= -\underline{L}_x \\ \frac{d\underline{\lambda}}{dt} &= -\underline{L}_\lambda \\ \frac{d\underline{\mu}}{dt} &= \begin{cases} 0 & \text{if } \underline{L}_\mu < 0 \text{ and } \underline{\mu} = \underline{0} \\ -\underline{L}_\mu & \text{otherwise} \end{cases} \end{aligned}\right\} \quad 3.3.14$$

This assures that $\underline{\mu}(t) \geqslant 0$ if $\mu(o) = 0$

$$\left.\begin{aligned} \underline{x}(t+1) &= \underline{x}(t) - K\,\underline{L}_x(t) \\ \underline{\lambda}(t+1) &= \underline{\lambda}(t) + K\,\underline{L}_\lambda(t) \\ \underline{\mu}(t+1) &= \max\left\{0,\ \underline{\mu}(t) + K\,\underline{L}_\mu(t)\right\} \end{aligned}\right\} \quad 3.3.15$$

K = iteration constant.

In order to provide a link between what we have done so far and the control problem that we will consider explicitly in parts 2, 3 and 4 of this book, let us use in this final section of this chapter, Mathematical programming techniques in order to solve optimal control problems.

3.4 APPLICATION OF MATHEMATICAL PROGRAMMING TO SOLVE CONTROL PROBLEMS

3.4.1 Theoretical optimality conditions

3.4.1.1 The problem

In the case of direct digital control, the process and the controller could be schematically described as in Fig. 3.22.

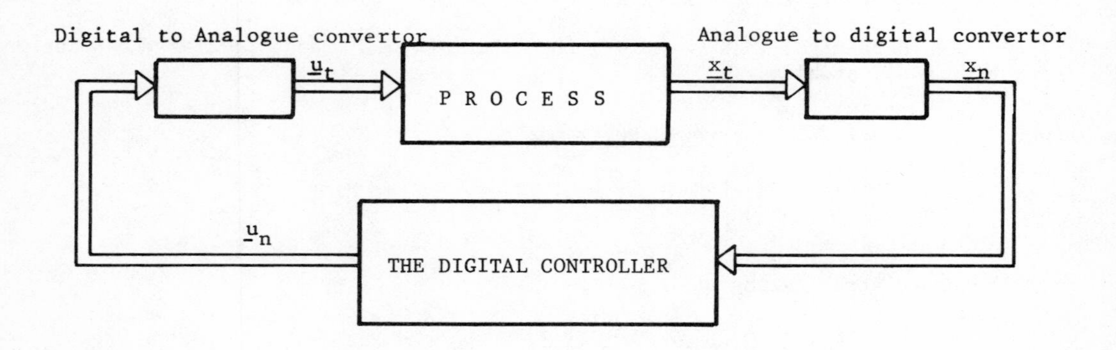

Fig. 3.22
Digital control structure

Here, the digital controller (computer) periodically (with a period T) provides a control $\underline{u}_n$ and it receives coded information $\underline{x}_n$ from the process.

The process is described by :

$\dot{\underline{x}} = F(\underline{x}, \underline{u})$ $\underline{x} \in R^m$ is the state vector, $\underline{u} \in R^{\ell}$ is the control vector. The equivalent discrete time equation obtained at the sampling instants n and n+1 is

$$\underline{x}_{n+1} = \underline{f}_n(\underline{x}_n, \underline{u}_n) \qquad 3.4.1$$

The problem is not only to determine the theoretical optimality conditions, but also the optimal control law which can be implemented, optimality here being with respect to some defined criterion.

The optimisation horizon is bounded and fixed and is given by $[1, N+1]$; starting from $\underline{x}_1$ the initial state, we wish to reach $\underline{x}_{N+1}$ by using the controls $\underline{u}_1, \underline{u}_2, \dots \underline{u}_N$.

* The state vector and the feasible controls will be defined by constraint relations of the type :

$$\text{- instantaneous :} \quad \underline{\gamma}_n(\underline{x}_n, \underline{u}_n) \leqslant \underline{0} \qquad 3.4.2$$

where $\underline{\gamma}_n$ is a vector function having p components and $1 \leqslant n \leqslant N$

(Ex : amplitude bound on the control).

$$\text{- global : } \underline{g}(\underline{x}_1, \underline{x}_2, \ldots, \underline{x}_N, \underline{u}_1, \ldots, \underline{u}_N) \leqslant \underline{0} \qquad 3.4.3$$

where $\underline{g}$ is a vector function with q components.

(If $\underline{x}_{N+1}$ appears in this constraint, we can eliminate it using 3.4.1. Example : limitation of energy).

* The initial and final states could be fixed, free or related by terminal constraints of the type :

$$n = 1 \quad \underline{h}'(\underline{x}_1) = \underline{0} \quad ; \quad n = N+1, \ \underline{h}''(\underline{x}_{N+1}) = \underline{0} \qquad 3.4.4$$

* The performance function will be :

$$R = r(\underline{x}_1, \ldots, \underline{x}_{N+1}, \underline{u}_1, \ldots, \underline{u}_N) \qquad 3.4.5$$

The problem is to find the optimal control law $\underline{u}_1, \underline{u}_2, \ldots, \underline{u}_N$ which takes the system from the initial state to the final state and which satisfies the constraints 3.4.1 to 3.4.4 and which minimises 3.4.5.

3.4.1.2 Necessary conditions for optimality

We assume that the functions that we use are continuous and continuously differentiable w.r.t. the appropriate variables. To the different constraints, we attach the following Lagrange or Kuhn-Tucker multipliers :

Equality constraints (n = 1 to N)

$$\underline{f}_n(\underline{x}_n, \underline{u}_n) - \underline{x}_{n+1} = \underline{0} \Longrightarrow \underline{\Psi}_n \in R^m$$

$$\underline{h}'(\underline{x}_1) = \underline{0} \Longrightarrow \underline{\xi}' \in R^{r'}$$

$$\underline{h}''(x_{N+1}) = \underline{0} \Longrightarrow \underline{\xi}'' \in R^{r''}$$

Inequality constraints (n = 1 to N)

$$\underline{\gamma}_n(\underline{x}_n, \underline{u}_n) \leqslant 0 \Longrightarrow \underline{\mu}_n \in R^p$$

$$\underline{g}(\underline{x}_1 \ldots \underline{x}_N, \underline{u}_1 \ldots \underline{u}_N) \leqslant 0 \Longrightarrow \underline{\lambda} \in R^q$$

For the overall problem we could write the Lagrangian

$$\phi = \tau + \sum_{n=1}^{N} \underline{\Psi}_n^T [\underline{f}_n - \underline{x}_{n+1}] + \sum_{n=1}^{N} \underline{\mu}_n^T \underline{\gamma}_n + \underline{\lambda}^T \underline{g} + \underline{\xi}^T \underline{h} + \underline{\xi}^T \underline{h}''$$

The first order conditions for optimality are :

a) current instant : $1 < n \leqslant N$

$$\underline{\phi}_{x_n} = \underline{0} = \underline{\tau}_{x_n} + F_{x_n}^T \underline{\Psi}_n - \underline{\Psi}_{n-1} + \Gamma_{x_n}^T \mu_n + G_{x_n}^T \underline{\lambda} = 0 \qquad 3.4.6$$

$$\underline{\phi}_{u_n} = \underline{0} = \underline{\tau}_{u_n} + F_{u_n}^T \underline{\Psi}_n + \Gamma_{u_n}^T \underline{\mu}_n + G_{u_n}^T \underline{\lambda} = \underline{0} \qquad 3.4.7$$

$$\underline{\phi}_{\Psi_n} = \underline{f}_n - \underline{x}_{n+1} = \underline{0}$$

$\gamma_n^i \mu_n^i = 0$ and $\mu_n^i \geqslant 0$ $\quad\left|\begin{array}{ll} \gamma_n^i < 0 & \mu_n^i = 0 \\ \gamma_n^i = 0 & \mu_n^i > 0 \end{array}\right.$

$g^i \lambda^i = 0$ and $\lambda^i \geqslant 0$ $\quad\left|\begin{array}{ll} g^i < 0 & \lambda^i = 0 \\ g^i = 0 & \lambda^i > 0 \end{array}\right.$

b) Initial instant : n = 1

$$\underline{\phi}_{x_1} = \underline{0} = \underline{\tau}_{x_1} + F_{x_1}^T \underline{\Psi}_1 + \Gamma_{x_1}^T + G_{x_1}^T \lambda + H'^T_{x_1} \underline{\xi}' = \underline{0} \qquad 3.4.8$$

$\underline{\phi}_{u_1} = \underline{0}$ (relation identical to 3.4.7)

If we want to use an identical form to 3.4.6 for 3.4.8, we should put :

$$\underline{\Psi}_0 = H'^T_{x_1} \underline{\xi} \qquad 3.4.9$$

This implies that Ψ_0 is orthogonal to the surface $\underline{h}'(\underline{x}_1) = \underline{0}$ at $\underline{x}_1$.

c) Final instant : n = N+1

$$\underline{\phi}_{x_{N+1}} = \underline{\tau}_{x_{N+1}} - \underline{\Psi}_N + H''^T_{x_{N+1}} \underline{\xi}'' = \underline{0} \qquad 3.4.10$$

This can be rewritten as :

$$\underline{\Psi}_N = \underline{\Psi}'_N + \underline{\Psi}''_N \quad \text{with} \quad \left|\begin{array}{l} \underline{\Psi}'_N = \underline{\tau}_{x_{N+1}} \\ \Psi''_N = H''^T_{x_{N+1}} \underline{\xi}'' \end{array}\right. \qquad 3.4.11$$

Ψ''_N is thus normal to the surface $\underline{h}''(\underline{x}_{N+1}) = \underline{0}$, at $\underline{x}_{N+1}$.

These initial and final conditions are the <u>transversality conditions</u>.

d) Conclusion : we can state the following necessary condition :

For $\underline{u}_n$ to be the optimal control which minimises r, it is necessary that there exist the multipliers $\underline{\Psi}_n$, $\underline{\mu}_n \geqslant 0$, and $\underline{\gamma} \geqslant 0$ such that :

$1 \leqslant n \leqslant N$

$$\underline{x}_{n+1} = \underline{f}(\underline{x}_n, \underline{u}_n) \qquad 3.4.12$$

$$\underline{\Psi}_{n-1} = F_{x_n}^T \underline{\Psi}_n + \underline{\tau}_{x_n} + \Gamma_{x_n}^T \underline{\mu}_n + G_{x_n}^T \underline{\lambda} \qquad 3.4.13$$

(dynamical equations)

$$\underline{\tau}_{u_n} + F_{u_n}^T \underline{\Psi}_n + \Gamma_{u_n}^T \underline{\mu}_n + G_{u_n}^T \underline{\lambda} = \underline{0} \qquad 3.4.14$$

$$\mu_n^i \gamma_n^i = 0 \qquad i = 1 \text{ to } p \qquad 3.4.15$$

(static equations)

$$\lambda^i g^i = 0 \qquad i = 1 \text{ to } q \qquad 3.4.16$$

(global static condition)

In addition, for :

* n = 1 $\underline{\Psi}_0$ is normal to the surface $\underline{h}'(\underline{x}) = \underline{0}$ at $\underline{x}_1$ 3.4.17

* n = N+1 $\underline{\Psi}_N = \underline{\Psi}'_N + \underline{\Psi}''_N$ $\underline{\Psi}'_N = \underline{\tau}_{x_{N+1}}$ 3.4.18

$\underline{\Psi}''_N$ normal to the surface $\underline{h}''(\underline{x}) = \underline{0}$ at $\underline{x}_{N+1}$

Remark : By passing from the discrete system to the continuous system, we can show that these equations are consistent with those given by the Maximum principle (cf. Chapter 5) for continuous systems.

3.4.2 The practical search for an optimal solution

The practical problem is to find the solution which satisfies the set of conditions 3.4.12 to 3.4.18. This will in general be the optimal solution to the practical problem (although these are only necessary conditions). However, this solution may not be unique so that the final choice will have to be made by examining the criterion.

Let us next count up the number of relations and variables that enter into our problem solution. We recall that $x_n \in R^m$, $u_n \in R^{\ell}$, h' : r' components, h" : r" components.

Number of relations

Equality relationships : 3.4.12 3.4.13 3.4.14 3.4.17 3.4.18

Number : $Nm + Nm + N\ell + \tau' + (m-\tau') + \tau'' + (m-\tau'')$

The reason is that (relation 3.4.13) $\Psi_n \in R^m$ and (relation 3.4.17 and 3.4.18) : $\underline{h}'(\underline{x}_1) = \underline{0}$ represent τ' relations, $\underline{h}''(\underline{x}_{N+1}) = \underline{0}$ represents τ'' relations, $\underline{\Psi}_0$ is normal to $\underline{h}'(\underline{x}_1) = \underline{0}$ (defining a subspace of dimension τ') i.e. it is in its complement of dimension $(m - \tau')$

(relation $\underline{\Psi}_0 = H'^T_{x_1} \underline{\xi}'$, $\Psi''_N = H''_{x_{N+1}} \underline{\xi}''$)

Inequality relationships : 3.4.15 3.4.16

Number : $Np + q$

As for the number of variables, we have :

$$\underbrace{(N+1)m}_{\underline{x}_1 \text{ to } \underline{x}_{N+1}} + \underbrace{(N+1)m}_{\underline{\Psi}_0 \text{ to } \underline{\Psi}_N} + \underbrace{N\ell}_{\underline{u}_1 \text{ to } \underline{u}_N} + \underbrace{Np}_{\underline{\mu}_1 \text{ to } \underline{\mu}_N} + \underbrace{q}_{\underline{\lambda}}$$

Although there are a sufficient number of equations here to enable us to find all the variables, in general it is not possible to solve these equations directly. It is the boundary conditions 3.4.17 and 3.4.18, and the global conditions 3.4.16 which create most of the difficulties.

Fortunately, the discrete equations 3.4.12 to 3.4.15 are recursive so that we attempt to satisfy 3.4.16 to 3.4.18 from the set of solutions which satisfy 3.4.12 to 3.4.15.

3.4.2.1 Solution of the Recursive Equations

To do this, we need to change 3.4.13 to

$$\underline{\Psi}_n = [F_{X_n}]^{-1} [\underline{\Psi}_{n-1} - \tau x_n - \Gamma_{x_n}^T \underline{\mu}_n - G_{x_n}^T \underline{\lambda}] \qquad 3.4.19$$

Let $[\mathcal{D}_n]$ be the set of data available at the nth instant, i.e.

$$[\mathcal{D}_n] = \begin{cases} \underline{\lambda} \\ \underline{x}_1 \text{ to } \underline{x}_n \\ \underline{\Psi}_0 \text{ to } \underline{\Psi}_{n-1} \\ \underline{\mu}_1 \text{ to } \underline{\mu}_{n-1} \end{cases}$$

Knowing $[\mathcal{D}_n]$, $\underline{\mu}_n$, $\underline{u}_n$ we can calculate $\underline{\Psi}_n$ from 3.4.19. But in order to compute $\underline{\Psi}_n$ in 3.4.19, we need also $\underline{\mu}_n$ from 3.4.14. Hence the implicit iterative nature of the procedure at each instant k.

Solution in reverse direction

In this case, it is necessary to change 3.4.12 to :

$$\underline{x}_n = f_n^{-1} [\underline{x}_{n+1}, \underline{\mu}_n] \qquad 3.4.20$$

The set $[\mathcal{D}_n]$ which is available at any instant n is in this case :

$$[\mathcal{D}_n] = \begin{cases} \underline{\lambda} \\ \underline{x}_{n+1}, \dots \underline{x}_{n+1} \\ \underline{\Psi}_{N+1}, \dots \underline{\Psi}_n \\ \underline{\mu}_N, \dots \underline{\mu}_{n+1} \end{cases}$$

Remark : The solution in the forward direction is not possible unless 3.4.12, 3.4.14, 3.4.15 and 3.4.19 do not require future information (w.r.t. $\underline{u}_n$, Ψ_n, $\underline{x}_n$).

The solution in the backward direction is not possible unless 3.4.13, 3.4.14, 3.4.15 and 3.4.20 do not require information previous to $\underline{u}_n$, Ψ_n, $\underline{x}_n$.

Combining these two restrictions, the integrations in the forward and backward directions will not be simultaneously possible unless

$$r(\underline{u}_1, \dots \underline{u}_N, \underline{x}_1, \dots \underline{x}_N) = \sum_{i=1}^{N} r_i(\underline{u}_i, \underline{x}_i)$$

$$\underline{g}(\underline{u}_1, \dots \quad \underline{x}_1, \dots) = \sum_{i=1}^{N} \underline{g}_i(\underline{u}_i, \underline{x}_i)$$

3.4.2.2. The boundary conditions

For the solution given by the 1st order conditions to be admissible, it must also satisfy the terminal conditions 3.4.16 to 3.4.18.

For example, the orthogonality of $\underline{\Psi}_0$ w.r.t. the corresponding surface can be written in two ways :

$$\underline{\Psi}_0 = H'^T_{x_1} \underline{\xi}'$$

or

$$(1 - Q')\underline{\Psi}_0 = 0$$

with $Q' = H'_{x_1} \left[H'_{x_1} H'^T_{x_1} \right]^{-1} H'_{x_1}$

since we have seen in the gradient method that $\left[1 - Q' \right]$ is the orthogonal projection operator on $\underline{h}'(x_1) = \underline{0}$.

Let us consider, for example, an integration in the backward direction. To solve these stationarity conditions, we must fix the values of the independent variables :

$$\begin{array}{ccc} \underline{x}_{N+1}, & \underline{\xi}', & \underline{\lambda} \\ \downarrow & \downarrow & \downarrow \\ m & r'' & q \end{array} \Longrightarrow (m+r''+q) \text{ variables}$$

$\underline{x}_{N+1}$ must satisfy the terminal conditions $\underline{h}''(\underline{x}_{N+1}) = \underline{0}$; these conditions often enable us to express certain components of $\underline{x}_{N+1}$ as a function of the rest of the components. In the best case, we could express r' components as a function of the remaining (m-r'). This reduces to (m+q) the number of independent variables which can be chosen arbitrarily.

To the vector $\underline{\xi}''$ corresponds

$$\underline{\Psi}_N = \underline{r}_{x_{N+1}} + H''^T_{x_{N+1}} \underline{\xi}''$$

and knowing $\underline{x}_{N+1}$, $\underline{\Psi}_N$ and $\underline{\lambda}$, we can integrate the recursive equations in the backward direction. At the end of the iteration, the quantities $\underline{x}_1$ and $\underline{\Psi}_0$ which are obtained must satisfy :

$$\begin{aligned} &\underline{h}'(\underline{x}_1) = \underline{0} \\ &(1 - Q')\underline{\Psi}_0 = 0 \qquad\qquad 3.4.21 \\ &\underline{g}(\underline{x}_1, \ldots, \underline{u}_1, \ldots) \leqslant \underline{0} \end{aligned}$$

The boundary value problem consists therefore in finding the independent variables $(\underline{x}_{N+1}, \underline{\xi}, \underline{\lambda})$ which at the end of the iteration enable us to satisfy the equations 3.4.21. In order to solve this problem, we must introduce in general, an iterative procedure since it is difficult, if not impossible, to relate $(\underline{x}_1, \underline{\Psi}_0)$ to $(\underline{x}_{N+1}, \underline{\Psi}_N)$.

In a case without global constraints, we could consider a function such as :

$$r = E\left[\underline{h}'(\underline{x}_1), \underline{h}''(\underline{x}_{N+1}), (1 - Q')\underline{\Psi}_0 \right]$$

For example :

$$E = \underline{h}'^T \underline{h} + \underline{h}''^T \underline{h}'' + \left\{ [1 - Q'] \underline{\Psi}_0 \right\}^T [1 - Q'] \underline{\Psi}_0$$

The extremalisation of E will give the (m+r") independent variables $\underline{x}_{N+1}$ and $\underline{\xi}''$. E is a direct or indirect function of these.

However, one difficulty remains in the fact that E is often multimodal with complicated forms.

In the case of a global constraint, we must add a term taking into account the non-satisfaction of $\underline{g}$. We must also iterate on $\underline{\lambda}$ with the conditions $\lambda_i \geq 0$, $\lambda_i g_i = 0$.

3.5 CONCLUSIONS

In this chapter we have extensively developed the theoretical optimality conditions (of the first and second order) and the two principal algorithms which arise from these, i.e. the gradient method and Newton's method. A certain number of other algorithms were also mentioned although our treatment was not exhaustive by any means, since our principal objective was to describe the main techniques of non-linear programming without doing a complete survey of the literature.

In the interest of homogenity of the book, we gave a well defined formulation of the optimisation problem and we ignored, more or less completely, "extremum seeking methods".

On the other hand, since the main theme of the book is the control of processes, we showed how mathematical programming techniques could be applied to such problems.

REFERENCES FOR CHAPTER 3

The literature on non-linear programming is quite vast. We have especially used the following major works in writing this chapter.

[1] BOUDAREL, R., DELMAS, J., GUICHET, P., 1968 : Commande optimale des processus Volume 2, Programmation non linéaire et ses applications. Dunod.

[2] PUN, L., 1972 : Introduction a la pratique de l'optimisation. Dunod, Paris.

[3] HADLEY, D., 1964 : Non linear and dynamic programming. Addison Wesley.

[4] ARROW, K., HURWICZ, L., UZAWA, H., 1968 : Studies in linear and non linear programming. Stanford University Press.

[5] LASDON, L.S., 1970 : Optimisation theory for large scale systems. Mac Millan series for operation research. New-York.

[6] VARAIYA, P.P., 1972 : Notes on optimisation. Van Nostrand Reinhold Company.

[7] COLLATZ, L., WETHERLING, W., 1975 : Optimisation problems. Springer Verlag.

PROBLEMS FOR CHAPTER 3

1. Consider the non linear programming problem :

$$\max Z = 2x_1 - x_1^2 + x_2$$

$$\text{subject to} \quad \left[\begin{array}{l} x_1 + 3x_2 \leqslant 6 \\ 2x_1 + x_2 \leqslant 4 \\ x_1 \geqslant 0 \quad x_2 \geqslant 0 \end{array} \right.$$

1°) Draw the feasible region
2°) Find the solution using the Kuhn-Tucker conditions
3°) Geometrically verify the results obtained

2. Consider the problem :

$$\text{minimize } Z = 10\,(x_1 - 3.5)^2 + 20(x_2 - 4)^2$$

$$\text{subject to} \quad \left[\begin{array}{l} x_1 + x_2 \leqslant 6 \\ x_1 - x_2 \leqslant 1 \\ 2x_1 + x_2 \geqslant 6 \\ 0.5\, x_1 - x_2 \geqslant -4 \\ \quad x_i \geqslant 1 \end{array} \right.$$

1°) Draw the feasible region of this problem and find the optimal solution using the Kuhn-Tucker conditions.

2°) Graphically verify the results obtained.

3°) What will the solution be if the criterion is now changed to

$$Z = 10(x_1 - 2)^2 + 20(x_2 - 3)^2$$

(please give the geometrical interpretation only).

3. Show that the control vector which minimizes the quadratic form (which is assumed to be non-negative definite i.e. $R > 0$, $Q \geqslant 0$) :

$$L = 1/2\, \underline{x}^T\, Q\, \underline{x} + 1/2\, \underline{u}^T\, R\, \underline{u}$$

subject to $\underline{f}(\underline{x}, \underline{u}) = \underline{x} + G\, \underline{u} + \underline{C} = \underline{0}$

is $\underline{u}^* = -(R + G^T\, Q\, G)^{-1}\, G^T\, Q\, \underline{C}$

Show also that the minimum of L is

$$L^* = 1/2\, \underline{C}^T \left[Q - Q\, G(R + G^T\, Q\, G)^{-1}\, G^T\, Q \right] \underline{C}$$

and $\underline{\lambda}^* = \left[Q - Q\, G(R + G^T\, Q\, G)^{-1}\, G^T\, Q \right] \underline{C}$

$= \left[Q^{-1} + G^T\, R^{-1}\, G^T \right]^{-1} \underline{C}$ if R^{-1} exists

($\underline{\lambda}$: Lagrange multiplier associated with the constraint $\underline{f}(\underline{x}, \underline{u}) = \underline{0}$).

$$\underline{x}^* = - \left[I - G(R + G^T\, Q\, G)^{-1}\, G^T\, Q \right] \underline{C}$$

(I : identity matrix).

4. Find the closest point to the origin on the lines

$$x + 2y + 2z = 10 \; ; \quad x - y + 2z = 1$$

in the space (x, y, z).

5. Solve the problem : $\quad \text{Min } Z = 2x_1 - x_1^2 + x^2$

subject to $\begin{cases} 2x_1^2 + 3x_2^2 \leqslant 6 \\ x_1 \geqslant 0 \;\; x_2 \geqslant 0 \end{cases}$

Verify geometrically the result obtained.

6. Show using the theory of mathematical programming that to the following primal linear programming problem :

$$\text{Min } \underline{C}^T\, \underline{x}$$

subject to $\begin{cases} A\, \underline{x} - \underline{b} \geqslant \underline{0} \\ \underline{x} \geqslant \underline{0} \end{cases}$

there corresponds the dual problem

$$\max \underline{y}^T\, \underline{b}$$

subject to $\begin{cases} A^T\, \underline{y} + \underline{C} \geqslant \underline{0} \\ \underline{y} \leqslant \underline{0} \end{cases}$

where $\underline{y}$ is the Kuhn-Tucker multiplier associated with $A\underline{x} - \underline{b} \geqslant \underline{0}$

7. Find the rectangle of maximum area which can be drawn inside the elipse

$$f(x, y) = \frac{x^2}{a^2} + \frac{y^2}{b^2} = \alpha$$

1°) What is the value of this maximum as a function of α , a, b.

2°) Verify and illustrate geometrically the solution using Lagrange multipliers.

8. We consider the system shown below and we assume that the initial conditions are zero. We wish to determine the values of the constants a and b such that the output at the instant T be maximal subject to the constraint :

$$x(t) \leqslant 1 \quad \text{for} \quad 0 \leqslant t \leqslant T$$

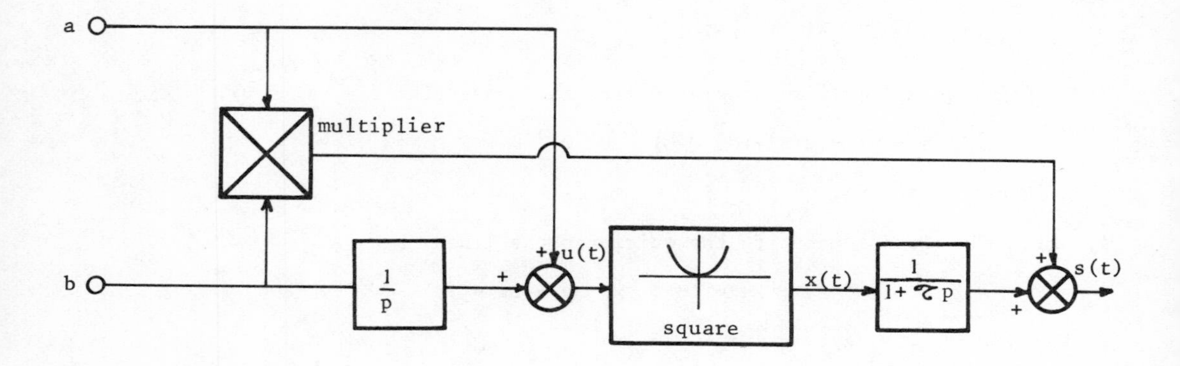

1°) Write the expressions for $x(t)$, $u(t)$, $y(t)$ and $s(t)$.

2°) Express $s(t)$ as a function of a and b and find the constraints on a and b which result from those on $x(t)$.

3°) For the special case where $\tau = T = 1$, verify that the problem reduces to the non-linear program :

$$\max_{a,b} \left[0.632\, a^2 + 1.736\, ab + 0.264\, b^2 \right]$$

$$\text{subject to} \quad \begin{cases} -1 \leqslant a \leqslant 1 \\ -1 \leqslant a+b \leqslant 1 \end{cases}$$

Find the optimal solution $\hat{a}$, $\hat{b}$.

9. We consider the general discrete time system :

$$\underline{x}_{k+1} = A\underline{x}_k + B\underline{u}_k, \qquad \underline{x}(0) = \underline{x}_0, \; k = 0, 1, \ldots k_{f-1}$$

where $\underline{x}$ is an m dimensional vector
$\underline{u}$ is an n dimensional vector
A and B are matrices of appropriate dimension and these could be functions of k.

The cost function is :

$$J = \frac{1}{2} \underline{x}_{k_f}^T S \underline{x}_{k_f} + \frac{1}{2} \sum_{k=0}^{k_f-1} \left[\underline{x}_k^T Q \underline{x}_k + \underline{u}_k^T R \underline{u}_k \right]$$

where the weighting matrices $Q \geqslant 0$, $R > 0$ could be functions of k. ($S \geqslant 0$)

1°) Write the optimality conditions for such a problem in the form of linear difference equations and boundary conditions (we will call $\underline{\lambda}(k+1)$ the Lagrange multiplier or adjoint vector associated with the dynamical constraints).

2°) We wish to find the closed loop solution for this problem by writing the adjoint vector as a linear function of the state $\underline{x}_k$. (we denote by P_k the matrix which relates $\underline{\lambda}_k$ to $\underline{x}_k$ at the instant k).
Give the difference equation (called the Riccati equation) that P_k should satisfy.

3°) Suppose we know P_k ; give an expression for $\underline{u}_k$ as a function of $\underline{x}_k$ and the block diagram for the resulting closed loop control structure.

4°) How could we obtain P_k ?

CHAPTER 4

Decomposition-Coordination Methods in Non-Linear Programming

4.1 INTRODUCTION

In this chapter, we introduce decomposition-coordination methods in mathematical programming for the solution of static optimisation problems. We basically treat the non-linear programming case since the linear case has already been treated in chapter 2. The solution methods will use a hierarchical calculation structure.

The function considered here (i.e. static optimisation) constitutes one of the levels of the functional hierarchy that we examined in chapter 1. Thus we will not discuss vertical hierarchies here but rather a "horizontal division" based on the decomposition of the process into interconnected subsystems for which a two level structure will be adequate.

In the context of a two level structure using a horizontal division of the static optimisation task, we will introduce the three basic methods of decomposition and we will study the stability of the corresponding coordinators.

In this chapter, for clarity of exposition, we treat separately the problems with and without inequality constraints.

4.2 DEFINITION OF THE SUBSYSTEM AND STATEMENT OF THE PROBLEM

The definition of the subsystem which leads to the definition of the subproblem can be imposed in certain cases using a purely physical reasoning although for the general case there exists a methodology for the partitioning of complex systems based on a graphical representation (cf. for example [1, 2]).

We assume here that the complex system under study has already been divided into subsystems of the type shown in Fig. 4.1.

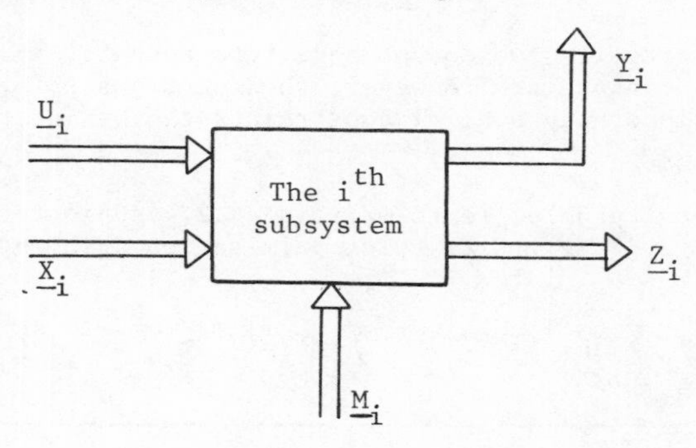

Fig. 4.1 : The ith subsystem

$\underline{U}_i$, $\underline{X}_i$, $\underline{M}_i$, $\underline{Z}_i$, $\underline{Y}_i$ are vectors having respectively m_{U_i}, m_{X_i}, m_{M_i}, m_{Z_i}, m_{Y_i} components. These vectors represent the following :

$\underline{U}_i$: inputs both to the global system and to the ith subsystem (disturbances or other fixed inputs which we can not influence).

$\underline{X}_i$: intermediate inputs supplied by the other subsystems.

$\underline{M}_i$: control variables for the ith subsystem.

$\underline{Y}_i$: outputs, both for the subsystem and for the global system. They could represent for example finished products in a production system.

$\underline{Z}_i$: outputs for subsystem i which act as inputs for other subsystems. In the case of production systems, these are intermediate products

For a given global input vector $\underline{U}$, the subsystem is completely described in the steady state by the vector equations :

$$\underline{Z}_i = \underline{T}_i\ (\underline{M}_i,\ \underline{X}_i) \qquad 4.2.1$$

$$\underline{Y}_i = \underline{S}_i\ (\underline{M}_i,\ \underline{X}_i) \qquad 4.2.2$$

The functions $\underline{T}_i$ and $\underline{S}_i$ are respectively of dimension m_{Z_i} and m_{Y_i}. The interconnection between the subsystems is represented by :

$$\underline{X}_i = \sum_{j=1}^{N} C_{ij}\ \underline{Z}_j \qquad i = 1 \text{ to } N \qquad 4.2.3$$

Here we assume that the decomposition is into N subsystems. C_{ij} are interconnection matrices with m_{X_i} rows and m_{Z_j} columns and provide a linear coupling between the subsystems.

The objective function of the system is assumed to be of the "additively separable" form :

$$F = \sum_{i=1}^{N} f_i(\underline{M}_i,\ \underline{X}_i) \qquad 4.2.4$$

Using, if necessary, equations 4.2.1 and 4.2.2, we could obviously express the criterion uniquely as a function of $\underline{M}_i$ and $\underline{X}_i$.

Here, initially, we do not take into account, in a direct way, any possible inequality constraints. However, the functions $f_i(\underline{M}_i,\ \underline{X}_i)$ could contain, using penalty functions, any special constraints that exist for any particular subsystem.

The global problem is to maximise 4.2.4 subject to the equality constraints 4.2.1 and 4.2.3 (*). With this optimisation problem we can associate the Lagrangian :

$$L = \sum_{i=1}^{N} f_i(\underline{M}_i,\ \underline{X}_i) + \sum_{i=1}^{N} \underline{\mu}_i^T(\underline{T}_i - \underline{Z}_i) + \sum_{i=1}^{N} \underline{\rho}_i^T(\underline{X}_i - \sum_{j=1}^{N} C_{ij}\ \underline{Z}_j) \qquad 4.2.5$$

(*) Since we do not assume that $\underline{Y}_i$ is constrained, we do not need to take 4.2.2 into account.

$\underline{\mu}_i$ (of dimension m_{Z_i}), $\underline{\rho}_i$ (of dimension m_{X_i}) being Lagrange multiplier vectors which have been introduced to take into account the equality constraints.

If we assume that the equality constraints are independent and the functions f_i and $\underline{T}_i$ (i = 1 to N) are continuous and continuously differentiable to first order, then the optimal solution must satisfy the stationarity conditions of the Lagrangian, i.e.

i=1 to N

$$\frac{\partial L}{\partial \underline{X}_i} = \frac{\partial f_i}{\partial \underline{X}_i} + \left(\frac{\partial \underline{T}_i}{\partial \underline{X}_i}\right)^T \underline{\mu}_i + \underline{\rho}_i = \underline{0} \qquad 4.2.6$$

$$\frac{\partial L}{\partial \underline{M}_i} = \frac{\partial f_i}{\partial \underline{M}_i} + \left(\frac{\partial \underline{T}_i}{\partial \underline{M}_i}\right)^T \underline{\mu}_i = \underline{0} \qquad 4.2.7$$

$$\frac{\partial L}{\partial \underline{Z}_i} = -\underline{\mu}_i - \sum_{j=1}^{N} C^T_{ji}\, \underline{\rho}_j = \underline{0} \qquad 4.2.8$$

$$\frac{\partial L}{\partial \underline{\mu}_i} = \underline{T}_i - \underline{Z}_i = \underline{0} \qquad 4.2.9$$

$$\frac{\partial L}{\partial \underline{\rho}_i} = \underline{X}_i - \sum_{j=1}^{N} C_{ij}\, \underline{Z}_j = \underline{0} \qquad 4.2.10$$

$\frac{\partial f_i}{\partial \underline{X}_i}$, $\frac{\partial f_i}{\partial \underline{M}_i}$, which are partial derivatives of a scalar f_i w.r.t. a vector, represent the gradient of f_i w.r.t. $\underline{X}_i$ and $\underline{M}_i$. The matrices $\left(\frac{\partial \underline{T}_i}{\partial \underline{X}_i}\right)$, $\left(\frac{\partial \underline{T}_i}{\partial \underline{M}_i}\right)$ which represent the partial derivative of a vector of functions w.r.t. a vector of variables could be written as :

$$\left(\frac{\partial \underline{T}_i}{\partial \underline{X}_i}\right) = \begin{bmatrix} \frac{\partial T_i^{1}}{\partial x_i^{1}} & \cdots\cdots\cdots & \frac{\partial T_i^{1}}{\partial x_i^{mX_i}} \\ & & \\ \frac{\partial T_i^{mZ_i}}{\partial x_i^{1}} & \cdots\cdots\cdots & \frac{\partial T_i^{mZ_i}}{\partial x_i^{mX_i}} \end{bmatrix}$$

4.2.11

$$\left(\frac{\partial \underline{T}_i}{\partial \underline{M}_i}\right) = \begin{bmatrix} \frac{\partial T_i^{1}}{\partial M_i^{1}} & \cdots\cdots\cdots & \frac{\partial T_i^{1}}{\partial M_i^{mM_i}} \\ & & \\ \frac{\partial T_i^{mZ_i}}{\partial M_i^{1}} & \cdots\cdots\cdots & \frac{\partial T_i^{mZ_i}}{\partial M_i^{mM_i}} \end{bmatrix}$$

We note that the global problem is a non-degenerate optimisation problem since for $\sum_i (m_{X_i} + m_{M_i} + m_{Z_i})$ independent variables, there are only $\sum_i (m_{X_i} + m_{Z_i})$ equality constraints and the solution of the system 4.2.6 to 4.2.10 should provide a local solution, if it exists. We will assume that such a solution does indeed exist.

In the case where N and m_w (for each vector) are large, it may be difficult to tackle directly, in a global way, the non-linear problem given by equations 4.2.6 to 4.2.10. It would be interesting to use decomposition-coordination methods in that case. We recall from chapter 1 that these methods involve solving the global problem P by solving a number of subproblems $P_i(\underline{\alpha})$ which are parameterised w.r.t. $\underline{\alpha}$ in order to satisfy

$$\text{Solution of } \left[P_1(\underline{\alpha}),\ \dots\ P_i(\underline{\alpha}),\ \dots\ P_N(\underline{\alpha})\right] \underset{\underline{\alpha}=\underline{\alpha}^*}{\Longrightarrow} \text{Solution of P}$$

The important steps here are :

- Definition of the subproblems $P_i(\underline{\alpha})$
- Evolution of $\underline{\alpha}$ from an initial value $\underline{\alpha}^o$ to a final value $\underline{\alpha}^*$ which solves the global problem (i.e. the coordination problem).

The decomposition coordination methods that have been proposed use basically two approaches [3], i.e.

1. Action on the objective functions of the subproblems (goal coordination).
2. Interaction prediction. In this case, we distinguish between :
   - a- prediction of the interaction outputs $\underline{Z}_i$ at the level of the subsystems (model coordination)
   - b- prediction of the interaction inputs $\underline{X}_i$ for the subsystems. In fact this latter mode is rarely applicable by itself (it involves basically neglecting the condition $L_{\underline{Z}} = \underline{0}$ [4] and we have to simultaneously do an interaction prediction and goal modification).

On the practical level this involves decoupling the subsystems, as we shall see, by acting on the coupling variables $\underline{X}$ or $\underline{Z}^*$ or the associated dual variables $\underline{P}$ and to distribute the treatment of the equations 4.2.6 to 4.2.10 between the two levels as shown in Table 4.1. This is not done in an arbitrary fashion but rather to obtain a "separable" problem at the lower level, thus providing a horizontal division of the tasks. This involves expressing 4.2.5 in the additive separable form :

$$L = \sum_{i=1}^{N} L_i(\underline{\beta}_i, \underline{\alpha}) \qquad 4.2.6$$

$\underline{\beta}_i$: local variables, $\underline{\alpha}$: coordination variables (Fig. 4.2).

---

* In fact, to reduce the inter-level information transfer (Fig. 4.2), we limit ourselves here to the output coupling variable $\underline{Z}$. Indeed, in general, for a coordination variable $\underline{\alpha}$, it is necessary to transfer to the higher level all the necessary information to evaluate $L_{\underline{\alpha}}$. Our choice is thus justified by an examination of equations 4.2.6 and 4.2.8.

	higher level	lower level	higher level	lower level	higher level	lower level
$L_{\underline{X}} = 0$		X		X		X
$L_{\underline{M}} = 0$		X		X		X
$L_{\underline{Z}} = 0$		X	X		X	
$L_{\underline{\mu}} = 0$		X		X		X
$L_{\underline{\rho}} = 0$	X			X	X	
Methods	"non feasible" (Goal coordination)		"feasible" method (model coordination)		Mixed method	

Table 4.1

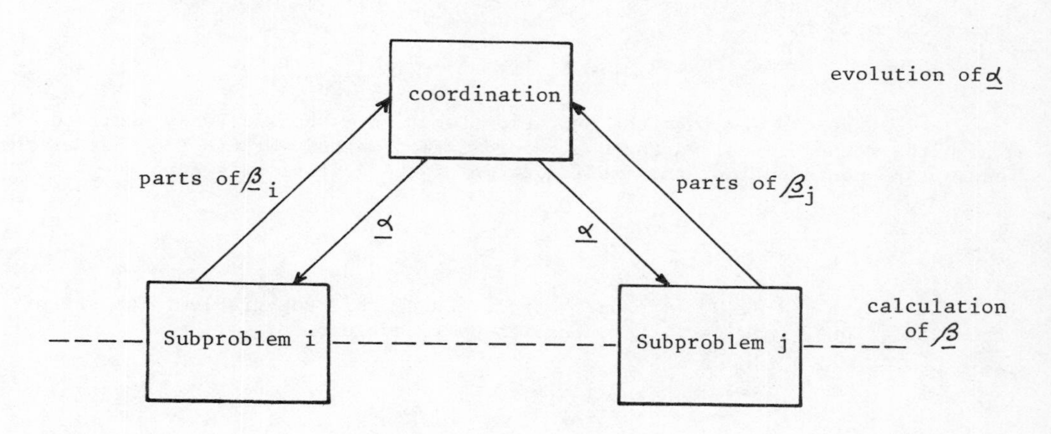

Fig. 4.2

4.3 THREE METHODS OF DECOMPOSITION

4.3.1 The "non feasible" or Goal coordination method

This method is characterized by the parameters $\underline{\rho}_j$ (j = 1 to N) which are fixed at the second level and therefore constitute the data for the first level.

Given that $\underline{\rho}_i$ is fixed, we can associate the term $\underline{\rho}_i{}^T \underline{X}_i$ in the Lagrangian, for the subsystem "i" and distribute the terms

$$\underline{\rho}_i{}^T \sum_j C_{ij} \underline{Z}_j$$

between all the other subsystems. Mathematically, this involves doing :

$$\sum_i \underline{\rho}_i^T \sum_j C_{ij} \underline{Z}_j = \sum_i \sum_j \rho_j^T C_{ji} \underline{Z}_i \qquad 4.3.1$$

and

$$L = \sum_{i=1}^{N} L_i = \sum_i \left[f_i(\underline{M}_i, \underline{X}_i) + \underline{\mu}_i^T (\underline{T}_i - \underline{Z}_i) \right. \qquad 4.3.2$$

$$+ \underline{\rho}_i^T \underline{X}_i - \sum_{j=1}^{N} \underline{\rho}_j^T C_{ji} \underline{Z}_i = \sum_{i=1}^{N} L_i(\underline{\beta}_i, \underline{\alpha}) \qquad 4.3.3$$

$(\underline{\beta}_i^T = (\underline{X}_i^T, \underline{M}_i^T, Z_i^T, \mu_i^T),\ \underline{\alpha} = \underline{\rho}\ ).$

The L function takes a "separable" form and each sub-problem at the first level is therefore defined by the Lagrangian L_i which is associated with it

sub-problem n° i
$$\begin{cases} \max \left[f_i(\underline{M}_i, \underline{X}_i) + \underline{\rho}_i^T \underline{X}_i - \sum_{j=1}^{N} \underline{\rho}_j^T C_{ji} \underline{Z}_i \right] & 4.3.4 \\ \text{subject to } \underline{Z}_i = T_i\ (\underline{M}_i, \underline{X}_i) & 4.3.5 \end{cases}$$
for $\underline{\rho}_i$, i = 1 to N, given

We can see that the criterion function is considerably modified. For this reason, the method is called the Goal coordination method. We can see that at the end of the coordination, this modification of the criterion is zero

$$\sum_i \Delta f_i = \sum_i \underline{\rho}_i^T (\underline{X}_i - \sum_j C_{ij} \underline{Z}_j) = 0 \qquad 4.3.6$$

Let $\underline{X}^*_i(\underline{\rho})$, $\underline{Z}^*_i(\underline{\rho})$, $\underline{M}^*_i(\underline{\rho})$ be the solution of this sub-problem. The set of these solutions, for i = 1 to N, enables us to evaluate

$$\underline{X}_i^*(\underline{\rho}) - \sum_{j=1}^{N} C_{ij}\, \underline{Z}^*_j(\rho) = \underline{\varepsilon}_i \qquad 4.3.7$$

If $\underline{\varepsilon}_i = \underline{0}\ \forall\ i$ we have the global solution ; if not, the interconnection constraint is not satisfied and after having taken into account the results of the first level, action at the second level is necessary in order to ensure the satisfaction of the constraint 4.3.7. The essential transfer of information for this case is shown schematically in Fig. 4.3. We see that it is quite limited

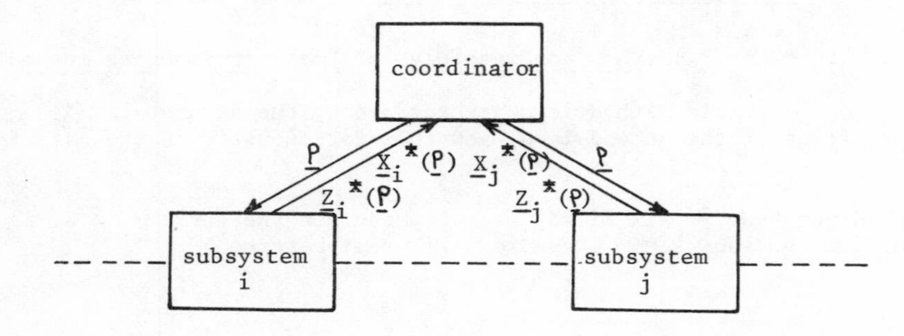

Fig. 4.3 : Information transfer in the "non-feasible" method

From an analytical point of view, the first level's task is to solve the system of equations 4.2.6, 4.2.7, 4.2.8 and 4.2.9 for given $\underline{\rho}$. At the second level, 4.2.10 is used to fix $\underline{\rho}$. However, $\underline{\rho}$ appears explicitly in this equation so that it is necessary to introduce an iterative coordinator algorithm.

But from equation 4.2.5, we have :

$$dL = \sum_i \left[L_{\underline{\rho}_i}^T \, d\underline{\rho}_i + L_{\underline{Z}_i}^T \, d\underline{Z}_i + L_{\underline{X}_i}^T \, d\underline{X}_i + L_{\underline{M}_i}^T \, d\underline{M}_i + L_{\underline{\mu}_i}^T \, d\underline{\mu}_i \right] \qquad 4.3.8$$

But when the 1st level's task is done, we have :

$$L_{\underline{Z}_i} = \underline{0} \; ; \; L_{\underline{M}_i} = \underline{0} \; ; \; L_{\underline{X}_i} = \underline{0} \; ; \; L_{\underline{\mu}_i} = \underline{0} \qquad 4.3.9$$

so that 4.3.8 reduces to

$$dL = \sum_i L_{\underline{\rho}_i}^T \, d\underline{\rho}_i \qquad 4.3.10$$

If we choose

$$d\underline{\rho}_i = -KL_{\underline{\rho}_i} \qquad K > 0 \qquad 4.3.11$$

we require $dL < 0$. This is necessary since we are seeking a solution, which according to the assumptions we have made, corresponds to a saddle point of L, i.e. which gives a maximum w.r.t. the variables and a minimum w.r.t. the Lagrange multipliers. Equation 4.3.11 leads after discretisation to the following coordinator algorithm of the gradient type :

$$\underline{\rho}_i(t+1) = \underline{\rho}_i(t) - KL_{\underline{\rho}_i}(t) \qquad 4.3.12$$

The method is shown in Fig. 4.4

It is also possible to use Newton's method directly to solve the equation $L_{\underline{\rho}} = \underline{0}$ which leads to the following coordinator algorithm :

$$\underline{\rho}(t+1) = \underline{\rho}(t) - C \left[\frac{dL_{\underline{\rho}}(t)}{d\underline{\rho}} \right]^{-1} L_{\underline{\rho}}(t) \qquad 0 < C < 2 \qquad 4.3.13$$

We have seen that

$$L_{\underline{\rho}_i} = \underline{X}_i - \sum_j C_{ij} \underline{Z}_j \implies L_{\underline{\rho}} = L_{\underline{\rho}}(\underline{X}, \underline{Z}) \qquad 4.3.14$$

and

$$\frac{dL_{\underline{\rho}}}{d\underline{\rho}} = L_{\underline{\rho}X} \, \underline{X}_{\underline{\rho}} + L_{\underline{\rho}Z} \, \underline{Z}_{\underline{\rho}} \qquad 4.3.15$$

This coordination algorithm will seek from each subsystem, not only $\underline{X}^*(\underline{\rho})$ and $\underline{Z}^*(\underline{\rho})$ but also $\underline{X}^*_{\underline{\rho}}(\underline{\rho})$ and $\underline{Z}^*_{\underline{\rho}}(\underline{\rho})$ and will also require the inversion of a matrix. Its advantage is that in this method the iteration constant is bounded from above and below whilst in the gradient method, the iteration constant is unbounded from above and this could cause problems. We recall also that Newton's method has good convergence (quadratic for $C = 1$).

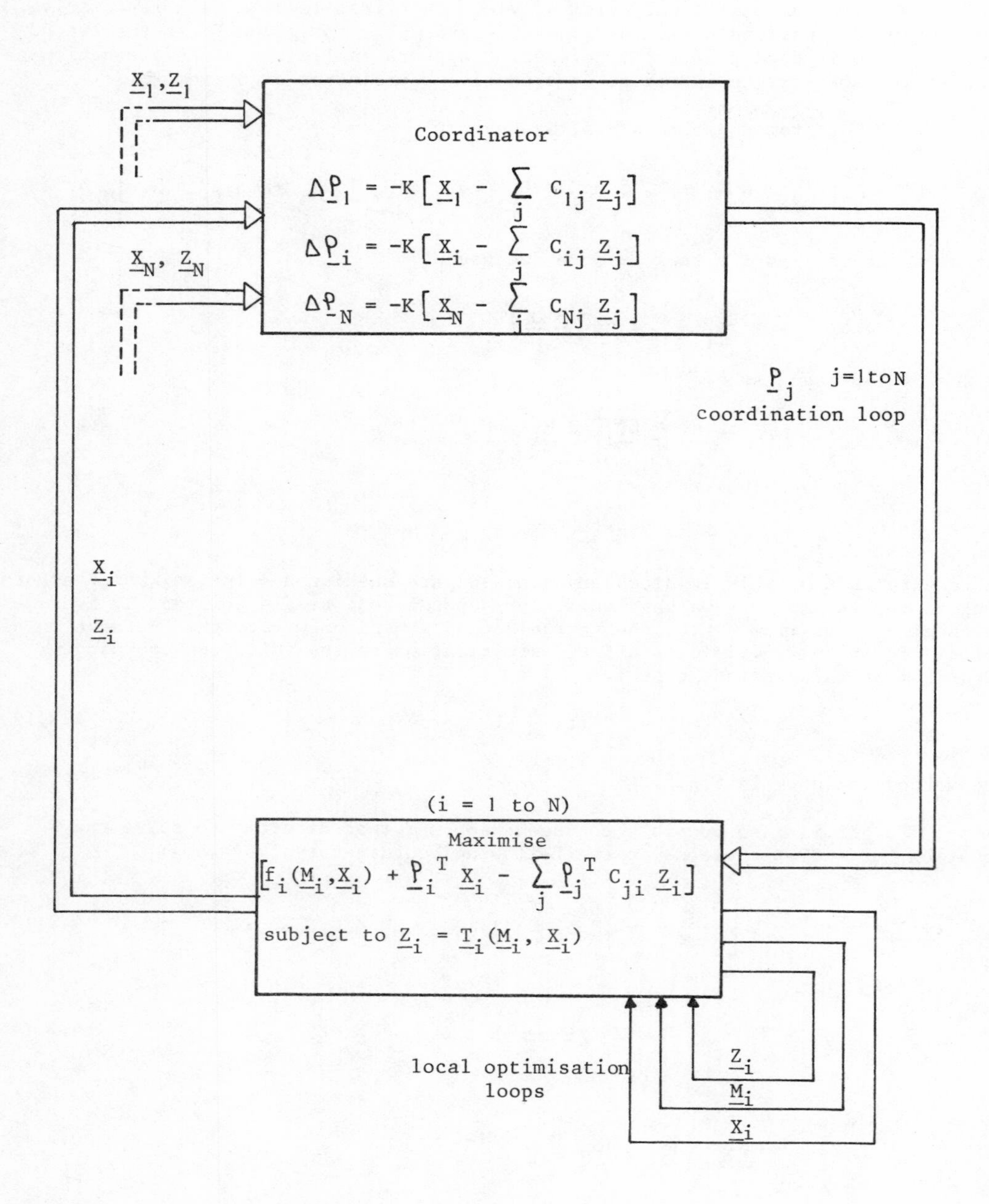

Fig. 4.4 : The "infeasible" method with the Gradient coordinator

Finally, we note that it may be desirable to use say a gradient method far from the optimum and Newton's method near the optimum. This has the advantage of providing both good initial convergence and good final convergence. We could also use other algorithms for example the conjugate gradient method which has good convergence and which requires little additional calculation as compared to the gradient method.

At the first level, the quantities to be determined for each "i" are : $\underline{M}_i$, $\underline{X}_i$, $\underline{Z}_i$ and $\underline{\mu}_i$, and the equations 4.2.6 to 4.2.9 provide us with an appropriate number of relations to do this.

We can find the different components of the unknown vectors using 4.2.6 to 4.2.9 step by step (and not necessarily globally) as indicated below :

a) $\underline{\rho}_j$, j = 1 to N being provided by the higher level, we can calculate $\underline{\mu}_i$ from equation 4.2.8.

b) $\underline{\rho}_i$, $\underline{\mu}_i$ being known, the solution of 4.2.6 + 4.2.7, which is a system of $(m_{X_i} + m_{M_i})$ equations in as many unknowns, enables us to compute $\underline{X}_i$ and $\underline{M}_i$.

c) Having evaluated $\underline{X}_i$ and $\underline{M}_i$, $\underline{Z}_i$ can be directly calculated from 4.2.9. Once this equation is solved, we send the results to level 2 where $\underline{\rho}$ is modified using one of the algorithms described above, provided of course that the overall optimum has not been reached.

We emphasise that in the "non-feasible" method, there are no requirements on the number of components of $\underline{M}_i$, $\underline{X}_i$, $\underline{Z}_i$ as in the case for the next method that we study.

4.3.2 Model coordination or the Feasible method [6]

In this method, the higher level fixes $\underline{Z}_j$ and provides it to the lower level. At the lower level, the Lagrangian can be written in the following "separable" form :

$$L = \sum_i L'_i(\underline{\beta}_i, \underline{\alpha}) = \sum_i \left[f_i(\underline{M}_i, \underline{X}_i) + \underline{\mu}_i^T(\underline{T}_i - \underline{Z}_i) + \underline{\rho}_i^T(\underline{X}_i - \sum_j C_{ij}\underline{Z}_j) \right] \qquad 4.3.16$$

$(\underline{\beta}_i^T = (\underline{X}_i^T, \underline{M}_i^T, \underline{\mu}_i^T, \rho_i^T), \underline{\alpha} = \underline{Z})$.

The form of L'_i shows that the ith subproblem is :

$$\left[\begin{array}{l} \max f_i(\underline{M}_i, \underline{X}_i) \\ \text{subject to} \left\{ \begin{array}{l} \underline{Z}_i = T_i(\underline{M}_i, \underline{X}_i) \\ \underline{X}_i = \sum_j C_{ij} \underline{Z}_j \end{array} \right. \quad \text{for given } \underline{Z}_j,\ j = 1 \text{ to } N \end{array} \right. \qquad 4.3.17$$

In this case, although the criterion is not changed, the variables $\underline{Z}_i$ and $\underline{X}_i$ are indirectly fixed in the model so we do indeed have model coordination.

The model constraints and the interconnection constraints are always satisfied. This is the reason why the method is also said to be feasible.

The information transfer between the two levels is still quite limited and this is shown in Fig. 4.5.

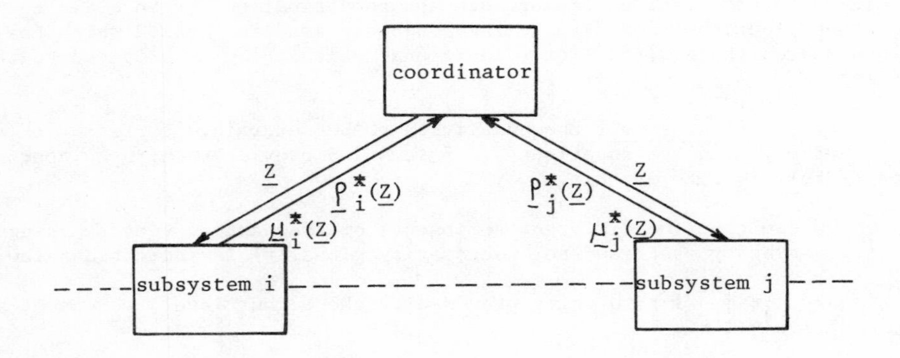

Fig. 4.5 : Information transfer in model coordination

$\underline{\mu}$ and $\underline{\rho}$ are necessary at level 2 since the only equation that remains to be treated at this level is :

$$L_{\underline{Z}} (\underline{\mu} , \underline{\rho}) = \underline{0} \qquad 4.3.18$$

The equations $L_{\underline{X}} = \underline{0}$, $L_{\underline{M}} = 0$, $L_{\underline{\rho}} = \underline{0}$, $L_{\underline{\mu}} = \underline{0}$ are satisfied at the local level in order to solve the set of problems 4.3.17. Thus, at the second level, $dL = L_{\underline{Z}}^T \, d\underline{Z}$. Since we need to maximise L w.r.t. $\underline{Z}$ in order to satisfy $dL > 0$ we could choose :

$$d\underline{Z} = K \, L_{\underline{Z}} \qquad K > 0 \qquad 4.3.19$$

which becomes after discretisation :

$$\underline{Z}(t+1) = \underline{Z}(t) + K \, L_{\underline{Z}}(t) \qquad 4.3.20$$

which is the gradient type coordination algorithm for model coordination.

The method is shown schematically in Fig. 4.6.

As for the Newton type algorithm, it can be written as :

$$\underline{Z}(t+1) = \underline{Z}(t) - C \left(\frac{dL_{\underline{Z}}}{d\underline{Z}}\right)^{-1} L_{\underline{Z}}(t) \qquad 0 < C < 2 \qquad 4.3.21$$

When all the subsystem controllers have performed their job, all the constraints are satisfied and we have :

$$L \equiv F \qquad 4.3.22$$

Thus the coordination algorithm will improve the value of the criterion at each iteration of the higher level and the results could be applied in "real time" since all the constraints are satisfied at each iteration. This is certainly one of the important advantages of the method.

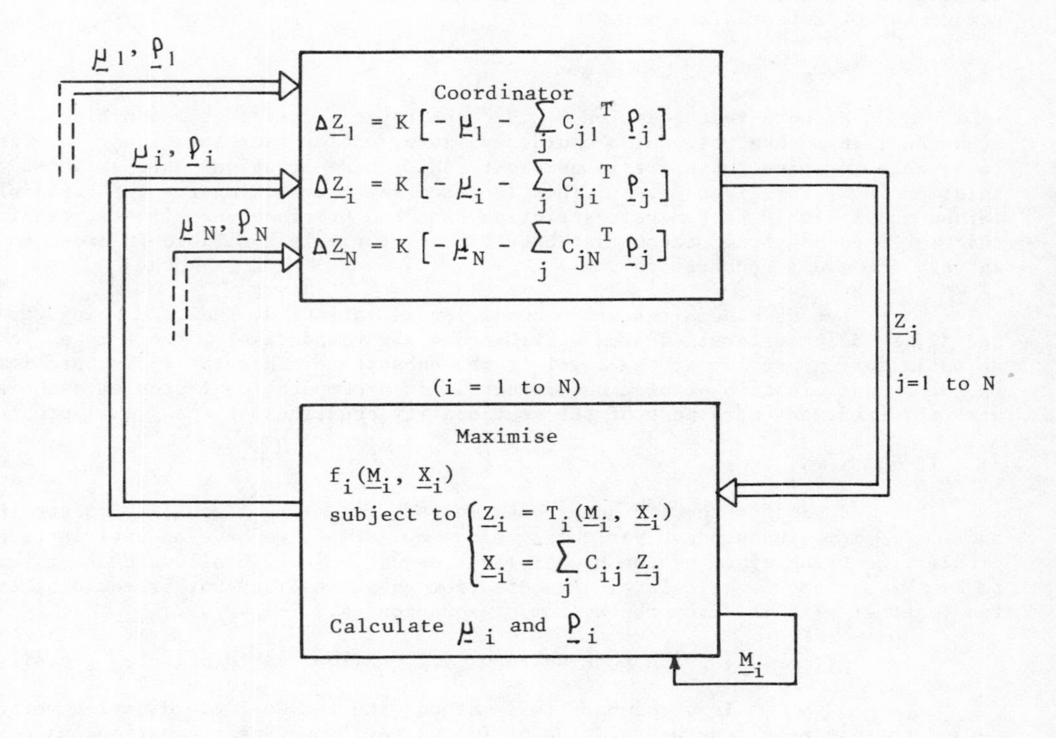

Fig. 4.6 : The model coordination method with gradient coordination

From an analytical point of view, the first levels job is to treat equations 4.2.6 to 4.2.10 in order to compute $\underline{X}_i$, $\underline{M}_i$, $\underline{\mu}_i$, $\underline{\rho}_i$ (i = 1 to N).

$\underline{X}_i$ can always be found easily. For the other variables, it is necessary to distinguish between the various cases which arise depending upon the number of components of each vector.

a) If $m_{M_i} < m_{Z_i}$:

In this case, equation 4.2.9 for $\underline{Z}$ is a system of m_{Z_i} equations with m_{M_i} unknowns. Since here we have more equations than unknowns, this system of equations does not have a solution and the method fails.

In order to solve such a problem, i.e. where $m_{M_i} < m_{Z_i}$, it is necessary to use some other decomposition technique (for example the non-feasible method is still valid) or use a global technique. The latter is a real possibility since here the number of controls is smaller than the number of intermediate outputs. Thus we can act directly and globally on $\underline{M}_i$. For this special case, therefore, all multi-level decomposition methods are less appealing.

Thus, for the "feasible" method of decomposition to be applicable, it is necessary that, for each subsystem, the number of controls is at least equal to the number of intermediate outputs.

b) If $m_{M_i} = m_{Z_i} = m$:

We note that equation 4.2.9 constitutes a system of m non-linear equations, in m unknowns, which could have zero, one or many solutions. In order to be able to solve these, it is necessary that these equations have at least one solution in $\underline{M}_i$ for given $\underline{Z}_i$, $\underline{X}_i$. This is a necessary condition for the "feasible" method and it is in fact more restrictive than the previous one since it requires that : "In each system, there must be a control variable available in order to satisfy the model constraints".

Let us assume that this condition is satisfied. Then, $\underline{Z}_i$ being given and $\underline{X}_i$, $\underline{M}_i$ being determined from 4.2.10 and 4.2.9 respectively, there is no longer an optimisation problem at the level of the subsystems. Thus the ith subproblem is not an optimisation problem but rather it is a computational problem associated with the treatment of a part of the stationarity equations of the global problem.

c) If $m_{M_i} > m_{Z_i}$:

Here, for the ith subproblem we have $(m_{X_i} + m_{Z_i})$ equality constraints and $(m_{M_i} + m_{X_i})$ independent variables ; since $m_{M_i} > m_{Z_i}$, we have an optimisation problem. $\underline{Z}_j$ being given by the higher level, equation 4.2.10 allows us to calculate $\underline{X}_i$; $\underline{M}_i$ can not be calculated any more from equation 4.2.9 but it could be computed together with $\underline{\mu}_i$ from the system of equations 4.2.7 + 4.2.9.

$\underline{M}_i$, $\underline{X}_i$, $\underline{\mu}_i$ being thus defined, 4.2.6 allows us to compute $\underline{\rho}_i$ easily.

If we compare this method with the Goal coordination method, we see that it requires the solution of $(m_{Z_i} + m_{M_i})$ non-linear equations as compared to $(m_{X_i} + m_{M_i})$ for the Goal coordination method.

The solution of these subproblems is repeated many times. It would therefore be useful to compare m_{X_i} with m_{Z_i} before we finally choose the method. Often, for physical systems, $m_{X_i} \gg m_{Z_i}$ which is an argument in favour of the model coordination method. But this method has an applicability condition $m_{M_i} \geq m_{Z_i}$.

4.3.3 The Mixed method [7]

In this method, $\underline{\rho}$ and $\underline{Z}$ are determined by the second level and are used at the first level. The Lagrangian thus takes on the following separable form :

$$L = \sum_{i=1}^{N} L_i''(\underline{\beta}_i, \underline{\alpha}) = \sum_{i=1}^{N} \Big[f_i(\underline{X}_i, \underline{M}_i) + \underline{\rho}_i^T (\underline{X}_i - \sum_{j=1}^{N} C_{ij}\underline{Z}_j) + \underline{\mu}_i^T [\underline{T}_i - \underline{Z}_i) \Big] \Big] \qquad 4.3.23$$

$(\underline{\beta}_i^T = (\underline{X}_i^T, \underline{M}_i^T, \underline{\mu}_i^T),\ \underline{\alpha}^T = (\underline{\rho}^T, \underline{Z}^T))$.

This allows us to decompose the problem such that each subproblem becomes :

$$\text{for given } \underline{\rho}_i \text{ and } \underline{Z}_i \begin{cases} \max \left[f_i(\underline{X}_i, \underline{M}_i) + \underline{\rho}_i^T \underline{X}_i + Ct \right] (- \sum_{j=1}^{N} \underline{\rho}_i^T C_{ij} \underline{Z}_j = \text{Constant}) \\ \text{subject to } \underline{Z}_i = \underline{T}_i(\underline{X}_i, \underline{M}_i) \end{cases} \qquad 4.3.24$$

In this case, we see that both the criterion function and the model are used in the coordination.

The solution of each subproblem corresponds to the treatment of equations 4.2.6, 4.2.7 and 4.2.9 which leaves for the second level :

$$L_{\underline{\rho}}\ (\underline{X}, \underline{Z}) = \underline{0}$$
$$L_{\underline{Z}}\ (\underline{\mu}, \underline{\rho}) = \underline{0} \qquad 4.3.25$$

Equation 4.2.9 is computable for given $\underline{\rho}$ and $\underline{Z}$, if :

$$m_{X_i} + m_{M_i} \geqslant m_{Z_i} \qquad i = 1 \text{ à } N \qquad 4.3.26$$

This is a less restrictive condition than the one required for the model coordination method.

The information transfer that is required between the levels in this method is shown in Fig. 4.7.

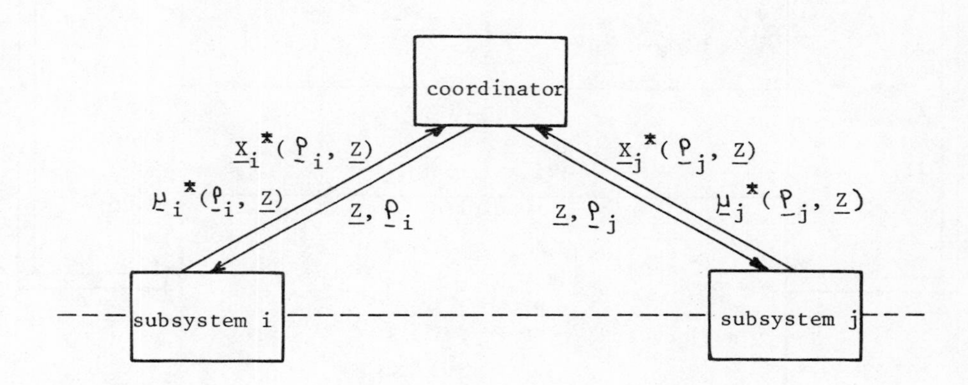

Fig. 4.7 : Information transfer in the mixed method

By analogy with the two previous methods as well as with the method of Arrow-Hurwicz [8] for finding a saddle point, we could use the gradient coordination algorithm :

$$\left.\begin{aligned} \frac{d\underline{\rho}}{dt} &= -L_{\underline{\rho}} \\ \frac{d\underline{Z}}{dt} &= L_{\underline{Z}} \end{aligned}\right| \Rightarrow \left[\begin{aligned} \underline{\rho}(t+1) &= \underline{\rho}(t) - KL_{\underline{\rho}} \qquad 4.3.27 \\ \underline{Z}(t+1) &= \underline{Z}(t) + KL_{\underline{Z}} \qquad 4.3.28 \end{aligned}\right. \qquad K > 0$$

Fig. 4.8 illustrates the application of this method.

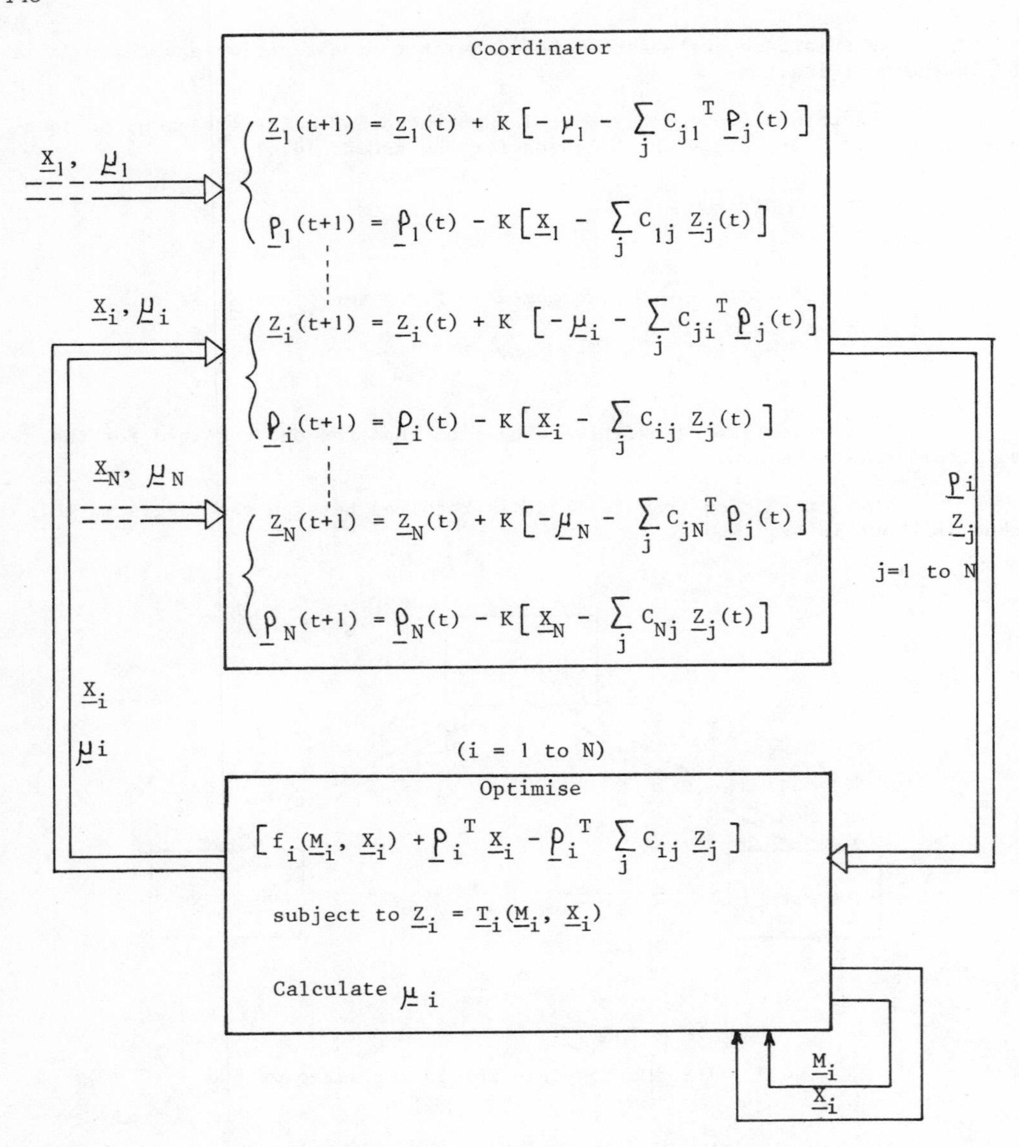

Fig. 4.8 : Mixed method with gradient coordination

Otherwise, we could use Newton's algorithm to solve the 2 vector equations :

$$L_{\underline{\rho}} = \underline{0} \qquad L_{\underline{Z}} = \underline{0} \tag{4.3.29}$$

by writing :

$$\underline{W}(t+1) = \underline{W}(t) - C\left[\frac{dL_{\underline{W}}}{d\underline{W}}\right]^{-1} L_{\underline{W}}$$

$$\underline{W} = \begin{bmatrix} \underline{z} \\ \underline{\rho} \end{bmatrix} \qquad 0 < C < 2 \qquad 4.3.30$$

or again, if certain conditions on the components are satisfied ($\sum m_{X_i} = \sum m_{Z_i}$), we could calculate $\underline{Z}$ using $L_{\underline{\rho}} = \underline{0}$ and $\underline{\rho}$ using $L_{\underline{Z}} = \underline{0}$ thus putting into practice a "direct iteration" method which has good convergence.

4.3.4 Comparison of the three methods

For comparison, we have grouped in Table 4.2 the essential results for the three methods.

We see from these results that the "non-feasible" method is always applicable but that we pay for this general applicability by having to solve more complex subproblems. The "feasible" method, which is only applicable under certain conditions, leads to simplifications and an easier implementation. As for the mixed method which represents a synthesis of the two previous techniques, it has a wider range of applicability than the feasible method since it only requires that $m_{M_i} + m_{X_i} \geqslant m_{Z_i}$. In addition, the mixed method simplifies the solution of each subproblem although this increases, slightly, the coordination task. In this way, the jobs are better distributed between the levels.

We note that we have assumed that we already have the division of the system into the subsystems. However, there does not exist, at the present time, any well established rule for dividing systems into subsystems. We note that the above results could perhaps be used as general guidlines, for this purpose.

Thus, if we want to use a "feasible" method so as to satisfy the interconnection constraints at each instant, we need to have a decomposition such that $m_{Z_i} \leqslant m_{M_i}$. On the other hand, the "non-feasible" or "mixed" methods allow us a bigger choice for the decomposition and we could take into account any physical peculiarities of the system.

Finally, we note that the decomposition of the system and the multilevel optimisation should not be done completely independently but rather they should complement each other in taking into account any special features of the system under study.

4.3.5 Extension to non-linear coupling between the subsystems

In this case, the basic problem remains the same as before except for the modification of the coupling constraint to the form :

$$\left.\begin{aligned} \underline{X}_i &= \underline{H}_i(\underline{Z}_j) && i = 1 \text{ to } N \\ &= H_i(\underline{Z}_1 \ldots \underline{Z}_N) && j = 1 \text{ to } N \end{aligned}\right\} \qquad 4.3.31$$

The advantage of having a non-linear coupling is two fold :

a - it allows us to limit the number of subsystems

b - it enables us to do coordinate transformations on the input-output relationships for some subsystems in order to simplify the treatment of the problem.

Unfeasible method	Feasible method	Mixed method
No condition on m_{X_i}, m_{M_i}, m_{Z_i} Analytical method : Solution of a system of $(m_{X_i} + m_{M_i})$ non-linear equations with as many unknowns Local optimisation : (always possible) $\max \left[f_i(\underline{M}_i, \underline{X}_i) + \underline{\rho}_i^T \underline{X}_i - \sum_j \underline{\rho}_j^T C_{ji} \underline{Z}_i \right]$ subject to $\underline{Z}_i = T_i(\underline{M}_i, \underline{X}_i)$ This is a problem with $(m_{X_i}+m_{M_i}+m_{Z_i})$ variables and m_{Z_i} constraints.	1. If $m_{M_i} < m_{Z_i}$ the method does not work. 2. If $m_{M_i} = m_{Z_i} = m$ there is no optimisation problem at the lower level Condition : It is also necessary that $\underline{Z}_i = \underline{T}_i(\underline{M}_i, \underline{X}_i)$ has at least a solution $\underline{M}_i$ for given $\underline{Z}_i$, $\underline{X}_i$. Thus we have to solve a non-linear system of equations with m variables. We need also to calculate $\underline{\mu}_i$ and $\underline{\rho}_i$. 3. If $m_{M_i} > m_{Z_i}$ the subproblem is an optimisation problem. 2 solutions are possible : - Analytical method : Solution of a non-linear system of $(m_{M_i} + m_{Z_i})$ equations. Calculation of $\underline{\rho}_i$ by a linear system. - Local optimisation : $\max f_i(\underline{M}_i, \underline{X}_i)$ subject to $\underline{Z}_i = \underline{T}_i(\underline{M}_i, \underline{X}_i)$ $\underline{X}_i = \sum_j C_{ij} \underline{Z}_j$ + calculation of $\underline{\mu}_i$ and $\underline{\rho}_i$	1. If $(m_{M_i} + m_{X_i}) < m_{Z_i}$ the method does not work 2. If $(m_{M_i}+m_{X_i}) = m_{Z_i} = n$ local optimisation is not possible. Condition : The equation $\underline{Z}_i = \underline{T}_i(\underline{X}_i, \underline{M}_i)$ should have at least one solution for given $\underline{Z}_i$. Solution of a system of n non-linear equations. Calculation of $\underline{\mu}_i$ by a linear system. 3. If $m_{M_i}+m_{X_i} > m_{Z_i}$ the subproblem is an optimisation problem. Two methods are possible : - Analytical method : Solution of a non-linear system of $(m_{X_i}+m_{M_i}+m_{Z_i})$ equations. - Local optimisation : $\max \left[f_i(\underline{M}_i, \underline{X}_i) + \underline{\rho}_i^T \underline{X}_i - \sum_j \underline{\rho}_i^T C_{ij} \underline{Z}_j \right]$ subject to $\underline{Z}_i = T_i(\underline{M}_i, \underline{X}_i)$ and calculate $\underline{\mu}_i$.

Table 4.2 : Comparison of the three methods

The introduction of the non-linear coupling modifies the stationarity conditions for the Lagrangian which now become :

$$L_{\underline{X}_i} = \underline{0} = \frac{\partial f_i}{\partial \underline{X}_i} + \left(\frac{\partial \underline{T}_i}{\partial \underline{X}_i}\right)^T \underline{\mu}_i + \underline{\rho}_i \qquad 4.3.32$$

$$L_{\underline{M}_i} = \underline{0} = \frac{\partial f_i}{\partial \underline{M}_i} + \left(\frac{\partial \underline{T}_i}{\partial \underline{M}_i}\right)^T \underline{\mu}_i \qquad 4.3.33$$

$$L_{\underline{Z}_i} = \underline{0} = -\underline{\mu}_i - \sum_{j=1}^{N} \left(\frac{\partial \underline{H}_j}{\partial \underline{Z}_i}\right)^T \underline{\rho}_j \qquad 4.3.34$$

$$L_{\underline{\mu}_i} = \underline{0} = \underline{T}_i - \underline{Z}_i \qquad 4.3.35$$

$$L_{\underline{\rho}_i} = \underline{0} = \underline{X}_i - \underline{H}_i(\underline{Z}_j) \qquad 4.3.36$$

since the Lagrangian of the problem could have been written as

$$L = \sum_{i=1}^{N} f_i(\underline{M}_i, \underline{X}_i) + \sum_{i=1}^{N} \underline{\mu}_i^T (\underline{T}_i - \underline{Z}_i) + \sum_{i=1}^{N} \underline{\rho}_i^T (\underline{X}_i - \underline{H}_i(\underline{Z}_i \ldots \underline{Z}_N)) \qquad 4.3.37$$

Non-feasible method

In this case, the distribution of the jobs is :

- first level : 4.3.32 to 4.3.35
- second level : 4.3.36

But we recall two basic notions in multi-level optimisation, i.e. distribution of the tasks between the levels with division of work at the lower level and coordination at the higher level.

Thus, in order that equations 4.3.32 to 4.3.35 ($i = 1$ to N) permit a division of work between the subsystems, it is necessary that for given i, the system of equations 4.3.32 to 4.3.35 be solvable to give $\underline{X}_i$, $\underline{M}_i$, $\underline{\mu}_i$, $\underline{Z}_i$ for given $\underline{\rho}_j$ ($j = 1$ to N).

This requires that in equation 4.3.34, the term $\frac{\partial \underline{H}_j}{\partial \underline{Z}_i}$ only be a function of $\underline{Z}_i$ which imposes on H_j the form :

$$\underline{H}_j(\underline{Z}_1 \ldots \underline{Z}_N) = \sum_{i=1}^{N} \underline{g}_{ji}(\underline{Z}_i) \qquad 4.3.38$$

Feasible method

Here, equation 4.3.34 is treated at the higher level for given $\underline{Z}$ and the distribution of the equations between the subsystems remains valid without any additional conditions on 4.3.31.

Mixed method

At the lower level, for given $\underline{\rho}$ and $\underline{Z}$, we treat equations 4.3.32, 4.3.33 and 4.3.35 and here again there are no problems about the division of work.

It is interesting to note that in the case of the last two methods, we do not have any longer the separability of the constraints whilst this notion of separability was the basis of all the decomposition methods that we have studied.

Next we will give an economic interpretation of these methods.

4.3.6 Economic interpretation [9]

The economic interpretation that we will give is macroscopic and it is certainly rather crude. However, it may provide valuable insights.

We assume that $f_i(\underline{X}_i, \underline{M}_i)$ is the production cost for each subproblem and the global problem is to minimise

$$\sum_{i=1}^{N} f_i(\underline{X}_i, \underline{M}_i)$$

subject to the constraints that we have previously stated.

In the goal coordination method, the subproblem is given by :

$$\text{for given } \underline{\rho} \quad \begin{cases} \min \left[f_i(\underline{X}_i, \underline{M}_i) + \underline{\rho}_i^T \underline{X}_i - \sum_{j=1}^{N} \underline{\rho}_j^T C_{ji} \underline{Z}_i \right] \\ \text{subject to } \underline{T}_i(\underline{X}_i, \underline{M}_i) - \underline{Z}_i = \underline{0} \end{cases} \qquad 4.3.39$$

If we give to the Lagrange multiplier $\underline{\rho}$ its traditional price interpretation, then $\underline{\rho}_i^T \underline{X}_i$ will correspond to the buying of raw materials for the subsystem and $\sum_{j=1}^{N} \underline{\rho}_j^T C_{ji} \underline{Z}_i$ the sale price of the finished products $\underline{Z}_i$ leaving the subsystem i. Thus the modified criterion can be viewed as the global financial balance at the level of the subsystems which try to minimise their spending.

The quantity $\Delta\underline{\rho}_i = \underline{X}_i - \sum_{j=1}^{N} C_{ij} \underline{Z}_j$ represents the difference between the supply ($\underline{Z}$) and the demand ($\underline{X}$). If this difference is > 0 ($\Delta\underline{\rho}_i < 0$), the prices decrease (and conversely). This is precisely what the gradient coordinator does using the relation :

$$\underline{\rho}_i(t+1) = \underline{\rho}_i(t) + K \, \Delta\underline{\rho}_i \qquad K > 0 \qquad 4.3.40$$

We thus conclude that the goal coordination method corresponds to a free enterprise economy where the coordinator acts as a "market".

In the model coordination method, the subproblem could be written as :

$$\left[\begin{array}{l} \min \left[f_i(\underline{X}_i, \underline{M}_i) \right] \\ \text{subject to } \begin{cases} \underline{T}_i - \underline{Z}_i = \underline{0} \\ \underline{X}_i - \sum_{j=1}^{N} C_{ij} \underline{Z}_j = \underline{0} \end{cases} \end{array} \right. \qquad \text{for given } \underline{Z} \qquad 4.3.41$$

By fixing $\underline{Z}$ the coordinator imposes the production objectives. It then remains only for the subsystem to minimise its direct production cost. Thus, this method is analogous to a planned economy.

The third method does not appear to have a direct economic interpretation. Perhaps it corresponds to certain cooperative structures ?

4.3.7 Examples

We will next tackle some simple academic problems with a view to illustrating the use of these methods. We will define in each case the subproblems, the coordination tasks, the information transfer between the levels, etc.

We note that the choice of a coordination variable $\underline{\alpha}$ (which is related to the coordination mode) fixes the transfer of the descending information and allows us to define the subproblems ; the choice of the coordination algorithm enables us to specify the information that goes to the coordinator.

Example 1

We consider two types of subsystems (Fig. 4.9, 4.10). In subsystems of type 1, we associate a criterion $Q_1 = a^2+d^2+e^2$ and with the subsystem of type 2 a criterion $Q_2 = f^2+g^2$.

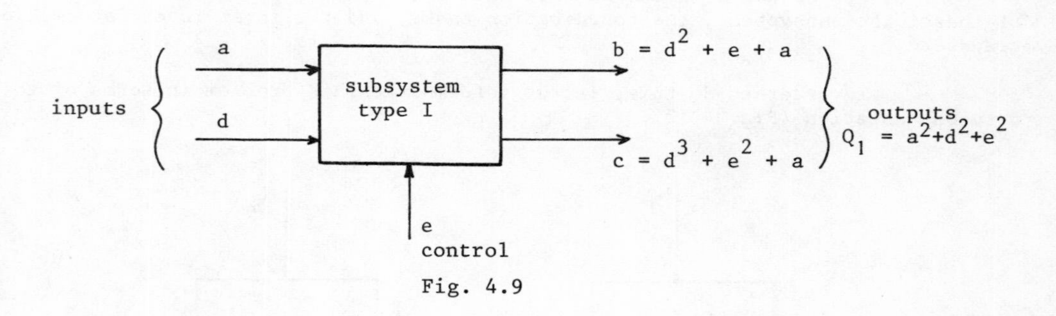

Fig. 4.9

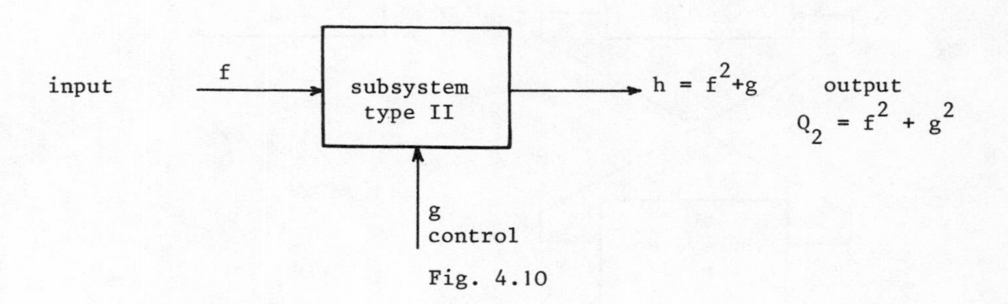

Fig. 4.10

These subsystems are interconnected as shown in Fig. 4.11. For given **a, we optimise the global system** . The global criterion is a sum of the local criteria and we can thus use the techniques of hierarchical optimisation.

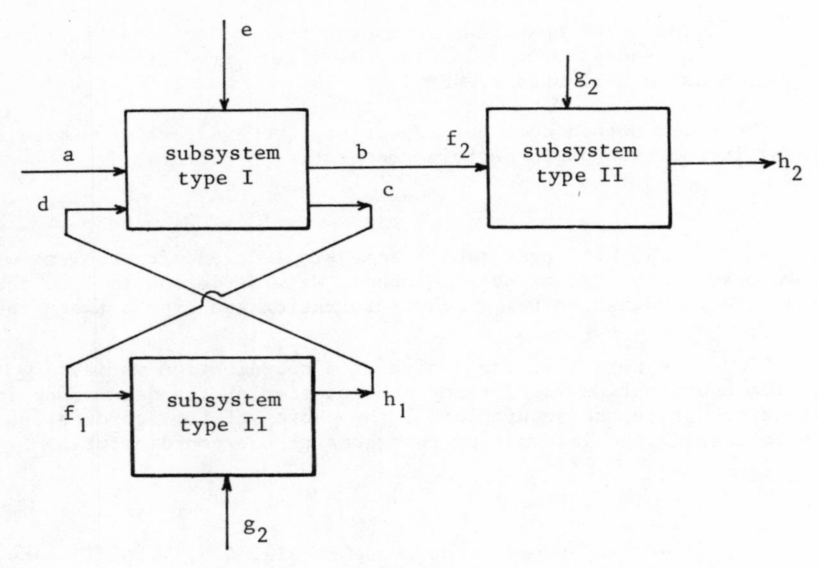

Fig. 4.11

Let us now find the methods that are applicable and formulate for each of these, the subsystems, the coordination tasks, and the inter-level information transfer.

In order to do this, let us reformulate this problem in terms of the previous notation (Fig. 4.12).

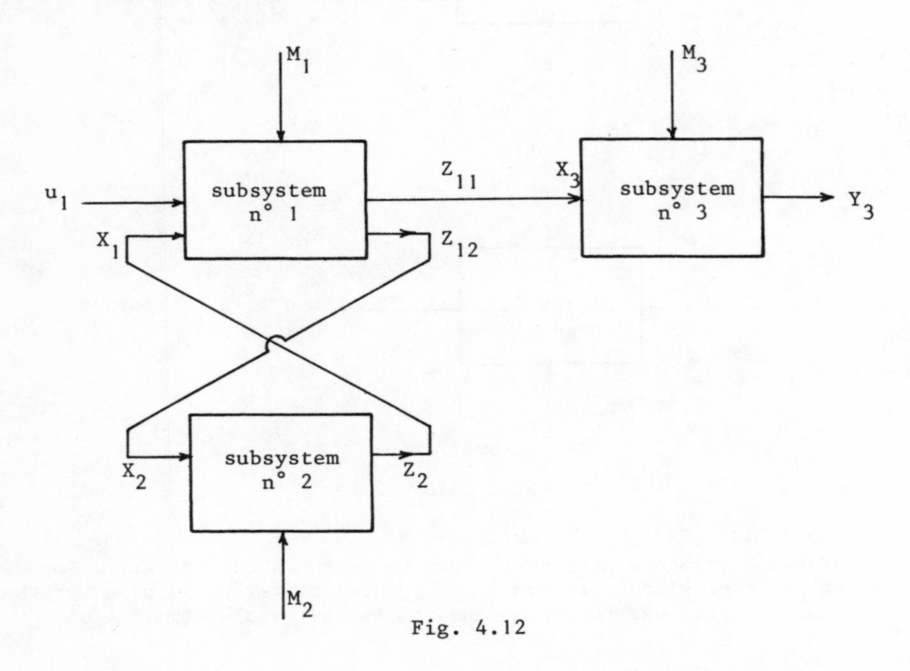

Fig. 4.12

$$\left[\begin{array}{l} \min \left[(U_1^{\,2} + X_1^{\,2} + M_1^{\,2}) + (X_2^{\,2} + M_2^{\,2}) + (X_3^{\,2} + M_3^{\,2})\right] \\ \text{subject to} \left[\begin{array}{l} Z_{11} = X_1^{\,2} + M_1 + U_1 \\ Z_{12} = X_1^{\,3} + M_1^{\,2} + U_1 \\ Z_2 = X_2^{\,2} + M_2 \\ X_1 = Z_2 \\ X_2 = Z_{12} \\ X_3 = Z_{11} \end{array}\right. \end{array}\right. \qquad 4.3.42$$

(Note : it is not necessary to take into account the model equation for Y_3).

If we examine the conditions of applicability of each method, we have :

- Goal coordination : none
- Model coordination : $m_{M_i} \geqslant m_{Z_i} \quad \forall\, i = 1$ to N
- Mixed : $m_{X_i} + m_{M_i} \geqslant m_{Z_i} \quad \forall\, i = 1$ to N

We see here that the condition $m_{M_i} \geqslant m_{Z_i}$ is not satisfied ($m_{M_1} = 1$; $m_{Z_1} = 2$) ; thus only the unfeasible or mixed methods are applicable.

Goal coordination : The Lagrangian in this case can be written as :

$$\begin{aligned} L = {} & U_1^{\,2} + X_1^{\,2} + M_1^{\,2} + \mu_{11}\left[X_1^{\,2} + M_1 + U_1 - Z_{11}\right] + \mu_{12}\left[X_1^{\,3} + M_1^{\,2} + U_1 - Z_{12}\right] \\ & + X_2^{\,2} + M_2^{\,2} + \mu_2\left[X_2^{\,2} + M_2 - Z_2\right] \\ & + X_3^{\,2} + M_3^{\,2} \\ & + \rho_1\left[X_1 - Z_2\right] + \rho_2\left[X_2 - Z_{12}\right] + \rho_3\left[X_3 - Z_{11}\right] \end{aligned} \qquad 4.3.43$$

($\mu_{11}, \mu_{12}, \mu_2, \rho_i$ = 1 to 3 : Lagrange multipliers).

For given $\underline{\rho}$, this Lagrangian can be written as :

$$\begin{aligned} L = \sum_{i=1}^{3} L_i = {} & \left.\begin{array}{l} U_1^{\,2} + X_1^{\,2} + M_1^{\,2} + \rho_1 X_1 - \rho_2 Z_{12} - \rho_3 Z_{11} \\ + \mu_{11}\left[X_1^{\,2} + M_1 + U_1 - Z_{11}\right] + \mu_{12}\left[X_1^{\,3} + M_1^{\,2} + U_1 - Z_{12}\right] \end{array}\right\} L_1 \\ & \left.\begin{array}{l} + X_2^{\,2} + M_2^{\,2} + \rho_2 X_2 - \rho_1 Z_2 \\ + \mu_2\left[X_2^{\,2} + M_2 - Z_2\right] \end{array}\right\} L_2 \\ & \left.\begin{array}{l} + X_3^{\,2} + M_3^{\,2} + \rho_3 X_3 \end{array}\right\} L_3 \end{aligned} \qquad 4.3.44$$

Examining each L_i, we can define the subproblems as :

Subproblem 1

$$\text{for given } \rho_1, \rho_2, \rho_3 \begin{cases} \min \left[U_1^2 + X_1^2 + M_1^2 + \rho_1 X_1 - \rho_2 Z_{12} - \rho_3 Z_{11} \right] \\ \text{subject to } \begin{cases} Z_{11} = X_1^2 + M_1 + U_1 \\ Z_{12} = X_1^3 + M_1^2 + U_1 \end{cases} \end{cases} \quad 4.3.45$$

Subproblem 2

$$\text{for given } \rho_1, \rho_2 \begin{cases} \min \left[X_2^2 + M_2^2 + \rho_2 X_2 - \rho_1 Z_2 \right] \\ \text{subject to } Z_2 = X_2^2 + M_2 \end{cases} \quad 4.3.46$$

Subproblem 3

$$\text{for given } \rho_3 \quad \left\{ \min \left[X_3^2 + M_3^2 + \rho_3 X_3 \right] \right. \quad 4.3.47$$

The resolution of these subproblems (which will not be done here) provides a solution which we will denote by $\underline{X}^*$, $\underline{M}^*$, $\underline{Z}^*$ which satisfies all the stationarity conditions of L except $L_{\underline{\rho}} = \underline{0}$ (which will be treated at the second level).

Gradient coordination : (t : iteration index for the coordinator)

$$\begin{aligned} \rho_1(t+1) &= \rho_1(t) + K L_{\rho_1} = \rho_1(t) + K \left[X_1^* - Z_2^* \right] \\ \rho_2(t+1) &= \rho_2(t) + K \left[X_2^* - Z_{12}^* \right] \qquad K > 0 \\ \rho_3(t+1) &= \rho_3(t) + K \left[X_3 - Z_{11}^* \right] \end{aligned} \quad 4.3.48$$

for X_1^*, X_2^*, X_3^*, Z_{11}^*, Z_{12}^*, Z_2^* provided by the lower level. We will assume that the global solution is achieved when

$$L_{\underline{\rho}}^T \underline{L}_{\rho} \leqslant \epsilon$$

where ϵ is a prechosen small number. Fig. 4.13 shows the calculation procedure. (Note that there is minimal information transfer between the levels).

Newton type coordination :

$$\underline{\rho}(t+1) = \begin{bmatrix} \rho_1(t+1) \\ \rho_2(t+1) \\ \rho_3(t+1) \end{bmatrix} = \underline{\rho}(t) - C \left[\frac{dL_{\underline{\rho}}}{d\underline{\rho}} \right]^{-1} L_{\underline{\rho}} \quad 4.3.49$$

with

$$\frac{dL_{\underline{\rho}}}{d\underline{\rho}} = \begin{bmatrix} \frac{\partial L_{\rho_1}}{\partial \rho_1} & \frac{\partial L_{\rho_1}}{\partial \rho_2} & \frac{\partial L_{\rho_1}}{\partial \rho_3} \\ \frac{\partial L_{\rho_2}}{\partial \rho_1} & \frac{\partial L_{\rho_2}}{\partial \rho_2} & \frac{\partial L_{\rho_2}}{\partial \rho_3} \\ \frac{\partial L_{\rho_3}}{\partial \rho_1} & \frac{\partial L_{\rho_3}}{\partial \rho_2} & \frac{\partial L_{\rho_3}}{\partial \rho_3} \end{bmatrix} \qquad 4.3.49$$

We note that this coordination mode could easily become difficult to implement.

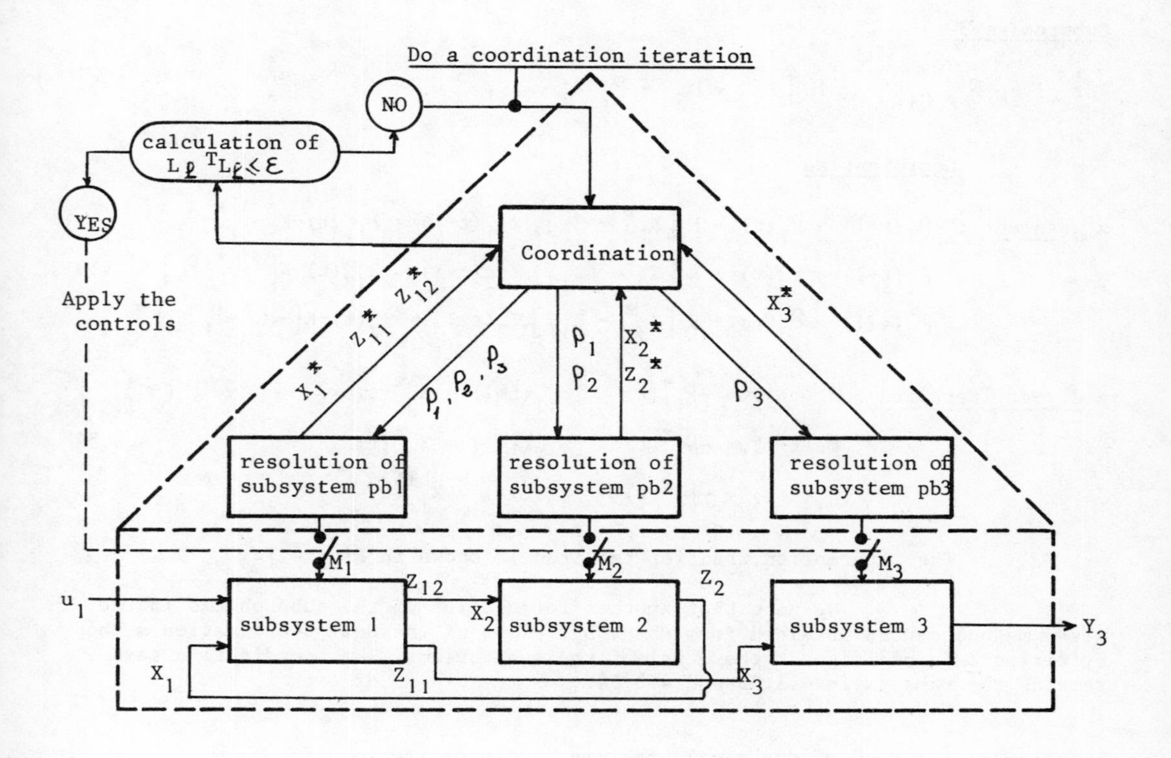

Fig. 4.13

<u>Mixed coordination</u> : The coordination variables are in this case $\underline{\rho}$ and $\underline{Z}$. For given $\underline{\rho}$ and $\underline{Z}$, the Lagrangian in equation 4.3.43 becomes separable, thus leading to the subproblems :

Subproblem 1

For given ρ_1, Z_{11}, Z_{12}, Z_2

$$\begin{cases} \min\left[U_1^2 + X_1^2 + M_1^2 + \rho_1[X_1 - Z_2]\right] \\ \text{subject to} \begin{cases} Z_{11} = X_1^2 + M_1 + U_1 \\ Z_{12} = X_1^3 + M_1^2 + U_1 \end{cases} \end{cases} \quad 4.3.50$$

Subproblem 2

For given ρ_2, Z_{12}, and Z_2

$$\begin{cases} \min\left[X_2^2 + M_2^2 + \rho_2[X_2 - Z_{12}]\right] \\ \text{subject to } Z_2 = X_2^2 + M_2 \end{cases} \quad 4.3.51$$

Subproblem 3

For given ρ_3, Z_{11}

$$\min\left[X_3^2 + M_3^2 + \rho_3 X_3 - Z_{11}\right] \quad 4.3.52$$

Coordination

* gradient :

$$\begin{aligned} \rho_1(t+1) &= \rho_1(t) + K\left[X_1^* - Z_2\right], & Z_{11}(t+1) &= Z_{11}(t) - K\left[-\mu_{11}^* - \rho_3\right] \\ \rho_2(t+1) &= \rho_2(t) + K\left[X_2^* - Z_{12}\right], & Z_{12}(t+1) &= Z_{12}(t) - K\left[-\mu_{12}^* - \rho_2\right] \\ \rho_3(t+1) &= \rho_3(t) + K\left[X_3^* - Z_{11}\right], & Z_2(t+1) &= Z_2(t) - K\left[-\mu_2^* - \rho_1\right] \end{aligned} \quad 4.3.53$$

* direct iteration :

$$\begin{aligned} \rho_3(t+1) &= -\mu_{11}^*(t) & Z_{11}(t+1) &= X_3^*(t) \\ \rho_2(t+1) &= -\mu_{12}^*(t) & Z_{12}(t+1) &= X_2^*(t) \\ \rho_1(t+1) &= -\mu_2^*(t) & Z_2(t+1) &= X_1^*(t) \end{aligned} \quad 4.3.54$$

The information transfer required is shown in Fig. 4.14.

We note that another formulation of the subproblems in the mixed method can be obtained from the subproblems of the Goal Coordination method by fixing $\underline{Z}$ in addition at the level of the subproblems. The coordination task remains the same as in 4.3.53 and 4.3.54.

Example 2 : Solution of dynamical problems by discretisation

Description of the static process obtained by discretisation

Consider the following dynamic optimisation problem :

$$\max_{\underline{M}(t)} R = \int_0^T r(\underline{X}, \underline{M})\, dt \quad 4.3.55$$

$$\text{subject to} \quad \underline{X}(0) = \underline{A}, \ \underline{X}(T) = \underline{B}$$
$$\dot{\underline{X}} = \underline{f}(\underline{X}, \underline{M}) \quad 4.3.56$$

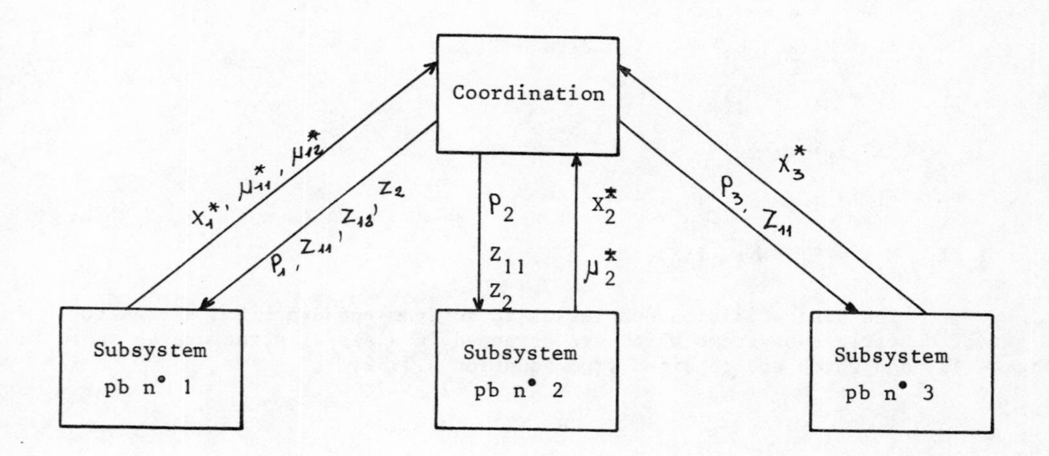

Fig. 4.14

$$\text{with} \quad \underline{X} = \begin{bmatrix} X^1 \\ \vdots \\ X^{m_X} \end{bmatrix} \quad \text{and} \quad \underline{M} = \begin{bmatrix} M^1 \\ \vdots \\ M^{m_M} \end{bmatrix} \qquad 4.3.57$$

Here $\underline{f}$ is an m_X dimensional vector function satisfying the usual Lipschitz condition which assures, for given $\underline{X}(0)$ and $\underline{M}(t)$, a unique solution for the differential equation which represents the dynamical system. These conditions are satisfied in particular when f^1 are continuous functions with continuous partial derivatives in the region defined by the constraints.

Let us decompose the finite optimisation interval (0, T) into N sub-intervals of length Δt. If Δt is sufficiently small, we can assume that the different quantities are constant with respect to time within each interval and we could write :

$$R = \Delta t \sum_{i=1}^{N} r(\underline{X}_i, \underline{M}_i)$$
$$\underline{X}_{i+1} = \underline{X}_i + \Delta t. \ \underline{f}(\underline{X}_i, \underline{M}_i) \qquad 4.3.58$$

The choice of the indices for the variable X is dictated by the need to find a formulation which is close to the one used previously :

$$R = \sum_{i=2}^{N} r_i(\underline{X}_i, \underline{M}_i) + r_1(\underline{A}_1, \underline{M}_1) \quad ; \quad r_j = r(\underline{X}_j, \underline{M}_j)\,\Delta t, \quad j = 1 \text{ to } N$$

$$\begin{aligned} \underline{Z}_i &= \underline{T}_i(\underline{X}_i,\ \underline{M}_i) \qquad i = 2 \text{ to } N-1 \\ \underline{Z}_1 &= \underline{T}_1(\underline{A},\ \underline{M}_1) \\ \underline{B} &= \underline{T}_N(\underline{X}_N,\ \underline{M}_N) \\ \underline{X}_i &= \underline{Z}_{i-1} \qquad i = 2 \text{ to } N \end{aligned} \qquad 4.3.59$$

with $\underline{T}_i(X_i,\ M_i) = \underline{X}_i + \Delta t \cdot \underline{f}(\underline{X}_i,\ \underline{M}_i)$

The discretisation enables us to go from one dynamical system to a set of N static subsystems which are arranged in a serial structure as shown in Fig. 4.15, and which are obtained from equation 4.3.59.

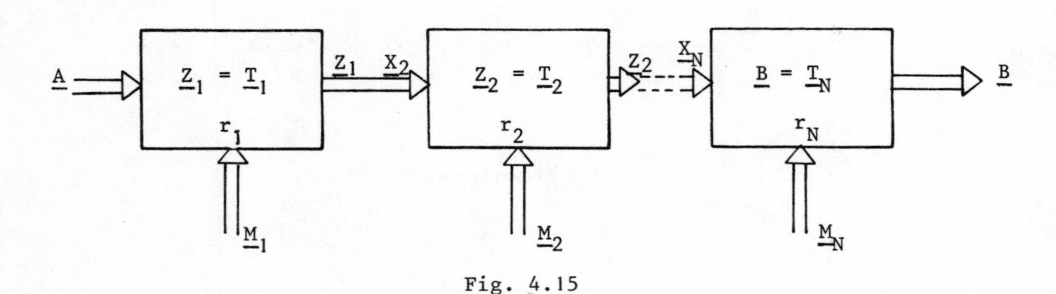

Fig. 4.15

However, this set could be high dimensional (which is an obvious function of the fineness of the discretisation), and it may be desirable to use multilevel techniques to solve such a high dimensional problem. Such solutions are in fact rather easy from a programming point of view, since each subproblem is of the same form. In order to solve it using hierarchical techniques, we start with the feasible method.

The feasible method

The coordination variable in this case is $\underline{Z}$ and the subproblem number i (i = 2 to N-1) could be written directly as :

$$\max_{\underline{X}_i, \underline{M}_i} \quad r_i(\underline{X}_i,\ \underline{M}_i)$$

$$\text{subject to} \begin{bmatrix} \underline{X}_i = \underline{Z}_{i-1} \\ \underline{Z}_i = \underline{T}_i(\underline{X}_i,\ \underline{M}_i) \end{bmatrix} \quad \text{for given } \underline{Z}_{i-1},\ \underline{Z}_i \qquad 4.3.60$$

To take into account the initial state, the first subproblem becomes :

$$\max_{\underline{M}_1} r_1(\underline{A}, \underline{M}_1)$$
$$\text{subject to } \underline{Z}_1 = \underline{T}_1(\underline{A}, \underline{M}_1) \qquad \text{for given } \underline{Z}_1 \qquad 4.3.61$$

The condition on the final state requires that subproblem number N be :

$$\max_{\underline{X}_N, \underline{M}_N} r_N(\underline{X}_N, \underline{M}_N)$$
$$\text{subject to } \underline{X}_N = \underline{Z}_{N-1}, \quad \underline{B} = \underline{T}_N(\underline{X}_N, \underline{M}_N) \qquad \text{for given } \underline{Z}_{N-1} \qquad 4.3.62$$

As for the coordinator, it could use the following algorithm (Fig. 4.16) :

$$\frac{d\underline{Z}_i}{dt} = L_{\underline{Z}_i} = -\underline{\mu}_i - \underline{\rho}_{i+1} \qquad i = 1 \text{ to } N-1 \qquad 4.3.63$$

since only $\underline{Z}_1, \ldots Z_{N-1}$ are necessary.

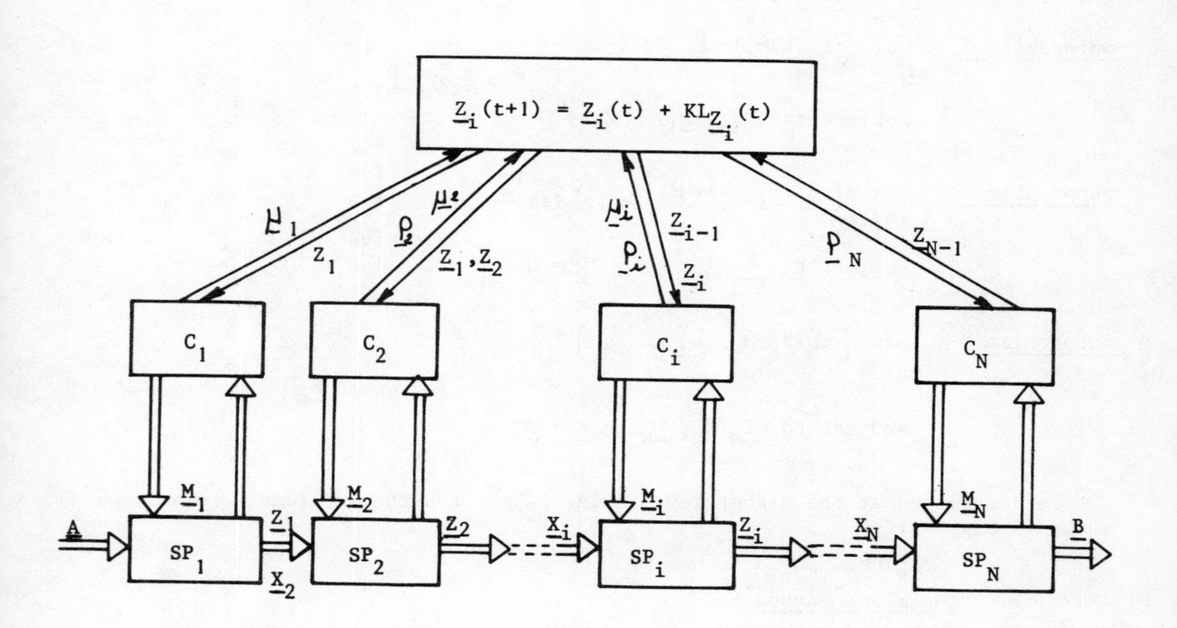

Fig. 4.16

Keeping in mind what we stated earlier, and knowing that each vector $\underline{Z}_i$ has m_X components, each vector $\underline{M}_i$, m_M components, the method is only applicable if

$$m_M \geqslant m_X$$

i.e. the number of controls of the original dynamical system is greater than the number of state variables.

In fact, in most cases, the inverse inequality is satisfied ($m_M \leqslant m_X$) and this limits the region of applicability of this method.

Non-feasible method

The Lagrangian of the global problem can be written as :

$$L = r_1(\underline{A}, \underline{M}_1) + \underline{\mu}_i^T(\underline{T}_1(\underline{A}, \underline{M}_1) - \underline{Z}_1) + \sum_{i=2}^{N-1} \left[r_i(\underline{X}_i, \underline{M}_i) + \underline{\mu}_1^T(\underline{T}_i(\underline{X}_i, \underline{M}_i) - \underline{Z}_i) \right.$$
$$\left. + \underline{\rho}_i^T(\underline{X}_i - \underline{Z}_{i-1}) \right] + r_N(\underline{X}_N, \underline{M}_N) + \underline{\mu}_N^T(\underline{T}_N(\underline{X}_N, \underline{M}_N) - \underline{B}) + \underline{\rho}_N^T(\underline{X}_N - Z_{N-1}) \qquad 4.3.64$$

Given L, it is possible to directly formulate the subproblems, knowing that for given $\underline{\rho}$, we wish to obtain a "separable" form in L at the lower level.

Subproblem 1 : $\max_{\underline{M}_1, \underline{Z}_1} \left[r_1(\underline{A}, \underline{M}_1) - \underline{\rho}_2^T \underline{Z}_1 \right]$ for given $\underline{\rho}_2$ 4.3.65

subject to $\underline{T}_1(\underline{A}, \underline{M}_1) - \underline{Z}_1 = \underline{0}$

Subproblem i : $\max_{\underline{X}_i, \underline{M}_i, \underline{Z}_i} \left[r_i(\underline{X}_i, \underline{M}_i) + \underline{\rho}_i^T \underline{X}_i - \rho_{i+1}^T \underline{Z}_i \right]$ for given $\underline{\rho}_i$, $\underline{\rho}_{i+1}$ 4.3.66

subject to $\underline{T}_i(\underline{X}_i, \underline{M}_i) - \underline{Z}_i = \underline{0}$

Subproblem N : $\max_{\underline{X}_N, \underline{M}_N} \left[r_N(\underline{X}_N, \underline{M}_N) + \underline{\rho}_N^T \underline{X}_N \right]$ for given $\underline{\rho}_N$ 4.3.67

subject to $\underline{T}_N(\underline{X}_N, \underline{M}_N) - \underline{B} = \underline{0}$

$\underline{\rho}$ is determined at the higher level using $L_{\underline{\rho}} = \underline{0}$ and the implemented structure is shown in Fig. 4.17.

The mixed method

The coordination variables are $\underline{\rho}$ and $\underline{Z}$ and the subproblems are directly obtained from those of the "non-feasible" method by fixing $\underline{Z}$ at the higher level.

The solution of the ith subproblem requires that :

$$m_X + m_M \geqslant m_Z \qquad \text{or} \qquad m_M \geqslant 0$$

which is obviously satisfied ; however, in order to solve subproblem 1 we require that : $$m_M \geqslant m_Z = m_X$$

because $\underline{X}_1 = \underline{A}$ and in this unfavourable case, we have the same conditions of utilisation as for the "feasible method".

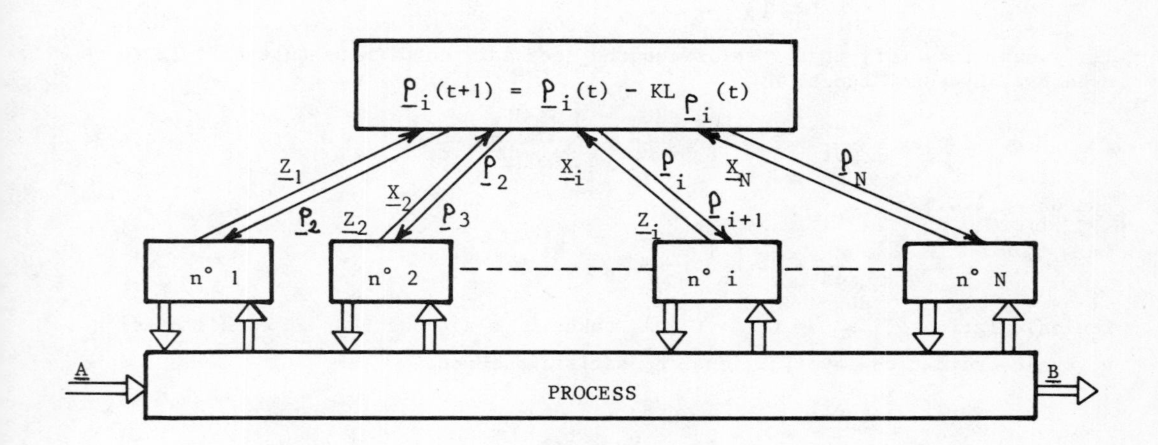

Fig. 4.17

Conclusions for these methods

Let us underline the advantages of these techniques which transform the constraints on the state variable into constraints on the independent variables. This allows us to consider more complex constraints on the initial and final states, than those required by simple modification of subproblems number 1 and N. We note that we can use finer discretisation by merely changing the function $T_i(X_i, M_i)$ within the sub-program since the serial structure presented here enables us to tackle the problem by successive calls of a single sub-program. However, the method runs into difficulty if the optimisation horizon is long and the discretisation is fine.

4.4 CONVERGENCE OF COORDINATION ALGORITHMS

4.4.1 Gradient coordination

4.4.1.1. Use of Lyapunov's direct method with the coordinator being considered in a continuous differential form :

The case of "feasible" and "mixed" methods :

Let $\underline{A}$ be the set of variables of the lower level (which thus has to solve $L_{\underline{A}} = \underline{0}$) and $\underline{B}$ the coordination variables. Then the coordinator equation can be written in the form of the differential equation :

$$\frac{d\underline{B}}{dt} = \mathcal{E}\, L_{\underline{B}} \qquad \mathcal{E} = \begin{cases} +1 & \text{if } \underline{B} \text{ is a physical variable} \\ -1 & \text{if } \underline{B} \text{ is a dual variable} \end{cases} \tag{4.4.1}$$

Let us consider the following Lyapunov function which suggests itself :

$$P = \frac{1}{2} L_{\underline{B}}^T L_{\underline{B}} \tag{4.4.2}$$

It is easy to verify that P satisfies the necessary conditions to enable it to be used as a Lyapunov function.

$$\frac{dP}{dt} = \mathring{P} = L_{\underline{B}}^T \frac{dL_{\underline{B}}}{dt} = L_{\underline{B}}^T \frac{dL_{\underline{B}}}{d\underline{B}} \frac{d\underline{B}}{dt} \tag{4.4.3}$$

and using 4.4.1 :

$$\frac{dP}{dt} = \mathcal{E}\, L_{\underline{B}}^T \frac{dL_{\underline{B}}}{d\underline{B}} L_{\underline{B}} \tag{4.4.4}$$

The calculation of $\frac{dL_{\underline{B}}}{d\underline{B}} = L_{\underline{BB}} + L_{\underline{BA}}\, \underline{A}_{\underline{B}}$ takes into account that at each iteration at the coordination level, $L_{\underline{A}}$ must be satisfied. Hence

$$\frac{dL_{\underline{A}}}{d\underline{B}} = \underline{0} = L_{\underline{AA}} \cdot \underline{A}_{\underline{B}} + L_{\underline{AB}} \tag{4.4.5}$$

$L_{\underline{AA}}$ is non singular if the set of subproblems has a solution (since this solution must satisfy $L_{\underline{A}} = \underline{0}$) which enables us to write :

$$\underline{A}_{\underline{B}} = - L_{\underline{AA}}^{-1} L_{\underline{AB}}$$

$$\text{and} \quad \frac{dL_{\underline{B}}}{d\underline{B}} = L_{\underline{BB}} - L_{\underline{BA}} L_{\underline{AA}}^{-1} L_{\underline{AB}} \tag{4.4.6}$$

In the case of linear coupling, $L_{\underline{BB}} = 0$ and the matrix to be tested is $\frac{dL_{\underline{B}}}{d\underline{B}} = - L_{\underline{BA}} L_{\underline{AA}}^{-1} L_{\underline{AB}}$. If in the case of the "non feasible" method ($\mathcal{E} = -1$) this matrix is positive definite, or if in the case of the "feasible" method ($\mathcal{E} = +1$) this matrix is negative definite, then the algorithm is asymptotically convergent in the sense of Lyapunov.

We can show [7] for both the cases that sufficient conditions for the conditions of Lyapunov to be satisfied are : $L_{\underline{VV}}(\underline{v} = \begin{bmatrix} \underline{X} \\ \underline{M} \end{bmatrix})$ negative definite which assures a maximum (a local solution) for each subproblem.

The case of the Mixed method :

The coordination algorithm is in this case :

$$\frac{d\underline{\rho}}{dt} = - L_{\underline{\rho}} ; \quad \frac{d\underline{Z}}{dt} = + L_{\underline{Z}} \tag{4.4.7}$$

Let $P = \frac{1}{2} \underline{L\rho}^T \underline{L\rho} + \frac{1}{2} \underline{L_Z}^T \underline{L_Z}$ be the Lyapunov function that is chosen. Let us put

$$\underline{A} = \begin{bmatrix} \underline{L_Z} \\ -\underline{L\rho} \end{bmatrix} \quad \text{and } \underline{W} = \begin{bmatrix} \underline{Z} \\ \underline{\rho} \end{bmatrix}$$

Then $\frac{d\underline{W}}{dt} = \underline{A}$ and $P = \frac{1}{2} \underline{A}^T \underline{A}$

$$\frac{dP}{dt} = \underline{A}^T \frac{d\underline{A}}{dt} = \underline{A}^T \frac{d\underline{A}}{d\underline{W}} \frac{d\underline{W}}{dt} = \underline{A}^T \frac{d\underline{A}}{d\underline{W}} \underline{A} \qquad 4.4.8$$

Thus for stability it is necessary that the matrix $\frac{d\underline{A}}{d\underline{W}} = \begin{bmatrix} A_{\underline{v}} & A_{\underline{\mu}} \end{bmatrix} \begin{bmatrix} \underline{v}_{\underline{W}} \\ \underline{\mu}_{\underline{W}} \end{bmatrix} + \underline{A}_{\underline{W}}$ be negative definite.

Taking into account the equations of the first level, we need to have

$$\begin{bmatrix} L_{\underline{vv}} & -L_{\underline{v\mu}} \\ L_{\underline{\mu v}} & 0 \end{bmatrix} \text{ negative definite} \qquad 4.4.9$$

4.4.1.2. <u>Use of Lyapunov's method with the coordination task being considered as one of optimisation</u>

<u>The unfeasible method</u>.

The global problem could be written as :

$$\max_{\underline{v},\underline{Z}} \ \min_{\underline{\mu},\underline{\rho}} \ L(\underline{v}, \underline{Z}, \underline{\mu}, \underline{\rho}) \qquad 4.4.10$$

Here, at the first level, we treat the equations $L_{\underline{v}} = \underline{0}$, $L_{\underline{Z}} = \underline{0}$, $L_{\underline{\mu}} = \underline{0}$. If a solution exists, we satisfy :

$$\max_{\underline{v},\underline{Z}} \ \min_{\underline{\mu},\underline{\rho}} \ L(\underline{v}, \underline{Z}, \underline{\mu}, \underline{\rho}) \qquad 4.4.11$$

At the second level, the gradient algorithm $\frac{d\underline{\rho}}{dt} = -L_{\underline{\rho}}$ is used to arrive at the satisfaction of $L_{\underline{\rho}} = \underline{0}$. This is an optimisation problem which is solved by a gradient algorithm. Let L^v be the solution of this problem and let $P = L - L^v$ be the appropriate Lyapunov function :

$$\frac{dP}{dt} = \frac{dL}{dt}$$

$$\frac{dL}{dt} = \left[(L_{\underline{v}}\, \underline{v}_{\underline{\rho}})^T + (L_{\underline{Z}}\, \underline{Z}_{\underline{\rho}})^T + (L_{\underline{\mu}}\, \underline{\mu}_{\underline{\rho}})^T + L_{\underline{\rho}} \right]^T \frac{d\underline{\rho}}{dt} = L_{\underline{\rho}}^T \frac{d\underline{\rho}}{dt} \qquad 4.4.12$$

because $L_{\underline{v}} = \underline{0}$, $L_{\underline{Z}} = \underline{0}$, $L_{\underline{\mu}} = \underline{0}$ by action of the lower level.

Knowing that $\frac{d\underline{\rho}}{dt} = -L_{\underline{\rho}}$, $\dot{P} = -L_{\underline{\rho}}^T L_{\underline{\rho}} < 0$. Thus the coordinator is asymptotically stable provided that P is continuous and bounded and the latter is verified if each subproblem has a unique solution.

<u>The feasible method</u>

Here the coordinator's job is :

$$\max_{\underline{Z}} L(\underline{v}, \underline{Z}, \underline{\mu}, \underline{\rho}) \qquad 4.4.13$$

and we seek the maximum by a gradient procedure $\frac{d\underline{Z}}{dt} = + L_{\underline{Z}}$. Let $\hat{L}$ be the solution and let $P = \hat{L} - L$ be the appropriate Lyapunov function that is chosen :

$$\dot{P} = -\frac{dL}{dt} = -L_{\underline{Z}}^{T}\frac{d\underline{Z}}{dt} = -L_{\underline{Z}}^{T} L_{\underline{Z}} < 0 \qquad 4.4.14$$

and this mode of coordination is asymptotically stable under the same conditions as above.

The mixed method

Here the coordination task is :

$$\max_{\underline{Z}} \min_{\underline{\rho}} L(\underline{v}, \underline{Z}, \underline{\mu}, \underline{\rho}) \qquad 4.4.15$$

Let $\underline{V}^*$, $\underline{Z}^*$, $\underline{\mu}^*$, $\underline{\rho}^*$ be the solution of the problem and let us put $\mathcal{E}_{\underline{\omega}} = \underline{\omega} - \underline{\omega}^*$ where $\underline{\omega} = \underline{v}$ or $\underline{Z}$ or $\underline{\rho}$ or $\underline{\mu}$.

Let us next formulate the error function at the coordinator level as :

$$D^2 = [\mathcal{E}_{\underline{\rho}}^{T}\mathcal{E}_{\underline{\rho}} + \mathcal{E}_{\underline{Z}}^{T}\mathcal{E}_{\underline{Z}}] \qquad 4.4.16$$

By linearising around the solution, we can show that :

$$D\dot{D} = [\mathcal{E}_{\underline{Z}}^{T}\ \mathcal{E}_{\underline{v}}^{T}]\begin{bmatrix} L^*_{\underline{ZZ}} & L^*_{\underline{Zv}} \\ L^*_{\underline{vZ}} & L^*_{\underline{vv}} \end{bmatrix}\begin{bmatrix} \mathcal{E}_{\underline{Z}} \\ \mathcal{E}_{\underline{v}} \end{bmatrix} \qquad 4.4.17$$

Hence for convergence : $\begin{bmatrix} L^*_{\underline{ZZ}} & L^*_{\underline{Zv}} \\ L^*_{\underline{vZ}} & L^*_{\underline{vv}} \end{bmatrix}$ must be negative definite 4.4.18

In conclusion, for all the three methods, the gradient type algorithm is always applicable (in the sense that it is easy to implement), but it has certain convergence restrictions.

4.4.2 Newton type coordinator

If $\underline{B}$ is the coordination variable, then the coordinator algorithm could be written as :

$$\frac{d\underline{B}}{dt} = -\left[\frac{dL_{\underline{B}}}{d\underline{B}}\right]^{-1} L_{\underline{B}} \qquad 4.4.19$$

Let us assume that the above matrix is non-singular and choosing as the Lyapunov function :

$$P = \frac{1}{2} L_{\underline{B}}^{T} L_{\underline{B}} \qquad 4.4.20$$

then :

$$\dot{P} = L_{\underline{B}}^{T} \frac{dL_{\underline{B}}}{d\underline{B}} \frac{d\underline{B}}{dt} = - L_{\underline{B}}^{T} L_{\underline{B}} \qquad 4.4.21$$

Hence the conclusion that Newton type algorithms are convergent provided the applicability condition, which is the non-singularity of $\frac{dL_{\underline{B}}}{d\underline{B}}$, is satisfied.

4.4.3 Direct study of iterative decomposition-coordination methods [7]

Consider Fig. 4.18.

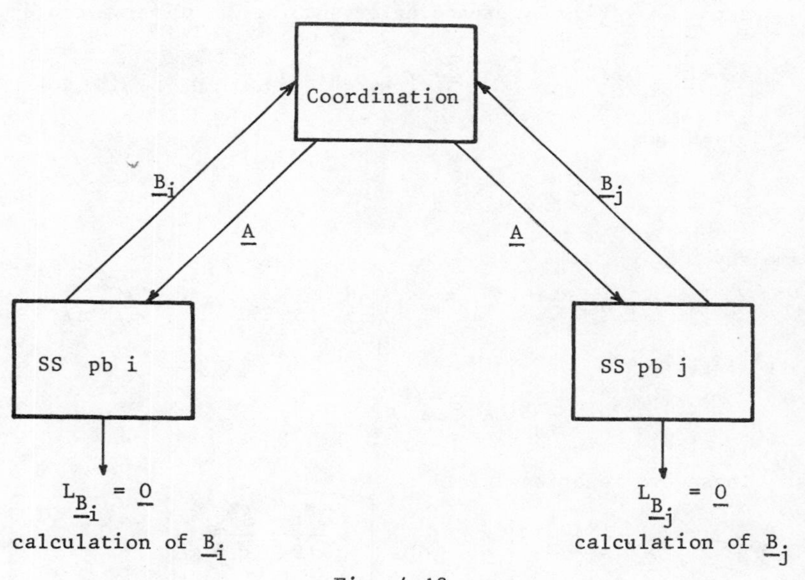

Fig. 4.18

It is interesting to inquire if we could, for whatever coordination method that we have chosen :

- from $L_{\underline{A}}(\underline{A}, \underline{B}) = \underline{0}$, calculate $\underline{A}$ for given $\underline{B}$, in the form

$$\underline{A}^{i} = g(\underline{B}^{i}) \qquad \text{(i : iteration index)}$$

- and from $L_{\underline{B}}(\underline{A}, \underline{B}) = \underline{0}$, calculate $\underline{B}$, for given $\underline{A}$, in the form

$$\underline{B}^{i+1} = f(\underline{A}^{i})$$

We have seen in the case of the feasible and non-feasible methods that $L_{\underline{A}}$ is only a function of $\underline{B}$ and not of $\underline{A}$. This makes this procedure impractical. On the other hand, for the mixed method, $L_{\underline{A}}$ is a function of $\underline{A}$ and $\underline{B}$ and $L_{\underline{A}}(\underline{A}, \underline{B}) = 0$ allows us in certain cases, to calculate $\underline{A}$ for $\underline{B}$ given by the lower level :

$$L_{\underline{A}} = \begin{bmatrix} L_{\underline{Z}} \\ L_{\underline{\rho}} \end{bmatrix} = \begin{bmatrix} -\underline{\mu}_i - \sum_j C_{ji}^T \underline{\rho}_j \\ \underline{X}_i - \sum_j C_{ij} \underline{Z}_j \end{bmatrix} \qquad 4.4.22$$

Here $L_{\underline{Z}}$ has $\sum_i m_{Z_i}$ equations and has $\sum_i m_{Z_i}$ unknowns $\underline{\rho}_j$

$L_{\underline{\rho}}$ has $\sum_i m_{X_i}$ equations and has $\sum_i m_{Z_i}$ unknowns $\underline{Z}_j$

If $\sum_i m_{X_i} = \sum_i m_{Z_i}$ then $L_{\underline{Z}} = \underline{0}$ allows us to calculate $\underline{\rho}$. $L_{\underline{\rho}} = \underline{0}$ allows us to calculate $\underline{Z}$.

The study of local convergence of such a scheme

Let $\underline{A}^i = \underline{g}(\underline{B}^i)$. By expanding around the solution we could write :

$$\underline{A}^* + \underline{\mathcal{E}}_{\underline{A}}^i = \underline{g}(\underline{B}^* + \underline{\mathcal{E}}_{\underline{B}}^1) \simeq \frac{\partial \underline{g}}{\partial \underline{B}} \underline{\mathcal{E}}_{\underline{B}}^i + \underline{g}(\underline{B}^*) \qquad 4.4.23$$

But $\underline{A}^* = \underline{g}(\underline{B}^*)$. Hence :

$$\underline{\mathcal{E}}_{\underline{A}}^i = \frac{\partial \underline{g}^*}{\partial \underline{B}} \underline{\mathcal{E}}_{\underline{B}}^i \qquad 4.4.24$$

In the same way :

$$\underline{B}^{i+1} = \underline{f}(\underline{A}^i) \Rightarrow \mathcal{E}_{\underline{B}}^{i+1} + \underline{B}^* = f(\underline{\mathcal{E}}_{\underline{A}}^i + \underline{A}) \simeq f(\underline{A}^*) + \frac{\partial \underline{f}^*}{\partial \underline{A}} \underline{\mathcal{E}}_{\underline{A}}^i \qquad 4.4.25$$

and since $\underline{B}^* = f(\underline{A}^*)$:

$$\underline{\mathcal{E}}_B^{i+1} = \frac{\partial^* \underline{f}}{\partial \underline{A}} \mathcal{E}_{\underline{A}}^i \qquad 4.4.26$$

These two relations yield :

$$\underline{\mathcal{E}}^{i+1} = \mathbb{M} \underline{\mathcal{E}}^i \ ; \ \underline{\mathcal{E}}^i = \begin{bmatrix} \underline{\mathcal{E}}_{\underline{B}}^i \\ \underline{\mathcal{E}}_{\underline{A}}^i \end{bmatrix}, \qquad \mathbb{M} = \begin{bmatrix} \frac{\partial^* \underline{f}}{\partial \underline{A}} \frac{\partial^* \underline{g}}{\partial \underline{B}} & 0 \\ 0 & \frac{\partial^* \underline{g}}{\partial \underline{B}} \frac{\partial^* \underline{f}}{\partial \underline{A}} \end{bmatrix} \qquad 4.4.27$$

and an examination of the eigenvalues of $\mathbb{M}$ (which is block diagonal) enables us to conclude about its stability.

4.5 EXTENSION TO THE CASE OF INEQUALITY CONSTRAINTS

We now consider the following problem where we have added on inequality constraints for each subsystem :

$$\max \sum_{i=1}^{N} f_i(\underline{X}_i, \underline{M}_i)$$

$$\text{subject to} \left\{ \begin{array}{l} \underline{Z}_i = \underline{T}_i(\underline{X}_i, \underline{M}_i) \\ \underline{X}_i = \sum_{j=1}^{N} C_{ij} \underline{Z}_j \\ \underline{h}_i(\underline{X}_i, \underline{M}_i, \underline{Z}_i) \geqslant \underline{0} \end{array} \right] \quad i = 1 \text{ to } N \qquad 4.5.1$$

and the corresponding Lagrangian is :

$$L = \sum_{i=1}^{N} f_i(\underline{X}_i,\underline{M}_i) + \sum_{i=1}^{N} \underline{\mu}_i^T[\underline{T}_i-\underline{Z}_i] + \sum_{i=1}^{N} \underline{\rho}_i^T[\underline{X}_i - \sum_{j=1}^{N} C_{ij}\,\underline{Z}_j]$$

$$+ \sum_{i=1}^{N} \underline{\gamma}_i^T\,\underline{h}_i\,(\underline{X}_i,\underline{M}_i,\underline{Z}_i) \qquad 4.5.2$$

Here, $\underline{\mu}_i$, $\underline{\rho}_i$ are Lagrange multipliers and $\underline{\gamma}_i$ are Kuhn-Tucker multipliers.

4.5.1 The feasible method

Since here $\underline{Z}$ is the coordination variable, the ith subproblem is :

$$\max\ f_i(\underline{X}_i,\ \underline{M}_i)$$

$$\text{subject to for given } \underline{Z} \quad \begin{cases} \underline{Z}_i - \underline{T}_i = 0 \\ \underline{X}_i - \sum_j C_{ij}\,\underline{Z}_j = \underline{0} \\ \underline{h}_i(\underline{X}_i,\ \underline{M}_i,\ \underline{Z}_i) \geqslant \underline{0} \end{cases} \qquad 4.5.3$$

(Verify that a solution exists for each subproblem as a function of the number of constraints). Then the coordination task for $\underline{Z}$ remains the same as before and the convergence study of section 4.4.1.2 remains valid.

4.5.2 The non-feasible method

In the first case, only the Lagrange multipliers $\underline{\rho}$ are the coordination variables and are thus located at the higher level. Since one of the disadvantages of the non-feasible method is that the interconnection constraints are only satisfied at the end of the coordination iteration, we do not lose anything by also considering the Kuhn-Tucker multipliers as coordination variables. The advantage of such a procedure is to eliminate at the lower level any optimisation problem with inequality constraints. In this case, the constraints $\underline{h}_i \geqslant \underline{0}$ as well as the constraints $\underline{X}_i - \sum_i C_{ij}\,\underline{Z}_j = \underline{0}$ are only satisfied at the optimum.

We will designate by the Type 1 non-feasible method, the first technique given below, and by Type 2 non-feasible method the second procedure given below.

The non-feasible method Type 1

For fixed $\underline{\rho}$, L given in 4.5.2 takes on the following separable form :

$$L = \sum_{i=1}^{N} L_i = \sum_{i=1}^{N} \Big[f_i(\underline{X}_i,\underline{M}_i) + \underline{\mu}_i^T\,(\underline{T}_i-\underline{Z}_i) + \underline{\rho}_i^T\,\underline{X}_i - \sum_{j=1}^{N} \underline{\rho}_j^T\,C_{ji}\,\underline{Z}_i$$

$$+ \underline{\gamma}_i^T\,\underline{h}_i(\underline{X}_i,\underline{M}_i,\underline{Z}_i)\Big] \qquad 4.5.4$$

This yields the ith subproblem :

$$\text{for given } \underline{\rho} \left\{ \begin{array}{l} \max \left[f_i(\underline{X}_i, \underline{M}_i) + \underline{\rho}_i^T \underline{X}_i - \sum_j \underline{\rho}_j^T C_{ji} \underline{Z}_i \right] \\ \text{subject to } \left\{ \begin{array}{l} \underline{T}_i - \underline{Z}_i = \underline{0} \\ \underline{h}_i(\underline{X}_i, \underline{M}_i, \underline{Z}_i) \geqslant \underline{0} \end{array} \right. \end{array} \right. \qquad 4.5.5$$

(We have to ensure the existence of a solution as a function of the number of constraints). The coordination task as far as $\underline{\rho}$ is concerned is unchanged and its convergence study could be done as in section 4.4.1.2.

<u>The non-feasible method Type 2</u>

The Lagrangian given in equation 4.5.4 obviously remains separable for fixed $\underline{\rho}$ and $\underline{\gamma}$ and we thus have the ith subproblem :

$$\text{for given } \underline{\rho} \text{ and } \underline{\gamma} \left\{ \begin{array}{l} \max \left[f_i(\underline{X}_i, \underline{M}_i) + \underline{\rho}_i^T \underline{X}_i - \sum_j \underline{\rho}_j^T C_{ji} \underline{Z}_i + \underline{\gamma}_i^T \underline{h}_i(\underline{X}_i, \underline{M}_i, \underline{Z}_i) \right. \\ \text{subject to } \underline{T}_i - \underline{Z}_i = \underline{0} \end{array} \right. \qquad 4.5.6$$

We note that :

a - No special conditions exist on the number of constraints required for the subproblem to have a solution.

b - The inequality constraint is used to modify the criterion. This modification is 0 at the solution since $\underline{\gamma}_i^T \underline{h}_i = 0$. The coordination task is :

$$\min_{\underline{\rho}, \underline{\gamma} \geqslant 0} L(\underline{\rho}, \underline{\gamma}) \qquad 4.5.7$$

In order to solve this problem, we can use the following algorithm [8] :

$$\frac{d\underline{\rho}}{dt} = -L_{\underline{\rho}} \qquad \text{where} \left[\begin{array}{l} \underline{\rho}(t+1) = \underline{\rho} - KL_{\underline{\rho}}(t) \\ \gamma^j(t+1) = \max\left\{0, \gamma^j(t) - KL_{\gamma^j}(t)\right\} \end{array} \right. \qquad 4.5.8$$

$$\frac{d\gamma^j}{dt} = \left[\begin{array}{l} 0 \text{ if } L_{\gamma}^j > 0 \text{ and } \gamma^j = 0 \\ -L_{\gamma}^j \text{ otherwise} \end{array} \right.$$

The continuous form of this algorithm is a differential system whose second term is discontinuous. The stability study of its solution using Lyapunov's method is rather difficult. We can however do a convergence study after discretisation. The proof in this case is rather long [7] and it shows that there always exist a $K > 0$ which assures convergence.

We note that we could do the same thing in the case of the mixed method but we will not do so here.

Let us next consider an example.

4.5.3 <u>Example</u>

We consider the process shown in Fig. 4.19 and its associated problem :

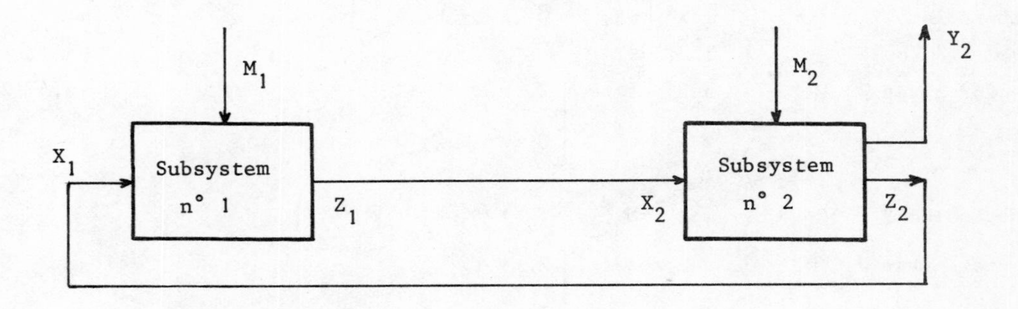

Fig. 4.19

$$\max \left[-M_1^{\,2} - M_2^{\,2} + 4\,Y_2\right] \text{ (with } Y_2 = S_2(X_2,\, M_2)) \qquad 4.5.9$$

$$\text{subject to} \quad \begin{cases} Z_1 = T_1(X_1,\, M_1) \\ Z_2 = T_2(X_2,\, M_2) \\ X_1 = Z_2 \\ X_2 = Z_1 \\ 0.8 - X_1 \geqslant 0 \\ 5 - X_2 - M_2 \geqslant 0 \end{cases} \qquad 4.5.10$$

Knowing that the constraint $0.8 - X_1 \geqslant 0$ is saturated at the solution, the feasible method leads to the following subproblems :

$$\text{Subsystem problem 1 :} \left[\begin{array}{l} \max\limits_{M_1} \left[-M_1^{\,2}\right] \\ \text{subject to} \quad \begin{cases} Z_1 = T_1 \\ X_1 = Z_2 \\ 0.8 - X_1 \geqslant 0 \end{cases} \end{array} \right.$$

$$\text{Subsystem problem 2 :} \left[\begin{array}{l} \max\limits_{M_2} \left[-M_2^{\,2} + 4Y_2\right] \qquad \text{for given } Z_1,\, Z_2 \\ \text{subject to} \quad \begin{cases} Z_2 = T_2 \\ X_2 = Z_1 \\ 5 - X_2 - M_2 \geqslant 0 \end{cases} \end{array} \right. \qquad 4.5.11$$

At the solution there may be incompatibility between $X_1 = Z_2$ fixed by the coordinator and $X_1 = 0.8$ fixed by the saturation of the inequality constraint. This method may not thus be applicable.

Non-feasible method Type 1

Subproblem 1

for given ρ_1, ρ_2

$$\left\{ \begin{array}{l} \max\limits_{X_1, M_1} \left[-M_1^2 + \rho_1 X_1 - \rho_2 Z_1 \right] \\ \text{subject to} \left[\begin{array}{l} Z_1 = T_1 \\ 0.8 - X_1 \geqslant 0 \end{array} \right. \end{array} \right. \qquad 4.5.12$$

Subproblem 2

for given ρ_1, ρ_2

$$\left\{ \begin{array}{l} \max \left[-M_2^2 + 4Y_2 + \rho_2 X_2 - \rho_1 Z_2 \right] \\ \text{subject to} \left[\begin{array}{l} Z_2 = T_2 \\ 5 - X_2 \geqslant 0 \end{array} \right. \end{array} \right. \qquad 4.5.13$$

Coordination : $\rho_1(t+1) = \rho_1(t) - K\left[X_1^* - X_2^*\right]$, $\rho_2(t+1) = \rho_2(t) - K\left[X_1^* - Z_1^*\right]$

Non-feasible method Type 2 :

Subproblem 1
for given ρ_1, ρ_2, γ_1

$$\left\{ \begin{array}{l} \max \left[-M_1^2 + \rho_1 X_1 - \rho_2 Z_1 + \gamma_1 \left[0.8 - X_1 \right] \right] \\ \text{subject to } Z_1 = T_1 \end{array} \right. \qquad 4.5.14$$

Subproblem 2
for given ρ_1, ρ_2, γ_1

$$\left\{ \begin{array}{l} \max \left[-M_2^2 + 4Y_2 + \rho_2 X_2 - \rho_1 Z_2 + \gamma_2 \left[5 - X_2 - M_2 \right] \right] \\ \text{subject to } Z_2 = T_2 \end{array} \right. \qquad 4.5.15$$

Coordination : On ρ as above and on γ :

$$\gamma_1(t+1) = \max\left[0, \gamma_1(t) - K\left[0.8 - X_1^*\right]\right] \qquad 4.5.16$$

$$\gamma_2(t+1) = \max\left[0, \gamma_2(t) - K\left[5 - X_2^* - M_2^*\right]\right]$$

4.6 NON-SEPARABLE PROBLEMS [10]

Non-separable problems constitute an important class of problems. In this case, the coupling relationship is implicit (in the model and/or in the criterion).

One possible approach to tackling such problems is by augmenting the coordination vector and we consider this next.

4.6.1 Augmentation of the coordination vector

The detailed study that we have just done of the different methods of decomposition-coordination shows that by choosing $\underline{\alpha}$ as the coordination variable (i.e. satisfying $L_{\underline{\alpha}} = \underline{0}$ at the higher level) allows us to take into account at the lower level all "non separable" forms in $\underline{\alpha}$ where $\underline{\alpha}$ provides the coupling in the criterion and/or in the model.

It is thus sufficient to detect such an $\underline{\alpha}$ and to add it on to the standard coordination variables $\underline{Z}$, $\underline{\rho}$ or $\underline{Z}$ and $\underline{\rho}$. Since $\underline{\alpha}$ is necessarily a physical variable, the non-feasible method(i.e. where the coordination variable is $\underline{\rho}$) becomes a mixed method (i.e. where the coordination variables are $\underline{\rho}$ and $\underline{\alpha}$).

Example : Consider the following simple problem :

$$\min F = X_1^2 + M_1^2 + M_1 M_2 + M_2^2 + X_2^2$$

$$\text{subject to} \begin{cases} Z_1 = T_1(X_1, M_1) \\ Z_2 = T_2(X_2, M_2) \\ X_1 = Z_2 \\ X_2 = Z_1 \end{cases} \qquad 4.6.1$$

in which an additional coupling appears in the criterion in the form of the term $M_1 \cdot M_2$.

Feasible method :

We choose M_1 (or M_2) in addition to Z_1 and Z_2 as the coordination variables. The subproblems thus become :

Subproblem 1

for given Z_1, Z_2 and M_1

$$\begin{cases} \min F_1 = X_1^2 + M_1^2 \\ \text{subject to} \begin{bmatrix} Z_1 = T_1(X_1, M_1) \\ X_1 = Z_2 \end{bmatrix} \end{cases} \qquad 4.6.2$$

Subproblem 2

for given Z_1, Z_2 and M_1

$$\begin{cases} \min F_2 = X_2^2 + M_2^2 + M_1 M_2 \\ \text{subject to} \begin{bmatrix} Z_2 = T_2(X_2, M_2) \\ X_2 = Z_1 \end{bmatrix} \end{cases} \qquad 4.6.3$$

The coordination task :

$$Z_1(t+1) = Z_1(t) - K\left[-\mu_1^* - \rho_2^*\right] = Z_1(t) - KL_{Z_1}$$

$$Z_2(t+1) = Z_2(t) - K\left[-\mu_2^* - \rho_2^*\right] = Z_2(t) - KL_{Z_2} \qquad 4.6.4$$

$$M_1(t+1) = M_1(t) - K\left[2M_1(t) + M_2^* + \frac{\partial T_1^*}{\partial M_1} \cdot \mu_1^*\right] = M_1(t) - KL_{M_1}$$

L being the Lagrangian associated with 4.6.1 (μ, ρ are the Lagrange multipliers). In fact, because of subproblem 1 which does not in general have a solution, this method is not applicable. To see this, we note that for given Z_1, M_1, and fixed X_1, there is no way to satisfy $Z_1 = T_1(X_1, M_1)$.

Non feasible method :

The coordination variables in this case are ρ_1, ρ_2 and M_1 (or M_2)

Subproblem 1

for given ρ_1, ρ_2 and M_1

$$\begin{cases} \min \left[X_1^2 + M_1^2 + \rho_1 X_1 - \rho_2 Z_1\right] \\ \text{subject to } Z_1 = T_1(X_1, M_1) \end{cases} \qquad 4.6.5$$

Subproblem 2 for given ρ_1, ρ_2 and M_1

$$\begin{cases} \min\left[X_2^{\;2} + M_2^{\;2} + M_1 M_2 + \rho_2 X_2 - \rho_1 Z_2\right] \\ \text{subject to } Z_2 = T_2(X_2, M_2) \end{cases} \qquad 4.6.6$$

The coordination task :

$$\rho_1(t+1) = \rho_1(t) + K\left[X_1^{*} - Z_2^{*}\right]$$
$$\rho_2(t+1) = \rho_2(t) + K\left[X_2^{*} - Z_1^{*}\right] \qquad 4.6.7$$
$$M_1(t+1) = M_1(t) - K\, L_{M_1}$$

We note that here by choosing M_1 as a coordination variable we again act on the model $Z_1 = T_1(X_1, M_1)$ by fixing M_1 so that the method becomes of the mixed type.

4.6.2 Introduction of pseudo-variables

The approach consists of introducing a number of pseudo-variables and with these pseudo-variables a corresponding number of coupling constraints. On this newly augmented problem, we can use standard decomposition techniques.

In fact, given the special form of the coupling constraints it would be advantageous to use a mixed method with direct iteration as we will show on the following simple example.

Example : Consider the problem :

$$\min F = X_1^{\;2} + M_1^{\;2} + M_1 M_2 + M_2^{\;2} + X_2^{\;2} \qquad 4.6.8$$

We can rewrite this as :

$$\min F = X_1^{\;2} + M_1^{\;2} + M_1 \alpha + M_2^{\;2} + X_2^{\;2}$$
$$\text{subject to } M_2 - \alpha = 0 \qquad 4.6.9$$

We can solve this using the mixed method (with coordination variables α and the Lagrange multiplier ρ associated with the constraint $M_2 - \alpha = 0$).

Subproblem 1 : $\min\ F_1 = \left[X_1^{\;2} + M_1^{\;2} + M_1 \alpha\right]$ for given α 4.6.10

Subproblem 2 : $\min\ F_2 = \left[M_2^{\;2} + Y_2^{\;2} + \rho\left[M_2 - \alpha\right]\right]$ for given ρ and α 4.6.11

Coordination :

$$\left.\begin{aligned} \rho(t+1) &= \rho(t) + K\left[M_2^{*} - \alpha(t)\right] \\ \alpha(t+1) &= \alpha(t) - K\left[M_1^{*} - \rho(t)\right] \end{aligned}\right\} \text{gradient type} \qquad 4.6.12$$

or better

$$\left.\begin{aligned} \rho(t+1) &= M_1^{*}(t) \\ \alpha(t+1) &= M_2^{*}(t) \end{aligned}\right\} \quad \text{direct iteration} \qquad 4.6.13$$

Next we examine applications.

4.7 APPLICATIONS

4.7.1 Stock building

The problem that we treat here is a simplified version of the one given in [11].

Statement of the problem

We need to take decisions at the instants $k = 1, 2, \ldots, K$ to build the stocks $x_i(k)$ $(i = 1, 2, \ldots N)$ of N products by placing the orders $u_i(k)$. In the interval $(k, k+1)$, sale and production considerations reduce the stocks by a known amount $s_i(k)$.

If we assume that delivery of orders is immediate, then the equation of evolution of stocks over the interval $(k, k+1)$ is :

$$x_i(k+1) = x_i(k) + u_i(k) - s_i(k) \qquad \text{with } x_i(1) = x_{i1} \tag{4.7.1}$$

Practical considerations require that $x_i(k)$ and $u_i(k)$ be bounded as :

$$\begin{aligned} x_{i\,min} &\leqslant x_i(k) \leqslant x_{i\,max} \\ u_{i\,min} &\leqslant u_i(k) \leqslant u_{i\,max} \end{aligned} \tag{4.7.2}$$

Besides, stock and/or production limitations impose the constraint :

$$r(x_1(k) \ldots x_N(k), u_1(k) \ldots u_N(k), k) = \sum_{i=1}^{N} r_i(x_i(k), u_i(k), k) \leqslant 0 \tag{4.7.3}$$

Let $C_i(u_i(k), k)$ be the cost of the order of $u_i(k)$ units at the kth instant and $h_i(x_i(k), k)$ the cost of storing $x_i(k)$ units. The total cost over the time horizon is given by :

$$J = \sum_{i=1}^{i=N} \left\{ h_i(x_i(K+1), K+1) + \sum_{k=1}^{k=K} \left[C_i(u_i(k), k) + h_i(x_i(k), k) \right] \right\} \tag{4.7.4}$$

Then the problem can be stated as :

1°) Knowing that N and K could be large so that this management problem could be tackled as a high order mathematical program, what decomposition-coordination method could we use ?

2°) What happens when we only have one product $(N = 1)$ and no global constraints $(r_i \equiv 0)$. In that case we will use as the subproblem goal :

$$J_1 = \left\{ \frac{1}{2} h_1 x_1^2(k+1) - h_{1L} x_1(k+1) + \sum_{k=1}^{k=K} \left(\frac{1}{2} h_1 x_1^2(k) - h_{1L} x_1(k) + C_1 u_1(k) \right\} \tag{4.7.5}$$

where h_1, C_1, h_{1L} are positive constants.

3°) We note that the problem is modified when we have a delivery delay of θ periods such that the dynamic equation now becomes :

$$x_1(k+1) = x_1(k) + u_1(k - \theta) - s_1(k) \qquad x_1(1) = x_{1\,1} \tag{4.7.6}$$

Solution

1°) The global problem could be written as :

$$\min J = \sum_i \left\{ J_i \right\} \qquad 4.7.7$$

subject to

$$\begin{cases} x_i(k) \in X_i(k) & \text{defined by the inequality constraints} \\ u_i(k) \in U_i(k) & \\ \sum_{i=1}^{N} r_i\,(x_i(k),\, u_i(k)) \leqslant 0 & i = 1 \text{ to } N \\ & k = 1 \text{ to } K \\ x_i(k+1) = x_i(k) + u_i(k) - s_i(k) & \end{cases} \qquad 4.7.8$$

The Lagrangian L is in this case :

$$L = \sum_{i=1}^{N} J_i + \sum_{k=1}^{K} \lambda(k) \sum_{i=1}^{N} r_i(x_i(k),\, u_i(k)) + \sum_{i=1}^{N}\sum_{k=1}^{K} P_i\,(k) \left[x_i(k+1) - x_i(k) - u_i(k) + s_i(k) \right] \qquad 4.7.9$$

Let us assume that $\lambda(k)$ $(k = 1, K)$ is given by the higher level (level 3). Then :

$$L = \sum_{i=1}^{N} \left[J_i + \sum_{k=1}^{K} \lambda(k)\, r_i(x_i(k),\, u_i(k)) + \sum_{k=1}^{K} p_i\,(k) \left[x_i(k+1) - x_i(k) - u_i(k) + s_i(k) \right] \right] \qquad 4.7.10$$

which corresponds to the definition of the N independent subproblems at the 2nd level. The ith subproblem can be written as :

$$\min \left[J_i + \sum_{k=1}^{K} \lambda(k)\, r_i(x_i(k),\, u_i(k)) \right]$$

for given $\lambda(k)$, $k = 1$ to K

$$\text{subject to } \begin{cases} x_i(k) \in X_i(k),\, u_i(k) \in U_i\,(k) & k = 1 \text{ to } K \\ x_i(k+1) = x_i(k) + u_i(k) - s_i(k) \end{cases} \qquad 4.7.11$$

The coordination of $\lambda(k)$ is done by maximising $\phi(\lambda)$ by using in a gradient type algorithm :

$$\phi_\lambda = \sum_{k=1}^{K} r_i(x_i(k),\, u_i(k)) \qquad 4.7.12$$

with

$$\phi(\lambda) = \min_{x,u} L(x,u,\lambda)$$
$$\text{subject to } x_i \in X_i,\; u_i \in U_i,\; x_i(k+1) = x_i(k) + u_i(k) - s_i(k) \qquad 4.7.13$$

The ith subprobmem itself can be decomposed by the index k :

$$L_i = \sum_{k=1}^{K} L_i(k) \quad \text{by fixing } P_i(k) \quad k = 1 \text{ to } K \tag{4.7.14}$$

$$L_i = J_i + \sum_{k=1}^{K} \lambda(k)\, r_i(x_i(k), u_i(k)) + \sum_{k=1}^{K} P_i(k)\left[x_i(k+1) - x_i(k) - u_i(k) + s_i(k)\right] \tag{4.7.15}$$

$$L_i = h_i(x_i(K+1), K+1) + \sum_{k=1}^{K} \Big[C_i(u_i(k),k) + h_i(x_i(k),k) + \lambda(k)\, r_i(x_i(k), u_i(k)) + P_i(k)\left[x_i(k+1) - x_i(k) - u_i(k) + s_i(k)\right]\Big]$$

or

$$L_i(k) = C_i + h_i + \lambda(k)\, r_i + P_i(k-1)\, x_i(k) - P_i(k)\, x_i(k) - P_i(k)\, u_i(k) + P_i(k)\, s_i(k) \tag{4.7.16}$$

Hence the (K+1) independent subproblems :

<u>k = K + 1</u>

for given $P_i(K)$

$$\min\left[h_i(x_i(K+1),K+1) + P_i(K)\, x_i(K+1)\right]$$
$$\text{subject to } x_i(K+1) \in X_i(K+1) \tag{4.7.17}$$

<u>k = K to 2</u>

for given $P_i(k)$, $P_i(k-1)$

$$\min\Big[C_i(u_i(k),k) + h_i(x_i(k),k) + \lambda(k)\, r_i(x_i(k), u_i(k)) - P_i(k)\, u_i(k) - \left[P_i(k) - P_i(k-1)\right] x_i(k) + \cancel{P_i(k)s_i(k)}\Big] \tag{4.7.18}$$

this is a constant term which can be eliminated

$$\text{subject to } x_i(k) \in X_i(k),\ u_i(k) \in U_i(k) \tag{4.7.19}$$

<u>k = 1</u>

for given $P_i(1)$

$$\min\Big[C_i(u_i(1),1) + h_i(x_i(1),1) + \lambda(1)\, r_i(x_i(1), u_i(1)) - \cancel{x(1)\, P_i(1)} - P_i(1)\, u_i(1)\Big],\ \cancel{x_i(1) \in X_i(1)},\ u_i(1) \in U_i(1) \tag{4.7.20}$$

But here $x_i(1)$ is given. Hence the elimination of the terms.

<u>Coordination</u> :

$$\max_P \phi_i \longrightarrow \text{price} : P_i^{\sigma+1}(k) = P_i^{\sigma}(k) + K\left[x_i^*(k+1) - x_i^*(k) - u_i^*(k) + s_i^*(k)\right] \tag{4.7.21}$$

where K is the iteration constant and σ is the coordination index.

The coordination structure is shown in Fig. 4.20.

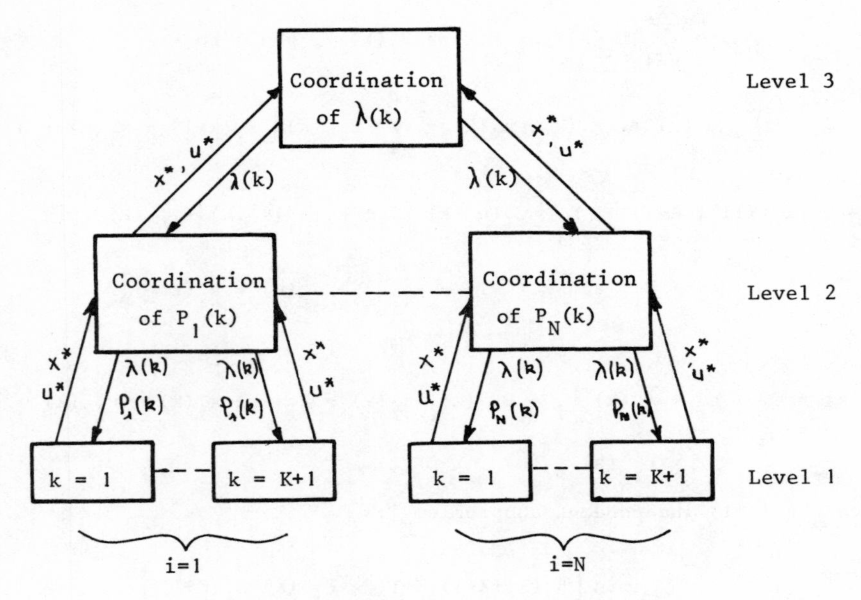

Fig. 4.20

2°) In the case of a single product, there are only two levels (the lowest two). The subproblems in this case are :

<u>k = 1</u>

$$\min \left[\frac{1}{2} h_1 x_1^{\,2}(1) - h_{1L} x_1(1) + C_1 u_1(1) - P_1(1) u_1(1) - P_1(1) x_1(1)\right] \qquad 4.7.22$$

for given P(1) $\quad\left|\begin{array}{l} x_1(1) = x_{11} \quad \text{(given)} \\ u_1(1) \in U_1(1) \rightarrow u_{1\,min}^{(1)} \leqslant u_1^{(1)} \leqslant u_{1\,max}^{(1)} \end{array}\right.$

The problem reduces to :

$$\min_{u_1(1)} \left\{ \left[C_1 - P_1(1) \right] u_1(1) \right\} = f \qquad 4.7.23$$

The solution is given by (Fig. 4.21) :

$$\begin{array}{ll} \cdot & u_1^*(1) = u_{1\,max} \quad \text{if } C_1 - P_1(1) < 0 \\ \cdot & u_1^*(1) = u_{1\,min} \quad \text{if } C_1 - P_1(1) > 0 \\ \cdot & u_1^*(1) \text{ arbitrary in } U_1(1) \text{ if } C_1 - P_1(1) = 0 \end{array} \qquad 4.7.24$$

<u>k = 2 to K</u>

$$\min \left[\frac{1}{2} h_1 x_1^{\,2}(k) - h_{1L} x_1(k) + C_1 u_1(k) + P_1(k-1) x_1(k) - P_1(k) x_1(k) - P_1(k) u_1(k)\right] \qquad 4.7.25$$

subject to $x_1(k) \in X_1(k)$, $u_1(k) \in U_1(k)$

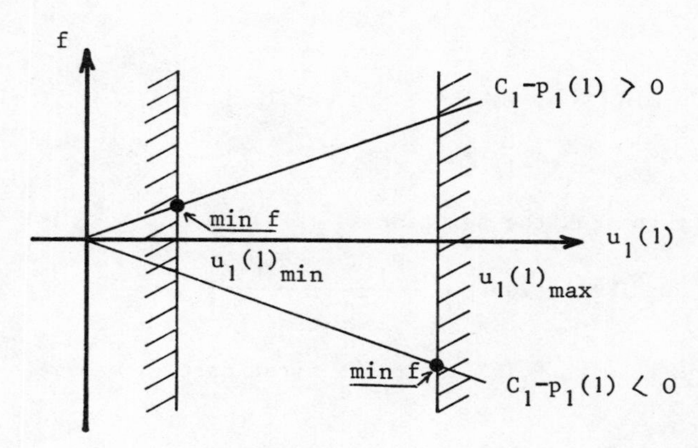

Fig. 4.21

The quadratic (in x) criterion enables us to calculate

$$x_1(k) = \frac{P_1(k) - P_1(k-1) + h_{1L}}{h_1} = f(P) \text{ if:} \qquad 4.7.26$$

$$\begin{bmatrix} f(P) < x_1(k)_{max} \\ f(P) > x_1(k)_{min} \end{bmatrix} \qquad 4.7.27$$

Else $x_1\ (k) = x_1(k)_{min}$ or $x_1(k)_{max}$

Let us define the function sat ξ by the graph in Fig. 4.22. Then :

$$x_1^*(k) = \text{sat}\left[\frac{P_1(k) - P_1(k-1)\ h_{1L}}{h_1}\right] \qquad 4.7.28$$

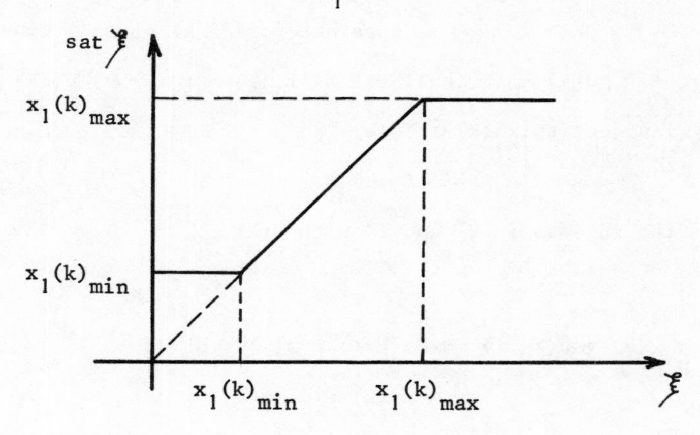

Fig. 4.22

The linear criterion (in $u_1(k)$) leads to the same type of solution as for k = 1.

k = K + 1

$$\min \left[\frac{1}{2} h_1 x_1^2(K+1) - h_{1L} x_1(K+1) + P_1(K) x_1(K+1)\right] \qquad 4.7.29$$

$$x_1(K+1) \in X_1(K+1)$$

We need only to seek the solution $x_1^*(k+1)$, i.e.

$$x_1^*(K+1) = \text{sat}\left[\frac{-P_1(K) + h_{1L}}{h_1}\right] \qquad 4.7.30$$

Coordination : with $P_1(k)$ as in the first part (1°) above.

3° The global problem

$$\min J = J(K+1) + \sum_{k=1}^{K} J(k) \qquad 4.7.31$$

$$\text{subject to} \left| \begin{array}{l} x_1(k+1) = x_1(k) + u_1(k-\theta) - s_1(k), x_1(1) = x_{11} \\ x_1(k) \in X_1(k) \\ u_1(k) \in U_1(k) \qquad k = 1 - K \end{array} \right. \qquad 4.7.32$$

The Lagrangian here is :

$$L = J(K+1) + \sum_{k=1}^{K} J(k) + \sum_{k=1}^{K} P_1(k)\left[x_1(k+1) - x_1(k) - u_1(k-\theta) + s_1(k)\right] \qquad 4.7.33$$

If we fix $P_1(k)$ $k = 1$ to K, $L = \sum_{k=1}^{K} L(k) + L(K+1)$ 4.7.34

$$L(k) = J(k) + P_1(k-1) x_1(k) - P_1(k) x_1(k) - P_1(k+\theta) u_1(k) + \underbrace{P_1(k) s_1(k)}_{\text{constant for } P_1(k) \text{ given : can be neglected}} \qquad 4.7.35$$

Subsystem problem : $\min \left[J(k) + (P_1(k-1) - P(k) x_1(k) - P_1(k+\theta) u_1(k)\right]$ 4.7.36

$P_1(k-1)\ldots P_1(k+\theta)$ given

$$\text{subject to} \left| \begin{array}{l} x_1(k) \in X_1(k) \\ u_1(k) \in U_1(k) \end{array} \right. \qquad 4.7.37$$

The coordination is the same as in 2° and 1° with P(k).

Standard approach :

$$x(k+1) = x(k) + u(k-\theta) \Rightarrow x(k+1) = x(k) + u_1(k)$$

$$u_1(k) = u_2(k-1) \qquad 4.7.38$$

$$u_2(k) = u_3(k-2)$$

$$\vdots$$

This leads to an augmentation of the dimension of the above problem and that is not very desirable.

As another application, we next consider the problem of management of hydroelectric systems.

4.7.2 Management of hydroelectric systems

4.7.2.1 Management strategies for production of electricity

We study here management strategies for production of electricity with a fixed amount of equipment over a time horizon of one year. The model for such a problem has the following general structure [12] :

$$\begin{cases} \min G(\underline{x}, \underline{y}) = G_o(\underline{x}_o, \underline{y}) + \sum_{j=1}^{m} G_j(\underline{x}_j) \\ \text{subject to} \begin{bmatrix} K_j(\underline{x}_j) \geqslant 0 & j = 0 \text{ to } m \\ \underline{y} = \sum_{j=1}^{m} H_j(\underline{x}_j) \end{bmatrix} \end{cases} \quad 4.7.39$$

In order to use decomposition-coordination techniques to solve this problem, let us rewrite it as :

$$\begin{cases} \min G\left[\underline{x}, H(\underline{x})\right] \\ \text{subject to } \underline{K}(\underline{x}) \geqslant \underline{0} \end{cases} \quad 4.7.40$$

or

$$\begin{cases} \min G(\underline{x}, \underline{y}) \\ \text{subject to} \begin{bmatrix} \underline{y} = \underline{H}(\underline{x}) \\ \underline{K}(\underline{x}) \geqslant \underline{0} \end{bmatrix} \end{cases} \quad 4.7.41$$

We assume that the standard differentiability assumptions are satisfied and we also assume that an optimal solution exists.

Let the Lagrangian for this problem be :

$$L = G(\underline{x}, \underline{y}) + \underline{\lambda}^T\left[\underline{y} - \underline{H}(\underline{x})\right] + \underline{\mu}^T \underline{K}(\underline{x}) \quad 4.7.42$$

where $\underline{\lambda}$ are the Lagrange multipliers and $\underline{\mu}$ are the Kuhn-Tucker multipliers.

Let us suppose that we wish to solve this problem by using a partially separable form so as to solve it by a mixed method. In that case the tasks of the two levels are :

Coordinator level :

This level fixes $\underline{\lambda}$ and $\underline{y}$ at each iteration and modifies it for the next iteration as a function of the information received from the lower levels so as to eventually satisfy

$$L_{\underline{y}} = \underline{0}, \qquad L_{\underline{\lambda}} = \underline{0}$$

Lower level :

Solve the problem :

$$\text{for fixed } \underline{\lambda} \text{ and } \underline{y} \quad \begin{cases} \min\left\{ G(\underline{x}, \underline{y}) + \underline{\lambda}^T \left[\underline{y} - \underline{H}(\underline{x}) \right] \right\} \\ \text{subject to } \underline{K}(\underline{x}) \geqslant \underline{0} \end{cases} \qquad 4.7.43$$

But since $\underline{\lambda}$ and $\underline{y}$ are fixed, this problem can be rewritten as :

$$\text{for fixed } \underline{\lambda} \text{ and } \underline{y} \quad \begin{cases} \min\left[G(\underline{x}, \underline{y}) - \underline{\lambda}^T \underline{H}(\underline{x}) \right] \\ \text{subject to } \underline{K}(\underline{x}) \geqslant \underline{0} \end{cases} \qquad 4.7.44$$

<u>Coordination task</u>

$$\begin{aligned} L_{\underline{y}} &= \underline{0} \Longrightarrow \frac{\partial G}{\partial \underline{y}} + \underline{\lambda} = \underline{0} \qquad \underline{\lambda} = -\frac{\partial G}{\partial \underline{y}} (\underline{x}, \underline{y}) \\ L_{\underline{\lambda}} &= \underline{0} \Longrightarrow \underline{y} - \underline{H}(\underline{x}) = \underline{0} \qquad \underline{y} = \underline{H}(\underline{x}) \end{aligned} \qquad 4.7.45$$

Hence we have the two level algorithm :

1° we guess $\underline{x}^S$

2° we calculate $\underline{y}^S = \underline{H}(\underline{x}^S)$ and $\underline{\lambda}^S = -\frac{\partial G}{\partial \underline{y}}(\underline{x}^S, \underline{y}^S)$

3° we solve for $\underline{x}$:

$$\begin{cases} \min\left[G(\underline{x}, \underline{y}^S) - \underline{\lambda}^{S^T} \underline{H}(\underline{x}) \right] \\ \text{subject to } \underline{K}(\underline{x}) \geqslant \underline{0} \end{cases} \qquad 4.7.46$$

Let $\underline{x}^{S+1}$ be this local solution.

4° we put $\underline{x}^S = \underline{x}^{S+1}$ and we go back to step 1 unless the termination criterion is satisfied.

Or again, using $\underline{\lambda} = -\frac{\partial G}{\partial \underline{y}}(\underline{x}, \underline{y})$ 4.7.47

1° we know $\underline{x}^S$

2° we can calculate $\underline{y}^S = \underline{H}(\underline{x}^S)$

3° we solve for $\underline{x}$

$$\begin{cases} \min\left[G(\underline{x}, \underline{y}^S) + \frac{\partial G^T}{\partial \underline{y}} (\underline{x}^S, \underline{y}^S) \underline{H}(\underline{x}) \right] \\ \text{subject to } \underline{K}(\underline{x}) \geqslant \underline{0} \end{cases} \qquad 4.7.48$$

Let $\underline{x}^{S+1}$ be the solution of this problem.

4° if $\underline{x}^{S+1}$ satisfies the termination criterion, we stop. Otherwise, we go back to 1°.

Let us put once again

$$G(\underline{x}, \underline{y}) = G_o(\underline{x}_o, \underline{y}) + \sum_{j=1}^{m} G_j(\underline{x}_j) \qquad 4.7.49$$

and let the problem be :

$$\text{Min } G(\underline{x}, \underline{y}) = G_o(\underline{x}_o, \underline{y}) + \sum_{j=1}^{m} G_j(\underline{x}_j)$$

$$\text{subject to} \left[\begin{array}{ll} \underline{K}_j(\underline{x}_j) \geq \underline{0} & j = 0 \text{ to } m \\ \underline{y} = \sum_{j=1}^{m} \underline{H}_j(\underline{x}_j) & \end{array} \right. \qquad 4.7.50$$

At each iteration of the decomposition algorithm, we will need to solve a problem of the type

$$\min \left[G_o(\underline{x}_o, \underline{y}^S) + \sum_{j=1}^{m} G_j(\underline{x}_j) + \frac{\partial G_o}{\partial \underline{y}} (\underline{x}^S, \underline{y}^S) \sum_{j=1}^{m} \underline{H}_j(\underline{x}_j) \right] \qquad 4.7.51$$

$$\text{subject to } \underline{K}_j(\underline{x}_j) \geq \underline{0}$$

which decomposes into the following (m+1) subproblems :

$$\text{Problem } P_o \left\{ \begin{array}{l} \min\limits_{\underline{x}_o} \; G_o(\underline{x}_o, \underline{y}^S) \\ \text{subject to } \underline{K}_o(\underline{x}_o) \geq \underline{0} \end{array} \right. \qquad 4.7.52$$

which corresponds to the thermal station problem of best placing of the thermal station for fixed demand.

$$\begin{array}{l} \text{Problem } P_j \\ j = 1 \text{ to } m \end{array} \left\{ \begin{array}{l} \min\limits_{\underline{x}_j} \left[G_j(\underline{x}_j) + \frac{\partial G_o}{\partial \underline{y}} (\underline{x}^S, \underline{y}^S) \, \underline{H}_j(\underline{x}_j) \right] \\ \text{subject to } \underline{K}_j(\underline{x}_j) \geq \underline{0} \end{array} \right. \qquad 4.7.53$$

which corresponds to the hydraulic problem for valley j, $\underline{\lambda}^S$ having the economic interpretation of the marginal value of the energy. The set of these hydraulic subproblems consists of maximising the hydraulic production of valley j for fixed demand.

We won't give the details of the utilisation of this algorithm. The interested reader is referred to [12]. The main point of the development here was to provide a theoretical justification of a decomposition algorithm which was introduced heuristically based on economic considerations (marginal value of electricity).

It is interesting to note that the algorithm converges in a fairly small number of iterations even for very high order problems.

Next we consider another hydroelectric problem.

4.7.2.2. Short term management of hydroelectric-thermal power systems

Basically, if we limit ourselves to a short-term time horizon, we can take advantage of the resulting simplifications to convert our problem to a high order linear programming problem which can subsequently be treated by decomposition coordination methods or by partitioning techniques which use the special structure of the problem.

Here, we retain the non-linearities in the problem and introduce other optimisation techniques to solve it.

Description of the system and of the problem

System

We will consider the production complex for producing electricity using hydro and thermal power plants, as shown in Fig. 4.23. The system comprises three hydroelectric stations in cascade on a river with a thermal power station (the methodology that we use here is obviously applicable for systems with an arbitrary structure but it is easier to understand the methodology on a simple example like the one chosen here).

We represent the system in discrete time form with the discretisation interval being equal to a unit of time. The variables in our problem are :

$Y_j(i)$ is the water used in the period i by the station j

$X_j(i)$ are the contents of the reservoir of the station j at the end of the period i

$q_j(i)$ is the inflow for the jth station in the period i

$V_j(i)$ is the spill for the jth station in the period i

$d_j(i)$ is the extra water that goes from the station j-1 to j

$H_j(i)$ is the hydroelectric power (station j, period i)

$S(i)$ is the thermal power in the period i

$D(i)$ is the global demand in the period i

$i \in T = (1, 2, \ldots, N)$ is the optimisation period.

Models and constraints

The time evolution of each reservoir is given by the following difference equation :

$$X_j(i) = X_j(i-1) + Y_j(i) - q_j(i) - V_j(i) \qquad j = 1 \text{ to } 3, \quad X_j(0) \text{ given} \tag{4.7.53}$$

In the case of the system shown in Fig. 4.23, the hydraulic coupling equations are

$$\begin{aligned} Y_2(i) &= V_1(i - \tau_1) + q_1(i - \tau_1) + d_2 \\ Y_3(i) &= V_2(i - \tau_2) + q_2(i - \tau_2) + d_3 \end{aligned} \tag{4.7.54}$$

where τ_1, τ_2 are transmission delays between the stations 1, 2 and 2, 3 respectively.

In addition, from physical considerations, we have the following inequality constraints :

$$0 < \underline{X}_j \leqslant X_j(i) \leqslant \overline{X}_j \qquad \forall\, i \in T \tag{4.7.55}$$

(bounds on the contents of the reservoirs)

$$0 \leqslant q_j(i) \leqslant \overline{Q}_j \qquad \forall\, i \in T \tag{4.7.56}$$

(lower and upper bounds on the flow rates)

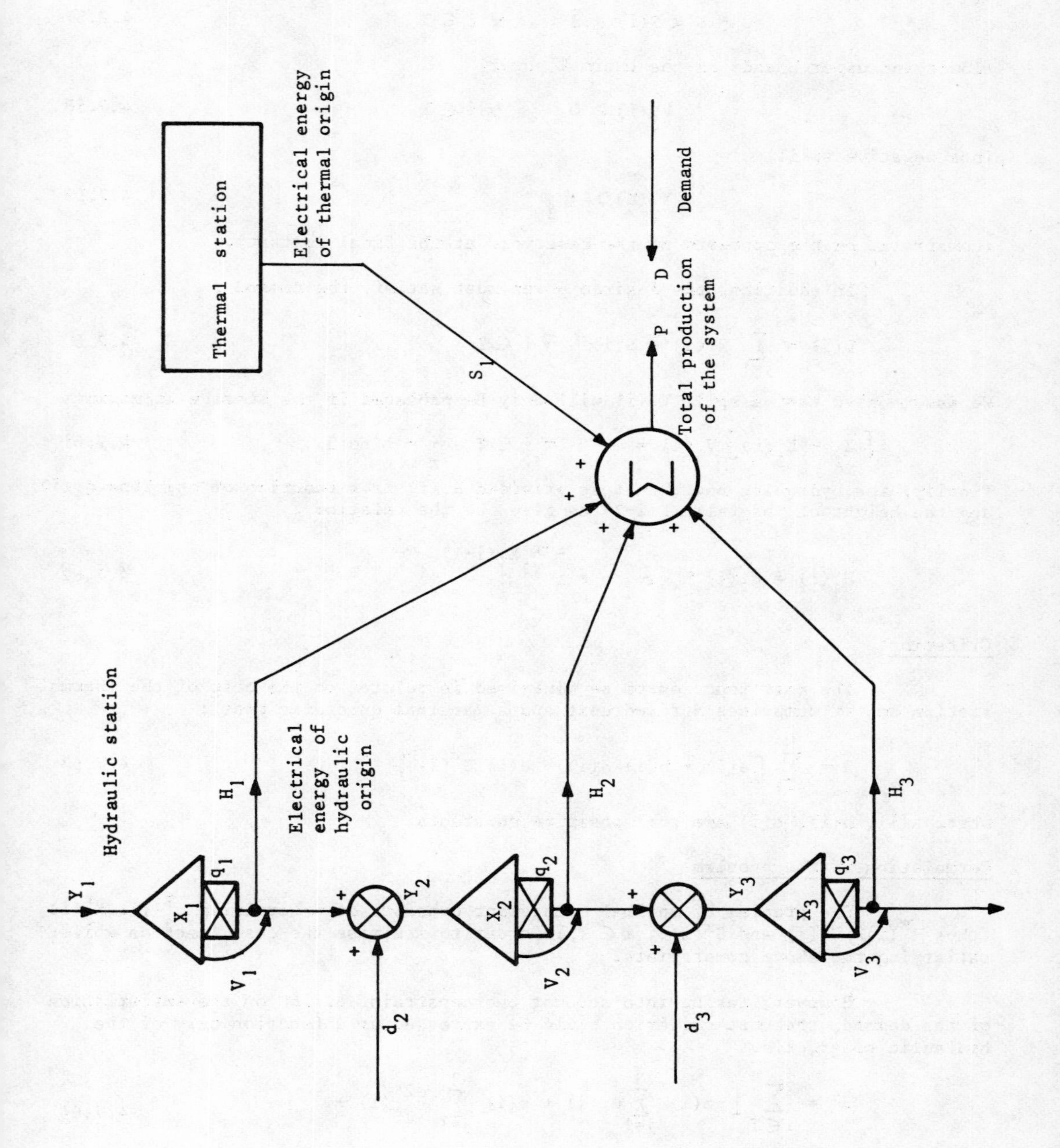

Fig. 4.23 : The hydro-thermal complex

$$0 < \underline{S} \leqslant S(i) \leqslant \overline{S} \qquad \forall\, i \in T \tag{4.7.57}$$

(lower and upper bounds on the thermal power)

$$V_j(i) \geqslant 0 \qquad \forall\, i \in T \tag{4.7.58}$$

(non negative spill)

$$X_j(N) \geqslant \gamma_j \tag{4.7.59}$$

(constraint on the contents of the reservoir at the final instant).

In addition, the desired power must satisfy the demand :

$$D(i) = \sum_{j=1}^{3} H_j(i) + S(i) \qquad \forall\, i \in T \tag{4.7.60}$$

We assume also that a spill $V_j(i)$ will only be produced if the storage is maximum

$$\left[\overline{X}_j - X_j(i)\right] V_j(i) = 0 \qquad \forall\, i \in T, \quad j = 1 \text{ to } 3 \tag{4.7.61}$$

Finally, the hydraulic power that is provided $H_j(i)$ is a function of the flow $q_j(i)$ and the height of the fall $X_j(i-1)$ as given by the relation :

$$H_j(i) = q_j(i)\, h_j(\delta_j - e^{-\alpha_j X_j(i-1)}) \tag{4.7.62}$$

Criterion

The cost function to be minimised is related to the cost of the thermal station and it comprises a fixed cost and a marginal quadratic cost :

$$J = \sum_{j=1}^{N} \left[a(i) + b(i)\, S(i) + c(i)\, S^2(i)\right] \tag{4.7.63}$$

where a(i), b(i), c(i) are real positive constants.

Formulation of the problem

The problem of optimal management consists of determining $q_j^*(i)$, $X_j^*(i)$, (thus $H_j^*(i)$), $V_j^*(i)$ and $S^*(i)$, $i \in T$, in order to minimise the cost function whilst satisfying the above constraints.

However, taking into account the constraint 4.7.60 on the satisfaction of the demand, the cost criterion could be expressed as a function only of the hydraulic power, i.e.

$$J' = \sum_{i \in T} \Big[-e(i) \sum_{j=1}^{3} H_j(i) + c(i) \sum_{j=1}^{3} H_j^2(i) + \underbrace{c(i)(H_1(i)\, H_2(i) + H_1(i) H_3(i) + H_2(i) H_3(i))}_{4.7.64'} \Big] \tag{4.7.64}$$

with e(i) = b(i) + 2c(i) D(i)

The original problem could thus be reformulated as :

$$\min J'$$

$$\text{subject to} \quad \underline{X}_j \leqslant X_j(i) \leqslant \overline{X}_j \qquad \text{a)}$$

$$\forall\, i \in T, \quad j = 1 \text{ to } 3 \quad \left\{ \begin{array}{ll} 0 \leqslant q_j(i) \leqslant \overline{Q}_j & \text{b)} \\ V_j(i) \geqslant 0 & \text{c)} \\ X_j(N) \geqslant \gamma_j & \text{d)} \\ X_j(i) = X_j(i-1) + Y_j(i) - q_j(i) - V_j(i) \quad X_j(0) \text{ given} & \text{e)} \\ Y_2(i) = V_1(i-\tau_1) + q_1(i-\tau_2) + d_2(i) & \text{f)} \\ Y_3(i) = V_2(i-\tau_2) + q_2(i-\tau_2) + d_3(i) & \text{g)} \\ \left[\overline{X}_j - X_j(i)\right] V_j(i) = 0 & \text{h)} \\ H_j(i) = q_j(i)\, h_j(\delta_j - e^{-\alpha_j X_j(i-1)}) & \text{i)} \\ \underline{H}_j(i) \leqslant H_j(i) \leqslant \overline{H}_j(i) & \text{j)} \end{array} \right. \qquad 4.7.65$$

The constraint $\underline{H}_j(i) \leqslant H_j(i) \leqslant \overline{H}_j(i)$ is a rewritten form of the constraint 4.7.57 (taking into account the demand satisfaction constraint 4.7.60).

$$\underline{S} \leqslant D(i) - \sum_{j=1}^{3} H_j(i) \leqslant \overline{S} \qquad 4.7.66$$

by putting for example

$$\underline{H}_j(i) = \max\ 0, \left[\frac{D(i) - \overline{S}}{3} \right]$$
$$\overline{H}_j(i) = \frac{D(i) - \underline{S}}{3} \qquad 4.7.67$$

Obviously, in this way, we choose an identical distribution for the potential production of the three hydraulic stations. Any other distribution which took account of the characteristics of the installations could also have been used by writing :

$$\left. \begin{array}{l} \underline{H}_j(i) = \max \left\{ 0, \left[\dfrac{D(i) - \overline{S}}{\varepsilon_j} \right] \right\} \\ \overline{H}_j(i) = \dfrac{D(i) - \underline{S}}{\varepsilon_j} \end{array} \right. \text{with } \sum_j \frac{1}{\varepsilon_j} = 1 \qquad 4.7.68$$

Thus, our final problem is :

Min J'
subject to 4.7.65, 4.7.67

In this set of constraints, the constraints (4.7.65, f,g) and the term 4.7.64' of the cost function 4.7.64 define the coupling between the stations (hydraulic and electrical coupling) whilst the other constraints define a feasible region $\Omega_j(i)$ for each station.

By taking into account only the coupling constraints, we could define the Lagrangian L for this optimisation problem as :

$$L = J' + \sum_{i \in T} \lambda_1(i) \left[Y_2(i) - V_1(i-\tau_1) - q_1(i-\tau_1) - d_2(i) \right] + \lambda_2(i) \left[Y_3(i) - V_2(i-\tau_2) - q_2(i-\tau_2) - d_3(i) \right] \quad 4.7.69$$

Given that we have to do a discretisation in time and given that the number of stations may be large, so that we might end up with a high order static optimisation problem, it is useful to be able to decompose it. We can do this by examining the Lagrangian L.

Decomposition of the problem

We basically wish to somehow modify the Lagrangian by fixing certain variables in order to obtain a separable form in L. We thus have :

$$L = \sum_j L_j(\underline{T}_j(i), \underline{V}_c(i)) \quad 4.7.70$$

where L_j is the Lagrangian of the jth subproblem which has the local variables $\underline{T}_j(i)$ and the coordination variables $\underline{V}_c(i)$. The latter are considered to be fixed as far as the lower level is concerned but which evolve at the higher level until the global solution is obtained (Fig. 4.24).

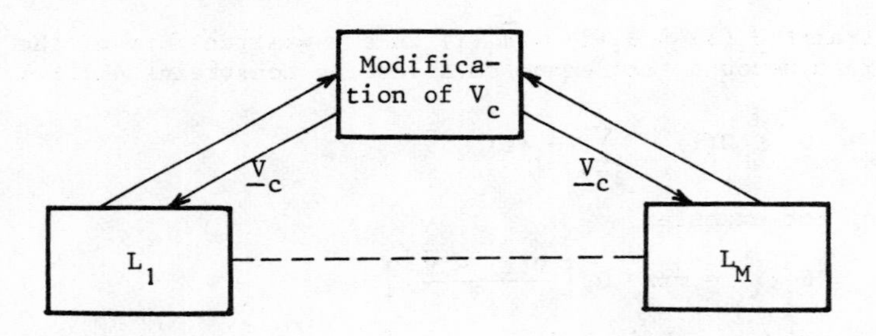

Fig. 4.24 : The two level structure

Decomposition of the non-separable term in the criterion 4.7.64'

We put :

$$H_1 H_2 + H_1 H_3 + H_2 H_3 = H_1 W_1 + H_1 W_2 + H_2 W_2 \quad 4.7.71$$

and we add in the equality constraints

$$W_1 - H_2 = 0 \longrightarrow \lambda_3$$
$$W_2 - H_3 = 0 \longrightarrow \lambda_4 \quad 4.7.72$$

and define the associated Lagrange multipliers λ_3 and λ_4. Thus for given W_1, W_2 (provided by the coordinator),

$H_1 W_1$ becomes a part of the criterion of subsystem 1

$H_1 W_2$ becomes a part of the criterion of subsystem 1

$H_2 W_2$ becomes a part of the criterion of subsystem 2.

Finally, in order to implement the efficient direct iteration coordinator [14], we choose, in addition, as the coordination variables, λ_3 and λ_4.

Decomposition of the equality constraint of the coupling relationship

From the equation 4.7.69, if we fix λ_1 and λ_2, we could distribute the coupling terms about the system. For example, for constraint 1 :

$\lambda_1 Y_2 \longrightarrow$ subsystem 2
$\lambda_1 V_1 \longrightarrow$ subsystem 1
$\lambda_1 q_1 \longrightarrow$ subsystem 1
$\lambda_1 d_2 \longrightarrow$ subsystem 2

Here also, in order to implement a direct iteration coordinator, we choose, in addition, as coordination variables, Y_2 and Y_3.

Additive separable form of the Lagrangian

With

$$V_c(i) = [\lambda_1^T(i)\ \lambda_2^T(i)\ \lambda_3^T(i)\ \lambda_4^T(i)\ W_1^T(i)\ W_2^T(i)\ Y_2^T(i)\ Y_3^T(i)]^T \qquad 4.7.73$$

as the coordination vector, we could write at the lower level :

$$\begin{aligned} L = \sum_{i \in T} & -e(i)\,H_1(i) + c(i)\left[H_1^2(i) + H_1(i)\,W_1(i) + H_1(i)\,W_2(i)\right. \\ & \left. + \lambda_3(i)\,W_1(i) + \lambda_4(i)\,W_2(i) - \lambda_1(i)\left[V_1(i-\tau_1) + q_1(i-\tau_1)\right]\right] \\ & + \left[-e(i)\,H_2(i) + c(i)\left[H_2^2(i) + H_2(i)\,W_2(i)\right]\right. \\ & \left. - \lambda_3(i)H_2(i) - \lambda_1(i)\left[Y_2(i) - d_2(i)\right] - \lambda_2(i)\left[V_2(i-\tau_2) + q_2(i-\tau_2)\right]\right] \\ & - \left[-e(i)H_3(i) + c(i)\left[H_3^2(i)\right] - \lambda_4(i)H_3(i) - \lambda_2(i)\left[Y_3(i) - d_3(i)\right]\right] \end{aligned} \qquad 4.7.74$$

or

$$L = L_1(\underline{T}_1(i), \underline{V}_c(i)) + L_2(\underline{T}_2(i), \underline{V}_c(i)) + L_3(\underline{T}_3(i)\ \underline{V}_c(i)) \qquad 4.7.75$$

Formulation of the subproblems

Examining $L_j(\underline{T}_j(i), \underline{V}_c(i))$ enables us to formulate the various subproblems as :

Subproblem 1

$$\min_{V_1, q_1, X_1, H_1} J'_1 = \sum_{i \in T} \left\{ -e(i)H_1(i) + c(i)H_1(i)\left[H_1(i) + W_1(i) + W_2(i)\right] + \lambda_3(i)W_1(i) + \lambda_4(i)W_2(i) - \lambda_1(i)\left[V_1(i-\tau_1) + q_1(i-\tau_1)\right]\right\} \qquad 4.7.76$$

subject to :

$$\forall i \in T \quad \begin{cases} H_1(i) = q_1(i)\, h_1\, (\delta_1 - e^{-\alpha_1 X_1(i-1)}) \\ X_1(i) = X_1(i-1) + Y_1(i) - q_1(i) - V_1(i) \\ \underline{X}_1 \leqslant Y_1(i) \leqslant \overline{X}_1 \\ X_1\,(N) \geqslant \gamma_1 \\ 0 \leqslant q_1(i) \leqslant \overline{Q}_1 \\ \underline{H}_1(i) \leqslant H_1(i) \leqslant \overline{H}_1(i) \text{ with } \begin{cases} \underline{H}_1(i) = \max\left\{0, \dfrac{D(i)-\overline{S}}{3}\right\} \\ \overline{H}_1(i) = \dfrac{D(i)-\underline{S}}{3} \end{cases} \\ \left[\overline{X}_1 - X_1(i)\right] V_1(i) = 0 \\ V_1(i) \geqslant 0 \end{cases} \qquad 4.7.77$$

This problem is to be solved for :

* $e(i)$, $c(i)$, τ_1, h_1, δ_1, α_1, $\underline{X}_1$, $\overline{X}_1$, γ_1, $\overline{Q}_1$ $D(i)$, $\overline{S}$, $\underline{S}$ given parameters.
* $W_1(i)$, $W_2(i)$, $\lambda_1(i)$, $\lambda_3(i)$, $\lambda_4(i)$, $Y_1(i)$, variables of coordination fixed by the higher level.

Subproblem 2

$$\underset{V_2,q_2,X_2^2,H_2}{\text{Min}}\ J'_2 = \sum_{i \in T} \Big\{ -e(i)H_2(i) + c(i)H_2(i)\left[H_2(i) + W_2(i)\right] - \lambda_3(i)H_2(i) - \lambda_2(i)\left[V_2(i-\tau_2) + q_2(i-\tau_2)\right] \Big\} \qquad 4.7.78$$

subject to :

$$\forall i \in T \quad \begin{cases} H_2(i) = q_2(i)h_2\,(\delta_2 - e^{-\alpha_2 X_2(i-1)}) \\ Y_2(i) = X_2(i-1) + Y_2(i) - q_2(i) - V_2(i) \\ \underline{X}_2 \leqslant Y_2(i) \leqslant \overline{X}_2 \\ X_2(N) \geqslant \gamma_2 \\ 0 \leqslant q_2(i) \leqslant \overline{Q}_2 \\ \underline{H}_2(i) \leqslant H_2(i) \leqslant \overline{H}_2(i) \quad \text{with } \begin{cases} \underline{H}_2(i) = \max\left\{0, \dfrac{D(i)-\overline{S}}{3}\right\} \\ H_2(i) = \dfrac{D(i)-\underline{S}}{3} \end{cases} \\ \left[\overline{X}_2 - X_2(i)\right] V_2(i) = 0 \\ V_2(i) \geqslant 0 \end{cases} \qquad 4.7.79$$

This problem is to be solved for :

* $e(i)$, $c(i)$, τ_2, h_2, δ_2, α_2, $\underline{X}_2$, $\overline{X}_2$, γ_2, $\overline{Q}_2$, $D(i)$, $\overline{S}$, $\underline{S}$, given parameters
* $W_2(i)$, $\lambda_2(i)$, $\lambda_3(i)$, $Y_2(i)$, coordination variables fixed by the coordinator.

This choice of coordination variables allows us, amongst other things, to eliminate from the cost function, (w.r.t. equation 4.7.74) the term $\lambda_2(i)\left[Y_3(i) - d_3(i)\right]$ which becomes constant for the optimisation problem.

Subproblem 3

$$\min_{V_3,q_3,X_3,H_3} J'_3 = \sum_{i\in T} \left\{ -e(i)H_3(i) + c(i)H_3^{\,2}(i) - \lambda_4(i)\, H_3(i) \right\} \qquad 4.7.80$$

subject to

$$\left\{\begin{array}{l} H_3(i) = q_3(i)\, h_3(\delta_3 - e^{-\alpha_3 X_3(i-1)}) \\ X_3(i) = X_3(i-1) + Y_3(i) - q_3(i) - V_3(i) \\ \underline{X}_3 \leqslant X_3(i) \leqslant \overline{X}_3 \\ X_3(N) \geqslant \gamma_3 \\ 0 \leqslant q_3(i) \leqslant \overline{Q}_3 \\ \underline{H}_3(i) \leqslant H_3(i) \leqslant \overline{H}_3(i) \quad \text{with} \left\{\begin{array}{l} \underline{H}_3(i) = \max\left\{0, \dfrac{D(i)-\overline{S}}{3}\right\} \\ H_3(i) = \dfrac{D(i)-\underline{S}}{3} \end{array}\right. \\ \left[\overline{X}_3 - X_3(i)\right] V_3(i) = 0 \\ V_3(i) \geqslant 0 \end{array}\right. \qquad 4.7.81$$

This problem is to be solved for :

* $e(i), c(i), h_3, \delta_3, \alpha_3, \underline{X}_3, \overline{X}_3, \gamma_3, \overline{Q}_3, D(i), \overline{S}, \underline{S}$, given parameters
* $\lambda_4(i)$ the coordination variable fixed by the coordination level.

Here again, since $\lambda_2(i)$, $Y_3(i)$, $d_3(i)$ are fixed, the corresponding terms could be eliminated from the criterion.

Definition of the coordination task

The solution of the subproblems enables us to satisfy the optimality conditions on $V_j(i)$, $q_j(i)$, $X_j(i)$, $H_j(i)$, $j = 1$ to 3, $i\in T$. It remains, therefore, to satisfy, at the second level, the optimality conditions which we write below. In these conditions, L' is the Lagrangian of the problem which takes into account all the constraints and which associates the multipliers $\mu_j(i)$ to the constraints 4.7.65.e.

$$i\in T \left\{\begin{array}{l} \dfrac{\partial L'}{\partial\lambda_1(i)} = L'_{\lambda_1(i)} = Y_2(i)-V_1(i-\tau_1)-q_1(i-\tau_1)-d_2(i) \\ L'_{\lambda_2(i)} = 0 = Y_3(i)-V_2(i-\tau_2)-q_2(i-\tau_2)-d_3(i) \\ L'_{\lambda_3(i)} = 0 = W_1(i) - H_2(i) \\ L'_{\lambda_4(i)} = 0 = W_2(i) - H_3(i) \\ L'_{W_1(i)} = 0 = \lambda_3(i) + c(i)\, H_1(i) \\ L'_{W_2(i)} = 0 = \lambda_4(i) + c(i)\left[H_2(i) + H_1(i)\right] \\ L'_{Y_2(i)} = 0 = \lambda_1(i) - \mu_2(i) \end{array}\right. \qquad 4.7.82$$

$$L'Y_3(i) = 0 = \lambda_2(i) - \lambda_3(i)$$

Let us denote by k the iteration index for the coordination level which at the kth iteration provides the coordination variables $\underline{V}_c(i)\,[k]$ to the lower level. Let us denote by $\underline{T}_j^*\,[k]$ the optimal solution of subproblem j. The coordinator will thus modify $\underline{V}_c(i)\,[k]$ to $\underline{V}_c(i)\,[k+1]$ using the relation :

$$\underline{V}_c(i)\,[k+1] = f\left[\underline{V}_c(i)\,[k],\ \underline{T}_j^*(i)\,[k]\right] \tag{4.7.83}$$

which is obtained by solving the linear system 4.7.82 (coordination by direct iteration) in a very simple way since each equation enables us to calculate one coordination variable :

$$\begin{aligned}
Y_2(i)\,[k+1] &= V_1^*(i-\tau_1)[k] + q_1^*(i-\tau_1)[k] + d_2(i) \\
Y_3(i)\,[k+1] &= V_2^*(i-\tau_2)[k] + q_2^*(i-\tau_2)[k] + d_3(i) \\
W_1(i)\,[k+1] &= H_3^*(i)\,[k] \\
W_2(i)\,[k+1] &= H_3^*(i)\,[k] \\
\lambda_1(i)\,[k+1] &= \mu_2^*(i)\,[k] \\
\lambda_2(i)\,[k+1] &= \mu_3^*(i)\,[k] \\
\lambda_3(i)\,[k+1] &= -c(i)\,H_1^*\,[k] \\
\lambda_4(i) &= -c(i)\left[H_2^*(i)\,[k] + H_1^*(i)\,[k]\right]
\end{aligned} \tag{4.7.84}$$

Having defined the subproblems and the coordination task, we show the information transfers required in Fig. 4.25.

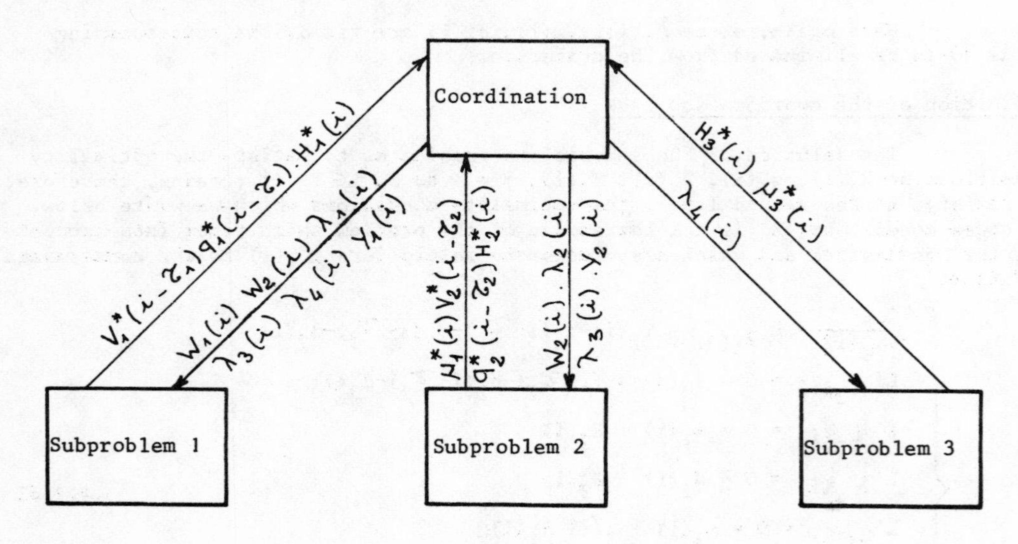

Fig. 4.25 : Transfer of information between the subproblems

Fig. 4.26 shows the flow-chart for solving this problem on a single computer. Note, however, that it might be more advantageous to use a multiprocessor here as we do in the next section.

The test for the end of coordination could be :

$$\| L'\underline{V}_c \| \leqslant \varepsilon_1 \tag{4.7.85}$$

or

$$\| \underline{V}_c[k+1] - \underline{V}_c[k] \| \leqslant \varepsilon_2 \tag{4.7.86}$$

Numerical results for this problem are given in [13].

As a final application, we consider a high order system obtained after discretisation of a distributed parameter problem.

4.7.3 Solution of a distributed parameter problem using a single processor and using a multi-processor [15]

4.7.3.1 The control of a distributed parameter system

The particular problem considered here was previously solved by Titli[7]. Here a brief description of the problem and hierarchical solution using a single computer will be given before tackling the implementation of the hierarchical structure on the multiprocessor.

Problem formulation

The problem is to minimise

$$\int_{x=0}^{1} \int_{t=0}^{\tau} u^2(x, t)\, dx\, dt$$

subject to the distributed parameter constraints

$$\frac{\partial \theta}{\partial t} = D^2 \frac{\partial^2 \theta}{\partial x^2} + u$$

with the boundary conditions : $\theta(0, t) = \theta(1, t) = 0$, i.e. it is desired to take the system state from $\theta(x, 0) = \theta_o$ to $\theta(x, \tau) = \theta_1$, whilst utilising minimal energy.

Assuming that a solution exists, discretise the system in space by putting $x_i = (i-1)\Delta x$ $(i = 1, \ldots, N_1)$; $\Delta x = \frac{1}{N_1 - 1}$ and in time by putting

$$t_j = (j-1)\Delta t \qquad (j = 1, \ldots, N_2)$$

$$\Delta t = \frac{\tau}{N_2 - 1}$$

Then, the problem becomes :

$$\underset{u(i,j)}{\text{Min}} \quad J = \left[\sum_{i=1}^{N_1} \sum_{i=1}^{N_2} u^2(i, j) \right] \Delta x\, \Delta t \tag{4.7.87}$$

subject to

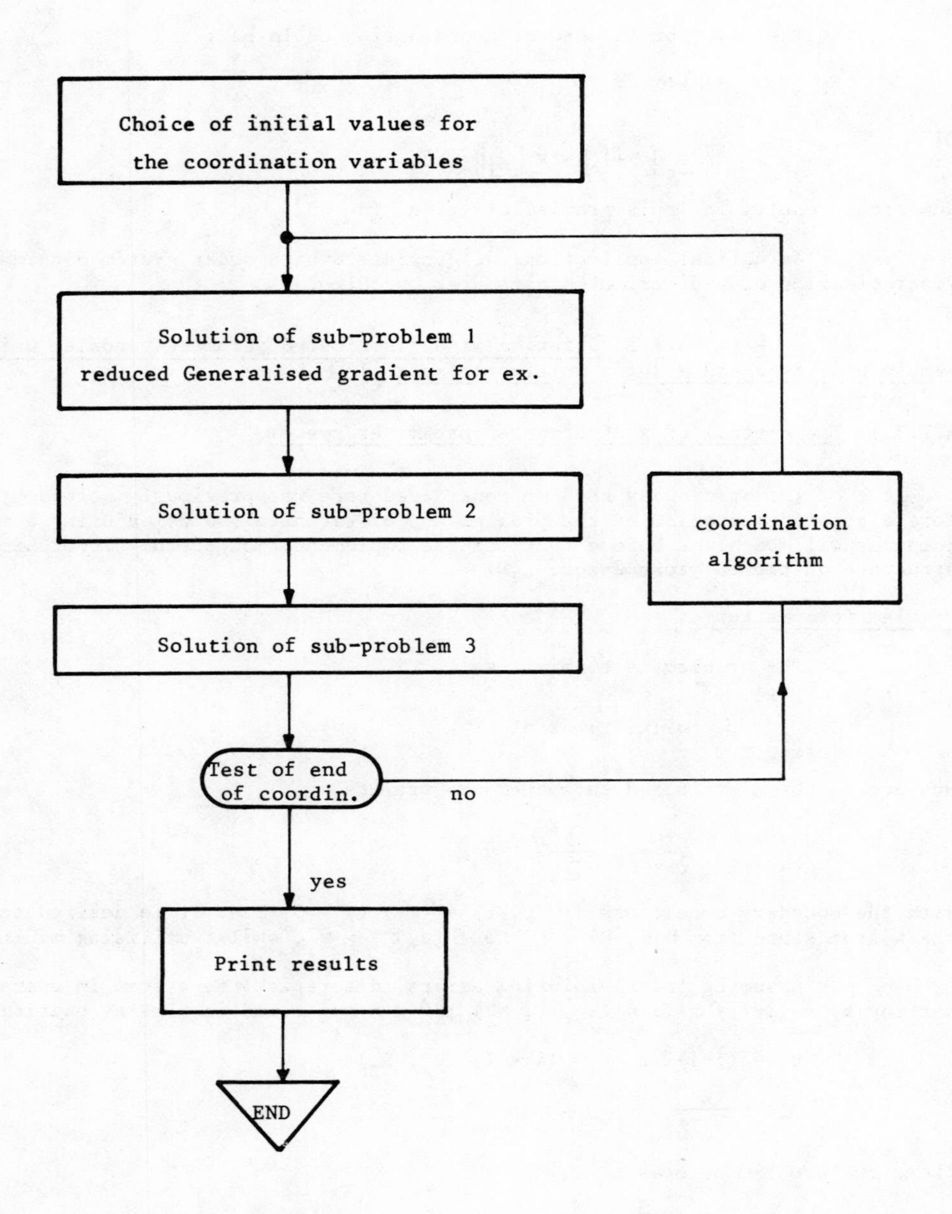

Fig. 4.26 : The flow-chart using a single computer

$$\frac{\theta(i,j) - \theta(i,j-1)}{\Delta t} = D^2 \left[\frac{\theta(i+1,j) - 2\,\theta(i,j) + \theta(i-1,j)}{(\Delta x)^2} \right] \quad 4.7.88$$

with

$$\theta(i,1) = \theta_{0i}, \theta(i, N_2) = \theta_{1i} \qquad i = 1, \ldots, N_1 \quad 4.7.89$$

which are the initial and final conditions, and

$$\theta(1,j) = 0, \ \theta(N_1,j) = 0 \qquad j = 1, 2, \ldots, N_2 \quad 4.7.90$$

which are the boundary conditions.

The constraint 4.7.88 can be rewritten as :

$$\theta(i,j) = \left[\frac{1}{\frac{1}{\Delta t} + 2(\frac{D}{\Delta x})^2} \right] \left[\frac{\theta(i,j-1)}{\Delta t} + (\frac{D}{\Delta x})^2 (\theta i+1,j) + \theta(i-1,j)) + u(i,j) \right] \qquad \begin{matrix} j = 2, \ldots N_2 \\ i = 2, \ldots N_1-1 \end{matrix} \quad 4.7.91$$

Suppose that it is possible to solve the problem :

$$\text{Min } J' = \sum_{i=2}^{N_1-1} \sum_{j=2}^{N_2-1} u^2(i, j) \Delta x \ \Delta t \quad 4.7.92$$

subject to

$$\theta(i,j) = \frac{1}{\frac{1}{\Delta t} + 2(\frac{D}{\Delta x})^2} \left[\frac{\theta(i, j-1)}{\Delta t} + (\frac{D}{\Delta x})^2 (\theta(i+1, j) + \theta(i-1, j)) + u(i, j)) \right] \qquad \begin{matrix} i = 2, \ldots N_1-1 \\ j = 2, \ldots N_2-1 \end{matrix} \quad 4.7.93$$

In that case, $\theta(i,j)$ ($j = 2$ and N_2-1) can be used to fix $u(i,j)$ for $j = 1$ and N_2.

Essentially, the equation $\frac{\partial \theta}{\theta t} = D^2 \frac{\partial^2 \theta}{\partial x^2} + u$ can be discretised in the following two ways :

$$\frac{\theta(i,j+1) - \theta(i,j)}{\Delta t} = D^2 \left[\frac{\theta(i+1,j) - 2\theta(i,j) + \theta(i-1,j)}{(\Delta x)^2} \right] + u(i,j) \quad 4.7.94$$

and

$$\frac{\theta(i,j) - \theta(i,j-1)}{\Delta t} = D^2 \left[\frac{\theta(i+1,j) - 2\theta(i,j) + \theta(i-1,j)}{(\Delta x)^2} \right] + u(i,j) \quad 4.7.95$$

Equation 4.7.94 enables one to calculate $u(i, 1)$ and 4.7.95 to calculate $u(i,N_2)$ since 4.7.94 is valid for $i = 2, \ldots N_1-1$, $j = 1, \ldots N_2-1$ and 4.7.95 for $i = 2, \ldots N_1-1$, $j = 2, \ldots N_2$.

Thus the problem can be written as :

$$\underset{U,\theta \in \overline{\Omega}}{\text{Min}} \ J = \underset{u,\theta \in \Omega_F}{\text{Min}} \ (J - J') + \underset{u,\theta \in \Omega}{\text{Min}} \ J'$$

where Ω is the feasible region of the problem, Ω_F the boundary of this region and $\overline{\Omega} = \Omega \cup \Omega_F$ (cf. Fig. 4.27).

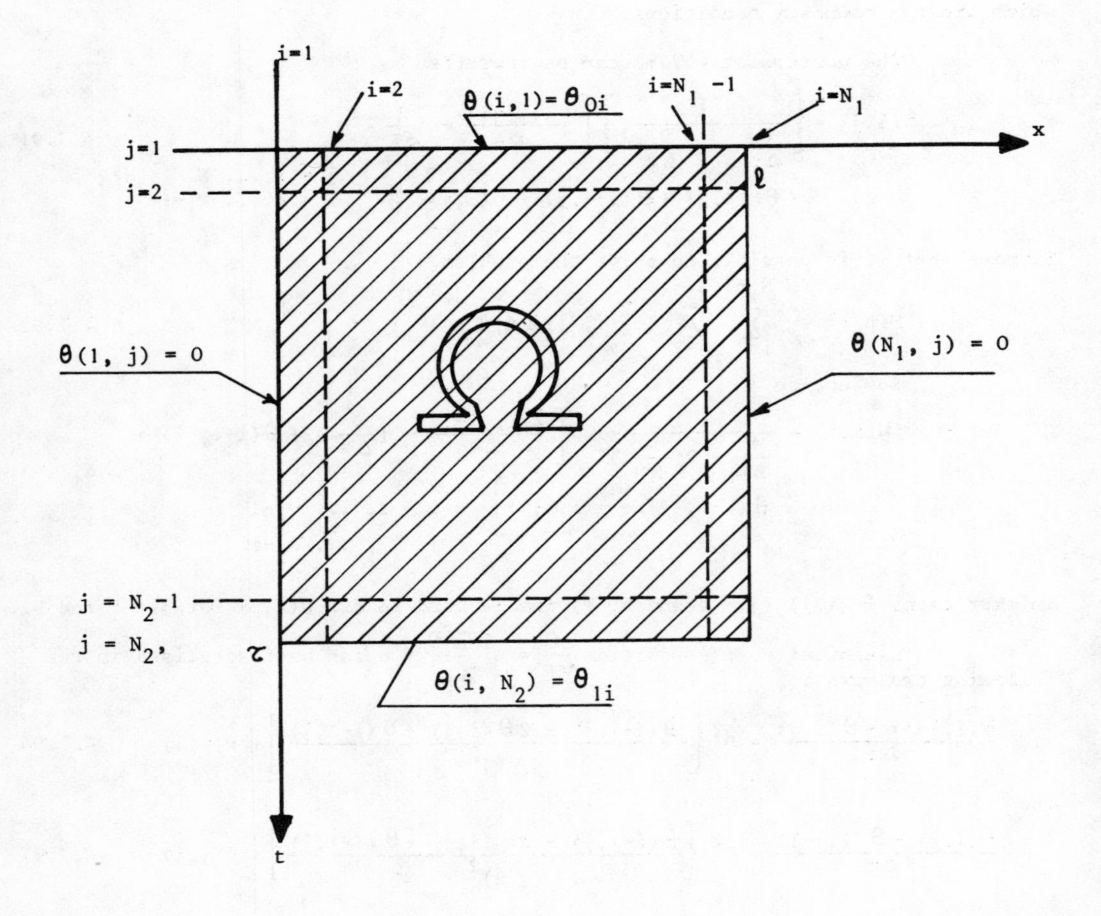

Fig. 4.27

Thus to minimise (J - J')

$$u, \theta \in \Omega_F$$

it is necessary to take into account the boundary conditions which fix and reduce the problem to

$$\text{Min } J'' = \sum_i u^2(i, j) \text{ where } i = 1 \text{ and } N_1, \ \theta(1, j) \text{ and } \theta(N_1, j) \text{ are given.}$$

The solution is obviously $u(i, j) = u(N_1, j) = 0 \quad \forall j = 1, \ldots N_2$

For the initial and final conditions, since u(i, j) is fixed, there is no optimisation problem. Thus, the problem reduces to

$$\min_{u, \theta \in \Omega} J'$$

which corresponds to equations 4.7.92, 4.7.93 and which can be written in terms of interconnected systems as :

$$\text{Min} \left[\sum_{i=2}^{N_1-1} \sum_{j=2}^{N_2-1} u^2(i, j) \right] \Delta x \ \Delta t \qquad 4.7.96$$

subject to

$$\theta(i, j) = \frac{1}{\frac{1}{\Delta t} + 2\left(\frac{D}{\Delta x}\right)^2} \left[W(i, j) + u(i, j) \right] \qquad 4.7.97$$

and the initial, final and boundary conditions where

$$W(i, j) = \frac{\theta(i, j-1)}{\Delta t} + \left(\frac{D}{\Delta x}\right)^2 \left[\theta(i+1, j) + \theta(i-1, j) \right] \qquad 4.7.98$$

Then the problem becomes one of $(N_1-2)(N_2-1)$ interconnected subsystems and $(N_1-2)(N_2-2)$ variables u(i, j) and θ (i, j). However, this is a large scale systems problem where solution by single level methods of non-linear programming is quite out of the question.

This problem can nevertheless be solved using a "feasible" hierarchical method.

4.7.3.2 Hierarchical solution using the feasible method

The interconnection variables θ(i, j) are given by the coordinator for use at the lower level. At the lower level, each subproblem becomes :

$$\text{Min } u^2 (i, j)$$

subject to

$$\begin{cases} \theta(i, j) = A \left[W(i, j) + u(i, j) \right] ; \ A = \dfrac{1}{\frac{1}{\Delta t} + 2\left(\frac{D}{\Delta x}\right)^2} \\ W(i, j) = \dfrac{\theta(i, j-1)}{\Delta t} + \left(\frac{D}{\Delta x}\right)^2 \left(\theta(i+1, j) + \theta(i-1, j) \right) \end{cases} \qquad 4.7.99$$

where $\theta(i, j)$ is given $\forall i, j$. Since this is a constrained optimisation problem, it can be solved by forming the Lagrangian

$$L(i,j) = u^2(i,j) + \lambda(i,j) \left[\theta(i,j) - A(W(i,j) + u(i,j)) \right]$$

$$+ \beta(i,j) \left[W(i,j) - \frac{\theta(i,j-1)}{\Delta t} - \left(\frac{D}{\Delta x}\right)^2 (\theta(i+1,j) + \theta(i-1,j)) \right]$$

Then from the necessary conditions for optimality

$$L(i,j)_{u(i,j)} = 0, \quad L(i,j)_{W(i,j)} = 0, \quad L(i,j)_{\lambda(i,j)} = 0$$
$$L(i,j)_{\beta(i,j)} = 0 \qquad 4.7.100$$

This gives :

$$W^*(i, j) = \frac{\theta(i,j-1)}{\Delta t} + \left(\frac{D}{\Delta x}\right)^2 \left[\theta(i+1,j) + \theta(i-1,j)\right]$$
$$u^*(i, j) = \frac{\theta(i,j)}{A} - W^*(i,j)$$
$$\lambda^*(i, j) = \frac{2u^*(i,j)}{A} \qquad 4.7.101$$
$$\beta^*(i, j) = 2u^*(i,j)$$

and

$$L = \sum_i \sum_j L(i, j) \qquad 4.7.102$$

The coordinator modifies $\theta(i, j)$ using a conjugate gradient method.

The calculations are terminated when $\left|\frac{\partial L}{\partial \theta}\right| \leqslant E$ where E is a prechosen small positive number.

Fig. 4.28 shows the flow-chart of this approach.

Computational results using a single computer

Using $\tau = 5$, $l = 1$, $D = \sqrt{0.0033}$, $\theta(i,j) = \theta(N_1,j) = 0$ $(j = 1,\dots N_2)$ $\theta(i,j)$ and $\theta(i,N_2)$ for $i = 1, \dots N_1$, i.e. initial values $\theta(i, j) = 50$ for $i = 2, \dots N_1-1$, $j = 2, \dots N_2-1$ and solving for $N_1 = 11$, $N_2 = 20$ gives a problem of 432 variables (of which 216 are coordination variables) and 216 subsystems (Let $\Delta t = 0.2$ $\Delta x = 0.1$).

The problem was programmed in the IBM 370/165 computer and the various operations were done serially. Convergence to an error E = 0.958 took place in 98 iterations which took 2.4 seconds to compute. The trajectories obtained were identical to those obtained by Wismer [16]. Further results on larger examples including that of a system with 1728 variables are given by Titli [7].

4.7.3.3. Hierarchical solution using the multiprocessor system

Description of the multiprocessor system

The studies described in this section were carried out on the "Closed Linked Union of Mini-Processors" (CLUMP) system in the Control and Management Systems Division of the Department of Engineering at Cambridge. Here, a very brief

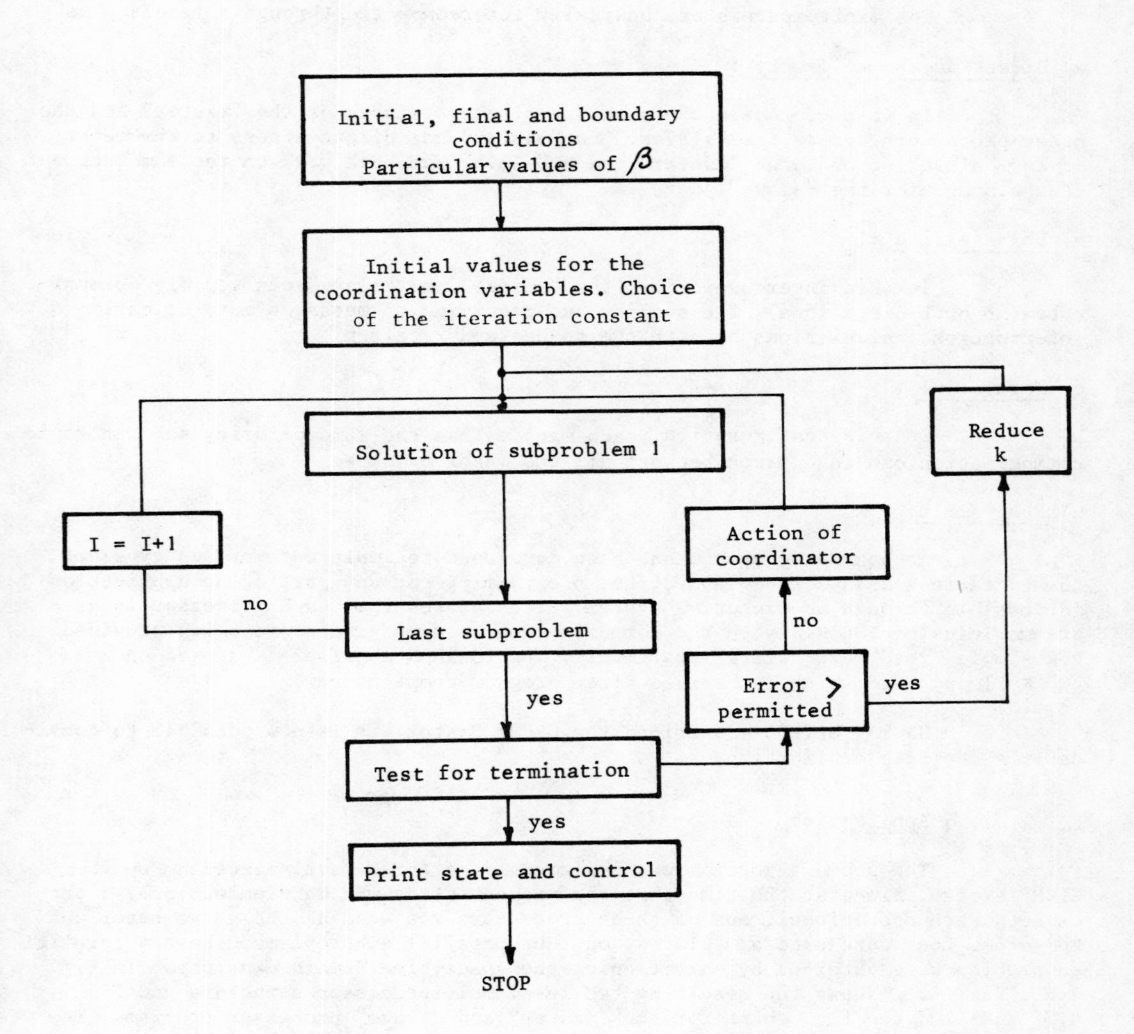

Initial, final and boundary conditions
Particular values of β
Initial values for the coordination variables. Choice of the iteration constant
Solution of subproblem 1
I = I+1
no
Last subproblem
yes
Test for termination
yes
Print state and control
STOP
Action of coordinator
no
Error > permitted
yes
Reduce k

Fig. 4.28

description of the system is given and the interested reader is referred to the paper of Billingsley et al.[19] for further details.

The system consists of 10 mini computers of which 5 were linked when the present study was carried out.

The minicomputers are basically interconnected through 3 levels, i.e.

a) Direct access to memory :

In this connection, one of the processors acts as the "master" and the other processors act as the "slaves. The "master" has direct access to the memory of any "slave" so that the "master" and a "slave" can both work on the same segment of the memory of the "slave".

b) Private line :

In this interconnection, the "master" and "slave" can rapidly communicate (in both directions). The sending and receiving of messages automatically interrupts the calculations of both the sender and receiver.

c) Party line :

In this configuration, each machine has the same priority so that rapid communication can take place between all the machines.

d) Programming :

Although the system has been developed to implement various types of communications as described above, the programming for any particular application is possible through an executive system which ensures that each processor is programmable using FORTRAN with the communication between processors being provided via simple "read" and "write" statements. The FORTRAN compiler is stored on a 400 K Floppy Disc which is removed after program compilation.

Having briefly described the CLUMP System, it is now possible to consider the concrete application.

Application

The above algorithm was implemented as a two level algorithm on the CLUMP system. Since at the time that the present study was carried out only 5 processors were operational, one of these processors was used as the coordinator and the other four were used as "slaves" on four parallel subproblems. These 4 parallel subproblems were obtained by partitioning the space-time domain Ω as shown in Fig. 4.29. Fig. 4.30 shows the resulting two level multiprocessor structure and Fig. 4.31 and 4.32 the flow chart for the "master" and "slave" processor programs. The actual "master" and "slave" programs used are listed in Titli et al. [15].

It should be noted that the partitioning of Ω was essentially dictated by the need to have 4 completely "identical" subproblems which being of the same dimension could be solved in parallel in an equal period of time.

The numerical values were chosen to be identical to the ones used previously for the single computer calculation. Here, however, convergence to an error E of 0.958 was obtained in 98 iterations on using a conjugate gradient method for the coordinator. This took 15 minutes on the multiprocessor system. Note that this

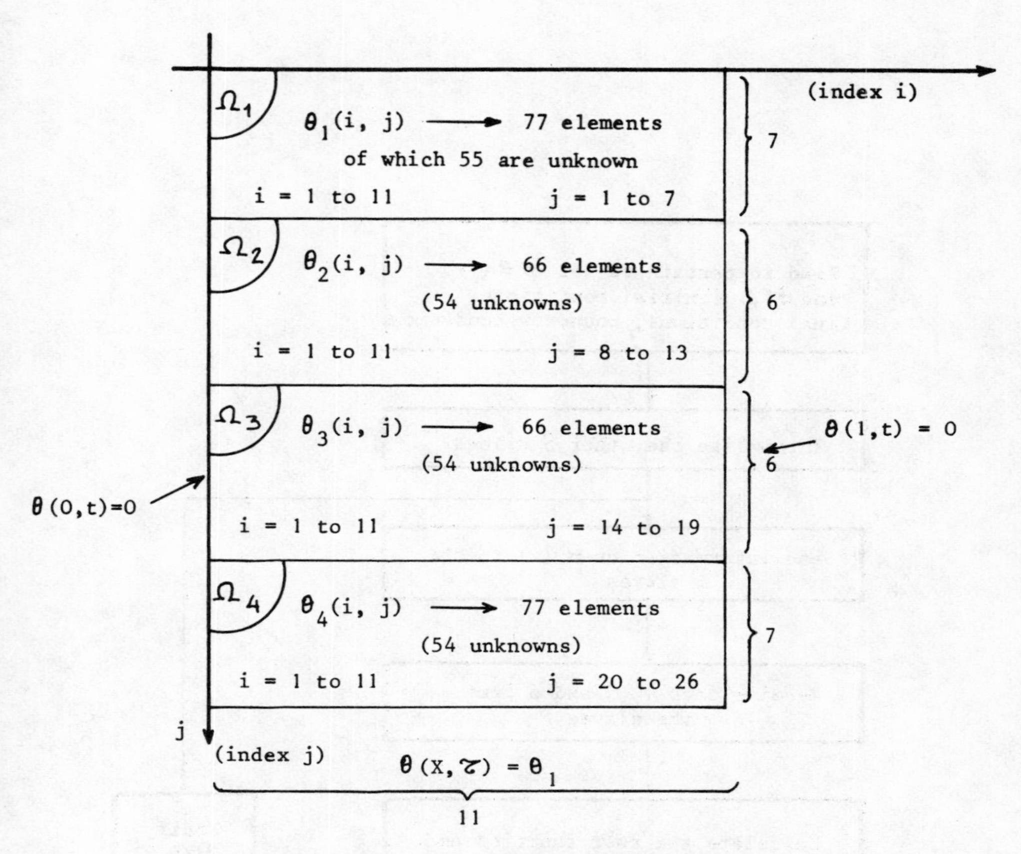

Fig. 4.29 : Partitions of the space-time domain

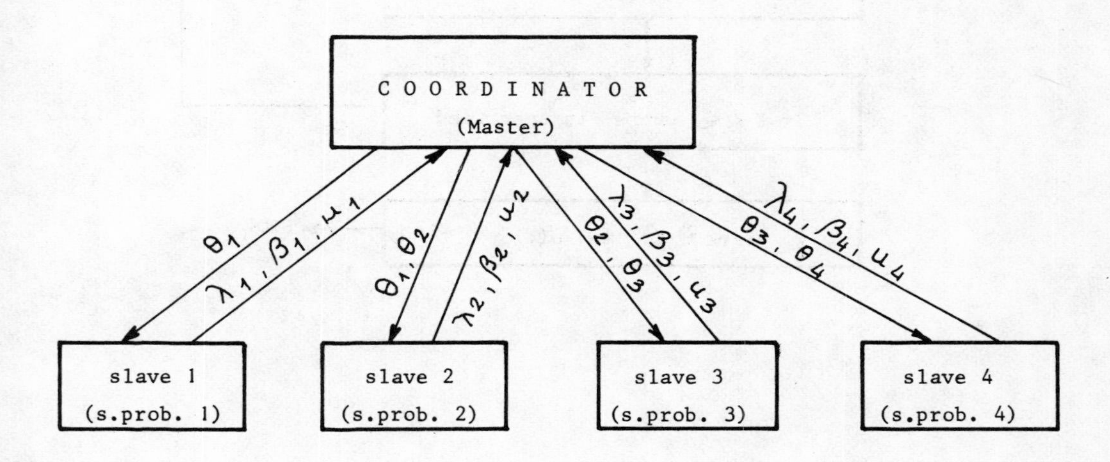

Fig. 4.30 : The 2 level multiprocessor structure

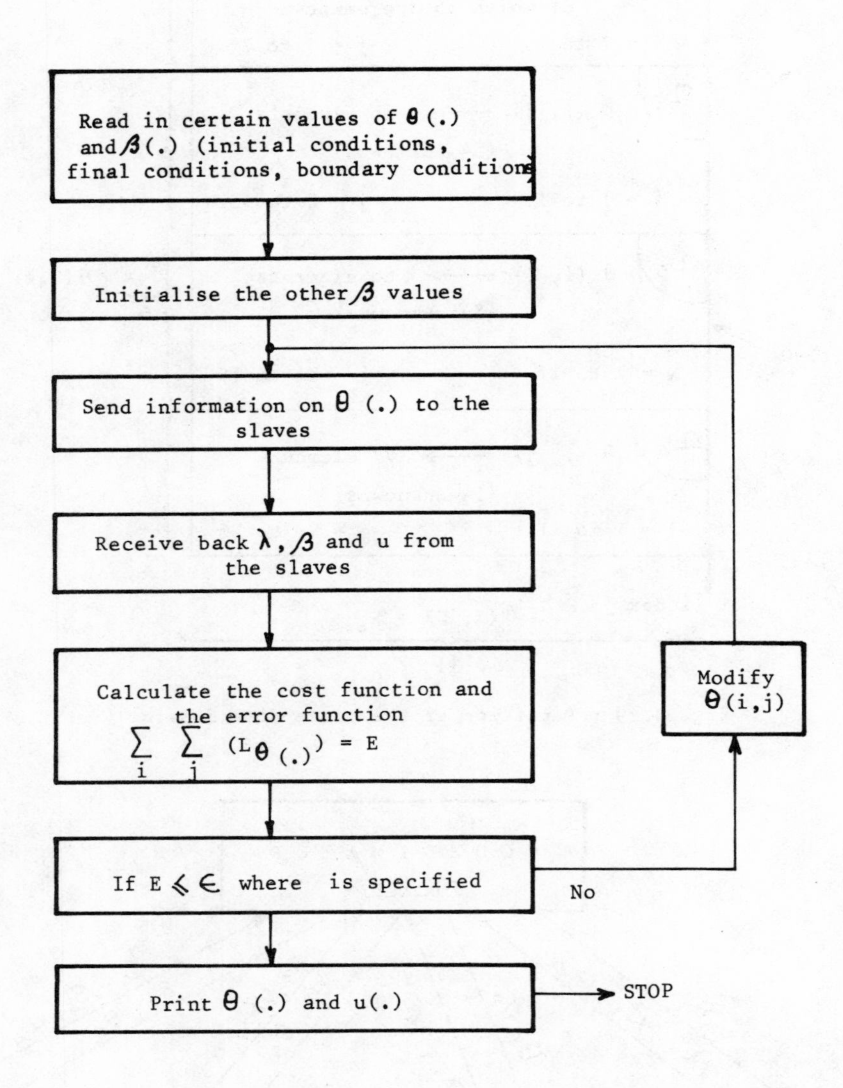

Fig. 4.31 : "Master" flow chart

would still appear to be smaller than the corresponding time of 2.4 seconds on the IBM 360/75 since each of the processors has some software arithmetic operations which make the calculation about 100 times slower than with the hardware functions on the IBM 360/75.

Comments

In implementing the hierarchical structure on a system of multi-processors it would appear that there are advantages both in computation speed and cost as compared to implementation on a single central computer like the IBM 360/75. The latter are particularly significant since minicomputers are now quite cheap and are getting cheaper every year. However, care must be taken in the implementation of the structure. For example, it is advantageous to partition the system into subsystems of roughly equal size so that the coordinator does not have to wait excessively for its next iteration.

Another important consideration is the one concerned with the efficiency of information transfer since the "master" is basically idle when the "slaves" are working and vice versa. In the present example, to speed up the information exchange, only vector information was transferred since this could be done without formatting.

4.8. CONCLUSIONS

We have presented the basic features of the theory of decomposition-coordination in Mathematical Programming in this chapter. We assumed largely throughout that our criterion was additively separable. This was done only for ease of exposition. In fact, this assumption is not strictly necessary and we can indeed solve problems with non-separable criteria [17, 18]. We did give some illustrative examples which showed the effectiveness of these procedures despite the effort required to put the problem into a desirable form (i.e. formulation in terms of interconnected subsystems).

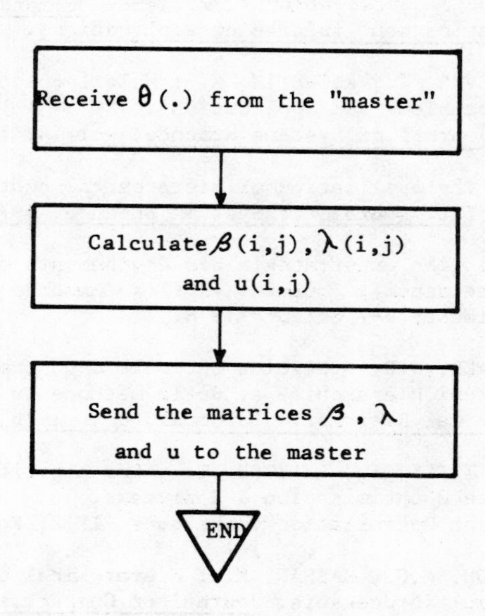

Fig. 4.32 : Flow chart for each "slave"

REFERENCES FOR CHAPTER 4

[1] HIMMELBLAU, D.M. (Editor), 1973 : Decomposition of large scale problems. North-Holland American elsevier.

[2] RICHETIN, M. : Analyse structurale des systemes complexes en vue d'une commande hiérarchisée.
Thèse de Doctorat d'Etat ès Sciences Physiques, U.P.S. Toulouse, 1975.

[3] BERNHARD, P., 1976 : Commande optimale, décentralisation et jeux dynamiques. Dunod Automatique, Paris.

[4] FOSSARD, A.J., CLIQUE, M., IMBERT, N. : Aperçus sur la commande hiérarchisée. Revue RAIRO, août 1972, I-3, p. 3-40.

[5] LASDON, L.S., SCHOEFFLER,J.D. : A multilevel technique for optimisation. J.A.C.C. 1965.

[6] BROSILOW, G.B., LASDON, L.S., PEARSON, J.D. : Feasible optimisation methods for interconnected systems.
J.A.C.C. 1965.

[7] TITLI, A., 1975 : Commande hiérarchisée et optimisation des processus complexes. Dunod Automatique, Paris.

[8] ARROW, K.J., HURWICZ, L., UZAWA, H., 1964 : Studies in linear and non linear programming. Stanford University, Press.

[9] BOUDAREL, R., DELMAS, J., GUICHET, P.,1968: Commande optimale des processus. T. 2. Programmation non linéaire et applications. Dunod, Paris.

[10] TITLI, A., LEFEVRE, T., RICHETIN, M. : Multilevel optimisation methods for non-separable problems and applications.
International Journal of Systems Science, vol. 4, n° 6, pp. 865-880, 1973.

[11] DREW, S.A.W. : The application of hierarchical control methods to a managerial problem. Int. Journal of Systems Science, 1975.

[12] BRETON, A. : Gestion en stratégie des équipements de production d'électricité dans le cadre annuel. Journées AFCET : Commande et gestion des systèmes complexes. Toulouse, 6-7 octobre 1976.

[13] TITLI, A., GODARD, J.P. : Gestion optimale d'un complexe hydro-thermique à l'aide du calcul hiérarchisé et de la méthode du gradient réduit généralisé. Canadian Electrical Engineering Journal, oct. 1976.

[14] GRATELOUP, G., TITLI, A., LEFEVRE, T. : Les algorithmes de coordination dans la méthode mixte d'optimisation à 2 niveaux.
5th Conference on Optimization techniques. IFIP, Rome, 7-11 March 1973.

[15] TITLI, A., SINGH, M.G., HASSAN, M. : Hierarchical optimisation of dynamical systems using multiprocessors. Journal of Computers and Electrical Engineering (to appear).

[16] WISMER, J.D.A. : Optimal control of distributed parameter systems using multilevel techniques. Ph. D. Los Angeles, 1966.

[17] FINDEISEN, W. : Parametric optimization by primal method in multilevel systems. IEEE Trans. on Syst. Sciences and Cybern. Vol. SSC, 4, n° 2, July 1968.

[18] MESAROVIC, M.D. : Role of separability of the performance (cost) function in the decomposition of decision problems, in D.M. Himmelblau (editor) 1973, Decomposition of large scale problems. North Holland American Elsevier.

[19] BILLINGSLEY, J. : Closed-linked union of mini-processors. Cambridge University Eng. Dept., Internal Note, November 1973.

PROBLEMS FOR CHAPTER 4

1. Solve by using pseudo variables and coordination by direct iteration the problem

$$\min_{x,y} F(x, y) = \min_{x,y} \left[x^2 - 12x + y^2/2 - yx \right]$$

2. Decomposition of a problem with a multiplicative criterion.

Consider the global problem :

$$\max \left[P_1(\underline{M}_1, \underline{M}_3) \cdot P_2(\underline{M}_2, M_3) \right]$$

$$\text{subject to} \begin{bmatrix} \underline{h}_1(\underline{M}_1, \underline{M}_3) \geqslant \underline{0} \\ \underline{h}_2(\underline{M}_2, \underline{M}_3) \geqslant \underline{0} \\ \underline{h}_3(\underline{M}_3) \geqslant \underline{0} \end{bmatrix}$$

Define in an intuitive way the 2 subproblems from which, after coordination, we could obtain the global solution.

Show that the Lagrangian of the global problem can be obtained in a partially multiplicative form. Define the conditions for this. Define also the coordination task.

3. Using the same reasoning, solve the problem

$$\min \left[(x_1^2 + x_2^2 + x_3^3 + x_4^2)(x_3^2 + x_4^2 + x_5^2 + x_6^2) \right]$$

$$\text{subject to} \quad \begin{bmatrix} x_1 + x_2 + x_3 - 2 = 0 \\ x_4 + x_5 + x_6 - 4 = 0 \\ x_3 - 6 = 0 \end{bmatrix}$$

4. Consider the problem :

$$\min \sum_{i=1}^{3} P_i(X_i)$$

$$\text{subject to} \sum_{i=1}^{3} g_i(X_i) = 0$$

Solve this problem using a multi-level optimisation technique by rewriting in the form :

$$\min \sum_{i=1}^{3} P_i(X_i)$$

$$\text{subject to} \begin{bmatrix} g_i(X_i) = \mu_i \\ \sum_{i=1}^{3} \mu_i = 0. \end{bmatrix}$$

Define the subproblems and the coordination task.

PART 2

Dynamic Optimisation

CHAPTER 5

Dynamic Optimisation for Low Order Systems

In this chapter, we introduce the basic notions of dynamic optimisation and treat in some detail a special case, i.e. systems with linear dynamics and quadratic cost functions. The computational techniques that we describe in this chapter are particularly applicable to low order systems. In the case of linear systems with quadratic performance indices, we could use these techniques to solve problems of 20th order, say, but as we discuss in the next chapter, it is more advantageous to use hierarchical techniques for systems of order greater than about 10. For non-linear problems, we will see in subsequent chapters that it is computationally attractive to use hierarchical techniques for even low order problems.

Let us begin by formulating the dynamic optimisation problem.

5.1 THE DYNAMIC OPTIMISATION PROBLEM

Let us suppose that we have a general non-linear dynamical system described by the state space equations

$$\dot{\underline{x}}(t) = \underline{f}(\underline{x}(t), \underline{u}(t), t) \qquad 5.1.1$$
$$\underline{x}(t_o) = \underline{x}_o$$

where $\underline{x}$ is an n dimensional state vector, $\underline{u}$ is an m dimensional control vector and $\underline{f}$ is a continuous doubly differentiable analytical vector function. The initial state $\underline{x}(t_o)$ is assumed known. Equation 5.1.1 could be used to describe many physical systems. In part 3 of this book, we will consider how we can construct a "model" of the type described by equation 5.1.1 for a given dynamical system. For the moment, let us assume that we have been able to construct such a model. Essentially our control problem is to choose the control trajectories $\underline{u}(t)$ ($t_o \leqslant t \leqslant t_f$), where t_f the final time may be fixed or free, so as to ensure that our system described by equation 5.1.1 has a "desirable" dynamical behaviour. For example, if our system is operating normally in the steady state when it receives an unknown disturbance which changes its state to some known (or measurable) state $\underline{x}_o$, the "desirable" dynamical behaviour could be represented by the control vector $\underline{u}$ which minimises some function of the states and controls all along the trajectory. A suitable way, thus, of choosing $\underline{u}$ may be to find the $\underline{u}$ which minimises the cost function.

$$J = h(\underline{x}(t_f), t_f) + \int_{t_o}^{t_f} g(\underline{x}(t), \underline{u}(t), t)\, dt \qquad 5.1.2$$

In this cost function, h, g are, in general, scalar non-linear functions in C^2. $h(\underline{x}(t_f), t_f)$ in the cost function allows us to ensure that at the final time t_f our state $\underline{x}$ will approach some target state whilst the integral ensures that over the optimisation interval, we are not utilising excessive control effort or allowing significant deviations from any given "desired" trajectories that our control system is required to follow.

In addition, since our model equation 5.1.1 represents a physical system and since we know that in physical systems we can not have infinite controls (which we can, in principle, in equation 5.1.1) or states, we must constrain the states and controls such that any control that we choose is a physically realisable one. Since the controls are related to the states, our controls must not be such that they cause violation of any bounds on the states. For an example of this idea of state and control bounds, consider a chemical reactor where the desired reaction occurs only if the temperature lies between certain limits $T \pm \Delta T$ so that if our state variable is the temperature then our state T must be bounded by

$$T - \Delta T \leqslant T \leqslant T + \Delta T$$

Again, physically, the control, which may be the rate of fuel flow which heats the reactor will also be bounded with the bounds depending on the size of the pipes, etc.

We shall call a state $\underline{x} \in X$ which satisfies the additional state constraints (aside from the dynamical equation 5.1.1) as an <u>admissible</u> state, and a control $\underline{u} \in U$ which satisfies the additional control constraints as an admissible control where X and U are respectively the sets of admissibles states and controls over the period (t_o, t_f).

The dynamical optimisation problem is to find an admissible control $\underline{u}^*$ which causes the dynamical system in equation 5.1.1 to follow an admissible trajectory $\underline{x}^*$ that minimises the performance measure J given in equation 5.1.2.

There are number of points which should be noted about the above optimal control problem. The first of these is the existence of the optimal control. In a number of important cases, the optimal control may not exist. Now, usually it is very difficult to tell a priori whether an optimal control will exist or not for a particular problem so that it is often easier to just try and solve the problem and if we are able to find the optimal control, it obviously exists.

The second important consideration is that except for a limited class of problems, the optimal control may be non-unique.

It should be emphasised that what we are seeking is basically the "global" optimum, i.e. $\underline{u}^*$ that is such that

$$J^* = h(\underline{x}^*(t_f), t_f) + \int_{t_o}^{t_f} g(\underline{x}^*(t), \underline{u}^*(t), t)\, dt$$

$$\leqslant h(\underline{x}(t_f), t_f) + \int_{t_o}^{t_f} g(\underline{x}(t), \underline{u}(t), t)\, dt$$

for all $\underline{u} \in U$ which make $\underline{x} \in X$.

In practice, except for the special case of linear-quadratic problems that we will discuss and with the method of Dynamic Programming, it is difficult to obtain the global optimal control numerically. We can however obtain local optima quite easily and in principle, if we obtain all the local optima, we could obtain the global optimum control by choosing from among the local optima the control that gives a minimal cost J. We next describe the most important approach to tackling such problems.

5.2 VARIATIONAL TECHNIQUES AND THE MAXIMUM PRINCIPLE

5.2.1 Necessary conditions of optimality

In this section, we will develop the necessary conditions for the solution of our dynamical optimisation problem using the calculus of variations. Satisfaction of these conditions will yield a local optimum.

Assume that the admissible state and control regions are not bounded and that the initial conditions $\underline{x}(t_o) = \underline{x}_o$ and the initial time t_o are specified. $\underline{x}$ is an n state vector and $\underline{u}$ is an m control vector.

Let us begin by noting that if in the cost function of equation 5.1.2 h is a differentiable function, then

$$h(\underline{x}(t_f), t_f) = \int_{t_o}^{t_f} \frac{d}{dt}\left[h(\underline{x}(t), t) \right] dt + h(\underline{x}(t_o), t_o) \qquad 5.2.1$$

so that we can rewrite equation 5.1.2 as :

$$J = \int_{t_o}^{t_f} \left\{ g(\underline{x}(t), \underline{u}(t), t) + \frac{d}{dt}\left[h(\underline{x}(t), t) \right] \right\} dt + h(\underline{x}(t_o), t_o) \qquad 5.2.2$$

Since $\underline{x}(t_o)$, t_o are fixed, these will not affect the minimisation of J in equation 5.2.2, so we need only consider the minimisation of

$$J(\underline{u}) = \int_{t_o}^{t_f} \left\{ g(\underline{x}(t), \underline{u}(t), t) + \frac{d}{dt}\left[h(\underline{x}(t), t) \right] \right\} dt \qquad 5.2.3$$

$$= \int_{t_o}^{t_f} \left\{ g(\underline{x}(t), \underline{u}(t), t) + \left[\frac{\partial h}{\partial \underline{x}} (\underline{x}(t), t) \right]^T \dot{\underline{x}}(t) + \frac{\partial h}{\partial t} (\underline{x}(t), t) \right\} dt \qquad 5.2.4$$

by the chain rule of differentiation.

Now if we include the differential equation constraints by introducing the Lagrange Multipliers $\lambda_1(t), \ldots, \lambda_n(t)$, then the cost function must be augmented to

$$L^* = \int_{t_o}^{t_f} \left\{ g(\underline{x}(t), u(t), t) + \left[\frac{\partial h}{\partial \underline{x}} (\underline{x}(t), t) \right]^T \dot{\underline{x}}(t) + \frac{\partial h}{\partial t} (\underline{x}(t), t) + \underline{\lambda}^T(t) \left[f(\underline{x}(t), \underline{u}(t), t) - \dot{\underline{x}}(t) \right] \right\} dt \qquad 5.2.5$$

Define $G\left[\underline{x}(t), \dot{\underline{x}}(t), \underline{u}(t), \underline{\lambda}(t), t \right]$ as the term inside the integral in the augmented cost function L^* so that :

$$L^* = \int_{t_o}^{t_f} \left\{ G\left[\underline{x}(t), \dot{\underline{x}}(t), \underline{u}(t), \underline{\lambda}(t), t \right] \right\} dt$$

Now, if we have small variations $\delta \underline{x}$, $\delta \underline{\dot{x}}$, $\delta \underline{u}$, $\delta \underline{\lambda}$ and δt_f, then on an extremal (i.e. on a local optimal solution), the variation of L^*, i.e. δL^* must vanish. It is easy to show, and we leave as an excercise, that δL^* is given by

$$\begin{aligned}
\delta L^* = 0 = & \left[\frac{\partial G}{\partial \underline{x}} (\underline{x}^*(t_f), \underline{\dot{x}}^*(t_f), \underline{u}^*(t_f), \underline{\lambda}^*(t_f), t_f) \right]^T \delta \underline{x}_f \\
& + \Big[G(\underline{x}^*(t_f), \underline{\dot{x}}^*(t_f), \underline{u}^*(t_f), \underline{\lambda}^*(t_f), t_f) \\
& - \left[\frac{\partial G}{\partial \underline{\dot{x}}} (\underline{x}^*(t_f), \underline{\dot{x}}^*(t_f), \underline{u}^*(t_f), \underline{\lambda}^*(t_f), t_f) \right]^T \underline{\dot{x}}^*(t_f) \Big] \delta t_f \\
& + \int_{t_o}^{t_f} \Bigg\{ \Bigg[\left[\frac{\partial G}{\partial \underline{x}} (\underline{x}^*(t), \underline{\dot{x}}^*(t), \underline{u}^*(t), \underline{\lambda}^*(t), t \right]^T \delta \underline{x}(t) \\
& - \frac{d}{dt} \left[\frac{\partial G}{\partial \underline{\dot{x}}} (\underline{x}^*(t), \underline{\dot{x}}^*(t), \underline{u}^*(t), \underline{\lambda}^*(t), t) \right]^T \Bigg] \delta \underline{\dot{x}}(t) \\
& + \left[\frac{\partial G}{\partial \underline{u}} (\underline{x}^*(t), \underline{\dot{x}}^*(t), \underline{u}^*(t), \underline{\lambda}^*(t), t) \right]^T \delta \underline{u}(t) \\
& + \left[\frac{\partial G}{\partial \underline{\lambda}} (\underline{x}^*(t), \underline{\dot{x}}^*(t), \underline{u}^*(t), \underline{\lambda}^*(t), t \right]^T \delta \underline{\lambda}(t) \Bigg\} \, dt
\end{aligned} \qquad 5.2.6$$

Consider next the terms inside the integral which involve the function h ; these terms contain

$$\begin{aligned}
& \frac{\partial}{\partial \underline{x}} \left[\left[\frac{\partial h}{\partial \underline{x}} (\underline{x}^*(t), t) \right]^T \underline{\dot{x}}^*(t) + \frac{\partial h}{\partial t} (\underline{x}^*(t), t) \right] \\
& \qquad - \frac{d}{dt} \left\{ \frac{\partial}{\partial \underline{\dot{x}}} \left[\left[\frac{\partial h}{\partial \underline{x}} (\underline{x}^*(t), t) \right]^T \underline{\dot{x}}^*(t) \right] \right\}
\end{aligned} \qquad 5.2.7$$

Writing out the above partial derivatives gives

$$\begin{aligned}
& \left[\frac{\partial^2 h}{\partial \underline{x}^2} (\underline{\dot{x}}^*(t), t) \right] \underline{\dot{x}}^*(t) + \left[\frac{\partial^2 h}{\partial t \partial \underline{x}} (\underline{x}^*(t), t) \right] \\
& \qquad - \frac{d}{dt} \left[\frac{\partial h}{\partial \underline{x}} (\underline{x}^*(t), t) \right]
\end{aligned} \qquad 5.2.8$$

Applying the chain rule to the last term

$$\begin{aligned}
& \left[\frac{\partial^2 h}{\partial \underline{x}^2} (\underline{\dot{x}}^*(t), t) \, \underline{\dot{x}}^*(t) \right] + \left[\frac{\partial^2 h}{\partial t \partial \underline{x}} (\underline{x}^*(t), t) \right] \\
& \qquad - \left[\frac{\partial^2 h}{\partial \underline{x}^2} (\underline{x}^*(t), t) \right] \underline{\dot{x}}^*(t) - \left[\frac{\partial^2 h}{\partial \underline{x} \partial t} (\underline{x}^*(t), t) \right]
\end{aligned}$$

Now if we assume that the second partial derivatives are continuous, the order of differentiation can be interchanged, and these terms add to zero. In the integral term we therefore have :

$$\int_{t_o}^{t_f} \left\{ \left[\left[\frac{\partial g}{\partial \underline{x}}(\underline{x}^*(t), \underline{u}^*(t), t \right]^T + \underline{\lambda}^{*T}(t) \left[\frac{\partial f^T}{\partial \underline{x}} (\underline{x}^*(t), \underline{u}^*(t), t) \right] - \frac{d}{dt}\left[- \underline{\lambda}^{*T}(t) \right] \right] \delta \underline{x}(t) + \left[\left[\frac{\partial g}{\partial \underline{u}} (\underline{x}^*(t), \underline{u}^*(t), t) \right]^T + \underline{\lambda}^{*T}(t) \left[\frac{\partial f^T}{\partial \underline{u}} (\underline{x}^*(t), \underline{u}^*(t), t) \right] \delta \underline{u}(t) + \left[\left[f(\underline{x}^*(t), \underline{u}^*(t), t) - \dot{\underline{x}}^*(t)^T \right] \delta \underline{\lambda}(t) \right] \right\} dt \qquad 5.2.9$$

Now this integral must vanish on an extremal trajectory. In addition, since the dynamic constraints

$$\dot{\underline{x}}^*(t) = f(\underline{x}^*(t), \underline{u}^*(t), t) \qquad 5.2.10$$

must be satisfied on an extremal, the coefficient of $\delta\underline{\lambda}$ must be zero.

Again, since the Lagrange multipliers are arbitrary, we can select them to make the coefficient of $\delta \underline{x}(t)$ zero, i.e.

$$\dot{\underline{\lambda}}^*(t) = - \left[\frac{\partial f^T}{\partial \underline{x}} (\underline{x}^*(t), \underline{u}^*(t), t) \right]^T \underline{\lambda}^*(t) - \frac{\partial g}{\partial \underline{x}} (\underline{x}^*(t), \underline{u}^*(t), t) \qquad 5.2.11$$

This equation is referred to as the costate equation and $\underline{\lambda}(t)$ as the costate vector.

Since the remaining variation $\underline{u}(t)$ is independent, its coefficient must be zero, i.e.

$$\frac{\partial g(\underline{x}^*(t), \underline{u}^*(t), t)}{\partial \underline{u}} + \left[\frac{\partial f^T(\underline{x}^*(t), \underline{u}^*(t), t)}{\partial \underline{u}} \right]^T \underline{\lambda}^*(t) = \underline{0} \qquad 5.2.12$$

For the terms outside the integral, since the variations must be zero, we have :

$$\left[\frac{\partial h}{\partial \underline{x}} (\underline{x}^*(t_f), t_f) - \underline{\lambda}^*(t_f) \right]^T \delta \underline{x}_f + \left[g(\underline{x}^*(t_f), \underline{u}^*(t_f), t_f) + \frac{\partial h}{\partial t} (\underline{x}^*(t_f), t_f) + \underline{\lambda}^*(t_f) \left[f(\underline{x}^*(t_f), \underline{u}^*(t_f), t_f) \right] \right] \delta t_f = 0 \qquad 5.2.13$$

where we have used the fact that $\dot{\underline{x}}^*(t) = f(\underline{x}^*(t_f), \underline{u}^*(t_f), t_f)$.

Equations 5.2.10, 5.2.11, 5.2.12 are the necessary conditions for optimality and consist of 2n first order differential equations (i.e. n state equations and n costate equations) and m algebraic relations (for the control equation 5.2.12) which need to be satisfied over the optimisation period $[t_o, t_f]$. To solve these equations, we require 2n boundary conditions. n of these are given by the initial conditions on the state, i.e. $\underline{x}(t_o) = \underline{x}_o$ and an additional n or (n+1) relationships depending on whether t_f is specified, are given by equation 5.2.13.

To write the necessary conditions for optimality in a more compact form, define the function H called the Hamiltonian as :

$$H(\underline{x}(t), \underline{u}(t), \underline{\lambda}(t), t) = g(\underline{x}(t), \underline{u}(t), t) + \underline{\lambda}^T[\underline{f}(\underline{x}(t), \underline{u}(t), t)]$$

Then the problem of minimising

$$\boxed{J = h(\underline{x}(t_f), t_f) + \int_{t_o}^{t_f} g(\underline{x}(t), \underline{u}(t), t) \quad dt} \qquad 5.2.14$$

subject to

$$\boxed{\begin{aligned} &\dot{\underline{x}}(t) = \underline{f}(\underline{x}(t), \underline{u}(t), t) \\ &\underline{x}(t_o) = \underline{x}_o \end{aligned}} \qquad 5.2.15$$

yields the necessary conditions

$$\boxed{\left.\begin{aligned} \dot{\underline{x}}^*(t) &= \frac{\partial H}{\partial \underline{\lambda}} (\underline{x}^*(t), \underline{u}^*(t), \underline{\lambda}^*(t), t) \\ \dot{\underline{\lambda}}^*(t) &= -\frac{\partial H}{\partial \underline{x}} (\underline{x}^*, t), \underline{u}^*(t), \underline{\lambda}^*(t), t) \\ \underline{0} &= \frac{\partial H}{\partial \underline{u}} (\underline{x}^*(t), \underline{u}^*(t), \underline{\lambda}^*(t), t) \end{aligned}\right] \forall\, t \in [t_o, t_f)]} \qquad 5.2.16$$

and

$$\boxed{\begin{aligned} &\left[\frac{\partial h}{\partial \underline{x}} (\underline{x}^*(t_f), t_f) - \underline{\lambda}^*(t_f)\right]^T \delta \underline{x}_f + \left[H(\underline{x}^*(t_f)\ \underline{u}^*(t_f), \underline{\lambda}^*(t_f), t_f)\right. \\ &\qquad \left. + \frac{\partial h}{\partial t} (\underline{x}^*(t_f), t_f)\right] \delta t_f = 0 \end{aligned}} \qquad 5.2.17$$

5.2.2 Boundary conditions

In a particular problem if g or h is missing, we can simply strike out the missing function. To determine the boundary conditions for a particular problem, it is necessary to make appropriate substitutions in equation 5.2.17. We assume that the initial state $\underline{x}(t_o) = \underline{x}_o$ is given. We will consider first of all the problems with a fixed terminal time since these are the problems which arise most often in practice.

5.2.2.1 Problems with fixed terminal time

If the terminal time t_f is specified, $\underline{x}(t_f)$ may be specified, free or required to lie on some surface in the state space.

If the final state is specified, $\underline{x}(t_f)$ and t_f being both specified means that $\delta \underline{x}(t_f) = \underline{0}$, $\delta t_f = 0$ in equation 5.2.17 so that the boundary condition becomes :

$$\boxed{\underline{x}(t_f) = \underline{x}_f} \qquad 5.2.18$$

If the final state is free, then only $\delta t_f = 0$ in equation 5.2.17 so that we must have :

$$\boxed{\frac{\partial h}{\partial \underline{x}} (\underline{x}^*(t_f)) - \underline{\lambda}^*(t_f) = \underline{0}}$$

Note here that we have been able to eliminate the term in h because the final time is fixed so that h will not depend on t_f.

<u>If the final state lies on a surface defined by $\underline{q}(\underline{x}(t)) = \underline{0}$</u> where $\underline{q}$ is a k vector, then since each component of $\underline{q}$ represent a hypersurface in n dimensional state space, the final state lies on the intersection of these k hypersurfaces and $\delta \underline{x}(t_f)$ is tangential to each of the hypersurface at the point $(\underline{x}^*(t_f), t_f)$. This implies that $\delta \underline{x}(t_f)$ is normal to each of the gradient vectors

$$\frac{\partial q_1}{\partial \underline{x}} (\underline{x}^*(t_f)), \ \dots \ \frac{\partial q_k}{\partial \underline{x}} (\underline{x}^*(t_f)) \qquad 5.2.19$$

where we assume that these gradient vectors are linearly independent.

From equation 5.2.17 since $\delta t_f = 0$

$$\left[\frac{\partial h}{\partial \underline{x}} (\underline{x}^*(t_f)) - \underline{\lambda}^*(t_f)\right]^T \delta \underline{x}(t_f) = \underline{\omega}^T \delta \underline{x}(t_f) = 0 \qquad 5.2.20$$

It is possible to show that this equation is satisfied if and only if the vector $\underline{\omega}$ is a linear combination of the gradient vectors in equation 5.2.19, i.e.

$$\frac{\partial h}{\partial \underline{x}} (\underline{x}^*(t_f)) - \underline{\lambda}^*(t_f) = d_1 \left[\frac{\partial q_1}{\partial \underline{x}} (\underline{x}^*(t_f))\right] + \dots$$
$$\dots + d_k \left[\frac{\partial q_k}{\partial \underline{x}} (\underline{x}^*(t_f))\right] \qquad 5.2.21$$

and in order to determine the 2n constants of integration in the solution of the state-costate equations, and $d_1, d_2, \dots d_k$ we have the n equations $\underline{x}^*(t_o) = \underline{x}_o$, the n equations given by equation 5.2.21 plus the k equations

$$\underline{q}(\underline{x}^*(t_f)) = \underline{0} \qquad 5.2.22$$

Next let us consider the case where the terminal time is free.

5.2.2.2 <u>Problems with free terminal time</u>

The first case of interest here is if the final state is fixed. In that case, $\delta \underline{x}_f = \underline{0}$ in equation 5.2.17 whilst δt_f is arbitrary so that in addition to the 2n state-costate equations we must have

$$H(\underline{x}^*(t_f), \underline{u}^*(t_f), \underline{\lambda}^*(t_f), t_f) + \frac{\partial h}{\partial t} (\underline{x}^*(t_f), t_f) = 0 \qquad 5.2.23$$

On the other hand if the final state is free, then $\delta \underline{x}_f$ and δt_f are arbitrary and independent so that their coefficients are zero, i.e. we obtain the n equations

$$\underline{\lambda}^*(t_f) = \frac{\partial h}{\partial \underline{x}} (\underline{x}^*(t_f), t_f) \qquad 5.2.24$$

and the additional equation

$$H(\underline{x}^*(t_f), \underline{u}^*(t_f), \underline{\lambda}^*(t_f), t_f) + \frac{\partial h}{\partial t} (\underline{x}^*(t_f), t_f) = 0 \qquad 5.2.25$$

Similarly it is possible to develop the boundary conditions for other situations by an appropriate substitution in equation 5.2.17.

Remark

We have only considered first order conditions (i.e. necessary conditions) of optimality which provide stationary points. In order to specify the nature of these points, we have to consider 2nd order conditions i.e. the condition of Weierstrass and of Legendre.

Thus we define, using 5.2.5 :

$$L^* = \int_{t_o}^{t_f} G \, dt$$

Then Legendre's condition could be stated as :

"For the solution of the 1st order conditions to correspond to a minimum, it is necessary that the matrix

$$\begin{bmatrix} G_{\underline{\mathring{x}}\underline{\mathring{x}}} & G_{\underline{\mathring{x}}\underline{u}} \\ G_{\underline{u}\underline{\mathring{x}}} & G_{\underline{u}\underline{u}} \end{bmatrix}_{\underline{x}^*, \underline{u}^*}$$

be non-negative definite".

From the formulation of the problem and the definition of G, this condition reduces to $G_{\underline{uu}} \geqslant 0$.

As for the condition of Weierstrass, it can be written as :

$$G(\underline{x}^*, \underline{\mathring{x}}, \underline{u}, t) - G(\underline{x}^*, \underline{\mathring{x}}^*, \underline{u}^*, t) - (\underline{\mathring{x}} - \underline{\mathring{x}}^*)^T \; G_{\underline{\mathring{x}}}(\underline{x}^*, \underline{\mathring{x}}^*, \underline{u}^*, t) \geqslant 0$$

$t \in [t_o, t_f]$, $\underline{\mathring{x}}$, $\underline{u}$ compatible with the constraints.

We can show that this condition is the same as the statement that the control should minimise $H(\underline{\mathring{x}}^*, \underline{u}, t)$ (which corresponds partially to the condition $\frac{\partial H}{\partial \underline{u}} = \underline{0}$) from which arises the name of the minimum principle (or maximum principle if we wish to maximise). In practice, to verify that a control is really a minimum, the easiest thing to do is to perturb it slightly and verify that this does indeed increase the cost.

5.2.3 Examples

Example 1 : To fix these ideas, consider the example of the sliding mass system shown in Fig. 5.1 (previously considered by Bauman). The state equations for this system can be written as :

$$\mathring{y}_1 = y_2$$

$$\mathring{y}_2 = m - y_2^2 \operatorname{sgn} y_2 \qquad 5.2.26$$

$$y_1(0) = y_1^*, \quad y_2(0) = y_2^*$$

where $\text{sgn } y_2 = \begin{cases} +1 & \text{if } y_2 > 0 \\ -1 & \text{if } y_2 < 0 \end{cases}$

and y_1^*, y_2^* are given initial conditions

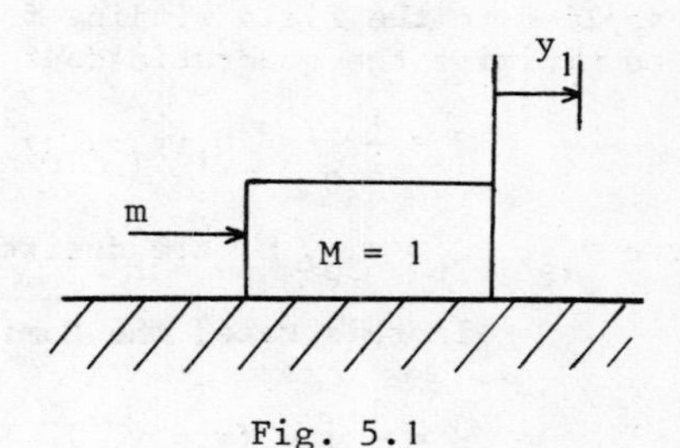

Fig. 5.1

and it is necessary to minimise

$$J = \frac{1}{2}\int_0^T (y_1^2 + y_2^2 + m^2)\, dt$$

Here, y_1, y_2 are the states, m is the control and T is the fixed final time.

In this case, the Hamiltonian H can be written as :

$$H = \frac{1}{2} y_1^2 + \frac{1}{2} y_2^2 + \frac{1}{2} m^2 + \lambda_1 y_2 + \lambda_2 \left[m - y_2^2 \text{ sgn } y_2 \right]$$

The necessary conditions in this case are :

$$\frac{\partial H}{\partial m} = 0 \; ; \quad \frac{\partial H}{\partial y_1} = -\dot{\lambda}_1, \quad \frac{\partial H}{\partial y_2} = -\dot{\lambda}_2$$

and these give

$$m = -\lambda_2 \qquad 5.2.27$$

$$-\dot{\lambda}_1 = y_1 \qquad 5.2.28$$
$$-\dot{\lambda}_2 = y_2 - 2\lambda_2 y_2 \text{ sgn } y_2$$

For the boundary conditions, since here we have a fixed terminal time, and "h" is missing,

$$\lambda_1(T) = \lambda_2(T) = 0 \qquad 5.2.29$$

Thus the problem reduces to solving equations 5.2.26 and 5.2.28 with the boundary conditions $y_1(0) = y_1^*$, $y_2(0) = y_2^*$, $\lambda_1(T) = \lambda_2(T) = 0$.

Example 2 : Consider the following synchronous machine excitation problem previously considered by Mukhopadhyay [7]. The model is of third order and is described by :

$$\left.\begin{aligned} \dot{y}_1 &= y_2 \\ \dot{y}_2 &= B_1 - A_1 y_2 - A_2 y_3 \sin y_1 - \frac{B_2}{2} \sin 2y_1 \\ \dot{y}_3 &= u - C_1 y_3 + C_2 \cos y_1 \end{aligned}\right] \qquad 5.2.30$$

with $y_1(0) = y_1^*$, $y_2(0) = y_2^*$, $y_3(0) = y_3^*$

where y_1 is the rotor angle (in radians).

y_2 is the speed deviation and y_3 is the field flux linkage. u the control variable is the voltage applied to the field winding of the synchronous machine and is assumed to be available for optimal manipulation.

The control problem is to determine the optimal voltage trajectory to be applied to the field winding following a transient disturbance in the system so as to minimise the quadratic cost function

$$J = \frac{1}{2}\int_0^T \left[Q_1(y_1-y_{1p})^2 + Q_2(y_2-y_{2p})^2 + Q_3(y_3-y_{3p})^2 + R(u-u_p)^2 \right] dt$$

where y_{1p}, y_{2p}, y_{3p}, u_p are desired steady state values.

In this case, the Hamiltonian can be written as :

$$H = \frac{1}{2}\left[Q_1(y_1 - y_{1p})^2 + Q_2(y_2 - y_{2p})^2 + Q_3(y_3-y_{3p})^2 + R(u - u_p)^2 \right] + \lambda_1[y_2] + \lambda_2\left[B_1 - A_1 y_2 - A_2 y_3 \sin y_1 - \frac{B_2}{2}\sin 2y_1\right] + \lambda_3\left[u - C_1 y_3 + C_2 \cos y_1\right]$$

Then the necessary conditions yield the following two point boundary value problem

$$\frac{\partial H}{\partial u} = 0 \quad \text{or} \quad u = u_p + \frac{\lambda_3}{R} \qquad 5.2.31$$

$$\frac{\partial H}{\partial y_1} = -\dot{\lambda}_1 = \lambda_2(-A_2 y_3 \cos y_1 - B_2 \cos 2y_1) - \lambda_3 C_2 \sin y_1 + Q_1(y_1-y_{1p})$$

$$\frac{\partial H}{\partial y_2} = -\dot{\lambda}_2 = \lambda_1 - \lambda_2 A_1 + Q_2(y_2-y_{2p}) \qquad 5.2.32$$

$$\frac{\partial H}{\partial y_3} = -\dot{\lambda}_3 = -\lambda_2 A_2 \sin y_1 - \lambda_3 C_1 + Q_3(y_3-y_{3p})$$

By substituting for the optimal control from equation 5.2.31 into the state equations 5.2.30 we have as our state costate equations 5.2.32 and 5.2.33, where equation 5.2.33 is :

$$\dot{y}_1 = y_2$$

$$\dot{y}_2 = B_1 - A_1 y_2 - A_2 y_3 \sin y_1 - \frac{B_2}{2}\sin 2y_1 \qquad 5.2.33$$

$$\dot{y}_3 = (u_p + \frac{\lambda_3}{R}) - C_1 y_3 + C_2 \cos y_1$$

with the two point boundary conditions :

$$y_1(0) = y_1^*, \; y_2(0) = y_2^*, \; y_3(0) = y_3^*$$

$$\lambda_1(T) = 0, \; \lambda_2(T) = 0, \; \lambda_3(T) = 0$$

We will see subsequently how we can solve such two point boundary value problems.

As a final example, let us consider a very special but important type of example, i.e. that of systems with linear dynamics and quadratic cost functions.

Example 3 : Consider the problem of minimising :

$$J = \frac{1}{2} x_1^2(T) + \frac{1}{2} \int_0^T (x_1^2 + x_2^2 + u^2)\, dt$$

$$\text{subject to} \quad \dot{x}_1 = x_1 + x_2 \qquad 5.2.34$$

$$\dot{x}_2 = x_2 + u$$

$$x_1(0) = x_1^*, \; x_2(0) = x_2^*$$

The Hamiltonian in this case can be written as :

$$H = \frac{1}{2} x_1^2 + \frac{1}{2} x_2^2 + \frac{1}{2} u^2 + \lambda_1(x_1+x_2) + \lambda_2(x_2+u)$$

and the necessary conditions yield

$$\frac{\partial H}{\partial u} = 0 \quad \text{or} \quad u = -\lambda_2$$

$$\frac{\partial H}{\partial x_1} = -\dot{\lambda}_1 = x_1 + \lambda_1 \qquad 5.2.35$$

$$\frac{\partial H}{\partial x_2} = -\dot{\lambda}_2 = x_2 + \lambda_1 + \lambda_2$$

with $\lambda_1(T) = x_1(T), \; \lambda_2(T) = 0$

and substituting for the control $u = -\lambda_2$ in equation 5.2.35, we obtain :

$$\dot{x}_1 = x_1 + x_2$$

$$\dot{x}_2 = x_2 - \lambda_2 \qquad 5.2.36$$

The optimisation problem therefore reduces to the solution of equations 5.2.35 and 5.2.36 with the split boundary conditions

$$x_1(0) = x_1^*, \; x_2(0) = x_2^*$$

$$\lambda_1(T) = x_1(T), \; \lambda_2(T) = 0$$

Remarks

We have considered three examples, above, two of which were for systems with non-linear dynamics whilst the third was for a system with linear dynamics and a quadratic cost function. In each case the necessary conditions for optimality led us to formulate a two point boundary value problem. Two point boundary value problems, in general, are extremely difficult to solve normally and we have to resort to iterative techniques. In Chapter 7, we study various numerical techniques for solving non-linear two point boundary value problems. Here, however, we will show that for the case where the dynamics are linear and the cost function is quadratic (as in the above example 3), the two point boundary value problem is linear. The linearity property can be used to superpose solutions and thus convert the problem to a single point boundary value problem. Single point boundary value problems are of course exceedingly easy to solve since they merely require the direct integration of the equations from the single point boundary conditions. The equation which needs to be integrated is a matrix equation of the Riccati type. Next we will examine this special case.

5.2.4 Linear-quadratic problems

5.2.4.1. The linear regulator

Although realistic descriptions of dynamical systems often require non linear differential equations, it is often possible to obtain a good approximation to the dynamic behaviour of the system by linearising the non-linear equations around a suitable trajectory or equilibrium point. Thus, many systems could be described by the state-space equations

$$\dot{\underline{x}}(t) = A(t)\,\underline{x}(t) + B(t)\,\underline{u}(t)$$
$$\underline{x}(t_o) = x_o \qquad 5.2.37$$

where $\underline{x}$ is the nth order state vector, $\underline{u}$ is the mth order control vector and A, B are respectively n x n and n x m time varying matrices.

The cost function to be minimised is assumed to be a weighted quadratic function of the states and controls, i.e.

$$J = \frac{1}{2}\,\underline{x}^T(t_f)\,S(t_f)\,\underline{x}(t_f) + \frac{1}{2}\int_{t_o}^{t_f}\left[\underline{x}(t)^T\,Q(t)\,\underline{x}(t) + \underline{u}(t)^T\,R(t)\,\underline{u}(t)\right]dt \qquad 5.2.38$$

Q, S, are assumed to be real symmetric positive semi-definite matrices, whilst R is a real symmetric positive definite matrix. We assume that the states and controls are not bounded and $\underline{x}(t_f)$ is free. The physical interpretation of this cost function is that we desire to maintain the state vector near the origin of the state space without utilising excessive control effort. The weighting matrices Q, R, S, enable us to define the relative importance of keeping the states near the origin, the expenditure of control effort and the need to ensure that at the final time t_f the states will be very close to the origin.

To solve this problem using the maximum principle, let us write the Hamiltonian as :

$$H = \frac{1}{2}\,\underline{x}(t)^T\,Q(t)\,\underline{x}(t) + \frac{1}{2}\,\underline{u}(t)^T R(t)\,\underline{u}(t) + \underline{\lambda}^T(t)A(t)\underline{x}(t) + \underline{\lambda}^T(t)B(t)\underline{u}(t)$$

The necessary conditions in this case yield :

$$\frac{\partial H}{\partial \underline{u}} = 0 \text{ or } \underline{u}(t) = -R^{-1}\,B^T\,\underline{\lambda}(t) \qquad 5.2.39$$

(R^{-1} exists since R is assumed to be positive definite)

$$\frac{\partial H}{\partial \underline{x}} = -\,\dot{\underline{\lambda}}(t) = Q(t)\,\underline{x}(t) + A^T(t)\,\underline{\lambda}(t) \qquad 5.2.40$$

with the terminal condition

$$\underline{\lambda}(t_f) = S(t)\,\underline{x}(t_f) \qquad 5.2.41$$

Substituting the control $\underline{u}$ from equation 5.2.39 into the state equation 5.2.37, we obtain

$$\dot{\underline{x}}(t) = A(t)\,\underline{x}(t) + B(t)\,R^{-1}(t)\,B^T(t)\,\underline{\lambda}(t) \qquad 5.2.42$$

with $$\underline{x}(t_o) = \underline{x}_o \qquad 5.2.43$$

Thus our two point boundary value problem becomes one of solving equations 5.2.40, 5.2.42, subject to the initial conditions 5.2.43 and terminal conditions 5.2.41.

Let us suppose that the solution for the costate $\underline{\lambda}$ is similar to equation 5.2.41, i.e. let us assume a solution of the form

$$\underline{\lambda}(t) = P(t)\ \underline{x}(t) \qquad 5.2.44$$

then

$$\dot{\underline{\lambda}}(t) = \dot{P}(t)\ \underline{x}(t) + P(t)\ \dot{\underline{x}}(t) \qquad 5.2.45$$

Now utilising equation 5.2.40 for $\dot{\underline{\lambda}}$ and equation 5.2.42 for $\dot{\underline{x}}$ in equation 5.2.45, we have :

$$\dot{\underline{\lambda}}(t) = -\ Q(t)\ \underline{x}(t) - A^T(t)\ P(t)\ \underline{x}(t) = \dot{P}(t)\underline{x}(t) + P(t)$$
$$\left[\ A(t)\underline{x}(t) + B(t)\ R^{-1}(t)\ B^T(t)\ P(t)\ \underline{x}(t)\ \right]$$

or

$$\left[\ \dot{P} + P(t)\ A(t) + A^T\ P(t) - P(t)\ B(t)\ R^{-1}(t)\ B^T(t)\ P(t) + Q(t)\right]\underline{x}(t) = \underline{0} \qquad 5.2.46$$

Since this must hold for arbitrary non zero $\underline{x}$, the terms premultiplying $\underline{x}(t)$ must be zero. Thus we have :

$$\boxed{\dot{P}(t) = -P(t)\ A(t) - A^T(t)\ P(t) + P(t)\ B(t)\ R^{-1}(t)\ B^T(t)\ P(t) - Q(t)} \qquad 5.2.47$$

with the terminal condition being given by equation 5.2.41 and 5.2.44 as

$$\boxed{P(t_f) = S} \qquad 5.2.48$$

Equation 5.2.47 is a matrix equation where P is an n x n symmetric matrix having n(n+1)/2 distinct elements. This equation is known as the matrix Riccati equation. It can be integrated backwards in time from the given terminal condition (equation 5.2.48).

If we set $G(t) = -R^{-1}(t)\ B^T(t)\ P(t)$ 5.2.49

then the control becomes :

$$\boxed{\underline{u}(t) = G(t)\ \underline{x}(t)} \qquad 5.2.50$$

where G is a time varying gain matrix.

It should be noted that the Riccati solution P(t) depends only upon the system matrices so that it can be pre-computed off-line and thus the gain G(t) can be pre-computed off-line and stored.

Note that we have a closed loop controller here, i.e. $\underline{u}(t)$ is a (linear) function of the current state $\underline{x}(t)$ and G(t) is independent of the initial conditions. This is a very desirable property because if we could measure all the states $\underline{x}(t)$, we merely have to multiply these states with the gain matrix G to obtain the optimal control. In fact, <u>if the final time is infinite</u> (which in practice means a few time constants of the system) and the matrices A, B, Q, R are time invariant, then we have the result that $\dot{P}(t)$ in equation 5.2.47 goes to zero as the time t approaches zero (since we are integrating $\dot{P}$ backwards in time) and P becomes a constant so that the gain G in equation 5.2.49 becomes time invariant.

Thus to summarise these results :

If we wish to minimise :

$$J = \frac{1}{2}\underline{x}^T(t_f)S(t_f)\underline{x}(t_f) + \frac{1}{2}\int_{t_o}^{t_f}\left[\underline{x}(t)^T\, Q(t)\underline{x}(t) + \underline{u}(t)^T\, R(t)\underline{u}(t)\right] dt \qquad 5.2.51$$

where Q, S are real symmetric positive semi-definite matrices whilst R is a real symmetric positive definite matrix, subject to the constraints

$$\begin{aligned} \mathring{\underline{x}}(t) &= A(t)\,\underline{x}(t) + B(t)\,\underline{u}(t) \\ \underline{x}(t_o) &= \underline{x}_o \end{aligned} \qquad 5.2.52$$

then the optimal control is given by

$$\underline{u}(t) = -R^{-1}\,B^T(t)\,P(t)\,\underline{x}(t) \qquad 5.2.53$$

where

$$\mathring{P}(t) = -P(t)A(t) - A^T(t)P(t) + P(t)B(t)\,R^{-1}(t)\,B^T(t)\,P(t) - Q(t) \qquad 5.2.54$$

with

$$P(t_f) = S \qquad 5.2.55$$

and when A, B, Q, R are time invariant, $t_f \rightarrow \infty$ and if the system is controllable, $\underline{u} = G\,\underline{x}$ where G is time invariant.

5.2.4.2 Examples of linear quadratic problems

Example 4 : Consider the scalar infinite time regulator

$$\mathring{x}_1 = x_1 + u_1$$

$$J = \frac{1}{2}\int_0^{\infty}(x_1^2 + u_1^2)\, dt$$

Here, A = 1, B = 1, Q = 1, R = 1 so that the Riccati equation in this case becomes

$$\mathring{P} = -2P + P^2 - 1 = 0$$

so that $P = \frac{2 \pm \sqrt{4+4}}{2} = 1 \pm \sqrt{2}$

Since P is required to be positive, we must have $P = 1 + \sqrt{2} \simeq 2.4$ so that the optimal control is :

$$u(t) = -2.4\, x_1(t)$$

which makes the dynamics $\mathring{x}_1 = -1.4\, x_1(t)$

or

$$x_1(t) = e^{-1.4\, t}x_1(0) + \text{constant}$$

Note that this kind of a constant feedback control is very cheap to implement since it merely requires an amplifier to produce 2.4 times the state.

Example 5 : Consider next another simple infinite time regulator, this time for a system of second order, i.e. the system is described by

$$\dot{x}_1 = -x_1 + x_2 + u_1$$

$$\dot{x}_2 = -x_2 + u_2$$

$$J = \frac{1}{2}\int_0^{\infty} (x_1^2 + 2x_2^2 + u_1^2 + u_2^2)\, dt$$

In this case, the Riccati equation can we written using the fact that

$$A = \begin{bmatrix} -1 & 1 \\ 0 & -1 \end{bmatrix} \quad ; \quad B = \begin{bmatrix} 1 & 0 \\ 0 & 1 \end{bmatrix} \quad ;$$

$$Q = \begin{bmatrix} 1 & 0 \\ 0 & 2 \end{bmatrix} \quad ; \quad R = \begin{bmatrix} 1 & 0 \\ 0 & 1 \end{bmatrix}$$

so that it becomes

$$\frac{d}{dt} = \begin{bmatrix} P_{11} & P_{12} \\ P_{12} & P_{22} \end{bmatrix} = - \begin{bmatrix} -1 & 0 \\ 1 & -1 \end{bmatrix} \begin{bmatrix} P_{11} & P_{12} \\ P_{12} & P_{22} \end{bmatrix} - \begin{bmatrix} P_{11} & P_{12} \\ P_{12} & P_{22} \end{bmatrix} \begin{bmatrix} -1 & 1 \\ 0 & -1 \end{bmatrix}$$

$$+ \begin{bmatrix} P_{11} & P_{12} \\ P_{12} & P_{22} \end{bmatrix} \begin{bmatrix} 1 & 0 \\ 0 & 1 \end{bmatrix} \begin{bmatrix} 1 & 0 \\ 0 & 1 \end{bmatrix} \begin{bmatrix} 1 & 0 \\ 0 & 1 \end{bmatrix} \begin{bmatrix} P_{11} & P_{12} \\ P_{12} & P_{22} \end{bmatrix} + \begin{bmatrix} 1 & 0 \\ 0 & 2 \end{bmatrix}$$

Since the final time $t_f \rightarrow \infty$, $\frac{d}{dt}\begin{bmatrix} P_{11} & P_{12} \\ P_{12} & P_{22} \end{bmatrix} \rightarrow 0$

so that we can either solve this by solving the non-linear algebraic equations for P_{11}, P_{12}, P_{22} iteratively or by taking t_f to be sufficiently long (about 3-4 time constants) for P_{11}, P_{12}, P_{22} to reach a constant steady state value if we integrate the Riccati equation backwards from

$$P_{11}(t_f) = 0,\ P_{12}(t_f) = 0\ ;\ P_{22}(t_f) = 0$$

We computed the gain matrix in this way for $t_f = 8$ and this gave the constant gain matrix :

$$G = - \begin{bmatrix} 0.4087 & 0.1274 \\ 0.1274 & 0.7995 \end{bmatrix}$$

Example 6 : As an example of a finite time regulator, consider the simple second order system

$$\dot{x}_1 = x_1 + u_1$$
$$\dot{x}_2 = x_1 + x_2 + u_2$$
$$x_1(0) = 1,\ x_2(0) = 1$$

and it is desired to minimise :

$$J = \int_0^4 (x_1^2 + x_2^2 + u_1^2 + u_2^2)\, dt$$

Here

$$A = \begin{bmatrix} 1 & 0 \\ 1 & 1 \end{bmatrix}, \quad B = \begin{bmatrix} 1 & 0 \\ 0 & 1 \end{bmatrix}, \quad Q = \begin{bmatrix} 1 & 0 \\ 0 & 1 \end{bmatrix}, \quad R = \begin{bmatrix} 1 & 0 \\ 0 & 1 \end{bmatrix}$$

For this problem, the Riccati equation can be integrated backwards from the final condition

$$\begin{bmatrix} P_{11} & P_{12} \\ P_{12} & P_{22} \end{bmatrix} = \begin{bmatrix} 0 & 0 \\ 0 & 0 \end{bmatrix}$$

Fig. 5.2 gives the optimal state and control trajectories obtained by integrating the Riccati equation backwards to the initial time and then using the initial state to compute recursively the controls and states.

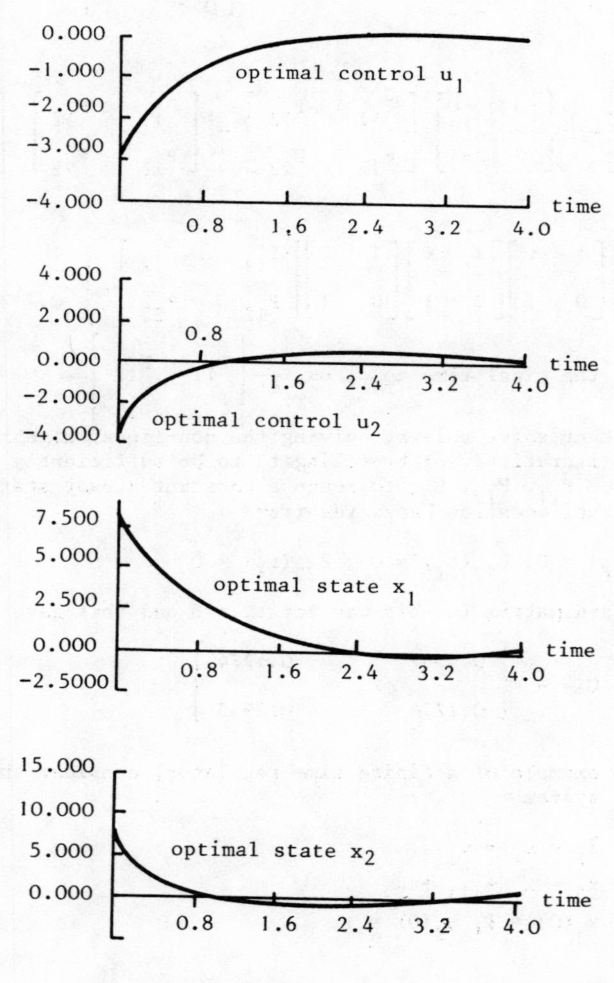

Fig. 5.2

Next let us consider a slightly more general version of this problem, i.e. the linear servomechanism problem.

5.2.4.3 The linear servomechanism problem

In the regulator problem that we examined above, we were essentially trying to find the controls which forced the states towards the origin. However, in many applications we would like our system outputs to follow given "desired" trajectories. In addition, our system may be subject to disturbances. Thus consider the problem of minimising

$$J = \frac{1}{2}\left[\underline{\eta}(t_f)-\underline{y}(t_f)\right]^T S \left[\underline{\eta}(t_f)-\underline{y}(t_f)\right] + \frac{1}{2}\int_{t_o}^{t_f}\left\{\left[\underline{\eta}(t)-\underline{y}(t)\right]^T Q(t) \left[\underline{\eta}(t)-\underline{y}(t)\right] + \underline{u}(t)^T R(t)\,\underline{u}(t)\right\} dt \qquad 5.2.56$$

where $\underline{y}$ are the outputs of the system whilst $\underline{\eta}$ are "desired" trajectories for the outputs that we would like our system to follow. For example $\underline{\eta}(t)$ could be a pre-computed desired trajectory for a rocket and the objective of our control may be to ensure that the actual rocket trajectory $\underline{y}(t)$ follows the desired pre-computed one whilst not utilising excessive control effort. In a chemical plant, $\underline{\eta}$ may be some constant steady state value that the plant normally operates at when it is subjected to some unknown and unmeasurable disturbance which changes the state to some value $\underline{x}_o$ and the object of the control is to bring the plant back to the desired steady state value $\underline{\eta}$ whilst ensuring a balance between control effort utilised and deviation of the outputs from $\underline{\eta}$.

The system dynamics here can be written as :

$$\dot{\underline{x}}(t) = A(t)\underline{x} + B(t)\underline{u}(t) + \underline{\omega}(t), \quad \underline{x}(t_o) = \underline{x}_o \qquad 5.2.57$$

$$\underline{y}(t) = C(t)\,\underline{x}(t) \qquad 5.2.58$$

where $\underline{\omega}(t)$ is an input disturbance which is assumed to be known.

In order to solve this problem, let us write our Hamiltonian as :

$$H = \frac{1}{2}\left\| \underline{\eta}(t) - \underline{y}(t)\right\|^2_{Q(t)} + \frac{1}{2}\left\|\underline{u}(t)\right\|^2_{R(t)} + \underline{\lambda}^T(t)\left[A(t)\underline{x}(t) + B(t)\underline{u}(t) + \underline{\omega}(t)\right]$$

where $\|\cdot\|^2_X = .^T X.$

From the necessary conditions for optimality we obtain the following equations :

$$\frac{\partial H}{\partial \underline{u}} = \underline{0} \text{ gives } \underline{u}(t) = -R^{-1}(t)\,B^T(t)\,\underline{\lambda}(t) \text{ so that}$$

$$\dot{\underline{x}}(t) = A\underline{x}(t) - B(t)\,R^{-1}(t)\,B(t)\,\underline{\lambda}^T(t) + \underline{\omega}(t) \qquad 5.2.59$$

$$\frac{\partial H}{\partial \underline{x}} = -\dot{\underline{\lambda}}(t) = C^T(t)\,Q(t)\left[C(t)\,\underline{x}(t) - \underline{\eta}(t)\right] + A^T(t)\,\underline{\lambda}(t) \qquad 5.2.60$$

with the terminal condition

$$\underline{\lambda}(t_f) = C^T(t_f)\,S\left[C(t_f)\,\underline{x}(t_f) - \underline{\eta}(t_f)\right] \qquad 5.2.61$$

Again, in order to solve this linear two point boundary value problem, let us assume a solution of the form

$$\underline{\lambda}(t) = P(t)\,\underline{x}(t) - \underline{\xi}(t)$$

Then $\dot{\underline{\lambda}}(t) = \mathring{P}(t)\,\underline{x}(t) + P(t)\,\mathring{\underline{x}}(t) - \dot{\underline{\xi}}(t)$

Substituting for $\mathring{P}$ and $\mathring{\underline{x}}$ in this equation, we obtain :

$$\left[\mathring{P}(t) + P(t)\,A(t) + A^T(t)\,P(t) - P(t)\,B(t)\,R(t)^{-1}\,B(t)^T\,P(t) - C(t)^T\,Q(t)\,C(t)\right]\underline{x}(t$$

$$+\left[\dot{\xi} + \left\{A(t) - B(t)\,R^{-1}(t)\,B^T(t)\,P(t)\right\}^T \underline{\xi} - P(t)\,\underline{\omega}(t) + C^T(t)\,Q(t)\,\underline{\eta}(t)\right] = 0$$

Since this equation is valid for arbitrary non zero $\underline{x}(t)$, we obtain the single point boundary value problem equations :

$$\mathring{P}(t) = -P(t)A(t) - A^T(t)P(t) + P(t)B(t)R^{-1}(t)B^T(t)P(t) - C^T(t)Q(t)C(t) \qquad 5.2.62$$

with $\qquad P(t_f) = C^T(t_f)\,S\,C(t_f) \qquad 5.2.63$

and

$$\dot{\underline{\xi}}(t) = -\left[A(t) - B(t)R^{-1}(t)B^T(t)P(t)\right]^T \underline{\xi} + P(t)\underline{\omega}(t) - C^T(t)Q(t)\underline{\eta}(t) \qquad 5.2.64$$

with $\quad \underline{\xi}(t_f) = C^T(t_f)\,S\,\underline{\eta}(t_f) \qquad 5.2.65$

and the control is given by :

$$\underline{u}(t) = -R^{-1}(t)\,B^T(t)\left[P(t)\,\underline{x}(t) - \underline{\xi}(t)\right] \qquad 5.2.66$$

We see that the linear servomechanism problem consists of two parts, one of which is a regulator part and the other is a pre-filter to determine the optimal driving function $R^{-1}(t)\,B^T(t)\,\underline{\xi}(t)$ from the desired value $\underline{\eta}(t)$, the disturbance $\underline{\omega}(t)$ and the system matrices.

Computation of the optimal control

The optimal control computation requires the off-line solution of the Riccati equation 5.2.62 backwards in time (from t_f to t_o) using the given final conditions 5.2.63 and the storage of P(t) ($t_o \leqslant t \leqslant t_f$), and if $\underline{\omega}(t)$, $\underline{\eta}(t)$ are also known, the off-line solution of the linear vector differential equation 5.2.64 backwards in time from t_f to t_o and the storage of this vector trajectory When the control is implemented on-line, it is necessary merely to multiply the pre-calculated time varying gain $G_1 = -R^{-1}(t)\,B^T(t)\,P(t)$ with the state $\underline{x}(t)$ and t_o add the pre-filter term $+R^{-1}(t)\,B^T(t)\,\underline{\xi}(t)$ to yield the optimal control from equation 5.2.66.

Thus, to summarise these results :

$$\text{Min } J = \frac{1}{2}\left\|\underline{\eta}(t_f) - \underline{y}(t_f)\right\|_S^2 + \frac{1}{2}\int_{t_o}^{t_f}\left(\left\|\underline{\eta}(t) - \underline{y}(t)\right\|_{Q(t)}^2 + \left\|\underline{u}(t)\right\|_{R(t)}^2\right)dt \qquad 5.2.67$$

subject to

$$\begin{aligned} \mathring{\underline{x}}(t) &= A(t)\ \underline{x}(t) + B(t)\ \underline{u}(t) + \underline{\omega}(t) \\ \underline{y}(t) &= C(t)\ \underline{x}(t) \end{aligned} \qquad 5.2.68$$

yields the optimal control

$$\underline{u}(t) = -R(t)^{-1}\ B(t)^T \left[P(t)\ \underline{x}(t) - \underline{\xi}(t) \right] \qquad 5.2.69$$

where

$$\mathring{P}(t) = -P(t)A(t) - A^T(t)P(t) + P(t)B(t)R^{-1}(t)B^T(t)P(t) - C^T(t)Q(t)C(t) \qquad 5.2.70$$

with

$$P(t_f) = C^T(t_f)\ S\ C(t_f) \qquad 5.2.71$$

and

$$\mathring{\underline{\xi}}(t) = -\left[A(t)-B(t)R^{-1}(t)B^T(t)P(t)\right]^T \underline{\xi}(t) + P(t)\underline{\omega}(t) - C^T(t)Q(t)\underline{\eta}(t) \qquad 5.2.72$$

with

$$\underline{\xi}(t_f) = C^T(t_f)\ S\ \underline{\eta}(t_f) \qquad 5.2.73$$

Example 7 : As a first example of the application of the servomechanism solution, let us consider a relatively practical example of the control of temperature and pressure in a boiler [6].

The optimisation problem can be written as :

$$\text{Min } J = \int_0^{t_f} \left[\underline{e}^T Q\ \underline{e} + \underline{u}^T R\ \underline{u} \right] dt$$

subject to the constraints

$$\mathring{\underline{y}} = A\ \underline{y} + B\ \underline{u} + \underline{C}$$

where

$$\underline{y} = \begin{bmatrix} P_b \\ q_c \\ t_s \\ x_3 \\ q_d \end{bmatrix}, \quad A = \begin{bmatrix} 0 & 396\text{x}10^{-6} & 0 & 0 & 325\text{x}10^{-7} \\ 0 & 0 & 0 & 0 & 0 \\ 0 & 1.25/154 & -\frac{1}{154} & -\frac{1}{154} & 0 \\ 0 & 0 & 0 & -\frac{1}{83} & 1.375/83 \\ 0 & 0 & & & \end{bmatrix}$$

$$\underline{u} = \begin{bmatrix} u_1 \\ u_2 \end{bmatrix}, \quad B = \begin{bmatrix} 0 & 0 \\ 1 & 0 \\ 0 & 0 \\ 0 & 0 \\ 0 & 1 \end{bmatrix}, \quad \underline{C} = \begin{bmatrix} -325\text{x}10^{-7} q_s \\ 0 \\ -649.35\text{x}10^{-6} q_s + 129.87\text{x}10^{-4} \\ -662.65\text{x}10^{-3} \\ 0 \end{bmatrix}$$

$\underline{e} = \underline{y} - \underline{a}$ where

$$\underline{a} = \begin{bmatrix} -1.225(\frac{q_s^2}{1000}) \\ 0 \\ 0 \\ 0 \\ 0 \end{bmatrix}$$

$$Q = \begin{bmatrix} 16 & 0 & 0 & 0 & 0 \\ 0 & 0 & 0 & 0 & 0 \\ 0 & 0 & 1 & 0 & 0 \\ 0 & 0 & 0 & 0 & 0 \\ 0 & 0 & 0 & 0 & 0 \end{bmatrix} ; \; R = \begin{bmatrix} 30 & 0 \\ 0 & 30 \end{bmatrix}$$

where

P_b is the deviation from a pressure of 20 bars in the boiler

t_s is the temperature deviation of the output from 300°C

x_3 is an intermediate variable

q_c is the fuel flow rate in Kg/h

q_d is the flow rate of the superheated steam

The two controls are the variations around the flow rates q_c (the fuel flow) and q_d (the flow rate of the superheated steam).

q_s is given whilst the final time t_f = 600 seconds is long enough for the system to reach a steady state.

For this problem, the optimal controls and states are shown in Figs. 5.3, 5.4, 5.5.

Example 8 : River pollution control

Since this is an example that we will treat subsequently also using hierarchical techniques, we will give a detailed description of the problem here.

In recent years there has been much interest in regulating the levels of pollution in rivers. A good measure of the quality of a stream is given by two main factors, i.e. (a) the instream biochemical oxygen demand (B.O.D.) * and (b) the Dissolved Oxygen (D.O.) in the stream. If the D.O. falls below certain levels or the B.O.D. rises above certain levels, fish die. In industrial societies, rivers are used as dumps for sewage except that the the sewage is treated at sewage stations prior to discharge into the stream. Sewage works in general operate a fixed level of treatment which could be determined by the level of B.O.D. in the sewage which can be safely absorbed in the stream. However, the ecological balance

* The Biochemical Oxygen Demand (B.O.D.) is a measure of the rate of absorption of oxygen by decomposing organic matter.

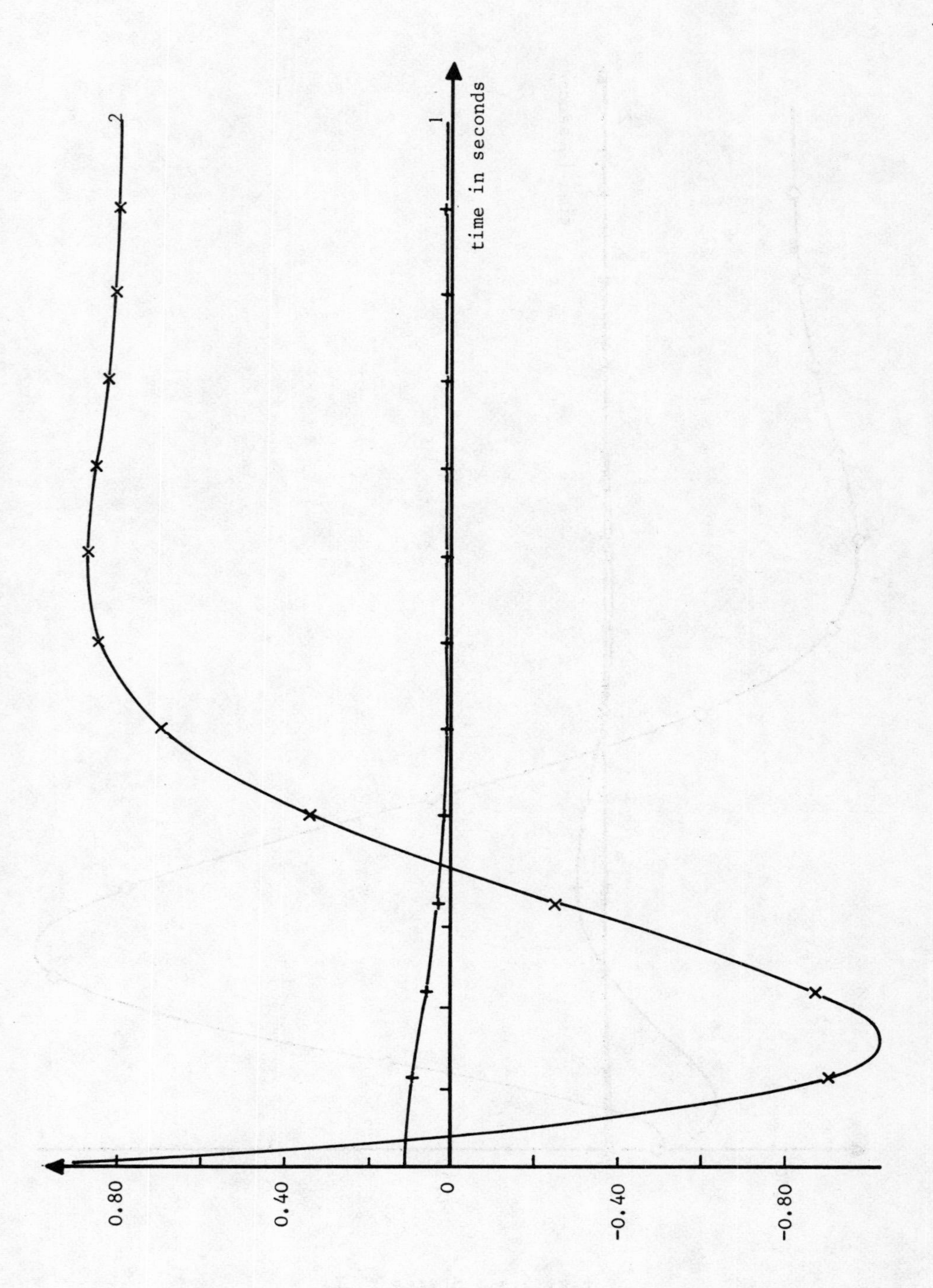

Fig. 5.3 : States Y_1 and Y_5

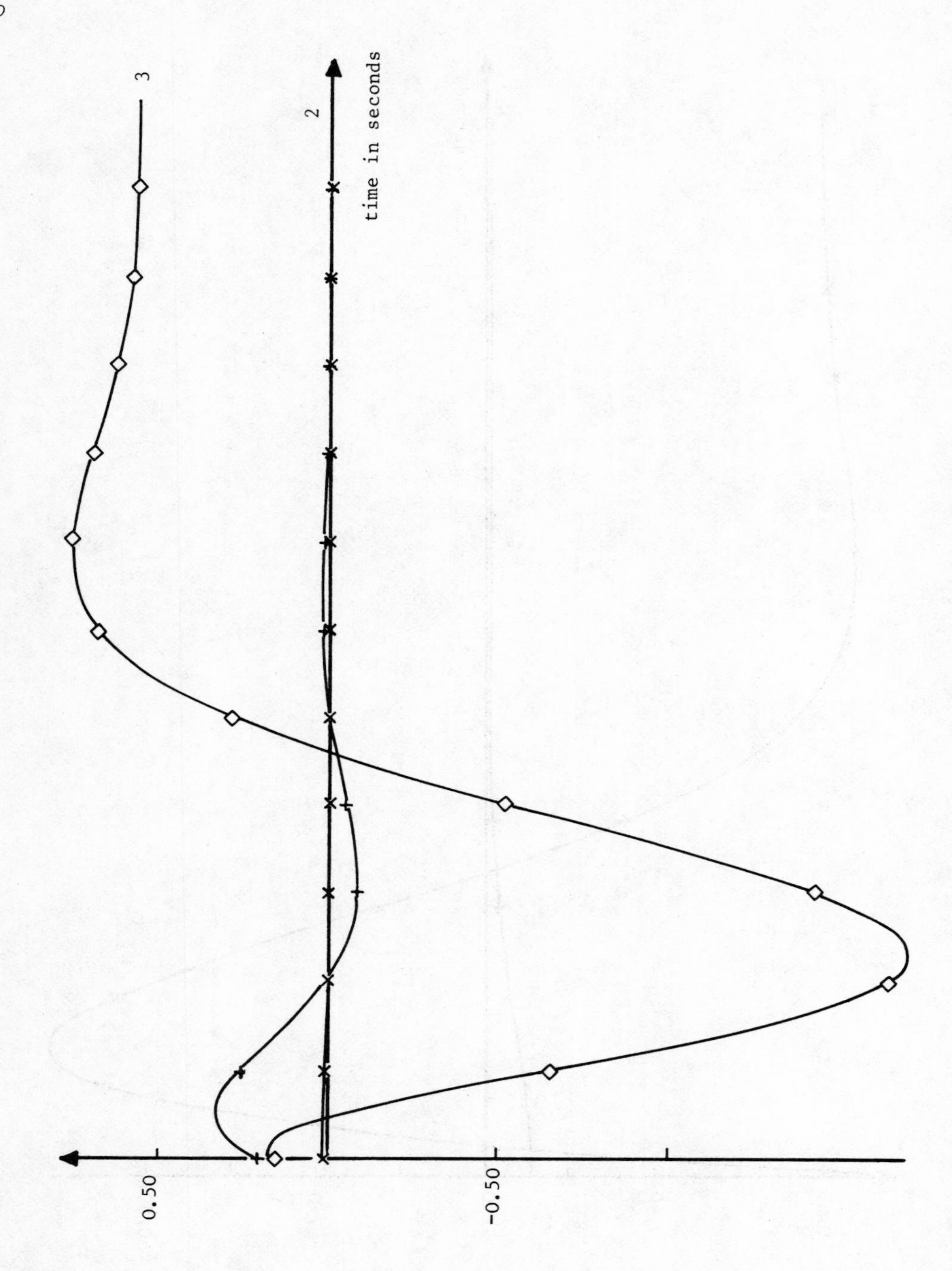

Fig. 5.4 : States Y_2 Y_4

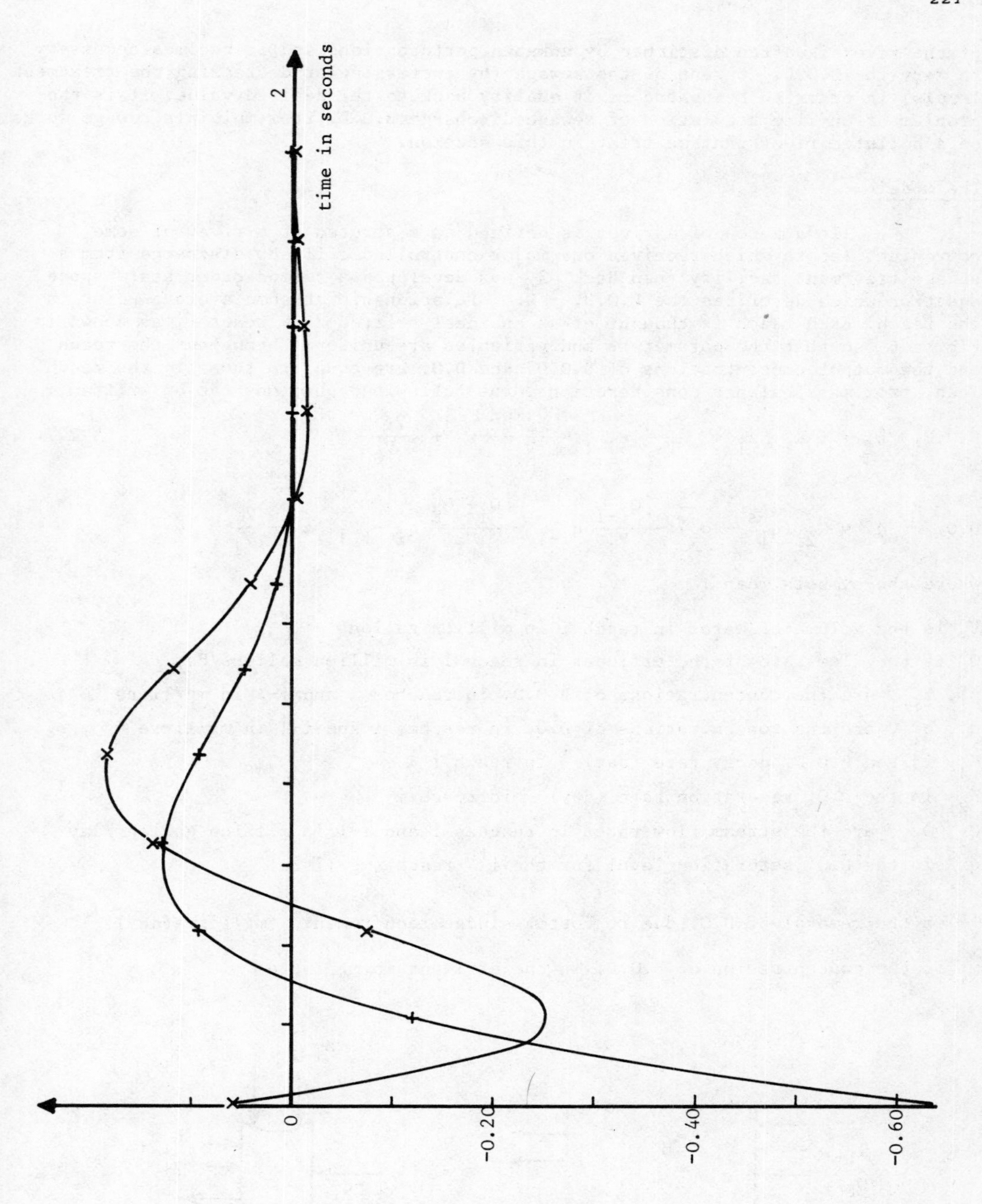

Fig. 5.5 : Optimal controls u_1, u_2

of the river is often disturbed by unknown perturbations and it becomes necessary to vary the B.O.D. content of the sewage (by increasing or decreasing the treatment levels) in order to bring the river quality back to the desired value. It is the problem of on-line regulation of sewage discharge B.O.D. from multiple sewage works on a polluted river that we treat in this section.

The model

If a reach of a river is defined as a stretch of a river of some convenient length which receives one major controlled effluent discharge from a sewage treatment facility then Beck [3] has developed a second order state space equation which describes the B.O.D. - D.O. relationship at some average point in the reach. Each reach is thought of as an ideal stirred tank reactor, as shown in Fig. 5.6 so that the parameters and variables are uniform throughout the reach and the output concentrations of B.O.D. and D.O. are equal to those in the reach. Then, from mass balance considerations, the following equations can be written :

$$\text{B.O.D.} \quad \mathring{z}_i = -K_{1i}\, z_i + \frac{Q_{i-1}}{V_i} z_{i-1} - \frac{Q_i + Q_E}{V_i} z_i + \frac{\eta_i Q_E}{V_i} \qquad 5.2.74$$

$$\text{D.O.} \quad \mathring{q}_i = K_{2i}(q_i^s - q_i) - \frac{Q_{i-1}}{V_i} q_{i-1} - \frac{Q_i + Q_E}{V_i} q_i - K_{1i}\, z_i - \frac{\eta_i}{V_i}$$

where the symbols mean :

V_i is the volume of water in reach i in million gallons

Q_E is the flow rate of the effluent in reach i in million gallons/day

z_i, z_{i-1} are the concentrations of B.O.D. in reaches i and i-1 in mg/litre

q_i, q_{i-1} are the concentrations of D.O. in reaches i and i-1 in mg/litre

K_{1i} is the B.O.D. decay rate $(\text{day})^{-1}$ in reach i

K_{2i} is the D.O. reaeration rate $(\text{day})^{-1}$ in reach i

Q_i, Q_{i-1} are the stream flow rates in reaches i and i-1 in million gallons/day

q_i^s is the D.O. saturation level for the ith reach (mg/litre)

$\frac{\eta_i}{V_i}$ is the removal of D.O. due to bottom sludge requirements (mg/litre$(\text{day})^{-1}$)

m_i is the concentration of B.O.D. in the effluent in mg/litre

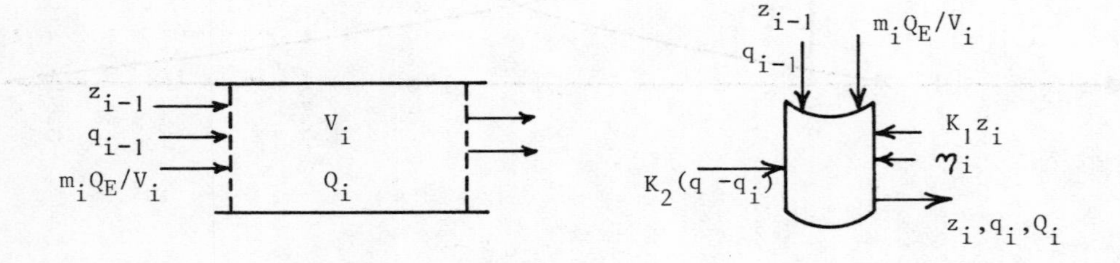

Fig. 5.6 : An ideal stirred tank reactor model for a reach of a river

Beck [3] found that for a section of the river Cam near Cambridge in England, the following values for the coefficients in equation 5.2.74 were appropriate :

$$K_{1i} = 0.32 \text{ day}^{-1}, \quad K_{2i} = 0.2 \text{ day}^{-1}$$

$$\frac{\eta_i}{V_i} = 0.1 \text{ mg/litre day}^{-1}, \quad q_i^s = 10 \text{ mg/litre}$$

$$\frac{Q_E}{V} = 0.1 \qquad \frac{Q}{V} = 0.9$$

Thus, for the ith reach, equation 5.2.74 could be rewritten as :

$$\frac{d}{dt}\begin{bmatrix} z_i \\ q_i \end{bmatrix} = \begin{bmatrix} -1.32 & 0 \\ -0.32 & -1.2 \end{bmatrix}\begin{bmatrix} z_i \\ q_i \end{bmatrix} + \begin{bmatrix} 0.1 \\ 0 \end{bmatrix} m_i + \begin{bmatrix} 0.9z_{i-1} \\ 0.9q_{i-1}+1.9 \end{bmatrix} \qquad 5.2.75$$

<u>Multiple Effluent inputs</u>

Beck's model [3] is based on a single reach of the river Cam which has only one effluent input. The interesting problem, however, is the one with multiple inputs. Since no additional data is available, we assume that there exists a system of many reaches each one of which has the properties of Beck's model.

The most realistic description of such a stream with multiple polluters was given by Tamura [4] who assumed that each reach was separated from the next by a distributed delay. This model is able to account for the dispersion of pollutants which actually occurs in rivers. In this model, for j = 1, 2, ... s, a fraction a_j of B.O.D. and D.O. in the $i-1^{th}$ reach at time $(t-\theta_j)$ arrives in the ith reach at time t, i.e. the transport delays are distributed in time between θ_1 and θ_s. Thus z_{i-1}, q_{i-1} are given by :

$$z_{i-1}(t) = \sum_{j=1}^{s} a_j\, z_{i-1}(t-\theta_j) \qquad 5.2.76$$

$$q_{i-1}(t) = \sum_{j=1}^{s} a_j\, q_{i-1}(t-\theta_j) \qquad 5.2.77$$

$$\sum_{j=1}^{s} a_j = 1 \text{ ; mean of } \theta_j = \theta_o \text{ ; } \theta_1 < \theta_2 < \dots \theta_s$$

Thus it is possible to write the state equations for a 3 reach system with distributed delays as :

$$\mathring{z}_1 = -1.32\, s_1 + 0.1\, m_1 + 0.9\, z_o + 5.35 \qquad 5.2.78$$

$$\mathring{q}_1 = -0.32\, z_1 + 1.2\, q_1 + 0.9\, q_o + 1.9 \qquad 5.2.79$$

$$\mathring{z}_2 = 0.9 \sum_{j=1}^{s} a_j z_1(t-\tau_j) - 1.32\, z_2 + 0.1\, m_2 + 4.19 \qquad 5.2.80$$

$$\mathring{q}_2 = 0.9 \sum_{j=1}^{s} a_j q_1(t-\tau_j) - 1.32\, z_2 - 1.29\, q_2 + 1.9 \qquad 5.2.81$$

$$\dot{z}_3 = 0.9 \sum_{j=1}^{s} a_j z_2(t-\tau_j) - 1.32\, z_3 + 0.1\, m_3 + 4.19 \qquad 5.2.82$$

$$\dot{q}_3 = 0.9 \sum_{j=1}^{s} a_j q_2(t-\tau_j) - 0.32\, z_3 - 1.29\, q_3 + 1.9 \qquad 5.2.83$$

Tamura [4] gives the following values for s, τ, a, etc. for each of the distributed delays :

$$s = 3,\ \tau_1 = 0,\ \tau_2 = \frac{1}{2}\ \text{day},\ \tau_3 = 1\ \text{day},\ z_o = 0,\ \ q_o = 10$$

$$a_1 = 0.15,\ a_2 = 0.7,\ a_3 = 0.15$$

The system as described by equations 5.2.78 - 5.2.83 is nominally of infinite dimension in the state space. It is possible to obtain a good finite dimensional approximation by expanding the delayed terms in a Taylor series and taking the first two terms. For example, consider equation 5.2.80 which can be rewritten as :

$$\dot{z}_2 = 0.9\,(0.15\, z_1(t) + 0.7\, z_1(t-0.5) + 0.15\, z_1(t-1))$$
$$- 1.32\, z_2 + 0.1\, m_2 + 4.19$$

Here, we have 2 delays so it is necessary to introduce four additional states. Let these be given by z_4, z_5, z_6, z_7. Let $z_4(t) = z_1(t-0.5)$; then $z_4(s) = z_1(s)e^{-0.5s}$. Now

$$z_1(s)\, e^{-0.5s} = z_1(s)\left[1 + 0.5s + \frac{0.25}{2} s^2 + \ldots\right]^{-1}$$

Taking only the first three terms :

$$z_1(t) = z_4(t) + 0.5\, \dot{z}_4(t) + 0.125\, \ddot{z}_4(t)$$

Let $\dot{z}_4 = z_5$ 5.2.84

then $\dot{z}_5 = 8\, z_1 - 8\, z_4 - 4\, z_5$ 5.2.85

Similarly for the other delay, we let :

$$z_1(t-1) = z_6 \qquad 5.2.86$$

then $\dot{z}_6 = z_7$ 5.2.87

$$\dot{z}_7 = 2\, z_1 - 2\, z_6 - 2\, z_7 \qquad 5.2.88$$

Thus we can rewrite equation 5.2.80 as

$$\dot{z}_2 = 0.135\, z_1 + 0.63\, z_4 + 0.135\, z_6 - 1.32\, z_2 + 0.1\, m_2 + 4.19 \qquad 5.2.89$$

Similarly, we can introduce four additional variables each for the delays in equations 5.2.78 - 5.2.83. This makes the overall system of order 22.

The servomechanism solution

Having described the dynamics of this 22nd order system, we are now in a position to examine the application of the servomechanism solution to this problem.

Let us assume that the problem is to minimise a cost function of the form

$$J = \int_0^8 \Big[2(z_1-z_1^*)^2 + (q_1-q_1^*)^2 + 2(z_2-z_2^*)^2 + (q_2-q_2^*)^2 + 2(z_3-z_3^*)^2 + (q_3-q_3^*)^2 + (m_1-m_1^*)^2 + (m_2-m_2^*) + (m_3-m_3^*)^2 \Big] dt \qquad 5.2.90$$

Minimising such a cost function ensures that if before time zero an unknown disturbance takes the system state to some value $\underline{x}_o$ where $\underline{x}_o$ is a 22nd order vector, then it will be possible to bring the system back to the steady state values of z_1^*, q_1^*, z_2^*, q_2^*, z_3^*, q_3^* and the controls back to the steady state controls m_1^*, m_2^*, m_3^* whilst ensuring that there are no unacceptably large deviations from the steady state B.O.D., D.O. or controls. Note that both positive and negative deviations are penalised. In the case of the control, this is meaningful since the treatment process is biological and it costs money to change the control in either direction. For the B.O.D. and D.O. such a penalty is not meaningful since we really want to ensure that B.O.D. does not exceed some value and D.O. is always greater than some value. It is nevertheless possible to interpret the quadratic form as follows : let us assume that there is some band $(z_i^* \pm z_i^{**})$ and $(q_i^* \pm q_i^{**})$ (i = 1, 2, 3) within which we should like our B.O.D.s and D.O.s to lie. Then the quadratic forms will ensure that this is so, provided adequate weights are used in the cost function. We have found after extensive experimentation that the weights given in equation 5.2.90 are adequate.

Simulation results

The problem of minimising J in equation 5.2.90 subject to the system dynamics given in equations 5.2.78 - 5.2.83 was solved using the IBM 370/165 digital computer. The initial conditions for B.O.D.s and D.O.s were taken to be :

$$z_1(0) = 10 \text{ mg/l}, \; z_2(0) = 5.94 \text{ mg/l}, \; z_3(0) = 5.237 \text{ mg/l},$$

$$q_1(0) = 7 \text{ mg/l}, \; q_2(0) = 6 \text{ mg/l}, \; q_3(0) = 4.69 \text{ mg/l}.$$

Such an initial condition implies that before t = 0, some disturbance affected reach one to make it very polluted whilst reaches 2 and 3 were in their desired steady state.

Figs. 5.7 to 5.9 show the resulting state and control trajectories. These figures show how the effect of the pollution load is gradually attenuated down the river. Note that the results are quite realistic with the distributed delay since the reaches downstream are only gradually affected as the pollutants are dispersed.

5.2.5 Singular solutions using the maximum principle

Basically, if we can solve dynamic optimisation problems, it is because the necessary conditions for optimality give us a relationship between the control $\underline{u}$ and the states and costates $\underline{x}$ and $\underline{\lambda}$. In certain special cases, such a relationship does not exist, i.e. the necessary conditions 5.2.16 do not provide any information between $\underline{u}^*(t)$, $\underline{x}^*(t)$ and $\underline{\lambda}^*(t)$. In this case, the problem is said to be singular. An important case where singularities arise is where the control $\underline{u}$ enters linearly into the Hamiltonian H. Consider for example the problem of minimising

$$J = \frac{1}{2} \left\| \underline{x}(t_f) \right\|_S^2 + \frac{1}{2} \int_{t_o}^{t_f} \left(\left\| \underline{x}(t) \right\|_Q^2 \right) dt \qquad 5.2.91$$

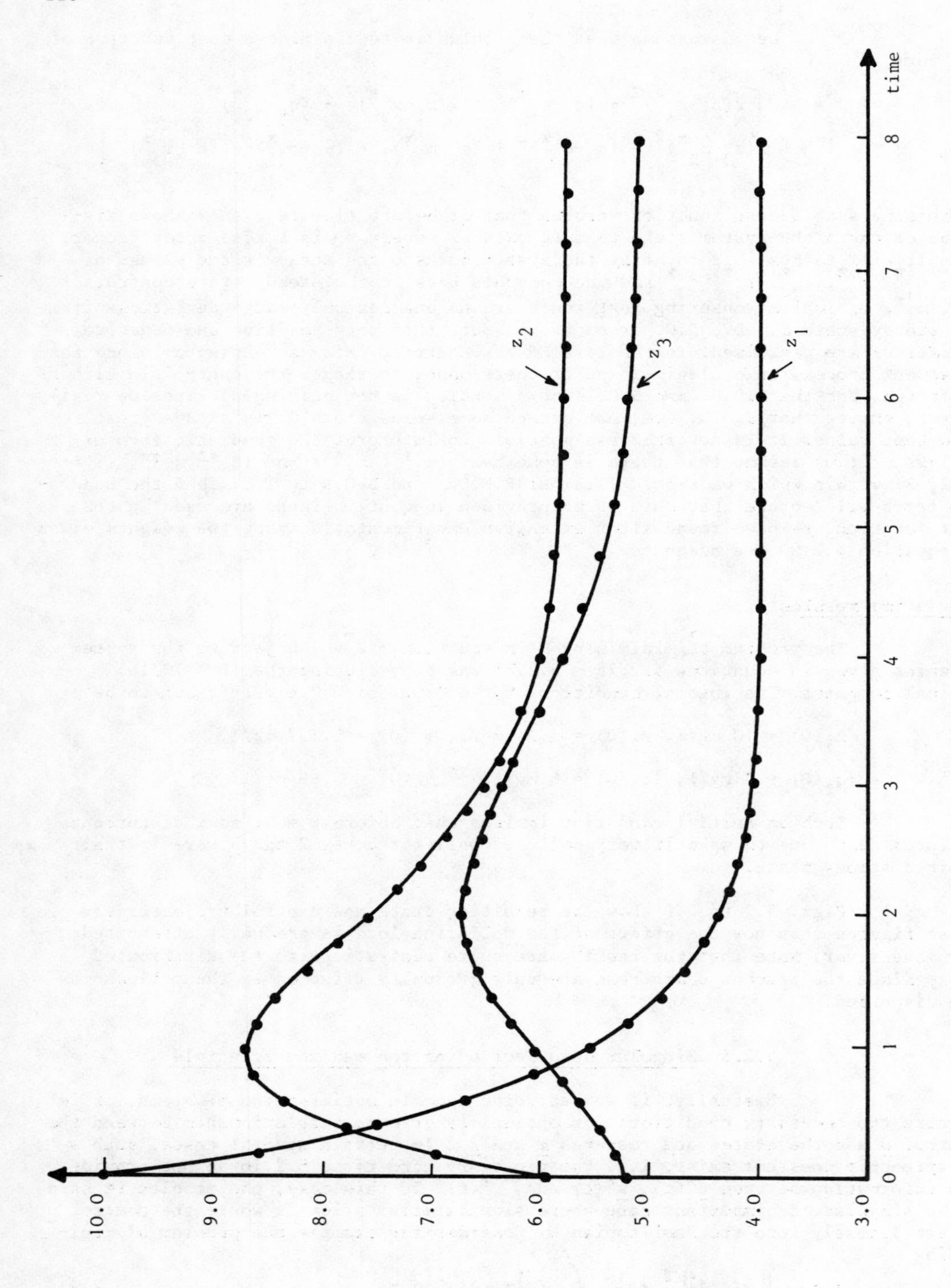

Fig. 5.7

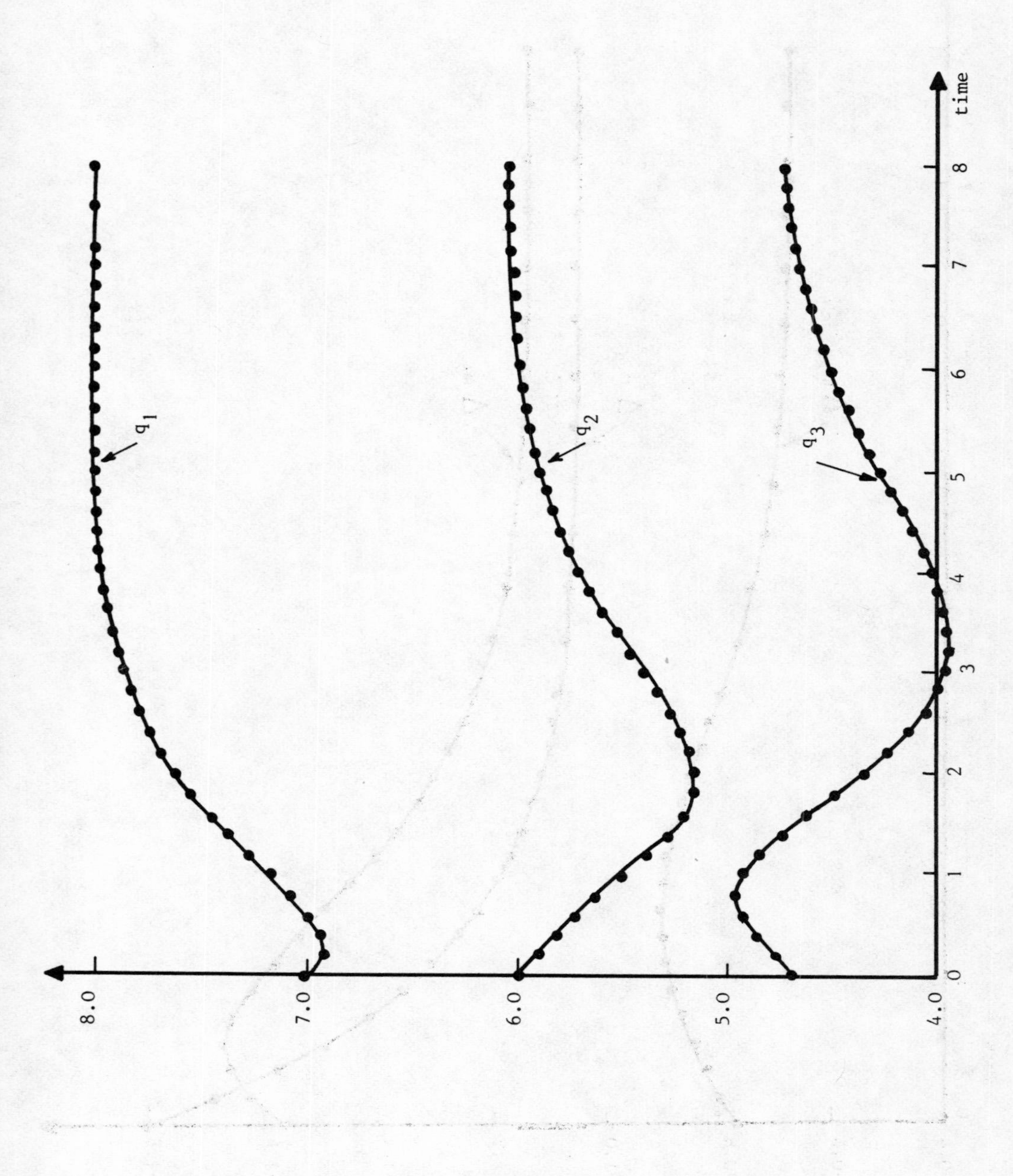

Fig. 5.8

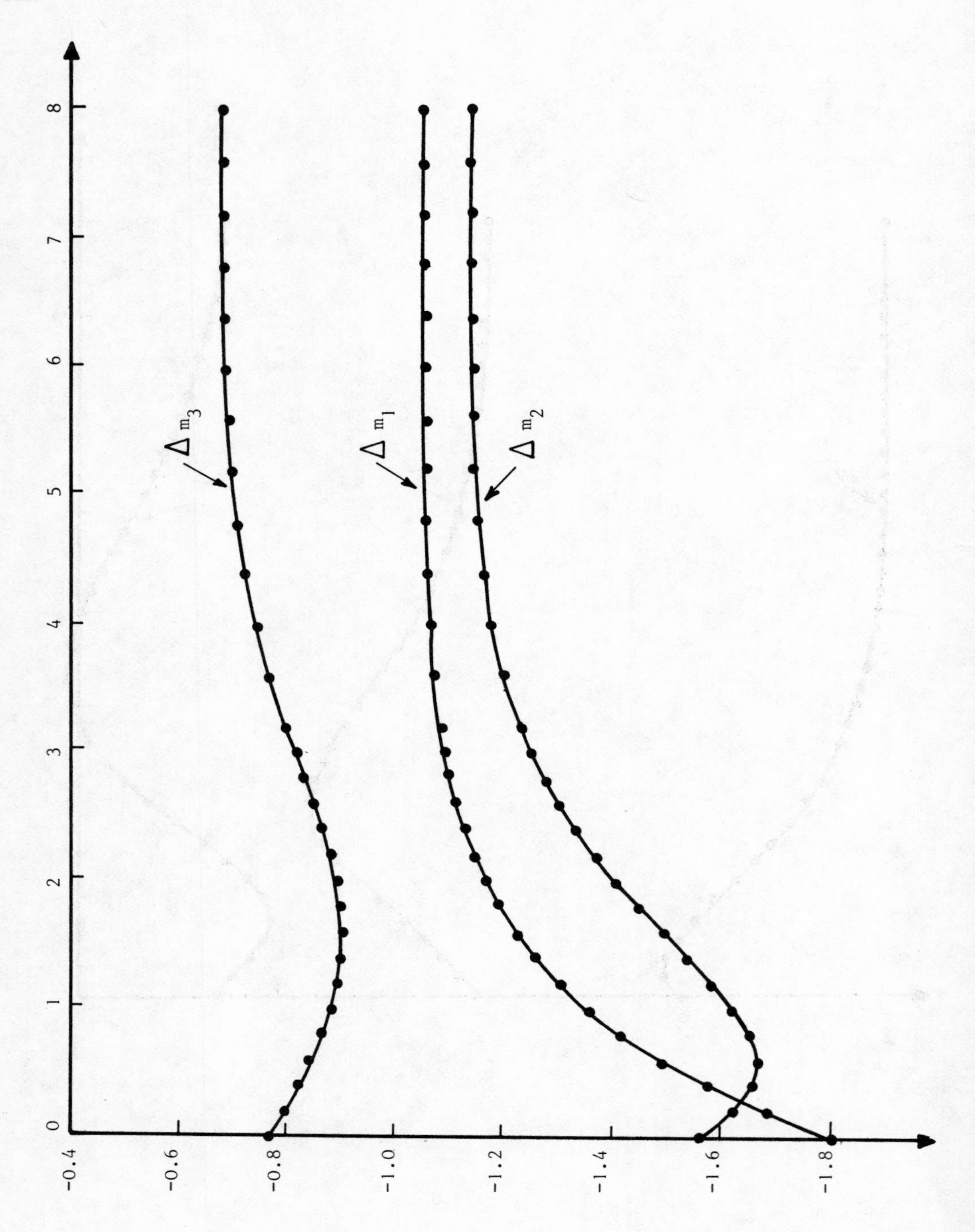

Fig. 5.9

subject to $\dot{\underline{x}}(t) = A(t)\,\underline{x}(t) + B(t)\,\underline{u}(t)$ 5.2.92

where t_f is fixed and the control vector $\underline{u}$ is bounded. The Hamiltonian in this case is

$$H = \frac{1}{2}\left\| \underline{x}(t) \right\|_Q^2 + \underline{\lambda}^T(t)\,A(t)\,\underline{x}(t) + \underline{\lambda}^T(t)\,B(t)\,\underline{u}(t) \qquad 5.2.93$$

For non-zero $\underline{\lambda}^T(t)\,B(t)$, $\underline{u}(t)$ here operates on the boundary of the admissible input set.

Now if we set $\frac{\partial H}{\partial \underline{u}} = \underline{0}$, we obtain $B^T(t)\,\underline{\lambda}(t) = \underline{0}$

as a solution which minimises the Hamiltonian. However, in this case we can't find the control $\underline{u}(t)$. If the Hamiltonian :

$$H = \frac{1}{2}\left\| \underline{x}(t) \right\|_Q^2 + \underline{\lambda}^T(t)\,A\underline{x}(t) + \underline{\lambda}^T(t)\,B(t)\,\underline{u}(t)$$

is not a non-linear function for $\underline{u}(t)$ for some finite range of t, we have a singular interval.

As an example, consider a problem treated previously by Johnson [8].

The system in this case is :

$$\dot{x}_1 = x_2 + u \qquad x_1(0) = x_{10}$$

$$\dot{x}_2 = -u \qquad x_2(0) = x_{20}$$

where we desire to drive the system to the origin at a fixed time, t_f, $(x_1(t_f) = x_2(t_f) = 0)$ with a bounded control $u(t) \leqslant K$

whilst minimising

$$J = \frac{1}{2}\int_{t_o}^{t_f} x_1^2\, dt$$

The Hamiltonian in this case is :

$$H = \lambda_1(t)\left[x_2(t) + u(t)\right] - \lambda_2(t)\,u(t) + \frac{1}{2}x_1^2(t)$$

Singular arcs for this system have the property that

$$\frac{\partial H}{\partial u} = 0 = \lambda_1(t) - \lambda_2(t)$$

for a finite time interval with the state-costate equations :

$$\dot{x}_1(t) = x_2(t) + u(t) \qquad x_1(0) = x_{10}$$

$$\dot{x}_2(t) = -u(t) \qquad x_2(0) = x_{20}$$

$$\dot{\lambda}_1(t) = -\frac{\partial H}{\partial x_1} = -x_1(t), \quad x_1(t_f) = 0$$

$$\dot{\lambda}_2(t) = -\frac{\partial H}{\partial x_2} = -\lambda_1(t), \quad x_2(t_f) = 0$$

For a singular solution, $\lambda_1(t) = \lambda_2(t)$ which from the state-costate equations yields

$$x_1(t) = \lambda_1(t)$$

Also, on a singular arc :

$$\frac{d^2}{dt^2}\frac{\partial H}{\partial u} = 0 = -\dot{x}_1 + \lambda_1 = -x_2(t) - u(t) - x_1(t)$$

Again, since the Hamiltonian depends only implicitly on time, it must be constant about an optimal trajectory and hence on a singular arc so that

$$H = x_1 x_2 + \frac{1}{2}x_1^2 = \text{constant}$$

so that, on the singular arc, the closed loop control is :

$$u(t) = -x_1(t) - x_2(t)$$

and the singular arcs are described by :

$$x_1 x_2 + \frac{1}{2}x_1^2 = \text{constant}$$

Remarks

The important point to note about solving singular problems is that each problem needs to be solved anew since there are no generally applicable methods for solving them. Thus in the case of large scale systems, as we will see in the next chapter, we try to avoid singular solutions if we possibly can.

Next, let us consider the discrete time problem.

5.3 THE DISCRETE MAXIMUM PRINCIPLE

5.3.1 Necessary conditions of optimality

Many systems particularly of the socio economic class are most naturally represented as discrete time processes. Econometric models for example which represent macro models of the economy receive data at discrete instants of time and are best modelled in discrete time. For such systems, it is possible to use a discrete version of the maximum principle that we derived in the beginning of this chapter. Here, we will present very briefly the equations of the discrete maximum principle.

Suppose that we have a discrete time non-linear dynamical system with state vector, $\underline{x}(k)$ and control vector, $\underline{u}(k)$, at the instant k. The state of the system at the instant k+1 is related to the state at the instant k by

$$\underline{x}_{k+1} = \underline{f}(\underline{x}_k, \underline{u}_k, k) \qquad 5.3.1$$

where f is a continuous doubly differentiable function, and we wish to minimise the cost function

$$J = \Big[g(\underline{x}_k, k) \Big]_{k=k_o}^{k=k_f} + \sum_{k=k_o}^{k_f-1} h(\underline{x}_k, \underline{u}_k, k) \qquad 5.3.2$$

subject to the constraints 5.3.1.

By defining the Lagrange Multipliers $\underline{\lambda}_k$, we can write the unconstrained cost function (Lagrange function) as :

$$J' = \Big[g(\underline{x}_k, k) \Big]_{k=k_o}^{k_f} + \sum_{k=k_o}^{k_f-1} h(\underline{x}_k, \underline{u}_k, k) - \underline{\lambda}_{k+1}^T \Big[-\underline{f}(\underline{x}_k, \underline{u}_k, k) + \underline{x}_{k+1} \Big] \qquad 5.3.3$$

Let us simplify J' by defining, as in the continuous time case treated previously, the Hamiltonian as :

$$H(\underline{x}_k, \underline{u}_k, \underline{\lambda}_{k+1}, k) = H_k = h(\underline{x}_k, \underline{u}_k, k) + \underline{\lambda}_{k+1}^T \underline{f}(\underline{x}_k, \underline{u}_k, k) \qquad 5.3.4$$

so that the unconstrained cost function becomes

$$J' = \Big[g(\underline{x}_k, k) \Big]_{k=k_o}^{k_f} + \sum_{k=k_o}^{k_f-1} \Big[H_k - \underline{\lambda}_{k+1}^T \underline{x}_{k+1} \Big] \qquad 5.3.5$$

In order to minimise this cost function w.r.t. $\underline{x}_k$ and $\underline{u}_k$ using variational techniques, let us introduce the independent perturbations

$$\underline{x}_k = \hat{\underline{x}}_k + \epsilon \underline{\eta}_k \qquad 5.3.6$$

$$\underline{x}_{k+1} = \hat{\underline{x}}_{k+1} + \epsilon \underline{\eta}_{k+1} \qquad 5.3.7$$

$$\underline{u}_k = \hat{\underline{u}}_k + \epsilon \underline{v}_k \qquad 5.3.8$$

It should be noted that the perturbations are independent at different instants of time so that $\underline{\eta}_k$, $\underline{\eta}_{k+1}$, $\underline{v}_k$ are all mutually independent. Introducing the perturbations into equation 5.3.5, we have :

$$J' = g(\hat{\underline{x}}_{k_f} + \epsilon \underline{\eta}_{k_f}, k_f) - g(\hat{\underline{x}}_{k_o} + \epsilon \underline{\eta}_{k_o}, k_o) + \sum_{k=k_o}^{k_f-1} \Big[H(\hat{\underline{x}}_k + \epsilon \underline{\eta}_k, \hat{\underline{u}}_k + \underline{v}_k, \underline{\lambda}_{k+1}, k) - \underline{\lambda}_{k+1}^T \big[\hat{\underline{x}}_{k+1} + \epsilon \underline{\eta}_{k+1} \big] \Big] \qquad 5.3.9$$

Now, the basic stationarity condition is that $\frac{\partial J'}{\partial \epsilon} = 0$ whilst for a local minimum we require that $\frac{\partial^2 J'}{\partial \epsilon^2} \geq 0$ for $\epsilon = 0$, independent of the variations.

Setting $\frac{\partial J'}{\partial \epsilon} = 0$, we obtain :

$$\left[\frac{\partial g_{kf}}{\partial \hat{\underline{x}}_{kf}}\right]^T \underline{\eta}_{kf} - \left[\frac{\partial g_{ko}}{\partial \hat{\underline{x}}_{ko}}\right]^T \underline{\eta}_{ko} + \sum_{k=k_o}^{k_f-1} \left(\frac{\partial H_k}{\partial \hat{\underline{x}}_k}\right)^T \underline{\eta}_k - \sum_{k=k_o}^{k_f-1} \underline{\lambda}_{k+1}^T \underline{\eta}_{k+1} + \sum_{k=k_o}^{k_f-1} \left(\frac{\partial H_k}{\partial \hat{\underline{u}}_k}\right)^T \underline{v}_k = 0 \qquad 5.3.10$$

Note that :

$$- \sum_{k=k_o}^{k_f-1} \underline{\lambda}^T_{k+1} \underline{\eta}_{k+1} = - \sum_{k=k_o+1}^{k_f} \underline{\lambda}^T_k \underline{\eta}_k = - \sum_{k=k_o}^{k_f-1} [\underline{\lambda}^T_k \underline{\eta}_k] - \underline{\lambda}^T_{k_f} \underline{\eta}_{k_f} + \underline{\lambda}^T_{k_o} \underline{\eta}_{k_o} \tag{5.3.11}$$

if we use a discrete version of integration by parts. If we now substitute equation 5.3.11 in equation 5.3.10, drop the (^) notation and drop the appropriate terms, we obtain :

$$\left[\left(\frac{\partial g_{k_f}}{\partial \underline{x}_{k_f}}\right)^T - \underline{\lambda}^T_{k_f} \right] \underline{\eta}_{k_f} - \left[\left(\frac{\partial g_{k_o}}{\partial \underline{x}_{k_o}}\right)^T - \underline{\lambda}^T_{k_o} \right] \underline{\eta}_{k_o} + \sum_{k=k_o}^{k_f-1} \left[\left(\frac{\partial H_k}{\partial \underline{x}_k}\right)^T - \underline{\lambda}^T_k \right] \underline{\eta}_k + \sum_{k=k_o}^{k_f-1} \left(\frac{\partial H_k}{\partial \underline{u}_k}\right)^T \underline{v}_k = 0 \tag{5.3.12}$$

Since the variations are mutually independent, the satisfaction of equation 5.3.12 requires that :

$$\underline{\lambda}_k = \frac{\partial H_k}{\partial \underline{x}_k} \tag{5.3.13}$$

$$\frac{\partial H_k}{\partial \underline{u}_k} = \underline{0} \tag{5.3.14}$$

$$\underline{\lambda}_{k_o} = \left(\frac{\partial g_{k_o}}{\partial \underline{x}_{k_o}}\right) \text{ or } \underline{\eta}^T_{k_o} \left[\lambda_{k_o} - \frac{\partial g_{k_o}}{\partial \underline{x}_{k_o}} \right] = 0 \tag{5.3.15}$$

$$\underline{\lambda}_{k_f} = \left(\frac{\partial g_{k_f}}{\partial \underline{x}_{k_f}}\right) \text{ or } \underline{\eta}^T_{k_f} \left[\underline{\lambda}_{k_f} - \frac{\partial g_{k_f}}{\partial \underline{x}_{k_f}} \right] = 0 \tag{5.3.16}$$

These equations are for the general case where there are no specified constraints on the variables. If any of the variables are specified, the corresponding variation vanishes and the corresponding equation among 5.3.13 to 5.3.16 does not apply.

Thus, for example, if $\underline{u}_k$ is a known, deterministic function, then $\underline{v}_k = 0$ and the requirement

$$\frac{\partial H_k}{\partial \underline{u}_k} = 0 \quad \text{does not hold.}$$

In addition to the requirements of equations 5.3.13 to 5.3.16, the optimal trajectory must also satisfy the state equation 5.3.1.

To summarise the discrete maximum principle, we have :

For the discrete non-linear dynamical system :

$$\underline{x}_{k+1} = \underline{f}(\underline{x}_k, \underline{u}_k, k) \tag{5.3.17}$$

$$\underline{x}_{ko} = \underline{x}_o \tag{5.3.18}$$

if it is desired to minimise the cost function :

$$J = \left[g_k(\underline{x}_k, k) \right]_{k=k_o}^{k=k_f} + \sum_{k=k_o}^{k_f-1} h(\underline{x}_k, \underline{u}_k, k) \qquad 5.3.19$$

the optimal trajectory must satisfy the following requirements :

$$H_k = h(\underline{x}_k, \underline{u}_k, k) + \underline{\lambda}^T_{k+1} \underline{f}(\underline{x}_k, \underline{u}_k, k) \qquad 5.3.20$$

$$\underline{x}_{k+1} = \frac{\partial H_k}{\partial \underline{\lambda}_{k+1}} \quad \text{or} \quad \underline{x}_{k+1} = \underline{f}(\underline{x}_k, \underline{u}_k, k) \qquad 5.3.21$$

$$\frac{\partial H}{\partial \underline{u}_k} = \underline{0} \quad \text{or} \quad \frac{\partial h_k}{\partial \underline{u}_k} + (\frac{\partial \underline{f}}{\partial \underline{u}_k}^T) \underline{\lambda}_{k+1} = \underline{0} \qquad 5.3.22$$

$$\underline{\lambda}_k = \frac{\partial H_k}{\partial \underline{x}_k} \quad \text{or} \quad \underline{\lambda}_k = (\frac{\partial f}{\partial \underline{x}_k})^T \underline{\lambda}_{k+1} + \frac{\partial h_k}{\partial \underline{x}_k} \qquad 5.3.23$$

$$\underline{\lambda}_{k+1} = (\frac{\partial \underline{f}^T}{\partial \underline{x}_k})^{-1} \left[\underline{\lambda}_k - \frac{\partial h_k}{\partial \underline{x}_k} \right] \qquad 5.3.23.a$$

(if $\frac{\partial f}{\partial \underline{x}_k}$ is a regular matrix).

$$\underline{\eta}^T_{k_o} \left[\underline{\lambda}_{k_o} - (\frac{\partial g_{k_o}}{\partial \underline{x}_{k_o}}) \right] = 0 \qquad 5.3.24$$

$$\underline{\eta}^T_{k_f} \left[\underline{\lambda}_{k_f} - (\frac{\partial g_{k_f}}{\partial \underline{x}_{k_f}}) \right] = 0 \qquad 5.3.25$$

Note that these equations are applicable only if the states and controls are not constrained by inequalities. The difficulty which occurs on constraining the states and controls by inequalities is due to the fact that unlike in the continuous case where a large variation of the control $\underline{u}(t)$ over a small enough time interval keeps the state perturbation small, here in the discrete case, a large variation of $\underline{u}_k$ at a single time instant necessarily results in a large perturbation of state.

The necessary conditions for optimality (equations 5.3.21 to 5.3.25) yield, in general, a non-linear two point boundary value problem since (a) we can use equation 5.3.22 to express $\underline{u}_k$ in terms of $\underline{\lambda}_{k+1}$, (b) the resulting expression can be substituted into equations 5.3.21 and 5.3.23 to eliminate $\underline{u}_k$ from the problem. The substitution of equation 5.3.21 may result in the variable $\underline{\lambda}_{k+1}$ appearing on the R.H.S. of the equation. In that case, equation 5.3.23 may, if desired, be solved for $\underline{\lambda}_{k+1}$ (5.3.23.a) and used in equation 5.3.21. Thus we obtain a two point boundary value problem of the form :

$$\begin{aligned} \underline{x}_{k+1} &= \underline{f}(\underline{x}_k, \underline{\lambda}_k, k) \\ \underline{\lambda}_{k+1} &= \underline{\phi}(\underline{x}_k, \underline{\lambda}_k, k) \end{aligned} \qquad 5.3.26$$

with $\underline{x}_{k_o}$ and $\underline{\lambda}_{k_f}$ given by equation 5.3.25. We will present in subsequent chapters various ways of obtaining numerical solutions to such non-linear discrete time two point boundary value problems.

5.3.2 Example

Consider the problem

$$\text{Min}\left\{ \underline{x}(k_f)^T Q\underline{x}(k_f) + \sum_{k=k_o}^{k_f-1} (\underline{x}(k)^T Q\underline{x}(k) + u^2(k)) \right. \qquad 5.3.27$$

where $\underline{x} = \begin{bmatrix} x_1 \\ x_2 \end{bmatrix}$, $Q = \begin{bmatrix} 1 & 0 \\ 0 & 1 \end{bmatrix}$

subject to

$$\begin{aligned} x_1(k+1) &= x_1^2(k) + x_2^2(k) + u(k) \; ; \; x_1(0) = 1 \\ x_2(k+1) &= - x_1(k) + x_2(k) \qquad ; \; x_2(0) = 2 \end{aligned} \qquad 5.3.28$$

The necessary conditions for optimality can be obtained by first writing the Hamiltonian function H as :

$$\begin{aligned} H = x_1^2(k) + x_2^2(k) + u^2(k) + \lambda_1(k+1) \left[x_1^2(k) + x_2^2(k) \right. \\ \left. + u(k) \right] + \lambda_2(k+1) \left[- x_1(k) + x_2(k) \right] \end{aligned} \qquad 5.3.29$$

Then $\frac{\partial H}{\partial x_1(k)} = \lambda_1(k)$ or $\lambda_1(k) = 2x_1(k) + 2x_1(k)$
$\lambda_1(k+1) - \lambda_2(k+1)$

$$\frac{\partial H}{\partial x_2(k)} = \lambda_2(k) = 2x_2(k) + 2x_2(k) \;\; \lambda_1(k+1) + \lambda_2(k+1)$$

$$\frac{\partial H}{\partial u} = 0 \quad \text{or} \quad 2u(k) + \lambda_1(k+1) = 0$$

$$\text{or} \quad u(k) = -\frac{1}{2}\lambda_1(k+1)$$

$$\frac{\partial H}{\partial \lambda_1(k+1)} = x_1(k+1) = x_1^2(k) + x_2^2(k) - \frac{1}{2}\lambda_1(k+1)$$

$$x_2(k+1) = - x_1(k) + x_2(k)$$

$$\lambda_1(k_f) = 2x_1(k_f)$$

$$\lambda_2(k_f) = 2x_2(k_f)$$

so that the non-linear two point boundary value problem becomes :

$$\begin{aligned} x_1(k+1) &= x_1^2(k) + x_2^2(k) - \frac{1}{2}\lambda_1(k+1) \\ x_2(k+1) &= - x_1(k) + x_2(k) \\ x_1(0) &= 1, \;\; x_2(0) = 2 \end{aligned} \qquad 5.3.30$$

and

$$\lambda_1(k) = 2x_1(k) + 2x_1(k)\ \lambda_1(k+1) - \lambda_2(k+1)$$
$$\lambda_2(k) = 2x_2(k) + 2x_2(k)\ \lambda_1(k+1) + \lambda_2(k+1) \qquad 5.3.31$$

$$\lambda_1(k_f) = 2x_1(k_f)$$
$$\lambda_2(k_f) = 2x_2(k_f) \qquad 5.3.32$$

In practice, before solving this two point boundary value problem, it will be necessary to ensure that terms of index k only appear on the right hand sides of the above equations. This can be done by first solving for $\lambda_1(k+1)\ \lambda_2(k+1)$ in terms of $\lambda_1(k)$, $\lambda_2(k)$, $x_2(k)$ from equations 5.3.31 and then substituting the expression for $\lambda_1(k+1)$ in equation 5.3.30.

Next let us consider the important case of linear-quadratic problems.

5.3.3 Discrete linear-quadratic problems

5.3.3.1 The regulator problem

Consider the regulator problem

$$\text{Min } J = \frac{1}{2}\left\|\underline{x}(k_f)\right\|^2_{S(k_f)} + \frac{1}{2}\sum_{k=0}^{k_f-1}\left\{\left\|\underline{x}(k)\right\|^2_{Q(k)} + \left\|\underline{u}(k)\right\|^2_{R(k)}\right\} \qquad 5.3.33$$

where Q, S are non-negative definite whilst R is positive definite

subject to the linear dynamical constraints

$$\underline{x}(k+1) = A(k)\ \underline{x}(k) + B(k)\ \underline{u}(k) \qquad 5.3.34$$

To solve this problem, let us form the Hamiltonian as

$$H = \frac{1}{2}\left\|\underline{x}(k)\right\|^2_{Q(k)} + \frac{1}{2}\left\|\underline{u}(k)\right\|^2_{R(k)} + \underline{\lambda}(k+1)^T\left[A(k)\underline{x}(k) + B(k)\underline{u}(k)\right] \qquad 5.3.35$$

Applying the necessary conditions for optimality, we have :

$$\frac{\partial H}{\partial \underline{x}(k)} = \underline{\lambda}(k) = Q(k)\ \underline{x}(k) + A^T(k)\ \underline{\lambda}(k+1) \qquad 5.3.36$$

Note that this equation can not be solved for $\underline{\lambda}(k+1)$ in terms of $\lambda(k)$ unless A is invertible. However, since A is a state transition matrix, A^{-1} exists.

Since the terminal state is unspecified, the boundary condition becomes

$$\underline{\lambda}(k_f) = S\ \underline{x}(k_f) \qquad 5.3.37$$

For the control

$$\frac{\partial H}{\partial \underline{u}(k)} = \underline{0} = R(k)\ \underline{u}(k) + B^T(k)\ \underline{\lambda}(k+1) \qquad 5.3.38$$

Thus our control can be obtained by solving the equations

$$\underline{x}(k+1) = A(k)\ \underline{x}(k)+B(k)\ R(k)^{-1}\ B(k)^{T}\ \underline{\lambda}(k+1) \qquad 5.3.39$$

$$\underline{x}(k_o) = \underline{x}_o \qquad 5.3.40$$

and

$$\underline{\lambda}(k) = Q(k)\ \underline{x}(k) + A^{T}(k)\ \underline{\lambda}(k+1) \qquad 5.3.41$$

with

$$\underline{\lambda}(k_f) = S(k_f)\ \underline{x}(k_f) \qquad 5.3.42$$

As in the continuous regulator treated previously, let us guess a solution of the form :

$$\underline{\lambda}(k) = P(k)\ \underline{x}(k) \qquad 5.3.43$$

Substituting in 5.3.39 and 5.3.41 in order to eliminate $\underline{\lambda}$, we have :

$$\underline{x}(k+1) = A(k)\ \underline{x}(k) - B(k)\ R(k)^{-1}\ B(k)^{T}\ P(k+1)\ \underline{x}(k+1) \qquad 5.3.44$$

$$P(k)\ \underline{x}(k) = Q(k)\ \underline{x}(k) + A^{T}(k)\ P(k+1)\ \underline{x}(k+1) \qquad 5.3.45$$

By solving for $\underline{x}(k+1)$ and eliminating it, we obtain :

$$P(k)\ \underline{x}(k) = Q(k)\ \underline{x}(k) + A^{T}(k)\ P(k+1)\left[I + B(k)\ R(k)^{-1}\ B(k)^{T}\ P(k+1)\right]^{-1} A(k)\ \underline{x}(k) \qquad 5.3.46$$

where I is the identity matrix.

This equation should hold for arbitrary $\underline{x}_k$ only if

$$P(k) = Q(k) + A^{T}(k)\ P(k+1)\left[I + B(k)\ R(k)^{-1}\ B(k)^{T}\ P(k+1)\right]^{-1} A(k) \qquad 5.3.47$$

$$= Q(k) + A^{T}(k)\left[P(k+1)^{-1} + B(k)\ R(k)^{-1}\ B(k)^{T}\right]^{-1} A(k) \qquad 5.3.48$$

with the condition at the final stage being

$$P_{k_f} = S \qquad 5.3.49$$

<u>Computational aspects</u>

The implementation of the control requires the solution of the matrix Riccati equation 5.3.47 or 5.3.48 backwards in time from $k = k_f$ to $k = k_o$ and then

$$\underline{u}(k) = -R^{-1}(k)\ B^{T}(k)\ A^{-1\ T}\left[P(k) - Q\right]\underline{x}(k) = G(k)\ \underline{x}(k) \qquad 5.3.50$$

where G(k) could be thought of as a "Gain". Thus using the P(k) sequence, we can obtain the G(k) sequence. This sequence can be stored and then applied to the system as it runs forward in real time. Thus we have obtained a closed loop discrete time system which is very similar to the continuous time regulator. The control is implemented as shown in Fig. 5.10.

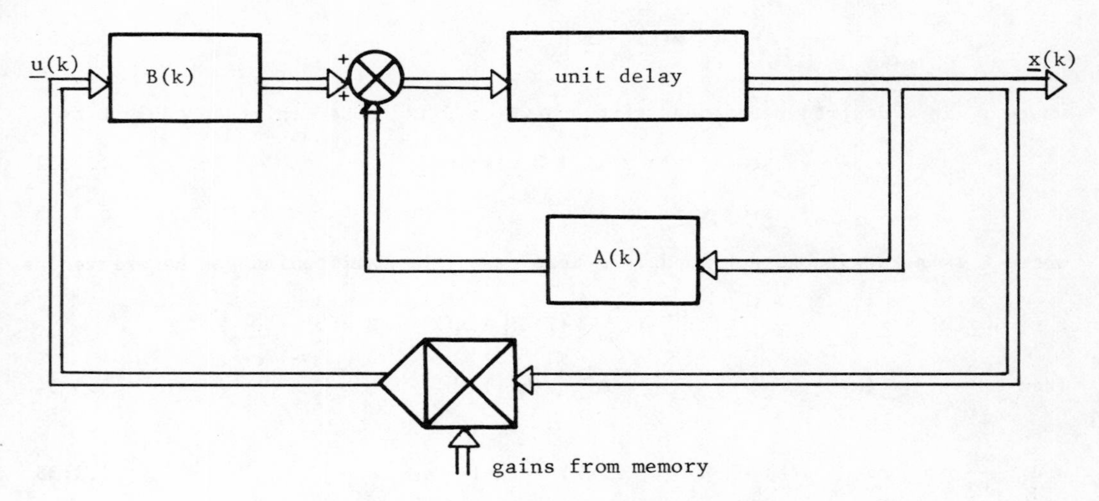

Fig. 5.10: The discrete regulator

Note that in our development we required Q, R, S to be positive semi-definite for the second variation to be positive (and thus gave us a minimum) whilst R is required to be positive definite for it to be invertible for the computation of $\underline{u}(k)$.

Infinite stage regulator

It is easy to show, and it is left as an excercise, that if A, B, Q, R, are time invariant, S = 0 and the system is controllable, then P(k) becomes constant as $k \longrightarrow \infty$ so that the gain G becomes constant and we can replace the prestored gains in the computer memory by a constant gain matrix.

Example

Consider the scalar problem :

$$\text{Min } J = \sum_{k=0}^{3} \frac{1}{2} x^2(k) + \frac{1}{2} u(k)^2$$

$$\text{subject to } x(k+1) = x(k) + u(k) \;;\; x(0) = 1$$

Here $\quad P(3) = 0 \;;\; A = 1,\ B = 1,\ Q = 1,\ R = 1$

Then $\quad P(2) = 1 \;;\; P(1) = 1 + (\frac{1}{2}) = 1.5 \;;\; P(0) = 1 + 1.5\ (\frac{1}{2.5}) = 1.6$

so that $G(0) = 1.6,\ G(1) = 1.5 \;;\; G(2) = 1,\ G(3) = 0$

and

$$u(k) = G(k)\ x(k)$$

5.3.3.2 : The servomechanism problem

Consider now the problem :

$$\underset{\underline{u}(k)}{\text{Min}}\ J = \frac{1}{2}\left[\sum_{k=0}^{k_f-1} \left(\left\|\underline{x}(k) - \underline{x}^d\right\|_Q^2 + \left\|\underline{u}(k)\right\|_R^2 \right)\right] \tag{5.3.51}$$

where $\underline{x}^d$ is a desired reference state and where J is to be minimised subject to

$$\underline{x}(k+1) = A\,\underline{x}(k) + B\,\underline{u}(k) + \underline{C} \tag{5.3.52}$$

$$\underline{x}(0) = \underline{x}_0 \tag{5.3.53}$$

where $\underline{C}$ is a known constant input. In this case, the Hamiltonian can be written as

$$H = \frac{1}{2}\left\| \underline{x}(k) - \underline{x}^d \right\|_Q^2 + \frac{1}{2}\left\|\underline{u}(k)\right\|_R^2 + \underline{\lambda}(k+1)^T \left[A\,\underline{x}(k) + B\,\underline{u}(k) + \underline{C}\right]$$

Then $$\frac{\partial H}{\partial \underline{x}(k)} = \underline{\lambda}(k) = Q(\underline{x}(k) - \underline{x}^d) + A^T\,\underline{\lambda}(k+1) \tag{5.3.54}$$

and $$\frac{\partial H}{\partial \underline{u}} = \underline{0} \quad \text{or} \quad \underline{u}(k) = -R^{-1}(k)\,B^T(k)\,\underline{\lambda}(k+1) \tag{5.3.55}$$

Let us assume a solution of the form

$$\underline{\lambda}(k) = P(k)\,\underline{x}(k) + \underline{s}(k) \tag{5.3.56}$$

Let us now eliminate $\underline{\lambda}$: then

$$\underline{x}(k+1) = A(k)\,\underline{x}(k) - B(k)\,R(k)^{-1}\,B(k)^T\,P(k+1)\,\underline{x}(k+1) - B(k)\,R(k)^{-1}\,B(k)^T\,\underline{s}(k+1)$$

and $P(k)\,x(k) + s(k)$

$$= Q(k)\left[\underline{x}(k) - \underline{x}^d\right] + A^T(k)\,P(k+1)\,\underline{x}(k+1) + A^T(k)\,\underline{s}(k+1)$$

By solving for $\underline{x}(k+1)$ and eliminating it, we obtain :

$$P(k)\,\underline{x}(k) + \underline{s}(k) = Q(k)\left[\underline{x}(k) - \underline{x}^d\right] + A^T(k)\,P(k+1)\Big\{\left[I + B(k)\,R(k)^{-1}\,B(k)^T P(k+1)\right]^{-1} A(k)\,\underline{x}(k) - \left[I + BR^{-1}\,B^T\,R(k+1)\right]^{-1} BR^{-1}\,B^T\,\underline{s}(k+1)\Big\} + A^T(k)\,\underline{s}(k+1)$$

For this equation to be valid for arbitrary $\underline{x}(k)$, we have :

$$P(k) = Q(k) + A^T(k)\,P(k+1)\left[I + B(k)\,R(k)^{-1}\,B(k)^T\,P(k+1)\right]^{-1} A(k) \tag{5.3.57}$$

and

$$\underline{s}(k) = -Q(k)\,\underline{x}^d - A^T(k)\,P(k+1)\left[I + B(k)\,R(k)^{-1}\,B(k)^T\,P(k+1)\right]^{-1} B(k)\,R(k)^{-1}\,B(k)^T\,\underline{s}(k+1) + A^T(k)\,\underline{s}(k+1) \tag{5.3.58}$$

with $P(k_f) = 0$

and $\underline{s}(k_f) = \underline{0}$

Example : River pollution control

To illustrate the servomechanism algorithm, consider a two reach river pollution control model which was simulated using some data from the river Cam near Cambridge. The model used was the two reach "no delay" model of Tamura [4].

The dynamic behaviour of a two reach system is given for the no delay model by Tamura [4] to be :

$$\underline{x}(k+1) = A\,\underline{x}(k) + B\,\underline{u}(k) + \underline{C}$$

where

$$\underline{x} = \begin{bmatrix} x_1 \\ x_2 \\ x_3 \\ x_4 \end{bmatrix}$$

is such that x_1, x_3 give the B.O.D. concentration (mg/1) in the stream and x_2, x_4 the D.O. concentration (mg/1). The control is given by :

$$\underline{u}(k) = \begin{bmatrix} u_1(k) \\ u_2(k) \end{bmatrix} \qquad k = 0, 1, \ldots, K-1$$

where u_1, u_2 are the maximum fraction of B.O.D. removed from the effluent in the reaches 1, 2. A, B, $\underline{C}$ from the Cam data are given by :

$$A = \begin{bmatrix} 0.18 & 0 & 0 & 0 \\ -0.25 & 0.27 & 0 & 0 \\ 0.55 & 0 & 0.18 & 0 \\ 0 & 0.55 & -0.25 & 0.27 \end{bmatrix}$$

$$B = \begin{bmatrix} 2.0 & 0 \\ 0 & 0 \\ 0 & 2.0 \\ 0 & 0 \end{bmatrix} \text{ and } \underline{C} = \begin{bmatrix} 4.5 \\ 6.15 \\ 2.0 \\ 2.65 \end{bmatrix}$$

A suitable cost function for this system is

$$J = \frac{1}{2}\left\| \underline{x}(K) \right\|^2_{I_4} + \sum_{k=0}^{K-1} \frac{1}{2}\left(\left\| \underline{x}(k) - \underline{x}^d \right\|^2_{I_4} + \left\| \underline{u}(k) \right\|^2_{100\, I_2} \right)$$

where I_4 is the fourth order identity matrix and I_2 is the second order identity matrix. The desired values $\underline{x}^d$ are :

$$\underline{x}^d = \begin{bmatrix} 5 \\ 7 \\ 5 \\ 7 \end{bmatrix}$$

This implies that it is desired to maintain the stream near B.O.D. values of 5 mg/1 and D.O. values of 7 mg/1 while minimising the treatment at the sewage works.

The above problem was solved using the servomechanism algorithm given above. The initial state was chosen to be $\underline{x}(0) = \underline{0}$ and K was chosen to be 23. The sampling interval is 0.5 days.

Here, K is long enough for the system to reach a steady state giving us a control law ofthe form

$$\underline{u} = G\,\underline{x} + \underline{d}$$

where $G = \begin{bmatrix} 0.0074 & -0.0011 & 0.0006 & -0.0001 \\ 0.0126 & -0.0015 & 0.0042 & -0.0004 \end{bmatrix}$

and $\underline{d} = \begin{bmatrix} 0.05449 \\ 0.00668 \end{bmatrix}$

using this control law, the system was simulated for the initial states

$$\underline{x} = \begin{bmatrix} 0 \\ 0 \\ 0 \\ 1 \end{bmatrix}$$

Figs. 5.11-5.12 show the B.O.D. and D.O. in reaches 1 and 2, and Figs. 5.13 - 5.14 the controls.

Let us next consider briefly the approach of dynamic programming to solving dynamic optimisation problems.

5.4 DYNAMIC PROGRAMMING AND HAMILTON-JACOBI EQUATION

5.4.1 Bellman's principle - Hamilton-Jacobi condition

The underlying idea behind the approach of dynamic programming is Bellman's principle of optimality which states that :

" An optimal policy has the property that whatever the initial state and decision are, the remaining decisions must constitute an optimal policy with regard to the state resulting from the first decision".

Let us see how we can apply this principle to the dynamic optimisation problem that we have been considering. For convenience, let us rewrite this problem as :

$$\text{Min } J = \int_{t_o}^{t_f} g(\underline{x}(t), \underline{u}(t), t)\, dt \qquad 5.4.1$$

subject to

$$\dot{\underline{x}}(t) = \underline{f}\,(\underline{x}(t), \underline{u}(t), t)$$
$$\underline{x}(0) = \underline{x}_0 \qquad 5.4.2$$

where $\underline{u}(t) \in U$ where U is possibly infinite or semi-infinite closed set. The admissible input set U may depend on $\underline{x}(t)$ and t. t_f is assumed fixed and $\underline{x}(t_f)$ is free. Suppose that we have calculated $\hat{\underline{u}}(t)$ as the optimal control and $\hat{\underline{x}}(t)$ as the resulting state trajectory. The cost function is then :

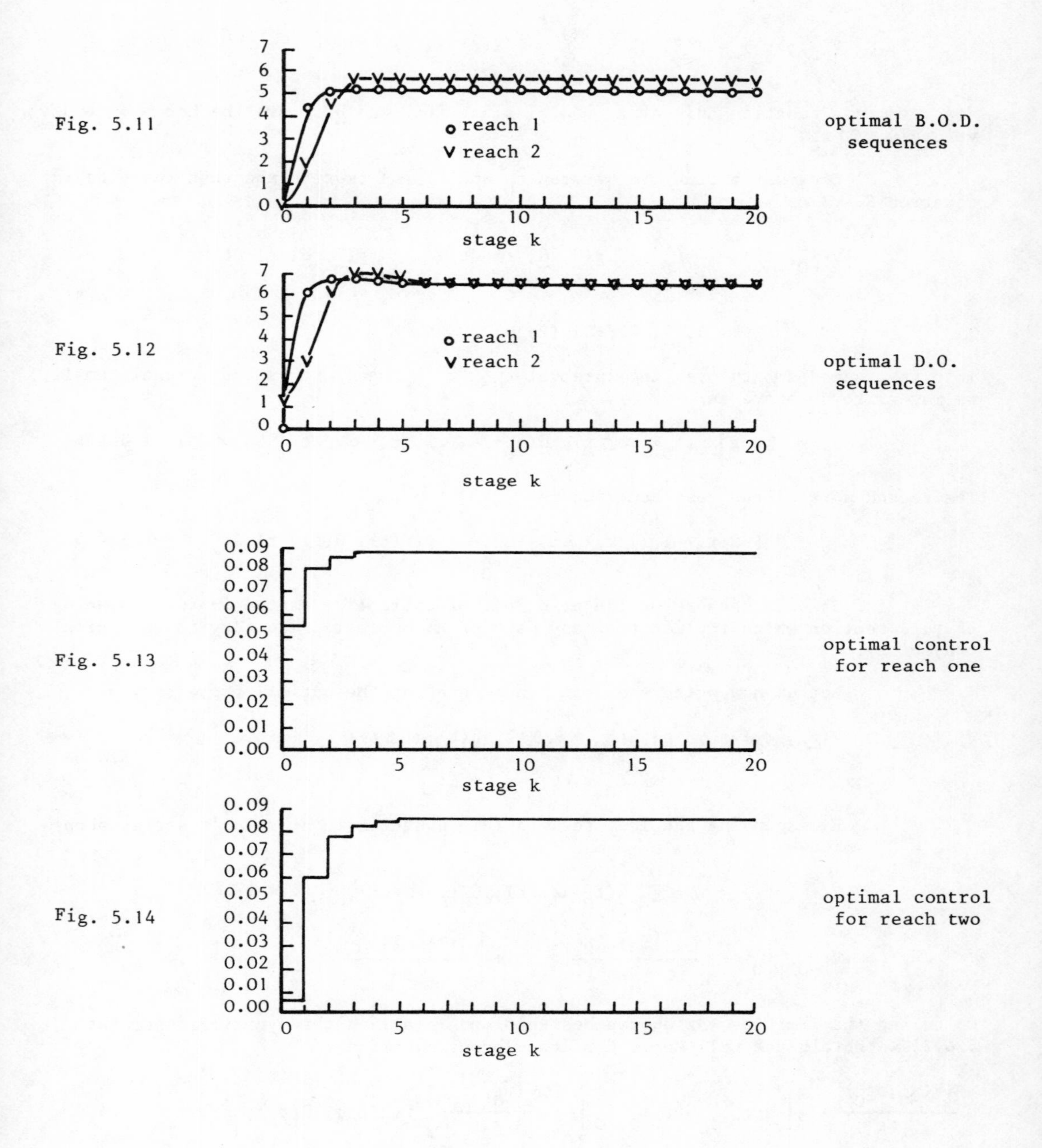

Fig. 5.11 optimal B.O.D. sequences

Fig. 5.12 optimal D.O. sequences

Fig. 5.13 optimal control for reach one

Fig. 5.14 optimal control for reach two

$$V(\underline{x}_0, t_0) = J(\hat{\underline{x}}, \hat{\underline{u}}) = \int_{t_0}^{t_f} g(\hat{\underline{x}}(t), \hat{\underline{u}}(t), t)\, dt$$

$V(\underline{x}_0, t_0)$ is a function only of $\underline{x}_0$ and t_0 since $\hat{\underline{x}}(t)$ and $\hat{\underline{u}}(t)$ are the known optimal values $\forall\, t \in [t_0, t_f]$.

Consider a time Δt between t_0 and t_f and rewrite the cost function in equation 5.4.1 as

$$V(\underline{x}_0, t_0) = \int_{t_0}^{t_0+\Delta t} g(\hat{\underline{x}}, \hat{\underline{u}}, t)\, dt + \int_{t_0+\Delta t}^{t_f} g(\hat{\underline{x}}, \hat{\underline{u}}, t)\, dt = J_1(\hat{\underline{x}}, \hat{\underline{u}}) + J_2(\hat{\underline{x}}, \hat{\underline{u}}) \qquad 5.4.3$$

If g is assumed smooth over the interval $t_0 \leqslant t \leqslant t$ and Δt is sufficiently small, we may write J_1 above as

$$J_1 = \Delta t\, g\left[\hat{\underline{x}}(t_0 + \alpha\Delta t), \hat{\underline{u}}(t_0 + \alpha\Delta t), t_0 + \alpha\Delta t\right] \quad 0 < \alpha < 1 \qquad 5.4.4$$

The second part of the cost function is

$$V_2 = V\left[\hat{\underline{x}}(t_0 + \Delta t), t_0 + \Delta t\right] = \int_{t_0+\Delta t}^{t_f} g(\hat{\underline{x}}(t), \hat{\underline{u}}(t), t)\, dt \qquad 5.4.5$$

This is because of the principle of optimality stated in the beginning of this section which implies that any part of an optimal trajectory is an optimal trajectory.

Let us now write the cost function along the optimal trajectory as

$$V(\underline{x}_0, t_0) = \Delta t\, g\left[\hat{\underline{x}}(t_0 + \alpha\Delta t), \hat{\underline{u}}(t_0 + \alpha\Delta t), t_0 + \alpha\Delta t\right] + V\left[\hat{\underline{x}}(t_0 + \Delta t), t_0 + \Delta t\right] \qquad 5.4.6$$

By expanding the last term in this equation in a Taylor's series about $t = 0$, we have

$$V(\underline{x}_0, t_0) = \Delta t\, g\left[\hat{\underline{x}}(t_0 + \alpha\Delta t), \hat{\underline{u}}(t_0 + \alpha\Delta t), t_0 + \alpha\Delta t\right] + V(\underline{x}_0, t_0) + \left[\frac{\partial V(\underline{x}_0, t_0)}{\partial t_0}\right]\Delta t + \left[\frac{\partial V(\underline{x}_0, t_0)}{\partial \underline{x}_0}\right]^T \dot{\underline{x}}_0\, \Delta t + \ldots \qquad 5.4.7$$

On taking the limit as Δt approaches zero and recalling the dynamic constraint 5.4.2, we obtain the well known Hamilton-Jacobi equation :

$$\frac{\partial V(\underline{x}_0, t_0)}{\partial t_0} + g\left[\hat{\underline{x}}(t_0), \hat{\underline{u}}(t_0), t_0\right] + \left[\frac{\partial V(\underline{x}_0, t_0)}{\partial \underline{x}_0}\right]^T \underline{f}(\hat{\underline{x}}(t_0), \hat{\underline{u}}(t_0), t_0) = 0 \qquad 5.4.8$$

In this expression we define

$$\underline{\lambda}(t_0) = \frac{\partial V(\underline{x}_0, t_0)}{\partial \underline{x}_0} \qquad 5.4.9$$

we may then rewrite the Hamilton-Jacobi equation, after dropping the subscript "0" for convenience as

$$\frac{\partial V(\underline{x}, t)}{\partial t} + H(\hat{\underline{x}}, \hat{\underline{u}}, \underline{\lambda}, t) = 0 \qquad 5.4.10$$

It should be noted that the Hamiltonian H in equation 5.4.10 is the Hamiltonian evaluated (at time t_0) for the optimal control $\hat{\underline{u}}(t)$ since it has been assumed that g was evaluated about the optimal control and state.

If we had used the cost function to be :

$$J = h(\underline{x}(t_f), t_f) + \int_{t_0}^{t_f} g(\underline{x}(t), \underline{u}(t), t)\, dt \qquad 5.4.11$$

we would have obtained the same Hamilton-Jacobi equation except that in this case the initial condition (at the terminal time) would have been

$$V\left[\underline{x}(t_f), t_f\right] = h(\underline{x}(t_f), t_f) \qquad 5.4.12$$

We note that :

1. The Hamilton-Jacobi equation cannot be easily solved in general. However, when it can, $\underline{u}(t)$ is determined as a function of $\underline{x}(t)$, i.e. we find a closed loop control.
2. The method can incorporate inequality constraints easily.
3. The solution obtained is the globally optimal solution.

As an example, let us consider the application of this approach to solving the linear-quadratic problem.

5.4.2 Example : Application of dynamic programming to the regulator problem.

In this section, we will consider one special case where it is possible to use the Hamilton-Jacobi equation.

The system is described by the linear vector differential equation

$$\dot{\underline{x}}(t) = A(t)\, \underline{x}(t) + B(t)\, \underline{u}(t) \qquad 5.4.13$$

and the cost function to be minimised is

$$J = \frac{1}{2} \underline{x}^T(t_f)\, S\underline{x}(t_f) + \int_{t_0}^{T_F} \frac{1}{2}\left[\underline{x}^T(t)\, Q(t)\, \underline{x}(t) + \underline{u}^T(t)\, R(t)\, \underline{u}(t)\right] dt \qquad 5.4.14$$

S, Q, are assumed to be real symmetric positive semi-definite matrices whilst R is assumed to be a real symmetric positive definite matrix. The initial and final times t_0 and t_f are assumed to be given and it is further assumed that $\underline{u}(t)$, $\underline{x}(t)$ are not constrained by any boundaries.

In order to use the Hamilton-Jacobi equation to solve this problem, let us write the Hamiltonian H as :

$$H(\underline{x}(t), \underline{u}(t), \underline{\lambda}(t), t) = \frac{1}{2}\|\underline{x}(t)\|_Q^2 + \frac{1}{2}\|\underline{u}(t)\|_R^2 + \underline{\lambda}(\underline{x}(t), t)^T [A(t)\ \underline{x}(t) + B(\underline{t})\ \underline{u}(t)]$$

5.4.15

Now, we know that a necessary condition for $\underline{u}(t)$ to minimise H is that $\frac{\partial H}{\partial \underline{u}} = \underline{0}$, i.e.

$$\frac{\partial H}{\partial \underline{u}} = R\ \underline{u} + B^T \underline{\lambda} = \underline{0}$$

or $$\underline{u}^*(t) = -R^{-1}(t)\ B^T(t)\ \underline{\lambda}^*(\underline{x}(t), t) \tag{5.4.16}$$

which yields for the Hamiltonian

$$H(\underline{x}(t), \underline{u}^*(t), \underline{\lambda}^*(\underline{x}(t), t), t) = \frac{1}{2}\|\underline{x}(t)\|_Q^2 + \frac{1}{2}\underline{\lambda}^{*T} B\ R^{-1} B^T \underline{\lambda}^* + \underline{\lambda}^{*T} A\underline{x} - \underline{\lambda}^{*T} B\ R^{-1}\ B^T\ \underline{\lambda}^* \tag{5.4.17}$$

$$= \frac{1}{2}\|\underline{x}\|_Q^2 - \frac{1}{2}\|\underline{\lambda}\|_{BR^{-1}B^T}^2 + \underline{\lambda}^T A\ \underline{x} \tag{5.4.18}$$

The Hamilton-Jacobi equation is then :

$$0 = \frac{\partial V}{\partial t} + \frac{1}{2}\|\underline{x}\|_Q^2 - \frac{1}{2}\|\underline{\lambda}\|_{BR^{-1}B^T}^2 + \underline{\lambda}^T A\underline{x} \tag{5.4.19}$$

with the boundary condition

$$\underline{\lambda}^*(\underline{x}(t_f), t_f) = \frac{1}{2}\ \|\underline{x}(t_f)\|_S^2 \tag{5.4.20}$$

From our study of the regulator problem in this chapter, it is not unreasonable to assume a solution of the form :

$$V(\underline{x}(t), t) = \frac{1}{2}\ \|\underline{x}(t)\|_P^2 \tag{5.4.21}$$

where P is a real symmetric positive definite matrix which needs to be determined. Substititing this assumed solution into the Hamilton-Jacobi equation 5.4.17, we have :

$$0 = \frac{1}{2}\|\underline{x}\|_{\dot{P}}^2 + \frac{1}{2}\|\underline{x}\|_Q^2 - \frac{1}{2}\|\underline{x}\|_{PBR^{-1}B^TP}^2 + \|\underline{x}\|_{PA}^2 \tag{5.4.22}$$

Now, let us write the matrix product PA as a sum of a symmetric part and an unsymmetric part, i.e.

$$PA = \frac{1}{2}[PA + (PA)^T] + \frac{1}{2}[PA - (PA)^T] \tag{5.4.23}$$

since for any two matrices F, G,

$$(FG)^T = F^T G^T$$

and since the transpose of a scalar equals itself, we can show that only the symmetric part of PA contributes to equation 5.4.22 so that equation 5.4.22 can be rewritten as :

$$0 = \frac{1}{2}\|\underline{x}(t)\|_{\dot{P}}^2 + \frac{1}{2}\|\underline{x}\|_Q^2 - \frac{1}{2}\|\underline{x}\|_{PBR^{-1}B^TP}^2 + \frac{1}{2}\|\underline{x}\|_{PA}^2 + \frac{1}{2}\|\underline{x}\|_{A^TP}^2 \tag{5.4.24}$$

Since this equation must hold for arbitrary $\underline{x}(t)$, we have :

$$\mathring{P}(t) + Q(t) - P(t)\ B(t)\ R^{-1}(t)\ B^T(t)\ P(t) + P(t)\ A(t) + A^T(t)\ P(t) = 0 \qquad 5.4.26$$

with

$$P(t_f) = S$$

Thus we see that the Hamilton-Jacobi partial differential equation reduces in this case to a set of ordinary non-linear differential equations, i.e. to the Riccati equation, and the optimal control is given by :

$$\underline{u}^*(t) = -R^{-1}(t)\ B^T(t)\ P(t)\ \underline{x}(t) \qquad 5.4.27$$

as before.

5.5 CONCLUSIONS

In this chapter, we have made the jump from static optimisation to dynamic optimisation. We have considered the necessary conditions for optimality (i.e. the maximum principle) for both discrete time and continuous time problems. We saw that this leads in general to the solution of a two point boundary value problem. We then examined the special case where these problems are linear and are thus easy to solve. This gave us a closed loop control. We also saw that as the dimensionality increases, even these linear-quadratic problems become difficult to solve so that in the next chapter we consider decomposition-coordination techniques for tackling such problems whilst in chapter 7 we tackle non-linear two point boundary value problems numerically.

REFERENCES FOR CHAPTER 5

[1] SAGE, A.P., 1968 : Optimum systems control". Prentice Hall.

[2] KIRK, D., 1970 : Optimal control theory". Prentice Hall.

[3] BECK, M.B. : "The application of control and systems theory to problems of river pollution control".
Cambridge University Ph. D. Thesis, 1974.

[4] TAMURA, H. : A discrete dynamical model with distributed transport delays and its hierarchical optimisation to preserve stream quality".
I.E.E.E. Trans. SMC 4, 424-429, 1974.

[5] BAUMAN, E.J., 1968 : Multi-level optimisation techniques with application to trajectory decomposition".
Advances in control systems, 6, p. 1 to 22.

[6] CHENEVEAUX, B. : "Contribution à l'optimisation hierarchisée des systèmes dynamiques".
Doctor Engineer thesis n° 4, Nantes, France, 1972.

[7] MUKHOPADHYAY, B.K. and MALIK, O.P. : "Optimal control of synchronous machine excitation by quasilinearisation."
Proc. IEE. vol. 119, 1, Jan. 1972.

[8.] JOHNSON, C.D. and GIBSON, J.E. : Singular solutions in problems of optimal control".
IEEE Trans. AC, 8, 4-14, 1963.

PROBLEMS FOR CHAPTER 5

1. Find the two point boundary value problem which when solved yields the control u(t) and trajectory x(t) which minimises

$$J = \frac{1}{2}x(2)^2 + \frac{1}{2}\int_0^2 (x^2 + u^2)\,dt$$

for the non-linear dynamical system

$$\dot{x}(t) = -x^2(t) + u(t) \quad \text{with } x(0) = 10$$

2. Find the two point boundary value problem for the optimal control problem of minimising

$$J = \int_0^1 (x_1 x_2 + x_1^2 + x_2^2 + u_1^2 + u_2^2)\,dt$$

subject to the constraints

$$\dot{x}_1 = x_1 + x_2^2 + u_1^2 + u_2 \qquad x_1(0) = 2$$

$$\dot{x}_2 = 2x_1 + x_2 + u_2 \qquad x_2(0) = -2$$

3. The optimal control problem for the ith subsystem controller in a hierarchical control structure is to minimise w.r.t. $\underline{u}_i$, $\underline{z}_i$

$$J_i = \frac{1}{2}\|\underline{x}_i(T)\|^2_{Q_i} + \int_0^T \Big\{\frac{1}{2}\Big[\|\underline{x}_i\|^2_{Q_i} + \|\underline{u}_i\|^2_{R_i} + \|\underline{z}_i\|^2_{S_i}\Big] + \underline{\lambda}_i^{*T}\underline{z}_i - \sum_j \underline{\lambda}_j^{*T} L_{ji}\,\underline{x}_i\Big\}\,dt$$

subject to

$$\dot{\underline{x}}_i = A_i\,\underline{x}_i + B_i\,\underline{u}_i + C_i\,\underline{z}_i$$

where $\underline{\lambda}_i^*$ $i = 1, 2, \ldots m$ are known trajectories.

Find in this case the optimal control $\underline{u}_i$ and the pseudo controls $\underline{z}_i$ as a function of $\underline{\lambda}^*$, P_i and the system parameters where P_i is the costate vector.

4. Find the optimal control trajectory u(t) which minimises

$$J = \frac{1}{2}\int_0^2 (x^2 + u^2)\,dt$$

subject to $\dot{x} = 2x + 3u$

How does this control differ from the one obtained when the optimisation horizon is infinite.

5. Find the optimal control trajectory which minimises

$$J = \frac{1}{2}\int_0^2 \left(\begin{bmatrix} x_1 \\ x_2 \end{bmatrix}^T \begin{bmatrix} 1 & 0 \\ 0 & 2 \end{bmatrix} \begin{bmatrix} x_1 \\ x_2 \end{bmatrix} + u^2 \right) dt$$

subject to $\dot{x}_1 = x_2$

$\dot{x}_2 = u$

6. Find the optimal control trajectory Δm_1 for the river system

$$\frac{d}{dt}\begin{bmatrix} z_1 \\ q_1 \end{bmatrix} = \begin{bmatrix} -1.32 & 0 \\ -0.32 & -1.2 \end{bmatrix} + \begin{bmatrix} 0.1 \\ 0 \end{bmatrix} \Delta m_1 + \begin{bmatrix} 0.9\, z_o + 5.35 \\ 0.9\, q_o + 1.9 \end{bmatrix}$$

with $q_o = 10$, $z_o = 0$, and the cost function to be minimised is

$$J = \int_0^8 \left\{ (z_1 - 4.06)^2 + 2(q - 8)^2 + \Delta m_1^2 \right\} dt$$

7. Write down the two point boundary value problem for the optimal control problem of minimising

$$J = \frac{1}{2}\begin{bmatrix} x_1(6) \\ x_2(6) \end{bmatrix}^T \begin{bmatrix} 1 & 0 \\ 0 & 2 \end{bmatrix} \begin{bmatrix} x_1(6) \\ x_2(6) \end{bmatrix}$$

$$+ \sum_{k=0}^{5} \begin{bmatrix} x_1(k) \\ x_2(k) \end{bmatrix}^T \begin{bmatrix} 1 & 0 \\ 0 & 3 \end{bmatrix} \begin{bmatrix} x_1(k) \\ x_2(k) \end{bmatrix} + \begin{bmatrix} u_1(k) \\ u_2(k) \end{bmatrix}^T \begin{bmatrix} 1 & 0 \\ 0 & 2 \end{bmatrix} \begin{bmatrix} u_1(k) \\ u_2(k) \end{bmatrix}$$

subject to the constraints

$$x_1(k+1) = 2x_1(k) + x_2^2(k)$$

$$x_2(k+1) = x_1(k) + x_2(k) + u_1(k) + u_2(k)$$

8. Find the optimal control for the system

$$\begin{bmatrix} x_1(k+1) \\ x_2(k+1) \end{bmatrix} = \begin{bmatrix} 1 & 0 \\ 2 & 2 \end{bmatrix} \begin{bmatrix} x_1(k) \\ x_2(k) \end{bmatrix} + \begin{bmatrix} 1 \\ 0 \end{bmatrix} u(k)$$

where the cost function to be minimised is

$$J = \sum_{k=0}^{\infty} (x_1(k)^2 + x_2(k)^2 + u^2)$$

CHAPTER 6

Hierarchical Optimisation and Control for Linear Systems with Quadratic Cost Functions

6.1 INTRODUCTION

We saw in the last chapter that we could solve both continuous time and discrete time linear-quadratic problems by solving an appropriate Riccati type equation. Now, the Riccati equation is a non-linear differential or difference equation with $\frac{n \times (n+1)}{2}$ elements for a system of order n so that its solution uses up a lot of core store and requires computation which increases rapidly with system order. Again we know that integration of high order equations involves the building up of rounding off errors in the computer which could make the numerical calculation unstable. In this chapter we will consider the hierarchical approach to solving such problems which enables us to get around some of these difficulties.

The main hierarchical methods that we will describe lead to open loop control, i.e. to a control which is initial state dependent and where the current control is not a direct function of the current state. We will show in later parts of this chapter how we can get around the difficulty and compute feedback control gains.

Let us begin by formulating the problem of optimisation and control of interconnected dynamical systems. Let us assume that the overall system comprises N subsystems which are interconnected together as shown, for instance, in fig. 6.1. For any subsystem i, $\underline{x}_i$ is the n_i dimensional state vector, $\underline{u}_i$ is the m_i dimensional control vector and $\underline{z}_i$ is an r_i dimensional vector of inputs which are generated by the states of the other subsystems. We assume that the subsystems themselves can be described by linear differential equations, i.e.

$$\dot{\underline{x}}_i(t) = A_i \underline{x}_i(t) + B_i \underline{u}_i(t) + C_i \underline{z}_i(t) \qquad 6.1.1$$

with

$$\underline{x}_i(0) = \underline{x}_{i0}$$

We assume also that the vector of inputs $\underline{z}_i$ is a linear combination of the states of the N subsystems, i.e.

$$\underline{z}_i = \sum_{j=1}^{N} L_{ij} \underline{x}_j \qquad 6.1.2$$

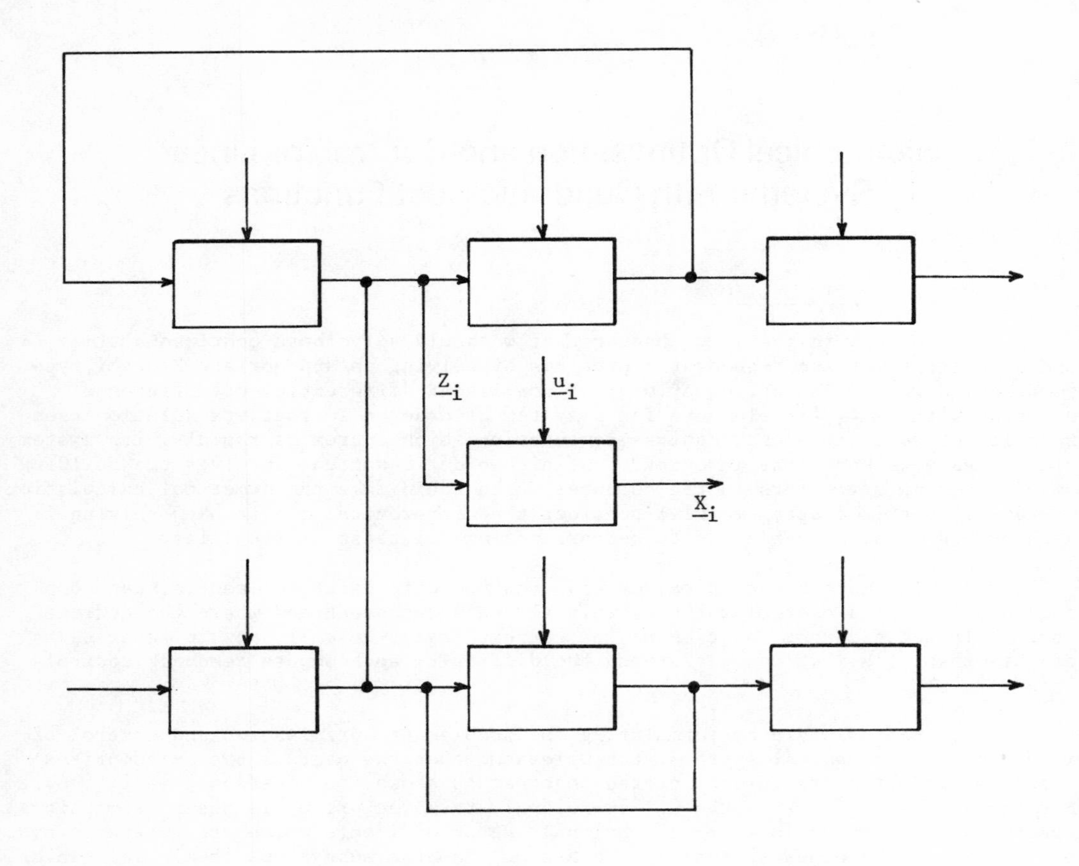

Fig. 6.1 : An example of an interconnected Dynamical System.

The optimisation problem of interest is to choose the controls $\underline{u}_1, \ldots \underline{u}_N$ in order to minimise a quadratic cost function of the kind :

$$J = \sum_{i=1}^{N} \left(\frac{1}{2}\left\|\underline{x}_i(T)\right\|^2_{P_i} + \int_0^T \frac{1}{2}\left[\left\|\underline{x}_i(t)\right\|^2_{Q_i} + \left\|\underline{u}_i\right\|^2_{R_i} + \left\|\underline{z}_i\right\|^2_{S_i}\right] dt\right) \qquad 6.1.3$$

subject to the constraints 6.1.1 and 6.1.2 where Q_i, P_i are positive semi-definite matrices and R_i, S_i are positive definite ($\left\|\underline{c}\right\|^2_L = \underline{c}^T L\underline{c}$.)

If the interconnection relationship 6.1.2 is substituted back into equation 6.1.1 it is possible to obtain a standard overall description of the form

$$\mathring{\underline{x}} = A\underline{x} + B\underline{u}$$

where $\underline{x} = \begin{bmatrix} \underline{x}_1 \\ \underline{x}_2 \\ \cdot \\ \cdot \\ \cdot \\ \underline{x}_N \end{bmatrix}$; $\underline{u} = \begin{bmatrix} \underline{u}_1 \\ \underline{u}_2 \\ \cdot \\ \cdot \\ \cdot \\ \underline{u}_N \end{bmatrix}$; A, B are full matrices

In the overall case it is difficult to interpret the physical significance of the quadratic term $\|\underline{z}_i\|^2_{S_i}$ since it can always be combined into the term $\|\underline{x}\|^2_Q$ in the case of the overall system. It is necessary to use this term, however, since otherwise the "pseudo" control $\underline{z}_i$ will enter linearly in the Hamiltonian and this will lead to singular solutions as we saw in the previous chapter. The present formulation is taken from Pearson [1] who introduced this term but in section 9 we show that it is often possible to avoid using it [3] . We begin our analysis with the "Goal Coordination" or "Interaction Balance" approach of Mesarovic [2] as specialised for "linear-quadratic" problems by Pearson [1].

6.2 THE GOAL COORDINATION APPROACH

6.2.1 Formulation

The basis of the approach is that it is possible to convert the original minimisation problem into a simpler maximisation problem and then solve this problem using a two level iterative calculation structure. To do this, define a dual function $\phi(\underline{\lambda})$ where

$$\phi(\underline{\lambda}) = \underset{\underline{x},\underline{u},\underline{z}}{\text{Min}} \left\{ L(\underline{x}, \underline{u}, \underline{z}, \underline{\lambda}) \text{ subject to equation 6.1.1} \right\} \qquad 6.2.1$$

where

$$L(\underline{x},\underline{u},\underline{z},\underline{\lambda}) = \sum_{i=1}^{N} \left\{ \frac{1}{2}\|\underline{x}_i(T)\|^2_{P_i} + \int_0^T \left(\frac{1}{2}\|\underline{x}_i\|^2_{Q_i} + \frac{1}{2}\|\underline{u}_i\|^2_{R_i} + \frac{1}{2}\|\underline{z}_i\|^2_{S_i} + \underline{\lambda}_i^T(\underline{z}_i - \sum_{j=1}^{N} L_{ij}\,\underline{x}_j)\right) dt \right\} \qquad 6.2.2$$

where $\underline{\lambda}$ is an r dimensional vector of Lagrange multipliers and $L(\underline{x},\underline{u},\underline{z},\underline{\lambda})$ is the Lagrangian which has been formed via the introduction of the Lagrange multipliers $\underline{\lambda}$. Now the theorem of strong Lagrange Duality [4] asserts that for cases like the one considered here where all the constraints are linear and the cost function is quadratic (i.e. the problem is convex) :

$$\underset{\underline{\lambda}}{\text{Max}} \quad (\underline{\lambda}) = \underset{\underline{u}}{\text{Min}}\ J \qquad 6.2.3$$

i.e. an equivalent way of solving the problem of minimising J in equation 6.1.3 subject to the linear equality constraints given by equations 6.1.1 and 6.1.2 is to maximise the dual function $\phi(\underline{\lambda})$ w.r.t. $\underline{\lambda}$. This can be done within a two level structure of the type considered in Chapter 4 since from equation 6.2.2, for given

$\underline{\lambda} = \underline{\lambda}^*$

$$L(\underline{x},\underline{u},\underline{z},\underline{\lambda}^*) = \sum_{i=1}^{N} \Big\{ \frac{1}{2}\|\underline{x}_i(T)\|^2_{P_i} + \int_0^T (\frac{1}{2}\|\underline{x}_i\|^2_{Q_i} + \frac{1}{2}\|\underline{u}_i\|^2_{R_i} +$$

$$\frac{1}{2}\|\underline{z}_i\|^2_{S_i} + \underline{\lambda}^{*T}_i \underline{z}_i - \sum_{j=1}^{N} \underline{\lambda}^{*T}_j L_{ji}\,\underline{x}_i)dt \Big\} = \sum_{i=1}^{N} L_i$$

i.e. the Lagrangian L is additively separable and can be decomposed into N independent sub Lagrangians, one for each subsystem. Thus for given $\underline{\lambda} = \underline{\lambda}^*$ which is treated as a known trajectory, it is possible to minimise the sub Lagrangian

$$L_i = \frac{1}{2}\|\underline{x}_i(T)\|^2_{P_i} + \int_0^T (\frac{1}{2}\|\underline{x}_i\|^2_{Q_i} + \frac{1}{2}\|\underline{u}_i\|^2_{R_i} + \frac{1}{2}\|\underline{z}_i\|^2_{S_i} + \underline{\lambda}^{*T}_i \underline{z}_i - \sum_{j=1}^{N} \underline{\lambda}^{*T}_j L_{ji}\,\underline{x}_i)\,dt \qquad 6.2.4$$

independently for the N subsystems where each subsystem's minimisation is subject to that subsystem's dynamical constraints given by equation 6.1.1. This enables us to obtain $\phi(\underline{\lambda}^*)$ in equation 6.2.1 and the $\phi(\underline{\lambda}^*)$ can be improved successively by an iterative exchange of information with a second level which improves $\phi(\underline{\lambda}^*)$ using the N independent first level minimisations. The actual mechanism for the improvement of $\phi(\underline{\lambda}^*)$ in order to maximise it relies on the fact that it is possible to write a simple expression for the gradient of $\phi(\underline{\lambda}^*)$ in terms of the solutions of the first level minimisations. In fact, the gradient is given by the error in the interconnection relationship, i.e.

$$\nabla\phi(\underline{\lambda})\Big|_{\underline{\lambda}=\underline{\lambda}^*} = \begin{bmatrix} \vdots \\ \underline{z}_i - \sum_{j=1}^{N} L_{ij}\,\underline{x}_j \\ \vdots \end{bmatrix} = \begin{bmatrix} \vdots \\ \underline{e}_i \\ \vdots \end{bmatrix} = \underline{e} \qquad 6.2.5$$

It is thus possible to envisage a two level hierarchical algorithm as shown in Fig. 6.2 where on level 1 for given $\underline{\lambda} = \underline{\lambda}^*$, supplied by the second level, L_i is minimised subject to the subsystem dynamic constraints and the resulting $\underline{x}_i$, $\underline{u}_i$ are sent back to level 2. At level 2, these vectors are collated and substituted into equation 6.2.5 to form the interconnection error.

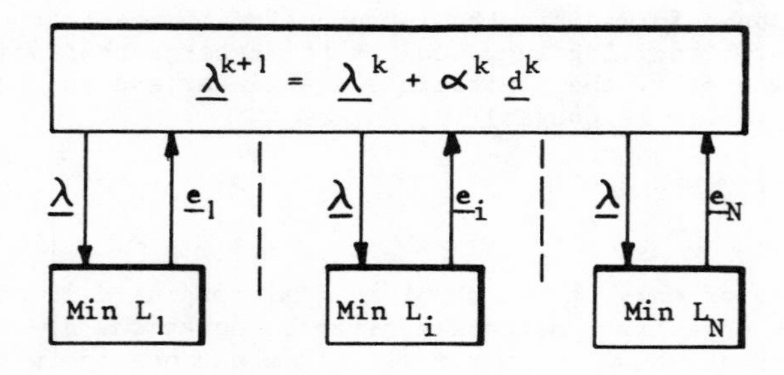

Fig. 6.2 : The two level Goal Coordination structure

This error vector is used in a gradient procedure to produce a new $\underline{\lambda}$. For example, from iteration k to k+1 :

$$\underline{\lambda}^{k+1}(t) = \underline{\lambda}^{k}(t) + \alpha^{k}\underline{d}^{k}(t) \qquad (0 \leqslant t \leqslant T) \qquad 6.2.6$$

where α is the step length and $\underline{d}^k$ is the search direction. If the steepest ascent method is used then $\underline{d}^k(t) = \underline{e}^k(t)$ $(0 \leqslant t \leqslant T)$. On the other hand, if the conjugate gradient method is used then

$$\underline{d}^{k+1}(t) = \underline{e}^{k+1}(t) + \beta^{k+1}\underline{d}^{k} \quad ; \quad 0 \leqslant t \leqslant T \qquad 6.2.7$$

where

$$\beta^{k+1} = \int_0^T (\underline{e}^{k+1^T} \underline{e}^{k+1})dt \Big/ \int_0^T (\underline{e}^{k^T} \underline{e}^{k})dt$$

with $\underline{d}^0 = \underline{e}^0$.

The overall optimum is achieved when $\underline{e}^k(t)$ $(0 \leqslant t \leqslant T)$ is sufficiently close to zero.

6.2.2. Comments

This method is a relatively classical one. Mesarovic et al. [2] call it the interaction balance method since at the optimum the interactions balance since the gradient is zero only when

$$\underline{z}_i = \sum_{j=1}^{N} L_{ij}\, \underline{x}_j$$

Pearson [1] calls it the Goal Coordination method since the goals (i.e. performance measures) of each subsystem are modified in each iteration. The main strengths of the method are :

a - It is possible to tackle large scale "linear-quadratic" problems with this approach since the computer storage requirements are no longer prohibitive. If parallel processors are used for the first level computation, it may be possible to tackle truly large problems.

b - It is easier than in a global approach, at least in principle, to take into account inequality constraints on the states and controls of the subsystems since only low order subproblems are solved.

However, the method also has certain drawbacks and these are perhaps the principal reason why the method has not, at least in this form, been extensively used to solve practical large scale problems. The main disadvantages of the method are :

a) Although the second level algorithm is attractive in principle since the gradient is easy to calculate, the choice of the step length causes some problems since one can either use a constant one or attempt to find an α at each iteration which gives the biggest increase in $\phi(\underline{\lambda})$. In the former cases no α is in general suitable for the whole of the convergence process since a large one may be desirable in the beginning when we are quite far from the optimum whilst a much smaller one may be most appropriate nearer the optimum. If a one dimensional search is made for the best α, a lot of additional computation is required since it will be necessary to solve the lower level problem a number of times to obtain the best step length. Given that one requires to do all this at each iteration and since the whole convergence process may require many iterations, the overall calculation although yielding substantial savings in computer storage (since only low order problems are solved) is unlikely to yield a saving in computation time if a single computer is used to perform the calculations of the two levels sequentially. It is argued that this may not be a real drawback since in future applications it is quite likely that parallel mini-computers will be used at the first level thus improving the computation time. However, that is not absolutely clear since parallel processing on minicomputers may lead to additional time consuming communication problems since with the present state of the technology, minicomputers calculate very much faster than they can communicate amongst themselves and it can be argued that a part of what one gains in parallel computation, one loses in slow communication of intermediate results.

b) The method is an infeasible one, i.e. it is only permissible to use the controls obtained at the end of the iterative process since otherwise the interconnection relationship $\underline{z}_i = \sum_{j=1}^{N} L_{ij}\,\underline{x}_j$ will not be satisfied.

c) Another disadvantage of the approach is that the inclusion of the term $\frac{1}{2}\|\underline{z}_i\|^2_{S_i}$ in the cost function does not correspond to a realistic physical situation and has been added purely to ensure that singular solutions do not arise at the first level. To see how such solutions arise in large scale problems, consider the lower level minimisation problem for the ith subsystem. The problem is to minimise L_i in equation 6.2.4 subject to equation 6.1.1. Write the Hamiltonian of the subsystem as

$$H_i = \frac{1}{2}\|\underline{x}_i\|^2_{Q_i} + \frac{1}{2}\|\underline{u}_i\|^2_{R_i} + \frac{1}{2}\|\underline{z}_i\|^2_{S_i} + \underline{\lambda}_i^{*T}\underline{z}_i - \sum_{j=1}^{N}\underline{\lambda}_j^{*T} L_{ji}\,\underline{x}_i + \underline{p}_i^T(A_i\underline{x}_i + B_i\underline{u}_i + C_i\underline{z}_i) \qquad 6.2.8$$

The singular solution arises if $\frac{1}{2}\|\underline{z}_i\|^2_{S_i}$ does not exist because it is necessary to minimise L_i w.r.t. $\underline{x}_i$, $\underline{u}_i$, $\underline{z}_i$ and the latter minimisation involves setting $\frac{\partial H_i}{\partial \underline{z}_i} = 0$ which in this case becomes

$$\underline{z}_i = -S_i^{-1}\left[C_i^T\underline{p}_i + \underline{\lambda}_i\right] \qquad 6.2.9$$

but which would be singular if no quadratic form in $\underline{z}_i$ existed in the Hamiltonian. But since the dynamics of the system are linear and the interconnection relationship is also linear, the quadratic form in $\underline{z}_i$ can only come from the cost function J. Thus one way of avoiding singularities is to add in the quadratic term in $\underline{z}$ in the cost function.

Now, although singular solutions are perfectly valid solutions to optimisation problems, they are certainly undesirable in our iterative hierarchical scheme since they complicate the lower level calculation enormously, as we saw in the previous chapter, whilst one of the main justifications of hierarchical optimisation is the ease of calculation achieved by decentralisation. Thus although the Goal Coordination method is attractive since it is able to solve large scale problems, it has not been extensively used for the two main reasons that the second level calculation gets complicated because of the need for finding a good step length which necessitates multiple solutions of the first level problems during a single iteration and the need to introduce terms which are not physically meaningful in order to avoid singularities. Nevertheless, the method is significant because of its conceptual simplicity.

6.2.3 Example

As the simplest possible example consider the problem of minimising

$$J = \int_0^T (x_1^2 + x_2^2 + u_1^2 + u_2^2)\, dt$$

subject to

$$\dot{x}_1 = x_1 - 2x_2 + u_1$$

$$\dot{x}_2 = -x_1 + x_2 + u_2$$

Let us rewrite the constraints as

$$\dot{x}_1 = x_1 + z_1 + u_1$$

$$\dot{x}_2 = -z_2 + x_2 + u_2$$

$$z_1 = -2x_2$$

$$z_2 = x_1$$

so that the Lagrangian becomes

$$L = \int_0^T \Big\{ x_1^2 + x_2^2 + z_1^2 + z_2^2 + u_1^2 + u_2^2 + p_1\left[-\dot{x}_1 + x_1 + z_1 + u_1\right] + p_2\left[-\dot{x}_2 - z_2 + x_2 + u_2\right] + \lambda_1\left[z_1 + 2x_2\right] + \lambda_2\left[z_2 - x_1\right]\Big\}\, dt$$

$= L_1 + L_2$ for given λ_1, λ_2 where

$$L_1 = \int_0^T (x_1^2 + z_1^2 + u_1^2 + p_1\left[-\dot{x}_1 + x_1 + z_1 + u_1\right] + \lambda_1 z_1 - \lambda_2 x_1)dt$$

$$L_2 = \int_0^T (x_2^2 + z_2^2 + u_2^2 + p_2\left[-\dot{x}_2 + x_2 + u_2 - z_2\right] + \lambda_2 z_2 + 2\lambda_1 x_2)dt$$

and the problem reduces to one of minimising independently L_1 w.r.t. x_1, z_1, u_1, and L_2 w.r.t. x_2, z_2, u_2 for given λ_1, λ_2 and at the second level of improving λ_1, λ_2 using the gradient

$$\nabla \phi(\underline{\lambda}) = \begin{bmatrix} z_1 - x_2 \\ z_2 - x_1 \end{bmatrix}$$

where z_1, z_2, x_1, x_2 are obtained from the independent first level solutions.

As a second more complicated example, consider the 12th order example of Pearson [1].

This example is sufficiently large to give one a "feel" for the computational approach whilst not being so excessively large that it cannot be solved by standard single level techniques.

6.3 EXAMPLE 2 : PEARSON'S 12TH ORDER EXAMPLE

In this case the overall system consists of four subsystems as shown in Fig. 6.3. Each subsystem is of order three. Here we use Pearson's [1] values for the parameters where available and reasonable values for the ones he does not give. The dynamic behaviour of the system is given by Pearson to be :

Subsystem 1 :

$$\begin{aligned} \mathring{x}_1 &= x_2 \\ \mathring{x}_2 &= x_3 \\ \mathring{x}_3 &= a_{31}x_1 + a_{32}x_2 + a_{33}x_3 + u_1 + z_1 \end{aligned} \qquad 6.3.1$$

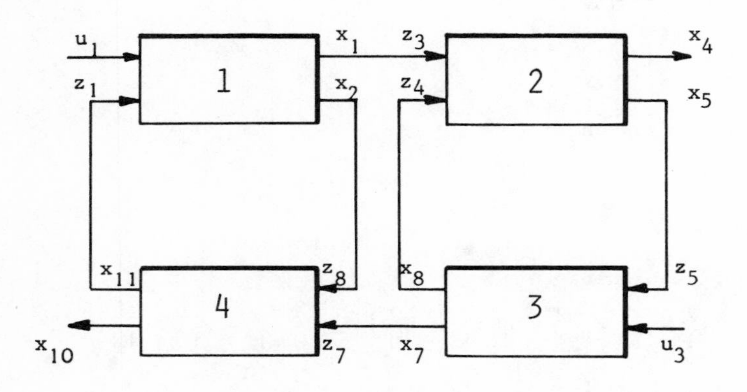

Fig. 6.3 : Subsystem structure for example 2

The output y are given by

$$\begin{bmatrix} y_1 \\ \\ y_2 \end{bmatrix} = \begin{bmatrix} 1 & 0 & 0 \\ & & \\ 0 & 1 & 0 \end{bmatrix} \begin{bmatrix} x_1 \\ x_2 \\ x_3 \end{bmatrix}$$

The interaction inputs are given by $z_1 = x_{11}$ where as shown in Fig. 6.3, x_{11} is a state of subsystem 4.

Subsystem 2 :

$$\begin{aligned} \mathring{x}_4 &= x_5 \\ \mathring{x}_5 &= x_6 \\ \mathring{x}_6 &= a_{64}x_4 + a_{65}x_5 + a_{66}x_6 + z_3 + z_4 \end{aligned} \qquad 6.3.2$$

$$\begin{bmatrix} y_3 \\ y_4 \end{bmatrix} = \begin{bmatrix} 1 & 0 & 0 \\ 0 & 1 & 0 \end{bmatrix} \begin{bmatrix} x_4 \\ x_5 \\ x_6 \end{bmatrix}$$

$z_3 = x_1, \quad z_4 = x_8$

<u>Subsystem 3</u> :

$$\begin{aligned} \mathring{x}_7 &= x_8 \\ \mathring{x}_8 &= x_9 \\ \mathring{x}_9 &= a_{97}x_7 + a_{98}x_8 + a_{99}x_9 + u_3 + z_5 \end{aligned} \tag{6.3.3}$$

with

$$\begin{bmatrix} y_5 \\ y_6 \end{bmatrix} = \begin{bmatrix} 1 & 0 & 0 \\ 0 & 1 & 0 \end{bmatrix} \begin{bmatrix} x_7 \\ x_8 \\ x_9 \end{bmatrix}$$

and $z_5 = x_5$

<u>Subsystem 4</u> :

$$\begin{aligned} \mathring{x}_{10} &= x_{11} \\ \mathring{x}_{11} &= x_{12} \\ \mathring{x}_{12} &= a_{12,10}\, x_{10} + a_{12,11}x_{11} + a_{12,12}x_{12} + z_7 + z_8 \end{aligned} \tag{6.3.4}$$

with

$$\begin{bmatrix} y_7 \\ y_8 \end{bmatrix} = \begin{bmatrix} 1 & 0 & 0 \\ 0 & 1 & 0 \end{bmatrix} \begin{bmatrix} x_{10} \\ x_{11} \\ x_{12} \end{bmatrix} \quad \text{and } z_7 = x_7, \; z_8 = x_2$$

Since no values were given by Pearson [1] for the 'a's, these were chosen to be

$a_{31} = -1, \; a_{32} = -2, \; a_{33} = -3, \; a_{64} = -2, \; a_{65} = -3$

$a_{66} = -1, \; a_{97} = -3, \; a_{98} = -2, \; a_{99} = -1, \; a_{12,10} = -1$

$a_{12,11} = -2, \; a_{12,12} = -3.$

Here N = 4 in the cost function of equation 6.1.3. The system weighting matrices were chosen to be block diagonal with

$$Q_{ii} = \begin{bmatrix} 1 & 0 & 0 \\ 0 & 1 & 0 \\ 0 & 0 & 0 \end{bmatrix} \quad ; \quad i = 1, 2, 3, 4$$

and the control weighting matrices were taken to be $[1]$. The interconnection weighting matrices were taken to be $S_{ii} = I_2$, the second order identify matrix, for subsystem 2, 4 and $[1]$ for subsystem 1, 3.

In order to solve this problem we note that the lower level problem is to minimise L_i for each of the four subsystems subject respectively to the dynamic constraints given by equations 6.3.1 to 6.3.4 given a Lagrange multiplier trajectory $\underline{\lambda} = \underline{\lambda}^*$ where $\underline{\lambda}$ and $\underline{\lambda}^*$ are of order 6 (since we have six interconnection constraints). The period of optimisation was chosen to be from zero to 10, with the time discretisation being in steps of 0.1. These lower level problems were solved sequentially (in the order of : Min L_1, Min L_2, Min L_3, Min L_4) on an IBM 370/165 digital computer by integrating for each subsystem a third order matrix Riccati equation and a third order vector differential equation and some simple multiplication and addition operations for the calculation of the control from these solutions.

On the second level, we used a conjugate gradient type of algorithm as described by equation 6.2.6 , 6.2.7 to improve the $\underline{\lambda}^*$ trajectories. Convergence was measured by the error criterion defined by Pearson [1], i.e.

$$\text{Error} = \sum_{i=1}^{4} \int_0^{10} \left\{ \underline{z}_i - \sum_{j=1}^{4} L_{ij}\underline{x}_j \right\}^T \left\{ \underline{z}_i - \sum_{j=1}^{4} L_{ij}\underline{x}_j \right\} dt/\Delta T$$

In this error measure when $\underline{z}_i = \sum_{j=1}^{4} L_{ij}\underline{x}_j$, then our original problem is solved, i.e. if the error approaches zero, the resulting $\underline{x}$, $\underline{u}$ are the optimal state and control trajectories.

Fig. 6.4 shows the decrease of this error measure at each iteration of the second level. The error stabilises around 10^{-5} because numerical inaccuracies build up around this point and this reduces the rate of convergence.

From fig. 6.4 we see that the error decreases rapidly, i.e. it takes only 6 iterations for the error to decrease to 10^{-5} thus providing a solution adequately close to the optimum of the original problem.

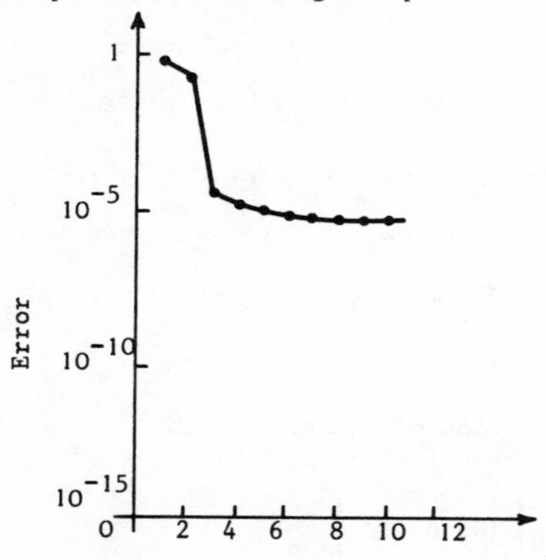

Fig. 6.4 : Convergence of the Goal Coordination method on Pearson's example.

In solving the above example we have seen that it is possible to solve large scale system problems using the method although it is not very attractive to do so unless a multiprocessor was used at least for the case where the overall system is small enough for it to be solvable using a single level method. Of course, if the system is of very high order, it will not be possible to store all the matrices required by the single level solution although it may well be possible to do so using the present approach. Even for relatively low order systems, an interesting modification by Tamura [5] makes the Goal Coordination approach more attractive. The basis of the modification is to convert the lower level problems from a functional to a parametric optimisation by adding an additional level to the hierarchy. We next describe this three level method of Tamura.

6.4 THE THREE LEVEL METHOD OF TAMURA

We develop first the Goal Coordination method in discrete time and then show how Tamura's modification leads to an attractive three level method.

6.4.1 The Goal Coordination method for Discrete Dynamical Systems

The problem is to minimise

$$J = \sum_{i=1}^{N}\left[\frac{1}{2}\|\underline{x}_i(K)\|^2_{P_i} + \sum_{k=0}^{K-1}\frac{1}{2}\left(\|\underline{x}_i(k)\|^2_{Q_i(k)} + \frac{1}{2}\|\underline{z}_i(k)\|^2_{S_i(k)} + \|\underline{u}_i(k)\|^2_{R_i(k)}\right)\right] \qquad 6.4.1$$

where $\frac{1}{2}\|\underline{x}_i(K)\|^2_{P_i}$ is the cost for the terminal interval and the terms within the inner summation represent the cost over the rest of the optimisation sequence, i.e. from k = 0 to K-1.

This minimisation is to be performed subject to the subsystem dynamic constraints, i.e.

$$\underline{x}_i(k+1) = A_i\,\underline{x}_i(k) + B_i\,\underline{u}_i(k) + C_i\,\underline{z}_i(k) \qquad 6.4.2$$

$$i = 1, 2, \ldots, N\ ;\ k = 0, 1, 2, \ldots K-1\ ;$$

and we assume that the initial state is known, i.e.

$$\underline{x}_i(0) = \underline{x}_{i0} \qquad 6.4.3$$

As in the continuous time case, $\underline{z}_i$ is the vector of interaction inputs coming in from the other subsystems, i.e.

$$\underline{z}_i(k) = \sum_{j=1}^{N} L_{ij}\,\underline{x}_j(k)\quad ;\ k = 0, 1, \ldots K-1;\quad i = 1, 2, \ldots, N \qquad 6.4.4$$

To solve this problem it is necessary as in the continuous time case described in the previous section to maximise a dual function $\phi(\underline{\lambda})$ w.r.t. $\underline{\lambda}$ where

$$\phi(\underline{\lambda}) = \min_{\underline{x},\underline{u},\underline{z}} L(\underline{x}, \underline{u}, \underline{z}, \underline{\lambda}) \qquad 6.4.5$$

subject to equations 6.4.2, 6.4.3 where

$$L(\underline{x},\underline{u},\underline{z},\underline{\lambda}) = \sum_{i=1}^{N}\left\{\frac{1}{2}\|\underline{x}_i(K)\|^2_{P_i} + \sum_{k=0}^{K-1} \frac{1}{2}\|\underline{x}_i(k)\|^2_{Q_i} + \frac{1}{2}\|\underline{z}_i(k)\|^2_{S_i} + \frac{1}{2}\|\underline{u}_i(k)\|^2_{R_i} + \right.$$

$$\left. \underline{\lambda}_i^T \underline{z}_i - \sum_{j=1}^{N} \underline{\lambda}_j^T L_{ji}\underline{x}_i(k) \right\} = \sum_{i=1}^{N} L_i \qquad 6.4.6$$

where

$$L_i = \frac{1}{2}\|\underline{x}_i(K)\|^2_{P_i} + \sum_{k=0}^{K-1} \frac{1}{2}\|\underline{x}_i(k)\|^2_{Q_i(k)} + \frac{1}{2}\|\underline{u}_i(k)\|^2_{R_i(k)} + \frac{1}{2}\|\underline{z}_i\|^2_{S_i(k)} +$$

$$\underline{\lambda}_i^T \underline{z}_i - \sum_{j=1}^{N} \underline{\lambda}_j^T L_{ji}\, \underline{x}_i(k) \qquad 6.4.7$$

Thus as in the continuous time case, it is possible to separate the problem of minimising the Lagrangian L into minimising N independent sub-Lagrangians L_i for given sequences $\underline{\lambda} = \underline{\lambda}^*$ supplied by a second level, each subject to equations 6.4.2, 6.4.3. The Lagrange Multiplier vector sequences can be improved at the second level by using a gradient type algorithm since

$$\nabla \phi(\underline{\lambda})\Big|_{\underline{\lambda}=\underline{\lambda}^*} = \left[\underline{z}_i(k) - \sum_{j=1}^{N} L_{ij}\, \underline{x}_j(k)\right] \qquad 6.4.8$$

$$i = 1, 2, \ldots N \quad ; \quad k = 0, 1, \ldots K-1$$

and this simple analytical expression for the gradient can be used in a steepest ascent or conjugate gradient type of algorithm as in the continuous time case. Having reformulated the Goal Coordination Method in discrete time, let us examine the modification of Tamura.

6.4.2 The modification of Tamura

Note that for a given trajectory $\underline{\lambda}^*(k)$, k = 0, 1,...K-1, the First Level problem of minimising L_i subject to the dynamic constraints given by equations 6.4.2, 6.4.3 can itself be treated by duality and decomposition, since, instead of decomposing the Lagrangian into the sub-Lagrangians for each subsystem, the subsystem Lagrangian itself can be decomposed by the index k leading at the lowest level to a parametric as opposed to a functional optimisation. Here we will consider a decomposition in time as opposed to the decomposition by subsystems considered previously.

Define the dual problem of minimising L_i in equation 6.4.7 subject to equation 6.4.2 as

$$\text{Maximise}_{\underline{p}}\ M(\underline{p})$$

where

$$M(\underline{p}) = \underset{\underline{x},\underline{u}}{\text{Min}} \left\{ \frac{1}{2} \| \underline{x}_i(K) \|^2_{P_i} + \sum_{k=0}^{K-1} \left(\frac{1}{2} \| \underline{x}_i(k) \|^2_{Q_i} + \frac{1}{2} \| \underline{u}_i(k) \|^2_{R_i} + \frac{1}{2} \| \underline{z}_i \|^2_{S_i} + \right.\right.$$

$$\underline{p}_i(k)^T \left[A_i \, \underline{x}_i(k) + B_i \, \underline{u}_i(k) + C_i \, \underline{z}_i(k) - \underline{x}_i(k+1) \right] + \qquad 6.4.9$$

$$\left.\left. \underline{\lambda}_i^{*T} \underline{z}_i - \sum_{j=1}^{N} \underline{\lambda}_j^{*T} L_{ji} \, \underline{x}_i \right) \right\}$$

subject to the known initial state $\underline{x}_i(0) = \underline{x}_{i0}$.

To solve this dual problem numerically, it is necessary to compute the value of the dual function $M(\underline{p})$ for given $\underline{p} = \underline{p}^*$ and then to maximise $M(\underline{p})$ w.r.t. $\underline{p}$ by some gradient technique. The gradient of $M(\underline{p})$ is given by

$$\nabla M(\underline{p}) \Big|_{\underline{p}=\underline{p}^*} = - \underline{x}_i(k+1) + A_i \underline{x}_i(k) + B_i \underline{u}_i(k) + C_i \underline{z}_i(k) \qquad 6.4.10$$

$$k = 0, 1, \ldots K-1 \; ; \qquad i = 1, 2, \ldots N$$

where $\underline{x}_i$, $\underline{u}_i$ are the solutions obtained after minimising L_i subject to equation 6.4.2 for given $\underline{p} = \underline{p}^*$. The computation of $M(\underline{p})$ for a fixed $\underline{p} = \underline{p}^*$ and $\underline{\lambda} = \underline{\lambda}^*$ can be performed by minimising the function independently for each time index k as follows ;

Define the Hamiltonian of the ith subsystem by

$$H_i(\underline{x}_i(k), \underline{u}_i(k), \underline{z}_i(k), k) = \frac{1}{2} \| \underline{x}_i(k) \|^2_{Q_i} + \frac{1}{2} \| \underline{u}_i(k) \|^2_{R_i} + \frac{1}{2} \| \underline{z}_i \|^2_{S_i} + \underline{\lambda}_i^{*T} \underline{z}_i$$

$$- \sum_{j=1}^{N} \underline{\lambda}_j^{*T} L_{ji} \, \underline{x}_i + \underline{p}_i^{*T}(k) \left[A_i \underline{x}_i(k) + B_i \underline{u}_i(k) + C_i \underline{z}_i(k) \right] \qquad 6.4.11$$

$$k = 0, 1, \ldots K-1 \; ; \qquad i = 1, 2, \ldots N$$

Then using equation 6.4.9 :

$$M(\underline{p}) = \frac{1}{2} \| \underline{x}_i(K) \|^2_{P_i} - p_i^{*T}(K-1) \, \underline{x}_i(K) + \sum_{k=0}^{K-1} \left(H_i(\underline{x}_i(k), \underline{u}_i(k), \underline{z}_i(k), k) \right.$$

$$\left. - \underline{p}_i^* (k-1)^T \, \underline{x}_i(k) \right)$$

where $\underline{p}(-1)$ is defined to be zero.

The minimisation problem for a fixed $\underline{p} = \underline{p}^*$ then becomes :

<u>for k = 0</u>

Minimise w.r.t. $\underline{u}_i(0)$, $\underline{z}_i(0)$: $\{ H_i(\underline{x}_i(0), \underline{u}_i(0), \underline{z}_i(0))$ subject to $\underline{x}_i(0) = \underline{x}_{i0} \}$. It is possible to obtain an explicit solution in this case by setting the partial derivative of H_i w.r.t. $\underline{u}_i(0)$, $\underline{z}_i(0)$ to zero to yield :

$$\underline{u}_i(0) = - R_i^{-1} \, B_i^T \, \underline{p}_i^* (0)$$

$$\underline{z}_i(0) = - S_i^{-1} \, (C_i^T \, \underline{p}_i^* (0) + \underline{\lambda}_i^*(0)) \qquad 6.4.12$$

For k = 1, 2, ... K-1

Minimise $\quad H_i(\underline{x}_i(k), \underline{u}_i(k), \underline{z}_i(k), k) - \underline{p}_i^*(k-1)^T \underline{x}_i(k)$

w.r.t. $\quad \underline{x}_i(k), \underline{u}_i(k), \underline{z}_i(k)$

The explicit solution in this case is

$$\underline{x}_i(k) = -Q_i(k)^{-1}\left[A_i^T \underline{p}_i^*(k) - \underline{p}_i^*(k-1) - \sum_{j=1}^{N}[\underline{\lambda}_j^{*T} L_{ji}]^T\right]$$

$$\underline{u}_i(k) = -R_i^{-1} B_i^T \underline{p}_i^*(k) \qquad 6.4.13$$

$$\underline{z}_i(k) = -S_i^{-1}(C_i^T \underline{p}_i^*(k) + \underline{\lambda}_i^*(k))$$

For k = K

Minimise w.r.t. $\underline{x}_i(K)$

$$\frac{1}{2}\|\underline{x}_i(K)\|^2_{P_i} - \underline{p}_i^{*T}(K-1)\,\underline{x}_i(K)$$

which gives

$$\underline{x}_i(K) = P_i^{-1}\underline{p}_i^*(K-1) \qquad 6.4.14$$

Thus the integrated problem of minimising J in equation 6.4.1 subject to the dynamics given by equations 6.4.2 - 6.4.4 can be solved by a three level algorithm where on level 1 for given $\underline{\lambda}^*(k)$, $\underline{p}^*(k)$ sequences, it is merely necessary to substitute into the explicit solutions given by equations 6.4.12 - 6.4.14 in order to obtain the optimal $\underline{x}$, $\underline{u}$, $\underline{z}$, which can be used at the second level to calculate the gradient of $M(\underline{p})$ from equation 6.4.10 and this gradient can be used to improve $\underline{p}$ in order to maximise $M(\underline{p})$. On the third level, the optimal $\underline{p}$ obtained from the second level optimisation can be used to iteratively improve $\phi(\underline{\lambda})$ in order to maximise it. The overall optimum is achieved when both $\nabla\phi(\underline{\lambda})$ and $\nabla M(\underline{p})$ go to zero. Fig. 6.5 shows the optimisation structure.

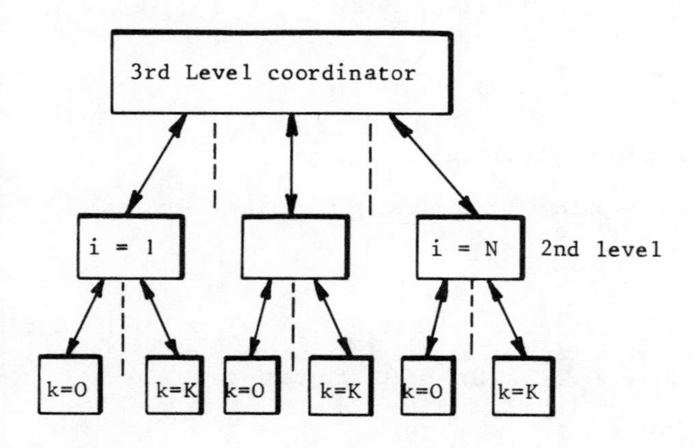

Fig. 6.5 : The three level method of Tamura

6.4.3 Remarks

1. The above method is attractive because an explicit solution is obtained at the lowest level making this level's problem trivial and ensuring that it is not necessary to solve complicated equations.

2. It is still necessary as in the standard Goal Coordination Method considered previously to introduce the term $\|\underline{z}_i\|^2_{S_i}$ in the cost function in order to avoid a singular solution.

6.4.4 Example

As a simple example of the application of the discrete minimum principle and Tamura's three level algorithm, consider the system :

$$x_1(k+1) = x_1(k) + u_1(k) + z_1(k) \qquad 6.4.15$$

$$x_2(k+1) = - x_2(k) + u_2(k) + z_2(k) \qquad 6.4.16$$

$$z_1(k) = x_2(k) \qquad 6.4.17$$

$$z_2(k) = x_1(k) \qquad 6.4.18$$

$$J = \sum_{k=0}^{3} \frac{1}{2} (x_1^2(k) + x_2(k)^2 + u_1^2(k) + u_2(k)^2 + z_1^2(k) + z_2^2(k))$$

Here the Lagrangian can be written as :

$$L(\underline{x}, \underline{u}, \underline{z}, \lambda) = \sum_{k=0}^{3} \frac{1}{2} (x_1(k)^2 + x_2(k)^2 + u_1(k)^2 + u_2(k)^2 + z_1(k)^2 + z_2(k)^2) + \lambda_1(k) \left[z_1(k) - x_2(k)\right] + \lambda_2(k) \left[z_2(k) - x_1(k)\right] \qquad 6.4.19$$

subject to equations 6.4.15 and 6.4.16.

The Lagrangian in equation 6.4.19 can be rewritten as :

$$\sum_{k=0}^{3} \left[\frac{1}{2} x_1(k)^2 + \frac{1}{2} u_1(k)^2 + \frac{1}{2} z_1(k)^2 + \lambda_1(k)\, z_1(k) - \lambda_2(k)\, x_1(k)\right] + \left[\frac{1}{2} x_2^2(k) + \frac{1}{2} u_2^2(k) + \frac{1}{2} z_2^2(k) - \lambda_1(k)\, x_2(k) + \lambda_2(k)\, z(k)\right]$$

$= L_1 + L_2$ for given $\underline{\lambda}$.

Thus in the Goal Coordination method for discrete dynamical systems, we obtain at level 1 two independent dynamic optimisation problems, i.e. Min L_1 s.t. equation 6.4.15 and Min L_2 s.t. equation 6.4.16, for given λ_1, λ_2, and at level two we use a gradient coordinator to improve λ_1, λ_2 using the fact that the gradient of the dual function is given by :

$$\begin{bmatrix} g_1 \\ g_2 \end{bmatrix} = \begin{bmatrix} z_1 - x_2 \\ z_2 - x_1 \end{bmatrix}$$

The minimisation of L_1 s.t. 6.4.15 and L_2 s.t. 6.4.16 can be done either by a straight application of the discrete minimum principle as outlined in the last chapter, or perhaps more efficiently by using the above three level method as follows :

Consider the minimisation of L_1 w.r.t. equation 6.4.15, i.e.

$$\text{Min}\left[\frac{1}{2}\,x_1(k)^2 + \frac{1}{2}\,u_1(k)^2 + \frac{1}{2}\,z_1(k)^2 + \lambda_1(k)\,z_1(k) - \lambda_2(k)\,x_1(k)\right]$$

subject to $\quad x_1(k+1) = x_1(k) + u_1(k) + z_1(k).$

This problem can be solved within a two level structure where for given p_1, $\underline{\lambda}$, we solve the 4 independent minimisation problems with the solutions given by

$$\begin{cases} u_1(0) = -\,p_1(0) \\ z_1(0) = -(p_1(0) + \lambda_1(0)) \end{cases}$$

$$\begin{cases} x_1(1) = -\left[p_1(1) - p_1(0) - \lambda_2(1)\right] \\ u_1(1) = -\,p_1(1) \\ z_1(1) = -\left[(p_1(1) + \lambda_1(1))\right] \end{cases}$$

$$\begin{cases} x_1(2) = -\,(p(2) - p(1) - \lambda_2(2)) \\ u_1(2) = -\,p_1(2) \\ z_1(2) = -\,(p_1(2) + \lambda_1(2)) \end{cases}$$

$$x_1(3) = p_1(2)$$

and on level 2 we improve p_1 by a gradient procedure using the fact that the gradient is given by

$$\text{grad} = -\,x_1(k+1) + x_1(k) + u_1(k) + z_1(k)$$

so that the whole problem can be solved in a three level structure.

Next we describe the important time delay method of Tamura.

6.5 THE TIME DELAY ALGORITHM OF TAMURA

This algorithm solves a class of problems which are of great practical importance. The overall system has multiple pure time delays in the state and control variables. In addition, the states and controls are bounded by hard inequality constraints. We will see how the method treats in a natural way these inequality constraints.

The system dynamics are assumed to be represented by a high order difference equation of the form :

$$\underline{x}(k+1) = A_0\underline{x}(k) + A_1\,\underline{x}(k-1) + \ldots + A_\theta\,\underline{x}(k-\theta) + B_0\,\underline{u}(k) +$$
$$B_1\,\underline{u}(k-1) + \ldots + B_\theta\,\underline{u}(k-\theta) \qquad 6.5.1$$

where A_i $(i = 0, 1, \ldots, \theta)$ are n x n matrices, $\underline{x}$ is a an n vector, $\underline{u}$ is an r vector, B_i $(i = 0, 1, \ldots \theta)$ are n x r matrices.

In addition it is assumed that $\underline{x}(k) = \underline{0}$; $\underline{u}(k) = \underline{0}$ for $k < 0$ and

$$\underline{x}(0) = \underline{x}_0 \qquad 6.5.2$$

We can interpret equation 6.5.2 as : we assume that the system is at some steady state operating point up to the instant k = 0 when it receives an unknown disturbance which takes the state of the system to a known value $\underline{x}_0$. Although we assume the steady state operating point to be zero here, it is just as easy to consider a non zero operating point by interpreting the sequences $\underline{x}(k)$, $\underline{u}(k)$, $k > 0$ to be varying about this actual steady state.

The state and control can be bounded by the inequality constraints

$$\underline{x}_{Min} \leqslant \underline{x}(k) \leqslant \underline{x}_{Max}, \quad k = 1, \ldots K$$
$$\underline{u}_{Min} \leqslant \underline{u}(k) \leqslant \underline{u}_{Max}, \quad k = 0, 1, \ldots K-1 \qquad 6.5.3$$

and it is desired to minimise

$$J = \frac{1}{2}\left\| \underline{x}(K) \right\|_P^2 + \sum_{k=0}^{K-1} \frac{1}{2}\left(\left\|\underline{x}(k)\right\|_{Q(k)}^2 + \left\|\underline{u}(k)\right\|_{R(k)}^2\right) \qquad 6.5.4$$

where P, Q and R are assumed to be positive definite diagonal matrices.

After our studies in the last chapter, even a cursory examination of this problem shows that it is very difficult to solve. The source of the difficulty is two fold : first of all the inequality constraints on the states and controls pose immense problems and then again the multiple time delays could lead to state space augmentation. For example if our system had been

$$x(k+1) = x(k) + x(k-1)$$

it would require the introduction of the additional variable $x(k-1) = x_1(k)$ to have the standard state space form

$$\begin{bmatrix} x(k+1) \\ x_1(k+1) \end{bmatrix} = \begin{bmatrix} 1 & 1 \\ 1 & 0 \end{bmatrix} \begin{bmatrix} x(k) \\ x_1(k) \end{bmatrix}$$

so that in the case of multiple time delays, we might end up with a very high order system. Tamura's method allows us to avoid both the difficulties of the inequality constraints as well as those of state space augmentation.

Write the Hamiltonian H of the overall system as

$$H(\underline{x}(k), \underline{u}(k), \underline{p}(k), k) = \frac{1}{2}\left(\left\|\underline{x}(k)\right\|_{Q(k)}^2 + \left\|\underline{u}(k)\right\|_{R(k)}^2\right) + \sum_{j=0}^{\theta} \underline{p}(k+j)^T (A_j\,\underline{x}(k) +$$
$$B_j\,\underline{u}(k)) \; ; \; k = 0, 1, \ldots, K-1 \qquad 6.5.5$$

where $\underline{p}(k)$ is defined to be zero for $k \geqslant K$.

For a fixed $\underline{p} = \underline{p}^* = [\underline{p}(0)^*, \ldots \underline{p}(K-1)^*]$ it is possible to write the Lagrangian as

$$L(\underline{x}, \underline{u}, \underline{p}^*, k) = \frac{1}{2} \| \underline{x}(K) \|^2_{P(K)} - \underline{p}^{*T}(K-1)\, \underline{x}(K) + \sum_{k=0}^{K-1} \Big\{ H(\underline{x}, \underline{u}, \underline{p}^*, k) - \underline{p}^*(k-1)\, \underline{x}(k) \Big\} \qquad 6.5.6$$

subject to

$$\underline{x}_{Min} \leqslant \underline{x}(k) \leqslant \underline{x}_{Max} \quad ; \quad k = 1, 2, \ldots K$$

$$\underline{u}_{Min} \leqslant \underline{u}(k) \leqslant \underline{u}_{Max} \quad ; \quad k = 0, 1, 2, \ldots, K-1$$

and as in the previous sections, in order to obtain the optimal control, it is necessary to maximise w.r.t. $\underline{p}$ the minimum w.r.t. $\underline{x}$, $\underline{u}$ of the Lagrangian $L(\underline{x}, \underline{u}, \underline{p}^*, k)$. One of the attractions of this formulation is that the Lagrange multiplier vector $\underline{p}$ is of the same dimension as $\underline{x}$ despite the existence of the delays.

As in the three level algorithm, on examining the expression for the Lagrangian in equation 6.5.6 it follows that the Lagrangian can be decomposed into the following (K+1) independent minimisation problems for fixed $\underline{p}^*$.

i) <u>for k = 0</u>

From equation 6.5.6 using the definition of H from equation 6.5.5, the first problem is :

$$\text{Min } H(\underline{x}(0), \underline{u}(0), \underline{p}(0)) = \min_{\underline{u}(0)} \Big\{ \frac{1}{2} \left(\| \underline{x}(0) \|^2_{Q(0)} + \| \underline{u}(0) \|^2_{R(0)} \right) + \sum_{j=0}^{\theta} \underline{p}^{*T}(j) \left(A_j\, \underline{x}(0) + B_j\, \underline{u}(0) \right) \Big\}$$

subject to

$$\underline{x}(0) = \underline{x}_0, \; \underline{u}_{Min} \leqslant \underline{u}(0) \leqslant \underline{u}_{Max}$$

The solution of this parametric optimisation problem is further simplified by the fact that here R(0) has been assumed to be a diagonal matrix so that the minimisation problem reduces to a set of r independent one variable minimisations. For a single variable minimisation it is of course easy to include the inequality constraints. The explicit solution is thus given by

$$\underline{u}^*(0) = \text{Sat}_2 \left[-R^{-1}(0) \sum_{j=0}^{\theta} B_j^T\, \underline{p}^*(j) \right] \qquad 6.5.7$$

where for i = 1, 2, ..., r, the ith element of $\text{Sat}_2(\eta)$ is given by

$$\text{Sat}_2(\eta) = \begin{cases} u_{Max,i} & \text{if } \eta_i > u_{Max,i} \\ \eta_i & \text{if } u_{Min,i} \leqslant \eta_i \leqslant u_{Max,i} \\ u_{Min,i} & \text{if } \eta_i < u_{Min,i} \end{cases} \qquad 6.5.8$$

The solution given in equation 6.5.7 is obtained from a trivial manipulation by setting $\partial H / \partial \underline{u}(0) = \underline{0}$ and then noting that since R is a diagonal matrix, this can be viewed as a set of r independent one variable solutions bounded by the limits.

ii) for $k = 1, 2, \ldots, K-1$

Similarly, for this case the minimisation problem is :

Minimise w.r.t. $\underline{x}(k)$, $\underline{u}(k)$

$$H(\underline{x}(k), \underline{u}(k), \underline{p}^*(k), k) - \underline{p}^*(k-1)^T \underline{x}(k)$$

subject to

$$\underline{x}_{Min} \leqslant \underline{x}(k) \leqslant \underline{x}_{Max}$$

$$\underline{u}_{Min} \leqslant \underline{u}(k) \leqslant \underline{u}_{Max}$$

Again since we have assumed that Q(k), R(k) are diagonal matrices, this becomes a set of n+r independent one variable minimisation with the solution given by

$$\underline{x}^*(k) = Sat_1\left\{ -Q^{-1}(k) \left[-p^*(k-1) + \sum_{j=0}^{\theta} A_j^T \underline{p}^*(k+j)\right]\right\}$$

$$\underline{u}^*(k) = Sat_2\left\{\left[-R^{-1}(k) \sum_{j=0}^{\theta} B_j^T \underline{p}^*(k+j)\right]\right\} \qquad 6.5.9$$

where for $i = 1, 2, \ldots n$, the ith element of $Sat_1(\xi_i)$ is

$$Sat_1(\xi_i) = \begin{cases} x_{Max,i} & \text{if} \quad \xi_i > x_{Max,i} \\ \xi_i & \text{if} \quad x_{Min,i} \leqslant \xi_i \leqslant x_{Max,i} \\ x_{Min,i} & \text{if} \quad \xi_i < x_{Min,i} \end{cases} \qquad 6.5.10$$

iii) for $k = K$

$$\underset{\underline{x}(K)}{\text{Minimise}} \ \frac{1}{2}\|\underline{x}(K)\|^2_{P(K)} - \underline{p}^{*T}(K-1)\,x(K)$$

subject to

$$\underline{x}_{Min} \leqslant x(K) \leqslant \underline{x}_{Max}$$

The solution is

$$\underline{x}^*(K) = Sat_1\left[P^{-1}\,\underline{p}^*(K-1)\right] \qquad 6.5.11$$

The solution given by equations 6.5.7 to 6.5.11 enables one to obtain the minimum of the Lagrangian analytically for a given $\underline{p} = \underline{p}^*$. On the second level it is necessary to improve $\underline{p}^*$ in order to maximise the dual function but this is easy to do since the gradient is given by the error in the system equation (i.e. R.H.S. of equation 6.5.1 minus $\underline{x}(k+1)$) so that a simple gradient method of the kind discussed previously could be used.

Comment

The Time Delay algorithm of Tamura is the only Goal Coordination type algorithm which has so far had much success. The main reasons for this are :

(a) There is a substantial computational saving since it is not necessary to augment the state space to account for the delays.

(b) The inequality constraints are treated in a very simple way.

(c) The cost function is meaningful since no additional terms have been introduced to avoid the singular solutions which occur in the standard Goal Coordination solution.

Thus the method is able to get around most of the disadvantages of the standard Goal Coordination approach (except that it is still necessary to perform a linear search for the second level gradient algorithm). However, the most important practical advantage of the approach is that time delay systems of the kind where the Tamura algorithm can be used have in general relatively slow dynamics (because of the delays). This means that given some disturbance before the origin of time which takes the system state to some known state $\underline{x}_0$ (or a state $\underline{x}_0$ which can be obtained by suitable measurement), it may well be possible to calculate the control sequences rapidly enough before the initial state changes significantly. This is one of the few cases where open-loop control methods of the kind developed here could well be used for on-line control. In the next section we apply the method to a significant practical problem which has slow dynamics, i.e. computer control of rush hour traffic.

6.6 EXAMPLE : CONTROL OF RUSH HOUR TRAFFIC

In this section we give a brief outline of the method of modelling oversaturated networks and then we apply the Tamura algorithm to the optimisation problem. Note that time delays arise in a natural way in urban road traffic networks and these ensure that the system has a slow dynamic response.

6.6.1 Model of an oversaturated traffic network [7]

Urban road traffic networks consist of junctions and interconnecting roads. We can model the dynamic behaviour of oversaturated networks by treating the queues at junction approaches as the state variables. Consider first a simple one-way no-turn intersection as shown in Fig. 6.6. Let $q_i(t)$ denote the arrival rates of vehicles in the direction i. Let i = 1 denote the horizontal traffic direction and i = 2 the vertical direction. Let s_i denote the saturation flow rates of vehicles in the direction i. This is the maximum number of vehicles which can pass through the intersection per cycle, in the direction i if this direction had all the available green signal. Let C be the duration of the cycle time and 1 the loss time due to the amber phase. Then $C = G_1 + G_2 + 1$ where G_i, i = 1, 2, is the duration of the green in direction i. Let $\overline{g}_i(t)$ be the average departure rate (number of vehicles/C). Define the control variables $u_i(t)$ to be the percentage of green in the direction i. Then it is easy to see that [7]

$$u_i(t) = \overline{g}_i(t) \quad / \quad s_i(t) = G_i/C \qquad i = 1,\ 2$$

In this formulation, the cycle time C is a known constant.

If the cycle time C is a constant, then it is necessary to have only one control $u = u_i$ and the other control is then $C - u_i - 1$.

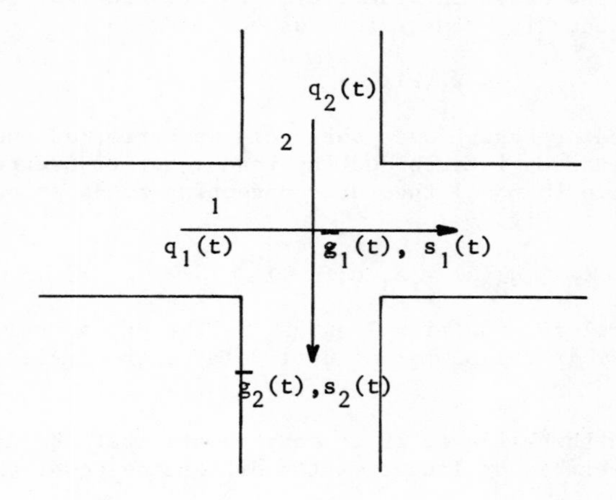

Fig. 6.6 : An oversaturated intersection

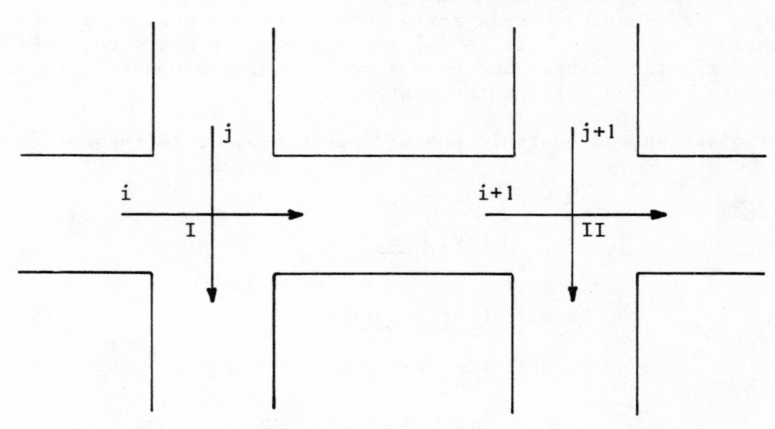

Fig. 6.7 : A model for the interconnecting road

The controls u must be subject to inequality constraints *

$$u_{min} \leqslant u \leqslant u_{max}$$

Now, define the state variables $x_i(t)$ as the instantaneous queue length in the direction i. Then, from one cycle to the next, the evolution of the queues can be described by

$$x_i(k+1) = x_i(k) + q_i(k) - \bar{g}_i(k) = x_i(k) + q_i(k) - s_i u_i(k)$$

---

* Too short a green is wasteful. On the other hand, if the green is too long, drivers tend to believe that the traffic signals have broken down. Hence the upper constraints.

Since queues exist in practice on interconnecting roads, such queues must be subject to inequality constraints of the form

$$0 \leqslant x_i(k) \leqslant x_{max,i}$$

Now, since averaging over one cycle is performed and over this period the diffusion phenomenon on interconnecting roads (due to overtaking) can be ignored, it is possible to model the interconnecting roads as pure delay elements so that, for example, in Fig. 6.7

$$q_3(k) = s_1 \; u_1(k-m)$$

where u_1 is the control for junction I and s_1 is the horizontal direction saturation flow from I and m is the number of unit delays, the cycles, between the two junctions.

Extending these ideas, it is easy to see that the dynamic behaviour of networks could be described by linear vector difference equations of the form

$$\underline{x}(k+1) = A\underline{x}(k) + B_0\underline{u}(k) + B_1\underline{u}(k-1) + \ldots B_m\underline{u}(k-m) + \underline{V}$$

where A is an n x n identity matrix for a network with n queues, $\underline{x}^T = (x_1, x_2 \ldots x_n)$ are the queues, B_j $(j = 0, 1, 2, \ldots, m)$ are appropriate matrices, $\underline{u}$ is a r vector of the r traffic signals and $\underline{V}$ is a vector which accounts for the external inputs which come in from outside the system.

The states and controls are of course subject to inequality constraints

$$\underline{0} \leqslant \underline{x}(k) \leqslant \underline{x}_{max}$$

$$\underline{u}_{min} \leqslant \underline{u}(k) \leqslant \underline{u}_{max}$$

6.6.2 Cost function

A suitable cost function for this system is

$$J = \min_{\underline{u}(k)} \sum_{k=1}^{K-1} \frac{1}{2}\| \underline{x}(k) \|_Q^2 + \frac{1}{2}\| \underline{u}(k) - \underline{u}^0 \|_R^2$$

where Q and R are diagonal weighting matrices and $\underline{u}^0$ is a nominal control vector chosen a priori to maximise the utilisation of the network. Note that the term $\frac{1}{2}\| \underline{u}(k) - \underline{u}^0 \|_R^2$ is not convincing from a practical point of view. It has been added essentially to fit the present problem into the standard form. For this reason R is usually chosen so as to make $\frac{1}{2}\| \underline{u}(k) - \underline{u}^0 \|_R^2$ a very small component of J. Basically this cost function minimises delays by penalising longer queues.

The traffic network optimisation problem is in a form where Tamura's Delay Algorithm can be used directly. As an example, consider the small network in London shown in Fig. 6.8.

6.6.3 An example

The network consists of three junctions and is a major trouble spot in the West London area. The difficulties arise because the junctions are very close, both to each other and to the neighbouring junctions, so that the storage available on the linking roads is fairly small. The state inequality constraints cannot therefore be relaxed. Based on the existing control structure, the system has the following three controls, i.e.

u_1 = fraction of the green associated with the flows 1 and 2,

u_2 = fraction of the green associated with the flows 6 and 7 and with a two-cycle delay with respect to u_1,

u_3 = fraction of the green associated with flows 10 and 12 and with a one-cycle delay with respect to u_2.

The cycle time for all the junctions is 1 min and the loss time is 6 sec. The time delays between junctions 236 and 234 in fig. 6.8 is 2 min and between junctions 233 and 234, the delay is 1 min.

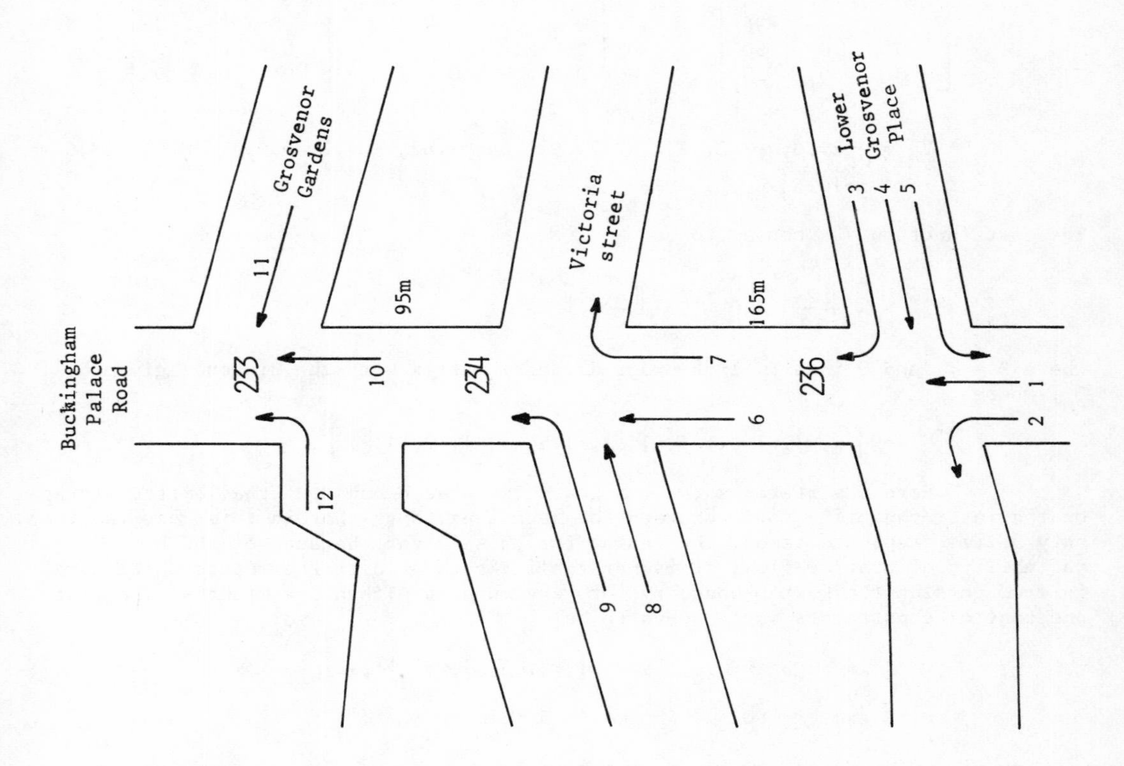

Fig. 6.8 : Junctions 233, 234 and 236 in the West London network

Based on saturation flows, desired values for the controls were chosen to be $u_1^0 = 0.45$, $u_2^0 = 0.30$, $u_3^0 = 0.50$. New control variables were then defined to be deviations from these values. Thus

$$\Delta u_1 = u_1 - 0.45, \quad \Delta u_2 = u_2 - 0.3, \quad \Delta u_3 = u_3 - 0.5$$

Using data on inflows into this network and saturation flows as measured between 17.00 and 18.00 hrs by the Greater London Council and provided to the authors, the state equations could be written in vector matrix form as

$$\underline{x}(k+1) = I_{12}\underline{x}(k) + B\,\Delta\underline{u}(k) + B_1\Delta\underline{u}(k-1) + B_2\Delta\underline{u}(k-2) + \underline{V}$$

where I_{12} is the twelfth-order identity matrix and where

$$B = \begin{bmatrix} -65 & 0 & 0 \\ -25 & 0 & 0 \\ 34 & 0 & 0 \\ 31 & 0 & 0 \\ 4 & 0 & 0 \\ 0 & -64 & 0 \\ 0 & -26 & 0 \\ 0 & 132 & 0 \\ 0 & 34 & 0 \\ 0 & 0 & -96 \\ 0 & 0 & 90 \\ 0 & 0 & 25 \end{bmatrix} \quad B_1 = \begin{bmatrix} 0 & 0 & 0 \\ 0 & 0 & 0 \\ 0 & 0 & 0 \\ 0 & 0 & 0 \\ 0 & 0 & 0 \\ 0 & 0 & 0 \\ 0 & 0 & 0 \\ 0 & 0 & 0 \\ 0 & 0 & 0 \\ 0 & 0 & 0 \\ 0 & 30 & 0 \\ 0 & 0 & 0 \end{bmatrix} \quad B_2 = \begin{bmatrix} 0 & 0 & 0 \\ 0 & 0 & 0 \\ 0 & 0 & 0 \\ 0 & 0 & 0 \\ 0 & 0 & 0 \\ 42.7 & 0 & 0 \\ 18.3 & 0 & 0 \\ 0 & 0 & 0 \\ 0 & 0 & 0 \\ 0 & 0 & 0 \\ 0 & 0 & 0 \\ 0 & 0 & 0 \end{bmatrix}$$

$$\underline{V}^T = \{-21.6,\ -8.2,\ 8.4,\ 7.7,\ 9.2.56,\ 1.62,\ -64.2,\ -16.4,\ -8.4,\ -33,6\ -10.5\}$$

The cost function was chosen to be

$$I = \sum_{k=0}^{3} \|\underline{x}(k)\|_Q^2 + 100\,\|\underline{u}(k)\|_R^2$$

where $R = I_3$ and Q is a twelfth-order diagonal matrix with the diagonal given by Q_D, where

$$Q_D = [1,\ 1,\ 1,\ 1,\ 1,\ 1,\ 5,\ 1.5,\ 1,\ 1,\ 2,\ 1,\ 1].$$

Here the states x_6, x_7, x_{10} are favoured because of the limited storage on the interconnecting roads between the junctions. Note that in this cost function only 3 time steps are taken. The reason for this is that because of the high variability of traffic flow, it is not realistic to predict the inputs which come in from outside the system boundary for very much more than 3-4 minutes. The state and control constraints were chosen to be

$$0 \leq x_i \leq 40, \qquad i = 1, 2, 3, 8, 9, 11, 12$$

$$0 \leq x_i \leq 80, \qquad i = 4, 5$$

$$0 \leq x_i \leq 50, \qquad i = 6, 7$$

$$0 \leq x_{10} \leq 25$$

$$-0.25 \leq \Delta u_1 \leq 0.25, \qquad -0.1 \leq \Delta u_2 \leq 0.4 \qquad -0.3 \leq \Delta u_3 \leq 0.2$$

For the initial conditions, a fairly loaded network was chosen with the following values :

$$x_i(0) = 30, \qquad i = 1, 2, 8, 9, 11, 12$$

$$x_i(0) = 70, \qquad i = 3, 4, 5$$

$$x_i(0) = 40 \qquad i = 6,\ 7$$

$$x_{10}(0) = 20$$

Note that this is a realistic problem of high dimension which is difficult to tackle without decomposition.

6.6.4 Simulation results

The optimisation problem for this oversaturated network was solved on an IBM 370/165 digital computer using Tamura's Time-Delay Goal Coordination algorithm. Convergence to the optimum took place in 153 iterations which required 2.73 min to execute. The optimal primal and dual costs were

$$\text{Primal cost} \ : 0.284 \times 10^5,$$

$$\text{Dual cost} \ : 0.284 \times 10^5$$

Table 6.1 shows the optimal control sequence and Fig. 6.9 the resulting state trajectories of the system as obtained from a simulation of the system with the optimal control sequences incorporated. Note that the queues x_3, x_4, x_5 are hardly reduced.

k	1	2	3
u_1	20 %	20 %	20 %
u_2	70 %	68 %	61 %
u_3	61 %	62 %	63 %

Table 6.1: Optimal control sequence for the network example

The reason for this is the high inflows q_3, q_4, q_5 compared to the corresponding saturation flows s_3, s_4, s_5, so that even though the control u_1 is such that the maximum permitted green is provided for these queues, the saturation flow is too small for the queues to be dissipated. This is the reason why the network is a trouble spot since unless a near optimal policy is implemented x_3, x_4, x_5 will build up and clog all the neighbouring junctions ; with the optimal controls calculated here on the other hand, the rest of the queues are all dissipated to low values and even the critical queues x_3, x_4, x_5 are not allowed to build up.

Comments

Applications of this method have also been made to water pollution and water resource problems. The main disadvantage of the approach is that the resulting control is still open loop. In a sense, that can't be helped, as long as we have inequality constraints on the states and controls. Later on in this chapter, we will see how for the case where there are no inequality constraints it is possible to achieve closed loop control.

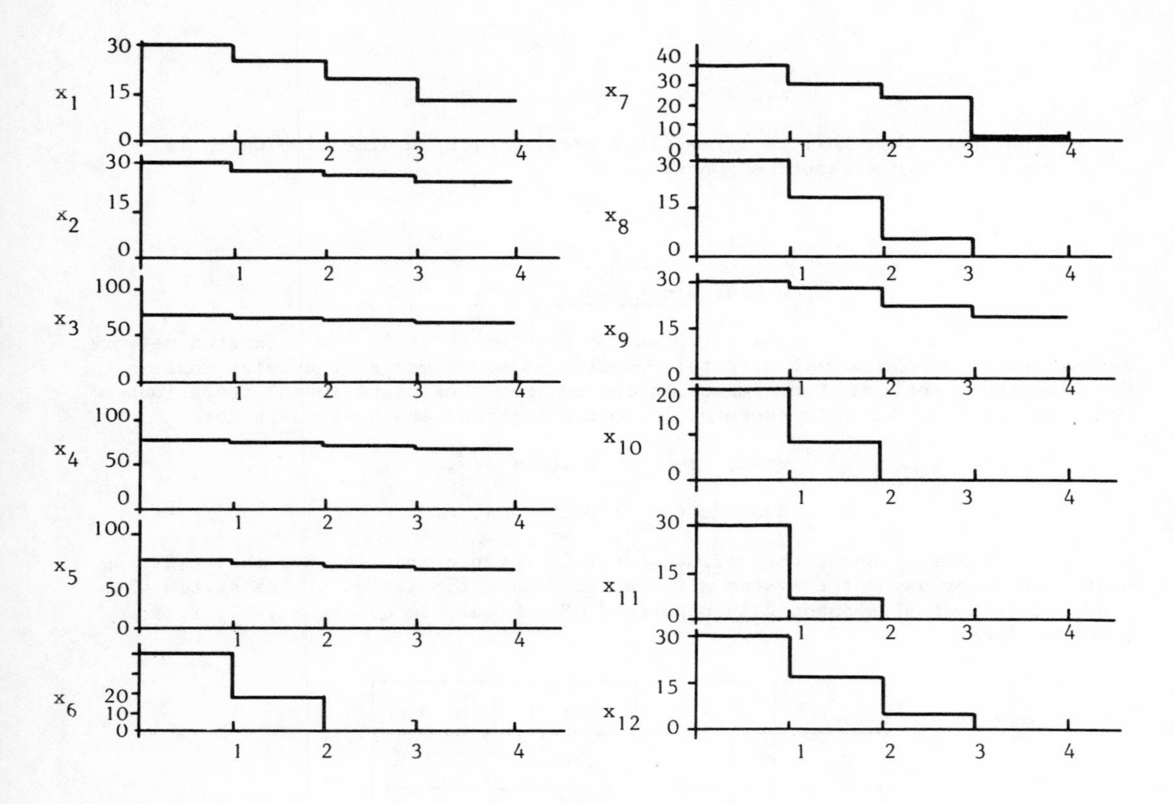

Fig. 6.9 : The optimal state trajectories for the London network.

Next let us consider another important approach to solving dynamic optimisation problems of the linear-quadratic type, i.e. the interaction prediction approach.

6.7 THE INTERACTION PREDICTION APPROACH

The overall system as usual consists of N linear interconnected subsystems described by

$$\dot{\underline{x}}_i = A_i \underline{x}_i + B_i \underline{u}_i + C_i \underline{z}_i \qquad 6.7.1$$

$$i = 1, 2, \ldots, N$$

where

$$\underline{z}_i = \sum_{j=1}^{N} L_{ij} \underline{x}_j \qquad 6.7.2$$

and it is desired to minimise

$$J = \sum_{i=1}^{N} \left\{ \frac{1}{2} \| \underline{x}_i(T) \|^2_{P_i} + \int_0^T \left(\frac{1}{2} \| \underline{x}_i(t) \|^2_{Q_i} + \frac{1}{2} \| \underline{u}_i(t) \|^2_{R_i} \right) dt \right\} \qquad 6.7.3$$

Note here that we do not have the quadratic term in $\underline{z}$ as in equation 6.2.3.

Let us write the Lagrangian as :

$$L = \sum_{i=1}^{N} \left\{ \frac{1}{2}\|\underline{x}_i(T)\|^2_{P_i} + \int_0^T \left(\frac{1}{2}\|\underline{x}_i(t)\|^2_{Q_i} + \frac{1}{2}\|\underline{u}_i(t)\|^2_{R_i} + \lambda_i^T\left[\underline{z}_i - \sum_{j=1}^{N} L_{ij}\underline{x}_j\right] + p_i^T\left[-\dot{\underline{x}}_i + A_i\underline{x}_i + B_i\underline{u}_i + C_i\underline{z}_i\right]\right) dt \right\} \quad 6.7.4$$

where $\underline{p}_i$ is the n_i dimensional adjoint vector and $\underline{\lambda}_i$ is the r_i dimensional vector of Lagrange multipliers. Now for given $\underline{\lambda}_i = \underline{\lambda}^*$, $\underline{z}_i = \underline{z}_i^*$, L in equation 6.7.4 is additively separable, i.e.

$$L = \sum_{i=1}^{N} L_i = \sum_{i=1}^{N} \left\{ \frac{1}{2}\|\underline{x}_i(T)\|^2_{P_i} + \int_0^T \left(\frac{1}{2}\|\underline{x}_i(t)\|^2_{Q_i} + \frac{1}{2}\|\underline{u}_i(t)\|^2_{R_i} + \underline{\lambda}_i^{*T}\underline{z}_i^* - \sum_{j=1}^{N}\underline{\lambda}_j^{*T} L_{ji}\underline{x}_i + \underline{p}_i^T\left[-\dot{\underline{x}}_i + A_i\underline{x}_i + B_i\underline{u}_i + C_i\underline{z}_i^*\right]\right) dt \right\}$$

where

$$L_i = \frac{1}{2}\|\underline{x}_i(T)\|^2_{P_i} + \int_0^T \left(\frac{1}{2}\|\underline{x}_i(t)\|^2_{Q_i} + \frac{1}{2}\|\underline{u}_i\|^2_{R_i} + \underline{\lambda}_i^{*T}\underline{z}_i^* - \sum_{j=1}^{N}\underline{\lambda}_j^{*T} L_{ji}\underline{x}_i + \underline{p}_i^T\left[A_i\underline{x}_i + B_i\underline{u}_i + C_i\underline{z}_i^* - \dot{\underline{x}}_i\right]\right) dt \quad 6.7.5$$

Here, unlike in the Goal Coordination case where the coordination vector was only the Lagrange multiplier vector, the coordination vector is $\begin{bmatrix}\underline{\lambda}\\ \underline{z}\end{bmatrix}$. This is of higher dimension than the coordination vector for the Goal Coordination case. However, the second level algorithm is exceedingly simple and this ensures that there is no disadvantage in using this more complex coordination vector. The second level algorithm provides an improvement of the coordination vector by reinjecting the value of the vector from the previous iteration into the stationarity conditions, i.e. from iteration k to k+1.

$$\begin{bmatrix}\underline{\lambda}^{*k+1}\\ \underline{z}^{*k+1}\end{bmatrix} = \begin{bmatrix}\underline{\lambda}^*(\underline{x}^k, \underline{u}^k, \underline{p}^k)\\ \underline{z}^*(\underline{x}^k, \underline{u}^k, \underline{p}^k)\end{bmatrix} \quad 6.7.6$$

where the expression on the R.H.S. of equation 6.7.6 is obtained by setting

$$\frac{\partial L}{\partial \underline{z}_i^*} = \underline{0} \quad \text{and} \quad \frac{\partial L}{\partial \underline{\lambda}_i^*} = \underline{0}$$

i.e.

$$\underline{\lambda}_i^* = C_i^T\underline{p}_i \quad \text{and} \quad \underline{z}_i^* = \sum_{j=1}^{N} L_{ij}\underline{x}_j$$

thus making the coordination rule

$$\begin{bmatrix} \underline{\lambda}_i^* \\ \\ \underline{z}_i^* \end{bmatrix}^{k+1} = \begin{bmatrix} C_i^T \underline{p}_i \\ \\ \sum_{j=1}^{N} L_{ij} \underline{x}_j \end{bmatrix}^k \qquad 6.7.7$$

The method therefore involves minimising the N independent sub-Lagrangians L_i for given $\underline{\lambda}_i^*$, $\underline{z}_i^*$ and then using the resultant $\underline{p}$ and $\underline{x}$ to calculate the new prediction for $\underline{\lambda}_i^*$, $\underline{z}_i^*$ by substituting into the R.H.S. of equation 6.7.7.

Comments

The interaction prediction approach is attractive for many reasons. For example here :

(a) The second level algorithm is very simple. It is not necessary to do the inefficient linear search as in the Goal Coordination method.

(b) The problem formulation is more meaningful since we do not have to include a quadratic term in $\underline{z}$ to avoid singularities as in the Goal Coordination approach.

(c) Experience shows that the method has extremely fast convergence.

Singh et al. [10] have made numerical comparisons of this approach with the Goal Coordination approach and their conclusion is that the present approach appears to be faster and requires less storage at least on the examples tested.

Next let us consider the river example of the previous chapter to test the approach.

6.8 RIVER POLLUTION CONTROL

As we recall from the previous chapter, the most realistic description of a stream with multiple polluters was given by Tamura [12] who assumed that each reach was separated from the next by a distributed delay. This model is able to account for the dispersion of pollutants which actually occurs in rivers. In this model, for j = 1, 2, ..., s, a fraction a_j of B.O.D. and D.O. in the (i-1)th reach at time $(t-\theta_j)$ arrives in the ith reach at time t, i.e. the transport delays are distributed in time between θ_1 and θ_s. Thus in equation 6.8.1, z_{i-1}, q_{i-1} are given by

$$z_{i-1}(t) = \sum_{j=1}^{s} a_j \, z_{i-1} \, (t-\theta_j)$$

$$q_{i-1}(t) = \sum_{j=1}^{s} a_j \, q_{i-1} \, (t-\theta_j) \qquad 6.8.1$$

$$\sum_{j=1}^{s} a_j = 1 \; ; \quad \text{mean of } \theta_j = \theta_0 \; ; \; \theta_1 < \theta_2 < \dots \; \theta_s .$$

$j = 1, 2, \ldots, s$ in the ith reach. Thus, it is possible to write the state equations for a 3 reach system with distributed delays as :

$$\dot{z}_1 = -1.32\, z_1 + 0.1\, \Delta m_1 + 0.9\, z_0 + 5.35 \qquad 6.8.2$$

$$\dot{q}_1 = -0.32\, z_1 - 1.2\, q_1 + 0.9\, q_0 + 1.9 \qquad 6.8.3$$

$$\dot{z}_2 = 0.9 \sum_{J=1}^{s} a_j z_1(t-\tau_j) - 1.32\, z_2 + 0.1\Delta m_2 + 4.19 \qquad 6.8.4$$

$$\dot{q}_2 = 0.9 \sum_{j=1}^{s} a_j q_1(t-\tau_j) - 1.32\, z_2 - 1.29\, q_2 + 1.9 \qquad 6.8.5$$

$$\dot{z}_3 = 0.9 \sum_{j=1}^{s} a_j z_2(t-\tau_j) - 1.32\, z_3 + 0.1\Delta m_3 + 1.59 \qquad 6.8.6$$

$$\dot{q}_3 = 0.9 \sum_{j=1}^{s} a_j q_2(t-\tau_j) - 0.32\, z_3 - 1.29\, q_3 + 1.9 \qquad 6.8.7$$

where $\Delta m = m - m^*$

Tamura gives the following values for s, τ, a, etc. for each of the distributed delays :

$$s = 3,\ \tau_1 = 0,\ \tau_2 = \tfrac{1}{2}\text{ day},\ \tau_3 = 1\text{ day},\ z_0 = 0,\ q_0 = 10$$

$$a_1 = 0.15,\quad a_2 = 0.7,\quad a_3 = 0.15$$

The system as described by equations 6.8.2 to 6.8.7 is nominally of infinite dimension in the state space. It is possible to obtain a good finite dimensional approximation by expanding the delayed terms in a Taylor series and taking the first two terms as we saw in the last chapter. For example, consider equation 6.8.4 which can be rewritten as :

$$\dot{z}_2 = 0.9\left[0.15\, z_1(t) + 0.7\, z_1(t-0.5) + 0.15\, z_1(t-1)\right] - 1.32\, z_2 + 0.1\Delta\, m_2 + 4.19$$

Here, we have 2 delays so it is necessary to introduce four additional states. Let these be given by z_4, z_5, z_6, z_7.

Let $z_4(t) = z_1(t-0.5)$; then $z_4(s) = z_1(s)\, e^{-0.5s}$.

Now

$$z_1(s)\, e^{-0.5s} = z_1(s)\left[1 + 0.5s + \frac{0.25}{2} s^2 + \ldots\right]^{-1}$$

Taking only the first three terms :

$$z_1(t) = z_4(t) + 0.5\, \dot{z}_4 + 0.125\, \ddot{z}_4(t) \qquad 6.8.8$$

Let $\dot{z}_4 = z_5$

then $\dot{z}_5 = 8\, z_1 - 8\, z_4 - 4\, z_5 \qquad 6.8.9$

Similarly for the other delay we let

$$z_1(t-1) = z_6$$

then $$\dot{z}_6 = z_7 \qquad 6.8.10$$

$$\dot{z}_7 = 2\, z_1 - 2\, z_6 - 2\, z_7 \qquad 6.8.11$$

Thus we can rewrite equation 6.8.4 as

$$\dot{z}_2 = 0.135\, z_1 + 0.63\, z_4 + 0.135\, z_6 - 1.32\, z_2 + 0.1\,\Delta m_2 + 4.19 \qquad 6.8.12$$

Similarly we can introduce four additional variables each for the delays in equations 6.8.5 to 6.8.7. This makes the overall system of order 22.

Having described the dynamics of this 22nd order system, we are now in a position to examine the application of the prediction principle to this problem. In order to do so, let us decompose our system into three subsystems where subsystem 1 consists of equations 6.8.2, 6.8.3, subsystem 2 consists of equations 6.8.4, 6.8.5, which ultimately yields 10 equations once the approximation to the delay is defined, and subsystem 3 consists of equations 6.8.6, 6.8.7, which again gives us a 10th order system once the delays are approximated by a second order Taylor series. Thus our system is decomposed into three parts respectively of order 2, 10 and 10. Clearly many other decompositions are also possible but the present one is the most convenient since it retains the subsystem structure. Other decompositions which consider for example 8, 7, 7 variables are almost certainly more efficient computationally but with such decompositions one does lose one's feel for the problem since the decomposition does not have a clear physical meaning. In any case, we are most concerned with the methodology since it will be possible to use the interaction prediction approach for systems with many more reaches and as the number of reaches increases, such a decomposition becomes computationally more attractive.

Let us assume that the problem is to minimise a cost function of the form

$$J = \int_0^8 \Big[2(z_1-z_1^*)^2 + (q_1-q_1^*)^2 + 2(z_2-z_2^*)^2 + (q_2-q_2^*)^2 + 2(z_3-z_3^*)^2 + (q_3-q_3^*)^2 + (m_1-m_1^*)^2 + (m_2-m_2^*)^2 + (m_3-m_3^*)^2 \Big]\, dt \qquad 6.8.13$$

Minimising such a cost function ensures that if before time zero an unknown disturbance takes the system state to some value $\underline{x}_0$ where $\underline{x}_0$ is a 22nd order vector then it will be possible to bring the system back to the steady state values of z_1^*, q_1^*, z_2^*, q_2^*, z_3^*, q_3^* and the controls back to the steady state controls m_1^*, m_2^*, m_3^* whilst ensuring that there are no unacceptably large deviations from the steady state B.O.D., D.O. or controls. Note that both positive and negative deviations are penalised. In the case of the control, this is meaningful since the treatment process is biological and it costs money to change the control in either direction. For the B.O.D. and D.O. such a penalty is not meaningful since we really want to ensure that B.O.D. does not exceed some value and D.O. is always greater than some value. It is nevertheless possible to interpret the quadratic form as follows : let us assume that there is some band ($z_1^* \pm z_1^{**}$) and $q_1^* \pm q_1^{**}$) (i = 1, 2, 3) within which we should like our B.O.D.s and D.O.s to lie. Then the quadratic forms will ensure that this is so provided adequate weights are used in the cost function. We have found after extensive experimentation that the weights given in equation 6.8.13 are adequate.

6.8.1 Simulation results

The problem of minimising J in equation 6.8.13 subject to the system dynamics given in equations 6.8.2 to 6.8.7 was solved using a hierarchical interaction prediction principle structure on an IBM 370/165 digital computer. The initial conditions for the B.O.D.s and D.O.s were taken to be :

$$z_1(0) = 10 \text{ mg/l}, \quad z_2(0) = 5.94 \text{ mg/l}, \quad z_3(0) = 5.237 \text{ mg/l},$$

$$q_1(0) = 7 \text{ mg/l}, \quad q_2(0) = 6 \text{ mg/l}, \quad q_3(0) = 4.69 \text{ mg/l}$$

Such an initial condition implies that before t = 0, some disturbance affected reach one to make it very polluted whilst reaches 2 and 3 were in their desired steady state.

Convergence to the optimum took place in 11 2nd level iterations. Figs. 6.10, 6.11 show the resulting state trajectories. These figures show how the effect of the pollution load is gradually attenuated down the river. Note that the results are quite realistic with the distributed delay since the reaches downstream are only gradually affected as the pollutants are dispersed.

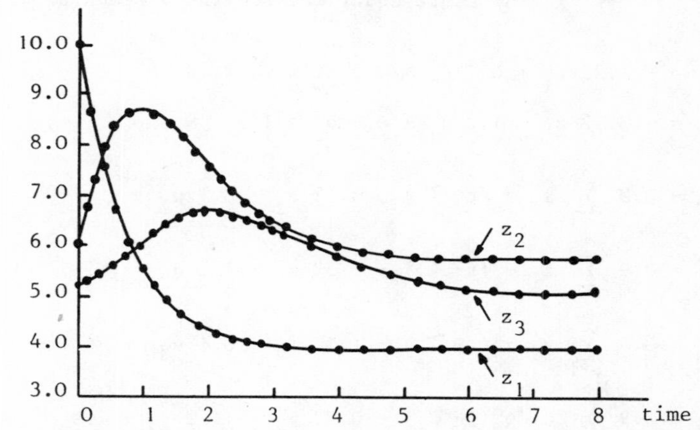

Fig. 6.10 : Optimal B.O.D. trajectories for the 3 reaches

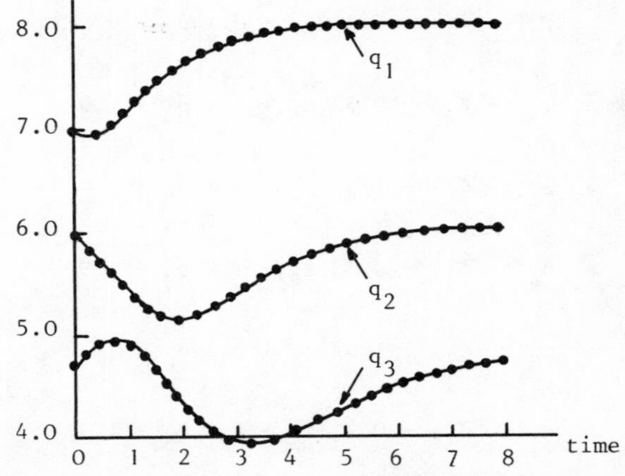

Fig. 6.11 : Optimal D.O. trajectories for the 3 reaches

Comments

In this section we have demonstrated the applicability of the prediction principle approach by solving a realistic large scale systems problem. We see that even for this very large system, convergence of the two level algorithm is very rapid even though our integration interval is very long. This rapid convergence property as well as the ease of programming makes this one of the most powerful approaches to hierarchical optimisation. In chapter 8, we will show how the method can be modified in order to apply it to the difficult non-linear case. Here we continue our discussion of the main hierarchical optimisation method which enables one to avoid in certain important cases the singularities which arise (section 6.9). However, before we do that, let us consider another example.

As a larger example where we can compare the results with the standard single level Riccati solution consider the 6 reach distributed delay problem which is of 52nd order.

6.8.2 The 6 reach river problem

The state equations for the 6 reach problem can be written as

$$\mathring{z}_1 = -1.32\, z_1 + 0.1\, q_1 + 0.9\, z_0$$

$$\mathring{q}_1 = -0.32\, z_1 - 1.2\, q_1 + 0.9\, q_0 + 1.9$$

$$\mathring{z}_2 = 0.9 \sum_{i=1}^{3} a_i\, z_n(t-\tau_i) - 0.32\, z_2 + 0.1\, u_2$$

$$\mathring{q}_2 = 0.9 \sum_{i=1}^{3} a_i\, q_1(t-\tau_i) - 0.32\, z_2 - 1.2\, q_2 + 1.9$$

$$\mathring{z}_3 = 0.9 \sum_{i=1}^{3} a_i\, z_2(t-\tau_i) - 1.32\, z_3 + 0.1\, u_3$$

$$\mathring{q}_3 = 0.9 \sum_{i=1}^{3} a_i\, q_2(t-\tau_i) - 0.32\, z_3 + 1.29\, q_3 + 1.9$$

$$\mathring{z}_4 = 0.9 \sum_{i=1}^{3} a_i\, z_3(t-\tau_i) - 1.32\, z_4 + 0.1\, u_4$$

$$\mathring{q}_4 = 0.9 \sum_{i=1}^{3} a_i\, q_3(t-\tau_i) - 0.32\, z_4 + 1.2\, q_4 + 1.9$$

$$\mathring{z}_5 = 0.9 \sum_{i=1}^{3} a_i\, z_4(t-\tau_i) - 1.35\, z_5 + 0.1\, u_5$$

$$\mathring{q}_5 = 0.9 \sum_{i=1}^{3} a_i\, q_4(t-\tau_i) - 0.32\, z_5 + 1.2\, q_5 + 1.9$$

$$\mathring{z}_6 = 0.9 \sum_{i=1}^{3} a_i\, z_5(t-\tau_i) - 1.32\, z_6 + 0.1\, u_6$$

$$\mathring{q}_6 = 0.9 \sum_{i=1}^{3} a_i\, q_5(t-\tau_i) - 0.32\, z_6 + 1.2\, q_5 + 1.9$$

As before, the 2nd order delay approximation makes this a 52nd order system with 6 subsystems, the first having 2 variables and subsystems 2 to 6 having 10 variables each. The control problem is to bring the system back to the steady state desired values (concentrations in mg/l) of

$$z_1^* = 2.10,\ q_1^* = 9.5,\ z_2^* = 2.47,\ q_2^* = 7.3,\ z_3^* = 2.47,\ q_3^* = 6.4,$$

$$z_4^* = 2.56,\ q_4^* = 5.7,\ z_5^* = 2.47,\ q_5^* = 5.2,\ z_6^* = 2.94,\ q_6^* = 4.7$$

whilst the control deviates as

$$u_1 = \Delta u_1 + 28.90,\ u_2 = \Delta u_2 + 12.90,\ u_3 = \Delta u_3 + 10.40,$$

$$u_4 = \Delta u_4 + 11.60,\ u_5 = \Delta u_5 + 9.50 \text{ and } u_6 = \Delta u_6 + 16.5$$

Here Δu_i, $i = 1$ to 6, is the deviation of the control around the nominal steady state value. The cost was chosen as

$$J = \sum_{i=1}^{6} \int_0^8 \left[(z_i - z_i^*)^2 Q_i + (q_i - q_i^*)^2 S_i + \Delta u_i^2 \right] dt$$

where $Q_i = 2$, $S_i = 1$.

RESULTS

This problem was solved using on the one hand the interaction prediction principle and on the other the standard single level Riccati equation solution as outlined in the last chapter. On the same computer system (IBM 370/165), the prediction principle approach allowed convergence in 80 seconds and used 186 K of core store. Compared to that, the global Riccati solution required 150 seconds of computation and 482 K of core store. This demonstrates that this hierarchical approach is computationally quite attractive for large scale problems.

6.9 AVOIDING SINGULARITIES IN THE GOAL COORDINATION METHOD

We saw that singularities at the first level arise because the local Hamiltonians are linear in $\underline{z}$. One way of avoiding singularities in the Goal Coordination method was proposed by Bauman [13] who suggested squaring the interconnection constraints since then $\underline{z}$ enters quadratically in the Hamiltonian.

Consider for example the system

$$\mathring{x}_1 = x_2 \tag{6.9.1}$$

$$\mathring{x}_2 = u - x_2 \tag{6.9.2}$$

and it is desired to minimise

$$J = \frac{1}{2} \int_0^1 (x_1^2 + x_2^2 + u^2)\, dt \tag{6.9.3}$$

Since the only coupling between the equations is through equation 6.9.1, it is possible to define an interconnection relationship of the form

$$z = x_2 \qquad 6.9.4$$

Then the Hamiltonian can be written as

$$H = \left[\frac{1}{2} x_1^2 - \mu z + \lambda_1 z\right] + \left[\frac{1}{2}(x_2^2 + u^2) + \mu x_2 + \lambda_2(u - x_2)\right] \qquad 6.9.5$$

But in this Hamiltonian, $\underline{z}$ enters linearly, so that its minimisation leads to a singular solution.

Bauman [13] suggested squaring the interconnection relationship to $x_2^2 - z^2 = 0$ to avoid singularity and then solved the problem using the Goal Coordination approach. However, this made convergence extremely slow. Moreover, for some initial conditions, the solution converged to $x_2 = - z$.

Another approach to avoiding singularities in a number of cases whilst at the same time avoiding the disadvantages of the approach of Bauman is to substitute for the interconnection vector $\underline{z}$ into the quadratic term for the state $\underline{x}$. The justification for this is that at the optimum, the Goal Coordination solution ensures that the interconnection relationship is satisfied so that we have solved the original problem whilst during the iterative process we have at the same time avoided the singular solutions at the first level.

To see this in more detail, consider the minimisation of J where

$$J = \int_{t_0}^{t_f} \left(\frac{1}{2}\|\underline{x}(t)\|_Q^2 + \frac{1}{2}\|\underline{u}(t)\|_R^2\right) dt \qquad 6.9.6$$

where as usual $\underline{x}$ is a $\sum_{i=1}^{N} n_i$ dimensional state vector consisting of N, n_i dimensional sub-state vector, $\underline{u}$ is a $\sum_{i=1}^{N} m_i$ dimensional control vector, Q, R are respectively $\sum_{i=1}^{N} n_i \times \sum_{i=1}^{N} n_i$ and $\sum_{i=1}^{N} m_i \times \sum_{i=1}^{N} m_i$ dimensional block diagonal matrices.

It is desired to minimise J in equation 6.9.6 subject to the system dynamics

$$\mathring{\underline{x}} = A\underline{x} + B\underline{u} + \underline{z} \qquad 6.9.7$$

where A is a $\sum_{i=1}^{N} n_i \times \sum_{i=1}^{N} n_i$ dimensional block diagonal matrix and $\underline{z}$ is a $\sum_{i=1}^{N} q_i$ dimensional vector of interaction inputs from other subsystems. B is an $\sum_{i=1}^{N} n_i \times \sum_{i=1}^{N} m_i$ dimensional block diagonal matrix. The interaction input $\underline{z}$ is given by

$$\underline{z} = M\,\underline{x} \qquad 6.9.8$$

We assume that $\sum_{i=1}^{N} n_i = \sum_{i=1}^{N} q_i$ and M^{-1} exists.

In order to solve this problem by the Goal Coordination approach, it is necessary to find the maximum w.r.t. $\underline{\lambda}$ of the minimum w.r.t. $\underline{x}$, $\underline{u}$, $\underline{z}$, of the Lagrangian L where

$$L = \int_{t_0}^{t_f} \left(\frac{1}{2}\|\underline{x}\|_Q^2 + \|\underline{u}\|_R^2 + \underline{\lambda}^T\left[\underline{z}-M\underline{x}\right] + \underline{p}^T\left[\underline{x}-A\underline{x}-B\underline{u}-\underline{z}\ \right)\right]dt \qquad 6.9.9$$

However, on setting $\frac{\partial L}{\partial \underline{z}} = \underline{0}$ singularities arise. Now, since at the optimum $\underline{z} = M\underline{x}$, substitute $\underline{x} = M^{-1}\underline{z}$ in the quadratic term. Then, in order to minimise L w.r.t. $\underline{x}$, $\underline{u}$, $\underline{z}$, for given $\underline{\lambda}$, the Hamiltonian can be written as

$$H = \frac{1}{2}\|M^{-1}\underline{z}\|_Q^2 + \frac{1}{2}\|\underline{u}\|_R^2 + \underline{\lambda}^T\underline{z} - \underline{\lambda}^T M\underline{x} + \underline{p}^T\left[A\underline{x}+B\underline{u}+\underline{z}\right]$$

This yields

$$\underline{z} = -MQ^{-1}M^T(\underline{\lambda} + \underline{p})$$

$$\underline{u} = -R^{-1}B^T\underline{p}$$

and the resulting optimal state equation is

$$\dot{\underline{x}} = A\underline{x} - BR^{-1}B^T\underline{p} - MQ^{-1}M^T(\underline{\lambda} + \underline{p}) \qquad 6.9.10$$

$$\underline{x}(t_0) = \underline{x}_0$$

with the costate equation

$$-\dot{\underline{p}} = -M^T\underline{\lambda} + A^T\underline{p}\ ; \quad \underline{p}(t_f) = \underline{0} \qquad 6.9.11$$

Note that since equation 6.9.10 does not have a term in $\underline{x}$, this equation can be integrated directly and the resulting $\underline{p}$ can be substituted in 6.9.11 to yield the optimum $\underline{x}$. This simplifies the computations considerably. Note also that because A, B, Q, R are block diagonal, equations 6.9.10, 6.9.11 can be solved independently subsystem by subsystem provided $\|M^{-1}z\|_Q^2$ remains separable in $\underline{z}$.

The above approach ensures that the subproblems are not singular, thus eliminating one of the main disadvantages of the standard Goal Coordination solution. However, it is necessary that M^{-1} exist and $\|M^{-1}\underline{z}\|_Q^2$ be separable and these restrictions limit the number of cases where the improved Goal Coordination method could be used. However, there are still a number of interesting problems where the approach works. Consider next an example where it does work.

EXAMPLE 1

Consider again Bauman's example which was formulated in the beginning of the section.The Hamiltonian of equation 6.9.7 can be rewritten as

$$H = \left\{\frac{1}{2}(z^2+x_1^2) + \lambda_1 z - \mu z\right\} + \left\{\frac{1}{2}u^2 - \lambda_2 x_2 + \lambda_2 u + \mu x_2\right\}$$

In this case, the subproblems become

Subproblem 1 $\qquad \frac{\partial H}{\partial x_1} = -\dot{\lambda}_1 = x_1$

$$\frac{\partial H}{\partial z} = 0 \qquad \text{or } z + \lambda_1 - \mu = 0$$

$$\mathring{x}_1 = z$$

Let $x_1(0) = 2$; also $\lambda_1(T) = 0$

Subproblem 2

$$\frac{\partial H}{\partial x_2} = -\dot{\lambda}_2 = -\lambda_2 + \mu$$

$$\frac{\partial H}{\partial u} = 0 \quad \text{or } u = -\lambda_2$$

$$\mathring{x}_2 = -x_2 + u \; ; \; \lambda_2(T) = 0$$

Let $x_2(0) = 2$

Second level : it is necessary to ensure that

$$z - x_2 = 0$$

This problem was solved on an IBM 370/165 digital computer. On the second level, a simple gradient algorithm was used. For T = 1, $\Delta T = 0.1$, convergence to the optimum took place in 11 second level iterations, yielding the true solution as opposed to the false solution found by Bauman using his modified approach. This solution is given in Fig. 6.12.

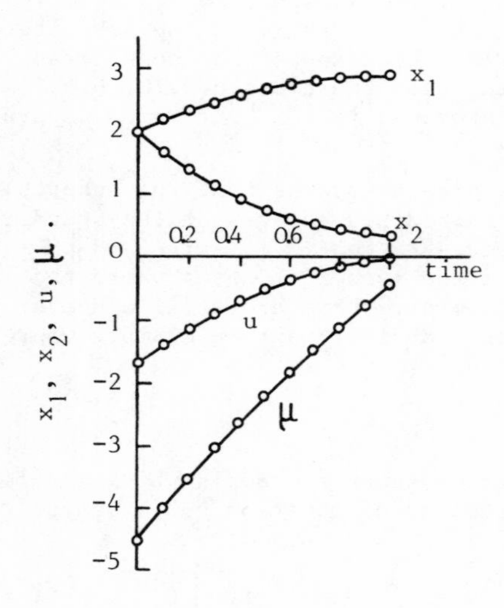

Fig. 6.12

EXAMPLE 2

Let us next consider as another example the problem :

$$\min_{\underline{u}} J = \frac{1}{2}\int_0^T (\underline{x}^T Q \underline{x} + \underline{u}^T R\underline{u})\, dt$$

subject to

$$\mathring{\underline{x}} = A\underline{x} + B\underline{u} + \underline{C} \quad ; \quad \underline{x}(0) = \underline{x}_0$$

where

$$\underline{x} = \begin{bmatrix} \underline{x}_1 \\ \hline \underline{x}_2 \end{bmatrix}. \quad A = \left[\begin{array}{c|c} A_{11} & A_{12} \\ \hline A_{21} & A_{22} \end{array}\right]. \quad \underline{u} = \begin{bmatrix} \underline{u}_1 \\ \hline \underline{u}_2 \end{bmatrix}. \quad B = \begin{bmatrix} B_1 \\ \hline B_2 \end{bmatrix}$$

$$C = \begin{bmatrix} C_1 \\ \hline C_2 \end{bmatrix}. \quad Q = \left[\begin{array}{c|c} Q_{11} & \mathbb{O} \\ \hline \mathbb{O} & Q_{22} \end{array}\right]. \quad R = \left[\begin{array}{c|c} R_{11} & \mathbb{O} \\ \hline \mathbb{O} & R_{22} \end{array}\right]$$

The problem can be written in partitioned form as :

$$\min_{\underline{u}} \frac{1}{2}\int_0^T \left(\begin{bmatrix} \underline{x}_1 \\ \hline \underline{x}_2 \end{bmatrix}^T \left[\begin{array}{c|c} Q_{11} & \mathbb{O} \\ \hline \mathbb{O} & Q_{22} \end{array}\right] \begin{bmatrix} \underline{x}_1 \\ \hline \underline{x}_2 \end{bmatrix} + \begin{bmatrix} \underline{u}_1 \\ \hline \underline{u}_2 \end{bmatrix}^T \left[\begin{array}{c|c} R_{11} & \mathbb{O} \\ \hline \mathbb{O} & R_{22} \end{array}\right] \begin{bmatrix} u_1 \\ \hline u_2 \end{bmatrix} \right) dt$$

subject to

$$\begin{bmatrix} \mathring{\underline{x}}_1 \\ \mathring{\underline{x}}_2 \end{bmatrix} = \begin{bmatrix} A_{11} & A_{12} \\ A_{21} & A_{22} \end{bmatrix} \begin{bmatrix} \underline{x}_1 \\ \underline{x}_2 \end{bmatrix} + \begin{bmatrix} B_1 \\ B_2 \end{bmatrix} \begin{bmatrix} \underline{u}_1 \\ \underline{u}_2 \end{bmatrix} + \begin{bmatrix} \underline{C}_1 \\ \underline{C}_2 \end{bmatrix}$$

or

$$\min_{\substack{\underline{u}_i \\ i = 1,\, 2}} \frac{1}{2}\int_0^T \sum_{i=1}^{2} \left[\underline{x}_i^T Q_{ii} \underline{x}_i + \underline{u}_i^T R_{ii} \underline{u}_i \right] dt$$

subject to

$$\dot{\underline{x}}_1 = A_{11}\underline{x}_1 + B_1\underline{u}_1 + \underline{C}_1 + A_{12}\underline{x}_2$$

$$\underline{x}_2 = A_{22}\underline{x}_2 + B_2\underline{u}_2 + \underline{C}_2 + A_{21}\underline{x}_1$$

In this formulation, let us put

$$\underline{z}_1 = A_{12}\underline{x}_2$$

$$\underline{z}_2 = A_{21}\underline{x}_1$$

as the interconnection variables

Then our problem becomes

$$\min_{\underline{u}_i} \sum_{i=1}^{2} \frac{1}{2}\int_0^T (\underline{x}_i^T Q_{ii}\underline{x}_i + \underline{u}_i^T R_{ii}\underline{u}_i)\, dt$$

subject to

$$\mathring{\underline{x}}_i = A_{ii}\underline{x}_i + B_i\underline{u}_i + \underline{C}_i + \underline{z}_i$$

$$\underline{x}_i(0) = \underline{x}_{i0}$$

$$\underline{z}_i = A_{ij}\,\underline{x}_j \qquad i = 1,\ j = 2;\quad i = 2,\ j = 1$$

The Goal coordination method :

Let $\underline{\lambda}_i$ be the Lagrange multiplier vector associated with :

$$\underline{z}_i - A_{ij}\,\underline{x}_j = \underline{0}$$

We can then write the Lagrangian as :

$$L = \sum_{i=1}^{2} \int_0^T \Big\{ (\underline{x}_i^T Q_{ii}\,\underline{x}_i + \underline{u}_i^T R_{ii}\,u_i) + \underline{p}_i^T [A_{ii}\,\underline{x}_i + B_i\,\underline{u}_i + \underline{C}_i + \underline{z}_i - \dot{\underline{x}}_i]$$
$$+ \underline{\lambda}_i^T (\underline{z}_i - A_{ij}\,\underline{x}_j) \Big\}\, dt \qquad i = 1,\ j = 2;\quad i = 2,\ j = 1$$

This Lagrangian is separable for a given $\underline{\lambda}_i$, i.e.

$$L = L_1 + L_2$$

This leads to the ith subproblem :

$$\min_{\underline{u}_i, \underline{z}_i} \frac{1}{2} \int_0^T (\underline{x}_i^T Q_{ii}\,\underline{x}_i + \underline{u}_i^T R_{ii}\,\underline{u}_i + \underline{\lambda}_i^T\,\underline{z}_i - \underline{\lambda}_j^T A_{ji}\,\underline{x}_i)\, dt$$

for given $\underline{\lambda}_i, \underline{\lambda}_j$

subject to

$$\dot{\underline{x}}_i = A_{ii}\,\underline{x}_i + B_i\,\underline{u}_i + \underline{C}_i + \underline{z}_i \qquad \underline{x}_i(0) = \underline{x}_{i0}$$

Here, the necessary conditions for optimality are :

$$\frac{\partial H_i}{\partial \underline{p}_i} = \dot{\underline{x}}_i \quad \underline{x}_i(0) = \underline{x}_{i0}, \quad \frac{\partial H_i}{\partial \underline{x}_i} = -\dot{\underline{p}}_i \qquad \underline{p}_i(T) = \underline{0}$$

$$\frac{\partial H_i}{\partial \underline{u}_i} = \underline{0} \quad \text{and} \quad \frac{\partial H_i}{\partial \underline{z}_i} = \underline{0}$$

where H_i is the Hamiltonian of the subproblem, i.e.

$$H_i = \frac{1}{2} (\underline{x}_i^T Q_{ii}\,\underline{x}_i + \underline{u}_i^T R_{ii}\,\underline{u}_i) + \underline{\lambda}_i^T\,\underline{z}_i - \underline{\lambda}_j^T A_{ji}\,\underline{x}_i + \underline{p}_i^T [A_{ii}\,\underline{x}_i + B_i\underline{u}_i + C_i + \underline{z}_i]$$

The coordination task is to modify $\underline{\lambda}_i$, i = 1, 2, ... using

$$L_{\underline{\lambda}_i} = [\underline{z}_i^* - A_{ij}\,\underline{x}_j^*]$$

$\underline{z}_i^*$, $\underline{x}_j^*$ being the solutions of the local subproblems.

However, we see using $\frac{\partial H}{\partial \underline{z}_i} = \underline{0}$ that singularities arise.

Different ways of avoiding the singularities

a) If $\dim \underline{x}_1 = \dim \underline{x}_2$, then A_{ij} and A_{ji} are square matrices and if these matrices are regular, then :

$$\underline{x}_j = A_{ij}^{-1} \underline{z}_i$$

The local problem then becomes :

$$\min_{\underline{u}_i, \underline{z}_i} \frac{1}{2} \int_0^T \left[(\underline{z}_i^T A_{ij}^{-1T} Q_{ii} A_{ij}^{-1} \underline{z}_i + \underline{u}_i^T R_{ii} \underline{u}_i + \underline{\lambda}_i^T \underline{z}_i - \underline{\lambda}_j^T A_{ji} \underline{x}_i) \right] dt$$

subject to $\dot{\underline{x}}_i = A_{ii} \underline{x}_i + B_i \underline{u}_i + \underline{C}_i + \underline{z}_i \qquad \underline{x}_i(0) = \underline{x}_{i0}$

and the singularity vanishes since $\frac{\partial H_i}{\partial \underline{z}_i}$ will now be a non linear function of the pseudo control $\underline{z}_i$.

b) If the coupling equation is a set of relation such that each element of $\underline{x}_j$ is linked only to one element of $\underline{z}_i$, then

$$\underline{z}_i^{\,l} = \underline{x}_j^{\,k}$$

Then these relations can be squared as $(z_i^{\,l})^2 = (x_j^{\,k})^2$ which makes $\underline{z}_i$ non linear. However it introduces the extra solution :

$$z_i^{\,l} = - x_j^{\,k}$$

c) Finally, since $\underline{z}_i$ which has to be computed at the lower level is the cause of the singularity, we could attempt to fix it at the higher level by considering it as a coordination variable and we thus arrive at the interaction prediction method where we have the sub-problem :

$$\min_{\underline{u}_i} \frac{1}{2} \int_0^T (\underline{x}_i^T Q_{ii} \underline{x}_i + \underline{u}_i^T R_{ii} \underline{u}_i - \underline{\lambda}_j^T A_{ji} \underline{x}_i) \, dt$$

for given $\underline{\lambda}_j$, $\underline{z}_i$

subject to $\dot{\underline{x}}_i = A_{ii} \underline{x}_i + B_i \underline{u}_i + \underline{C}_i + \underline{z}_i \qquad \underline{x}_i(0) = \underline{x}_{i0}$

($\underline{\lambda}_i^T \underline{z}_i$ being constant can be eliminated from the cost function).

Coordination : From $\frac{\partial H_i}{\partial \underline{\lambda}_i} = 0$, $\frac{\partial H_i}{\partial \underline{z}_i} = 0$ we have

$$\underline{z}_i = A_{ij} \underline{x}_j$$

$$\underline{\lambda}_i = - \underline{p}_i^*$$

Comment

We have so far examined most of the currently available methods for the hierarchical optimisation for linear-quadratic problems which lead to open loop control. Now, if we were faced with an actual practical problem, which one would we use ? There are in fact no hard and fast rules although the authors experience would indicate that in general the prediction principle approach is very attractive except for cases where there are time delays and/or inequality constraints on the states and controls. In this latter case, it might be more appropriate to use the Tamura method.

6.10 HIERARCHICAL FEEDBACK CONTROL : MOTIVATION

Thus far we considered the optimisation problem to arise from the following kind of situation : we have some large scale system operating in a steady state when some unknown disturbance occurs at time t_0 which changes the state vector $\underline{x}$ from the given steady state value $\underline{x}_s$ to some known value $\underline{x}_0$. It is then necessary to calculate some control over the fixed period $(t_f - t_0)$ to bring the system back to steady state $\underline{x}_s$ whilst minimising some integral function of the states and controls. In that case, it was assumed that we could perform our hierarchical calculation sufficiently rapidly and apply the control before the state had significantly changed. However, this is not a realistic situation unless the system has particularly slow dynamics (as, for example, in the case of the river) or if we know a priori precisely what the initial state value $\underline{x}_0$ will be (as, for example, in the case of start up or shut down of plants or change of product, etc.). In the general case, although we may be able to measure the initial state $\underline{x}_0$, it will almost certainly have changed by the time we have calculated our control and applying the open loop control for the actual state $\underline{x}$ instead of for the initial state $\underline{x}_0$ for which the hierarchical calculation was done could prove to be disastrous. It is therefore highly desirable to be able to calculate a feedback control which is independent of initial conditions since then we could apply the control as soon as a measurement was available. In such cases, it would also not be necessary to calculate the controls repeatedly for differing initial conditions since the same control law would be valid for all initial conditions. In this chapter we develop such control laws for the only case where such laws exist for the integrated problems, i.e. for systems comprising "linear-quadratic" subsystems.

Here, we will develop a scheme for solving this problem such that the whole hierarchical calculation is performed off-line and the resulting feedback gains are implemented on-line. The resulting controller having "gains" independent of the initial conditions will bring the system back to the steady state "optimally" from any state that an unknown disturbance takes the system to. Such a scheme is ideal for on-line implementation since the actual on-line calculation is quite minimal.

6.11 THE INTERACTION PREDICTION APPROACH TO DECENTRALISED CONTROL

To begin with the interaction prediction principle approach of Takahara [8], as formulated earlier on, is used as the vehicle for the development of the closed loop decentralised control. The main reason for this is that the prediction method requires very little second level calculation. This is important since for all multi-level methods, the overall system variables are manipulated at the second level and for truly large systems computational difficulties could arise with other methods. Nevertheless it is easy to do a similar analysis for the other hierarchical optimisation approaches although this will not be done here except for the discrete time case where we use the Tamura method for the off-line optimisation. We leave as an exercise the development of feedback control using other hierarchical methods.

The dynamic optimisation problem for an interconnected system can be written as before as :

$$\text{Minimise } J = \sum_{i=1}^{N} \int_{t_0}^{t_f} \left(\frac{1}{2}\|\underline{x}_i\|^2_{Q_i} + \frac{1}{2}\|\underline{u}_i\|^2_{R_i}\right) dt \qquad 6.11.1$$

where Q_i is positive semi-definite and R_i is positive definite,

subject to

$$\mathring{\underline{x}}_i = A_i\,\underline{x}_i + B_i\,\underline{u}_i + C_i\,\underline{z}_i \qquad i = 1, \ldots, N \qquad 6.11.2$$

$$\underline{z}_i = \sum_{j=1}^{N} L_{ij}\,\underline{x}_j \qquad i = 1, \ldots, N \qquad 6.11.3$$

Here, $\underline{x}_i$ is the n_i dimensional state vector of the ith subsystem, $\underline{u}_i$ is the m_i dimensional control vector and $\underline{z}_i$ is the q_i dimensional vector of interconnections from the other subsystems. Such a problem can be viewed in the integrated or composite sense as a large multivariable regulator but with a special structure. In order to utilise this structure to reduce the computer storage, we will decompose the necessary conditions for optimality within a two level structure. Thus, let us write the Lagrangian as

$$L(\underline{x}_i, \underline{u}_i, \underline{\lambda}, \underline{z}_i, \underline{p}_i) = \sum_{i=1}^{N} \int_{t_0}^{t_f} \left(\frac{1}{2}\|\underline{x}_i\|^2_{Q_i} + \frac{1}{2}\|\underline{u}_i\|^2_{R_i} + \underline{\lambda}_i^T\left[\underline{z}_i - \sum_{j=1}^{N} L_{ij}\underline{x}_j\right] + \underline{p}_i^T\left[-\dot{\underline{x}}_i + A_i\underline{x}_i + B_i\underline{u}_i + C_i\underline{z}_i\right]\right) dt \qquad 6.11.4$$

where $\underline{\lambda}$ is a $\sum_{i=1}^{N} q_i$ dimensional vector of Lagrange multipliers and $\underline{p}_i$ is an n_i dimensional vector of adjoint variables.

Now, in this linear case, convexity assures that the necessary conditions for optimality are also sufficient so that by satisfying the necessary conditions we can obtain the global optimum. To satisfy the necessary conditions within a two level computational structure, let us rewrite the Lagrangian L as

$$L = \sum_{i=1}^{N} L_i = \sum_{i=1}^{N} \left\{ \int_{t_0}^{t_f} \left\{ \frac{1}{2}\|\underline{x}_i\|^2_{Q_i} + \frac{1}{2}\|\underline{u}_i\|^2_{R_i} + \underline{\lambda}_i^T\,\underline{z}_i - \sum_{j=1}^{N} \underline{\lambda}_j^T\,L_{ji}\underline{x}_i + \underline{p}_i^T\left[-\mathring{\underline{x}}_i + A_i\underline{x}_i + B_i\underline{u}_i + C_i\underline{z}_i\right]\right\} dt \right\} \qquad 6.11.5$$

From equation 6.11.5 we see that the Lagrangian L is additively separable for any given $\underline{z}_i$ and $\underline{\lambda}_i$ trajectory. This implies that for any given $\underline{z}_i$, $\underline{\lambda}_i$, the optimality with respect to the other variables can be obtained by N independent minimisations. This leaves the problem of improving $\underline{z}_i$, $\underline{\lambda}_i$ such that the global optimum is achieved. A necessary condition for doing this is that :

$$\frac{\partial L}{\partial \underline{z}_i} = \underline{0} \qquad 6.11.6$$

$$\frac{\partial L}{\partial \underline{\lambda}} = \underline{0} \qquad 6.11.7$$

which gives $\underline{\lambda}_i = -C_i^T\,\underline{p}_i$ and $\underline{z}_i = \sum_{j=1}^{N} L_{ij}\,\underline{x}_j$ $\qquad$ 6.11.8

The coordination rule using the interaction prediction principle is from iteration k to k+1

$$\begin{bmatrix} \underline{\lambda}_i \\ \\ \underline{z}_i \end{bmatrix}^{k+1} = \begin{bmatrix} -C_i^T \underline{p}_i \\ \\ \sum_{j=1}^{N} L_{ij}\underline{x}_j \end{bmatrix}^{k} \qquad 6.11.9$$

i.e. the $\underline{x}_i$, $\underline{p}_i$ which are obtained from the independent minimisations in equation 6.11.5 are used in equation 6.11.8 to get the values of $\underline{\lambda}_i$, $\underline{z}_i$ ($i = 1, \ldots, N$) which are subsequently used as the new prediction of $\underline{\lambda}_i$ and $\underline{z}_i$. The important point to note is that at the second level it is necessary to do very little work, i.e. the 2nd level only evaluates the R.H.S. of equation 6.11.9 which involves a few multiplications. The lower level also does very little work since only low order problems are solved here. Also, the convergence is extremely rapid.

6.11.1 Modification to give partial feedback control

Consider the lower level problem. The Hamiltonian for the ith independent subproblem can be written as :

$$H_i = \frac{1}{2}\|\underline{x}_i\|^2_{Q_i} + \frac{1}{2}\|\underline{u}_i\|^2_{R_i} + \underline{\lambda}_i^T \underline{z}_i - \sum_{j=1}^{N} \underline{\lambda}_j^T L_{ji}\underline{x}_i + \underline{p}_i^T\left[A_i\underline{x}_i + B_i\underline{u}_i + C_i\underline{z}_i\right] \qquad 6.11.10$$

Then from the necessary conditions

$$\overset{\circ}{\underline{p}}_i = -Q_i\underline{x}_i - A_i^T\underline{p}_i + \sum_{j=1}^{N}\left[\underline{\lambda}_j^T L_{ji}\right]^T \qquad 6.11.11$$

with

$$\underline{p}_i(t_f) = \underline{0} \qquad 6.11.12$$

$$\underline{u}_i = -R_i^{-1}B_i^T\underline{p}_i \qquad 6.11.13$$

Let

$$\underline{p}_i = K_i\underline{x}_i + \underline{s}_i \qquad 6.11.14$$

where $\underline{s}_i$ is the open-loop compensation vector.

Then

$$\overset{\circ}{\underline{p}} = K_i\overset{\circ}{\underline{x}}_i + \overset{\circ}{K}_i\underline{x}_i + \overset{\circ}{\underline{s}}_i \qquad 6.11.15$$

substituting into equation 6.11.2 gives

$$\overset{\circ}{\underline{x}}_i = A_i\underline{x}_i - B_iR_i^{-1}B_i^TK_i\underline{x}_i - B_iR_i^{-1}B_i^T\underline{s}_i + C_i\underline{z}_i \qquad 6.11.16$$

using equation 6.11.15

$$\left[\overset{\circ}{K}_i + A_i^TK_i + K_iA_i - K_iB_iR_i^{-1}K_i + Q_i\right]\underline{x}_i + \left[\overset{\circ}{\underline{s}}_i + A_i^T\underline{s}_i - K_iB_iR_i^{-1}B_i^T\underline{s}_i + K_iC_i\underline{z}_i - \sum_{j=1}^{N}\left[\underline{\lambda}_j^T L_{ji}\right]^T\right] = 0 \qquad 6.11.17$$

Since this equation is valid for arbitrary $\underline{x}_i$

$$\overset{\circ}{K}_i + K_i A_i + A_i^T K_i - K_i B_i R_i^{-1} B_i^T K_i + Q_i = 0 \qquad 6.11.18$$

with $\qquad K_i(t_f) = 0$

and $$\overset{\circ}{\underline{s}}_i = \left[K_i B_i R_i^{-1} B_i^T - A_i^T \right] \underline{s}_i - K_i C_i \underline{z}_i + \sum_{j=1}^{N} \left[\underline{\lambda}_j^T L_{ji} \right]^T \qquad 6.11.19$$

with $\qquad \underline{s}_i(t_f) = \underline{0}$

and the local control $\underline{u}_i$ is given by

$$\underline{u}_i = - R_i^{-1} B_i^T K_i \underline{x}_i - R_i^{-1} B_i^T \underline{s}_i \qquad 6.11.20$$

6.11.2 Remarks

(i) The K_i in equation 6.11.18 is independent of the initial state $\underline{x}(0)$. Thus the N matrix Riccati equations each involving $\frac{n_i \times (n_i+1)}{2}$ non-linear differential equations can be solved independently from the given final condition $K_i(t_f) = 0$. These give a partial feedback control. It can be argued that this feedback around each subsystem does provide some degree of stabilisation against small disturbance and moreover allows one to correct the control based on the current state as opposed to the initial condition.

(ii) $\underline{s}_i$ in equation 6.11.19 is not independent of the initial state $\underline{x}_i(t_0)$. Thus the second term in equation 6.11.20 provides open loop compensation. To see this using equation 6.11.8, at the optimum, $\underline{s}_i$ can be written as

$$\overset{\circ}{\underline{s}}_i = \left[- A_i^T + K_i B_i R_i^{-1} \right] \underline{s}_i - K_i C_i \sum_{j=1}^{N} \left[L_{ij} \underline{x}_j \right]$$
$$+ \sum_{j=1}^{N} L_{ji}^T \left[- C_j^T K_j \underline{x}_j - C_j^T \underline{s}_j \right] \qquad 6.11.21$$

i.e. $\underline{s}_i$ is a function of the states of all the other subsystems, and is thus dependent on the initial state $\underline{x}(t_0)$ of the overall system.

6.12 THE CLOSED LOOP CONTROLLER

Let $\underline{x}$ be the overall state vector of the system, $\underline{u}$ the overall control vector, $\underline{s}$ the overall open loop part of the compensator and A, B, C, L, Q, R, K etc. the matrices for the overall system. Then it is easy to show that the open loop compensation vector $\underline{s}$ and the state $\underline{x}$ are related by a transformation Y

i.e. $\qquad \underline{s} = Y \underline{x}$

where Y is an n x n matrix [3] [15] .

For the infinite time regulator $Y(t_f, t)$ is time invariant.

As a simple intuitive proof of this we note that the global control for the composite system can be written as :

$$\underline{u} = -R^{-1} B^T K' \underline{x} = -G \underline{x}$$

where K' is the solution of the global Riccati equation.

But from 6.11.20 we can also write

$$\underline{u} = -G_d \underline{x} + \underline{V}$$

where G_d is a block diagonal matrix with $+R_i^{-1} B_i^T K_i$ as the block elements.

Comparing these two expressions for the control, we obtain :

$$\underline{V} = -(G - G_d) \underline{x}$$

where G is a full matrix and G_d is a block diagonal matrix. Then $\underline{V}$, the open-loop part of the control, is linearly dependent on the state vector and $\underline{s}(t) = Y \underline{x}(t)$.

6.12.1 The regulator solution

As seen in equation 6.11.21, $\underline{s}$ provides the open-loop part of the controller and the above remark shows that $\underline{s}$ is related very simply to the overall state vector. However $\underline{s}$ is not easy to obtain directly.

For the infinite time regulator Y is particularly easy to compute since near $t=t_o$, Y is constant whereas $\underline{x}$ and $\underline{s}$ are not. Thus if the values of $\underline{x}$ and $\underline{s}$ are recorded at the first $n = \sum_{i=1}^{N} n_i$ time points, very close to $t = t_0$, $\underline{Y}$ can be determined as follows

Form the matrix $S = \left[\underline{s}(t_0)\ \underline{s}(t_1) \ldots \underline{s}(t_n) \right]$ and

$X = \left[\underline{x}(t_0), \underline{x}(t_1) \ldots \underline{x}(t_n) \right]$ and then

$S = Y X$

or $\quad Y = S X^{-1}$

This inversion of X should not pose much of a problem for even large systems since it is to be done off-line.*

If it is desired to calculate the time varying Y (i.e. if the horizon must be considered finite) it is possible to do so by solving the problem n times for n different initial conditions and then forming the n x n matrices :

---

* It should be noted that the approach hinges on one's ability to invert X which basically depends on the linear independence of the chosen record. In practical cases, if a large enough perturbation is given, it will be possible to obtain an invertible X. Otherwise, one can solve the problem off-line n times successively for the initial conditions

$$\underline{x}(t_0) = \begin{bmatrix} 1 \\ 0 \\ \cdot \\ \cdot \\ 0 \end{bmatrix}, \begin{bmatrix} 0 \\ 1 \\ \cdot \\ \cdot \\ 0 \end{bmatrix}, \ldots, \begin{bmatrix} 0 \\ 0 \\ \cdot \\ \cdot \\ 1 \end{bmatrix}$$

Then Y = S and it is not necessary to invert X. This of course requires more computation but it is off-line and in a decentralised way.

$$S = \left[\underline{s}^1 (t),\ \underline{s}^2(t) \ \ldots\ \underline{s}^n(t) \right]$$

$$X = \left[\underline{x}^1 (t),\ \underline{x}^2(t) \ \ldots\ \underline{x}^n(t) \right]$$

for each integration point and determining each value of Y by the relationship

$$Y = S\ X^{-1}$$

6.12.2 <u>Remarks</u>

(1) The above method enables one to solve the large interconnected systems regulator problem within a decentralised calculation structure.

(2) The resulting gains are independent of the initial conditions since using equation 6.11.20 in the composite case

$$\underline{u} = -\ R^{-1}\ B^T\ K\ \underline{x} - R^{-1}\ B^T\ \underline{s} \qquad 6.12.1$$

where K is block diagonal. Substituting for $\underline{s}$ from the equation $\underline{s} = Y\underline{x}$,

$$\underline{u} = -\ R^{-1}B^TK\underline{x} - R^{-1}B^TY\underline{x} = -\left[R^{-1}B^TK+R^{-1}\ B^TY\right]\underline{x} = -\ G\underline{x} \qquad 6.12.2$$

where none of the terms in the gain matrix G are dependent on $\underline{x}_0$. Thus this gain will bring the system back to the steady state optimally from any initial condition.

(3) Against the above advantages, there is the difficulty that a large amount of off-line calculation needs to be done for the finite time case. However, even here, all this off-line computation is within a decentralised structure so that its storage requirements are minimal. Also, for large systems, the case of most practical interest is the one where the period of optimisation is infinite (or large with respect to the dynamics of the system) and for this important case, the off-line computational requirements are very small.

The method also has the advantage that once the gains have been calculated, the on-line calculation is minimal. It may not even be necessary to implement the controller using a digital computer -a few hardware components like amplifiers is all that is required for the infinite time case. The implementation can be done in a global approach (Fig. 6.13) or in a hierarchical way (Fig. 6.14).

6.12.3 <u>Example</u>

As a very simple illustrative example, consider the minimisation of J where

$$J = \frac{1}{2} \int_0^{8} (x_1^2 + 2\ x_2^2 + u_1^2 + u_2^2)\ dt$$

subject to

$$\mathring{x}_1 = -\ x_1 + x_2 + u_1$$

$$\mathring{x}_2 = -\ x_2 + u_2$$

This can be broken into 2 subproblems by defining the interaction variable $Z = x_2$.

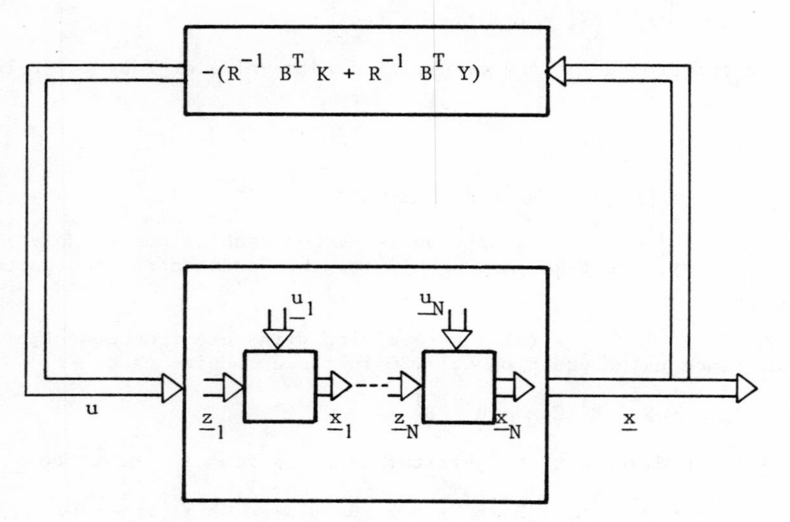

Fig. 6.13 : Global approach

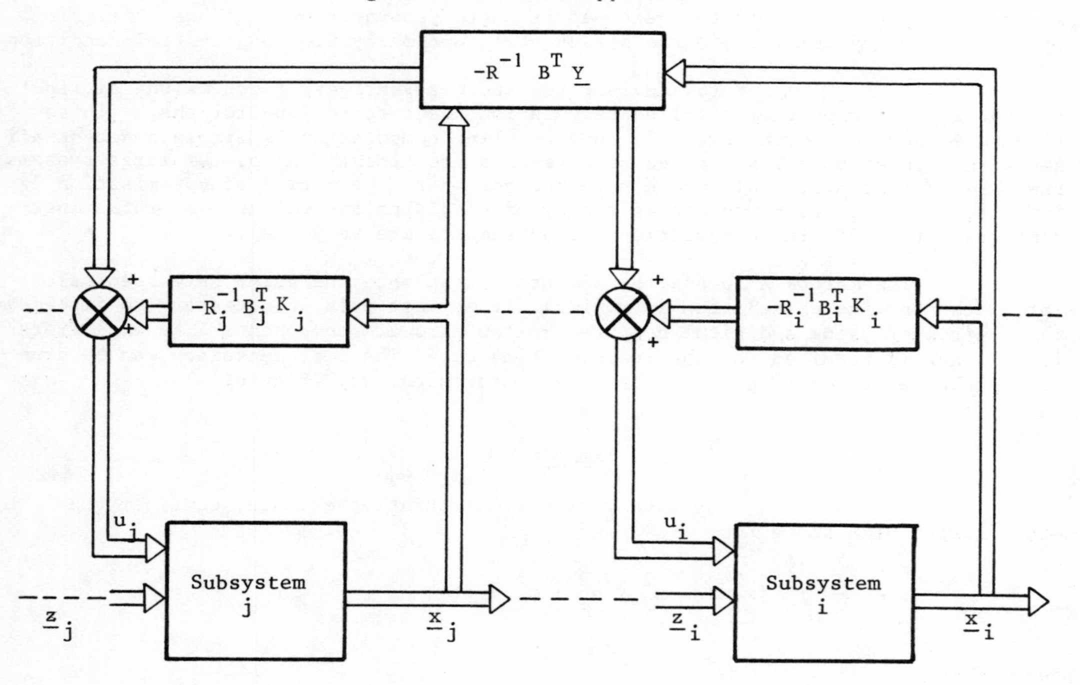

Fig. 6.14 : Hierarchical implementation

Then the first level problem becomes

$$\text{Min } J_1 = \int_0^8 \left[\frac{1}{2} x_1^2 + \frac{1}{2} u_1^2 + \lambda z \right]$$

subject to

$$\overset{\circ}{x}_1 = - x_1 + z + u_1$$

and

$$\text{Min } J_2 = \int_0^8 \left[x_2^2 + \frac{1}{2} u_2^2 - \lambda x_2 \right]$$

subject to

$$\overset{\circ}{x}_2 = - x_2 + u_2$$

This problem was solved on an IBM 370/165 digital computer for initial conditions (1.5, 5). Convergence to the optimum took place in 10 second level iterations. Here the period of optimisation is long enough for it to be considered infinite and the constant partial feedback gains were found to be $K_1 = 0.4142$ and $K_2 = 0.732$ respectively. Using the values of $\underline{x}$ and $\underline{s}$ at time t = 0,1, Y was evaluated by inverting a 2 x 2 matrix and multiplying with a 2 x 2 matrix:

$$R^{-1} B^T Y = \begin{bmatrix} -0.0055 & 0.1274 \\ 0.1274 & 0.0675 \end{bmatrix}$$

and

$$R^{-1} B^T K = \begin{bmatrix} 0.4142 & 0 \\ 0 & 0.732 \end{bmatrix}$$

This gave the overall gain to be

$$G = \begin{bmatrix} 0.4087 & 0.1274 \\ 0.1274 & 0.7995 \end{bmatrix}$$

which is identical to that obtained by solving the Riccati equation for the composite system.

6.13 EXTENSION TO THE SERVOMECHANISM CASE

Consider now the problem

$$\text{Min } J = \sum_{i=1}^{N} \frac{1}{2} \int_{t_0}^{t_f} \left[\| \underline{x}_i - \underline{x}_i^* \|^2_{Q_i} + \| \underline{u}_i \|^2_{R_i} \right] dt \qquad 6.13.1$$

subject to

$$\dot{\underline{x}}_1 = A_i \underline{x}_i + B_i \underline{u}_i + C_i \underline{z}_i + \underline{D}_i \qquad 6.13.2$$

and

$$\underline{z}_i = \sum_{j=1}^{N} L_{ij} \underline{x}_j \qquad 6.13.3$$

Here $\underline{x}_i^*$ is a constant known desired trajectory for the ith subsystem and $\underline{D}_i$ is a vector of constant known inputs which come into the ith subsystem.

Using a similar development as in section 6.12, it is easy to show that the control is given by

$$\underline{u} = - R^{-1} B^T K \underline{x} - R^{-1} B^T \underline{\xi} \qquad 6.13.4$$

where R, B, K are block diagonal matrices. Similarly, it is also easy to show that

$$\underline{\xi} = T_1 \underline{x} + \underline{T}_2 \qquad 6.13.5$$

where for the infinite time case for A, B, C, D, etc. time invariant, T_1 is an $\sum_{i=1}^{N} n_i \times \sum_{i=1}^{N} n_i$ constant matrix and T_2 is a $\sum_{i=1}^{N} n_i$ dimensional constant vector.

Thus

$$\underline{u} = - \left[R^{-1} B^T K + R^{-1} B^T T_1\right] \underline{x} - R^{-1} B^T \underline{T}_2$$

or

$$\underline{u} = - G\underline{x} - R^{-1} B^T \underline{T}_2 \qquad 6.13.6$$

Here again, as for the regulator case, for $t_f - t_0 \rightarrow \infty$, T_1 and T_2 can be obtained from the $(\sum_{i=1}^{N} n_i + 1)$ time points by inverting a $(\sum_{i=1}^{N} n_i \times \sum_{i=1}^{N} n_i)$ matrix where the points are obtained from an off-line decentralised calculation using the interaction prediction principle.

Note that for the servomechanism case, the form of the overall optimal control law for the composite system is the same as that given in equation 6.13.6.

6.14 EXAMPLE : RIVER POLLUTION CONTROL [19]

As a practical example of the application of the approach, let us reconsider our river pollution control problem. We will begin by recapitulating the various river pollution models that have been suggested and then apply the approach to each model.

6.14.1 The river pollution control models

As we saw before, the basic elements of the models are the dynamics of Biochemical Oxygen Demand (B.O.D.) and Dissolved Oxygen (D.O.) in the stream. For a reach which is defined as a section of the river of some convenient length having one polluter, these can be represented by the state space equations

$$\dot{\underline{x}}_i(t) = A \underline{x}_i(t) + B \underline{u}_i(t) + \underline{C}_i \qquad 6.14.1$$

where for the ith reach

$$\underline{x}_i = \begin{bmatrix} z_i \\ q_i \end{bmatrix}$$

where z_i is the concentration of B.O.D. in reach i (mg/l) and q_i is the concentration of D.O. in reach i. u_i is the concentration of B.O.D. in the effluent discharged into the stream.

From the River Cam model [11]

$$A = \begin{bmatrix} -1.32 & 0 \\ -0.32 & -1.2 \end{bmatrix} \qquad B = \begin{bmatrix} 0.1 \\ 0 \end{bmatrix} \qquad C_i = \begin{bmatrix} 0.9\, z_{i-1} \\ 0.9\, q_{i-1} + 1.0 \end{bmatrix}$$

In equation 6.14.1, C_i can be written in several ways to account for the different transport delays between the reaches as Tamura [12] points out.

1) No delay model

Here

$$\begin{bmatrix} z_{i-1}(t) \\ q_{i-1}(t) \end{bmatrix} = \begin{bmatrix} z_{i-1}(t) \\ q_{i-1}(t) \end{bmatrix} \qquad 6.14.2$$

2) Pure delay model

Here

$$\begin{bmatrix} z_{i-1}(t) \\ q_{i-1}(t) \end{bmatrix} = \begin{bmatrix} z_{i-1}(t-\theta_0) \\ q_{i-1}(t-\theta_0) \end{bmatrix} \qquad 6.14.3$$

$$\theta_0 = \frac{V_i}{q_{i-1}} = 1$$

where V_i is volume of water (m.gl/d) in reach i and Q_{i-1} is the volume of water which flows from the (i-1)th reach to the ith reach in one day (mg/l). θ_0 is the pure delay.

3) Distributed delay model

$$z_{i-1}(t) = \sum_{j=1}^{m} a_j\, z_{i-1}(t-\theta_j)$$

$$q_{i-1}(t) = \sum_{j=1}^{m} a_j\, q_{i-1}(t-\theta_j) \qquad 6.14.4$$

$$\sum_{j=1}^{m} a_j = 1 \text{ ; mean of } \theta_j = \theta_0 \text{ ; } \theta_1 < \theta_2 \ldots < \theta_m$$

The "no-delay" model shows that the state of the (i-1)th reach affects the state of the ith reach immediately. Equations 6.14.1 and 6.14.2 give the Kendrick model [20]. The "pure delay" model assumes that there exists a pure time delay between adjacent reaches, i.e. the state of the (i-1)th reach affects the state of the ith reach after θ_0 days. The "distributed delay" model shows that for j = 1, 2, ..., m, a fraction a_j of B.O.D. and D.O. in the (i-1)th reach at time $(t-\theta_j)$ arrives in the ith reach at time t ; i.e. the transport delays are distributed in time between θ_1 and θ_m. Fig. 6.15 shows the distributed phenomena whereby B.O.D. (or D.O.) is discharged at time t = 0 in the (i-1)th reach and fractions a_j arrive at $t = \theta_j$, j=1, ..., m in the ith reach. We know that this last model is particularly realistic since it takes into account the dispersion of pollutants in the river.

6.14.2 The optimisation problem

A realistic optimisation problem for an N reach system is to minimise

$$J = \sum_{i=1}^{N} \int_0^\infty \left[(\underline{x}_i - \underline{x}_i^*)^T Q_i (\underline{x}_i - \underline{x}_i^*) + \underline{u}_i^T R_i \underline{u}_i \right] dt \qquad 6.14.5$$

where $\underline{x}_i^* = \begin{bmatrix} z_i^* \\ q_i^* \end{bmatrix}$ are the desired levels of B.O.D. and D.O. in the stream, Q_i is an appropriate 2 x 2 constant positive definite weighting matrix and R_i is a positive scalar. The period of optimisation here is taken to be long enough for the system to return to the steady state from any initial condition that an unknown disturbance may take the system to.

J in equation 6.14.5 is to be minimised subject to the system dynamics 6.14.1 and successively to the three models given by equations 6.14.2, 6.14.3 and 6.14.4 for the delays.

6.14.3 Feedback control for a two reach river system

6.14.3.1 No delay model

The "no delay" model for the Cam river of two reaches is already in the form given by equations 6.14.1 and 6.14.2 . This problem was solved on an IBM 370/165 for the initial state $z_1(0) = 10$ mg/l, $q_1(0) = 7$ mg/l, $z_2(0) = 5$ mg/l, $q_2(0) = 7$ mg/l. These initial conditions imply that the second reach is initially "clean" while the first reach suddenly receives a pollution load. The cost function was chosen to be such that

$$Q_i = \begin{bmatrix} 2 & 0 \\ 0 & 1 \end{bmatrix} \qquad i = 1, 2$$

and

$$R_i = 1 \qquad i = 1, 2$$

The desired states were chosen to be consistent with the steady state of the system as $z_1^* = 4.06$ mg/l ; $q_1^* = 8$ mg/l ; $z_2^* = 5.94$ mg/l ; $q_2^* = 6$ mg/l ; $u_1^* = 53.3$ mg/l ; $u_2^* = 41.9$ mg/l.

The subsystem Riccati equations, etc., were integrated over a period of 8 days. This period is certainly sufficiently long for the system to reach a steady state. From the open loop control which was obtained after 6 second level iterations, the states $\underline{x}$ and $\underline{u}$ were recorded at the first 5 sampling points and from this the gain matrix G was calculated to be

$$G = \begin{bmatrix} 0.0960 & -0.0095 & 0.0270 & -0.0038 \\ 0.0270 & -0.0038 & 0.0768 & -0.0053 \end{bmatrix}$$

and ξ to be

$$\underline{\xi} = \begin{bmatrix} -.4295 \\ -.2693 \end{bmatrix}$$

and the optimal feedback control was

$$\underline{u} = G\,\underline{x} + \underline{\xi}$$

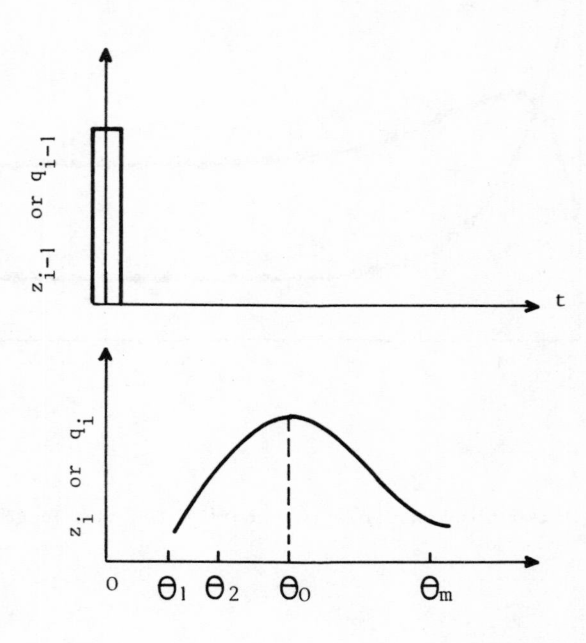

Fig. 6. 15 : The distributed Delay Phenomena

These feedback parameters were found to be identical with those obtained from a globally optimal solution calculated from the overall Riccati equation. Fig.6.16-17 show the trajectories of B.O.D. and D.O. for the two reaches using the above control law and Fig. 6.18 the resulting control deviations Δu_1, Δu_2 from the steady state control of 53.3 mg/l for reach 1 and 41.9 for reach 2.

These trajectories show that the B.O.D. and D.O. are forced towards their desired values exponentially in reach 1 whereas in reach 2, they get the effect of the pollution in reach 1. The responses are not too realistic since the effect of the pollution is apparent immediately in reach 2 whereas in practice it should occur after some time.

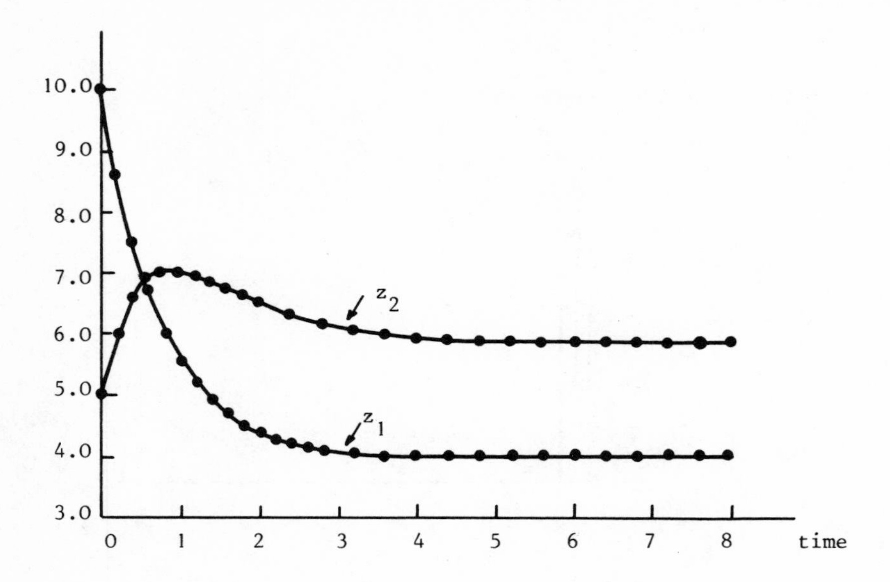

Fig. 6.16 : Optimal B.O.D. trajectories for the no-delay model

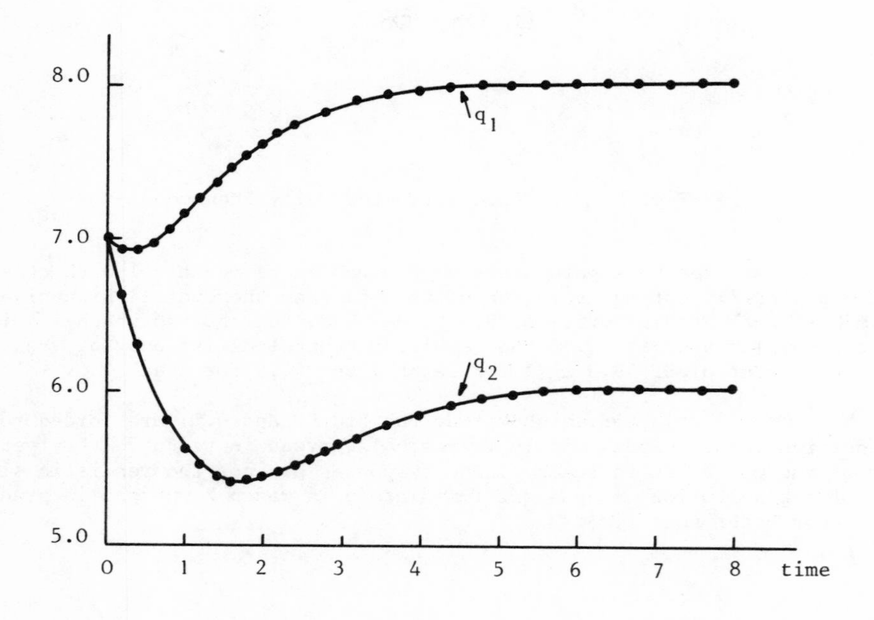

Fig. 6.17 : Optimal D.O. trajectories for the no-delay model

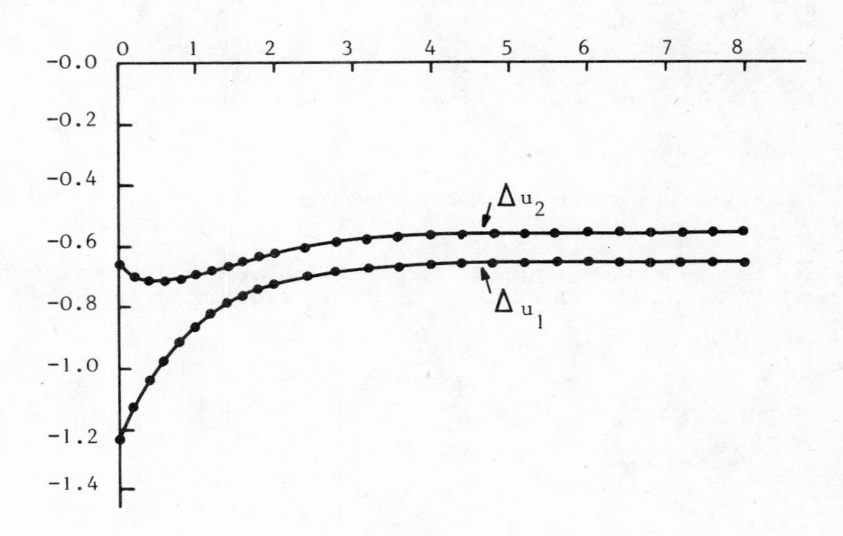

Fig. 6.18 : Optimal control trajectories for the no-delay model

6.14.3.2 Pure delay model

For the two reach pure delay model, the delays were approximated by a 2nd order Taylor series expansion, i.e. in this case the state equations were

$$\dot{z}_1 = -1.32\ z_1 + 0.1\Delta u_1 + 0.9\ z_0 + 5.35 \qquad 6.14.6$$

$$\dot{q}_1 = -0.32\ z_1 - 1.2\ q_1 + 0.9\ q_0 + 1.9 \qquad 6.14.7$$

$$\dot{z}_2 = 0.9\ z_1\ (t-\tau) - 1.32\ z_2 + 0.1\Delta u_2 + 4.19 \qquad 6.14.8$$

$$\dot{q}_2 = 0.9\ q_1\ (t-\tau) - 0.32\ z_2 - 1.2\ q_2 + 1.9 \qquad 6.14.9$$

Let $z_3(t) = z_1(t-\tau)$ then $z_3(s) = z_1(s)\ e^{-s\tau}$ where s is the Laplace variable.

Now $z_1(s)\ e^{-s\tau} = z_1(s)\left[1 + s\tau + s^2\frac{\tau^2}{2}\ \ldots\right]^{-1}$ taking only the first three terms

$$z_1(t) = z_3(t) + \tau\ z_3(t) + \frac{\tau^2}{2}\ z_3(t)$$

This requires the introduction of two additional variables z_3, z_4 such that

$$\dot{z}_3 = z_4 \qquad 6.14.10$$

$$\dot{z}_4 = \frac{2}{\ _2}\ z_1 - \frac{2}{\tau^2}\ z_3 - \frac{2}{\tau}\ z_4 \qquad 6.14.11$$

Similarly for q,

$$\mathring{q}_3 = q_4 \qquad 6.14.12$$

$$\mathring{q}_4 = \frac{2}{\tau^2} q_1 - \frac{2}{\tau^2} q_3 - \frac{2}{\tau} q_4 \qquad 6.14.13$$

Then the state equations are given by equations 6.14.6 to 6.14.13.

In this analysis, τ was taken to be 1/2 day.

The system was divided into two subsystems with (z_1, q_1) as the states of subsystem 1 and $(z_2, q_2, z_3, q_3, z_4, q_4)$ as the states of subsystem 2.

For the optimisation, Q_1, R_1 were the same as for the no-delay case for subsystem 1. However, for subsystem 2, Q_2 and R_2 were chosen to be

$$Q_2 = \begin{bmatrix} 2 & 0 & 0 & 0 & 0 & 0 \\ 0 & 0 & 0 & 0 & 0 & 0 \\ 0 & 0 & 0 & 0 & 0 & 0 \\ 0 & 0 & 0 & 1 & 0 & 0 \\ 0 & 0 & 0 & 0 & 0 & 0 \\ 0 & 0 & 0 & 0 & 0 & 0 \end{bmatrix}$$

and $R_2 = 1$.

This problem was again solved on an IBM 370/165 computer and the resulting gain matrix G and disturbance vector $\underline{\xi}$ are given in Table 6.2.

These were again found to be identical to the global solution obtained by integrating the 8th order Riccati equation for the overall system. Using these gains, the system was simulated for initial conditions

$$z_1(0) = 10 \text{ mg/l},\ z_2(0) = 5.94 \text{ mg/l},\ q_1(0) = 7 \text{ mg/l},\ q_3(0) = 6 \text{ mg/l}$$

Figs. 6.19 - 6.20 show the B.O.D. and D.O. trajectories and Fig. 6.21 the corresponding control sequences. On comparing Figs. 6.19 - 6.20 with Figs. 6.16 6.17, the response of the "pure delay" model appears to be more realistic than that of the "no delay" model.

The Gain G for 2 reaches with pure delay :

The first row :

+0.094848156 -0.009303570 +0.014567137 +0.009602249 +0.002856791
-0.002140164 -0.001366377 -0.000450552

The second row :

+0.014566869 -0.003238410 +0.076799244 +0.024565300 +0.004629463
-0.005270749 -0.002777129 -0.000679225

The disturbance vector $\underline{\xi}$:

-0.471007645
-0.238365799

TABLE 6.2

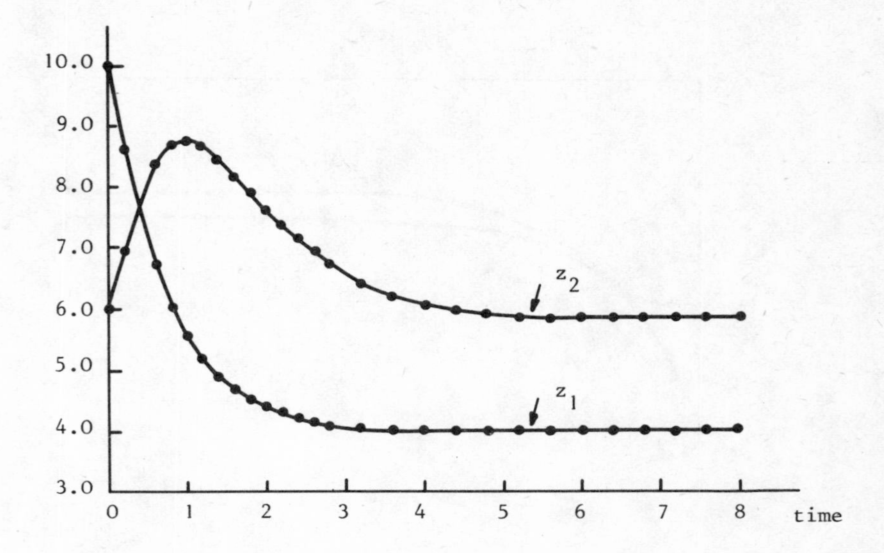

Fig. 6. 19: Optimal B.O.D. trajectories for the pure delay model

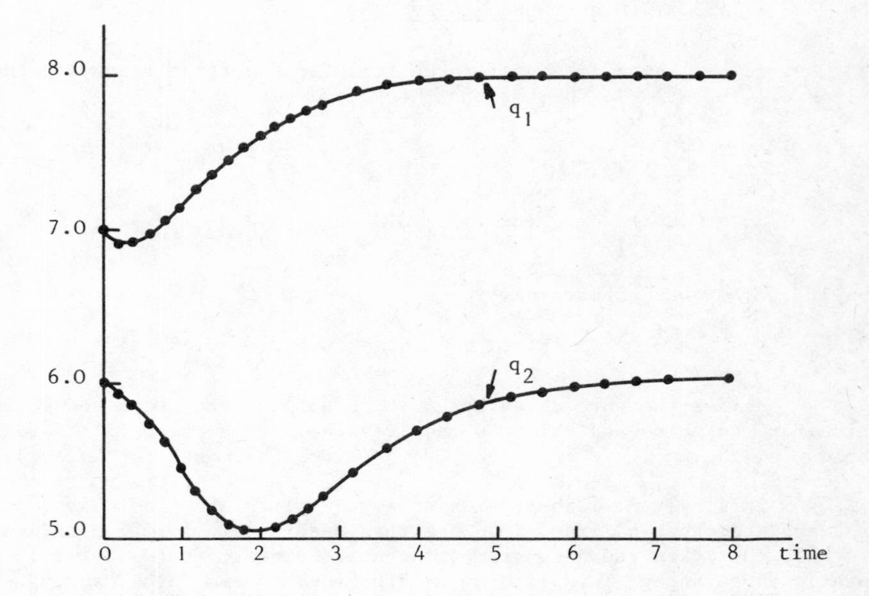

Fig. 6. 20: Optimal D.O. trajectories for the pure delay model

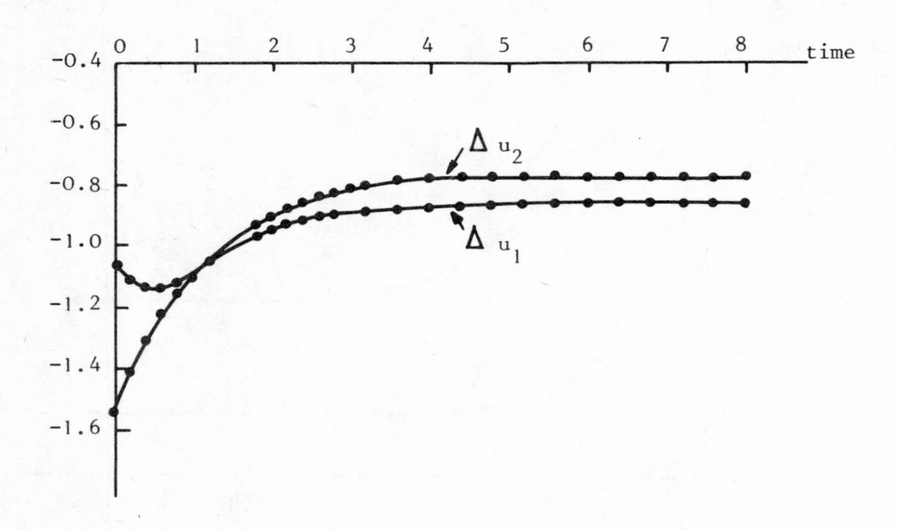

Fig. 6. 21 : Optimal control trajectories for the pure delay model

6.14.3.3. Distributed delay model

In this case, the state equations for the first reach are the same as equations 6.14.6, 6.14.7, whereas for the second reach

$$z_2 = 0.9 \sum_{i=1}^{m} a_i \, z_1(t-\tau_i) - 1.32\, z_2 + 0.1\Delta u_2 + 4.19 \qquad 6.14.14$$

$$q_2 = 0.9 \sum_{i=1}^{m} a_i \, q_1(t-\tau_i) - 0.32\, z_2 - 1.2\, q_2 + 1.9$$

As in [12], m, τ_i were chosen as :

$m = 3, \tau_1 = 0, \tau_2 = \frac{1}{2}, \tau_3 = 1$; $z_0 = 0$, $q_0 = 10$; $a_1 = 0.15$, $a_2 = 0.7$, $a_3 = 0.15$

As for the pure delay case, it is possible to use a second order approximation for the delay and this now requires the introduction of the additional variables (z_3, q_3, z_4, q_4, z_5, q_5, z_6, q_6) (since there are two delays).

This overall 12th order system was split into two subsystems, one of order 2 and the other of order 10. Note that such a decomposition is useful in that we are able to retain the subsystem structure. However, there is little if any computational advantage in splitting a 12th order system into 2 subsystem of order 10 and 2. But here we are only demonstrating the methodology. As the number of subsystems increases, the decomposition becomes more computationally viable as we see in the next example where a three reach system is considered.

The gain G matrix for 2 reaches with distributed delay

The first row

+0.093525320	-0.009112746	+0.015547663	+0.006788820	+0.001955301	+0.003402621
+0.001717001	-0.002268046	-0.001000434	-0.000320464	-0.000610203	-0.000370532

The second row :

+0.015546888	-0.003200263	+0.076798409	+0.017195493	+0.003240615	+0.005426496
+0.001640528	-0.005270392	-0.001943916	-0.000475317	-0.000784606	-0.000327557

The disturbance vector $\underline{E}$

-0.468761355
-0.241522819

TABLE 6.3

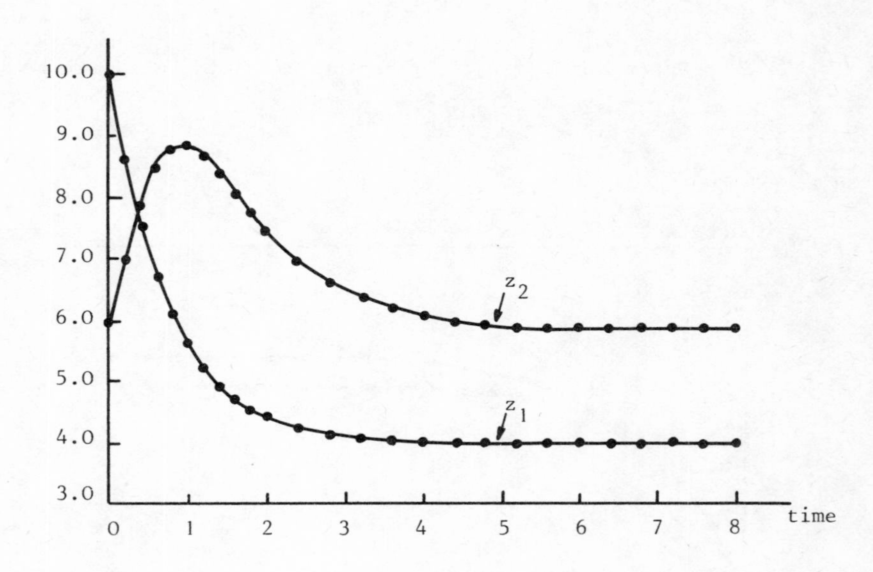

Fig. 6.22 : Optimal B.O.D. trajectories for the distributed delay model

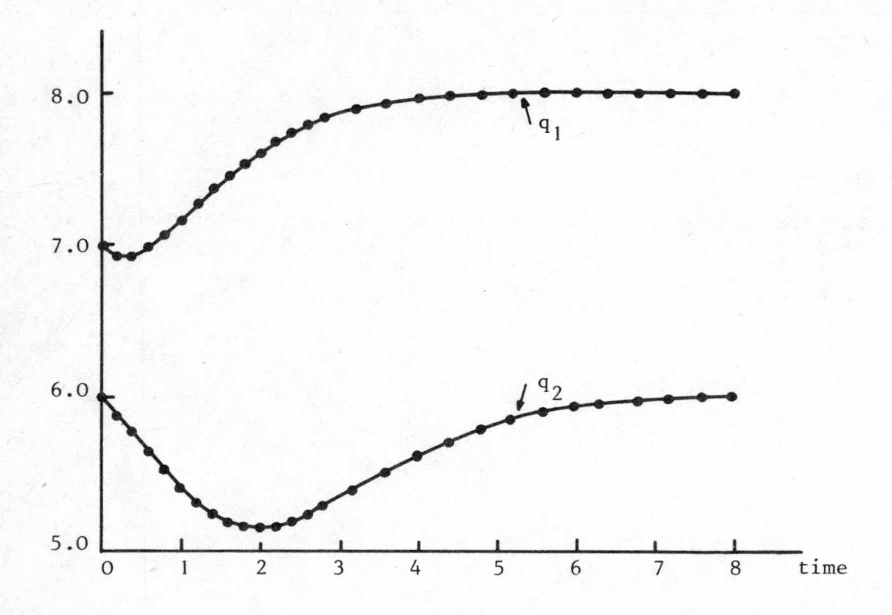

Fig. 6.23 : Optimal D.O. trajectories for the distributed delay model.

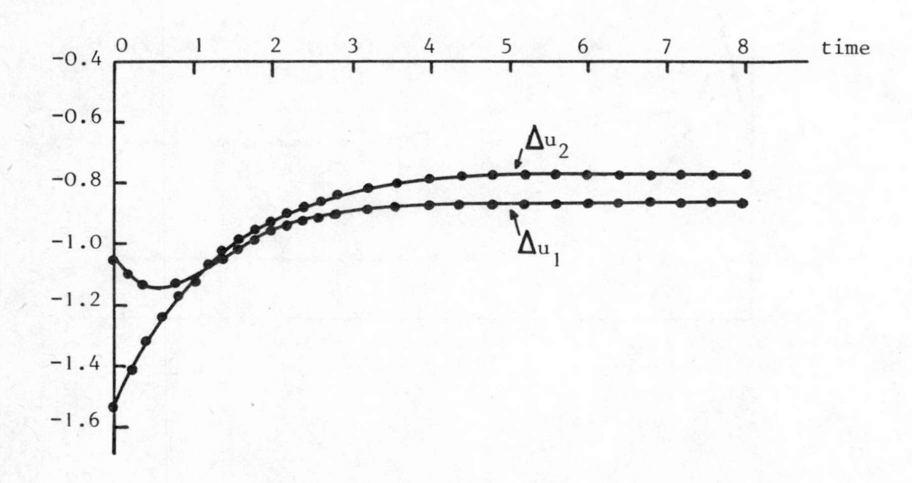

Fig. 6.24 : Optimal control trajectories for the distributed delay model.

<u>The gain matrix G for the 3 reaches with distributed delays</u>

The first row :

+0.099062085 -0.010958791 +0.022846639 +0.009379625 +0.002637625 +0.004595816
+0.002281010 -0.004275560 -0.001736045 -0.000522673 -0.000974357 -0.000555217
+0.003174365 +0.001376390 +0.000480294 +0.000990987 +0.000697553 -0.000700831
-0.000285327 -0.000102699 -0.000220478 -0.000178099

The second row :

+0.022842050 -0.005539298 +0.093502045 +0.022300422 +0.004415929 +0.007374823
+0.002398074 -0.009108484 -0.003236697 -0.000795007 -0.001329780 -0.000566363
+0.015544653 +0.006787241 +0.001954496 +0.003401399 +0.001715660 -0.002267599
-0.001000464 -0.000320613 -0.000610054 -0.000370383

The third row :

+0.003168851 -0.001074106 +0.015543908 +0.003468424 +0.000657767 +0.001095563
+0.000330418 -0.003199846 -0.000876218 -0.000188380 -0.000311464 -0.000111490
+0.076797992 +0.017195195 +0.003240436 +0.005426317 +0.001640469 -0.005270451
+0.003240436 +0.005426317 -0.000784665 -0.000327677 -0.000327677 -0.000475317

The disturbance vector $\underline{\xi}$:

-0.588996768
-0.410924196
-0.112484366

TABLE 6.4

Using similar weighting matrices as for the no delay and pure delay cases, the gain matrix G and the vector $\underline{\xi}$ were calculated using the hierarchical approach. Table 6.3 gives G and $\underline{\xi}$. Using these, the system was simulated. Figs.6.22 6.23 give the trajectories for B.O.D. and D.O. and Fig. 6.24 the corresponding control trajectories for the same initial conditions as for the pure delay case. These were found to be more spread out showing the effects of the dispersion of the pollutants.

We next apply the method to the "distributed delay" model for 3 reaches.

6.14.4 Control of the 3 reaches distributed delay model

In this case, the system is of order 22 which can be split up into 3 subsystems of order 2, 10 and 10. From a steady state analysis, the desired values for the 3rd reach are :

$$z_3^* = 5.237 \quad ; \quad q_3^* = 4.69$$

and the control for reach 3 was taken to be a deviation of the B.O.D. in the effluent discharge from 15.91 mg/l. All the other parameters were taken to be the same as for the 2 reach distributed delay model.

Results

This 3 reach 22 order system was solved using the decentralised 3 subsystem calculation structure on the IBM 370/165. For the off-line calculation, convergence took place in 11 second level iterations. From this, the control gain matrix was calculated to be as shown in Table 6.4. Fig. 6.26 and Fig. 6.27 show the trajectories of B.O.D. and D.O. for the 3 reaches from initial conditions

$$z_1(0) = 10 \text{ mg/l}, \quad z_2(0) = 5.94 \text{ mg/l}, \quad z_3(0) = 5.237 \text{ mg/l}$$

$$q_1(0) = 7 \text{ mg/l}, \quad q_2(0) = 6 \text{ mg/l}, \quad q_3(0) = 4.69 \text{ mg/l}$$

Physically, these imply that the 2nd and 3rd reaches are in the steady state when a large pollution load comes into reach 1. Figs. 6.26, 6.27 show the attenuation of this pollution down the river. Fig. 6.25 shows the control trajectories.

Remark

A method of providing optimal constant feedback control has been demonstrated for river pollution control using some data from the river Cam near Cambridge. The method is advantageous since the resulting control is closed loop and the gain matrix is constant so that a computer is not necessary for its implementation. It is also optimal for any initial condition.

Our various simulation studies on the river pollution problem have shown the efficacy of the method for river pollution control.

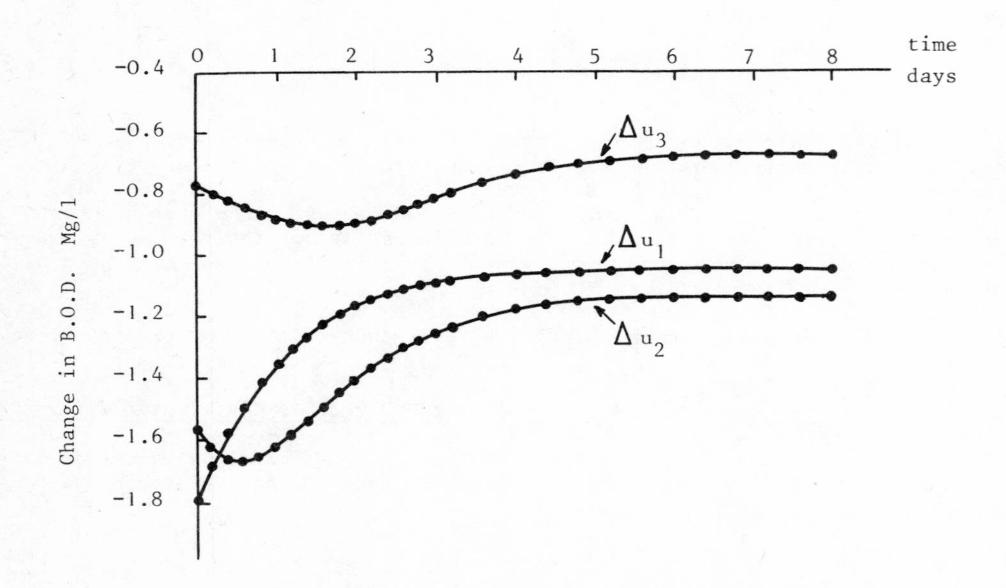

Fig. 6.25 : Optimal control trajectories for the 3 reach distributed delay model.

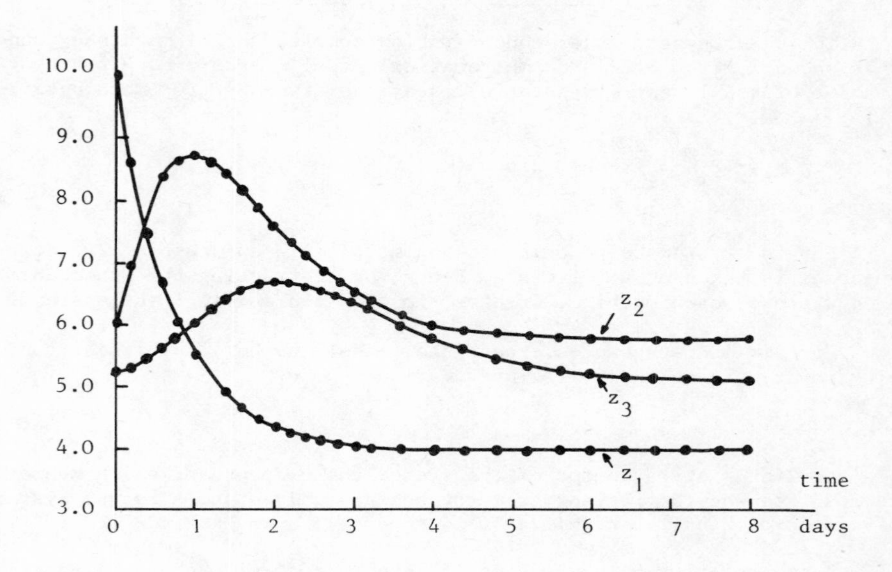

Fig. 6.26 : Optimal B.O.D. trajectories for the three reach distributed delay model

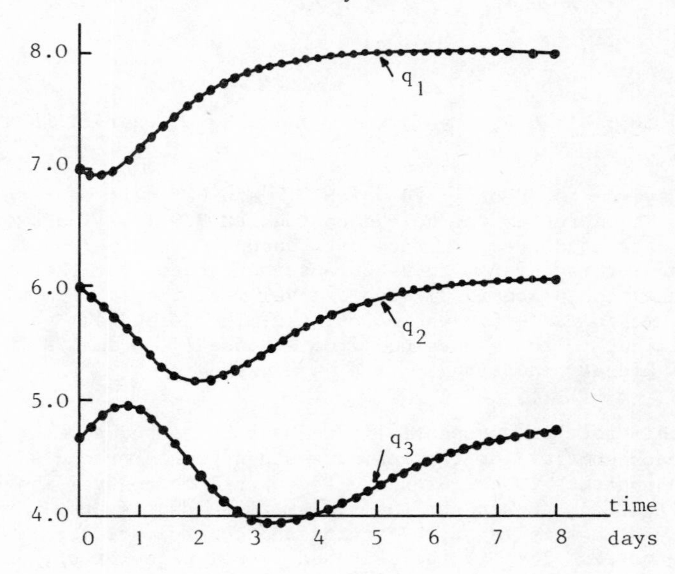

Fig. 6.27 : Optimal D.O. trajectories for the three reach distributed delay model

6.15 EXAMPLE : FEEDBACK CONTROL FOR POWER SYSTEM [17]

The power system under consideration consists of 11 machines connected together as shown in Fig. 6.28. The mathematical model of the n-machines system consists of a set of non-linear differential equations given by (for more details, see [18]) :

$$M_i \ddot{\delta}_i = P_i - \sum_{\substack{j=1 \\ j \neq i}}^{n} b_{ij} \sin \delta_{ij} \qquad i = 1, \ldots, n \qquad 6.15.1$$

For the system to be completely controllable and completely observable, the nth machine is taken as a reference. Hence, by subtracting the nth equation from the equation of each machine, we have the required model of the system.

For small disturbances, the system model can be linearized around the equilibrium point, giving the following relation :

$$\underline{\dot{x}} = A \, \underline{x} \qquad 6.15.2$$

where $\underline{x}$ is the 2(n-1) state vector of the system the elements of which represent the angle and frequency deviations from the nominal values and A is the system matrix of dimension 2(n-1) x 2(n-1).

Using relation 6.15.2, we would like to control each state variable of our system having 11 machines such that a quadratic cost function is minimized. Therefore, we have the following 20th order power system control problem :

$$\min J = \frac{1}{2} \int_0^4 (\underline{x}^T Q \, \underline{x} + \underline{u}^T R \, \underline{u}) \, dt \qquad 6.15.3$$

subject to : $\underline{\dot{x}} = A \, \underline{x} + B \, \underline{u}$

where A is given in Table 6.5, B = I, Q = 0.1 and R = I where I is the identity matrix.

The system is of order 20 which can be split up into 5 subsystems each of order 4. This problem was solved on the IBM 370/165. The system Riccati equations were integrated over a period of 4 seconds. This period is certainly sufficiently long for the system to reach a steady state. For the off-line calculations, convergence took place in 31 second level iterations. From this, the control gain matrix G given in Table 6.6 was calculated. These feedback parameters were found to be identical to those obtained from a globally optimal solution calculated from the overall Riccati equation.

In this section, essentially we have developed a method of providing a constant feedback control for the infinite stage regulator and servomechanism. The basis of the approach is to calculate the open loop control and state trajectories for a period long enough for the system to reach its steady state. Then for a system of order n, n samples of the states and control are used for the calculation of the gain matrix. This is the only additional calculation, and since it is done off-line, it should not pose much of a problem.

We next develop a feedback structure for discrete time systems.

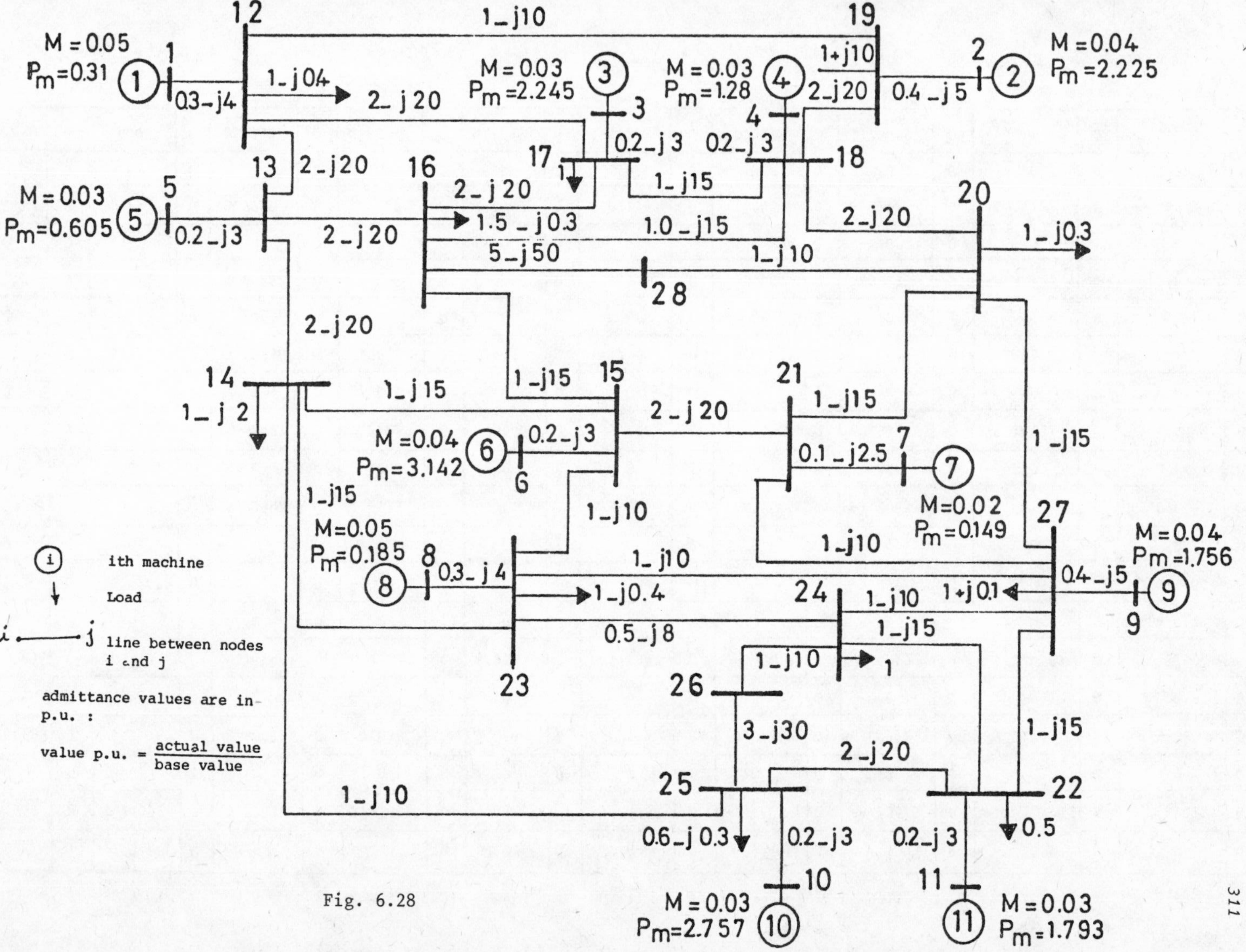

M = 0.05
Pm=0.31
M = 0.03
Pm=0.605
M = 0.03
Pm =2.245
M = 0.03
Pm = 1.28
M = 0.04
Pm =2.225
M =0.04
Pm=3.142
M=0.05
Pm=0.185
M=0.02
Pm=0.149
M = 0.04
Pm =1.756
M = 0.03
Pm=2.757
M = 0.03
Pm=1.793
1_j10
1+j10
2_j20
0.4_j5
0.3-j4
1_j04
2_j20
2_j20
2_j20
1.5_j0.3
2_j20
0.2_j3
0.2_j3
0.2_j3
1_j15
1.0_j15
5_j50
1_j10
2_j20
1_j0.3
2_j20
1_j15
1_j15
1_j2
0.2_j3
2_j20
1_j15
0.1_j2.5
1_j15
1_j15
1_j10
1_j10
0.3_j4
1_j10
1_j0.4
0.5_j8
1_j10
1+j0.1
0.4_j5
1_j15
1_j10
3_j30
2_j20
1_j15
1_j10
0.6_j0.3
0.2_j3
0.2_j3
0.5
ith machine
Load
line between nodes i and j
admittance values are in p.u. :
value p.u. = actual value / base value

Fig. 6.28

	1	2	3	4	5	6	7	8	9	10	11	12	13	14	15	16	17	18	19	20
1	0	1	0	0	0	0	0	0	0	0	0	0	0	0	0	0	0	0	0	0
2	-93.931	0	8.74383	0	5.37437	0	2.62699	0	4.05603	0	-1.64764	0	-2.7157	0	-6.33619	0	-14.37576	0	-12.59014	0
3	0	0	0	1	0	0	0	0	0	0	0	0	0	0	0	0	0	0	0	0
4	15.84999	0	-140.367	0	9.60086	0	14.026	0	6.684	0	2.67515	0	1.51862	0	-1.99922	0	-7.02907	0	-9.53736	0
5	0	0	0	0	0	1	0	0	0	0	0	0	0	0	0	0	0	0	0	0
6	10.6983	0	10.258	0	-116.061	0	6.21808	0	4.3133	0	0.41594	0	-1.06214	0	-4.6919	0	-11.74903	0	-11.52972	0
7	0	0	0	0	0	0	0	1	0	0	0	0	0	0	0	0	0	0	0	0
8	8.51556	0	18.07436	0	7.65584	0	-13.972	0	4.075	0	1.7033	0	1.16265	0	-3.055	0	-7.66032	0	-10.25109	0
9	0	0	0	0	0	0	0	0	0	1	0	0	0	0	0	0	0	0	0	0
10	11.29531	0	8.60465	0	5.98992	0	4.3135	0	-124.466	0	2.5982	0	-0.0644	0	-1.3065	0	-9.4617	0	-9.0164	0
11	0	0	0	0	0	0	0	0	0	0	0	1	0	0	0	0	0	0	0	0
12	1.96937	0	1.96213	0	1.68471	0	1.2125	0	1.6449	0	-99.694	0	1.9443	0	-0.933	0	-9.1423	0	-9.939	0
13	0	0	0	0	0	0	0	0	0	0	0	0	0	1	0	0	0	0	0	0
14	7.0993	0	10.2119	0	5.69536	0	7.5948	0	5.5155	0	11.7763	0	-166.755	0	5.63471	0	7.2399	0	-4.4116	0
15	0	0	0	0	0	0	0	0	0	0	0	0	0	0	0	1	0	0	0	0
16	0.19444	0	-0.31102	0	0.05862	0	-0.397	0	0.4133	0	1.255	0	-0.689	0	-89.399	0	-6.811	0	-8.578	0
17	0	0	0	0	0	0	0	0	0	0	0	0	0	0	0	0	0	1	0	0
18	5.084	0	7.478	0	3.9759	0	5.6047	0	4.0148	0	6.0777	0	7.286	0	8.009	0	-141.137	0	-0.465	0
19	0	0	0	0	0	0	0	0	0	0	0	0	0	0	0	0	0	0	0	1
20	1.532	0	1.400	0	0.9652	0	0.8061	0	1.8019	0	1.577	0	0.5298	0	1.739	0	-1.011	0	-110.087	0

TABLE 6.5

	1	2	3	4	5	6	7	8	9	10	11	12	13	14	15	16	17	18	19	
1	-3.3572	0.043	0.1698	-0.0353	0.0609	-0.0257	0.0985	-0.0273	0.1448	-0.0305	-0.08	-0.0257	0.2734	-0.0216	-0.0344	-0.0431	0.05583	-0.0418	0.4468	-0.047
2	0.0430	-0.0364	0.0297	-0.0012	0.0191	-0.0012	0.0229	-0.0002	0.0260	-0.0003	0.0196	0.0	0.0243	0.0022	0.0389	0.0017	0.0513	0.0066	0.0555	0.0062
3	0.1698	0.0297	-3.7532	0.0456	0.1176	-0.0015	0.1993	-0.0147	0.1107	-0.0068	-0.0230	-0.0007	0.2249	-0.0228	-0.0722	-0.014	0.2479	-0.045	0.0642	-0.0456
4	-0.0353	-0.0012	0.0456	-0.0281	-0.0042	-0.0011	0.0096	-0.0016	0.0018	-0.0006	-0.0055	-0.0004	0.022	0.0007	0.0073	0.0001	0.0456	0.002	0.0446	0.0015
5	0.0609	0.0191	0.1176	-0.0042	-3.3533	0.0454	0.1326	-0.0084	0.1039	-0.0086	-0.0504	-0.0123	0.1761	-0.017	-0.1239	-0.0284	0.2675	-0.0456	0.1043	-0.0492
6	0.0257	-0.0012	-0.0015	-0.0011	0.0454	-0.0301	0.0033	-0.0004	0.0035	-0.0002	0.0061	-0.0003	0.017	0.001	0.022	-0.0001	0.0482	0.0031	0.05	0.0022
7	0.0985	0.0229	0.1993	0.0096	0.1326	0.0033	-3.5331	0.0461	0.0930	-0.0025	-0.0283	-0.0018	0.1887	-0.0168	-0.114	-0.0148	0.1786	-0.0422	-0.0185	-0.0443
8	-0.0273	-0.0002	-0.0147	-0.0016	-0.0084	-0.0004	0.0461	-0.0281	-0.0021	-0.0001	-0.0037	-0.0003	0.0157	0.0006	0.0084	-0.0003	0.0418	0.0015	0.0423	0.0008
9	0.1448	0.026	0.1107	0.0018	0.1039	0.0035	0.0930	-0.0021	-3.4488	0.0461	-0.0056	-0.0013	0.1505	-0.0147	-0.0713	-0.0129	0.1649	-0.0424	0.0083	-0.0432
10	-0.0305	-0.0003	-0.0068	-0.0006	-0.0086	-0.0002	-0.0025	-0.0001	0.461	-0.0289	-0.0042	-0.0003	0.0137	0.0006	0.0065	-0.0003	0.0420	0.0018	0.0412	0.0008
11	-0.08	0.0196	-0.023	-0.0055	-0.0504	0.0061	-0.0283	-0.0037	-0.0056	-0.0042	-3.2311	0.0448	0.2797	-0.0257	-0.1210	-0.0154	0.2545	-0.0526	0.0577	-0.0547
12	-0.0257	0.0	-0.0007	-0.0004	-0.0123	-0.0003	-0.0018	-0.0003	-0.0013	-0.0003	0.0448	-0.0334	0.0249	0.0008	0.0077	-0.0009	0.0532	0.0024	0.0534	0.0014
13	0.2734	0.0243	0.2249	0.0220	0.1761	0.017	0.1687	0.0157	0.1505	0.0137	0.2797	0.0249	-3.9382	0.0471	0.1317	0.017	0.1012	-0.0027	-0.1721	-0.0136
14	-0.0216	0.0022	-0.0228	0.0007	-0.0170	0.0010	-0.0168	0.0006	-0.0147	0.0006	-0.0257	0.0008	0.0471	-0.025	-0.0197	0.0004	-0.0036	-0.0012	0.007	-0.0016
15	-0.0344	0.0389	-0.0722	0.0073	-0.1239	0.022	-0.1140	0.0084	-0.0713	0.0065	-0.1210	0.0077	0.1317	-0.0197	-3.1540	0.0433	0.2022	-0.0620	-0.0922	-0.0609
16	-0.0431	0.0017	-0.014	0.0001	-0.0284	-0.0001	-0.0148	-0.0003	-0.0129	-0.0003	-0.0154	-0.0009	0.0170	0.0004	0.0433	-0.0366	0.0582	0.0010	0.0547	-0.0006
17	0.5583	0.0513	0.2479	0.0456	02675	0.0482	0.1786	0.0418	0.1649	0.0420	0.2545	0.0532	0.1012	-0.0036	0.02021	0.0582	-4.1375	0.0402	-0.4702	0.011
18	-0.0418	0.0066	-0.045	0.0020	-0.0456	0.0031	-0.0422	0.0015	-0.0424	0.0018	-0.0526	0.0024	-0.0027	-0.0012	-0.0620	0.001	0.0402	-0.0325	-0.0274	-0.0053
19	0.4468	0.0555	0.0642	0.0446	0.1043	0.050	-0.0185	0.0423	0.0083	0.0412	0.0577	0.0534	-0.1721	0.0070	-0.0922	0.0547	-0.4702	-0.0274	-3.9216	0.0396
20	-0.0475	0.0062	-0.0456	0.0015	-0.0492	0.0022	-0.0443	0.0008	-0.0432	0.0008	-0.0547	0.0014	-0.0136	-0.0016	-0.0609	-0.0006	0.011	-0.0053	0.0396	-0.0385

TABLE 6.6

6.16 OPEN LOOP HIERARCHICAL OPTIMISATION BY DUALITY AND DECOMPOSITION

6.16.1 Problem formulation

The problem to be solved could be written as

Minimise $J(\underline{x}(k), \underline{u}(k), k)$

w.r.t. $\underline{u}(k)$

where $k = 0, 1, \ldots, K-1$

and

$$J = \frac{1}{2}\|\underline{x}(K)\|_P^2 + \frac{1}{2}\sum_{k=0}^{K-1}\left(\|\underline{x}(k)\|_Q^2 + \|\underline{u}(k)\|_R^2\right) \quad 6.16.1$$

where $\underline{x}$ is an n dimensional state vector, $\underline{u}$ is an r dimensional control vector, Q, R, are respectively n x n and r x r positive definite diagonal matrices. K is assumed to be "very" large.

J in equation 6.16.1 is to be minimised subject to the system dynamics

$$\underline{x}(k+1) = A\,\underline{x}(k) + B\,\underline{u}(k) \quad 6.16.2$$

$$\underline{x}(0) = \underline{x}_0 \quad 6.16.3$$

where A is an n x n time invariant matrix and B is an n x r time invariant matrix.

6.16.2 Open loop hierarchical optimisation structure

As we saw before, the approach of Tamura to tackling this problem is to solve instead the dual maximisation problem since from the theorem of strong Lagrange duality, Geoffrion [16], the solution of the two problems is equivalent at the extremum.

In order to formulate the dual problem, it is necessary to define the dual function $\phi(\underline{p}) = \phi[\underline{p}(0), \underline{p}(1), \ldots, \underline{p}(K-1)]$ which is to be maximised where $\underline{p}(k) \in E^n$ are the dual variables. The dual function ϕ is given by

$$\phi(\underline{p}) = \min_{\underline{x},\underline{u}} L(\underline{x}, \underline{u}, \underline{p}) \quad 6.16.4$$

where $L(\underline{x}, \underline{u}, \underline{p})$ is the Lagrangian function defined as

$$L(x,u,p) = \frac{1}{2}\|\underline{x}(K)\|_P^2 + \sum_{k=0}^{K-1} \frac{1}{2}\|\underline{x}(k)\|_Q^2 + \frac{1}{2}\|\underline{u}(k)\|_R^2 - \underline{p}(k)^T \left[\underline{x}(k+1) - A\underline{x}(k) - B\,\underline{u}(k)\right] \quad 6.16.5$$

The Lagrangian can be rewritten as

$$L(\underline{x},\underline{u},\underline{p}) = \frac{1}{2}\|\underline{x}(K)\|_P^2 - \underline{p}(K-1)^T\,\underline{x}(K) + \sum_{k=0}^{K-1}\Big(\frac{1}{2}\|\underline{x}(k)\|_Q^2 + \frac{1}{2}\|\underline{u}(k)\|_R^2 - \underline{p}(k-1)^T\,\underline{x}(k) + \underline{p}(k)^T\left[A\underline{x}(k) + B\underline{u}(k)\right]\Big) \quad 6.16.6$$

Define the Hamiltonian function H by

$$H(\underline{x}(k),\ \underline{u}(k),\ \underline{p}(k),\ k) = \frac{1}{2}\|\underline{x}(k)\|_Q^2 + \frac{1}{2}\|\underline{u}(k)\|_R^2 + \underline{p}(k)^T\ (A\underline{x}(k) + B\underline{u}(k)) \qquad 6.16.7$$

Then the Lagrangian function can be written using the Hamiltonian as

$$L(\underline{x},\underline{u},\underline{p}) = -\underline{p}(K-1)^T\ \underline{x}(K) + \frac{1}{2}\|\underline{x}(K)\|_P^2 + \sum_{k=0}^{K-1}\left[H(\underline{x},\underline{u},\underline{p},k) - \underline{p}(k-1)^T\ \underline{x}(k)\right] \qquad 6.16.8$$

where we assume that $\underline{p}(-1) = \underline{0}$; $\underline{p}(k) = \underline{0}$ for $k \geqslant K$.

To evaluate the value of the dual function in equation 6.16.4 for fixed $\underline{p} = \underline{p}^* = (\underline{p}^*(0)^T, \underline{p}^*(1)^T, \ldots, \underline{p}^*(K-1)^T)^T$, it is necessary to minimise the Lagrangian function in equation 6.16.8, w.r.t. $\underline{x}$, $\underline{u}$. However, for fixed $\underline{p}$, this functional minimisation problem can be separated into K + 1 independent parametric minimisation problems.

These are given by

(i) For k = 0

$$\text{Min } H(\underline{x}(0),\ \underline{u}(0),\ \underline{p}^*(0)) = \frac{1}{2}\|\underline{x}(0)\|_Q^2 + \frac{1}{2}\|\underline{u}(0)\|_R^2 + \underline{p}^*(0)^T\ \left[A\underline{x}(0) + B\underline{u}(0)\right]$$

subject to $\underline{x}(0) = \underline{x}_0$

Since R is assumed to be a positive definite diagonal matrix, this minimisation problem splits up into r independent one variable minimisation problems whose analytical solution is given by

$$\underline{u}^*(0) = -\ R^{-1}\ B^T\ \underline{p}^*(0) \qquad 6.16.9$$

(ii) For k = 1, 2, ..., K-1

$$\underset{\text{w.r.t.}\underline{x}(k),\underline{u}(k)}{\text{Min}}\quad H(\underline{x}(k),\ \underline{u}(k),\ \underline{p}^*(k),\ k) - \underline{p}^*(k-1)\ \ \underline{x}(k)$$

$$= \frac{1}{2}(\|\underline{x}(k)\|_Q^2 + \|\underline{u}(k)\|_R^2) - \underline{p}^*(k-1)^T\ \underline{x}(k) + \underline{p}^{*T}(k)\left[A\underline{x}(k) + B\underline{u}(k)\right]$$

Since this is a set of (n+r) independent one variable minimisation problems, the solution is given by

$$\begin{aligned}\underline{x}^*(k) &= -\ Q^{-1}\left[-\ \underline{p}^*(k-1) + A^T\ \underline{p}^*(k)\right]\\ \underline{u}^*(k) &= -\ R^{-1}\ B^T\ \underline{p}^*(k)\end{aligned} \qquad 6.16.10$$

(iii) For k = K

$$\underset{\underline{x}(K)}{\text{Min}}\quad -\ \underline{p}^{*T}(K-1)\,\underline{x}(K) + \frac{1}{2}\|\underline{x}(K)\|_P^2$$

The solution is

$$\underline{x}^{*}(K) = P^{-1}\ \underline{p}^{*}(K-1) \tag{6.16.11}$$

Equations 6.16.9 to 6.16.11 give the solutions to the first level problems and thus enable the value of the dual function $\emptyset$ to be evaluated for a fixed $\underline{p} = \underline{p}^{*}$. It merely remains to find a $\underline{p} = \underline{p}^{**}$ which maximises $\emptyset$. This is easy to do since the gradient of $\emptyset$ ($\underline{p}$) at $\underline{p} = \underline{p}^{*}$ is given by

$$\nabla \emptyset(\underline{p})\Big|_{\underline{p}=\underline{p}^{*}} = -\ \underline{x}^{*}(k+1) + A\underline{x}^{*}(k) + B\underline{u}^{*}(k) \tag{6.16.12}$$

Using the value of the dual function calculated by substituting the solutions given in equations 6.16.9 to 6.16.11 into the Lagrangian in equation 6.16.5 as well as the resulting gradient given in equation 6.16.12, it is possible to maximise $\emptyset(\underline{p})$ by a gradient technique such as the conjugate gradient method.

The complete algorithm is given below.

6.16.3 The Algorithm

Step 1 : Choose arbitrary initial values of $\underline{p}^{1} = (\underline{p}^{1}(0)^{T}, \underline{p}^{1}(1)^{T} \ldots \underline{p}^{1}(K-1)^{T})^{T}$. Set i = 1. One could use $\underline{p}^{1} = \underline{0}$ for i = 1.

Step 2 : Substitute $\underline{p} = \underline{p}^{i}$ in equations 6.16.10 to 6.16.12 to obtain the local optimal solutions

$$\begin{aligned} \underline{x}^{i} &= (\underline{x}^{i}(1)^{T}, \underline{x}^{i}(2)^{T}, \ldots, \underline{x}^{i}(K)^{T})^{T} \\ \underline{u}^{i} &= (\underline{u}^{i}(0)^{T}, \underline{u}^{i}(1)^{5}, \ldots, \underline{u}^{i}(K-1)^{T})^{T} \end{aligned} \tag{6.16.13}$$

Then, using equation 6.16.12, compute the n x K dimensional error vector (gradient of $\emptyset$) which comprises the error in each system equation, i.e.

$$\underline{e}(\underline{p}^{i}) = \nabla \emptyset(\underline{p})\Big|_{\underline{p}=\underline{p}^{i}} = \begin{bmatrix} \nabla\phi\,(\underline{p},\ k = 0) \\ \nabla\phi\,(\underline{p},\ k = 1) \\ ------- \\ \nabla\phi\,(\underline{p},\ k = K-1) \end{bmatrix}_{\underline{p}=\underline{p}_{i}} \tag{6.16.14}$$

if $\|\underline{e}(\underline{p}^{i})\|^{2} < \mathcal{E}$ for a small positive $\mathcal{E}$, record $\underline{x}^{i}$, $\underline{u}^{i}$ as the optimal solution. Else :

Step 3 : Set $\underline{p}^{i+1}$ according to

$$\underline{p}^{i+1} = \underline{p}^{i} + \alpha_{i}\ \underline{d}^{i} \tag{6.16.15}$$

where

$$\underline{d}^{i} = \underline{e}(\underline{p}^{i}) + \beta_{i-1}\ \underline{d}^{i-1}\ \ ;\ \underline{d}^{1} = \underline{e}(\underline{p}^{1}) \tag{6.16.16}$$

$$\beta_{i-1} \equiv \frac{\sum_{k=0}^{K-1} \underline{e}(\underline{p}^{i}(k))^T \underline{e}(\underline{p}^{i}(k))}{\sum_{k=0}^{K-1} \underline{e}(\underline{p}^{i-1}(k))^T \underline{e}(\underline{p}^{i-1}(k))} \qquad 6.16.17$$

$\alpha_i > 0$ is the step length and $\underline{d}^i$ is the search direction in the ith iteration. Set $i + 1 \longrightarrow i$ and go back to step 2.

6.16.4 Remarks

(1) The above two level algorithm is attractive because at the lowest level only a one dimensional problem has to be solved and even for this one dimensional problem, an analytical solution is available.

(2) Convergence of the second level algorithm is assured since the dual function to be maximised on the second level is convex.

(3) Against these advantages, there is the difficulty that the resulting control is open loop. This means that it is necessary to recalculate the controls every time the initial state changes. Moreover, the control is very sensitive to small parametric variations. Also, the control trajectory needs to be stored before open loop implementation. A constant feedback control matrix is clearly desirable. In the next Section, a simple way of calculating such a constant matrix is described.

6.17 A MULTILEVEL SOLUTION OF THE INFINITE STAGE REGULATOR

The feedback solution of equation 6.16.1 subject to equations 6.16.2 and 6.16.3 is given as in the last chapter by

$$\underline{u}(k) = - R^{-1} B^T A^{-1} \left[P(k) - Q \right] \underline{x}(k) \qquad 6.17.1$$

$$= G\, \underline{x}(k) \qquad 6.17.2$$

where $P(k)$ is the solution of the discrete matrix Riccati equation. For the infinite stage case ($K \longrightarrow \infty$), one way of obtaining a solution is to solve the Riccati equation for K large enough such that $P(k) = P(k+1)$, and to take this value of $P(k)$ to calculate the constant feedback matrix G in equation 6.17.2. In general, if K is several time periods of the system, P does reach a steady state.

Now, although P can be obtained from the Riccati equation, it requires the storage of large matrices and this may not be feasible for large scale systems. However, it is feasible to solve the problem in a hierarchical fashion as described in Section 16. In that case, at the optimum, when the dual function ϕ is maximised, equation 6.16.9 and 6.16.10 give the optimal sequences of $\underline{x}$ and $\underline{u}$. It is possible to obtain G in equation 6.17.2 from these sequences.

In order to do this, form the matrix

$$X = \left[\underline{x}(0) \quad \underline{x}(1) \ \ldots \ \underline{x}(n) \right]$$

and

$$U = \left[\underline{u}(0) \quad \underline{u}(1) \ \ldots \ \underline{u}(n) \right]$$

Then

$$G = U X^{-1} \qquad 6.17.3$$

This involves the additional calculation of inverting X but this can be done off-line.

Extension to the servomechanism case

$$\underset{\underline{u}(k)}{\text{Min}} \quad J = \frac{1}{2}\left[\sum_{k=0}^{K-1} \left\| \underline{x}(k) - \underline{x}^d \right\|_Q^2 + \left\| \underline{u}(k) \right\|_R^2 \right] \qquad 6.17.4$$

where $\underline{x}^d$ is a desired reference state and where J is to be minimised subject to

$$\underline{x}(k+1) = A\underline{x}(k) + B\underline{u}(k) + \underline{C} \qquad 6.17.5$$

$$\underline{x}(0) = \underline{x}_0 \qquad 6.17.6$$

where $\underline{C}$ is a known constant input. It is possible to obtain a similar solution by noting that the standard single level solution is now of the form

$$\underline{u}(k) = G\,\underline{x}(k) + \underline{d} \qquad 6.17.7$$

The two level off-line algorithm is changed slightly at the first level where the analytical solution can now be written as (equation 6.16.9 remains unchanged but 6.16.10 changes) :

$$\begin{aligned} \underline{u}^*(0) &= -R^{-1} B^T \underline{p}^*(0) \\ \underline{x}^*(k) &= \underline{x}^d - Q^{-1}\left[-\underline{p}^*(k-1) + A^T \underline{p}^*(k) \right] \\ \underline{u}^*(k) &= -R^{-1} B^T \underline{p}^*(k) \end{aligned} \qquad 6.17.8$$

$$k = 0, 1, \ldots, K-1$$

At the second level, the gradient becomes

$$\nabla_{\underline{p}(k)} \phi(\underline{p}) \Big|_{\underline{p}=\underline{p}^*} = -\underline{x}^*(k-1) + A\underline{x}^*(k) + B\underline{u}^*(k) + \underline{C}(k)$$

The method of obtaining G and $\underline{d}$ is to solve the problem within a two level structure by duality and decomposition using the initial condition $\underline{x}(0) = \underline{x}_0$ for K large. Then, the first n values of $\underline{x}$ and $\underline{u}$ can be recorded. From these, form the matrices

$$X = \left\{ \left[\underline{x}(0) - \underline{x}(n) \right], \left[\underline{x}(1) - x(n) \right] \ldots \left[x(n-1) - x(n) \right] \right\}$$

and

$$U = \left\{ \left[\underline{u}(0) - \underline{u}(n) \right], \left[\underline{u}(1) - \underline{u}(n) \right] \ldots \left[\underline{u}(n-1) - \underline{u}(n) \right] \right\}$$

Then

$$G = U X^{-1}$$

and

$$\underline{d} = \underline{u}(0) - G\,\underline{x}(0)$$

6.18 SIMULATION EXAMPLE

To illustrate the above approach, the two reach river pollution control model was simulated. The model used was the two reach "no delay" model, i.e.

$$\underline{x}(k+1) = A\,\underline{x}(k) + B\,\underline{u}(k) + \underline{C}$$

where

$$\underline{x} = \begin{bmatrix} x_1 \\ x_2 \\ x_3 \\ x_4 \end{bmatrix}$$

is such that x_1, x_3 give the B.O.D. concentration (mg/l) in the stream and x_2, x_4 the D.O. concentration (mg/l). The control is given by

$$\underline{u}(k) = \begin{bmatrix} \pi_1(k) \\ \pi_2(k) \end{bmatrix} \qquad k = 0, 1, \ldots, kK-1$$

where π_1, π_2 are the maximum fraction of B.O.D. removed from the effluent in the reaches 1, 2. A, B, $\underline{C}$ from the Cam data are given by

$$A = \begin{bmatrix} 0.18 & 0 & 0 & 0 \\ -0.25 & 0.27 & 0 & 0 \\ 0.55 & 0 & 0.18 & 0 \\ 0 & 0.55 & -0.25 & 0.27 \end{bmatrix}$$

$$B = \begin{bmatrix} -2.0 & 0 \\ 0 & 0 \\ 0 & -2.0 \\ 0 & 0 \end{bmatrix} \qquad \text{and } \underline{C} = \begin{bmatrix} 4.5 \\ 6.15 \\ 2.0 \\ 2.65 \end{bmatrix}$$

A suitable cost function for this system is

$$J = \frac{1}{2}\|\underline{x}(K)\|^2_{I_4} + \sum_{k=0}^{K-1} \frac{1}{2}\left(\|\underline{x}(k) - \underline{x}^d\|^2_{I_4} + \|\underline{u}(k)\|^2_{100\ I_2}\right.$$

where I_4 is the fourth order identity matrix and I_2 is the second order identity matrix. The desired values $\underline{x}^d$ are

$$\underline{x}^d = \begin{bmatrix} 5 \\ 7 \\ 5 \\ 7 \end{bmatrix}$$

This implies that it is desired to maintain the stream near B.O.D. values of 5 mg/l and D.O. values of 7 mg/l while minimising the treatment at the sewage works.

Simulation results

The above problem was solved using the hierarchical structure given in Section 6.17. The initial state was chosen to be $\underline{x}(0) = \underline{0}$ and K was chosen to be 23. The sampling interval here is 0.5 days. K is certainly sufficiently long for the system to settle to a steady state. Convergence to the optimum took place in 89 iterations of the second level which took 30.6 seconds to execute. At this point

$$\emptyset = J = 1607$$

The first 5 values were then taken of $\underline{x}$ and $\underline{u}$. From this, by inverting a 4 x 4 matrix, the control was found to be $\underline{u} = G\,\underline{x} + \underline{d}$ where

$$G = \begin{bmatrix} 0.0074 & -0.0011 & 0.0006 & -0.0001 \\ 0.0126 & -0.0015 & 0.0042 & -0.0004 \end{bmatrix}$$

and $\underline{d} = \begin{bmatrix} 0.05449 \\ 0.00668 \end{bmatrix}$

using this control law, the system was simulated for the initial states

$$\underline{x}(0) = \begin{bmatrix} 0 \\ 0 \\ 0 \\ 1 \end{bmatrix}$$

Figs. 6.29, 6.30 show the B.O.D. and D.O. in reaches 1 and 2 and Fig. 6.31 the controls.

These trajectories are identical to those found from a global solution.

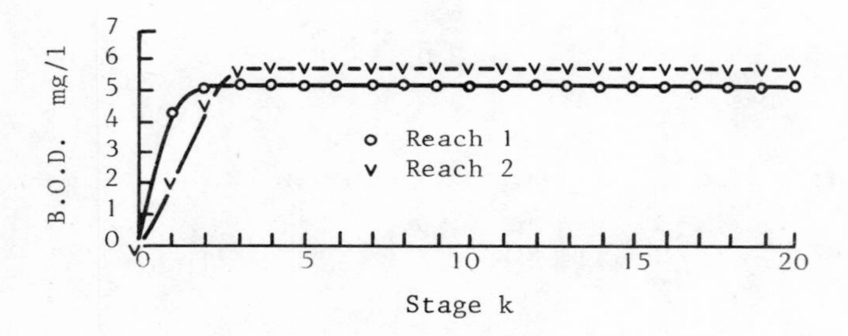

Fig. 6.29 : Optimal B.O.D. sequences

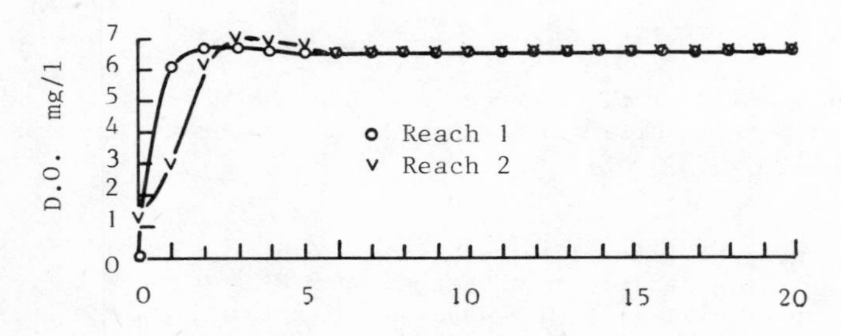

Fig. 6.30 : Optimal D.O. sequences

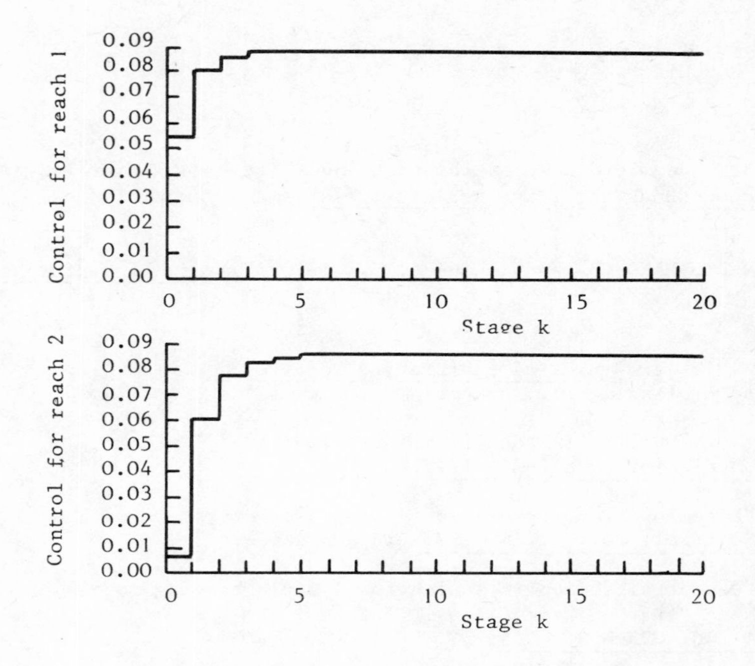

Fig. 6.31 : Optimal control for reaches one and two.

6.19 CONCLUSION

In this chapter, we have considered how we could use decomposition-coordination techniques in order to solve high order Linear-Quadratic problems. We have considered both open loop control as well as closed loop control. In each case, we have developed methods for tackling both continuous time and discrete time problems. We have seen that the methods are certainly very practical and we hope that the readers will use them to solve many other large scale problems.

In the next chapter, we return to the general non-linear dynamic optimisation problem formulated in chapter 5 and develop iterative techniques for solving it.

REFERENCES

[1] PEARSON, J.D. "Dynamic decomposition techniques" in D.A. Wismer (editor), 1971 : "Optimisation methods for large scale systems". Mac Graw Hill.

[2] MESAROVIC, M.D., MAKO, D. and TAKAHARA, Y. 1970 :"Theory of hierarchical multi-level systems".
Academic Press.

[3] SINGH, M.G. 1977 : "Dynamical hierarchical control".
North Holland Publishing Co., Amsterdam.

[4] GEOFFRION, A.M. "Duality in non-linear programming".
SIAM Review, vol. 13, 1, p.1-37, 1971.

[5] TAMURA, H. "Decomposition techniques in large scale systems with applications".
Systems and Control, vol. 17, 6, 1973 (in Japanese).

[6] TAMURA, H. "Decentralised optimisation for distributed lag models of discrete systems".
Automatica 11, 6, 593-602, 1975.

[7] SINGH, M.G. and TAMURA, H. "Modelling and hierarchical optimisation for oversaturated urban road traffic networks".
Int. J. Control. 20, 913-934, 1974.

[8] TAKAHARA, Y. "A multi-level structure for a class of dynamical optimisation problems".
M.S. Thesis, Case Western Reserve University, Cleveland, USA, 1965.

[9] COHEN, G., BENVENISTE, A. and BERNHARD, P. "Coordination algorithms for optimal control problems. Part. 1". Report A 157, Centre d'Automatique, Ecole des Mines, Paris, 1974.

[10] SINGH, M.G. and HASSAN, M. "A comparison of two hierarchical optimisation methods".
Int. J. Syst. Sci., 7, 6, 603-611, 1976.

[11] BECK, M.B. "The application of control and systems theory to problems of river pollution".
University of Cambridge, Ph. D. Thesis, 1973.

[12] TAMURA, H. "Decentralised optimisation for distributed-lag models of discrete systems".
I.E.E.E. Trans. SMC 4, 424-429, 1974.

[13] BAUMAN, E. "Multi-level optimisation techniques with application to trajectory decomposition".
Advances in control systems, 6, 160-222, 1968.

[14] FALLSIDE, F. and PERRY, P. "Hierarchical optimisation of a water supply network".
Proc. I.E.E. vol. 122, 2, 202-208, 1975.

[15] SINGH, M.G., HASSAN, M. and A. TITLI, A. "Multi-level feedback control for interconnected dynamical systems using the prediction principle".
I.E.E.E. Trans. SMC 6, 233-239, 1976.

[16] GEOFFRION, A.M. "Duality in non-linear programming".
SIAM review, 13-1, 1971.

[17] HASSAN, M. and TITLI, A. "Closed-loop hierarchical control for practical large scale systems using the prediction principle".
12th Conference on Statistics and Computer Science, 5-6 April 1976, Cairo.

[18] DARWISH, M. and FANTIN, J. "The application of Lyapunov method to large power systems using decomposition and aggregation techniques".
To appear Int. J. Control.

[19] SINGH, M.G. and HASSAN, M. "A closed loop hierarchical solution for the continuous time river pollution control problem".
Automatica 12, 261-264, 1976.

[20] KENDRICK, D.A., RAO, H.S. and WELLS, C.H. "Optimal operation of a system of waste water facilities".
Proc. IEEE Symposium on Adaptive processes, Austin, Texas, 1970.

[21] CHENEVEAUX, B. "Contribution à l'optimisation hiérarchisée des systèmes dynamiques".
Thèse de Docteur-Ingénieur, Université de Nantes, 1972.

PROBLEMS FOR CHAPTER 6

The authors beleive that in the area of large scale systems it is extremely important for anybody trying to master the subject to write computer programs for the various algorithms. For this reason, most of the excercises given below involve writing a program.

1. We would like to control the temperature and pressure in a boiler system. This control problem could be formulated as *

$$\text{Min } J = \frac{1}{2}\int_0^{t_f} \left[\|\underline{e}\|_Q^2 + \|\underline{u}\|_R^2 \right] dt$$

subject to $\mathring{\underline{y}} = A\underline{y} + B\underline{u} + \underline{C}$

where

$$\underline{y} = \begin{bmatrix} P_u \\ q_c \\ t_s \\ x_3 \\ q_d \end{bmatrix} ; \quad A = \begin{bmatrix} 0 & 396\text{x}10^{-6} & 0 & 0 & 325\text{x}10^{-7} \\ 0 & 0 & 0 & 0 & 0 \\ 0 & 1.25/154 & -1/154 & -1/154 & 0 \\ 0 & 0 & 0 & -1/83 & 1.375/83 \\ 0 & 0 & 0 & 0 & 0 \end{bmatrix}$$

$$\underline{u} = \begin{bmatrix} u_1 \\ u_2 \end{bmatrix} ; \quad B = \begin{bmatrix} 0 & 0 \\ 1 & 0 \\ 0 & 0 \\ 0 & 0 \\ 0 & 1 \end{bmatrix} ; \quad \underline{C} = \begin{bmatrix} -325\text{x}10^{-7}\, q_s \\ 0 \\ -649.35\text{x}10^{-6}\, q_s + 129.87\text{x}10^{-4} \\ -662.65\text{x}10^{-3} \\ 0 \end{bmatrix}$$

$$\underline{e} = \underline{y} - \underline{a} \quad \text{where } \underline{a} = \begin{bmatrix} -1.225\text{x}\left(\frac{q_s}{1000}\right)^2 \\ 0 \\ 0 \\ 0 \\ 0 \end{bmatrix}$$

$$Q = \begin{bmatrix} 16 & 0 & 0 & 0 & 0 \\ 0 & 0 & 0 & 0 & 0 \\ 0 & 0 & 1 & 0 & 1 \\ 0 & 0 & 0 & 0 & 0 \\ 0 & 0 & 0 & 0 & 0 \end{bmatrix} ; \quad R = \begin{bmatrix} 30 & 0 \\ 0 & 30 \end{bmatrix}$$

---

* Numerical values for the problem are taken from B. CHENEVEAUX [21]

t_f = 600 seconds, q_s is given and known.

The various variables are :

P_u is the deviation of the pressure around 20 bars in the boiler.

t_s is the deviation of the output temperature from 300°C.

x_3 is an intermediate variable.

q_c is the flow rate of the fuel (KG/h).

q_d is the flow rate of the superheated steam (KG/h).

q_s is the flow rate of the steam.

The criterion function J implies that we would like to minimise the deviations of pressure, temperature and the speed of variations of the controls.

For this system :

(a) write the problem in the form of interconnected subsystems. Write the Lagrangian and Hamiltonian.

(b) what choice of coordination variables and/or methods of coordination will avoid singular subproblems.

(c) write down the precise tasks of the two levels for solving this problem.

(d) How could we compute optimal feedback control here.

(e) write a computer program for implementing (c) above.

(f) write a computer program for computing (d) above.

2. We consider the global problem :

$$\min_{\underline{u}} \frac{1}{2} \int_0^T (y^T Q\underline{y} + y^T R\underline{u})dt$$

subject to $\mathring{\underline{x}} = A\underline{x} + B\underline{u}$ $\qquad \underline{x}(0) = \underline{x}_o$, $\quad \underline{x} \in R^n$

and the corresponding decomposed problem :

$$\min_{\underline{u}} \sum_{i=1}^{N} \frac{1}{2} \int_0^T \left[\underline{x}_i^T C_i^T Q_i C_i \underline{x}_i + \underline{u}_i^T R_i \underline{u}_i\right] dt$$

subject to $\mathring{\underline{x}}_i = \sum_{j=1}^{N} A_{ij}\, \underline{x}_j + B_i\, \underline{u}_i \qquad \underline{x}_i(0) = \underline{x}_{i0} \qquad i = 1$ à N

with $\underline{x}^T = \left[\underline{x}_1^T \ldots \underline{x}_N^T\right]$ etc...

and where A, B, C, Q, R, A_i to R_i are matrices of appropriate dimension and Q_i, R_i

satisfy the standard assumptions.

We introduce an additional variable which relates the state of subsystem i with the interaction for this subsystem, i.e.

$$\underline{\beta}_i = \sum_{j=1}^{N} A_{ij}\, \underline{x}_j \qquad i = 1 \text{ to } N$$

Let $\underline{\lambda}_i$ be the Lagrange multiplier for this new constraint. Show that we can use $\underline{\lambda}_i$ as a coordination variable in the goal coordination method.

Define thus the subproblem and the coordination task. Show that this method leads to performing differentiation operations, i.e. given a vector $\underline{v}$ we need to calculate $\overset{\circ}{\underline{v}}(t) = \frac{d\underline{v}(t)}{dt}$ which causes problems.

Define the conditions for the applicability of such a method (we assume that $C_i^T Q_i C_i$ is regular).

3. Starting from the same formulation as above, we now put

$$\underline{\alpha}_i = \underline{x}_i \qquad i = 1 \text{ à } N$$

Show that $\underline{\alpha} = \begin{bmatrix} \underline{\alpha}_1 \\ \vdots \\ \underline{\alpha}_N \end{bmatrix}$ could be used as a coordination variable. Define thus the subproblem and the coordination task. Show that this method is the dual of the previous one.

4. We consider the distributed parameter linear-quadratic problem :

$$\min_{\underline{u}} \int_0^T \int_{\Omega} \sum_{i=1}^{2} \left[y_i^2 + k_3\, u_i^2 \right] dx\, dt$$

subject to
$$\frac{\partial y_1}{\partial t} = k_1 \frac{\partial^2 y_1}{\partial x^2} + y_1 - y_2 + u_1$$

$$\frac{\partial y_2}{\partial t} = k_2 \frac{\partial^2 y_2}{\partial x^2} + y_2 - y_1 + u_2$$

$$\left.\begin{array}{ll} y_i(x, 0) = y_{io}(x) & \text{initial condition} \\ \dfrac{\partial y_1(0, t)}{\partial x} = \dfrac{\partial y_i(1,t)}{\partial x} = 0 & \text{boundary conditions} \end{array}\right\} i=1,\ 2$$

Ω : Space domain $]0, 1[$, x : space variable

y_i: state, u_i control, k_i given, $i = 1, 2, 3$

Rewrite this problem in the form of interconnected systems. Show that the Goal coordination method is applicable here. Define the subproblems and the coordination tasks. What are the difficulties of the subproblems and how can these be avoided ? Apply also the interaction prediction method.

5. For Pearson's 12th order example in section 63 write the tasks for the two levels. Write a general purpose computer program for the solution of L - Q problems using the Goal coordination approach. Test the program on Pearson's example.

6. Write down for the discrete river problem in section 6.18 the tasks for the 3 levels in Tamura's three level method of section 6.4. Write a program to solve the above problem.

7. Write down the tasks for the 2 levels for the problem in excercise 3 using the discrete goal coordination method. Write a computer program to solve this problem and verify that the solution is identical to that found in excercise 3. Compare computation times and storage requirements.

8. Write a general purpose program for Tamura's time delay algorithm. Verify the traffic example in section 6.8 using this program.

9. Write a general purpose program for the interaction prediction principle algorithm. Test it on the example in section 6.3. Verify that the river solution in section 6.9 is correct.

10. Compute feedback controls for Pearson's example.

CHAPTER 7

Dynamical Optimisation for Non-Linear Systems

In the last two chapters, we have studied in some detail how we can solve problems of optimisation and control in the case of linear systems with quadratic cost functions. We saw in chapter 5 that this was the special case where the two point boundary value problem which results from the application of the maximum principle is linear. In this chapter we will examine various ways of tackling the general non-linear two point boundary value problem * and try to compare the numerical efficiency of the various methods. We will examine gradient methods, quasilinearisation, variation of extremals, and the invariant imbedding method.

7.1 FORMULATION OF NON-LINEAR TWO POINT BOUNDARY VALUE PROBLEMS

Let us review first of all how the two point boundary value problem arises. We are concerned with the problem of minimising a non-linear cost function of the form

$$\text{Min } J = h(\underline{x}(t_f)) + \int_{t_o}^{t_f} g(\underline{x}, \underline{u}, t)\, dt \tag{7.1.1}$$

subject to the dynamical constraints

$$\dot{\underline{x}}(t) = \underline{f}(\underline{x}, \underline{u}, t) \tag{7.1.2}$$

where the states and controls are not constrained by any boundaries, the final time t_f is fixed and the final state $\underline{x}(t_f)$ is free.

From chapter 5, we know that in order to solve this problem, we should find the states and controls which satisfy the necessary conditions for optimality. These necessary conditions can be written in terms of the Hamiltonian function H where

$$H = g(\underline{x}, \underline{u}, t) + \underline{\lambda}^T(t)\, \underline{f}(\underline{x}, \underline{u}, t) \tag{7.1.3}$$

and the first order necessary conditions are given by

$$\dot{\underline{x}}(t) = \frac{\partial H}{\partial \underline{\lambda}} = \underline{f}(\underline{x}, \underline{u}, t) \tag{7.1.4}$$

$$\dot{\underline{\lambda}}(t) = -\frac{\partial H}{\partial \underline{x}} = -\left[\frac{\partial \underline{f}}{\partial \underline{x}}(\underline{x}, \underline{u}, t)\right]^T \underline{\lambda}(t) - \frac{\partial g}{\partial \underline{x}}(\underline{x}, \underline{u}, t) \tag{7.1.5}$$

---

* The solution of this problem will give us an open loop control. It is only for very special cases that we can obtain a closed loop control for non-linear systems.

$$\frac{\partial H}{\partial \underline{u}} = \underline{0}$$

$$\text{or} \quad \left[\frac{\partial \underline{f}}{\partial \underline{u}}(\underline{x}, \underline{u}, t)\right]^T \underline{\lambda}(t) + \frac{\partial g}{\partial \underline{u}}(\underline{x}, \underline{u}, t) = \underline{0} \qquad 7.1.6$$

$$\underline{x}(t_o) = \underline{x}_o \qquad 7.1.7$$

$$\underline{\lambda}(t_f) = \frac{\partial h}{\partial \underline{x}}(\underline{x}(t_f)) \qquad 7.1.8$$

Note that the optimal $\underline{x}(t)$ and $\underline{u}(t)$ that we obtain will depend explicitely on the initial state $\underline{x}_o$, i.e. we will in general only be able to obtain an open loop control.

Let us assume that we can solve the vector algebraic equation 7.1.6 to obtain an explicit relationship between $\underline{u}(t)$ on the one hand and $\underline{x}(t)$, $\underline{\lambda}(t)$ on the other, i.e.

$$\underline{u}(t) = \underline{a}(\underline{x}(t), \underline{\lambda}(t), t) \qquad 7.1.9$$

If this expression is substituted into equations 7.1.4 and 7.1.5, we have a set of 2n first order ordinary differential equations involving only $\underline{x}$, $\underline{\lambda}$, t. Now, if the boundary conditions on $\underline{x}$ and $\underline{\lambda}$ were known at t_o or t_f, we could have integrated the 2n reduced equations to obtain $\underline{x}(t)$, $\underline{\lambda}(t)$, $t \in [t_o, t_f]$ and from these solutions we could have obtained the optimal control trajectories by substituting for $\underline{x}$, $\underline{\lambda}$, in equation 7.1.9. However, because the boundary conditions are split, we can't apply this method. We saw in chapter 5 that if the reduced equations are linear, then by superposition we can obtain the Riccati equation which needs to be integrated numerically from a single point boundary condition. For the non-linear case, superposition of solutions is not valid and we have to resort to iterative techniques to satisfy the five necessary conditions for optimality given by equations 7.1.4 to 7.1.8.

We will next examine three iterative schemes for solving non-linear two point boundary value problems.

Each of these techniques is based on guessing certain variables or trajectories and obtaining a solution of the equations, such that one or more of the necessary conditions for optimality 7.1.4 to 7.1.8 is not satisfied whilst all the others are satisfied. This solution is used to adjust the initial guess in an attempt to make the next solution come "closer" to satisfying all the necessary conditions. If these steps are repeated and the iterative procedure converges, the necessary conditions 7.1.4 to 7.1.8 will eventually be satisfied.

7.2 THE GRADIENT METHOD

We examined in the first part of the book gradient methods for solving non-linear programming problems and here we will extend the method to the case of functional optimisation. However, we will start by recapitulating the gradient methods for parametric optimisation problems.

7.2.1 The gradient method for parametric optimisation

The gradient method in the control context can be formulated as the following non-linear programming problem :

$$\text{Min } J = g(\underline{x}, \underline{u}) \tag{7.2.1}$$

where by a suitable choice of the control vector $\underline{u}$ we wish to minimise J in 7.2.1 subject to the equality constraints

$$\underline{f}(\underline{x}, \underline{u}) = \underline{0} \tag{7.2.2}$$

Here, $\underline{x}$, $\underline{f}$, are n vectors, g is a scalar and $\underline{u}$ is an m vector. Such problems arise in the control context in the steady state optimisation of many industrial processes where we desire to find the optimum steady state values that we would like our system to maintain. Subsequently, these "desired" values are used in the dynamic optimisation stage where we maintain our states (and controls) near these desired values even if our system is subjected to disturbances.

From part I of this book, we know that we can adjoin the constraint equation 7.2.2 to the cost function J in 7.2.1, using a suitable n dimensional Lagrange multiplier vector, i.e.

$$L = g(\underline{x}, \underline{u}) + \underline{\lambda}^T(\underline{f}(\underline{x}, \underline{u})) \tag{7.2.3}$$

Then the necessary conditions for a stationary point require that

$$\frac{\partial L}{\partial \underline{x}} = \underline{0} \quad \text{and} \quad \frac{\partial L}{\partial \underline{u}} = \underline{0}$$

i.e.

$$\frac{\partial g(\underline{x}, \underline{u})}{\partial \underline{u}} + \left[\frac{\partial \underline{f}(\underline{x}, \underline{u})}{\partial \underline{u}}\right]^T \underline{\lambda} = \underline{0} \tag{7.2.4}$$

and

$$\frac{\partial g(\underline{x}, \underline{u})}{\partial \underline{x}} + \left[\frac{\partial \underline{f}(\underline{x}, \underline{u})}{\partial \underline{x}}\right]^T \underline{\lambda} = \underline{0} \tag{7.2.5}$$

Now, if the constraint equation $\underline{f}(\underline{x}, \underline{u}) = \underline{0}$ is satisfied, then we have approximately

$$\Delta J \simeq \left[\frac{\partial L}{\partial \underline{u}}\right]^T \Delta \underline{u} \tag{7.2.6}$$

so that if we want to make the biggest change in J, i.e. if we want to make ΔJ large, we could perhaps calculate the gradient $\frac{\partial L}{\partial \underline{u}}$ and make a change in J by making $\Delta \underline{u}$ directed opposite to the gradient

i.e.

$$\Delta \underline{u} = -K\left[\frac{\partial L}{\partial \underline{u}}\right] \tag{7.2.7}$$

so that from 7.2.6 in order to make J smaller we have

$$\Delta J = -K\left[\frac{\partial L}{\partial \underline{u}}\right]^T \left[\frac{\partial L}{\partial \underline{u}}\right] \tag{7.2.8}$$

Thus, to solve the problem iteratively, we will guess some control $\underline{u} = \underline{u}^{\alpha}$ and use this to determine the $\underline{x} = \underline{x}^{\alpha}$ which satisfies the equality constraint

$$\underline{f}(\underline{x}^{\alpha}, \underline{u}^{\alpha}) = \underline{0} \tag{7.2.9}$$

Then we find the value of the Lagrange multiplier vector $\underline{\lambda}$ in order to ensure that

$\frac{\partial L}{\partial \underline{x}^{\alpha}} = 0$ from equation 7.2.5

i.e. $$\underline{\lambda}^{\alpha} = - \left[\frac{\partial \underline{f}(\underline{x}^{\alpha}, \underline{u}^{\alpha})^T}{\partial \underline{x}^{\alpha}} \right]^{-1} \left[\frac{\partial g(\underline{x}^{\alpha}, \underline{u}^{\alpha})}{\partial \underline{x}^{\alpha}} \right] \quad 7.2.10$$

Next we compute the gradient vector from equation 7.2.3

i.e. $$\frac{\partial L}{\partial \underline{u}^{\alpha}} = \frac{\partial g(\underline{x}^{\alpha}, \underline{u}^{\alpha})}{\partial \underline{u}^{\alpha}} + \left[\frac{\partial \underline{f}(\underline{x}^{\alpha}, \underline{u}^{\alpha})}{\partial \underline{u}^{\alpha}} \right]^T \underline{\lambda}^{\alpha} \quad 7.2.11$$

This vector will be zero only when we satisfy the necessary conditions for a stationary point. We use the gradient vector to compute the steepest descent, i.e.

$$\Delta \underline{u}^{\alpha} = - K \frac{\partial L(\underline{x}^{\alpha}, \underline{u}^{\alpha}, \underline{\lambda}^{\alpha})}{\partial \underline{u}^{\alpha}} \quad 7.2.12$$

and we use this to improve $\underline{u}$ for the next iteration as

$$\underline{u}^{\alpha+1} = \underline{u}^{\alpha} + \Delta \underline{u}^{\alpha}$$

and we restart. We repeat this process until the change in the cost function

$$\Delta J^{\alpha} = - K \left[\frac{\partial L(\underline{x}^{\alpha}, \underline{u}^{\alpha}, \underline{\lambda}^{\alpha})}{\partial \underline{u}^{\alpha}} \right]^T \frac{\partial L(\underline{x}^{\alpha}, \underline{u}^{\alpha}, \underline{\lambda}^{\alpha})}{\partial \underline{u}^{\alpha}} \quad 7.2.13$$

is smaller than some particular prechosen small number.

Next, let us extend this procedure to the case of functional optimisation which is the main problem of interest in this chapter.

7.2.2 The gradient method for functional optimisation

Here we come back to the non-linear two point boundary value problem of interest in this chapter and use the basic idea of the gradient method to solve the problem iteratively. We saw that the necessary conditions for optimality which must be satisfied are

$$\dot{\underline{x}}(t) = \frac{\partial H}{\partial \underline{\lambda}} = \underline{f}(\underline{x}, \underline{u}, t) \quad 7.2.14$$

$$\dot{\underline{\lambda}}(t) = -\frac{\partial H}{\partial \underline{x}} = - \left[\frac{\partial \underline{f}(\underline{x}, \underline{u}, t)}{\partial \underline{x}} \right]^T \underline{\lambda}(t) - \frac{\partial g}{\partial \underline{x}}(\underline{x}, \underline{u}, t) \quad 7.2.15$$

$$\frac{\partial H}{\partial \underline{u}} = \underline{0} = \left[\frac{\partial \underline{f}}{\partial \underline{u}} (\underline{x}, \underline{u}, t) \right]^T \underline{\lambda}(t) + \frac{\partial g}{\partial \underline{u}} (\underline{x}, \underline{u}, t) \quad 7.2.16$$

$$\underline{x}(t_o) = \underline{x}_o$$

$$\underline{\lambda}(t_f) = \frac{\partial h}{\partial \underline{x}} (\underline{x}(t_f))$$

In the steepest ascent method for functional optimisation, as for parametric optimisation we start with a nominal control trajectory $\underline{u}^{\alpha}(t)$, $t \in [t_o, t_f]$ and use this to solve the differential equations 7.2.14 and 7.2.15 such that the nominal state-costate trajectories $\underline{x}^{\alpha}$, $\underline{\lambda}^{\alpha}$ satisfy the boundary conditions

$$\underline{x}^{\alpha}(t_o) = \underline{x}_o, \quad \underline{\lambda}(t_f) = \frac{\partial h}{\partial \underline{x}}(\underline{x}^{\alpha}(t_f)) \qquad 7.2.17$$

If the nominal control trajectory $\underline{u}^{\alpha}(t)$ also satisfies

$$\frac{\partial H}{\partial \underline{u}}(\underline{x}^{\alpha}(t), \underline{u}^{\alpha}(t), \underline{\lambda}^{\alpha}(t), t) = \underline{0} \qquad t \in [t_o, t_f] \qquad 7.2.18$$

then $\underline{u}^{\alpha}(t)$, $\underline{x}^{\alpha}(t)$, $\underline{\lambda}^{\alpha}(t)$ are extremal. In actual fact, for an arbitrary choice of $\underline{u}^{\alpha}(t)$, equation 7.2.18 will not in general be satisfied. In that case the variation of the augmented function J^* on the nominal state-costate trajectory is (cf.chap.5)

$$\delta J^* = \left[\frac{\partial h}{\partial \underline{x}}(\underline{x}^{\alpha}(t_f)) - \underline{\lambda}^{\alpha}(t_f)\right]^T \delta \underline{x}(t_f) + \int_{t_o}^{t_f} \left\{ \left[\dot{\underline{\lambda}}^{\alpha}(t) + \frac{\partial H}{\partial \underline{x}}(\underline{x}^{\alpha}(t), \underline{u}^{\alpha}(t), \underline{\lambda}^{\alpha}(t), t)\right]^T \right.$$

$$\left. \delta \underline{x}(t) + \left[\frac{\partial H}{\partial \underline{u}}(\underline{x}^{\alpha}(t), \underline{u}^{\alpha}(t), \underline{\lambda}^{\alpha}(t), t)\right]^T \delta \underline{u}(t) + \left[\underline{f}(\underline{x}^{\alpha}(t), \underline{u}^{\alpha}(t), t) - \dot{\underline{x}}^{\alpha}(t)\right]^T \delta(t) \right\} dt \qquad 7.2.19$$

where $\delta \underline{x}(t) = \underline{x}^{\alpha+1}(t) - \underline{x}^{\alpha}(t)$

$\delta \underline{u}(t) = \underline{u}^{\alpha+1}(t) - \underline{u}^{\alpha}(t)$

and $\delta \underline{\lambda}(t) = \underline{\lambda}^{\alpha+1}(t) - \underline{\lambda}^{\alpha}(t)$

Now if equations 7.2.14, 7.2.15 and 7.2.17 are satisfied, then

$$\delta J^* = \int_{t_o}^{t_f} \left[\frac{\partial H}{\partial \underline{u}}(\underline{x}^{\alpha}(t), \underline{u}^{\alpha}(t), \underline{\lambda}^{\alpha}(t), t)\right]^T \delta \underline{u}(t)\, dt \qquad 7.2.20$$

It should be noted that δJ^* is the linear part of the increment $\Delta J = J(\underline{u}^{\alpha+1}) - J(\underline{u}^{\alpha})$ and that if the norm of $\delta \underline{u}$, $\|\underline{u}^{\alpha+1} - \underline{u}^{\alpha}\|$ is small, the sign of ΔJ will be determined by the sign of δJ^*. Since we wish to minimise J^*, we would like to make ΔJ negative. Then a good choice of the change in $\underline{u}$, $\delta \underline{u}$ is

$$\delta \underline{u}(t) = \underline{u}^{\alpha+1}(t) - \underline{u}^{\alpha}(t) = -\tau \frac{\partial H^{\alpha}}{\partial \underline{u}}(t) \qquad t \in [t_o, t_f] \qquad 7.2.21$$

with $\tau > 0$.

Then
$$\delta J^* = -\tau \int_{t_o}^{t_f} \left[\frac{\partial H^{\alpha}}{\partial \underline{u}}(t)\right]^T \left[\frac{\partial H}{\partial \underline{u}}(t)\right] dt \leq 0 \qquad 7.2.22$$

The integrand is non-negative for all $t \in [t_o, t_f]$. The equality holds if and only if

$$\frac{\partial H^{\alpha}}{\partial \underline{u}}(t) = \underline{0} \qquad \text{for all } t \in [t_o, t_f] \qquad 7.2.23$$

Thus if we choose $\delta \underline{u}$ as in equation 7.2.21, with $\|\delta \underline{u}\|$ sufficiently small, then each value of the cost function will be at least as small as the preceding value. Eventually, when J^* reaches a relative minimum, the vector $\frac{\partial H}{\partial \underline{u}}$ will be zero throughout the time interval $[t_o, t_f]$.

This algorithm can be programmed on a digital computer using the following steps.

The gradient algorithm

Step 1

The first step in the algorithm is to choose a nominal control trajectory over the period t_o to t_f. Since the calculations are to be performed on a digital computer, it is necessary to discretise this trajectory and store the discrete time points in the memory of the computer. The discretisation can be done, for example, by subdividing the interval $[t_o, t_f]$ into N subintervals which are usually of equal length and considering the control $\underline{u}^{(o)}$ as being piecewise-constant during each of the subintervals

i.e. $\underline{u}^{(o)}(t) = \underline{u}^{(o)}(t_\beta) \qquad t \in [t_\beta, t_{\beta+1}[, \beta = 0, 1, \ldots N-1$

Let the iteration index be i.

Step 2

Using the nominal control trajectory $\underline{u}^i$, integrate the state equations 7.2.14 from t_o to t_f with initial conditions $\underline{x}(t_o) = \underline{x}_o$ and store the resulting state trajectory $\underline{x}^i(t)$ as a piecewise-constant vector function.

Step 3

Calculate $\underline{\lambda}^{(i)}(t_f)$ by substituting $\underline{x}^{(i)}(t_f)$ from step 2 into the R.H.S. of the equation $\underline{\lambda}(t_f) = \frac{\partial h}{\partial \underline{x}}(\underline{x}(t_f))$. Using this value of $\underline{\lambda}^{(i)}(t_f)$ as the terminal condition and the piecewise constant values of $\underline{x}^i$ stored in step 2, integrate the co-state equation backwards in time from t_f to t_o. Calculate $\frac{\partial H^{(i)}}{\partial \underline{u}}(t)$, $t \in [t_o, t_f]$ and store this function in a piecewise-constant fashion. We do not need to store the costate trajectory.

Step 4

$$\text{If } \left\| \frac{\partial H}{\partial \underline{u}} \right\| \leqslant \gamma \qquad 7.2.24$$

where γ is a pre-selected small positive constant and

$$\left\| \frac{\partial H^{(i)}}{\partial \underline{u}} \right\| = \int_{t_o}^{t_f} \left[\frac{\partial H^{(i)}}{\partial \underline{u}}(t) \right]^T \left[\frac{\partial H^{(i)}}{\partial \underline{u}}(t) \right] dt \qquad 7.2.25$$

stop and record the $\underline{x}^{(i)}$, $\underline{u}^{(i)}$ as the optimal state and control trajectories. If the termination criterion 7.2.24 is not satisfied, generate a new piecewise-constant control function given by

$$\underline{u}^{(i+1)}(t_\beta) = \underline{u}^{(i)}(t_\beta) - \tau \frac{\partial H^{(i)}}{\partial \underline{u}}(t_\beta) \qquad \beta = 0, 1, \ldots, N-1 \qquad 7.2.26$$

where

$$\underline{u}^{(i)}(t) = \underline{u}^{(i)}(t_\beta) \text{ for } \beta \in [t_\beta, t_{\beta+1}[\; \beta = 0, 1, \ldots, N-1 \qquad 7.2.27$$

Replace $\underline{u}^{(i)}(t_\beta)$ by $\underline{u}^{(i+1)}(t_\beta)$, $\beta = 0, \ldots N-1$

and return to step 2.

In the above algorithm, we require τ and γ. The step size τ is usually determined by a single variable search. For example, we could start with an arbitrary value of τ, compute $\frac{\partial H^{(i)}}{\partial u}$ and find $\underline{u}^{(i+1)}$ using equation 7.2.26. Then a search among values of $\tau > 0$ is carried out until the smallest value of J^* is obtained and this is the value of τ that is used.

The termination constant γ depends mainly on the accuracy of the solution required. Thus it might be desirable to do several trial runs for any problem before fixing γ.

Illustrative examples

Example 1

To illustrate the mechanics of the above algorithm, consider the simple scalar problem

$$\dot{x} = x + u \quad ; \quad x(0) = 2$$

and

$$\text{Min } J = \frac{1}{2}\int_0^1 (x^2 + u^2)\, dt$$

Here the Hamiltonian is :

$$H = \frac{1}{2}x^2 + \frac{1}{2}u^2 + \lambda(x+u)$$

so that

$$\frac{\partial H}{\partial u} = u + \lambda$$

$$\frac{\partial H}{\partial x} = -\dot{\lambda} = x + \lambda \qquad \lambda(1) = 0$$

Now, let us choose a discretisation interval of 0.1 so that our control is sampled at ten discrete points. As an initial guess, let us choose a constant initial control trajectory

$$u^{(1)}(o) = 1, \quad u^{(1)}(.1) = 1, \; u^{(1)}(.2) = 1 \ldots \; u^{(1)}(1) = 1.$$

Now the first step is to integrate the equation

$$\dot{x}^{(1)} = x^{(1)} + u^{(1)} \quad ; \; x(o) = 2$$

using this control. Let us suppose that the trajectory is given by $x^{(1)}$. Using this $x^{(1)}$, we integrate $\dot{\lambda}^{(1)} = -x^{(1)} - \lambda^{(1)}$ backwards and substitute the $\lambda^{(1)}$ trajectory obtained to calculate

$$\frac{\partial H}{\partial u} = u^{(1)} + \lambda^{(1)}$$

Now we calculate the norm of $\frac{\partial H}{\partial u}$ and if it is sufficiently small, we stop. Else, we choose the new control

$$u^{(2)} = u^{(1)} - \tau \frac{\partial H}{\partial u}$$

and restart. We continue this process until our algorithm converges and we obtain a relative minimum.

Remark : This problem is a L-Q one and the relative minimum that is obtained is also the global minimum. For this reason, also, the two point boundary value problem is linear and can be solved directly without using iterative techniques as we do here.

Example 2 : Synchronous machine control

As a second more significant example which we will treat in detail using the different algorithms in this chapter and in the next, consider the problem of optimal control of a synchronous machine. A model for this system is given by Mukhopadhyay [1] as :

$$\dot{y}_1 = y_2$$
$$\dot{y}_2 = B_1 - A_1 y_2 - A_2 y_3 \sin y_1 - \frac{B_2}{2} \sin 2y_1$$
$$\dot{y}_3 = u_1 - C_1 y_3 + C_2 \cos y_1$$

where y_1 is the rotor angle (radians)

y_2 is the speed deviation ($\frac{d}{dt} y_1$ rad/s)

y_3 is the field flux linkage

u_1 the control variable is the voltage applied to the machine and is assumed to be available for optimal manipulation.

The parameter values that we will use in this example are due to Mukhopadhyay [1] and these are given by

$$A_1 = 0.2703,\ A_2 = 12.012,\ B_1 = 39.1892,\ B_2 = -48.048$$
$$C_1 = 0.3222,\ C_2 = 1.9$$

It is desired to minimise

$$J = \frac{1}{2} \int_0^2 \left[Q_1 (y_1 - y_{1p})^2 + Q_2 (y_2 - y_{2p})^2 + Q_3 (y_3 - y_{3p})^2 + R(u_1 - u_p)^2 \right] dt$$

Then the Hamiltonian can be written as :

$$H = \Big\{ \tfrac{1}{2} \left[Q_1 (y_1 - y_{1p})^2 + Q_2 (y_2 - y_{2p})^2 + Q_3 (y_3 - y_{3p})^2 + R(u_1 - u_p)^2 \right] + \lambda_1 [y_2] +$$
$$\lambda_2 \left[B_1 - A_1 y_2 - A_2 y_3 \sin y_1 - \frac{B_2}{2} \sin 2y_1 \right] + \lambda_3 \left[u_1 - C_1 y_3 + C_2 \cos y_1 \right] \Big\}$$

The costate system of equations can therefore be written as :

$$\frac{\partial H}{\partial y_1} = -\dot{\lambda}_1 = \lambda_2 (-A_2 y_3 \cos y_1 - B_2 \cos 2y_1) - \lambda_3 C_2 \sin y_1 + Q_1 (y_1 - y_{1p})$$

$$\frac{\partial H}{\partial y_2} = -\mathring{\lambda}_2 = \lambda_1 - \lambda_2 A_1 + Q_2(y_2 - y_{2p})$$

$$\frac{\partial H}{\partial y_3} = -\mathring{\lambda}_3 = -\lambda_2 A_2 \sin y_1 - \lambda_3 C_1 + Q_3(y_3 - y_{3p})$$

and for the control we have :

$$\frac{\partial H}{\partial u_1} = 0 = \lambda_3 + R(u_1 - u_p)$$

or

$$u_1 = -\frac{\lambda_3}{R} + u_p$$

and substituting this optimal control into the state equations we have

$$\dot{y}_1 = y_2$$

$$\dot{y}_2 = B_1 - A_1 y_2 - A_2 y_3 \sin y_1 - \frac{B_2}{2} \sin 2y_1$$

$$\dot{y}_3 = u_p - \frac{\lambda_3}{R} - C_1 y_3 + C_2 \cos y_1$$

The initial and final conditions were

$$\underline{y}(0) = \begin{bmatrix} 0.734 \\ 0.215 \\ 7.7443 \end{bmatrix} \quad \text{and } \underline{\lambda}(2) = \begin{bmatrix} 0 \\ 0 \\ 0 \end{bmatrix}$$

We choose as termination criterion the norm of $\frac{\partial H}{\partial u_1}$, i.e.

$$\left\| \frac{\partial H}{\partial u_1} \right\|^2 = \int_0^2 \left[\frac{\partial H}{\partial u_1} \right]^2 dt$$

In detail , this problem could be solved using the flow chart 7.1

Flow chart 7.1

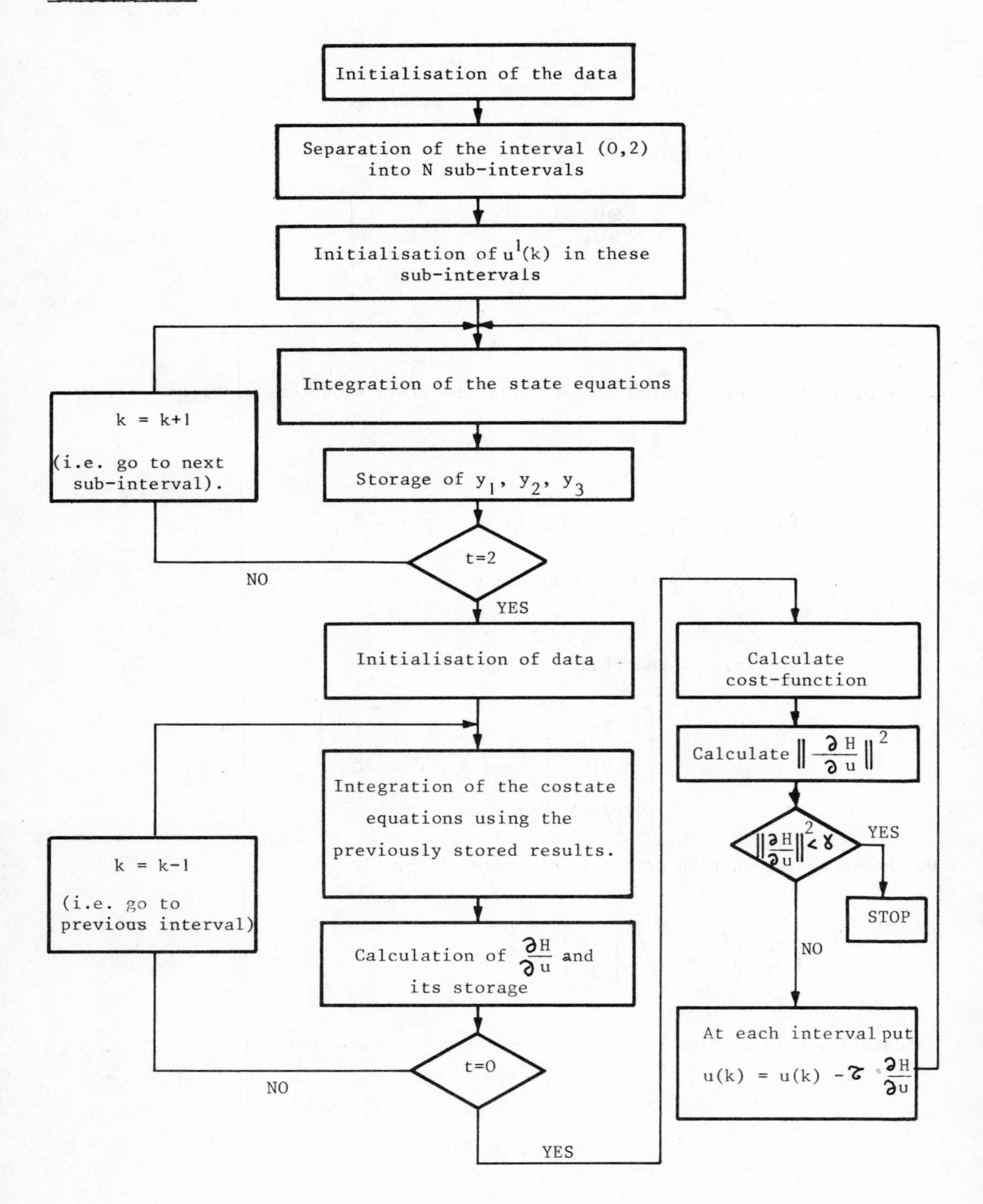
Initialisation of the data
Separation of the interval (0,2) into N sub-intervals
Initialisation of $u^1(k)$ in these sub-intervals
Integration of the state equations
k = k+1
(i.e. go to next sub-interval).
Storage of y_1, y_2, y_3
t=2
NO
YES
Initialisation of data
Integration of the costate equations using the previously stored results.
k = k-1
(i.e. go to previous interval)
Calculation of $\frac{\partial H}{\partial u}$ and its storage
t=0
NO
YES
Calculate cost-function
Calculate $\left\| \frac{\partial H}{\partial u} \right\|^2$
$\left\| \frac{\partial H}{\partial u} \right\|^2 < \gamma$
YES
STOP
NO
At each interval put
$u(k) = u(k) - \tau \frac{\partial H}{\partial u}$

By using the above flow chart a program can be written. We wrote such a program using the 4th order Runge-Kutta method for the integrations. All the calculations were done in double precision. Figs. 7.1 to 7.3 show the optimal state trajectories and Fig. 7.4 the optimal control. The minimum cost was found to be 1.807×10^{-2}.

τ was choosen arbitrarily. We found that for $\tau = 0.1$, the solution diverged. For $\tau = 0.0025$, the cost J decreased slowly. Indeed, after 25 iterations, J was 1.819×10^{-2} as opposed to the final value of 1.807×10^{-2}. For $\tau = 0.005$, the decrease in the cost function value was much more rapid and we obtained an accuracy of 10^{-6} in 15 iterations. For $\tau = 0.01$, the decrease of J is extremely rapid so that we obtain the optimum by the third iteration.

Fig. 7.5 shows the decrease of J for varying τ . The important thing to note here is that the initial decrease is very rapid but as in the gradient method for parametric optimisation in part I of the book, the convergence slows down as the gradient becomes small.

We also tested different initial guesses for the control u_1 and found that (a) when the initial control chosen was far from the optimum one, convergence was slightly slower than when the initial control was quite near the optimal one, and (b) in any case, <u>for virtually any initial choice of the control, we obtained convergence</u>. This is one of the significant advantages of the present method, i.e. it is not very critical to the choice of initial guess. However, although its initial convergence can be rapid, its final convergence is usually slow.

Let us summarise the important features of the gradient algorithm :

a) <u>Computational requirements</u>

These can be split up into storage requirements and computation time requirements with a trade off between the two.

<u>Storage</u> :

The current trial control $\underline{u}^{(i)}$, the corresponding state trajectory $\underline{x}^{(i)}$ and the gradient trajectory $\partial H^{(i)}/\partial \underline{u}$ need to be stored. If the computer used is small with very limited storage, the state values needed to determine $\partial H^{(i)}/\partial \underline{u}$ can be obtained by reintegrating the state equations with the costate equations. If this is done then $\underline{x}^{(i)}$ need not be stored ; however, there is a trade off here with computation time since the latter will increase. However, accuracy of the computation is improved in this case because we do not need to use the piecewise-constant approximation for $\underline{x}^{(i)}$.

<u>Computation time</u> :

In each iteration it is necessary to integrate numerically 2n first order ordinary differential equations. In addition, the trajectory of $\partial H^{(i)}/\partial \underline{u}$ at the sampled times t_s, s = 0, 1, ..., N-1, must be computed. To speed up the iterative process, a single variable search may be used to determine the step size for the change in the trial control.

b) <u>Importance of initial guess and convergence characteristics</u>

A nominal control trajectory $\underline{u}^{o}(t)$, $t \in [t_o, t_f]$ must be chosen to start the numerical procedure. In selecting this nominal control we utilise

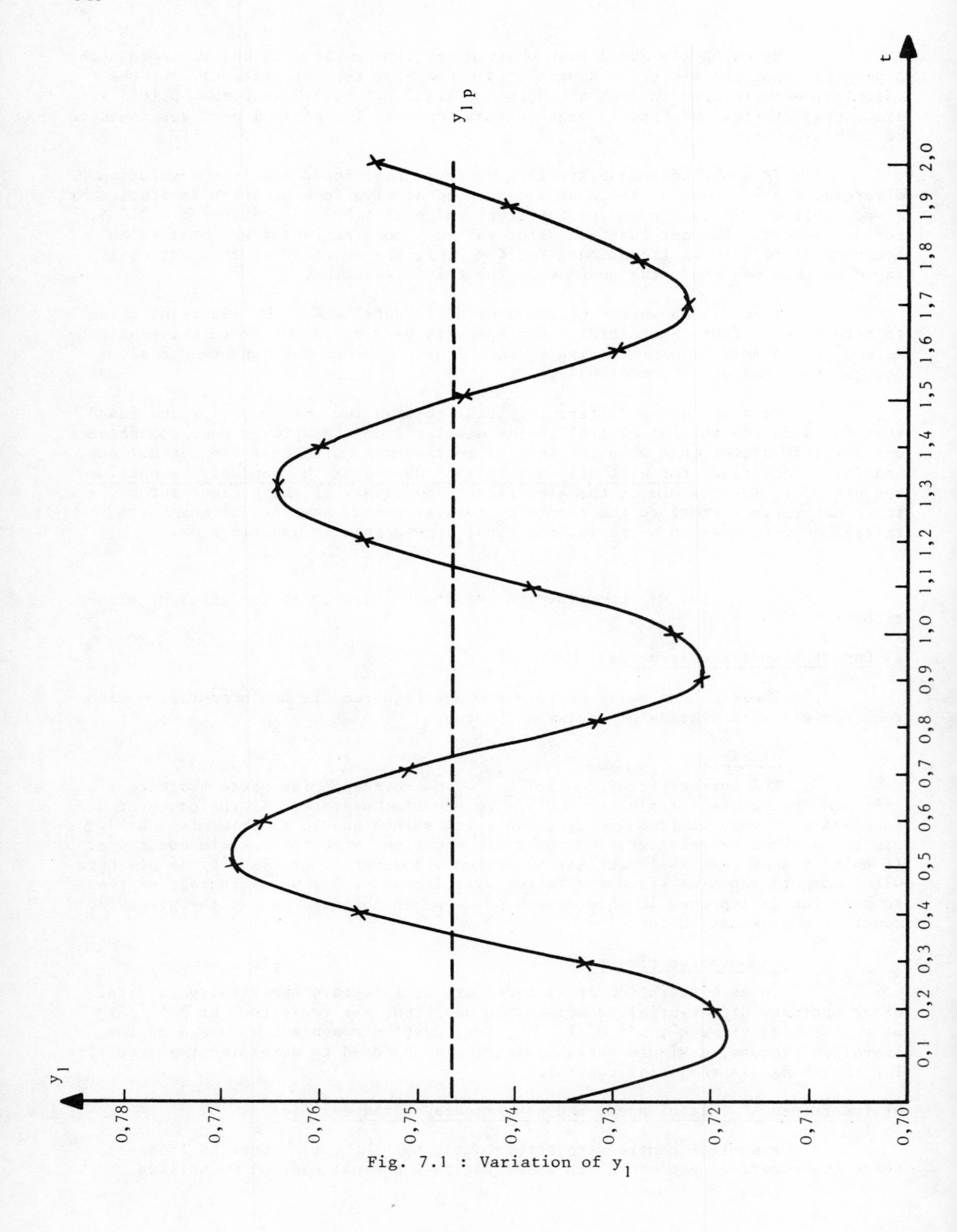

Fig. 7.1 : Variation of y_1

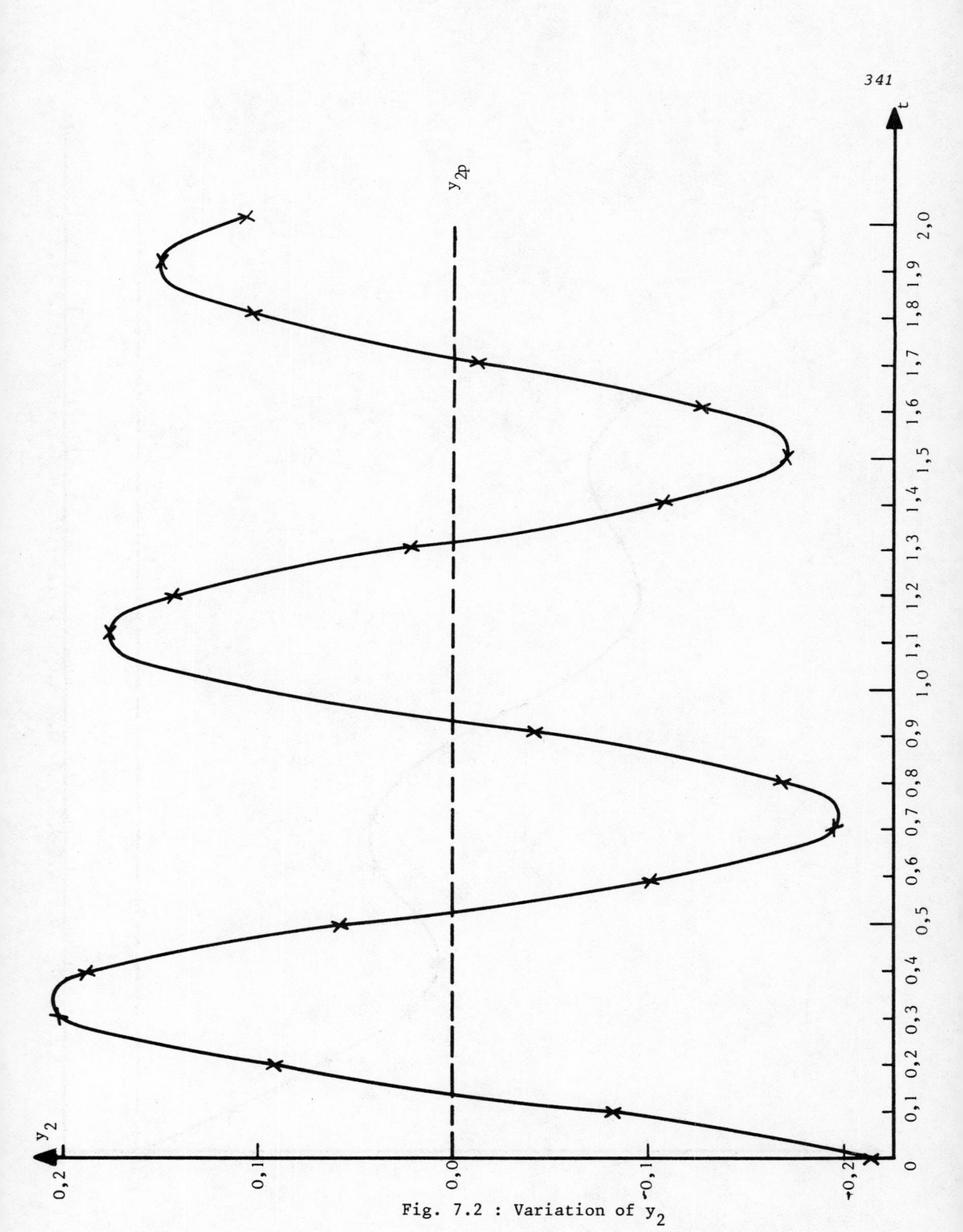

Fig. 7.2 : Variation of y_2

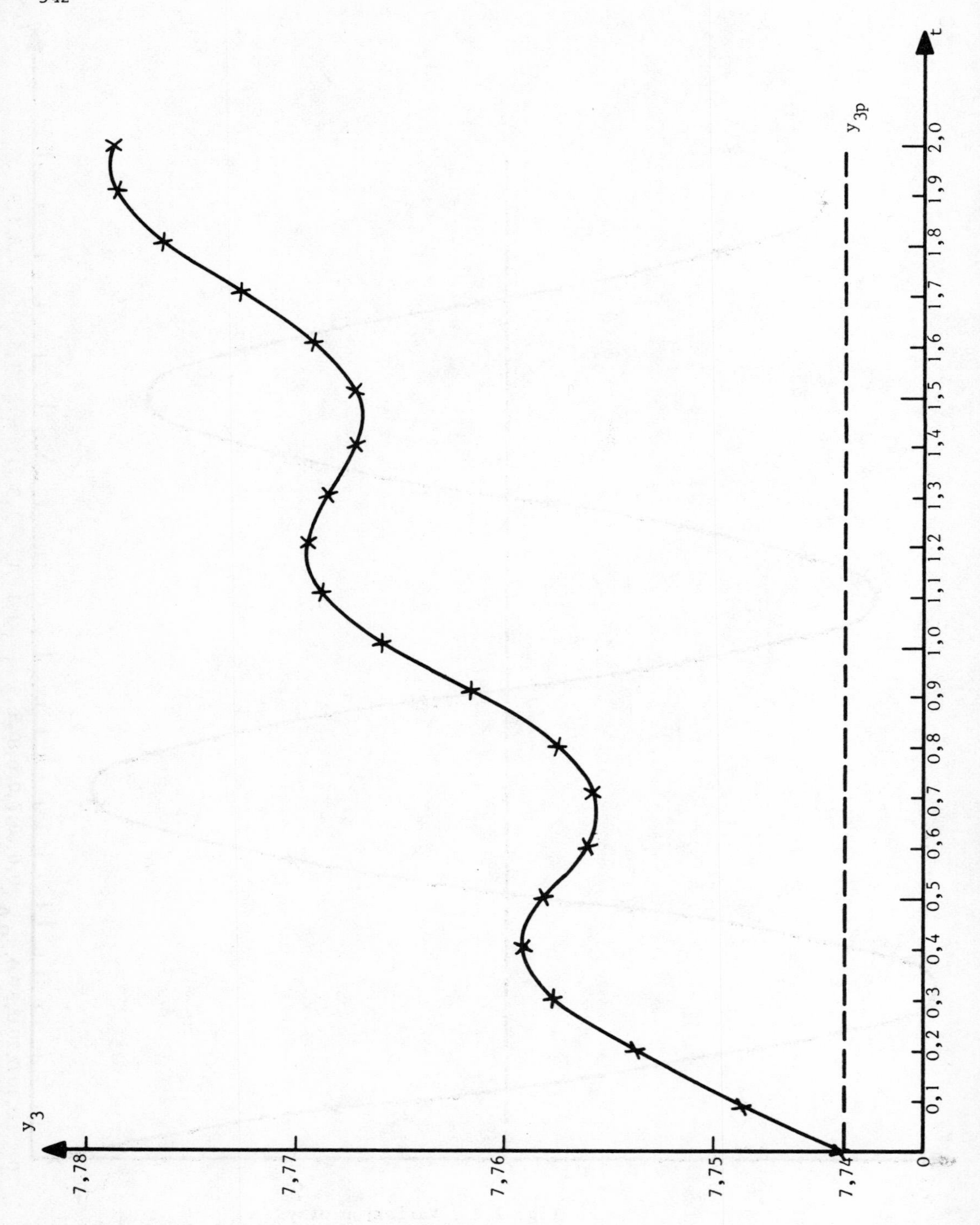

Fig. 7.3 : Variation of y_3

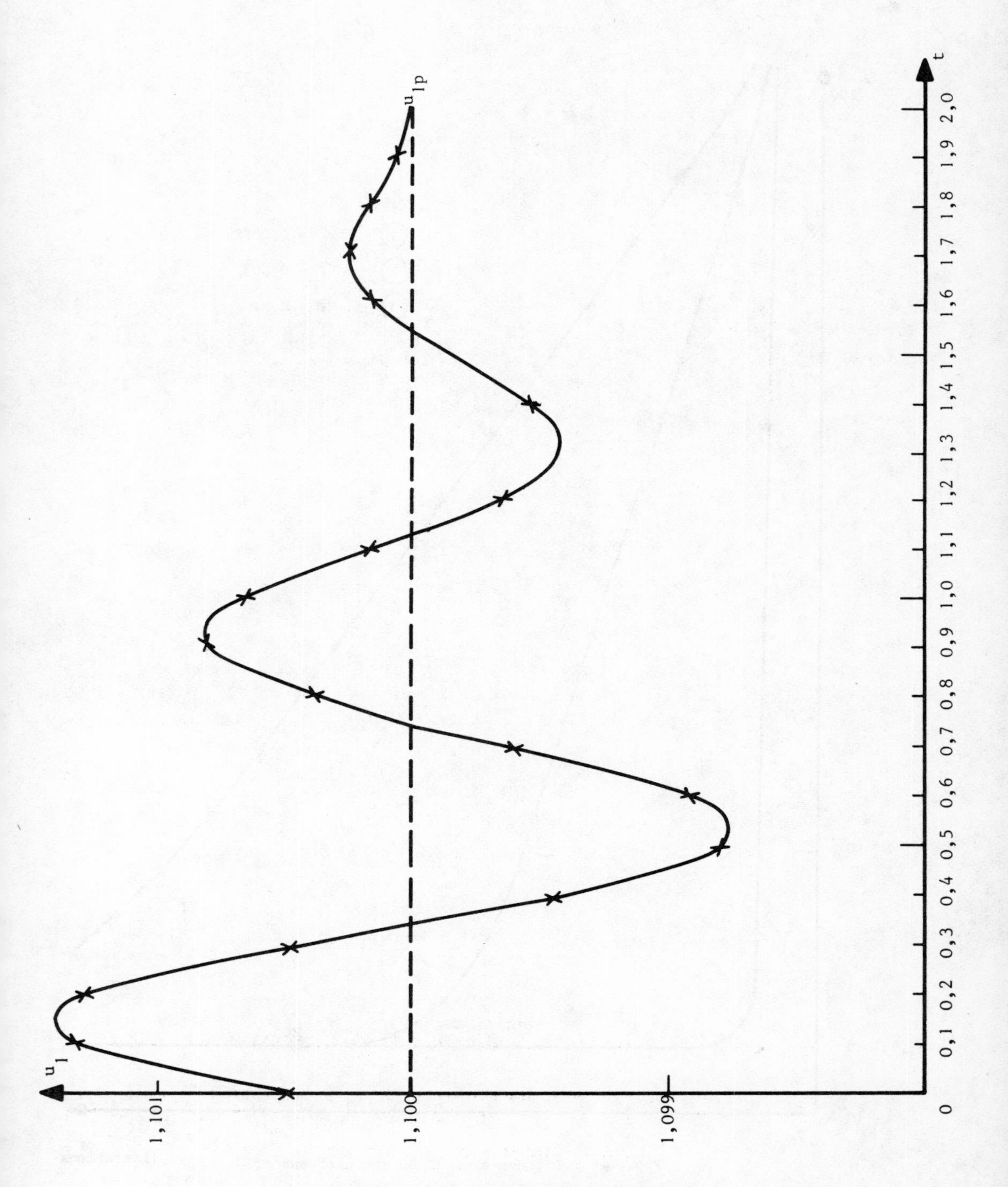

Fig. 7.4 : The optimal control

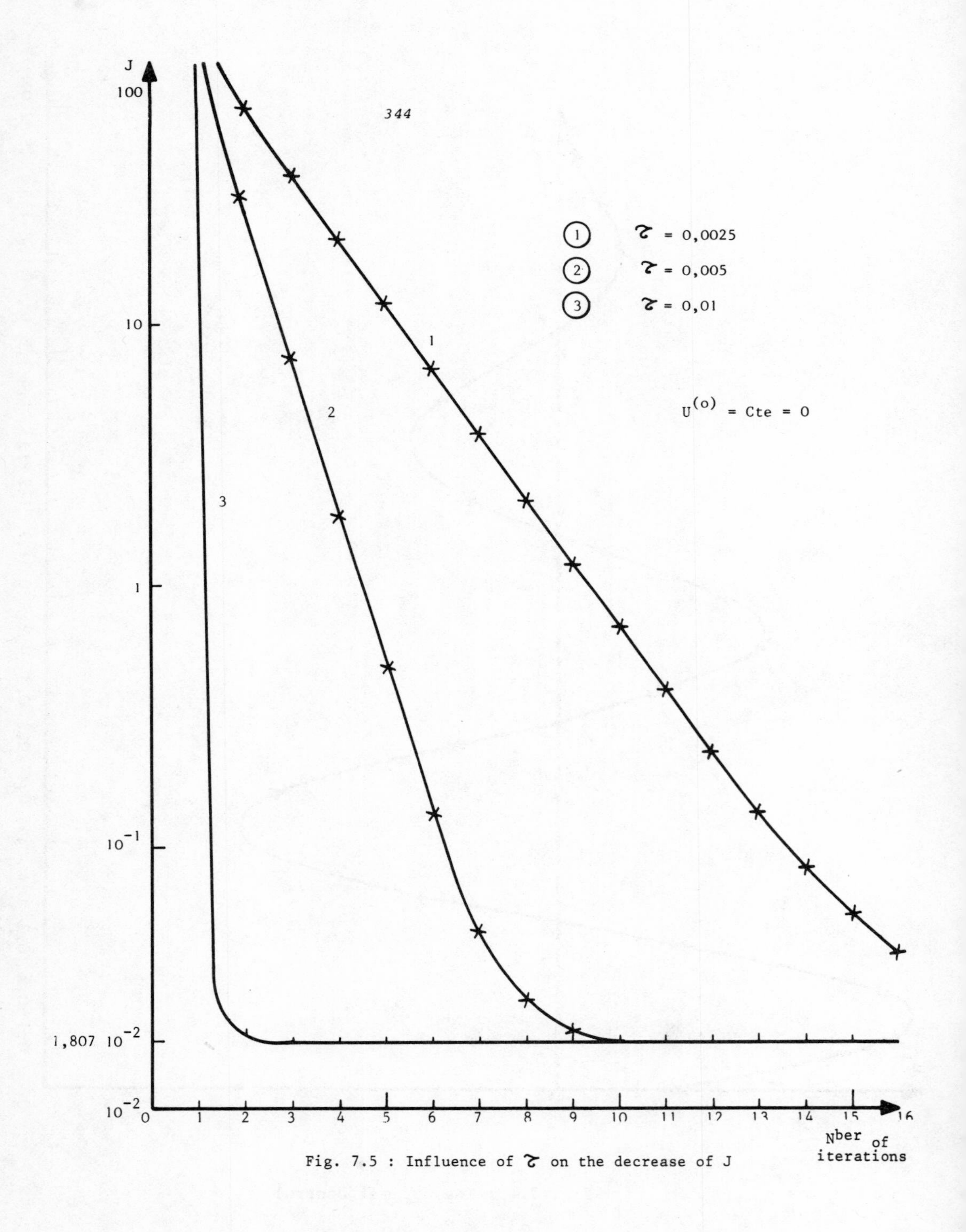

Fig. 7.5 : Influence of τ on the decrease of J

whatever physical insight we have about the problem. The actual initial guess chosen is not usually crucial. This makes the method rather easy to use. However, as with the parametric gradient methods, near the minimum, the gradient becomes small and the method has a tendency to converge very slowly.

c) Stopping criterion

The iterative procedure can be terminated when a criterion such as the norm of the gradient becomes smaller than some pre-selected positive number or when say the difference in the cost function value between two iterations is sufficiently small.

d) Modifications required for fixed end point problems

One way to handle the case where some of the final states are fixed is to use the penalty function approach. For example, if the desired final state is $\underline{x}_f(t_f)$, we can add an extra term to the cost function of the form

$$\frac{1}{2}\left\|\underline{x}_f(t_f) - \underline{x}(t_f)\right\|_S^2$$

where S is a diagonal matrix with large positive elements. In this case, the only change in the algorithm is that $\underline{\lambda}(t_f)$ is now given by $-S\left[\underline{x}(t_f) - \underline{x}_f(t_f)\right]$.

Having discussed an important approach to the numerical solution of two point boundary value problems, let us go on to another approach : the quasilinearisation method.

7.3 QUASILINEARISATION

The quasilinearisation approach to the solution of non-linear dynamic optimisation problems is based on solving the non-linear two point boundary value problem iteratively as a series of linear two point boundary value problems. As we saw in chapter 5, it is possible to solve a linear two point boundary value problem non-iteratively since for the case of linear differential equations it is possible to superpose solutions.

The stationarity conditions for the problem yield the following non-linear two point boundary value problem

$$\dot{\underline{x}} = \underline{a}(\underline{x}(t),\ \underline{\lambda}(t),\ \underline{u}(t),\ t) \qquad 7.3.1$$

$$\dot{\underline{\lambda}} = \underline{d}(\underline{x}(t),\ \underline{\lambda}(t),\ \underline{u}(t),\ t) \qquad 7.3.2$$

where $\underline{a}$ and $\underline{d}$ are non-linear functions of $\underline{x}(t)$, $\underline{\lambda}(t)$ and $\underline{u}(t)$; the boundary conditions are

$$\underline{x}(t_o) = \underline{x}_o \text{ and } \underline{\lambda}(t_f) = \underline{\lambda}_f \qquad 7.3.3$$

Let us rewrite these equations by putting

$$\underline{Z} = \begin{bmatrix} \underline{x} \\ \underline{\lambda} \end{bmatrix} \qquad \text{and} \qquad \underline{F} = \begin{bmatrix} \underline{a} \\ \underline{d} \end{bmatrix} \qquad 7.3.4$$

We will assume for the moment that the vector function $\underline{F}$ is independent of the control $\underline{u}$. We will see later how the equations change when $\underline{u}$ is included.

With the transformations given by 7.3.4, it is now necessary to solve the homogenous differential system

$$\frac{d\underline{Z}}{dt} = \underline{F}(\underline{Z}) \qquad 7.3.5$$

where $\underline{Z}$ is a non-explicit vector function of the independent variable t over the interval $[t_o, t_f]$ and $\underline{F}$ is a functional (of dimension m_z) with components Z_i and F_i ($i = 1, \ldots, m_z$) respectively.

The functional F is assummed to be differentiable with respect to $\underline{Z}$ and each component Z_i should be a strictly convex function of $\underline{Z}$ in the interval $[t_o, t_f]$ i.e. the matrix F_{iZZ} must be positive definite ($i = 1, \ldots, m_z$).

We assume also that

$$\text{Max} \left[F_i(\underline{Z}),\ F_{iZ_j}(\underline{Z}),\ \frac{1}{2} F_{iZ_jZ_k}(\underline{Z}) \right] \qquad t \in [t_o, t_f] \qquad 7.3.6$$

with $i = 1, \ldots m_z$
$j = 1, \ldots m_z$
$k = 1, \ldots m_z$

exists.

The boundary conditions become

$$\left.\begin{aligned} Z_j(t_o) &= Z_{jo} \qquad j = 1, 2, \ldots m \\ Z_j(t_f) &= Z_{jf} \qquad j = m+1, \ldots m_z \end{aligned}\right\} \qquad 7.3.7$$

Having formulated the problem, we next consider its linearisation.

7.3.1 Linearisation

As we said before, the quasi-linearisation method is based on successive linearisation of dynamical equations around different solutions. To start off such a method, we need to guess an initial approximation to the trajectory that we are seeking which we call $\overset{\circ}{\underline{Z}}(t)$. This initial guess must be such that $\underline{Z}^{\circ}(t)$ satisfies the boundary conditions, i.e.

$$\left.\begin{aligned} \overset{\circ}{Z}_j(t_o) &= Z_{jo} \qquad j = 1, \ldots, m \\ \overset{\circ}{Z}_j(t_f) &= Z_{jf} \qquad j = m+1, \ldots, m_z \end{aligned}\right\} \qquad 7.3.8$$

The linearisation of the dynamical equations i.e. of the vector function $\underline{F}$, around the initial guess $\overset{\circ}{\underline{Z}}(t)$ can be done by expanding in a Taylor series up to first order, i.e.

$$\frac{d\underline{Z}}{dt} = \underline{F}(\overset{\circ}{\underline{Z}}) + J(\underline{Z}^{\circ})\ (\underline{Z} - \underline{Z}^{\circ}) + \Phi \qquad 7.3.9$$

Φ in 7.3.9 represents terms of order higher than one and $J(\underline{Z}^{\circ})$ is the Jacobian of the function $\underline{F}(\underline{Z}, t)$ at $\underline{Z}^{\circ}(t)$, i.e.

$$J(\underline{Z}^\circ) = \begin{bmatrix} \frac{\partial F_1}{\partial Z_1} & \cdots\cdots\cdots & \frac{\partial F_1}{\partial Z_{m_z}} \\ \vdots & & \vdots \\ \vdots & \cdots \frac{\partial F_i}{\partial Z_j} \cdots & \vdots \\ \vdots & & \vdots \\ \frac{\partial F_{m_z}}{\partial Z_1} & \cdots\cdots\cdots & \frac{\partial F_{m_z}}{\partial Z_{m_z}} \end{bmatrix}_{\underline{Z}=\underline{Z}^\circ} \qquad 7.3.10$$

Note that in order to be able to neglect the terms represented by Φ, it is necessary for the initial approximation $\underline{Z}^\circ(t)$ of $\underline{Z}(t)$ to be quite close to the solution. In practice, this is a difficult problem to which we will return when we discuss the convergence of the method.

Thus, neglecting the higher order terms, we obtain :

$$\frac{dZ}{dt} = \underline{F}(\underline{Z}^\circ) + J(\underline{Z}^\circ)\ (\underline{Z} - \underline{Z}^\circ) \qquad 7.3.11$$

The solution of 7.3.11 may not be identical to that of the system 7.3.5 but it satisfies the boundary conditions 7.3.6. Since $\underline{Z}^\circ(t)$ is a known trajectory, equation 7.3.11 can be solved. If the solution of 7.3.11, $\underline{Z}^1(t)$ is not also a solution to 7.3.5, $\underline{Z}^1(t)$ is used as the new approximation for the solution and the linearisation is performed around it.

Using this reasoning on successive solutions, we obtain the following recursive formulae :

$$\frac{dZ}{dt}(\underline{Z}^{n+1}) = \underline{F}(\underline{Z}^n) + J(\underline{Z}^n)\ (\underline{Z}^{n+1} - \underline{Z}^n) \qquad 7.3.12$$

where

$$J(\underline{Z}^n) = \begin{bmatrix} & \vdots & \\ \cdots & \frac{\partial F_i}{\partial Z_j} & \cdots \\ & \vdots & \end{bmatrix}_{\underline{Z} = \underline{Z}^n} \qquad 7.3.13$$

n being in this case the iteration number.

The boundary conditions corresponding to 7.3.12 could be written as

$$\begin{aligned} Z_j^{n+1}(t_o) &= Z_{jo} \qquad & j = 1, \ldots, m \\ Z_j^{n+1}(t_f) &= Z_{jf} \qquad & j = m+1, \ldots, m_z \end{aligned} \qquad 7.3.14$$

Here $\underline{Z}^n(t)$ represents the $(n+1)^{th}$ approximation to $\underline{Z}(t)$. We can thus solve the linear differential equation 7.3.12 subject to the boundary conditions 7.3.14 in order to obtain a solution which is closer to $\underline{Z}(t)$.

We thus need to solve at each iteration a linear vector differential equation which satisfies the boundary conditions. But does this sequence of trajectories indeed converge to the solution of the system of equations 7.3.5 ? We next study the convergence of the solution sequence.

7.3.2 Convergence

Bellman and Kalaba [5] have studied the convergence of the algorithm and the uniqueness of the solution on the following problem :

$$\ddot{u} = f(u) \qquad \text{with } u(t_o) = u(t_f) = 0 \tag{7.3.15}$$

The quasilinearisation approach gives the following recursive formulae for this case :

$$\ddot{u}^{n+1} = f(u^n) + (u^{n+1} - u^n)\,\dot{f}(u^n) \tag{7.3.16}$$

with

$$u(t_o)^{n+1} = u(t_f)^{n+1} = 0 \tag{7.3.17}$$

On an interval (t_o, t_f) sufficiently small, Bellman and Kalaba [5] have shown that the solution is unique and that convergence is monotonic and quadratic, provided that

a) $\max \left[\max |f(u)|, \max |\dot{f}(u)| \right] = m_1 \leqslant \infty$

and b) $|u^\circ(t)| \leqslant 1$

Obviously, the upper bound is inversely proportional to the width of the interval $[t_o, t_f]$ and this dictates the choice of the initial approximation $u^\circ(t)$. Indeed, in this special case, convergence occurs provided the following inequality is satisfied [5]

$$\frac{(t_f - t_o)^2\, k/8}{1 - (t_f - t_o)^2 \,\frac{m_2}{4}} \max\,(u^1 - u^\circ) \leqslant 1 \tag{7.3.18}$$

where $k = \max |\ddot{f}(u)|$

$m_2 = \max |\dot{f}(u)|$

$|u| \leqslant 1$

If m_2 is small, $(t_f - t_o)$ can be large so that the choice of u° is relatively insensitive. On the other hand if m_2 is large, i.e. for a monotone function, $(t_f - t_o)$ should be small or otherwise u° should be judiciously chosen.

It is interesting to note that it is sufficient for convergence that $\max_t |u^{n+1} - u^n|$ be adequately small for some value of n.

Thus, even if the interval (t_o, t_f) appears initially to be too big, we would hope that a good initial approximation to u(t) will ensure that $|u^1(t) - u^\circ(t)|$ is sufficiently small. Bellman and Kalaba also show that

$$\max |u - u_n| \leqslant k_2 \max |u - u_{n-1}|^2$$

where

$$k_2 = \frac{(t_f - t_o)^2\, k / 8}{1 - (t_f - t_o)^2 m_1 / 4}$$

These convergence results can be generalised to make them applicable to first order differential equations in which case the following algorithm converges quadratically

$$Z^{n+1} - Z^n = K(Z^n - Z^{n-1})\ (Z^n - Z^{n-1}) \qquad 7.3.19$$

Next let us study the case of the optimisation problem.

7.3.3 The case of the optimisation problem

Up to now, we have only studied solutions of differential equations. However, our main interest is in optimisation problems and we are only interested in solving differential equations if they enable us to solve our non-linear dynamical optimisation problem. In that case, the variables of the dynamical system are not only the state variables $\underline{x}$ and the costate variables $\boldsymbol{\lambda}$ but also the control variables $\underline{u}$ and as we will see in the case of hierarchical control in the next chapter, the coupling variables. Here we will only consider the standard case of control variables so that our system 7.3.5 becomes

$$\frac{d\underline{Z}}{dt} = \underline{F}(\underline{Z}, \underline{u}) \qquad 7.3.20$$

where $\underline{u}$ is an m_u vector. Now, two possibilities exist, i.e.

a) the controls are free or

b) the controls are bounded.

If the controls are free, then we can replace $\underline{u}$ by its expression in $\underline{Z}$ given by the optimality condition ($\frac{\partial H}{\partial \underline{u}} = 0$), i.e.

$$\underline{u} = \mathcal{F}(\underline{Z}) \qquad 7.3.21$$

on substituting for $\underline{u}$ from 7.3.21 into 7.3.20, we can eliminate $\underline{u}$ in 7.3.20 so that we find again equation 7.3.5 and we can apply the above quasilinearisation method to it directly.

If this substitution is not feasible, the terms of the Jacobian will be given by the formulae 7.3.23 given below.

In the case where the controls are bounded, i.e. of the form :

$$|\underline{u}| \leqslant \underline{u}_{max}$$

we can't directly replace $\underline{u}$ by $\mathcal{F}(\underline{Z})$ in equation 7.3.20 and the equation 7.3.12 is modified to

$$\frac{d}{dt}(\underline{Z}^{n+1}) = F(\underline{Z}^n, \underline{u}) + J(\underline{Z}^n)\ (\underline{Z}^{n+1} - \underline{Z}^n) \qquad 7.3.22$$

where $J(\underline{Z}^n)$ represents the Jacobian at $\underline{Z}^n$ and is formed by terms J_{ij} given by

$$J_{ij} = \frac{\partial F_i}{\partial Z_j^n} + \sum_{i=1}^{m_u} \frac{\partial F_i}{\partial u_i} \frac{\partial u_i}{\partial Z_j^n} \qquad \begin{matrix} i = 1, 2, \ldots m_z \\ j = 1, 2, \ldots m_z \end{matrix} \qquad 7.3.23$$

On the boundaries, we have :

$$\frac{\partial u_i}{\partial Z_j^n} = 0 \qquad \text{and} \qquad u_i = \pm u_{imax}$$

whilst in the interior of the constrained region, u_i and its derivatives could be obtained using the optimality conditions.

It should be noted that in the above analysis only autonomous systems have been considered. In the case of non-autonomous systems of the form

$$\underline{Z} = \underline{F}(\underline{Z}, t) \tag{7.3.24}$$

we can convert to the autonomous case by adding another component to the $\underline{Z}$ vector, i.e.

$$Z_{m_{z+1}} = t \quad \text{and} \quad \frac{d\, Z_{m_{z+1}}}{dt} = 1 \tag{7.3.25}$$

7.3.4 The algorithm

In the previous parts of this section, we have briefly described the quasilinearisation approach and we now go on to develop an algorithm to enable us to compute the optimal control using this approach.

Let us begin by considering the set of given conditions 7.3.7 which could be rewritten as

$$\begin{aligned} (\underline{Z}(t_o), a_k) &= b_k & k &= 1, \ldots, m \\ (\underline{Z}(t_f), a_\ell) &= b_\ell & \ell &= m+1, \ldots, m_z \end{aligned} \tag{7.3.26}$$

where a_i and b_i represent respectively the given vectors and scalars and (u, v) represents the usual scalar product. The set of equations 7.3.26 can be regrouped into the matrix form

$$\begin{bmatrix} \ldots & Z_i(t_o) & \ldots \end{bmatrix} \begin{bmatrix} & \vdots & \\ \ldots & a_{ik} & \ldots \\ & \vdots & \end{bmatrix} = \begin{bmatrix} \ldots & b_k & \ldots \end{bmatrix} \tag{7.3.27}$$

$$\begin{bmatrix} \ldots & Z_i(t_f) & \ldots \end{bmatrix} \begin{bmatrix} & \vdots & \\ \ldots & a_{i\ell} & \ldots \\ & \vdots & \end{bmatrix} = \begin{bmatrix} \ldots & b_\ell & \ldots \end{bmatrix} \tag{7.3.28}$$

with $i = 1, \ldots, m_z$

$k = 1, \ldots, m$

$\ell = m+1, \ldots, m_z$

Let us put $A_o = \begin{bmatrix} a_{ik} \end{bmatrix}$ and $A_f = \begin{bmatrix} a_{i\ell} \end{bmatrix}$

$\underline{b}_o = \begin{bmatrix} b_k \end{bmatrix}$ and $\underline{b}_f = \begin{bmatrix} b_\ell \end{bmatrix}$

The solutions 7.3.27 and 7.3.28 become thus

$$\left.\begin{aligned} \underline{Z}(t_o)^T A_o &= \underline{b}_o^T \\ \underline{Z}(t_f)^T A_f &= \underline{b}_f^T \end{aligned}\right\} \tag{7.3.29}$$

or again

$$\left.\begin{array}{l} A_o^T \cdot \underline{Z}(t_o) = \underline{b}_o \\ A_f^T \cdot \underline{Z}(t_f) = \underline{b}_f \end{array}\right\} \quad 7.3.30$$

Again the recursive relation 7.3.12 can not be treated numerically as it is written but it should be modified to :

$$\frac{d}{dt}(\underline{Z}^{n+1}) = J(\underline{Z}^n)\,\underline{Z}^{n+1} + \underline{F}(\underline{Z}^n) - J(\underline{Z}^n)\,\underline{Z}^n \quad 7.3.31$$

Let $X^{n+1}(t)$ be the matrix solution of a homogeneous form of 7.3.31 (i.e. without the second term on the right hand side), i.e. $X^{n+1}(t)$ is the matrix solution of

$$\frac{d}{dt}(X^{n+1}) = J(\underline{Z}^n)\,X^{n+1} \quad 7.3.32$$

with $X^{n+1}(t_o) = I$ 7.3.33

where I is the identity matrix.

Let $\underline{P}^{n+1}$ be the particular solution of the vector equation (i.e. from 7.3.31)

$$\frac{d}{dt}(\underline{P}^{n+1}) = J(\underline{Z}^n)\,\underline{P}^{n+1} + \underline{F}(Z^n) - J(\underline{Z}^n)\,\underline{Z}^n \quad 7.3.34$$

with

$$\underline{P}(t_o)^{n+1} = \underline{0} \quad 7.3.35$$

The complete solution of 7.3.31 could thus be written as

$$\underline{Z}^{n+1} = X^{n+1} \cdot \underline{C}^{n+1} + \underline{P}^{n+1} \quad 7.3.36$$

where $\underline{C}^{n+1}$ is a vector of integration constants. $\underline{C}^{n+1}$ can be evaluated from the boundary conditions by solving $(\underline{C}^{n+1}, a_i) = b_i$, $i = 1, \ldots, m$

$$(X^{n+1}(t_f) \cdot \underline{C}^{n+1} + \underline{P}^{n+1}(t_f), a_j) = b_j \qquad j = m+1, \ldots m_z \quad 7.3.37$$

Using equation 7.3.30, this system of algebraic equations could be rewritten as

$$\left.\begin{array}{l} A_o^T\,\underline{C}^{n+1} = \underline{b}_o \\ A_f^T\left[X^{n+1}(t_f) \cdot \underline{C}^{n+1} + \underline{P}^{n+1}(t_f)\right] = \underline{b}_f \end{array}\right\} \quad 7.3.38$$

expanding the second equation 7.3.38 we have

$$A_f^T\,X^{n+1}(t_f)\,\underline{C}^{n+1} + A_f^T\,\underline{P}^{n+1}(t_f) = \underline{b}_f \quad 7.3.39$$

The system of equations 7.3.38 becomes thus

$$\begin{array}{l} A_o^T\,\underline{C}^{n+1} = \underline{b}_o \\ A_f^T \cdot X^{n+1}(t_f)\,\underline{C}^{n+1} = \underline{b}_f - A_f^T\,\underline{P}^{n+1}(t_f) \end{array} \quad 7.3.40$$

On replacing A_o, A_f by their values, we obtain the matrix equation

$$\begin{bmatrix} \begin{bmatrix} & \vdots & \\ \cdots & a_{ki} & \cdots \\ & \vdots & \end{bmatrix} & \\ \begin{bmatrix} & \vdots & \\ \cdots & a_{\ell i} & \cdots \\ & \vdots & \end{bmatrix} & \begin{bmatrix} & \vdots & \\ \cdots & X_{ij} & \cdots \\ & \vdots & \end{bmatrix} \end{bmatrix}_{t=t_f} \begin{bmatrix} \vdots \\ C_j^{n+1} \\ \vdots \end{bmatrix} = \begin{bmatrix} \begin{bmatrix} \vdots \\ b_k \\ \vdots \end{bmatrix} \\ \begin{bmatrix} \vdots \\ b_\ell \\ \vdots \end{bmatrix} - \begin{bmatrix} \vdots \\ \cdots a_{\ell i} \\ \vdots \end{bmatrix} \begin{bmatrix} \vdots \\ P_i^{n+1} \\ \vdots \end{bmatrix} \end{bmatrix}_{t=t_f} \qquad 7.3.41$$

with $i = 1, \ldots m_z$

$j = 1, \ldots m_z$

$k = 1, \ldots m$

$\ell = m+1, \ldots m_z$

We can rewrite 7.3.41 in the more compact form :

$$\begin{bmatrix} A_o^T \\ \\ A_f^T X \end{bmatrix} \begin{bmatrix} \underline{C}^{n+1} \end{bmatrix} = \begin{bmatrix} \underline{b}_o \\ \\ \underline{b}_f - A_f^T \underline{P}^{n+1} \end{bmatrix} \qquad 7.3.42$$

Equation 7.3.42 can be easily solved to obtain the integration constants $\underline{C}$. The general solution $\underline{Z}^{n+1}(t)$ is thus completely specified by the equation

$$Z_i(t)^{n+1} = P_i^{n+1}(t) + \sum_{j=1}^{m_z} X_{ij}^{n+1}(t)\, C_j^{n+1} \qquad i = 1, \ldots m_z \qquad 7.3.43$$

Finally, we need a test for the end of the calculation. For this, it is sufficient to compare the $(n+1)^{th}$ and n^{th} trajectories, i.e.

$$\left| Z_i(t)^{n+1} - Z_i(t)^n \right| \leqslant \epsilon_i \qquad i = 1, 2, \ldots m_z \qquad 7.3.44$$

Here, ϵ_i are components of a prechosen vector $\underline{\epsilon}$ which defines the accuracy of the desired results.

Summary of the algorithm

The algorithm can be summarised by the following steps :

Step 1

Enter the initial approximation $\underline{Z}^\circ(t_o)$, the matrix A and the vector $\underline{b}$ where

$$A = \begin{bmatrix} A_o^T \\ A_f^T \end{bmatrix}, \qquad \underline{b} = \begin{bmatrix} \underline{b}_o \\ \underline{b}_f \end{bmatrix}$$

Enter also the accuracy of the final result desired, i.e. ϵ .

Step 2

Integrate simultaneously the equations

$$\dot{X}^1 = J(\underline{Z}^\circ)\, X^1$$

$$\underline{\dot{P}}^1 = J(\underline{Z}^\circ)\, \underline{P}^1 + \underline{F}(\underline{Z}^\circ) - J(\underline{Z}^\circ)\, Z^\circ$$

with $X^1(t_o) = I$, $\underline{P}^1(t_o) = \underline{0}$

Step 3

Form the matrices A_1 and the vector $\underline{b}_1$ where

$$A_1 = \begin{bmatrix} A_o^T \\ A_f^T\, \underline{Z}^1(t_f) \end{bmatrix} \qquad \underline{b}_1 = \begin{bmatrix} \underline{b}_o \\ \underline{b}_f - A_f^T \,.\, \underline{P}^1(t_f) \end{bmatrix}$$

Step 4

Find the integration constants $\underline{C}^1$ using equation 7.3.41.

Step 5

Compute the solution $\qquad \underline{Z}^1 = X^1 \,.\, C^1 + \underline{P}^1$

Step 6

Test for convergence, i.e.

$$|\underline{Z}^1 - \underline{Z}^o| \leqslant \epsilon$$

If this inequality is satisfied, verify that the solution obtained, i.e. $\underline{Z}^1$ also satisfies the non-linear equation 7.3.5. If so, stop. If the inequality is not satisfied, use Z^1 as the new Z^o in step 1 and start again.

7.3.5 Examples

Next, let us consider some illustrative examples.

Example 1

As a very simple example, consider the minimisation of

$$J = \frac{1}{2}\int_0^1 (x^2 + u^2)\, dt$$

subject to $\qquad \dot{x} = x^2 + u \; ; \; x(o) = x(o).$

Here the Hamiltonian is :

$$\frac{1}{2}\, x^2 + \frac{1}{2}\, u^2 + \lambda\,(x^2 + u)$$

The necessary conditions for optimality yield

$$\frac{\partial H}{\partial u} = 0 \qquad \text{or} \qquad u = -\lambda$$

$$\frac{\partial H}{\partial x} = -\dot{\lambda} = x + 2\lambda x$$

The reduced state costate equations therefore become

$$\dot{x} = x^2 - \lambda$$

$$\dot{\lambda} = -(1 + 2\lambda)x$$

with $x(o) = x_o$ and $\lambda(1) = 0$

Let us assume we have a nominal trajectory $\binom{x^o}{\lambda^o}$. Expanding the state-costate equations in a Taylor's series about this trajectory up to the first order we obtain the linearised state-costate equations as

$$\dot{x}^{(i)}(t) = \dot{x}^{(o)}(t) + \left[2x^{(o)}\right]\left[x^{(i)}(t) - x^{(o)}(t)\right] + \left[-1\right]\left[\lambda^{(i)}(t) - \lambda^{(o)}(t)\right]$$

$$\dot{\lambda}^{(i)}(t) = \dot{\lambda}^{(o)}(t) + \left[(-1-2\lambda^{(o)})(x^{(i)}(t) - x^{(o)}(t)\right] + \left[-2x^{(o)}\,(\lambda^{(i)}(t) - \lambda^{(o)}(t))\right]$$

which becomes on substituting for $\dot{x}^{(o)}(t)$ and $\lambda^{(o)}(t)$ and simplifying

$$\begin{bmatrix} \dot{x}^{(i)}(t) \\ \dot{\lambda}^{(i)}(t) \end{bmatrix} = \begin{bmatrix} A_{11} & A_{12} \\ A_{21} & A_{22} \end{bmatrix} \begin{bmatrix} x^{(i)}(t) \\ \lambda^{(i)}(t) \end{bmatrix} + \begin{bmatrix} e_1(t) \\ e_2(t) \end{bmatrix}$$

where

$$A_{11} = 2x^{(o)}$$

$$A_{12} = -1$$

$$A_{21} = (-1-2\lambda^{(o)})$$

$$A_{22} = (-2x^{(o)})$$

$$e_1 = x^{(o)2} - \lambda^{(o)} - 2x^{(o)2} + \lambda^{(o)} = -x^{(o)2}$$

$$e_2 = -(1+2\lambda^{(o)})x^{(o)} - (-1-2\lambda^{(o)})x^{(o)} - \left[-2x^{(o)}\right]\lambda^{(o)} = 2x^{(o)}\lambda^{o}.$$

Next we integrate the homogeneous equations

$$\begin{bmatrix} \dot{x}^{(i)}(t) \\ \dot{\lambda}^{(i)}(t) \end{bmatrix} = \begin{bmatrix} A_{11} & A_{12} \\ A_{21} & A_{22} \end{bmatrix} \begin{bmatrix} x^{(i)}(t) \\ \lambda^{(i)}(t) \end{bmatrix}$$

from the initial conditions

$$x^h(o) = 0 \quad ; \quad \lambda^h(o) = 1$$

to obtain the trajectories $x^h(t)$, $\lambda^h(t)$. We obtain also a particular solution by integrating the equation

$$\begin{bmatrix} \dot{x}^{(i)} \\ \dot{\lambda}^{(i)} \end{bmatrix} \quad \begin{bmatrix} A_{11} & A_{12} \\ A_{21} & A_{22} \end{bmatrix} \begin{bmatrix} x^{(i)} \\ \lambda^{(i)} \end{bmatrix} + \begin{bmatrix} e_1^{(i)} \\ e_2^{(i)} \end{bmatrix}$$

from the initial conditions $x(t_o) = x_o$, $\lambda(o) = 0$, over the period (0 - 1).

Thus we obtain the trajectories $\begin{bmatrix} x^P \\ \lambda^P \end{bmatrix}$

Then the complete solution of the linearised equations is given by

$$x^{(i)}(t) = \alpha \; x^h(t) + x^P(t)$$

$$\lambda^{(i)}(t) = \alpha \; \lambda^h(t) + \lambda^P(t)$$

Using $\lambda(1) = 0$, we have

$$\alpha = \frac{\lambda^P(1)}{\lambda^h(1)}$$

Using this value of α in the above solution completes one iteration of the quasilinearisation procedure. Now we can test if we have reached the optimum and if not, we restart using the $\begin{bmatrix} x^{(1)} \\ \lambda^{(1)} \end{bmatrix}$ as the new nominal trajectories.

<u>Example 2</u>

As a second example, consider the sliding mass system shown in Fig. 7.6

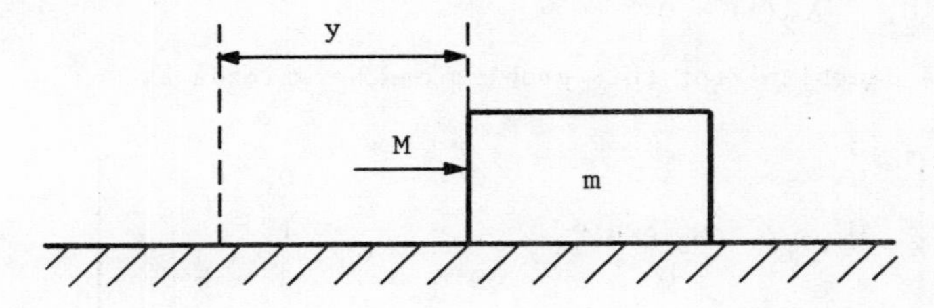

Fig. 7.6

This system can be described by the second order differential equation

$$m\ddot{y} + C\,\dot{y}^2 \quad \mathrm{Sgn}\,\dot{y} = M$$

$$\mathrm{Sgn}\,\dot{y} = 1 \text{ if } \dot{y} > 0$$
$$= -1 \text{ if } \dot{y} < 0$$

where m is the mass of the sliding solid
C is the coefficient of friction
M is the controlled force applied and
y is the displacement.

To simplify the calculations, let us put m = 1 ; C = 1. Then we can rewrite the above second order equation in the state space form as

$$\dot{y}_1 = y_2$$

$$\dot{y}_2 = M - y_2^{\,2} \text{ Sgn } y_2$$

with the initial conditions $y_1(o) = 2$, $y_2(o) = -2$. We would like to minimise the cost function

$$J = \underset{M}{\text{Min}} \int_0^1 \frac{1}{2} (y_1^{\,2} + y_2^{\,2} + M^2) \, dt$$

This example was previously solved by Bauman [2] using a hierarchical technique, and subsequently by Galy [8].

To solve this problem using the quasilinearisation approach, let us write the Hamiltonian as

$$H = \frac{1}{2} (y_1^{\,2} + y_2^{\,2} + M^2) + \lambda_1 y_2 + \lambda_2 \left[M - y_2^{\,2} \text{ Sgn } y_2 \right]$$

Then the conditions of optimality yield the two point boundary value problem :

$$\dot{\lambda}_1 = y_1$$

$$\dot{\lambda}_2 = y_2 - \lambda_1 + 2 \lambda_2 y_2 \text{ Sgn } y_2$$

$$\dot{y}_1 = y_2$$

$$\dot{y}_2 = M - y_2^{\,2} \text{ Sgn } y_2 = \lambda_2 - y_2^{\,2} \text{ Sgn } y_2$$

$$y_1(o) = 2, \qquad y_2(o) = -2$$

$$\lambda_1(1) = 0, \qquad \lambda_2(1) = 0$$

The Jacobian for this problem can be written as

$$J(\underline{Z}) = \begin{bmatrix} 0 & 1 & 0 & 0 \\ 0 & -2y_2 \text{ Sgn } y_2 & 0 & 1 \\ 1 & 0 & 0 & 0 \\ 0 & 1+2\lambda_2 \text{ Sgn } y_2 & -1 & 2y_2 \text{ Sgn } y_2 \end{bmatrix}$$

with $\underline{Z}^T = \begin{bmatrix} y_1 & y_2 & \lambda_1 & \lambda_2 \end{bmatrix}$

We wrote a computer program based on the quasilinearisation algorithm and solved the problem on an IBM 370/165 digital computer.

Using a termination error of 10^{-3} and an integration interval of 0.02 seconds in the Runge-Kutta procedure for integrating the differential equations, we were able to obtain convergence in 4 iterations which took 1.33 seconds to execute. Table 7.1 gives the decrease of the norm of each component trajectory between two iterations.

The state and control trajectories are given in Fig. 7.7.

	y_1	y_2	λ_1	λ_2
$\|z^1 - z^0\|$	1.220	1.227	1.322	0.137
$\|z^2 - z^1\|$	0.131	0.319	0.035	0.026
$\|z^3 - z^2\|$	5×10^{-4}	0.019	1×10^{-3}	1×10^{-3}
$\|z^4 - z^3\|$	1×10^{-4}	5×10^{-4}	0	0

Table 7.1

As a final example, consider the synchronous machine example that we treated previously.

Example 3 : Synchronous machine control.

As we saw before, the model in this case is given by the non-linear state space equations

$$\dot{y}_1 = y_2$$

$$\dot{y}_2 = B_1 - A_1 y_2 - A_2 y_3 \sin y_1 - \frac{B_2}{2} \sin 2y_1$$

$$\dot{y}_3 = M - C_1 y_3 + C_2 \cos y_1$$

where y_1 is the rotor angle

y_2 is the speed deviation

and y_3 is the flux variation

M is the excitation voltage which is assumed to be available for optimal manipulation. A_1, A_2, B_1, B_2, C_1, C_2 are constants which characterise the machine and the network.

The cost function is given by

$$J = \underset{M}{\text{Min}} \int_0^T \left[A_1^*(y_1-y_{1p})^2 + A_2^*(y_2-y_{2p})^2 + A_3^*(y_3-y_{3p})^2 + A_M(M - M_p)^2 \right] dt$$

where y_{1p}, y_{2p}, y_{3p}, M_p are desired steady state values and A_1^*, A_2^*, A_3^*, A_M are weighting factors.

As initial conditions, we have

$$y_1(o) = 0.7347, \; y_2(o) = -0.2151, \; y_3(o) = 7.7443$$

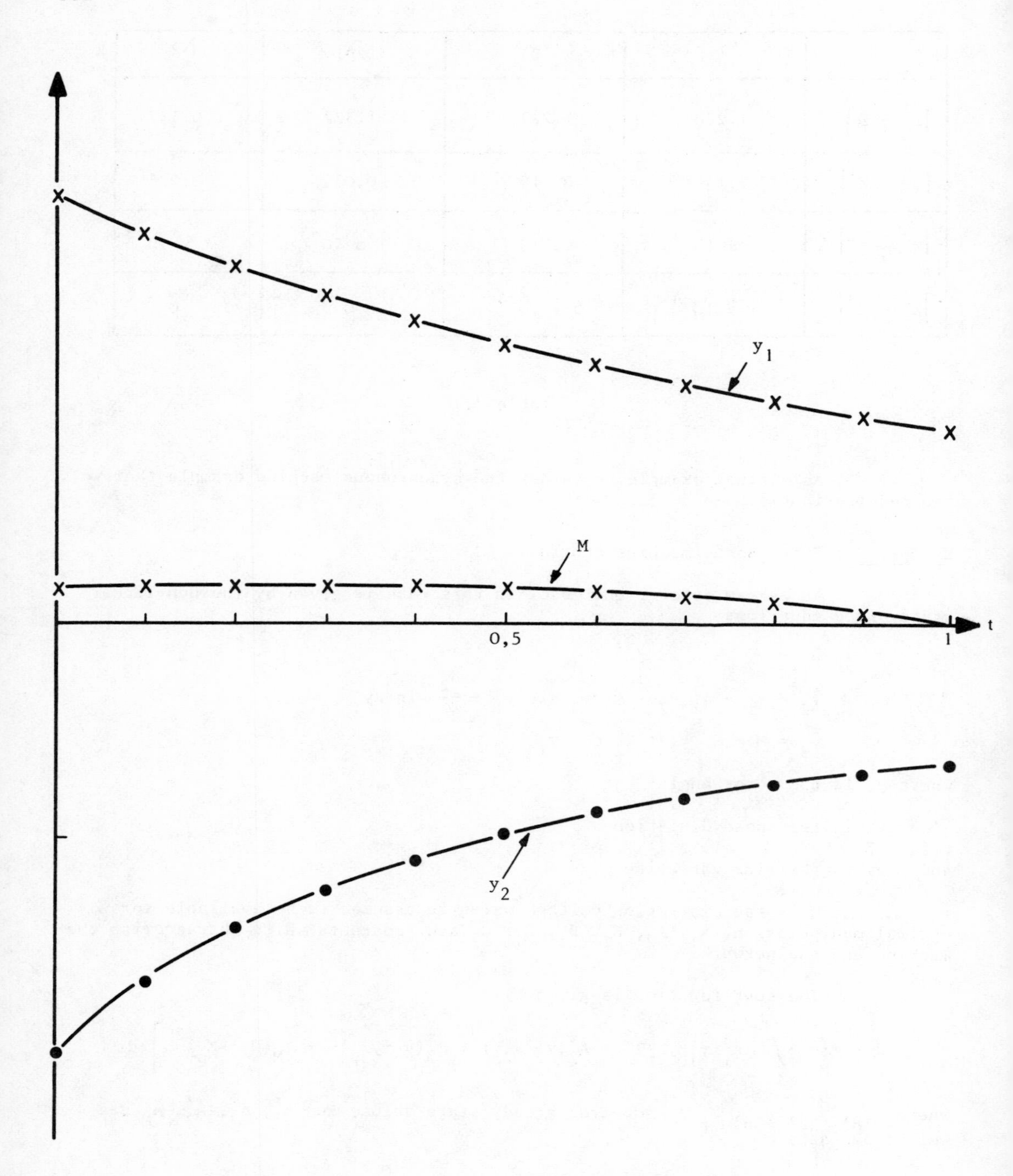

Fig. 7.7. Optimal trajectories for the sliding mass system

To solve this problem, let us write the Hamiltonian as

$$H = A_1^*(y_1-y_{1p})^2 + A_2^*(y_2-y_{2p})^2 + A_3^*(y_3-y_{3p})^2 + A_M(M-M_p)^2 + \lambda_1 y_2 +$$

$$\lambda_2 \left[B_1 - A_1 y_2 - A_2 y_3 \sin y_1 - \frac{B_2}{2}\sin 2y_1)\right] + \lambda_3\left[M - C_1 y_3 + C_2 \cos y_1\right]$$

The conditions of optimality in this case yield the following reduced two point boundary value problem (after substituting for the optimal control $M = (Mp - \frac{1}{2A_M}\lambda_3)$

$$\dot{y}_1 = y_2$$

$$\dot{y}_2 = B_1 - A_1 y_2 - A_2 y_3 \sin y_1 - \frac{B_2}{2}\sin 2y_1$$

$$\dot{y}_3 = M_p - \frac{1}{2A_M}\lambda_3 - C_1 y_3 + C_2 \cos y_1$$

$$\dot{\lambda}_1 = -2A_1^*(y_1 - y_{1p}) + A_2\lambda_2 y_3 \cos y_1 + B_2\lambda_2 \cos 2y_1 + C_2\lambda_3 \sin y_1$$

$$\dot{\lambda}_2 = -2A_2^* y_2 - \lambda_1 + A_1\lambda_2$$

$$\dot{\lambda}_3 = -2A_3^*(y_3 - y_{3p}) + A_2\lambda_2 \sin y_1 + C_1\lambda_3$$

$y_1(0) = 0.7437 \qquad \lambda_1(T) = 0$

$y_2(0) = -0.2151 \qquad \lambda_2(T) = 0$

$y_3(0) = 7.7443 \qquad \lambda_3(T) = 0$

The Jacobian $J(\underline{Z})$ of the system can be written as

$$\begin{bmatrix} 0 & 1 & 0 & 0 & 0 & 0 \\ \begin{bmatrix}-A_2y_3\cos y_1\\ -B_2\cos 2y_1\end{bmatrix} & -A_1 & -A_2\sin y_1 & 0 & 0 & 0 \\ -C_2\sin y_1 & 0 & -C_1 & 0 & 0 & \frac{1}{2A_M} \\ \begin{bmatrix}2A_1^*-A_2\lambda_2y_3\sin y_1\\ -B_2\lambda_2\sin 2y_1+C_2\lambda_3\cos y_1\end{bmatrix} & 0 & \begin{bmatrix}A_2\lambda_2\cos y_1\end{bmatrix} & 0 & \begin{bmatrix}A_2\lambda_3\cos y_1\\ +B_2\cos 2y_1\end{bmatrix} & C_2\sin y_1 \\ 0 & -2A_2^* & 0 & -1 & A_1 & 0 \\ A_2\lambda_2\cos y_1 & 0 & -2A_3^* & 0 & A_2\sin y_1 & C_1 \end{bmatrix}$$

where $\underline{Z}^T = \left[y_1 \quad y_2 \quad y_3 \quad \lambda_1 \quad \lambda_2 \quad \lambda_3\right]$

For this problem also, a computer program was written [8] and the problem was tackled on an IBM 370/165 digital computer. The convergence test was again 10^{-3}. In a first test the weighting coefficients used were the same as those of Mukhopadhyay [1], i.e.

$$A_1^* = 10 \qquad A_2^* = 1 \qquad A_3^* = 10 \qquad A_M = 0.5$$

The values of the states and control in the steady state were taken to be

$y_{1p} = 0.7461$ radians, $y_{2p} = 0$ radian/sec., $y_{3p} = 7.7438$, $M_p = 1.1$

The optimisation interval was taken to be 2 seconds. We used the values of Mukhopadhyay [1] for $A_1, A_2, \ldots$ etc., i.e.

$$A_1 = 0.27 \qquad B_1 = -4.804 \qquad A_2 = 12.01 \qquad C_1 = 0.32 \qquad B_1 = 39.18 \qquad C_2 = 1.9$$

The initial approximation to the state-costate trajectories was taken to be a constant at

$$y_1^{(o)}(t) = 0.7347 \qquad \lambda_1(t) = 0 \qquad t \in [\, 0,\ 2\,]$$

$$y_2^{(o)}(t) = -0.2151 \qquad \lambda_2(t) = 0$$

$$y_3^{(o)}(t) = 7.7443 \qquad \lambda_3(t) = 0$$

The problem converged in only 3 iterations giving results which are identical to those of Mukhopadhyay [1]. Fig. 7.8, 7.9, 7.10 and 7.11 give respectively the optimal states y_1, y_2, y_3 and the optimal control M.

In another test, the same cost function weights as those used in the example in section 7.2.2 were used to provide a numerical comparison with the previous method. Here, convergence to the optimum took place in 15 seconds on the IBM 370/165 digital computer and gave the same trajectories as previously.

To conclude our discussion of the quasilinearisation approach, let us summarise the important features of the method.

a) Computational requirements

As in the case of the gradient method, we split these into storage and computations required.

Storage requirements : It is only necessary to store the linearising state-costate trajectories, the specific value of $\underline{x}(t_o)$, the value of $\underline{C}$, $\underline{x}^P(t_f)$, $\underline{\lambda}^P(t_f)$ and $\underline{x}^j(t_f)$, $\underline{\lambda}^j(t_f)$, $j = 1, 2, \ldots, n$. The ith trajectory can be generated by reintegrating equation 7.3.5 with the initial conditions $\underline{x}^{(i)}(t_o) = \underline{x}_o$, $\underline{\lambda}^{(i)}(t_o) = \underline{C}$.

Computations required : It is necessary to integrate 2n(n+1) first order linear differential equations and invert an n x n matrix at each iteration. If, to save storage, we generate the ith trajectory by reintegration, it is necessary to integrate another 2n linear differential equations.

b) Importance of initial guess and convergence characteristics

In this method, the initial guess is rather more important than in the

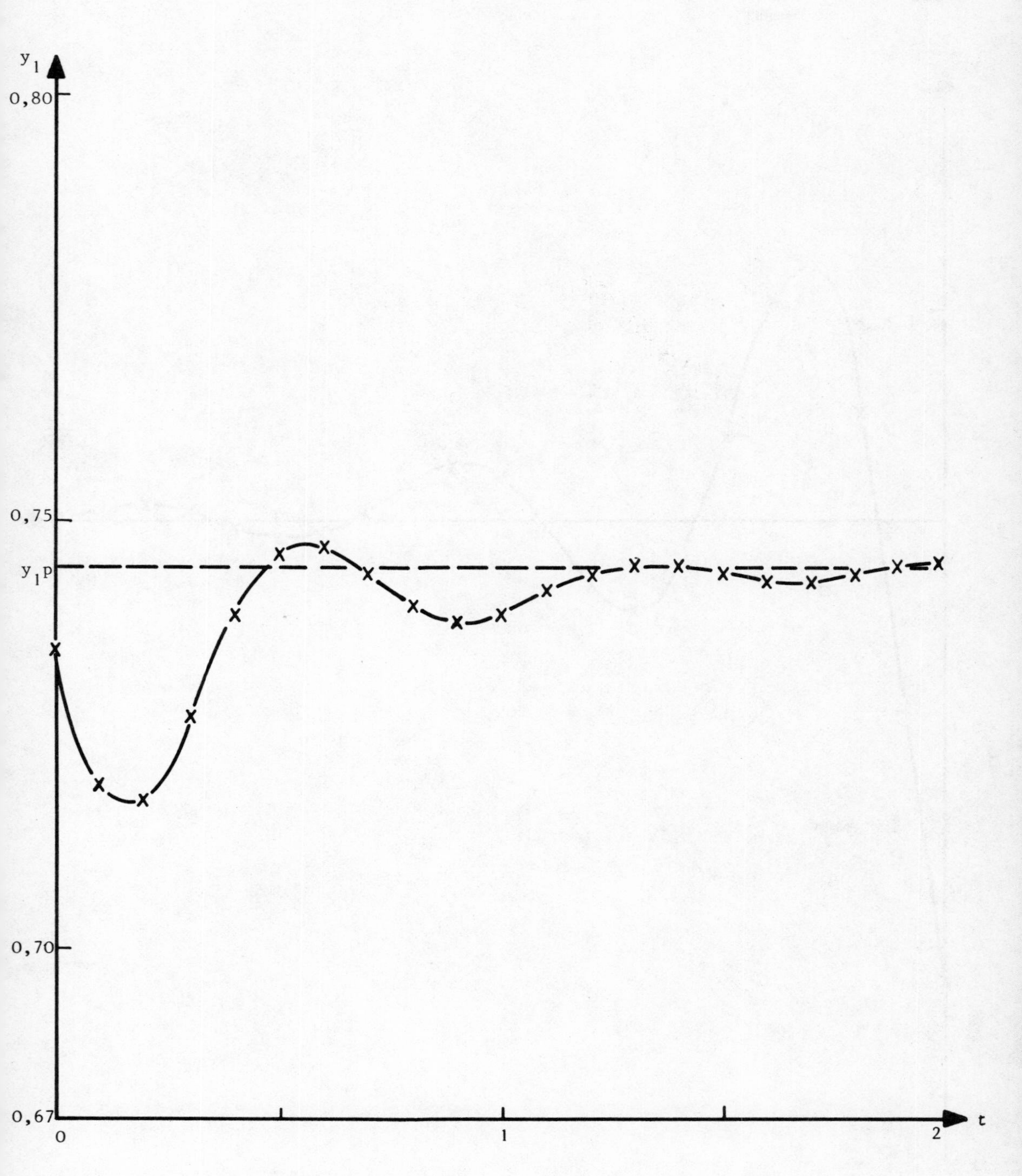

Fig. 7.8 : The optimal state y_1 for the synchronous machine

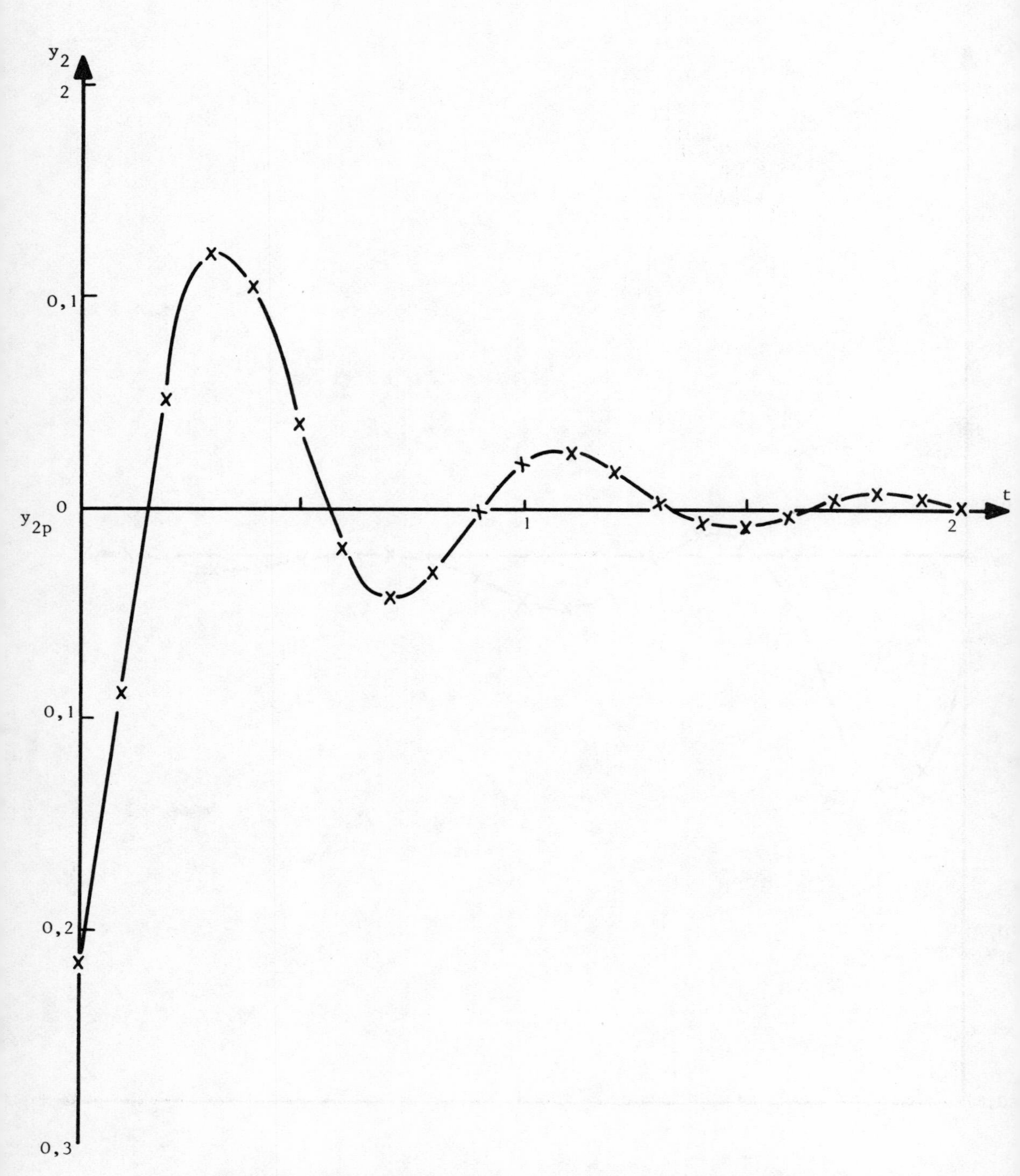

Fig. 7.9 : The optimal state y_2 for the synchronous machine

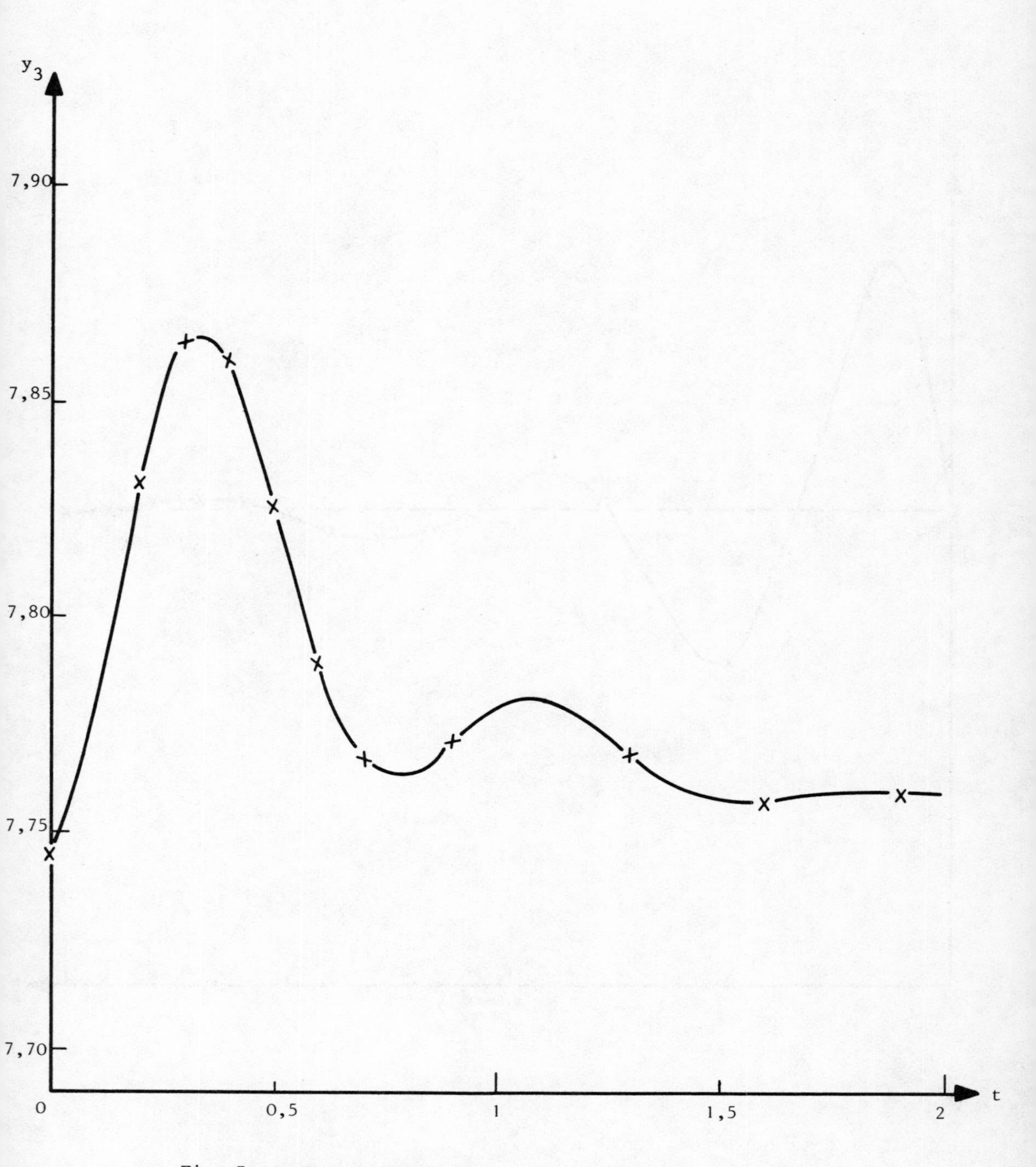

Fig. 7.10: The optimal state y_3 for the synchronous machine

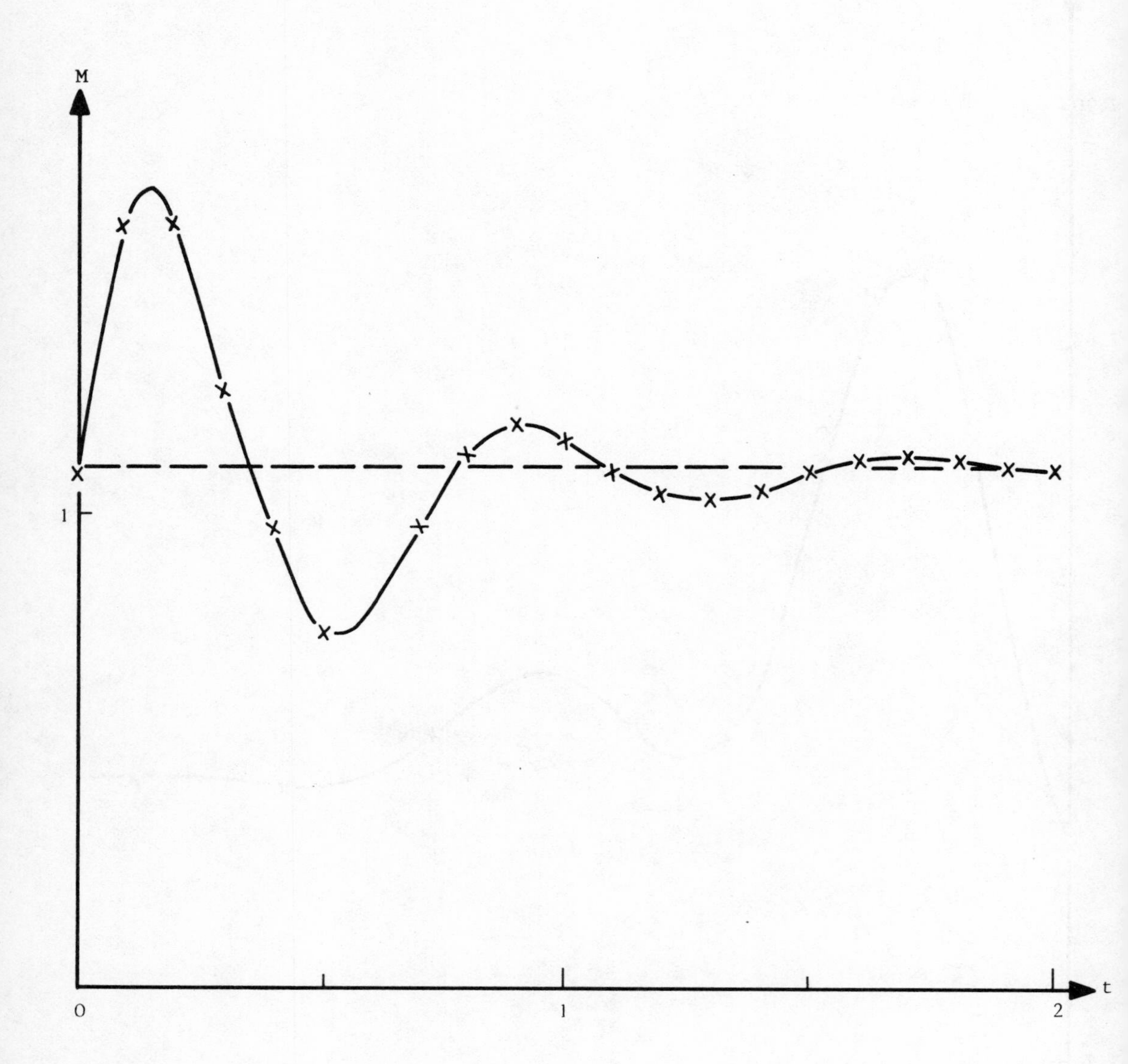

Fig. 7.11 : The optimal control for the synchronous machine

previous method. An initial state-costate trajectory $\begin{bmatrix} \underline{x}^{(o)}(t) \\ \underline{\lambda}^{(o)}(t) \end{bmatrix}; t \in [t_o, t_f]$ needs to be selected to linearise the non-linear reduced differential equations. This nominal trajectory need not necessarily satisfy the specified boundary conditions although the procedure ensures that all subsequent iterates will do so. If the nominal trajectory chosen is too far from the optimum, the algorithm may diverge. The initial guess is made on whatever physical information is available. Although much physical information may be available on the state trajectories, usually little information is available about the costate trajectories.

The convergence of this method has been extensively studied by Bellman and Kalaba [5], Mc Gill and Kenneth [3]. They have proved that the sequence of solutions to the linearised equations 7.3.7 converges quadratically to the solution of the reduced non-linear differential equations provided that :

1. The functions $\underline{a}$, $\frac{\partial H}{\partial \underline{x}}$ are continuous
2. The partial derivatives $\frac{\partial \underline{a}}{\partial \underline{x}}$, $\frac{\partial \underline{a}}{\partial \underline{\lambda}}$, $\frac{\partial^2 H}{\partial \underline{x}^2}$ and $\frac{\partial^2 H}{\partial \underline{x} \partial \underline{\lambda}}$ exist and are continuous.
3. The partial derivates in 2 above satisfy a Lipschitz condition w.r.t. $[\underline{x}(t)^T \vdots \underline{\lambda}^T(t)]^T$.
4. The norm of the initial guess from the solutions of the reduced non-linear differential equations is sufficiently small.

c) Termination criterion

We have seen that the quasilinearisation method produces a series of solutions of linear differential equations each one of which is "closer" to the solution of the original non-linear differential equation. Thus it is natural that to test whether the method has converged, we use a measure of the deviation of adjacent members of the sequence. For example, Mc Gill and Kenneth [3] use the norm

$$\gamma = \sum_{j=1}^{n} \left\{ \max_t \left| x_j^{i+1}(t) - x_j^{(i)}(t) \right| + \max_t \left| \lambda_j^{(i+1)}(t) - \lambda_j^{(i)}(t) \right| \right\}$$

where $x_j^{(i+1)}$ is the jth component of the state vector generated in the (i+1)st iteration. When two successive trajectories yield a value of γ which is smaller than some pre-selected positive number β, the iterations are stopped. It is prudent at this point to verify that the solution obtained indeed solves the original non-linear two point boundary value problem.

d) Modifications for fixed end point problems

Although the approach outlined in this section only deals explicitly with problems where the final states are free, it is easy to modify the problem to the case of fixed final states. For example, if $\underline{x}(t_f)$ is specified. Then in order to determine $\underline{C}$, we solve equation 7.3.36 with $t = t_f$.

If some of the components of $\underline{x}(t_f)$ are fixed and others free, we could select the appropriate equations among equations 7.3.36, let $t = t_f$ and solve for $\underline{C}$.

Next we will consider another important method for solving two point boundary value problems, i.e. the variation of extremals method.

7.4 THE VARIATION OF EXTREMALS METHODS

Let us reconsider our problem of minimising

$$J = \int_{t_o}^{t_f} g(\underline{x}, \underline{u}, t)\, dt \qquad 7.4.1$$

subject to the dynamical constraints

$$\dot{\underline{x}} = \underline{f}(\underline{x}, \underline{u}, t) \qquad 7.4.2$$

where the states and controls are not constrained by any boundaries, the final time t_f is fixed and the final state $\underline{x}(t_f)$ is free.

To write the necessary conditions for optimality, let us define as usual the Hamiltonian function

$$H = \underline{f}(\underline{x}, \underline{u}, t) + \underline{\lambda}^T(t)\, g(\underline{x}, \underline{u}, t) \qquad 7.4.3$$

so that the necessary conditions become

$$\dot{\underline{x}}(t) = \frac{\partial H}{\partial \underline{\lambda}} = \underline{f}(\underline{x}, \underline{u}, t) \qquad 7.4.4$$

$$\dot{\underline{\lambda}}(t) = -\frac{\partial H}{\partial \underline{x}} = -\left[\frac{\partial \underline{f}(\underline{x}, \underline{u}, t)}{\partial \underline{x}}\right]^T \underline{\lambda}(t) - \frac{\partial g}{\partial \underline{x}}(\underline{x}, \underline{u}, t) \qquad 7.4.5$$

$\frac{\partial H}{\partial \underline{u}} = \underline{0}$ or

$$\left[\frac{\partial \underline{f}(\underline{x}, \underline{u}, t)}{\partial \underline{u}}\right]^T \underline{\lambda}(t) + \frac{\partial g(\underline{x}, \underline{u}, t)}{\partial \underline{u}} = \underline{0} \qquad 7.4.6$$

$$\underline{x}(t_o) = \underline{x}_o \qquad 7.4.7$$

$$\underline{\lambda}(t_f) = \underline{0} \qquad 7.4.8$$

Now, if we can solve the equation 7.4.6 for the control $\underline{u}$ in terms of $\underline{x}$, $\underline{\lambda}$, and substitute this $\underline{u}$ into equation 7.4.4 and 7.4.5, we obtain reduced differential equations of the form

$$\left.\begin{aligned} \dot{\underline{x}}(t) &= \underline{a}(\underline{x}(t), \underline{\lambda}(t), t) \\ \dot{\underline{\lambda}}(t) &= \underline{d}(\underline{x}(t), \underline{\lambda}(t), t) \end{aligned}\right\} \qquad 7.4.9$$

Thus to solve our optimisation problem, it is necessary to solve the reduced non-linear differential equations 7.4.9 subject to the split boundary conditions given by equations 7.4.7 and 7.4.8. We recall that in the quasilinearisation method considered previously we linearised equations 7.4.9 so that in any intermediate iteration the non-linear differential equations 7.4.9 were not satisfied, but the initial and final conditions were satisfied. In the variation of extremals method, the reduced differential equations 7.4.9 are satisfied in every iteration. This is the reason why the method is called the variation of extremals method since each trajectory generated by the algorithm satisfies equation 7.4.9 and is hence an extremal. We vary the extremals from one iteration to the next by guessing either the terminal state $\underline{x}(t_f)$ or the initial costate $\underline{\lambda}(t_o)$. For example, if we guess $\underline{\lambda}(t_o)$, we can use $\underline{x}(t_o)$, $\underline{\lambda}(t_o)$ to integrate numerically the reduced state-costate equations 7.4.9 from t_o to t_f. If the resulting final costate $\underline{\lambda}(t_f) = \underline{0}$, obviously our guess of the initial co-state was correct and we have solved the problem. Otherwise, we must improve the guess of $\underline{\lambda}(t_o)$ in some way.

One possible way of improving the guess in the variation of extremals approach is to use Newton's method [4] for finding the roots of non-linear equations. The idea here is quite easy to understand in the scalar case. Basically since $\lambda(t_f)$ is some unknown non-linear function of $\lambda(t_o)$, we could use any given value of $\lambda(t_o)$ to compute the tangent to the unknown $\lambda(t_f)$ versus $\lambda(t_o)$ relationship and use as the next guess for $\lambda(t_o)$ the extrapolated point where the tangent intersects the $\lambda(t_f)$ axis (i.e. where $\lambda(t_f)$ is zero). Fig. 7.11 shows a typical relationship between $\lambda(t_f)$ and $\lambda(t_o)$.

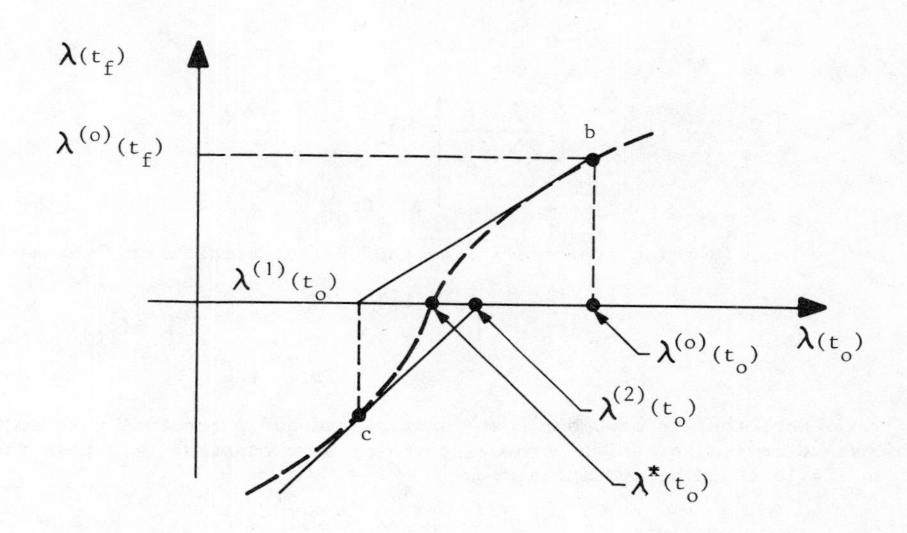

Fig. 7.11

Suppose we guess $\lambda^{(o)}(t_o)$ which corresponds to point b on the curve. Suppose we draw the tangent at b and extrapolate it to intersect the desired value of $\lambda(t_f) = 0$ ($\lambda^{(1)}(t_o)$ in the figure). Then $\lambda^{(1)}(t_o)$ is the new guess. This corresponds to point c on the $\lambda(t_f)$ vs $\lambda(t_o)$ curve. At c we determine the tangent and see that the tangent intersects the $\lambda(t_o)$ axis at $\lambda^{(2)}(t_o)$ and this then is the new guess in the next iteration. We continue the iterative procedure until the value of $\lambda^{(i)}(t_f)$ obtained is sufficiently close to zero to satisfy the specified termination criterion.

The slope of the curve at the point b which is needed to determine the equation of the tangent can be found approximately by perturbing the value of $\lambda^{(o)}(t_o)$ and evaluating the perturbed value of the final costate $\lambda^{(o)}(t_f) + \delta\lambda^{(o)}(t_f)$ by integration of the state-costate equations

i.e. $$\lambda(t_f) = \beta\,\lambda(t_o) + \alpha \qquad 7.4.10$$

where β is the slope at b and it is given by

$$\beta = \left.\frac{d\lambda(t_f)}{d\lambda(t_o)}\right|_{\lambda^{(o)}(t_o)} \simeq \frac{\delta\lambda^{(o)}(t_f)}{\delta\lambda^{(o)}(t_o)} \qquad 7.4.11$$

and α , the intercept on the $\lambda(t_f)$ axis, is given by

$$\alpha = \lambda^{(o)}(t_f) - \beta\,\lambda^{(o)}(t_o) \qquad 7.4.12$$

so that

$$\lambda(t_f) = \beta\,\lambda(t_o) + \left[\lambda^{(o)}(t_f) - \beta\,\lambda^{(o)}(t_o)\right] \qquad 7.4.13$$

Thus the new value $\lambda^{(1)}(t_o)$ is given by

$$\lambda^{(1)}(t_o) = \lambda^{(o)}(t_o) - \left[\left.\frac{d\lambda(t_f)}{d\lambda(t_o)}\right|_{\lambda^{(o)}(t_o)}\right]^{-1}\lambda^{(o)}(t_f) \qquad 7.4.14$$

Thus, in going from the ith to the (i+1)st trial value, we use the equation

$$\lambda^{(i+1)}(t_o) = \lambda^{(i)}(t_o) - \left[\left.\frac{d\lambda(t_f)}{d\lambda(t_o)}\right|_{\lambda^{(i)}(t_o)}\right]^{-1}\lambda^{(i)}(t_f) \qquad 7.4.15$$

Note that in case our cost function had had a terminal cost so that the final costate instead of being zero was given by some constant λ_f, then the above equation 7.4.15 could be rewritten as

$$\lambda^{i+1}(t_o) = \lambda^{(i)}(t_o) - \left[\left.\frac{d\lambda(t_f)}{d\lambda(t_o)}\right|_{\lambda^{(i)}(t_o)}\right]^{-1}\left[\lambda^{(i)}(t_f) - \lambda_f\right] \qquad 7.4.16$$

The equation 7.4.15 is of course only for the scalar case. For the general case of n state equations and n costate equations, the matrix generalisation of equation 7.4.15 is

$$\underline{\lambda}^{(i+1)}(t_o) = \underline{\lambda}^{(i)}(t_o) - \left[P_\lambda(\underline{\lambda}^{(i)}(t_o), t_f)\right]^{-1}\underline{\lambda}^{(i)}(t_f) \qquad 7.4.17$$

where $P_\lambda(\underline{\lambda}^{(i)}(t_o), t)$ is the n x n matrix of partial derivatives of the components of $\underline{\lambda}(t)$, w.r.t. each of the components of $\underline{\lambda}(t_o)$ evaluated at $\underline{\lambda}^{(i)}(t_o)$, i.e

$$P_\lambda(\underline{\lambda}^{(i)}(t_o), t) = \begin{bmatrix} \frac{\partial\lambda_1(t)}{\partial\lambda_1(t_o)} & \frac{\partial\lambda_1(t)}{\partial\lambda_2(t_o)} & \cdots & \frac{\partial\lambda_1(t)}{\partial\lambda_n(t_o)} \\ \vdots & \vdots & & \\ \frac{\partial\lambda_n(t)}{\partial\lambda_1(t_o)} & \frac{\partial\lambda_n(t)}{\partial\lambda_2(t_o)} & \cdots & \frac{\partial\lambda_n(t)}{\partial\lambda_n(t_o)} \end{bmatrix}_{\underline{\lambda}^{(i)}(t_o)} \qquad 7.4.18$$

P_λ is called the costate influence function matrix since it indicates the influence of changes in the initial costate on the costate trajectory at time t. It should be noted that in equation 7.4.17, we need to know P_λ only at the terminal time t_f.

Note also that equation 7.4.18 is appropriate only for the case where there is no termination time weighting term in the cost function as in equation 7.4.1. If we were to include a termination time term, i.e. if the cost function became

$$J = h(\underline{x}(t_f)) + \int_{t_o}^{t_f} g(\underline{x}, \underline{u}, t)\, dt \qquad 7.4.19$$

then it is left as an exercise for the reader to show that the initial costate $\underline{\lambda}(t_o)$ should be improved using the relation (cf. problem 8)

$$\underline{\lambda}^{(i+1)}(t_o) = \underline{\lambda}^{(i)}(t_o) + \left\{\left[\left[\frac{\partial^2 h}{\partial \underline{x}^2}(\underline{x}(t_f))\right] P_x(\underline{\lambda}(t_o), t_f) - P_\lambda(\lambda(t_o), t_f)\right]_i\right\}^{-1} \cdot \left[\underline{\lambda}(t_f) - \frac{\partial h}{\partial \underline{x}}(\underline{x}(t_f))\right]_i \qquad 7.4.20$$

$[\cdot]_i$ implies that the enclosed terms are evaluated on the ith trajectories and $P_x(\lambda^{(i)}(t_o), t_f)$ is the n x n state influence function matrix

$$P_x(\underline{\lambda}^{(i)}(t_o), t) = \begin{bmatrix} \frac{\partial x_1(t)}{\partial \lambda_1(t_o)} & \frac{\partial x_1(t)}{\partial \lambda_2(t_o)} & \cdots & \frac{\partial x_1(t)}{\partial \lambda_n(t_o)} \\ \vdots & & & \\ \frac{\partial x_n(t)}{\partial \lambda_1(t_o)} & \frac{\partial x_n(t)}{\partial \lambda_2(t_o)} & \cdots & \frac{\partial x_n(t)}{\partial \lambda_n(t_o)} \end{bmatrix}_{\underline{\lambda}^{(i)}(t_o)} \qquad 7.4.21$$

$\frac{\partial^2 h}{\partial \underline{x}^2}(x(t_f))$ is the matrix whose jk^{th} element is

$$\left[\frac{\partial^2 h}{\partial \underline{x}^2}(\underline{x}(t_f))\right]_{jk} = \frac{\partial^2 h}{\partial x_j \partial x_k}(\underline{x}(t_f)) \qquad 7.4.22$$

Note that if h is zero, equation 7.4.20 reduces to equation 7.4.18.

7.4.1 Determination of influence function matrices

We have seen that the most important step in the variation of extremals method is the computation of the influence function matrices P_x and P_λ. Next we will describe an easy way of computing these on a digital computer. Basically, we will develop a set of differential equations, the integration of which will yield P_x and P_λ .

Let us assume that $\frac{\partial H}{\partial \underline{u}}$ has been solved for $\underline{u}(t)$ and used to obtain the reduced state and costate differential equations

$$\dot{\underline{x}}(t) = \frac{\partial H}{\partial \underline{\lambda}} (\underline{x}(t), \underline{\lambda}(t), t)$$
$$\dot{\underline{\lambda}}(t) = - \frac{\partial H}{\partial \underline{x}} (\underline{x}(t), \underline{\lambda}(t), t) \qquad 7.4.23$$

Taking the partial derivatives of these equations, w.r.t. the initial value of the costate vector, we obtain

$$\frac{\partial}{\partial \underline{\lambda}(t_o)} [\dot{\underline{x}}(t)] = \frac{\partial}{\partial \underline{\lambda}(t_o)} \left[\frac{\partial H}{\partial \underline{\lambda}} (\underline{x}(t), \underline{\lambda}(t), t) \right]$$
$$\frac{\partial}{\partial \underline{\lambda}(t_o)} [\dot{\underline{\lambda}}(t)] = \frac{\partial}{\partial \underline{\lambda}(t_o)} \left[-\frac{\partial H}{\partial \underline{x}} (\underline{x}(t), \underline{\lambda}(t), t) \right] \qquad 7.4.24$$

Now, if we assume that $\partial \left[\frac{d\underline{x}}{dt} \right] / \partial \underline{\lambda} \;(t_o)$ and $\partial \left[\frac{d\underline{\lambda}}{dt} \right] / \;\partial \underline{\lambda}(t_o)$ are continuous w.r.t. $\underline{\lambda}\,(t_o)$ and t, the order of differentiation can be interchanged on the left hand side of equation 7.4.24 so that if we do this and use the chain rule of differentiation on the R.H.S., we obtain

$$\frac{d}{dt}\left[\frac{\partial \underline{x}(t)}{\partial \underline{\lambda}(t_o)}\right] = \left[\frac{\partial^2 H}{\partial \underline{\lambda} \partial \underline{x}} (\underline{x}(t), \underline{\lambda}(t), t)\right] \frac{\partial \underline{x}(t)}{\partial \underline{\lambda}(t_o)} + \left[\frac{\partial^2 H}{\partial \underline{\lambda}^2} (\underline{x}(t), \underline{\lambda}(t), t)\right] \frac{\partial \underline{\lambda}(t)}{\partial \underline{\lambda}(t_o)}$$

$$\frac{d}{dt}\left[\frac{\partial \underline{\lambda}(t)}{\partial \underline{\lambda}(t_o)}\right] = \left[-\frac{\partial^2 H}{\partial \underline{x}^2} (\underline{x}(t), \underline{\lambda}(t), t)\right] \frac{\partial \underline{x}(t)}{\partial \underline{\lambda}(t_o)} - \left[\frac{\partial^2 H}{\partial \underline{x} \partial \underline{\lambda}} (\underline{x}(t), \underline{\lambda}(t), t)\right] \frac{\partial \underline{\lambda}(t)}{\partial \underline{\lambda}(t_o)} \qquad 7.4.25$$

where the above partial derivatives are n x n matrices whose jk^{th} elements are given by

$$\left[\frac{\partial^2 H}{\partial \underline{\lambda} \partial \underline{x}}\right] = \frac{\partial^2 H}{\partial \lambda_j \partial x_k} \; ; \left[\frac{\partial \underline{x}(t)}{\partial \underline{\lambda}(t_o)}\right]_{jk} = \frac{\partial x_j(t)}{\partial \lambda_k(t_o)} \qquad \text{etc.}$$

Now if we use the definitions of the influence function matrices P_x and P_λ from equations 7.4.21 and 7.4.18, we obtain the required differential equations i.e.

$$\frac{d}{dt}\left[P_x(\underline{\lambda}^{(i)}(t_o), t)\right] = \left[\frac{\partial^2 H}{\partial \underline{\lambda} \partial \underline{x}}(t)\right]_i P_x(\underline{\lambda}^{(i)}(t_o), t) + \left[\frac{\partial^2 H}{\partial \underline{\lambda}^2}(t)\right]_i P_\lambda(\underline{\lambda}^{(i)}(t_o), t)$$

$$\frac{d}{dt}\left[P_\lambda(\underline{\lambda}^{(i)}(t_o), t)\right] = \left[-\frac{\partial^2 H}{\partial \underline{x}^2}(t)\right]_i P_x(\underline{\lambda}^{(i)}(t_o), t) + \left[-\frac{\partial^2 H}{\partial \underline{x} \partial \underline{\lambda}}(t)\right]_i P_\lambda(\underline{\lambda}^{(i)}(t_o), t) \qquad 7.4.26$$

where as usual $[\cdot]_i$ indicates that the enclosed matrices are evaluated on the trajectory $\underline{x}^{(i)}$, $\underline{\lambda}^{(i)}$ obtained by integrating the reduced state-costate equations with initial conditions $\underline{x}(t_o) = \underline{x}_o$, $\underline{\lambda}(t_o) = \underline{\lambda}^{(i)}(t_o)$.

Thus to evaluate the influence function matrices P_x, P_λ, we can integrate the set of $2n^2$ first order differential equations 7.4.26 using the initial conditions

$$P_x(\underline{\lambda}^{(i)}(t_o), t_o) = \left.\frac{\partial \underline{x}(t_o)}{\partial \underline{\lambda}(t_o)}\right|_{\underline{\lambda}^{(i)}(t_o)} = 0 \quad \text{(the n x n zero matrix)} \qquad 7.4.27$$

$$P_\lambda(\underline{\lambda}^{(i)}(t_o), t_o) = \left.\frac{\partial \underline{\lambda}(t_o)}{\partial \underline{\lambda}(t_o)}\right|_{\underline{\lambda}^{(i)}(t_o)} = I \quad \text{(the n x n identity matrix)} \qquad 7.4.28$$

These initial conditions 7.4.27, 7.4.28 on the P_x, P_λ equations arise from the facts that (a) a change in any of the components of $\underline{\lambda}(t_o)$ does not affect the value of $\underline{x}(t_o)$ since $\underline{x}(t_o)$ is specified, (b) a change in the jth component of $\underline{\lambda}(t_o)$ changes only $\lambda_j(t_o)$. Hence equation 7.4.28.

We are now in a position to write down the precise steps necessary in solving practical problems using the variation of extremals method.

7.4.2 The variation of extremals algorithm

Step 1

Form the reduced differential equations by solving $\frac{\partial H}{\partial \underline{u}} = \underline{0}$ for $\underline{u}(t)$ in terms of $\underline{x}(t)$, $\underline{\lambda}(t)$ and substituting in the state-costate equations to obtain the equivalent of equations 7.4.23 which only contains terms in $\underline{x}(t)$, $\underline{\lambda}(t)$, t.

Step 2

Guess the initial value of the costate vector $\underline{\lambda}^{(o)}(t_o)$ and set the iteration index i to zero.

Step 3

Using $\underline{\lambda}(t_o) = \underline{\lambda}^{(i)}(t_o)$ and $\underline{x}(t_o) = \underline{x}_o$ as initial conditions, integrate the reduced state-costate equations and the influence function equations 7.4.26 (with the initial conditions given by equation 7.4.27) from t_o to t_f. Store only the values $\underline{\lambda}^{(i)}(t_f)$, $\underline{x}^{(i)}(t_f)$ and the n x n matrices $P_\lambda(\underline{\lambda}^{(i)}(t_o), t_f)$, $P_x(\underline{\lambda}^{(i)}(t_o), t_f)$.

Step 4

Check to see if the termination criterion $\|\underline{\lambda}^{(i)}(t_f) - \partial h(\underline{x}^{(i)}(t_f))/\partial \underline{x}\| < \gamma$ is satisfied where γ is a small prechosen positive constant. If it is, use the final iterate $\underline{\lambda}^*(t_o)$ to reintegrate the state and costate equations and print out the optimal states and controls. Otherwise find new value for $\underline{\lambda}(t_o)$, i.e $\underline{\lambda}^{(i+1)}(t_o)$ using equation 7.4.20. Set i to i+1 and return to step 3.

It should be noted that steps 1 and 2 are performed off-line by the user whilst steps 3 and 4 are performed on a digital computer.

Next, let us illustrate the variation of extremals method on our synchronous machine example.

The synchronous machine example

We recall that our state equations were

$$\dot{y}_1 = y_2$$
$$\dot{y}_2 = B_1 - A_1 y_2 - A_2 y_3 \sin y_1 - \frac{B_2}{2} \sin 2y_1$$
$$\dot{y}_3 = u - C_1 y_3 + C_2 \cos y_1$$

where y_1 is the rotor angle (in radians)

y_2 is the speed (rad./ sec.)

y_3 is the flux linkage.

We would like to minimise

$$J = \frac{1}{2} \int_0^2 (Q_1(y_1 - y_{1p})^2 + Q_2(y_2 - y_{2p})^2 + Q_3(y_3 - y_{3p})^2 + R(u - u_p)^2) \, dt$$

The Hamiltonian is as before

$$H = \frac{1}{2} \left\{ Q_1(y_1 - y_{1p})^2 + Q_2(y_2 - y_{2p})^2 + Q_3(y_3 - y_{3p})^2 + R(u - u_p)^2 \right\}$$
$$+ \lambda_1 y_2 + \lambda_2 \left[B_1 - A_1 y_2 - A_2 y_3 \sin y_1 - \frac{B_2}{2} \sin 2y_1 \right] + \lambda_3 \left[u - C_1 y_3 + C_2 \cos y_1 \right]$$

The costate equations become

$$\frac{\partial H}{\partial y_1} = -\dot{\lambda}_1 = \lambda_2 \left[- A_2 y_3 \cos y_1 - B_2 \cos 2y_1 \right] - \lambda_3 C_2 \sin y_1 + Q_1(y_1 - y_{1p})$$

$$\frac{\partial H}{\partial y_2} = -\dot{\lambda}_2 = \lambda_1 - \lambda_2 A_1 + Q_2(y_2 - y_{2p}) \qquad 7.4.29$$

$$\frac{\partial H}{\partial y_3} = -\dot{\lambda}_3 = -\lambda_2 A_2 \sin y_1 - \lambda_3 C_1 + Q_3(y_3 - y_{3p})$$

The optimal control is given by

$$\frac{\partial H}{\partial u} = 0 = \lambda_3 + R(u - u_p) \text{ or } u = u_p - \frac{\lambda_3}{R} \qquad 7.4.30$$

Using this expression in the state equations, we obtain the reduced state equations

$$\dot{y}_1 = y_2$$
$$\dot{y}_2 = B_1 - Ay_2 - A_2 y_3 \sin y_1 - \frac{B_2}{2} \sin 2y_1 \qquad 7.4.31$$
$$\dot{y}_3 = u_p - \frac{\lambda_3}{R} - C_1 y_3 + C_2 \cos y_1$$

and the reduced Hamiltonian as

$$H = \lambda_1 y_2 + \lambda_2(B_1 - A_1 y_2 - A_2 y_3 \sin y_1 - \frac{B_2}{2} \sin 2y_1) + \lambda_3(u_p - \frac{\lambda_3}{R} - C_1 y_3 + C_2 \cos y_1)$$

$$+ \frac{1}{2} \left[(y_1 - y_{1p})^2 Q + Q_2(y_2 - y_{2p})^2 + Q_3(y_3 - y_{3p})^2 - \frac{\lambda_3^2}{R} \right]$$

Next let us write down the influence function equations in the form

$$\frac{d}{dt} [P_x] = \left[\frac{\partial^2 H}{\partial \underline{\lambda} \partial \underline{y}}\right] P_x + \left[\frac{\partial^2 H}{\partial \underline{\lambda}^2}\right] P_\lambda$$

$$\frac{d}{dt} [P_\lambda] = -\left[\frac{\partial^2 H}{\partial \underline{y}^2}\right] P_x - \left[\frac{\partial^2 H}{\partial \underline{y} \partial \underline{\lambda}}\right] P_\lambda$$

i.e.

$$\frac{d}{dt}[P_x] = \begin{bmatrix} 0 & 1 & 0 \\ [-A_2 y_3 \cos y_1 - B_2 \cos 2y_1] & -A_1 & -A_2 \sin y_1 \\ -C_2 \sin y_1 & 0 & -C_1 \end{bmatrix} P_x + \begin{bmatrix} 0 & 0 & 0 \\ 0 & 0 & 0 \\ 0 & 0 & -\frac{1}{R} \end{bmatrix} P_x \qquad 7.4.32$$

and

$$\frac{d}{dt}[P_\lambda] = -\begin{bmatrix} [\lambda_2(A_2 y_3 \sin y_1 + 2B_2 \sin 2y_1 \quad \lambda_3 C_2 \cos y_1 + Q_1] & 0 & [-\lambda_2 A_2 \cos y_1] \\ 0 & Q_2 & 0 \\ -\lambda_2 A_2 \cos y_1 & 0 & Q_3 \end{bmatrix} . P_x$$

$$- \begin{bmatrix} 0 & [-A_2 y_3 \cos y_1 - B_2 \cos 2y_1] & -C_2 \sin y_1 \\ 1 & -A_1 & 0 \\ 0 & -A_2 \sin y_1 & 0 \end{bmatrix} P_\lambda \qquad 7.4.33$$

It is thus necessary to integrate these $2n+2n^2 = 24$ (for $n = 3$) equations using $P_x = 0$; $P_p = I$. Note that we have taken

$$\underline{y}(t_o) = \begin{bmatrix} y_1(t_o) \\ y_2(t_o) \\ y_3(t_o) \end{bmatrix} = \begin{bmatrix} 0.7347 \\ -0.215 \\ 7.7443 \end{bmatrix}$$

Thus starting from an initial guessed $\underline{\lambda}(t_o)$, we compute $\underline{\lambda}(t_f)$ (by integrating the state-costate equations) and P_λ (by integrating equations 7.4.33). If $\underline{\lambda}(t_f) \simeq \underline{0}$ we stop and record the optimal control. Otherwise we improve $\underline{\lambda}(t_o)$ from iteration i to i+1 using the formula

$$\underline{\lambda}^{(i)}(t_o) = \underline{\lambda}^{(i-1)}(t_o) - \left[P_\lambda(t_f)\right]^{-1}_{(i-1)} \underline{\lambda}^{(i-1)}(t_f) \qquad 7.4.34$$

Thus the problem could be solved using the flow chart 7.2.

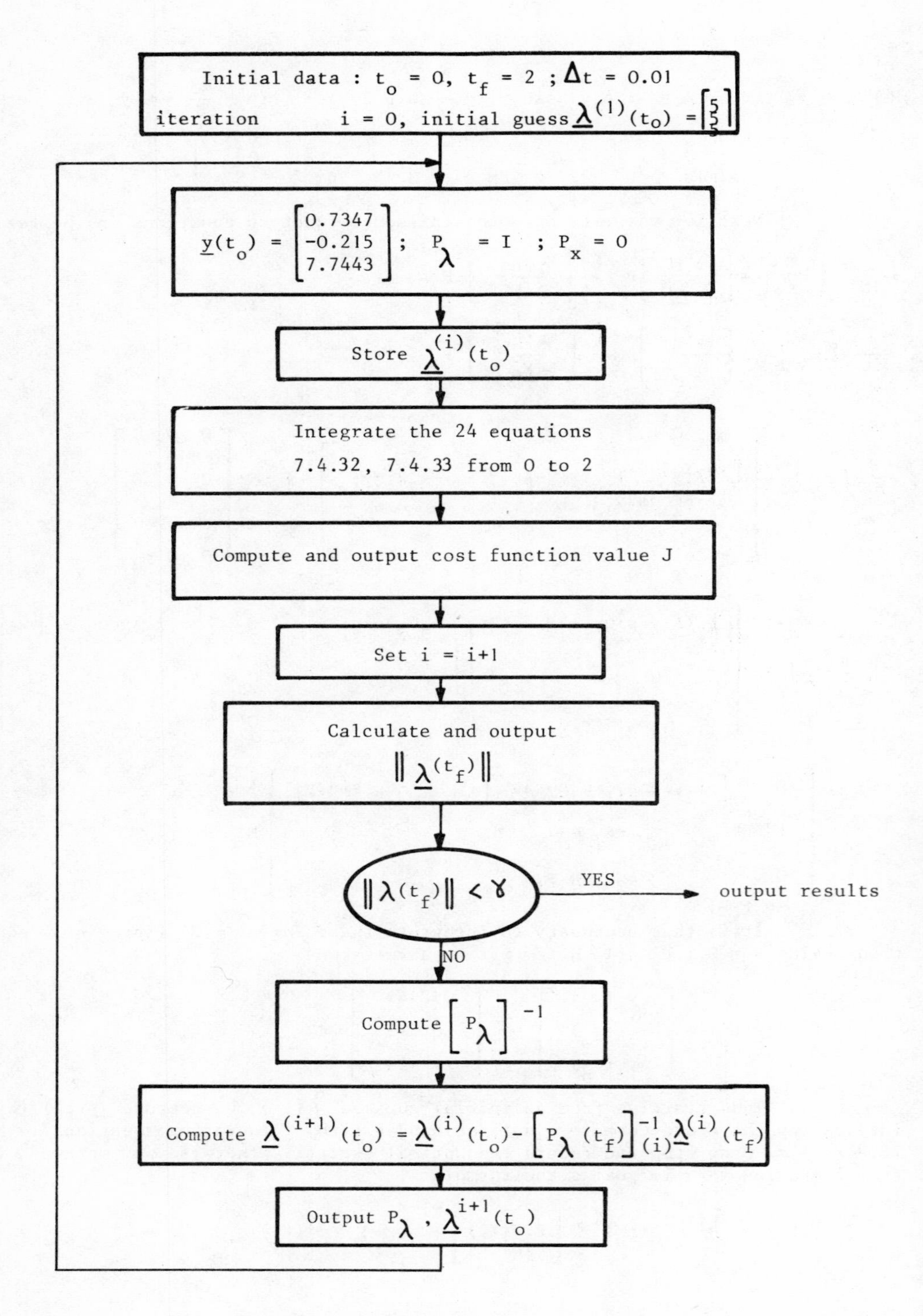

Flow chart 7.2

This calculation was performed on the IBM 370/165 digital computer. The optimal trajectories obtained were identical to those obtained previously. The minimal cost was found to be J = 0.01797.

We found that the results obtained were highly sensitive to the length of the integration step. Indeed, the smaller the step length, the more precise the final result. However, it is not possible to reduce the step length infinitely because first of all truncation errors eventually become significant even in double precision and calculation time increases proportionally to the reduction of step length.

Let us next write down the important features of this method.

a) Computational requirements

As usual we split this up into the storage and computation time requirements.

Storage : The storage requirements of this method are quite small as we saw in our flow chart 7.2 since it is not necessary to store trajectories. Only the values of the influence function matrices at $t = t_f$, the value of $\underline{\lambda}^{(i)}(t_o)$ the given initial state value and the appropriate desired boundary conditions are retained in the computer memory.

Computations required : It is necessary to integrate 2n(n+1) first order differential equations numerically and to integrate an n x n matrix at each iteration.

b) Importance of initial guess and convergence characteristics

As in the quasilinearisation method, the initial guess is important. If the initial guess (in this case of $\underline{\lambda}^{(i)}(t_o)$) is sufficiently close to the optimal $\underline{\lambda}(t_o)$, the method converges extremely rapidly. However, if the initial guess for $\underline{\lambda}(t_o)$ is poor, the method may not converge at all.

c) Stopping criterion

The iterative procedure is terminated when

$$\left\| \underline{\lambda}^{(i)}(t_f) - \frac{\partial h}{\partial \underline{x}}(\underline{x}(t_f)) \right\| < \gamma$$

where γ is a preselected small positive number.

7.5 COMPARISON OF THE THREE METHODS

So far in this chapter we have studied in detail three important methods for solving non-linear dynamical optimisation problems and we are now in a position to compare and contrast their strengths and weaknesses.

In all the three methods, it is necessary to satisfy the necessary conditions for optimality iteratively. All the methods require, at least in principle, an infinite number of iterations, although in practice iterations are terminated when certain norms become smaller than a certain predefined number. Thus in the gradient method, we obtain convergence when $\frac{\partial H}{\partial \underline{u}} \simeq \underline{0}$ over the entire time interval, the other optimality conditions having already been satisfied throughout the iterative procedure. In the quasilinearisation method, we obtain trajectories which satisfy the boundary conditions and when from among the successive linearised trajectories we find a trajectory which also satisfies the reduced state-costate equations to an acceptable degree, we have solved our problem. In the variation of extremals method, convergence is obtained when the boundary condition

$$\underline{\lambda}(t_f) \simeq \partial h(\underline{x}(t_f)) \;/\; \partial \underline{x}$$

is satisfied.

After the description of each method, we summarised the principle characteristics of each method in terms of the (a) computational requirements (storage and computations required), (b) importance of initial guess and convergence characteristics, and (c) termination criterion, and table 7.2 summarises these aspects of the three methods.

Characteristic	Gradient	Quasilinearisation	Variation of extremals
Computations required	Integration of 2n differential equations, calculation of $\frac{\partial H}{\partial \underline{u}}$, step length τ.	Integration of 2n(n+1) first order differential equations. Inversion of an n x n matrix	Integration of 2n(n+1) first order differential equations. Inversion of an n x n matrix
Storage requirements	$\underline{u}^{(i)}(t)$, $\underline{x}^{(i)}(t)$ and $\frac{\partial H}{\partial \underline{u}}^{(i)}(t), t \in [t_o, t_f]$	$\underline{x}^{(i)}(t)$, $\underline{\lambda}^{(i)}(t)$ $t \in [t_o, t_f]$. n x n matrix, boundary conditions, integration constants $\underline{C}$.	2 [n x n] matrices boundary conditions.
Must provide initial guess of	$\underline{u}(t)$ $t \in [t_o, t_f]$	$\underline{\lambda}(t_o)$ or $\underline{x}(t_f)$	$\underline{x}(t)$, $\underline{\lambda}(t)$ $t \in (t_o, t_f)$
Termination criterion	$\frac{\partial H}{\partial \underline{u}} \simeq \underline{0}$	$\underline{\lambda}(t_f) \simeq \frac{\partial h}{\partial \underline{x}}(\underline{x}(t_f))$	$\underline{\lambda}^{(i)}$, $\underline{x}^{(i)}$ satisfy reduced state-costate equations
Importance of good initial guess for convergence	Not important	Very important - Divergence may occur if guess is poor	Very important - Divergence may occur if guess is poor.
Convergence characteristics	Very fast initially and then very slow near the optimum	Quadratic near the optimum.	Fast if it converges.

Table 7.2

Having summarised the characteristics of the various methods, let us examine how we would go about choosing a method for a particular problem.

The gradient method requires the least amount of calculation per iteration although its storage requirements are quite comparable with those of the quasilinearisation method. Its main strength is the ease of starting since virtually any initial guess will do. Its main drawback is poor convergence characteristics near the optimum. The quasilinearisation method requires the most computation and storage but this is offset by the fact that convergence is very fast. However, the method is critically dependent on the initial guess for $\underline{\lambda}^{(o)}(t)$, $\underline{x}^{(o)}(t)$, $t \in [t_o, t_f]$ and there is no easy way of providing a good initial guess. The variation of extremals method requires a similar quantity of computation per iteration as for the quasilinearisation method although its storage requirements are virtually negligible compared to the other 2 methods since here 2 $[n \times n]$ matrices are stored plus the boundary conditions whereas the other methods require storage of entire trajectories.

In general, because of the difficulty of starting the quasilinearisation or variation of extremals procedure, it is often more attractive to use the gradient method to begin with and then one or the other of the 2 methods as we approach the optimum.

It must be emphasised that all the three iterative methods that we have so far outlined for solving non-linear dynamic optimisation problems essentially ensure the satisfaction of the necessary conditions for optimality. Thus we are assured of obtaining a local minimum. It would therefore be desirable in any practical case to resolve the problem starting from different initial guesses to see if we obtain convergence to a better local optimum.

Finally, let us study an entirely different approach to solving the non-linear two point boundary value problem, i.e. the invariant imbedding approach.

7.6 THE INVARIANT IMBEDDING METHOD

We have seen that the difficulty in solving non-linear two point boundary value problems arises due to the iterative nature of the solution procedure. However, we saw in chapter 5 that for the case of linear two point boundary value problems, it is possible to obtain a solution non-iteratively since we can convert our problem to a single point boundary value problem and integrate an equation of the Riccati type. In this section, we describe the invariant imbedding procedure which allows solution of non-linear two point boundary value problems directly. In this method, the original two point boundary value problem is imbedded in a class of more general single point boundary value problems. The terminal conditions are held invariant and thus solving the more general problem ultimately provides a solution for the original two point boundary value problem.

Let us suppose that the reduced differential equations obtained after substituting for $\frac{\partial H}{\partial \underline{u}} = \underline{0}$ in the necessary conditions for optimality are (as in equation 7.31)

$$\dot{\underline{x}}(t) = \underline{a}(\underline{x}(t), \underline{\lambda}(t), t)$$
$$\dot{\underline{\lambda}}(t) = \underline{d}(\underline{x}(t), \underline{\lambda}(t), t) \qquad 7.6.1$$

Let us assume that the boundary conditions for the equation 7.6.1 are given by

$$\underline{x}(t_o) = \underline{b} \qquad \underline{\lambda}(t_f) = \underline{c} \qquad 7.6.2$$

One way of solving this problem non-iteratively is to solve the problem as an initial value problem by finding the initial condition vector $\underline{\lambda}(t_o)$. Now $\underline{\lambda}(t_o)$ is clearly a function of $\underline{x}(t_o)$ and t_o. So we could write

$$\underline{x}(t_o) = \underline{D} \qquad 7.6.3$$

$$\underline{\lambda}(t_o) = \underline{r}(\underline{D}, t_o) \qquad 7.6.4$$

In other words, we imbed the given condition $\underline{x}(t_o) = \underline{b}$ in the more general condition $\underline{x}(t_o) = \underline{D}$ which includes the former. We are letting $\underline{r}(\underline{D}, t_o)$ denote the missing condition on $\underline{\lambda}(t)$ for the process which starts at t_o and ends at t_f and also satisfies $\underline{\lambda}(t_f) = \underline{c}$, $\underline{x}(t_o) = \underline{D}$ with $\underline{D}$ and t_f being regarded as independent variables. Let us consider a perturbation Δt_o to obtain

$$\underline{\lambda}(t_o + \Delta t_o) = \underline{\lambda}(t_o) + \dot{\underline{\lambda}}(t_o)\,\Delta t_o + \underline{O}(\Delta^2) \qquad 7.6.5$$

where $O(\Delta^2)$ are higher order terms such that

$$\lim_{\Delta t_o \to 0} \left[\underline{O}(\Delta^2)\right] / \Delta t_o = \underline{0}$$

Using equation 7.6.4, this becomes

$$\underline{r}(\underline{D} + \Delta\underline{D},\ t_o + \Delta t_o) = \underline{r}(\underline{D}, t_o) + \underline{d}(\underline{D}, \underline{r}, t_o)\Delta t_o + \underline{O}(\Delta^2) \qquad 7.6.6$$

Let us now expand the left hand side of equation 7.6.6 to obtain

$$\underline{r}(\underline{D} + \Delta\underline{D},\ t_o + \Delta t_o) = \underline{r}(\underline{D}, t_o) + \frac{\partial \underline{r}}{\partial \underline{D}}\Delta\underline{D} + \frac{\partial \underline{r}}{\partial t_o}\Delta t_o + \underline{O}(\Delta^2) \qquad 7.6.7$$

Equating equations 7.6.6 and 7.6.7 and noting that

$$\Delta\underline{D} = \underline{a}(\underline{D}, \underline{r}, t_o)\Delta t_o + \underline{O}(\Delta^2)$$

we obtain the invariant imbedding equation

$$\frac{\partial \underline{r}}{\partial t_o} + \left[\frac{\partial \underline{r}}{\partial \underline{D}}\right] \underline{a}(\underline{D}, \underline{r}, t_o) = \underline{d}(\underline{D}, \underline{r}, t_o) \qquad 7.6.8$$

This equation can also be rewritten in terms of the reduced Hamiltonian H^* obtained on substituting for $\frac{\partial H}{\partial \underline{u}} = \underline{0}$ in the standard Hamiltonian H as

$$\frac{\partial \underline{r}}{\partial t_o} + \left[\frac{\partial \underline{r}}{\partial \underline{D}}\right] \frac{\partial H^*}{\partial \underline{r}} = -\frac{\partial H^*}{\partial \underline{D}} \qquad 7.6.9$$

Note that if the terminal conditions on the two point boundary value problem change then it is possible to obtain an appropriate invariant imbedding equation for the new case. For example, in estimation theory, as we will see in the next part of the book, we often obtain a two point boundary value problem with the boundary conditions

$$\underline{\lambda}(t_o) = \underline{b}\,, \quad \underline{\lambda}(t_f) = \underline{c} \qquad 7.6.10$$

In this case, we can imbed the problem by letting

$$\underline{x}(t_f) = \underline{r}(\underline{D}, t_f) \qquad 7.6.11$$

where we imbed the final condition $\underline{\lambda}(t_f) = \underline{c}$ in a more general class of final conditions such that

$$\underline{\lambda}(t_f) = \underline{D} \tag{7.6.12}$$

which includes the given boundary conditions 7.6.10. With $\underline{D}$ and t_f regarded as independent, we expand 7.6.1 as

$$\underline{x}(t_f + \Delta t_f) = \underline{x}(t_f) + \dot{\underline{x}}(t_f)\,\Delta t_f + \underline{O}(\Delta^2) \tag{7.6.13}$$

so that we have

$$\underline{r}(\underline{D} + \Delta\underline{D},\ t_f, \Delta t_f) = \underline{r}(\underline{D},\ t_f) + \underline{a}(\underline{r}, \underline{D}, t_f)\,\Delta t_f + \underline{O}(\Delta^2) \tag{7.6.14}$$

Expanding the left hand side of 7.6.14 in a Taylor series

$$\underline{r}(\underline{D} + \Delta\underline{D},\ t_f + \Delta t_f) = \underline{r}(\underline{D},\ t_f) + \frac{\partial r}{\partial \underline{D}}\,\Delta\underline{D} + \frac{\partial \underline{r}}{\partial t_f}\,\Delta t_f + \underline{O}(\Delta^2) \tag{7.6.15}$$

From equations 7.6.1 and 7.6.12, we may write

$$\Delta\underline{D} = \underline{d}(\underline{r},\ \underline{D},\ t_f)\,\Delta t_f + \underline{O}(\Delta^2) \tag{7.6.16}$$

Then equating the right hand sides of equation 7.6.14 and 7.6.15, and substituting for $\Delta\underline{D}$ from equation 7.6.16, we have

$$\frac{\partial \underline{r}}{\partial t_f} + \left[\frac{\partial \underline{\underline{r}}}{\partial \underline{D}}\right] \underline{d}(\underline{r},\ \underline{D},\ t_f) = \underline{a}(\underline{r},\ \underline{D},\ t_f) \tag{7.6.17}$$

In terms of the reduced Hamiltonian H^*, this can be written as

$$\frac{\partial \underline{r}}{\partial t_f} - \left[\frac{\partial \underline{\underline{r}}}{\partial \underline{D}}\right] \frac{\partial H^*}{\partial \underline{r}} = \frac{\partial H^*}{\partial \underline{D}} \tag{7.6.18}$$

Having briefly described the invariant imbedding approach, let us illustrate it on an example.

Example

The example considered is that of the classical regulator problem, i.e

Minimise $$J = \frac{1}{2}\,\|\underline{x}(t_f)\|^2_S + \frac{1}{2}\int_{t_o}^{t_f} \left[\|\underline{x}(t)\|^2_{Q(t)} + \|\underline{u}(t)\|^2_{R(t)}\right] dt$$

subject to

$$\dot{\underline{x}}(t) = A(t)\,\underline{x}(t) + B(t)\,\underline{u}(t) \qquad \underline{x}(t_o) = \underline{x}_o$$

Writing the Hamiltonian as

$$H = \frac{1}{2}\|\underline{x}(t)\|^2_{Q(t)} + \frac{1}{2}\|\underline{u}(t)\|^2_{R(t)} + \underline{\lambda}(t)^T\left[A(t)\,\underline{x}(t) + B(t)\,\underline{u}(t)\right]$$

Then from the necessary conditions

$$\frac{\partial H}{\partial \underline{u}} = \underline{0} \qquad \text{or } \underline{u} = -R^{-1}\,B^T\,\underline{\lambda}$$

$$\frac{\partial H}{\partial \underline{\lambda}} = \overset{\circ}{\underline{x}} = A\underline{x} + B\underline{u} = A\underline{x} - B R^{-1} B^T \underline{\lambda}$$

$$\frac{\partial H}{\partial \underline{x}} = - \dot{\underline{\lambda}} = - Q\underline{x} - A^T\underline{\lambda}$$

The boundary conditions are

$$\underline{x}(t_o) = \underline{x}_o \quad ; \quad \underline{\lambda}(t_f) = S\ \underline{x}(t_f)$$

For this problem we use the invariant imbedding equation 7.6.8 or 7.6.9 where the reduced Hamiltonian H^* is

$$H^* = \frac{1}{2} \| \underline{x}(t) \|_Q^2 + \underline{\lambda}^T Ax - \frac{1}{2} \underline{\lambda}^T B R^{-1} B^T \underline{\lambda}$$

we use $\underline{x} = \underline{D}$, $\underline{\lambda} = \underline{r}$

Then

$$\frac{\partial \underline{r}}{\partial t_o} + \frac{\partial \underline{r}}{\partial \underline{D}} \left[A\ \underline{D} - B R^{-1} B^T \underline{r} \right] = - Q\ \underline{D} - A^T \underline{r}$$

Since the equations are linear, it is possible to assume a solution valid for arbitrary $\underline{D}$ and arbitrary running time $t = t_1$, of the form

$$\underline{\lambda}(t_1) = \underline{r}(\underline{D}, t_1) = P(t_1)\ \underline{D} + \underline{N}(t_1)$$

By substituting in the invariant imbedding equation, we have

$$\dot{\underline{N}} + \dot{P}\underline{D} + P\left[A\underline{D} - BR^{-1}B^T P\ \underline{D} - BR^{-1}B^T \underline{N}\right] = - Q\ \underline{D} - A^T P\ \underline{D} - A^T \underline{N}$$

This can be a solution for arbitrary $\underline{D}$ only if

$$\dot{P} = - P A - A^T P + P B R^{-1} B^T P - Q$$

$$\dot{\underline{N}} = \left[P B R^{-1} B^T - A^T \right] \underline{N}$$

which must satisfy at the start of the computation where $t = t_f$

$$\underline{\lambda}(t_f) = P\left[t_f \right] \underline{D} + \underline{N}(t_f) = \underline{\lambda}(t_f) = S\ \underline{x}(t_f) = S\ \underline{D}$$

Since C is arbitrary, this gives

$$P(t_f) = S, \quad \underline{N}(t_f) = \underline{0}$$

i.e. we must integrate the matrix Riccati equation and the linear vector equation backwards from t_f to t_o.

Note that this result is identical to that obtained in chapter 5 using other methods.

7.7 CONCLUSIONS

In this chapter we have examined the four principal methods of solving non-linear dynamic optimisation problems using global techniques. In the next chapter, we study how we can use decomposition-coordination techniques to tackle such problems for large scale systems.

REFERENCES

[1] MUKHOPADHYAY, B.R. and MALIK, O.P. : "Optimal control of synchronous machine excitation by quasilinearisation techniques".
Proc. IEE, vol. 119, 1, Jan. 1972.

[2] BAUMAN, E.J., 1968 : Multilevel optimisation techniques with application to trajectory decomposition.
Advances in control systems, vol. 6,C.T. Leondes (editor),Academic Press,New-Yor

[3] KENNETH, P. and Mc GILL, R., 1966 : Two point boundary value problem techniques.
Advances in control systems, C.T. Leondes (editor), Academic Press, New York.

[4] FOX, L., 1962 : Numerical solution of ordinary and partial differential equations. Addison Wesley Publishing Co., Inc.

[5] BELLMAN, R. and KALABA, R.E., 1965 : Quasilinearisation and non-linear boundary value problems", in "Modern Analytical and computational methods in Science and Mathematics" series, American Elsevier.

[6] SAGE, A.P., 1969 : Optimum systems control.
Prentice Hall.

[7] KIRK, D.E., 1970 : Optimal control theory - An introduction.
Prentice Hall.

[8] GALY, J. : "Optimisation dynamique par quasilinéarisation et commande hiérarchisée".
Thèse de Doctorat de 3e cycle, Université Paul Sabatier, Toulouse, 1973.

[9] CALVET, J.L. : "Optimisation par calcul hiérarchisé et coordination en ligne des systèmes dynamiques de grande dimension".
Thèse de Doctorat de 3e cycle, Université Paul Sabatier, Toulouse, 1976.

In writing this chapter, we have made extensive use of the excellent texts of SAGE [6] and KIRK [7].

PROBLEMS FOR CHAPTER 7

1. Minimise $J = \frac{1}{2} \int_0^1 (x^2 + u^2)\, dt$

for the system $\mathring{x} = -x^2 + u \qquad x(0) = 1$

using the gradient method.

2. Use the gradient method to solve the sliding mass problem of minimising

$$J = \underset{u}{\text{Min}} \int_0^1 \frac{1}{2} (y_1^2 + y_2^2 + u^2)\, dt$$

subject to

$$\dot{y}_1 = y_2 \qquad\qquad y_1(0) = 2$$

$$\dot{y}_2 = u - y_2^2 \text{ sgn } y_2 \qquad y_2(0) = -2$$

where $\text{sgn}(y) = \begin{cases} 1 & \text{if } y > 0 \\ 0 & \text{if } y < 0 \end{cases}$

3. Write down the various steps to do one iteration of the gradient procedure for the problem of a continuous stirred tank reactor

$$\mathring{x}_1(t) = -2\left[x_1(t) + 0.25\right] + \left[x_2(t) + 0.5\right] \exp\left[\frac{25x_1(t)}{x_1(t)+2}\right]$$

$$\mathring{x}_2(t) = 0.5 - x_2(t) - \left[x_2(t) + 0.5\right] \exp\left[\frac{25x_1(t)}{x_1(t)+2}\right]$$

where the cost function is

$$J = \int_0^{0.78} \left[x_1^2(t) + x_2^2(t) + u^2(t) \right] dt$$

4. Write a computer program to solve problem 3. Use a step length of 0.25.

5. Solve problem 1 using the quasilinearisation approach.

6. Solve problem 2 using the quasilinearisation approach for the initial states $y_1(0) = 3$, $y_2(0) = -3$.

7. Write down the various steps to do one iteration of the quasilinearisation method for problem 3.

8. Show that for the cost function of equation 7.4.19, the initial costate $\underline{\lambda}(t_o)$ should be improved from the ith to the (i+1)st iteration using the formula

$$\underline{\lambda}^{i+1}(t_o) = \underline{\lambda}^{(i)}(t_o) + \left[\left[\frac{\partial^2 h}{\partial \underline{x}^2}(x(t_f))\right] P_x(\underline{\lambda}(t_o), t_f - P_\lambda(\underline{\lambda}(t_o), t_f)\right]_i^{-1} \cdot \left[\underline{\lambda}(t_f) - \frac{\partial h}{\partial \underline{x}}(\underline{x}(t_f))\right]_i$$

where $[\cdot]_i$ implies that the enclosed terms are evaluated on the ith trajectory and $P_x(\underline{\lambda}^{(i)}(t_o^i), t_f)$ is the n x n state influence matrix given by equation 7.4.21, $\frac{\partial^2 h}{\partial \underline{x}^2}(\underline{x}(t_f))$ is given by equation 7.4.22 and P_λ is given by equation 7.4.18.

9. Solve problem 3 using the variation of extremals method.

10. Obtain the invariant imbedding equation for the two point boundary value problem

$$\dot{\underline{x}}(t) = \underline{f}(\underline{x}(t), \underline{\lambda}(t), t)$$

$$\dot{\underline{\lambda}}(t) = \underline{g}(\underline{x}(t), \underline{\lambda}(t), t)$$

$$\underline{x}(t_o) = \underline{a}, \; \underline{x}(t_f) = b$$

11. Obtain the invariant imbedding equations for the problem of minimising

$$J = \| \underline{x}(t_f) \|_S^2 + \int_{t_o}^{t_f} (\| \underline{x} - \underline{x}^* \|_Q^2 + \| \underline{u} - \underline{u}^* \|_R^2) \, dt$$

subject to

$$\dot{\underline{x}} = A\underline{x} + B\underline{u} + C\underline{Z}$$

$$\underline{Z} = L\underline{x}$$

where $\underline{x}^*$, $\underline{u}^*$ are given trajectories.

CHAPTER 8

Dynamic Optimisation for Large Scale Non-Linear Systems

In this chapter we continue our study of non-linear systems by going from the small decomposed subsystem considered in the previous chapter to the interconnection of such subsystems leading to a large scale non-linear interconnected dynamical system. As in the previous chapter, the methods for the optimisation of such systems are necessarily iterative since ultimately they all involve solving in some form a non-linear two point boundary value problem. We will see that the hierarchical methods developed in this chapter are basically more attractive than the global methods of the previous chapter since at any stage we only manipulate low order subproblems. This makes the calculations more accurate since manipulations on low order subproblems ensure that the trunction and rounding off errors are smaller. We will see that in fact most of the hierarchical methods are, where they work, also more attractive computationally than the methods of chapter 7. We should note this area of hierarchical optimisation for non-linear systems is currently being investigated by many researchers so that the methods outlined here represent essentially the author's view of the subject as it exists today. The chapter does represent, however, most of the published work in this area up to the year 1977.

The plan of this chapter is as follows. We begin by describing the goal coordination method for non-linear systems. We then go on to describe the prediction approaches of Hassan and Singh. We subsequently treat the three level costate prediction method as well as the discrete time costate coordination method. We also outline the practical hierarchical model follower. Finally, we consider ways of obtaining near optimal feedback control.

8.1 THE GOAL COORDINATION METHOD

This method is a direct extension to the non-linear case of the goal coordination method considered in chapter 7.

The problem is to minimise

$$J = \sum_{i=1}^{N} \pi_i(\underline{x}_i(t_f)) + \int_{t_o}^{t_f} g_i(\underline{x}_i(t), \underline{u}_i(t), \underline{z}_i(t), t)\, dt \qquad 8.1.1$$

where π_i is the terminal cost for the ith subsystem, $g_i(\,.\,)$ is the cost at time t for the ith subsystem ($t \in [t_o, t_f]$ where t_o, t_f are known). The above functional J is to be minimised subject to the constraints which define the subsystem dynamics, i.e.

$$\dot{\underline{x}}_i(t) = \underline{f}_i(\underline{x}_i(t), \underline{u}_i(t), \underline{z}_i(t), t) \qquad 8.1.2$$

$$\underline{x}_i(t_o) = \underline{x}_{io} \qquad i = 1, 2, \ldots, N$$

where $\underline{z}_i(t)$ is the ith subsystem's interaction input. Also

$$\underline{G}_i^{*}(\underline{x}_i(t), \underline{z}_j(t)) = \underline{x}_i - \sum_{j=1}^{N} \underline{g}_{ij}(\underline{z}_j(t)) \qquad i = 1 \text{ à } N \qquad 8.1.3$$

Equation 8.1.3 represents the interconnection constraints.

Such an interconnection structure is clearly very general.

$$\text{Let us define } \underline{x} = \begin{bmatrix} \underline{x}_1 \\ \vdots \\ \underline{x}_N \end{bmatrix}, \quad \underline{u} = \begin{bmatrix} \underline{u}_1 \\ \vdots \\ \underline{u}_N \end{bmatrix}, \quad \underline{z} = \begin{bmatrix} \underline{z}_1 \\ \vdots \\ \underline{z}_N \end{bmatrix}, \quad \underline{\lambda} = \begin{bmatrix} \underline{\lambda}_1 \\ \vdots \\ \underline{\lambda}_N \end{bmatrix}$$

where $\underline{\lambda}$ is a Lagrange multiplier vector which arises in the definition of

$$L(\underline{x},\underline{u},\underline{z},\underline{\lambda}) = \sum_{i=1}^{N} \pi_i(\underline{x}_i(t_f)) + \int_{t_o}^{t_f} g_i(\underline{x}_i,\underline{u}_i,\underline{z}_i,t)\,dt + \int_{t_o}^{t_f} \underline{\lambda}_i^T G_i^*(\underline{x}_i,\underline{z}_j,t)\,dt \qquad 8.1.4$$

Under certain assumptions which we discuss later (existence of a saddle point for L), the solution of the problem is given by :

$$\max_{\underline{\lambda}} \; \min_{\underline{u},\underline{x},\underline{z}} \; L(\underline{x}, \underline{u}, \underline{z}, \underline{\lambda}) = \max_{\underline{\lambda}} \; \phi(\underline{\lambda}) \qquad 8.1.5$$

$(\phi(\underline{\lambda}) : \min_{\underline{u},\underline{x},\underline{z}} L(\underline{x},\underline{u},\underline{z},\underline{\lambda})$ subject to equation 8.1.2)

Given the special structure of L, $\phi(\underline{\lambda})$ can be obtained by the solution of N independent subproblems. Thus let us rewrite as :

$$L = \sum_{i=1}^{N} L_i = \sum_{i=1}^{N} \pi(\underline{x}_i(t_f)) + \int_{t_o}^{t_f} \Big[g_i(\underline{x}_i,\underline{u}_i,\underline{z}_i,t) + \underline{\lambda}_i^T \underline{x}_i - \sum_{j=1}^{N} \underline{\lambda}_j^T g_{ji}(\underline{z}_i(t)) \Big] dt \qquad 8.1.6$$

Thus consider a two level optimisation structure where on level 1, for given $\underline{\lambda}$, the following N independent minimisation problems are solved, i.e.

$$\min_{\underline{x}_i,\underline{u}_i,\underline{z}_i} \; L_i = \left\{ \pi_i(\underline{x}_i(t_f)) + \int_{t_o}^{t_f} \Big[g_i(\underline{x}_i,\underline{u}_i,\underline{z}_i,t) + \underline{\lambda}^T \underline{G}_i(\underline{x}_i,\underline{z}_i,t) \Big] dt \right\} \qquad 8.1.7$$

$$\text{subject to } \overset{\circ}{\underline{x}}_i = \underline{f}_i(\underline{x}_i, \underline{u}_i, \underline{z}_i, t) \qquad t_o \leqslant t \leqslant t_f \qquad 8.1.8$$

$$\underline{x}_i(t_o) = \underline{x}_{io}$$

and on level 2, the $\underline{\lambda}$ trajectory is improved in order to maximise $\phi(\underline{\lambda})$. This can be done using say the steepest ascent method, i.e. from iteration j to j+1

$$\underline{\lambda}(t)^{j+1} = \underline{\lambda}(t)^j + \alpha^j \underline{d}^j(t) \qquad t_o \leqslant t \leqslant t_f \qquad 8.1.9$$

$$\text{where} \quad \underline{d} = \nabla \phi(\underline{\lambda}) \Big|_{\underline{\lambda}=\underline{\lambda}^*} = \sum_{i=1}^{N} \underline{G}_i(\underline{x}_i, \underline{z}_i) \qquad 8.1.10$$

where $\nabla\phi(\underline{\lambda})$ is the gradient of $\phi(\underline{\lambda})$ for a fixed $\lambda = \lambda^*$ trajectory, $\underline{x}_i$, $\underline{z}_i$ are the values of $\underline{x}_i$, $\underline{z}_i$ which minimise L_i in equation 8.1.7 whilst using $\underline{\lambda} = \underline{\lambda}^*$. $\alpha^j > 0$ is the step length and $\underline{d}^j$ is the steepest ascent search direction. At the optimum $\underline{d}^j \to 0$ and the appropriate Lagrange multiplier trajectory is the optimum one.

Discussion

The whole argument hinges on the validity of the assertion Max $\phi(\underline{\lambda})$ = Min J and this may in fact not be valid since $\underline{G}_i^*$, $\underline{f}_i$ need to be linear for the constrained domain to be convex and the convexity of the constrained domain is necessary to prove the assertion [1]. This means that in general, equation 8.1.5 may not be valid so that maximising $\phi(\underline{\lambda})$ on the R.H.S. of the equation using a hierarchical structure may not give the optimal control. Unfortunately, at the present time, we cannot say a priori for any given problem if there will be a difference between the values on the L.H.S. and R.H.S. of equation 8.1.5 (called a duality gap). This duality gap can be detected by evaluating the dual and primal functions ϕ and J. Nevertheless, the method has an intrinsic simplicity which makes it attractive. Another attraction of the approach is that the dual function is still concave for this non-linear case. This ensures that if the duality assertion is valid, the optimum obtained is the Global Optimum. This is clearly a very useful property for non-linear systems.

We will now look at an example where the method does work.

Example : Control of two coupled synchronous machines

The problem is to control the excitation voltages of two coupled synchronous machines optimally. A model for a single machine is given by Mukhopadhayay [2] which is of order 3. In this case, for the two machines, the model has 6 coupled non-linear differential equations given by :

$$\begin{aligned}
\dot{y}_1 &= y_2 \\
\dot{y}_2 &= B_1 - A_1 y_2 - A_2 y_3 \sin y_1 - \frac{B_2}{2} \sin 2\, y_1 \\
\dot{y}_3 &= u_1 - C_1 y_3 + C_2 \cos y_1 \\
\dot{y}_4 &= y_5 \\
\dot{y}_5 &= B_4 - A_4\, y_5 - A_5\, y_6 \sin y_4 - \frac{B_5}{2} \sin 2y_4 \\
\dot{y}_6 &= u_2 - C_4\, y_6 - C_5 \cos y_4
\end{aligned}$$

where $y_1, y_2, \ldots, y_6$ are the six state variables and u_1, u_2 are the two controls. The parameters are given by Mukhopadhayay :

$$\begin{array}{lll}
A_1 = A_4 = 0.2703, & A_2 = 12.012, & A_5 = 14.4144 \\
B_1 = B_4 = 39.1892, & B_2 = -48.048, & B_5 = -57.6576 \\
C_1 = C_4 = 0.3222, & C_2 = 1.9, & C_5 = 2.28
\end{array}$$

and the system comprises two subsystems coupled by $y_1 = y_4 = x$.

It is desired to minimise J where

$$J = \int_0^2 \frac{1}{2} \left[\| \underline{y} - \underline{y}_p \|_Q^2 + \| \underline{u} - \underline{u}_p \|_R^2 \right] dt$$

where $\underline{y}_p^T = \begin{bmatrix} 0.7461 & 0 & 7.7438 & 0.7461 & 0 & 7.7438 \end{bmatrix}$

$\underline{u}_p^T = \begin{bmatrix} 1.1 & 1.1 \end{bmatrix}$

are desired steady state values for the states and controls.

$R = I_2$ the second order identity matrix and Q is a diagonal matrix with the diagonal given by Q_D where

$$Q_D = \begin{bmatrix} 20 & 20 & 2 & 20 & 20 & 2 \end{bmatrix}$$

Let us put $\underline{y}_1^T = [y_1, y_2, y_3]$ and $\underline{y}_2^T = [y_4, y_5, y_6]$. We can then rewrite the state equations in the form :

$$\dot{\underline{y}}_1 = \underline{f}_1(\underline{y}_1, u_1)$$
$$\dot{\underline{y}}_2 = \underline{f}_2(\underline{y}_2, u_2)$$

with the coupling constraint $x = y_4 = y_1$

Let us rewrite the cost function also as :

$$\min_{u_1,u_2} \left\{ \frac{1}{2} \int_0^2 \left[\| \underline{y}_1 - \underline{y}_{p1} \|^2_{Q_1} + R_1 (u_1 - u_{P1})^2 \right] dt + \frac{1}{2} \int_0^2 \left[\| \underline{y}_2 - \underline{y}_{p2} \|^2_{Q_2} + R_2 (u_2 - u_{p2})^2 \right] dt \right]$$

with $Q_1 = \begin{bmatrix} 20 & & 0 \\ & 20 & \\ 0 & & 2 \end{bmatrix}$ $\quad Q_2 = \begin{bmatrix} 20 & & 0 \\ & 20 & \\ 0 & & 2 \end{bmatrix}$ $\quad R_1 = 1, \; R_2 = 1$

Let us write the Lagrangian as :

$$L = \int_0^2 \left\{ \frac{1}{2} \left[\| \underline{y}_1 - \underline{y}_{p1} \|^2_{Q_1} + R_1 (u_1 - u_{p1})^2 + \| \underline{y}_2 - \underline{y}_{p2} \|^2_{Q_2} + R_2 (u_2 - u_{p2})^2 \right] + \underline{\lambda}_1^T \left[\dot{\underline{y}}_1 - \underline{f}_1(\underline{y}_1, u_1) \right] + \underline{\lambda}_2^T \left[\dot{\underline{y}}_2 - \underline{f}_2(\underline{y}_2, u_2) \right] + \gamma (y_1 - x) + \beta (x - y_4) \right\} dt$$

where $\underline{\lambda}_1 = \begin{bmatrix} \lambda_1 \\ \lambda_2 \\ \lambda_3 \end{bmatrix}$ $\quad \underline{\lambda}_2 = \begin{bmatrix} \lambda_4 \\ \lambda_5 \\ \lambda_6 \end{bmatrix}$

In the solution procedure we will use a gradient method at level 2 to force $\frac{\partial L}{\partial \beta} = 0 = x - y_4$ and at level 1 we perform

$$\left[\min_{u_1, u_2, x} L \right] \quad \text{for given } \beta \,.$$

The Lagrangian can be decomposed at level 1 into two sub Lagrangians

$$L = L_1 + L_2 \qquad \text{where}$$

$$L_1 = \int_0^2 \left[\frac{1}{2} \| \underline{y}_1 - \underline{y}_{p1} \|^2_{Q_1} + \frac{1}{2} R_1 (u_1 - u_{p1})^2 + \underline{\lambda}_1^T \left[\dot{\underline{y}}_1 - \underline{f}_1(\underline{y}_1, u_1) \right] + \gamma (y_1 - x) + \beta x \right] dt$$

$$L_2 = \int_0^2 \frac{1}{2} \left[\| \underline{y}_2 - \underline{y}_{p2} \|^2_{Q_2} + \frac{1}{2} R_2 (u_2 - u_{p2})^2 + \underline{\lambda}_2^T \left[\dot{\underline{y}}_2 - \underline{f}_2(\underline{y}_2, u_2) \right] - \beta y_4 \right] dt$$

Thus at level 1 we perform independent minimisations of L_1 and L_2 for given $\underline{\lambda}_1$, $\underline{\lambda}_2$ supplied by the second level.

The independent minimisation of L_1 leads to the following two point boundary value problem :

$$\dot{y}_1 = y_2$$

$$\dot{y}_2 = B_1 - A_1 y_2 - A_2 y_3 \sin y_1 - \frac{B_2}{2} \sin 2y_1$$

$$\dot{y}_3 = u_{p_1} + \frac{\lambda 3}{R_1} - C_1 y_3 + C_2 \cos y_1$$

$$\dot{\lambda}_1 = 20(y_1 - y_{p1}) + \lambda_2(A_2 y_3 \cos y_1 + B_2 \cos 2y_1) + C_2 \lambda_3 \sin y_1 + \beta$$

$$\dot{\lambda}_2 = 20(y_2 - y_{p2}) - \lambda_1 + A_1 \lambda_2$$

$$\dot{\lambda}_3 = 2(y_3 - y_{p3}) + A_2 \lambda_2 \sin y_1 + C_1 \lambda_3$$

with $y_i(0) = y_{i0}$ and $\lambda_i(2) = 0$ $i = 1, 2, 3$

and the control is :

$$u_1 = u_{p1} + \frac{\lambda 3}{R_1} \quad \text{with } x = y_1$$

Similarly for the second subproblem, we need to solve the two point boundary value problem

$$\dot{y}_4 = y_5$$

$$\dot{y}_5 = B_4 - A_1 y_5 - A_5 y_6 \sin y_4 - \frac{B_5}{2} \sin 2y_4$$

$$\dot{y}_6 = u_{p2} + \frac{\lambda 6}{R_2} - C_4 y_6 + C_5 \cos y_4$$

$$\dot{\lambda}_4 = 20(y_4 - y_{p4}) + \lambda_5(A_5 y_6 \cos y_4 + B_5 \cos 2y_4) + C_5 \lambda_6 \sin y_4 - \beta$$

$$\dot{\lambda}_5 = 20(y_5 - y_{p5}) - \lambda_4 + A_4 \lambda_5$$

$$\dot{\lambda}_6 = 2(y_6 - y_{p6}) + A_5 \lambda_5 \sin y_4 + C_4 \lambda_6$$

with $y_i(0) = y_{i0}$ $\lambda_i(2) = 0$ $i = 4,5,6$

These two subproblems were solved on the I.B.M. 370 / 165 digital computer using the quasilinearisation approach and at the second level a simple gradient coordinator was used.

The initial condition used was

$$\underline{y}^T(0) = \left[0.7347 \quad -0.2151 \quad 7.7443 \quad 0.7347 \quad -0.2151 \quad 6.9483\right]$$

and the optimum was reached in 10 second level iterations. At this point the primal and dual costs were found to be identical, i.e.

$$J = \emptyset = 1.1658$$

In another test, the initial conditions were changed to

$$\underline{y}^T(0) = [1 \quad 0.1 \quad 10 \quad 1 \quad 0.1 \quad 8.69]$$

The problem was resolved in 10 second level iterations giving in this case

$$J = \phi = 78.5048$$

In this Section we have obtained the solution of a practical problem using the Goal Coordination method. There are many other cases where the method works (cf. Smith and Sage [3]). However, there are still others where it does not but one is not able to know this prior to simulation. This is obviously a serious drawback for the practical optimisation of large scale systems since for such systems even the simulation is costly.

Having considered the Goal Coordination approach we next examine the new prediction approach of Hassan and Singh [4] which ensures that the necessary conditions for optimality are satisfied whilst calculating the control within a decentralised computational structure.

8.2 THE NEW PREDICTION METHOD OF HASSAN AND SINGH

The basis of the approach is that it is possible to use the equilibrium point* of the system to expand the dynamic equations in a Taylor series and then to "fix" the second and higher order terms by predicting the states and controls which arise in these terms. This also enables us to decompose the optimisation problem into independent "linear-quadratic" subproblems for given states and controls to be provided by a second level. On the second level, a prediction algorithm of the type discussed in Chapter 6 can be used.

The algorithm, aside from being a viable hierarchical optimisation method for non-linear systems, has the additional advantage that only "linear-quadratic" problems are solved at the first level and trivial updating is done at the second. So in many cases there are substantial computational savings compared to the global single level solution, making the method suitable for solving even low order non-linear problems. Against this, there is the disadvantage that in the present version of the method, the cost function is restricted to a quadratic form in the states and controls. We will show subsequently how this restriction can be lifted.

Consider the non linear system given by

$$\dot{\underline{x}} = \underline{f}(\underline{x}, \underline{u}, t) \qquad 8.2.1$$

where $\underline{x}$ is an n dimensional state vector and $\underline{u}$ is an m dimensional control vector. Let the equilibrium point of the system be at the origin**

---

* For most practical plants the equilibrium point is taken to be the steady state operating point of the plant obtained a priori by suitable steady state analysis or experimentation. The need to know the equilibrium point of the plant is one of the failings of the method. Later on in this chapter, we will see how we can get around this.

** This assumption is made for the sake of convenience only. It is possible to derive all the results of this section without this assumption.

i.e. $$\underline{f}(\underline{x}, \underline{u}, t) = \underline{0} \text{ at } \underline{x} = \begin{bmatrix} 0 \\ 0 \\ \cdot \\ \cdot \\ \cdot \\ 0 \end{bmatrix}, \quad \underline{u} = \begin{bmatrix} 0 \\ 0 \\ \cdot \\ \cdot \\ \cdot \\ 0 \end{bmatrix}$$

expanding 8.2.1 by a Taylor series about this point

$$\dot{\underline{x}} = \left.\frac{\partial \underline{f}}{\partial \underline{x}}\right|_{\underline{x}=\underline{u}=0} \underline{x} + \left.\frac{\partial \underline{f}}{\partial \underline{u}}\right|_{\underline{x}=\underline{u}=0} \underline{u} + \frac{1}{2} \underline{x}^T \frac{\partial \underline{f}^2}{\partial \underline{x}^2} \underline{x} + \frac{1}{2} \underline{u}^T \frac{\partial^2 \underline{f}}{\partial \underline{u}^2} \underline{u} + \quad \cdots$$

This equation can be rewritten as

$$\dot{\underline{x}} = A^* \underline{x} + B^* \underline{u} + \underline{f}(\underline{x}, \underline{u}, t) - A^* \underline{x} - B^* \underline{u}$$

where $$A^* = \left.\frac{\partial \underline{f}}{\partial \underline{x}}\right|_{\underline{x}=\underline{u}=0}, \quad B^* = \left.\frac{\partial \underline{f}}{\partial \underline{u}}\right|_{\underline{x}=\underline{u}=0}$$

or $$\dot{\underline{x}} = A\underline{x} + B\underline{u} + \left[C_1\underline{x} + C_2\underline{u} + \underline{f}(\underline{x}, \underline{u}, t) - A^*\underline{x} - B^*\underline{u} \right]$$

where A, B are respectively the diagonal (or block diagonal) parts of A^*, B^* and C_1, C_2 are the off diagonal parts of A^*, B^*.

Then

$$\dot{\underline{x}} = A\underline{x} + B\underline{u} + \underline{D}(x, \underline{u}) \qquad 8.2.2$$

where $$\underline{D}(\underline{x}, \underline{u}) = C_1\underline{x} + C_2\underline{u} + \underline{f}(\underline{x}, \underline{u}, t) - A^*\underline{x} - B^*\underline{u} \qquad 8.2.3$$

Let it be desired to minimise the additively separable cost function

$$J = \sum_{i=1}^{N} \int_0^T \frac{1}{2} \left(\| \underline{x}_i \|^2_{Q_i} + \| \underline{u}_i \|^2_{R_i} \right) dt \qquad 8.2.4$$

where it is assumed that there are N blocks in A and B and corresponding to each of these blocks there is a component of the summation on the right hand side of equation 8.2.4. In this equation, Q_i are assumed to be positive semi-definite and R_i to be positive definite.

To solve this problem of minimising J in equation 8.2.4 subject to equations 8.2.2 and 8.2.3, let us add the additional constraints

$$\underline{x} = \underline{x}^\circ$$

$$\underline{u} = \underline{u}^\circ$$

in order to fix the non-linearities and off diagonal terms $\underline{D}(\underline{x}_i, \underline{u}_i)$, so that our optimisation problem is decomposed into N independent minimisation problems the ith of which is given by

$$\text{Min } J_i = \int_0^T \frac{1}{2} \left(\| \underline{x}_i \|^2_{Q_i} + \| \underline{u}_i \|^2_{R_i} \right) dt$$

subject to $$\dot{\underline{x}}_i = A_i \underline{x}_i + B_i \underline{u}_i + \underline{D}_i (\underline{x}^\circ, \underline{u}^\circ)$$

$$\underline{x}_i^o = \underline{x}_i$$
$$\underline{u}_i^o = \underline{u}_i$$

In order to aid the convergence of the two level algorithm that we will develop to solve this modified problem, let us add quadratic penalty terms in the cost function such that it becomes

$$J^1 = \frac{1}{2}\int_0^T (\|\underline{x}\|_Q^2 + \|\underline{u}\|_R^2 + \|\underline{x}-\underline{x}^o\|_S^2 + \|\underline{u}-\underline{u}^o\|_H^2)\, dt \qquad 8.2.5$$

where S, H are block diagonal matrices with the blocks S_i, H_i corresponding respectively to Q_i, R_i. Note that if the constraints $\underline{x}_i^o = \underline{x}_i$, $\underline{u}_i^o = \underline{u}_i$ are strictly satisfied, the above problem is identical to our original problem. To solve this new problem of minimising J^1 in 8.2.5, subject to the dynamics

$$\dot{\underline{x}} = A\underline{x} + B\underline{u} + \underline{D}(\underline{x}^o, \underline{u}^o) \text{ and the constraints } \underline{u} - \underline{u}^o = \underline{0}, \quad \underline{x} - \underline{x}^o = \underline{0}$$

let us write the Hamiltonian as

$$H^o = \frac{1}{2}\|\underline{x}\|_Q^2 + \frac{1}{2}\|\underline{u}\|_R^2 + \frac{1}{2}\|\underline{x}-\underline{x}^o\|_S^2 + \frac{1}{2}\|\underline{u}-\underline{u}^o\|_H^2 + \underline{\lambda}^T\left[A\underline{x} + B\underline{u} + D(\underline{x}^o, \underline{u}^o)\right] + \underline{\pi}^T\left[\underline{x}-\underline{x}^o\right] + \underline{\beta}^T\left[\underline{u}-\underline{u}^o\right] \qquad 8.2.6$$

where

$\underline{\lambda}$ is a $\sum_{i=1}^N n_i$ dimensional co-state vector and $\underline{\pi}$, $\underline{\beta}$, are respectively $\sum_{i=1}^N n_i$ and $\sum_{i=1}^N m_i$ dimensional vectors of Lagrange multipliers.

To obtain the first level solutions, let us set $\frac{\partial H^o}{\partial \underline{u}} = \underline{0}$ to obtain

$$\underline{u} = R^{*-1}\left[-B^T\underline{\lambda} + H\underline{u}^o - \underline{\beta}\right] \qquad 8.2.7$$

where $\qquad R^* = R + H$

Also, setting $\frac{\partial H^o}{\partial \underline{\lambda}} = \dot{\underline{x}}$ gives

$$\dot{\underline{x}} = A\underline{x} + BR^{*-1}\left[-B^T\underline{\lambda} + H\underline{u}^o - \underline{\beta}\right] + \underline{D}(\underline{x}^o, \underline{u}^o)$$

i.e.
$$\dot{\underline{x}} = A\underline{x} + BR^{*-1}\left[-B^T\underline{\lambda} + H\underline{u}^o - \underline{\beta}\right] + C_1\underline{x}^o + C_2\underline{u}^o + \underline{f}(\underline{x}^o, \underline{u}^o, t) - A^*\underline{x}^o - B^*\underline{u}^o \qquad 8.2.8$$

with $\underline{x}(0) = \underline{x}_o$

Also
$$\dot{\underline{\lambda}} + \frac{\partial H^o}{\partial \underline{x}} = \underline{0} \quad \text{i.e. } \dot{\underline{\lambda}} = -Q^*\underline{x} - A^T\underline{\lambda} + S\underline{x}^o - \underline{\pi} \qquad 8.2.9$$

where $\quad Q^* = Q + S$

Note that from the transversality conditions

$$\underline{\lambda}(T) = \underline{0}$$

On the first level it is merely necessary to solve the two point boundary value problem given by equation 8.2.8 and 8.2.9 for given $\underline{x}^o$, $\underline{u}^o$, $\underline{\pi}$, $\underline{\beta}$. The interesting thing is that these equations are linear so that it is possible to write a Riccati type equation with a single point boundary value condition to solve it. We will use such an equation when we look at a practical example in a subsequent Section.

For the second level, it is necessary to force $\underline{x}^o$, $\underline{u}^o$ towards, $\underline{x}$, $\underline{u}$. We envisage a prediction type algorithm to do this. We know that at the optimum, we must have, in addition to

$$\frac{\partial H^o}{\partial \underline{u}} = \underline{0} \quad , \quad \frac{\partial H^o}{\partial \underline{\lambda}} = \underline{0} \quad ; \quad \frac{\partial H^o}{\partial \underline{x}} = -\dot{\underline{\lambda}} \quad \text{also}$$

$$\frac{\partial H^o}{\partial \underline{\pi}} = \underline{0} \quad , \quad \frac{\partial H^o}{\partial \underline{\beta}} = \underline{0} \quad , \quad \frac{\partial H^o}{\partial \underline{x}^o} = \underline{0} \quad , \quad \frac{\partial H^o}{\partial \underline{u}^o} = \underline{0}$$

As in the prediction algorithm in Chapter 6 for the linear quadratic case, we use these stationarity conditions to generate new values for these last four variables. For example :

using $\quad \dfrac{\partial H^o}{\partial \underline{\pi}} = \underline{0}$ we obtain $\underline{x}^o = \underline{x}$ $\qquad$ 8.2.10

similarly $\quad \dfrac{\partial H^o}{\partial \underline{\beta}} = \underline{0}$ gives $\underline{u}^o = \underline{u}$ $\qquad$ 8.2.11

$$\frac{\partial H^o}{\partial \underline{x}^o} = \underline{0} \text{ gives } \underline{\pi} = \left[-S(\underline{x}-\underline{x}^o) + \left[C_1^T + \frac{\partial \underline{f}^T}{\partial \underline{x}}(\underline{x}^o,\underline{u}^o,t)\bigg|_{\substack{\underline{x}=\underline{x}^o \\ \underline{u}=\underline{u}^o}} - A^{*T} \right] \underline{\lambda}(t) \right] \qquad 8.2.12$$

$$\frac{\partial H^o}{\partial \underline{u}^o} = \underline{0} \text{ gives } \underline{\beta} = \left[H\underline{u}^o - H(-R^{*-1}B^T\underline{\lambda} + R^{*-1}H\underline{u}^o - R^{*-1}\underline{\beta}) + \left[C_2^T + \frac{\partial \underline{f}^T}{\partial \underline{u}}\bigg|_{\substack{\underline{x}=\underline{x}^o \\ \underline{u}=\underline{u}^o}} - B^{*T} \right] \underline{\lambda}(t) \right] \qquad 8.2.13$$

Thus the second level algorithm from iteration k to k+1 is :

$$\begin{bmatrix} \underline{x}^o \\ \underline{\pi} \\ \underline{u}^o \\ \underline{\beta} \end{bmatrix}^{k+1} = \begin{bmatrix} \underline{x} \\ \underline{\pi}\ (\underline{x}^k, \underline{\lambda}^k) \\ \underline{u} \\ \underline{\beta}\ (\underline{\lambda}^k) \end{bmatrix}^k$$

where the expression on the R.H.S. of the above equation is obtained by substituting the optimal $\underline{x}$, $\underline{u}$, from the independent first level minimisation into the R.H.S. of equation 8.2.10 to 8.2.13

To sum up, the various steps in the algorithm are :

Step 1 : Start with some guessed values for the trajectories $\underline{x}^o$, $\underline{u}^o$, $\underline{\pi}$, $\underline{\beta}$ and send to level 1. It is possible for example to start with the initial guess $\underline{x}^o = \underline{\pi} = 0$; $\underline{u}^o = \underline{\beta} = \underline{0}$.

Step 2 : At level 1, solve the N independent minimisation problems for the given $\underline{x}^o$, $\underline{u}^o$, π, β by noting that (a) the two point boundary value problem of equations 8.2.8 - 8.2.9 splits into N independent blocks because of the block diagonal nature of all the relevant matrices and (b) each of these low order blocks represents a linear two point boundary value problem which can be solved easily by integrating an appropriate Riccati type equation.

The resulting locally optimal trajectories for the states, controls and adjoint vectors are sent to level one where they are collated as $\underline{x}$, $\underline{u}$, $\underline{\lambda}$.

Step 3 : Produce new prediction for $\underline{x}^o$, $\underline{u}^o$, π, β by substituting the $\underline{x}$, $\underline{u}$, into the R.H.S. of equations 8.2.10 to 8.2.13.

Step 4 : If the new prediction is "closer" to the old one (in the sense of an appropriate norm) than some predefined small number ϵ , stop and record the $\underline{u}$ as the optimal control. Else go back to Step 2

The convergence of this method is studied by Singh [5]. Essentially, Hassan and Singh [4] derived a convergence condition the satisfaction of which guarantees convergence. The condition can be used to provide qualitative guidelines. For example, convergence can be assured for any problem where the functions are continuous and bounded by manipulating (a) the optimisation interval length and (b) the Q and R matrices, and (c) the S and H matrices.

Basically if the R matrix has sufficiently large elements and/or the period of optimisation is sufficiently small, the method converges.

In this section, we have developed a powerful method for the optimisation of large scale non-linear systems. The method is particularly attractive for the following reasons :

a - We usually do know a priori that our algorithm will converge. We can, moreover, increase the period of optimisation (which defines the period over which it will converge), in certain cases, by manipulating the matrices H and S which are at our disposal.

b - The jobs of level one are particularly simple since we merely have to solve a Riccati type equation for a low order system. On level 2, simple substitution is all that is required. Thus the algorithm is very simple and could prove to be a good alternative to traditional single level methods for optimising even low order non-linear problems.

Against these, we have the disadvantages that :

a - Unlike the case where the Goal Coordination method works, here only a locally optimal solution is obtained. Note that in this sense it is no worse than standard single level methods like quasi-linearisation, gradient methods, etc. of Chapter 7.

b - In this version of the method we are limited to a quadratic cost function.

c - For any particular problem, we have to verify that we have indeed obtained the minimum by, say, examining either a small variation from the optimum or by checking if the second order conditions of optimality are satisfied.

Experience shows that the algorithm converges very fast and this is another attraction of the approach. We next show a practical application of the approach on an example of a single synchronous machine of the type studied in Chapter 7. In view of the importance of the approach, we will go through the example in sufficient detail to give the reader a "feel" for the application of the method.

EXAMPLE : CONTROL OF A SYNCHRONOUS MACHINE

A model for this system is given by Mukhopadhayay et al. [2] as :

$$\begin{aligned} \dot{y}_1 &= y_2 \\ \dot{y}_2 &= B_1 - A_1 y_2 - A_2 y_3 \sin y_1 - \frac{B_2}{2} \sin 2y_1 \\ \dot{y}_3 &= u_1 - C_1 y_3 + C_2 \cos y_1 \end{aligned} \tag{8.2.14}$$

where y_1 is the rotor angle (radians)

y_2 is the speed deviation ($\frac{d}{dt} y_1$ rad/s)

y_3 is the field flux linkage

u_1 the control variable is the voltage applied to the machine and it is assumed to be available for optimal manipulation.

Mukhopadhayay gives the values of the parameters to be

$A_1 = 0.2703$, $A_2 = 12.012$, $B_1 = 39.1892$, $B_2 = -48.048$

$C_1 = 0.3222$, $C_2 = 1.9$

It is desired to minimise

$$J = \frac{1}{2} \int_0^2 ((y_1 - 0.7461)^2 + y_2^2 + 0.1\,(y_3 - 7.7438)^2 + 100(u_1 - 1.1)^2)\, dt$$

The equilibrium point of the system is

$$\underline{y}_e^T = \begin{bmatrix} 0.7461 & 0.0 & 7.7438 \end{bmatrix}^T$$

$$u_e = 1.1$$

then

$$A^* = \left[\frac{\partial \underline{f}^T}{\partial \underline{y}}\right]^T_{\underline{y}_e} = \begin{bmatrix} \frac{\partial f_1}{\partial y_1} & \frac{\partial f_1}{\partial y_2} & \frac{\partial f_1}{\partial y_3} \\ \frac{\partial f_2}{\partial y_1} & \frac{\partial f_2}{\partial y_2} & \frac{\partial f_2}{\partial y_3} \\ \frac{\partial f_3}{\partial y_1} & \frac{\partial f_3}{\partial y_2} & \frac{\partial f_3}{\partial y_3} \end{bmatrix}_{\underline{y}_e} \tag{8.2.15}$$

$$= \begin{bmatrix} 0 & 1 & 0 \\ -64.5348 & -0.2703 & -8.1533 \\ -1.2896 & 0.0 & -0.3222 \end{bmatrix}$$

$$B^* = \begin{bmatrix} 0 & 0 & 1 \end{bmatrix}^T$$

let $\underline{y} - \underline{y}_e = \underline{x}$ and $u = u_1 - u_e$

then

$$\mathring{\underline{x}} = A^* \underline{x} + B^* \underline{u} + \underline{f}(\underline{y}, \underline{u}_e, t) - A^* \underline{x}$$

Note that for this problem, the term in u is linear so that it is not necessary to linearise for it and not in fact necessary to add an extra term $\|\underline{u} - \underline{u}^o\|_H^2$ in the cost function to iterate on $\underline{u}^o$ at the second level. However, it does mean that if the system is split into subsystems then all but one of them will not have a control, making the R matrix impossible to invert. To get around this, define a fictitious control so that

$$\frac{d}{dt}\begin{bmatrix} x_1 \\ x_2 \\ x_3 \end{bmatrix} = \begin{bmatrix} 0 & 1 & 0 \\ -64.5348 & 02703 & 0 \\ 0 & 0 & -0.3222 \end{bmatrix}\begin{bmatrix} x_1 \\ x_2 \\ x_3 \end{bmatrix} + \begin{bmatrix} 0 & 0 \\ 0 & 0 \\ 0 & 1 \end{bmatrix}\begin{bmatrix} u_1 \\ u_2 \end{bmatrix}$$

$$+ \left\{ \underline{f}(\underline{y},t) + \begin{bmatrix} 0 & 0 & 0 \\ 0 & 0 & -8.1533 \\ -1.2846 & 0 & 0 \end{bmatrix}\begin{bmatrix} x_1 \\ x_2 \\ x_3 \end{bmatrix} - A^* \underline{x} \right\}$$

Now let $\underline{x} = \underline{x}^o$ for the term in the bracket and minimise

$$J = \frac{1}{2}\int_0^2 \left(\|\underline{x}\|_Q^2 + \|\underline{u}\|_R^2 + \|\underline{x} - \underline{x}^o\|_S^2 \right) dt$$

where

$$Q = \begin{bmatrix} 1 & 0 & 0 \\ 0 & 1 & 0 \\ 0 & 0 & 0.1 \end{bmatrix} \qquad R = \begin{bmatrix} 100 & 0 \\ 0 & 100 \end{bmatrix} \qquad S = \begin{bmatrix} 1 & 0 & 0 \\ 0 & 1 & 0 \\ 0 & 0 & 1 \end{bmatrix}$$

s.t. $\dot{\underline{x}} = A\underline{x} + B\underline{u} + \underline{f}(\underline{y}_e + \underline{x}^o, \underline{u}_e, t) + C_1 \underline{x}^o - A^* \underline{x}^o$

This can be separated into the following two subproblems :

$$\text{I : Min } J = \frac{1}{2}\int_0^T (\underline{x}_1^T Q_1 \underline{x}_1 + (\underline{x}_1 - \underline{x}_1^o)^T S_1 (\underline{x}_1 - \underline{x}_1^T) + u_1^T R_1 u_1)\, dt$$

s.t.

$$\dot{\underline{x}}_1 = \frac{d}{dt}\begin{bmatrix} x_1 \\ x_2 \end{bmatrix} = \begin{bmatrix} 0 & 1 \\ -64.5348 & 0.2703 \end{bmatrix}\begin{bmatrix} x_1 \\ x_2 \end{bmatrix} + \begin{bmatrix} 0 \\ 0 \end{bmatrix} u_1 + \left\{ \begin{bmatrix} f_1(\underline{y}_e + \underline{x}^o, t) \\ f_2(\underline{y}_e + \underline{x}^o, t) \end{bmatrix} \right.$$

$$\left. + \begin{bmatrix} \sum_{i=1}^3 C_{1i} & x_1^o \\ \sum_{i=1}^3 C_{2i} & x_1^o \end{bmatrix} - \begin{bmatrix} \sum_{i=1}^3 a_{1i} & x_1^o \\ \sum_{i=1}^3 a_{2i} & x_1^o \end{bmatrix} \right\}$$

where $Q_1 = \begin{bmatrix} 1 & 0 \\ 0 & 1 \end{bmatrix}$ $S = \begin{bmatrix} 1 & 0 \\ 0 & 1 \end{bmatrix}$ $R = 100$

which is now in the required form

$$\dot{\underline{x}}_1 = A_1 \underline{x}_1 + B_1 \underline{u}_1 + \underline{D}(\underline{x}_1^o, t)$$

and II : $\text{Min } J = \frac{1}{2} \int_0^T (x_3^2 Q_2 + (x_3 - x_3^o)^2 S_2 + u_2^2 R)\, dt$

s.t. $\dot{x}_3 = -0.3222\, x_3 + u_2 + (f_3(\underline{y}_e + \underline{x}^o, u_e, t) - \sum_{i=1}^{3} a_{3i}\, x_1^o + \sum_{i=1}^{3} c_{3i}\, x_1^o.$

The first level problem

It is left as an exercise for the reader to show that the first level problems are to find $\underline{x}_i$, $\underline{u}_i$ independently for the 2 subsystems by solving the following Riccati type equations

i.e. $\dot{P}_i = -P_i A_i - A_i^T P_i + P_i B_i R_i^{-1} B_i^T P_i - (Q_i + S_i)$

with $P_i(T) = 0$

and the disturbance equations

$$\dot{\underline{\xi}}_i = - A_i^T \underline{\xi}_i + P_i B_i R_i^{-1} \underline{\xi}_i - \underline{D}_i + S_i \underline{x}_i^o - \underline{\pi}_i$$

with $\dot{\underline{\xi}}_i(T) = \underline{0}$

where $\underline{x}_i^o, \underline{\pi}_i$ are given by the second level and where the $P_i, \underline{\xi}_i$ are used to give

$$\underline{u}_i = - R_i^{-1} B_i^T \left[P_i \underline{x}_i + \underline{\xi}_i \right]$$

and

$$\dot{\underline{x}}_i = A_i \underline{x}_i - B_i R_i^{-1} B_i^T P_i \underline{x}_i - B_i R_i^{-1} B_i^T \underline{\xi}_i + \underline{D}_i$$

Level II

From the optimality conditions, it is easy to see that the updating at the second level is given by

$$\underline{x}^{ok+1} = \underline{x}^k$$

and

$$\underline{\pi}^{k+1} = - S(\underline{x}^k - \underline{x}^{ok}) + \frac{\partial \underline{D}^{kT}}{\partial \underline{x}^{ok}} \underline{\lambda}^k$$

where $\underline{\lambda}_i^k = P_i \underline{x}_i^k + \underline{\xi}_i^k$

and $\underline{D}^k = \underline{f}(\underline{x}^{ok}) + C_1 \underline{x}^{ok} - A^* \underline{x}^{ok}$

$$\underline{\pi}^{k+1} = - S(\underline{x}^k - \underline{x}^{ok}) - A^* \underline{\lambda}^k + C_1^T + \frac{\partial \underline{f}^{kT}}{\partial_x{}^{ok}}$$

where

$$\frac{\partial \underline{f}^T}{\partial \underline{x}^o} = \begin{bmatrix} 0 & A_2x_2 \cos x_1^o + B_2 \cos 2x_1^o C_2 \sin x_1^o \\ -1 & A_1 & 0 \\ 0 & A_2 \sin x_1^o & C_1 \end{bmatrix}$$

Results

The above algorithm was programmed on the I.B.M. 370 / 165 digital computer. Convergence to the optimum where

$$x_{error} = \sqrt{\sum_{i=1}^{3} \int_0^2 (x_i^{ok+1} - x_i^{ok})^2 \, dt} < 10^{-5}$$

and

$$\pi_{error} = \sqrt{\sum_{i=1}^{3} \int_0^2 (\pi_i^{k+1} - \pi_i^{k})^2 \, dt} < 10^{-5}$$

took place in 13 second level iterations which took 7.63 seconds to execute.

The convergence has been tested also for

$$S = \begin{bmatrix} 0.10 & 0 & 0 \\ 0 & 0.10 & 0 \\ 0 & 0 & 0.10 \end{bmatrix}$$

and convergence took place in 11 second level iterations which were executed in 6.26 seconds. Thus S could be used to improve convergence.

Figs. 8.1 to 8.3 show the optimal state trajectories y_1, y_2, y_3. On these figures, the initial guessed trajectories which were taken to be $\underline{y}_e$ are also drawn, fig. 8.4 shows the control. Note that because of the cost function weights chosen, the control gets close to the desired control quite rapidly whilst the states are damped in more slowly.

For comparison, a solution of the overall problems was available using quasilinearisation. The optimal trajectories in this case were identical to those in figs. 8.1 to 8.4. However, their convergence to the optimum took place in 15 seconds, i.e. in roughly twice the time of the hierarchical solution.

Remark

We have demonstrated a powerful method for the optimisation of non-linear systems. We have seen that it is able to solve non-linear problems rapidly and efficiently. In the original method we included penalty terms in the modified cost function for the intuitive reason that we felt it would aid the speed of convergence. We will next examine if these terms are really necessary. In other words, can we obtain convergence without these terms ?

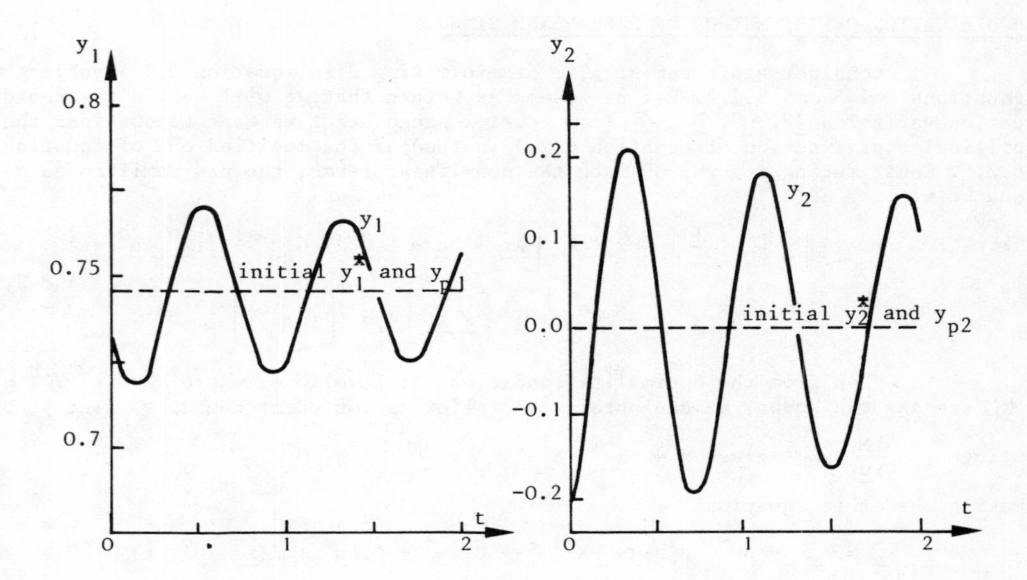

Fig. 8.1 : The optimal state y_1

Fig. 8.2 : The optimal state y_2

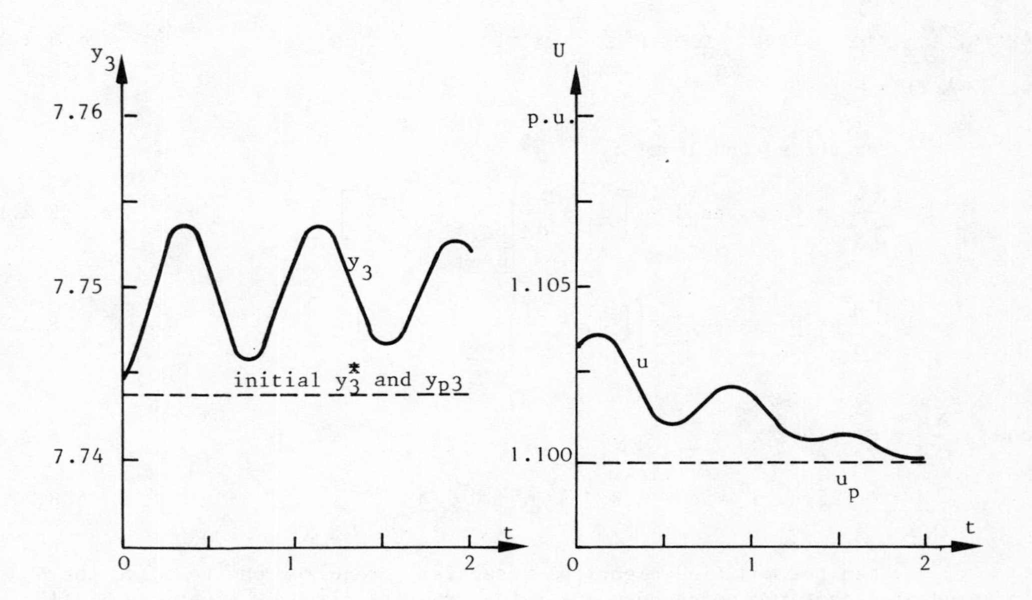

Fig. 8.3 : The optimal state y_3

Fig. 8.4 : The optimal control

MODIFICATION OF THE METHOD OF HASSAN AND SINGH

Consider again the problem of minimising J in equation 8.2.4 subject to equations 8.2.3 and 8.2.2. Let us assume as before that we will obtain the coordination variables $\underline{x}^o$, $\underline{u}^o$, $\underline{\pi}$, $\underline{\beta}$ from level 2 except that we want to consider the original cost function of equation 8.2.4 instead of the modified one of equation 8.2.5. Substituting for $\underline{x}^o$, $\underline{u}^o$ into the non-linear terms, the new Hamiltonian M can be written as

$$M = \frac{1}{2}\|\underline{x}\|_Q^2 + \frac{1}{2}\|\underline{u}\|_R^2 + \underline{\lambda}^T\Big[A\underline{x} + B\underline{u} + C_1\underline{x}^o + C_2\underline{u}^o + \underline{f}(\underline{x}^o, \underline{u}^o, t) - A^* \underline{x}^o - B^* \underline{u}^o\Big] + \underline{\pi}^T\Big[\underline{x}-\underline{x}^o\Big] + \underline{\beta}^T\Big[(\underline{u}-\underline{u}^o)\Big] \qquad 8.2.16$$

Then from the optimality conditions at level one, where $\ell^T = (\underline{x}^{oT}, \underline{\pi}^T, \underline{u}^{oT}, \underline{\beta}^T)^T$ are assumed known, we can obtain the following two point boundary value problem:

using $\dfrac{\partial M}{\partial \underline{u}} = 0$ gives $\underline{u} = -R^{-1}\Big[B^T \underline{\lambda} + \underline{\beta}\Big]$

making the state equation

$$\dot{\underline{x}} = A\underline{x} - BR^{-1}B^T\underline{\lambda} - BR^{-1}\underline{\beta} + C_1\underline{x}^o + C_2\underline{u}^o + \underline{f}(\underline{x}^o, \underline{u}^o, t) - A^*\underline{x}^o - B^*\underline{u}^o \qquad 8.2.17$$

with $\underline{x}(0) = \underline{x}_0$.

Again, $\dfrac{\partial M}{\partial \underline{x}} = -\dot{\underline{\lambda}}$ gives $\dot{\underline{\lambda}} = -A^T\underline{\lambda} - Q\underline{x} - \underline{\pi}$ $\qquad$ 8.2.18

with $\underline{\lambda}(T) = \underline{0}$

For the second level :

$$\frac{\partial M}{\partial \underline{x}^o} = \underline{0} \text{ gives } \underline{\pi} = \Big[\underline{C}_1^T + \frac{\partial \underline{f}}{\partial \underline{x}}\Big|_{\substack{\underline{x}=\underline{x}^o\\ \underline{u}=\underline{u}^o}}^T - A^{*T}\Big]\underline{\lambda} \qquad 8.2.19$$

$$\frac{\partial M}{\partial \underline{u}^o} = \underline{0} \text{ gives } \underline{\beta} = \Big[\Big[C_2^T + \frac{\partial \underline{f}}{\partial \underline{u}}\Big|_{\substack{\underline{x}=\underline{x}^o\\ \underline{u}=\underline{u}^o}}^T - B^{*T}\Big]\underline{\lambda}\Big] \qquad 8.2.20$$

and $\dfrac{\partial M}{\partial \underline{\pi}} = \underline{0}$ gives $\underline{x}^o = \underline{x}$ $\qquad$ 8.2.21

$\dfrac{\partial M}{\partial \underline{\beta}} = \underline{0}$ gives $\underline{u}^o = \underline{u} = -R^{-1}B^T\underline{\lambda} - R^{-1}\underline{\beta}$ $\qquad$ 8.2.22

Then the modified algorithm essentially requires one to solve the N independent linear two point boundary value problems given by equations 8.2.17, 8.2.18 at level 1 for given $\underline{\ell}^T = [\underline{x}^{oT}, \underline{\pi}^T, \underline{u}^{oT}, \underline{\beta}^T]^T$ and at level 2 to improve from iteration k to k+1 by using the simple prediction algorithm

$$\underline{\ell}^{k+1} = \underline{\ell}^k \qquad 8.2.23$$

where the $\underline{\ell}^k$ in 8.2.23 is given by the R.H.S. of equations 8.2.19 to 8.2.21.

The convergence of this method is studied by Singh [5].

Discussion

In the case of this modified algorithm as well, we are able to prove convergence. The assumptions are slightly weaker. Thus, we can use the above algorithm in the place of the one of Hassan and Singh. It remains to be seen whether there is any computational advantage in using the penalty terms in the algorithm of Hassan and Singh [4]. We next investigate this question on our numerical example of the synchronous machine.

The machine example

The problem of controlling the excitation of the synchronous machine is reconsidered using the modified algorithm.

Since in that example the control was linear, H = 0 and the new algorithm can be obtained by setting the matrix S = 0. The problem was solved with the new algorithm such that convergence of x_{error} and π_{error} was obtained to less than 10^{-5}. This took 11 second level iterations to execute.

This is the same number of iterations that were required for the original algorithm. However, the value of x_{error} and π_{error} was smaller than for the original algorithm of Hassan and Singh. Thus the modified algorithm would appear to have slightly faster convergence. In addition the period of optimisation was increased to (0, 6) and for this case whilst the new algorithm converged, the previous one diverged. Thus the new method appears to converge over a longer time horizon [6].

Next, let us see how the basic method can be extended to the more general case of non-linear non separable cost functions.

8.3 EXTENSION TO THE CASE OF NON-LINEAR NON SEPARABLE PROBLEMS

The modified algorithm uses the same sort of a basic idea as the previous method. In the previous method, "desired" state and control trajectories were used to linearise the non-linear dynamics. At the second level the "desired" control and state trajectories were forced towards the "optimal" states and controls using a prediction type method. In the new method we use the given desired "state" and control trajectories to make the cost function a separable quadratic function (by adding and subtracting certain terms) and the dynamics linear and then, as in the previous method, we improve these "desired" trajectories at the second level in order to force them towards the optimal states and controls.

Formulation of the problem and the new algorithm

The optimisation problem is to minimise :

$$J = \int_0^T g(\underline{x}, \underline{u}, t)\, dt \qquad 8.3.1$$

subject to the non-linear dynamical constraints

$$\dot{\underline{x}} = \underline{f}(\underline{x}, \underline{u}, t) \qquad 8.3.2$$

It is assumed that the dynamical system given by equation 8.3.2 is asymptotically stable and that it has an equilibrium point at the origin.

i.e. $\underline{f}(\underline{x}, \underline{u}, t) = \underline{0}$

For the sake of convenience let the equilibrium point obtained be

$$\underline{x}^o = \underline{0} \qquad \underline{u}^o = \underline{0}$$

Expanding $\underline{f}(\underline{x}, \underline{u}, t)$ in a Taylor series about this point up to first order, we have :

$$\underline{f}(\underline{x}, \underline{u}, t) = \underline{f}(\underline{x}^o, \underline{u}^o, t) + \left[\frac{\partial \underline{f}^T}{\partial \underline{x}}\right]^T_{\underline{x}^o, \underline{u}^o} \underline{x} + \left[\frac{\partial \underline{f}^T}{\partial \underline{u}}\right]^T_{\underline{x}^o, \underline{u}^o} \underline{u} \qquad 8.3.3$$

Substituting for $\underline{f}$ in 8.3.2 from 8.3.3

$$\dot{\underline{x}} = A^* \underline{x} + B^* \underline{u} + \underline{f}(\underline{x}, \underline{u}, t) - A^* \underline{x} - B^* \underline{u} \qquad 8.3.4$$

where $A^* = \left[(\frac{\partial \underline{f}^T}{\partial \underline{x}})^T \right]$, $B^* = \left[(\frac{\partial \underline{f}^T}{\partial \underline{u}})^T \right]$

or $\dot{\underline{x}} = A\underline{x} + B\underline{u} + \underline{D}(\underline{x}, \underline{u}, t)$ 8.3.5

where

$$\underline{D}(\underline{x}, \underline{u}, t) = C_1 \underline{x} + C_2 \underline{u} + \underline{f}(\underline{x}, \underline{u}, t) - A^* \underline{x} - B^* \underline{u}$$

or $\underline{D}(\underline{x}, \underline{u}, t) = \underline{f}(\underline{x}, \underline{u}, t) - A\underline{x} - B\underline{u}$ 8.3.6

where C_1, C_2 are the off diagonal parts of A^*, B^* and A, B are block diagonal. Note that the size of the blocks in A and B is essentially chosen such that each block is of roughly similar dimension to the other block since this will minimise the computational load.

Next let us make the cost function separable quadratic in $\underline{x}$ and $\underline{u}$ by adding and subtracting certain terms i.e.

$$J = \int_0^T \left[\frac{1}{2} \| \underline{x} \|^2_Q + \frac{1}{2} \| \underline{u} \|^2_R + G(\underline{x}, \underline{u}, t) \right] dt \qquad 8.3.7$$

where $G(\underline{x}, \underline{u}, t) = g(\underline{x}, \underline{u}, t) - \frac{1}{2} \| \underline{x} \|^2_Q - \frac{1}{2} \| \underline{u} \|^2_R$

where Q, R are block diagonal matrices with blocks corresponding to those of A and B respectively. In this algorithm, "desired" states $\underline{x}^*$ and controls $\underline{u}^*$ will be used to fix $G(\underline{x}, \underline{u}, t)$ in the cost function and $\underline{D}(\underline{x}, \underline{u}, t)$ in the dynamics by the introduction of additional constraints $\underline{x} = \underline{x}^*$, $\underline{u} = \underline{u}^*$ which will be satisfied at the second level. Also, an additional penalty term $1/2 \| \underline{u} - \underline{u}^* \|^2_H$ can be added into the cost function to aid convergence. Thus the new problem becomes :

$$\text{Min } J' = \int_0^T \left[\frac{1}{2} (\| \underline{x} \|^2_Q + \| \underline{u} \|^2_R) + G(\underline{x}^*, \underline{u}^*, t) + \frac{1}{2} \| \underline{u} - \underline{u}^* \|^2_H \right] dt \qquad 8.3.8$$

$$\text{s.t } \dot{\underline{x}} = A\,\underline{x}(t) + B\,\underline{u}(t) + \underline{D}(\underline{x}^*, \underline{u}^*, t) \qquad 8.3.9$$

$$\underline{x}^* = \underline{x} \qquad 8.3.10$$

$$\underline{u}^* = \underline{u} \qquad 8.3.11$$

Clearly, if $\underline{x} \simeq \underline{x}^*$, $\underline{u} \simeq \underline{u}^*$, the original non-linear optimisation problem is identical to the above problem. In order to force $\underline{x}^*$ towards $\underline{x}$ and $\underline{u}^*$ towards $\underline{u}$ whilst satisfying the dynamic constraints and minimising J' in equation 8.3.8 write the Hamiltonian of the overall system as

$$H = \frac{1}{2}\|\underline{x}\|_Q^2 + \frac{1}{2}\|\underline{u}\|_R^2 + G(\underline{x}^*, \underline{u}^*, t) + \frac{1}{2}\|\underline{u}-\underline{u}^*\|_H^2 + \underline{\lambda}^T(A\underline{x} + B\underline{u} + \underline{D}(\underline{x}^*, \underline{u}^*, t))$$

$$+ \underline{\beta}^T(\underline{x}-\underline{x}^*) + \underline{\nu}^T(\underline{u}-\underline{u}^*)$$

where $\underline{\lambda}$ is the adjoint vector whilst $\underline{\beta}$ and $\underline{\nu}$ are Lagrange multipliers associated with the constraints 8.3.10 and 8.3.11.

The necessary conditions for optimality are :

$$\frac{\partial H}{\partial \underline{u}} = \underline{0},\quad \frac{\partial H}{\partial \underline{\lambda}} = \dot{\underline{x}},\quad \frac{\partial H}{\partial \underline{x}} = -\dot{\underline{\lambda}},\quad \frac{\partial H}{\partial \underline{\beta}} = \underline{0},\quad \frac{\partial H}{\partial \underline{\nu}} = \underline{0},\quad \frac{\partial H}{\partial \underline{x}^*} = \underline{0},\quad \frac{\partial H}{\partial \underline{u}^*} = \underline{0}$$

and these give

$$\underline{u}(t) = -R^{*-1}\left[B^T\underline{\lambda}(t) + \underline{\nu}(t) - H\underline{u}^*(t)\right] \qquad 8.3.12$$

where $R^* = R + H$.

$$\dot{\underline{x}}(t) = A\underline{x} - BR^{*-1}\left[B^T\underline{\lambda}(t) + \underline{\nu}(t) - H\underline{u}^*(t)\right] + \underline{D}(\underline{x}^*, \underline{u}^*, t) \qquad 8.3.13$$

with $\underline{x}(0) = \underline{x}_0$

$$\dot{\underline{\lambda}} = -Q\underline{x}(t) - A^T\underline{\lambda}(t) - \underline{\beta}(t) \qquad 8.3.14$$

$$\underline{x}^*(t) = \underline{x}(t) \qquad 8.3.15$$

$$\underline{u}^*(t) = \underline{u}(t) \qquad 8.3.16$$

$$\underline{\beta}(t) = \frac{\partial G(\underline{x}^*, \underline{u}^*, t)}{\partial \underline{x}^*} + \frac{\partial \underline{D}^T(\underline{x}, \underline{u}, t)}{\partial \underline{x}^*}\underline{\lambda}(t)$$

where $\dfrac{\partial G}{\partial \underline{x}^*} = \dfrac{\partial g(\underline{x}, \underline{u}, t)}{\partial \underline{x}^*} - Q\underline{x}^*$

$$\frac{\partial \underline{D}^T}{\partial \underline{x}^*} = \frac{\partial \underline{f}^*(\underline{x}^*, \underline{u}^*, t)}{\partial \underline{x}^*} - A^T$$

$$\underline{\beta}(t) = \frac{\partial g(\underline{x}, \underline{u}, t)}{\partial \underline{x}^*} - Q\underline{x}^* + \left[\frac{\partial \underline{f}^T(\underline{x}^*, \underline{u}^*, t)}{\partial \underline{x}^*} - A^T\right]\underline{\lambda}(t) \qquad 8.3.17$$

$$\underline{\nu}(t) = \frac{\partial g}{\partial \underline{u}^*} - R\underline{u}^* + H\underline{u}^* - H\left[-R^{*-1}B^T\underline{\lambda} - R^{*-1}\underline{\nu} + R^{*-1}H\underline{u}^*\right] + \left[\frac{\partial \underline{f}^T}{\partial \underline{u}^*} - B^T\right]\underline{\lambda}(t) \qquad 8.3.18$$

Essentially, the hierarchical algorithm iteratively satisfies equations 8.3.12 to 8.3.18 within a two level structure as follows :

Level 1

At the first level the following N independent* two point boundary value problems are solved i.e.

$$\dot{\underline{x}}(t) = A\underline{x} - BR^{*-1}\left[B^T\underline{\lambda}(t) + \underline{\nu}(t) - H\underline{u}^*(t)\right] + \underline{D}(\underline{x}^*, \underline{u}^*, t)$$

with $\underline{x}(0) = \underline{x}_o$

$$\dot{\underline{\lambda}} = -Q\underline{x} - A^T\underline{\lambda} - \underline{\beta}(t) \quad \text{with} \quad \underline{\lambda}(T) = \underline{0}.$$

These constitute N linear two point boundary value problems for given $\underline{x}^*$, $\underline{u}^*$, $\underline{\beta}$, $\underline{\nu}$ provided by the second level. In order to solve them, let $\underline{\lambda} = K\underline{x} + \underline{\xi}$ which on substituting and manipulating leads to N independent low order Riccati equations which need to be solved only once and a vector linear differential equation in $\underline{\xi}$ which is solved subsystem by subsystem at each iteration (i.e. for each trajectory of $\underline{x}^*$, $\underline{u}^*$, $\underline{\beta}$, $\underline{\nu}$ provided by level 2).

Level 2

At level 2, new improved values for $\underline{x}^*$, $\underline{u}^*$, $\underline{\beta}$, $\underline{\nu}$ are provided by substituting the $\underline{x}$, $\underline{u}$ obtained at level one into the right hand sides of equations 8.3.15 to 8.3.18 respectively.

We can show that the present two level algorithm is a fixed point algorithm if a certain condition is satisfied.

Singh and Hassan [6] have shown that under certain conditions, the algorithm converges uniformly to the optimum.

Next, we illustrate the kind of situation where the new algorithm will be particularly useful.

Example

Whilst previous algorithms for non-linear systems could cope with the non-linear dynamics, the cost function had to be a separable quadratic form. Let us consider as a very simple example a non-separable cost function in order to illustrate the application of the last algorithm**

$$\text{Min } J = \frac{1}{2}\int_0^T (x_1 x_2 + x_2^2 + u_1^2 + u_2^2)\, dt \qquad 8.3.19$$

where we have the non-separable term $x_1 x_2$: the dynamics are :

---

* These first level problems are independent because A, B, R^*, H, Q are all block diagonal matrices so the equations for $\underline{x}, \underline{\lambda}$, split up into N blocks of lower order equations.

** Another standard way to solve the problem is to write :

$$\min J = \frac{1}{2}\int_0^T (x_1 W_1 + u_1^2) + (x_2^2 + u_2^2)\, dt$$

$$\text{s.t. } W_1 - x_2 = 0$$

in place of 8.3.19. The same procedure is used for 8.3.20, but this leaves a difficult non-linear problem to be solved at the lower level.

$$\dot{x}_1 = -x_2 + x_1 \sin x_2 + u_1 \qquad 8.3.20$$
$$\dot{x}_2 = \cos x_2 + x_1 x_2 + u_2 + C$$

Let us develop these dynamic equations around the equilibrium point as

$$\dot{\underline{x}} = A\underline{x} + B\underline{u} + \underline{f}(\underline{x}, t) - A\underline{x} + \underline{C}$$

or

$$\dot{\underline{x}} = A_b \underline{x} + B\underline{u} + \underline{D}(x, t) + \underline{C}$$

where A_b, B are diagonal 2 x 2 matrices.

Let us modify the problem to the form

$$\text{Min } J = \frac{1}{2} \int_0^T (x_1^2 + x_2^2 + u_1^2 + u_2^2 + x_1^* x_2^* - x_1^{*2})\, dt$$

subject to

$$\dot{x}_1 = a_{11} x_1 + u_1 + (x_1^* \sin x_2^* - x_2^* - a_{11} x_1^*)$$
$$\dot{x}_2 = a_{22} x_2 + u_2 + C + (\cos x_2^* + x_1^* x_2^* - a_{22} x_2^*)$$
$$x_1^* = x_1$$
$$x_2^* = x_2$$

In order to solve this problem, write the Hamiltonian as

$$H = \frac{1}{2} (x_1^2 + x_2^2 + u_1^2 + u_2^2 - x_1^{*2} + x_1^* x_2^*) +$$
$$\lambda_1(a_{11}x_1 + u_1 + x_1^* \sin x_2^* - x_2^* - a_{11}^* x_1^*) + \lambda_2(a_{22}x_2 + u_2 + C + \cos x_2^* + x_1^* x_2^* - a_{22} x_2^*) +$$
$$\nu_1(x_1 - x_1^*) + \nu_2 (x_2 - x_2^*)$$

Then from the necessary conditions for optimality we have :

$$\frac{\partial H}{\partial u_1} = 0 \qquad \text{or } u_1 = -\lambda_1 \qquad 8.3.21$$

$$\frac{\partial H}{\partial u_2} = 0 \qquad \text{or } u_2 = -\lambda_2 \qquad 8.3.22$$

$$\frac{\partial H}{\partial \lambda_1} = \dot{x}_1 = a_{11} x_1 - \lambda_1 + x_1^* \sin x_2^* - x_2^* - a_{11} x_1^* \qquad 8.3.23$$
$$x_1(0) = x_{10}$$

$$\frac{\partial H}{\partial \lambda_2} = \dot{x}_2 = a_{22} x_2 - \lambda_2 + C + \cos x_2^* + x_1^* x_2^* - a_{22} x_2^* \qquad 8.3.24$$
$$x_2(0) = x_{20}$$

$$\frac{\partial H}{\partial x_1} = -\dot{\lambda}_1 = x_1 + a_{11} + \nu_1 \qquad ; \lambda_1 (T) = 0 \qquad 8.3.25$$

$$\frac{\partial H}{\partial x_2} = -\dot{\lambda}_2 = x_2 + a_{22} + \nu_2 \quad ; \lambda_2 (T) = 0 \qquad 8.3.26$$

$$\frac{\partial H}{\partial \nu_1} = 0 \qquad x_1^* = x_1 \qquad 8.3.27$$

$$\frac{\partial H}{\partial \nu_2} = 0 \qquad x_2^* = x_2 \qquad 8.3.28$$

$$\frac{\partial H}{\partial x_1^*} = 0 \qquad \nu_1 = -x_1^* + x_2^* + \lambda_1 \sin x_2^* - a_1^* \lambda_1 + \lambda_2 x_2^* \qquad 8.3.29$$

$$\frac{\partial H}{\partial x_2^*} = 0 \qquad \nu_2 = x_1^* + \lambda_1 x_1^* \cos x_2^* - \lambda_1 - \lambda_2 \sin x_2^* + \lambda_2 x_1^* - a_{22} \lambda_2$$

8.3.30

In this algorithm $x_1^*, x_2^*, \nu_1, \nu_2$ are supplied by the second level. Using these fixed trajectories, the two independent linear two point boundary value problems given by equations 8.3.23-25and 8.3.24-26 are solved by integrating in each case a scaler Riccati equation and a scaler linear equation. At level 2, the x_1^*, x_2^*, ν_1, ν_2 trajectories are improved by substituting the values obtained at level 1 into the right hand sides of equations 8.3.27 to 8.3.30.

Next, let us consider the costate coordination method for discrete systems. The original costate coordination method was developed by Mahmoud et al. [8] . Here we present a modification of their method which has attractive computational properties.

8.4 THE COSTATE PREDICTION METHOD

Problem formulation and the two-level costate prediction solution structure

The problem is to minimise

$$J = \frac{1}{2} \sum_{k=k_o}^{k_f-1} \left(\| \underline{x}(k) \|_Q^2 + \| \underline{u}(k) \|_R^2 \right) \qquad 8.4.1$$

subject to the discrete dynamical constraints

$$\underline{x}(k+1) = \underline{f} \left[\underline{x}, \underline{u}, (k+1, k) \right] \qquad 8.4.2$$

Rewrite equation 8.4.2 in the form

$$\underline{x}(k+1) = \phi\left[\underline{x},\underline{u},(k+1,k)\right] \underline{x}(k) + \Gamma\left[\underline{x},\underline{u},(k+1,k)\right] \underline{u}(k) + \underline{D}\left[\underline{x},\underline{u},(k+1,k)\right] \qquad 8.4.3$$

where ϕ and Γ are block diagonal matrices with

$$\underline{D}(\underline{x},\underline{u},(k+1,k)) = C_1 \left[\underline{x},\underline{u},(k+1,k) \right] \underline{x}(k) + C_2 \left[\underline{x},\underline{u},(k+1,k) \right] \underline{u}(k) \qquad 8.4.4$$

Note that ϕ , Γ , C_1, C_2 now contain non-linear terms in $\underline{x}$, $\underline{u}$, multiplicative and other non-linearities etc. and that it is always possible to write the non-linear function $\underline{f}$ in equation 8.4.2 in the form of equations 8.4.3 and 8.4.4. If there are

n_1 blocks in ϕ and n_1 blocks in Γ, then it is assumed that Q and R are also block diagonal with n_1 blocks in Q corresponding to the blocks in ϕ and n_1 in R corresponding to those in Γ.

The first main idea in this method for solving the problem is to provide 'predicted' vectors x^* and u^* and to use these for $\underline{x}$ and $\underline{u}$ in equations 8.4.3 and 8.4.4 to 'fix' the non-linearities in ϕ, Γ and $\underline{D}$. Thus, rewrite the problem as :

$$\text{Min } J' = \frac{1}{2} \sum_{k=k_o}^{k_f-1} \left(\| \underline{x}(k) \|_Q^2 + \| \underline{u}(k) \|_Q^2 \right) \qquad 8.4.5$$

subject to

$$\underline{x}(k+1) = \phi\left[\underline{x}^*, \underline{u}^*, (k+1,k)\right] \underline{x}(k) + \Gamma\left[\underline{x}^*, \underline{u}^*, (k+1,k)\right] \underline{u}(k) + \underline{D}\left[\underline{x}^*, \underline{u}^*(k+1,k)\right] \quad 8.4.6$$

$$\underline{x}^*(k) = \underline{x}(k) \qquad 8.4.7$$

$$\underline{u}^*(k) = \underline{u}(k) \qquad 8.4.8$$

To solve this new problem, write the Hamiltonian as

$$H = \frac{1}{2} \| \underline{x}(k) \|_Q^2 + \frac{1}{2} \| \underline{u}(k) \|_Q^2 + \underline{\lambda}^T(k+1) \Big[\phi\left[\underline{x}^*, \underline{u}^*, (k+1,k)\right] \underline{x}(k)$$
$$+ \Gamma\left[\underline{x}^*, \underline{u}^*, (k+1,k)\right] \underline{u}(k) + \underline{D}\left[\underline{x}^*, \underline{u}^*, (k+1,k)\right] \Big] \qquad 8.4.9$$
$$+ \underline{\beta}^T(k) \left[\underline{x}(k) - \underline{x}^*(k) \right] + \underline{\gamma}(k)^T \left[\underline{u}(k) - \underline{u}^*(k) \right]$$

Note that the Hamiltonian H in equation 8.4.9 is additively separable for given $\underline{x}^*$, $\underline{u}^*$, i.e.

$$H = \sum_{i=1}^{n_1} H_i$$

where

$$H_i = \frac{1}{2} \| \underline{x}_i(k) \|_{Q_i}^2 + \frac{1}{2} \| \underline{u}_i(k) \|_{R_i}^2 + \underline{\lambda}_i^T(k+1) \Big[\phi_i \underline{x}_i(k) + \Gamma_i \underline{u}_i(k)$$
$$+ \underline{D}_i \left[\underline{x}^*, \underline{u}^*, (k+1,k) \right] \Big] + \underline{\beta}_i^T(k) \left[\underline{x}_i(k) - \underline{x}_i^*(k) \right] + \underline{\gamma}_i^T(k) \left[\underline{u}_i(k) - \underline{u}_i^*(k) \right]$$

Then from the necessary conditions for optimality, i.e. $\frac{\partial H_i}{\partial \underline{u}_i} = \underline{0}$, gives

$$\underline{u}_i(k) = -R_i^{-1} \left[\Gamma_i^T \left[\underline{x}^*, \underline{u}^*, (k+1,k) \right] \underline{\lambda}_i(k+1) + \underline{\gamma}_i(k) \right] \qquad 8.4.10$$

Using

$$\underline{x}_i(k+1) = \frac{\partial H_i}{\partial \underline{\lambda}_i(k+1)}$$

gives

$$\underline{x}_i(k+1) = \phi_i\left[\underline{x}^*, \underline{u}^*, (k+1,k)\right]\underline{x}_i(k) - \Gamma_i\left[\underline{x}^*, \underline{u}^*, (k+1)\right]R_i^{-1}$$
$$\left\{\Gamma_i^T\left[\underline{x}^*, \underline{u}^*, (k+1,k)\right]\underline{\lambda}_i(k+1) + \underline{\gamma}_i(k)\right\} + \underline{D}_i\left[\underline{x}^*, \underline{u}^*, (k+1,k)\right] \qquad 8.4.11$$

with
$$\underline{x}(k_o) = \underline{x}_o$$

Using
$$\frac{\partial H_i}{\partial \underline{x}_i(k)} = -\underline{\lambda}_i(k)$$

gives
$$\underline{\lambda}_i(k) = Q_i\underline{x}_i(k) + \phi_i^T\left[\underline{x}^*, \underline{u}^*, (k+1,k)\right]\underline{\lambda}_i(k+1) + \underline{\beta}_i(k) \qquad 8.4.12$$

with
$$\underline{\lambda}_i(k_f) = \underline{0}$$

Using
$$\frac{\partial H}{\partial \underline{\beta}} = \underline{0}$$

gives
$$\underline{x}^*(k) = \underline{x}(k) \qquad 8.4.13$$

and using
$$\frac{\partial H}{\partial \underline{\gamma}} = \underline{0}$$

gives
$$\underline{u}^*(k) = \underline{u}(k) \qquad 8.4.14$$

Using
$$\frac{\partial H}{\partial \underline{x}^*(k)} = 0$$

$$\underline{\beta}(k) = \left[\frac{\partial \underline{z}^T}{\partial \underline{x}^*}\left[\underline{x}^*, \underline{u}^*, \underline{x}, (k+1,k)\right] + \frac{\partial \underline{y}^T}{\partial \underline{x}^*}\left[\underline{x}^*, \underline{u}^*, \underline{u}, (k+1,k)\right] + \frac{\partial \underline{D}^T}{\partial \underline{x}^*}\left[\underline{x}^*, \underline{u}^*, (k+1,k)\right]\right]\underline{\lambda}(k+1) \qquad 8.4.15$$

where
$$\underline{z}\left[\underline{x}^*, \underline{u}^*, \underline{x}, (k+1,k)\right] = \phi\left[\underline{x}^*, \underline{u}^*, (k+1,k)\right]\underline{x}(k)$$
$$\underline{y}\left[\underline{x}^*, \underline{u}^*, \underline{u}, (k+1,k)\right] = \Gamma\left[\underline{x}^*, \underline{u}^*, (k+1,k)\right]\underline{u}(k)$$

Also
$$\frac{\partial H}{\partial \underline{u}^*(k)} = \underline{0}$$

so that
$$\underline{\gamma}(k) = \frac{\partial \underline{z}^T}{\partial \underline{u}^*}\left[\underline{x}^*, \underline{u}^*, \underline{x}, (k+1,k)\right] + \frac{\partial \underline{y}^T}{\partial \underline{u}^*}\left[\underline{x}^*, \underline{u}^*, (k+1,k)\right] + \frac{\partial \underline{D}^T}{\partial \underline{u}^*}\left[\underline{x}^*, \underline{u}^*, (k+1,k)\right]\underline{\lambda}(k+1) \qquad 8.4.16$$

Thus to solve this problem it is necessary to satisfy equations 8.4.10 to 8.4.16. It is possible to do so using the following two level algorithm :

Step 1

Guess the vector sequences $\underline{\lambda}$, $\underline{x}^*$, $\underline{u}^*$, $\underline{\gamma}$. Set the iteration index L=1.

Step 2

Substitute the sequences $\underline{\lambda}^L(k)$, $\underline{x}^{*L}(k)$, $\underline{u}^{*L}(k)$, $\underline{\gamma}(k)$ $(k_o \leqslant k \leqslant k_{f-1})$ into the right hand sides of equations 8.4.11 and 8.4.10 to obtain the sequences $\underline{x}_i^L(k)$, $\underline{u}_i^L(k)$ $(i = 1, 2, \ldots, n_1)$ $(k_o \leqslant k \leqslant k_{f-1})$ and then into 8.4.15 to obtain $\underline{\beta}^L(k)$.

Step 3

Update $\underline{\ell}^L = \left[\underline{\lambda}_i^L, \underline{x}^{*L}, \underline{u}^{*L}, \underline{\gamma}^L\right]^T$ by substituting $\underline{x}_i^L$, $\underline{u}_i^L$ directly into the equations 8.4.12 to 8.4.16., i.e.

$$\underline{\lambda}_i^{L+1}(k) = Q_i\underline{x}_i^L(k) + \underline{\phi}_i^T\left[\underline{x}^*, \underline{u}^*, (k+1,k)\right]\underline{\lambda}_i^{L+1}(k+1) + \underline{\beta}_i^L(k)$$

$$\underline{x}^{*L+1}(k) = \underline{x}^L(k)$$

$$\underline{u}^{*L+1}(k) = \underline{u}^L(k)$$

$$\underline{\gamma}^{L+1}(k) = \left\{ \frac{\partial \underline{z}^T}{\partial \underline{u}^*}\left[\underline{x}^*, \underline{u}^*, \underline{x}, (k+1, k)\right]\Bigg|_{(\underline{x}^{*L}, \underline{u}^{*L}, \underline{x}^L)} + \frac{\partial \underline{y}^T}{\partial \underline{u}^*}\left[\underline{x}^*, \underline{u}^*, \underline{u}, (k+1)\right]\Bigg|_{(\underline{x}^{*L}, \underline{u}^{*L}, \underline{x}^L)} + \frac{\partial \underline{D}^T}{\partial \underline{u}^*}\left[\underline{x}^*, \underline{u}^*, (k+1, k)\right]\Bigg|_{(\underline{x}^{*L}, \underline{u}^{*L})} \right\} \underline{\lambda}(k+1)$$

If $\underline{\ell}^{L+1} \simeq \underline{\ell}^L$, stop and record $\underline{u}$ as the optimal control, or else go back to Step 2.

Hassan and Singh [9] have shown that there exists an interval (k_o, k_{f-1}) within which the above algorithm converges uniformly thus satisfying equations 8.4.10 to 8.4.16 provided that

(a) $\underline{\ell}$, $\underline{y}$, $\underline{z}$ and $\underline{D}$ are bounded functions of the stage index k.

(b) $\underline{y}$, $\underline{z}$ and $\underline{D}$ are differentiable w.r.t. $\underline{x}^*$ and $\underline{u}^*$ at each stage k $(k_o \leqslant k \leqslant k_{f-1})$ and their derivatives are bounded.

Remarks

1. The first level algorithm is trivially simple since it merely requires substitution into the right hand side of equations 8.4.14 and 8.4.10 using the given vector $\underline{\ell}$ obtained from the second level and the resulting $\underline{x}$, $\underline{u}$ are collated and substituted into the right hand side of equation 8.4.15 to obtain $\underline{\beta}$. This level's task is much simpler than in previous algorithms since it is not necessary to solve either a two point boundary value problem (as in the goal coordination method of section 8.1.), or even a Riccati type equation for the

subproblems as in the previous method of Hassan and Singh of section 8.2. This has been achieved by using the costates as a part of the coordination vector.

2. The second level algorithm is also trivially simple since it also requires mere substitution of the $\underline{x}$, $\underline{u}$, $\underline{\beta}$ vectors that have been obtained in the previous iteration to yield the new coordination vector $\underline{\ell}$. The second level task is of similar magnitude to the task in the algorithm of section 8.2. and since the lower level task is easier than in that method, one would expect the approach to give even faster convergence than in that method. Convergence of the present approach will be compared with that of the method of section 8.2. on the machine example.

3. One of the difficulties of the algorithm of section 8.2. was that since the non-linear dynamic equality constraints were expanded in a Taylor series about the fixed point of the system, it was necessary to determine the fixed point of the system by a priori experimentation. This difficulty does not arise in the present algorithm.

4. From the convergence condition of Hassan and Singh [9] it would appear that the number of stages in the multistage optimisation problem could be small. It may be possible to increase this by

(a) using a form of smoothing, i.e. instead of taking $\underline{\ell}^{L+1}$ as the next iterate, use the relaxation procedure :

$$\bar{\underline{\ell}}^{L+1} = \varepsilon\, \underline{\ell}^{L+1} + (1-\varepsilon)\, \underline{\ell}^{L} \qquad 0<\varepsilon<1$$

(b) Alternatively, from the convergence condition, it is not too difficult to show that if R is "large" and Q is "small", the number of stages can be large. This fact can be used to add an additional term of the form

$$\frac{1}{2}\, \| \underline{u} - \underline{u}^* \|^2_{R_A}$$

into the cost function of equation 8.4.1. Then the number of stages could be increased by choosing R_A to be sufficiently "large". Note that this trick does not affect the final result since the additional term goes to zero at the optimum.

The method is next demonstrated on the two examples.

Example 1

The first example is the one previously treated by Mahmoud et al. [8]. The problem is to minimise

$$J = \sum_{k=0}^{50} \left[0.05\, x_1^2(k) + 0.05\, x_2^2(k) + 0.1\, u_1^2(k) + 0.05\, u_2^2(k) \right]$$

subject to

$$\begin{bmatrix} x_1(k+1) \\ x_2(k+1) \end{bmatrix} = \begin{bmatrix} 0.9 & 0.1 \\ 0.2 & (0.1 - 0.1\ x_2(k)) \end{bmatrix} \begin{bmatrix} x_1(k) \\ x_2(k) \end{bmatrix} + \begin{bmatrix} 0.1 & 0 \\ 0 & 0.1 \end{bmatrix} \begin{vmatrix} u_1(k) \\ u_2(k) \end{vmatrix}$$

with $x_1(0) = 10.0$, $x_2(0) = 4.5$.

The only non-linearity arises in the term $-0.1\ x_2^2$ in the equation for x_2.

The Hamiltonian for this problem can be written as

$$H = \Big\{ 0.05\, x_1^2(k) + 0.1\, u_1^2(k) + 0.1\, \lambda_1(k+1) \big[0.9\, x_1(k) + 0.1\, x_2^*(k) + 0.1\, u_1(k)\big] + 0.1 \beta_1 \big[x_1(k) - x_1^*(k)\big] \Big\} + \Big\{ 0.05\, x_2^2(k) + 0.05\, u_2^2(k) + 0.1 \lambda_2(k+1) \big[0.1\, x_2(k) + 0.2\, x_1^*(k) - 0.1\, x_2^*(k)^2\, x_2(k) + 0.1\, u_2(k)\big] + 0.1\, \beta_2 \big[x_2(k) - x_2^*(k)\big] \Big\}$$

Here $x_1^*(k) = x_1(k)$, $x_2^*(k) = x_2(k)$ have been included as extra constraints and β_1, β_2 are the multipliers associated with these constraints.

The coordination vector for this problem is given by

$$\underline{\ell} = \begin{bmatrix} \lambda_1 \\ \lambda_2 \\ x_1^* \\ x_2^* \end{bmatrix}$$

and at the first level it is necessary to calculate x_1, x_2 and then use these to calculate β_1, β_2. From the necessary conditions for optimality, the first level formulae are :

For given $\underline{\ell}$,

$$\frac{\partial H}{\partial u_1} = 0 \text{ gives } u_1(k) = -0.05\, \lambda_1(k+1)$$

$$\frac{\partial H}{\partial u_2} = 0 \text{ gives } u_2(k) = -0.1\, \lambda_2(k+1)$$

$$\frac{\partial H}{\partial \lambda_1(k+1)} = 0 \text{ gives } x_1(k+1) = 0.9\, x_1(k) + 0.1\, x_2^*(k) + 0.1\, u_1(k)$$

$$\frac{\partial H}{\partial \lambda_2(k+1)} = 0 \text{ gives } x_2(k+1) = 0.1\, x_2(k) + 0.2\, x_1^*(k) - 0.1\, x_2^*(k)\, x_2(k) + 0.1\, u_2(k)$$

with $x_1(0) = 10$, $x_2(0) = 4.5$.

At the second level $\underline{x}^*$, $\underline{\lambda}$ are improved by the algorithm :

From iteration L to L+1

$$\begin{bmatrix} \lambda_1(k) \\ \lambda_2(k) \\ x_1^*(k) \\ x_2^*(k) \end{bmatrix}^{L+1} = \begin{bmatrix} x_1(k) + 0.9\, \lambda_1(k+1) + \beta_1(k) \\ x_2(k) + 0.1\, \lambda_2(k+1) + \beta_2(k) \\ x_1 \\ x_2 \end{bmatrix}^{L}$$

with $\lambda_1(51) = \lambda_2(51) = 0$.

The above formulae were programmed on the I.B.M. 370/165 digital computer. Convergence to the optimum within the tolerance of 0.001 was achieved in 13 second level iterations. At the optimum, the minimum cost was found to be J = 33.7373. Figs. 8.5 and 8.6 show the optimal trajectories of x_1, x_2, u_1, u_2.

Example 2

As a second example, consider the synchronous machine excitation problem. The problem is to minimise the cost function

$$J = \int_0^2 \left[(y_1 - 0.7461)^2 + y_2^2 + 0.1(y_3 - 7.7438)^2 + 100(u_1 - 1.1)^2 \right] dt$$

subject to the non-linear dynamical constraints :

$$\dot{y}_1 = y_2 \; ; \; \dot{y}_2 = 39.1892 - 0.2703\, y_2 - 12.012\, y_3 \sin y_1 + 24.024 \sin 2y_1$$

$$\dot{y}_3 = u_1 - 0.3222\, y_3 - 1.9 \cos y_1$$

with $\underline{y}(0)^T = \left[0.500 \quad -0.3999 \quad 8.000 \right]$

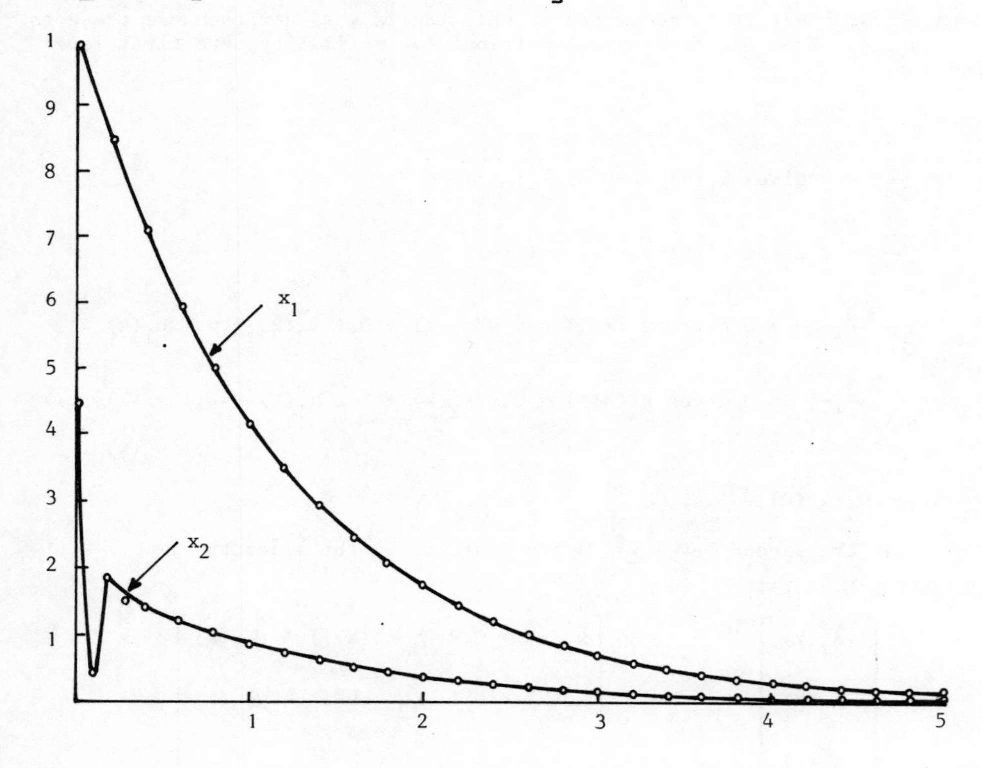

Fig. 8.5 : States for example 1 of section 8.4

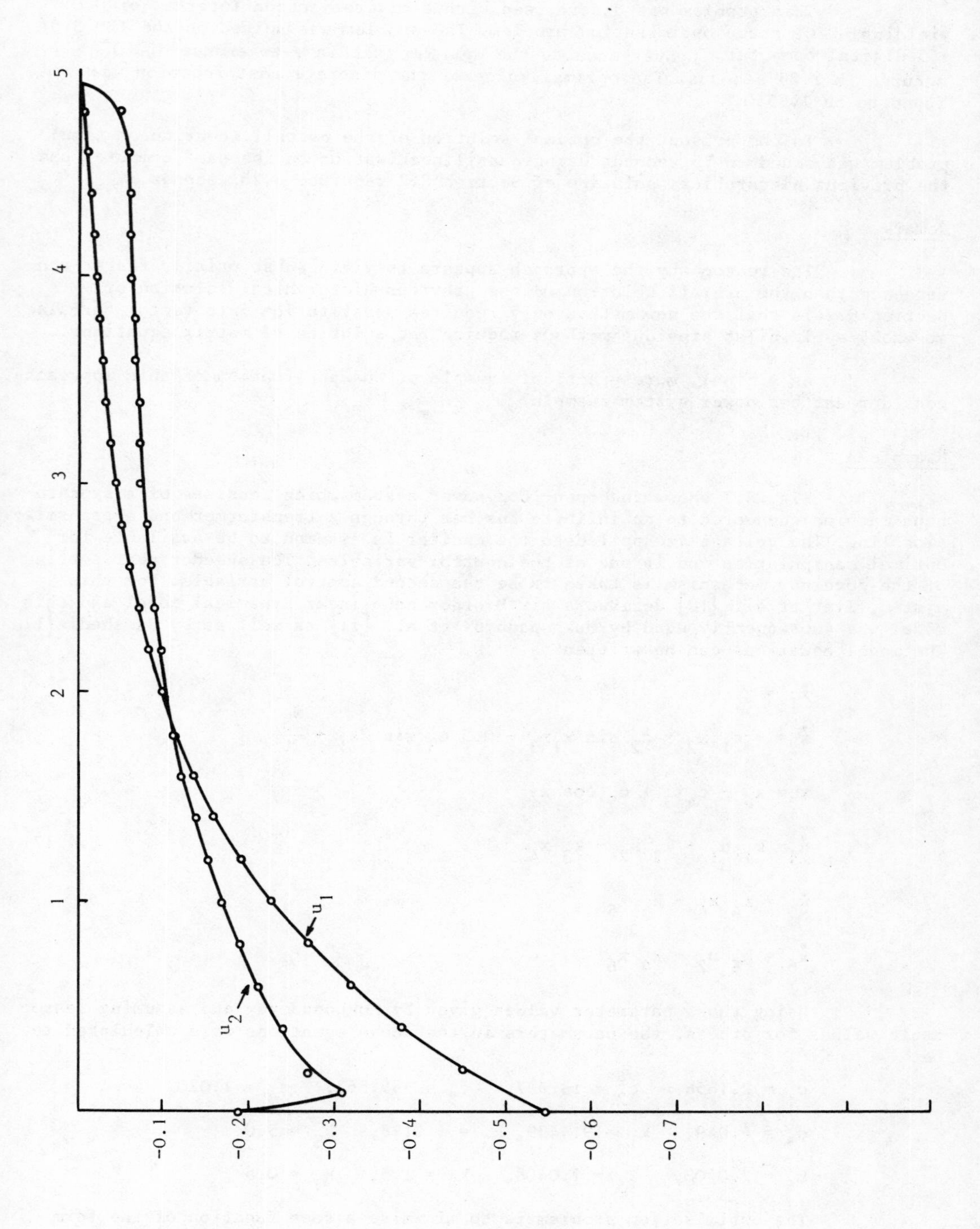

Fig. 8.6 : Control for example 1 of section 8.4

This problem was discretised with a discretisation interval of 0.01 yielding a 201 stage optimisation problem. The problem was solved on the IBM 370/165 digital computer. Convergence to the optimum (within a tolerance of 10^{-4}) occured in 2.23 seconds. The optimal value of the discrete cost function was found to be 1.9550.

In comparison, the optimal solution of the overall (continuous time) problem was found in 15 seconds using quasilinearisation on the same computer and the previous hierarchical solution of Section 8.2 required 6.26 seconds.

Remark

The reason why the approach appears to yield substantially faster convergence than the overall solution and the previous hierarchical solution of Section 8.2 is that the new method only requires substitution into vector formulae at each level whilst previous methods require the solution of matrix equations.

As a final, more practical example of the application of this approach consider another power system example.

Example 3

Fig. 8.7 shows the open-loop power system which consists of a synchronous machine connected to an infinite bus bar through a transformer and a transmission line. The voltage Ve applied to the exciter is assumed to be available for optimal manipulation and is one of the control variables. The speeder gear setting of the governor mechanism is taken to be the second control variable. For this system, Iyer et al. [10] derived a sixth order non-linear dynamical model and this model was subsequently used by Mukhopadhyay et al. [11] as well as by Jamshedi [12]. The model equations can be written as

$$\begin{aligned}
\dot{x}_1 &= x_2 \\
\dot{x}_2 &= - c_1 x_2 - c_2 \sin x_1 x_3 - 0.5\, c_3 \sin 2x_1 + \frac{x_5}{M} \\
\dot{x}_3 &= x_6 - c_4 x_3 + c_5 \cos x_1 \\
\dot{x}_4 &= k_1 u_1 + k_2 x_2 - k_3 x_4 \\
\dot{x}_5 &= k_4 x_4 - k_5 x_5 \\
\dot{x}_6 &= k_6 u_2 - k_4 x_6
\end{aligned} \qquad 8.4.17$$

Using those parameter values given by Mukhopadhyay and assuming reasonable values for others, the parameters in the above equations were calculated to be

$$c_1 = 2.1656, \quad c_2 = 13.997, \quad c_3 = -55.565, \quad c_4 = 1.020,$$

$$c_5 = 4.049, \quad k_1 = 9.4429, \; k_2 = 1.0198, \quad k_3 = 5.0$$

$$k_4 = 2.0408, \quad k_5 = 2.0408, \quad k_6 = 1.5, \quad k_7 = 0.5$$

The optimisation problem is to minimise a cost function of the form

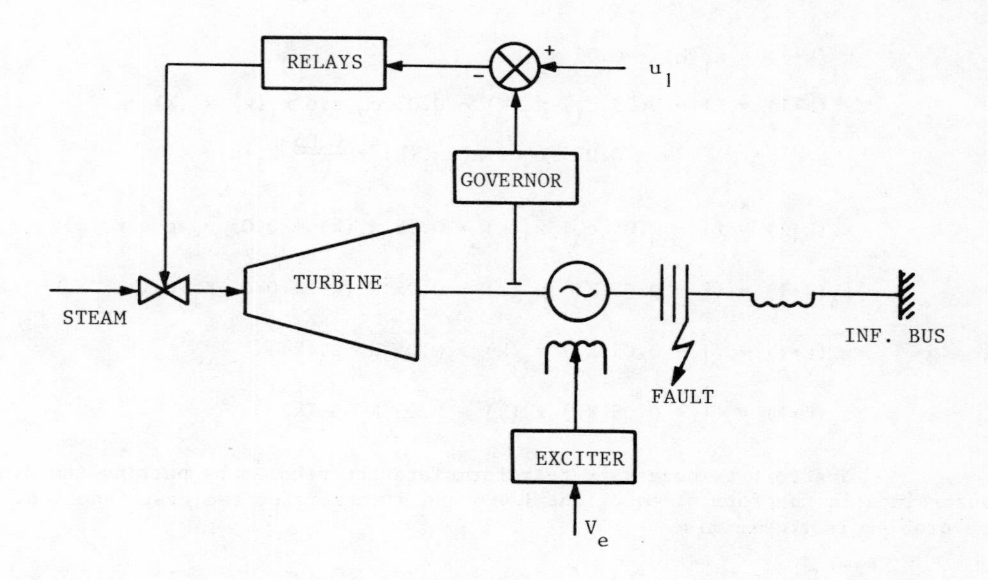

Fig. 8.7

$$J = \int_0^t \left\{ Q_{11}\left[x_1(t) - x_{f1}(t)\right]^2 + Q_{33}\left[x_3(t) - x_{f3}(t)\right]^2 + R_{11}\left[u_1(t) - u_{f1}(t)^2\right] + R_{22}\left[u_2(t) - u_{f2}(t)\right]^2 \right\} dt \qquad 8.4.18$$

subject to the dynamic constraints of equation 8.4.17. The physical significance of the problem is that it is assumed that a three phase to ground fault of a short duration occurs on the line side of the transformer and it is desired to compute the optimal synchronous machine excitation u_1 and the speeder governor setting u_2 to bring the system back to normal operation whilst minimising J in equation 8.4.18.

Single level methods of solving this problem involve the use of the maximum principle to derive a set of 12 coupled non-linear differential equations with split boundary conditions. This problem has previously been solved by Mukhopadhayay et al. [11] by a hybrid quasilinearisation technique for $t_f = 2$ seconds by first solving it from 0 to 1.25 using the terminal and initial conditions on the states and costates as constant initial guesses for these trajectories and then over the period 0 to 2 seconds using the optimal values calculated over 0 to 1.25 and fixing the values at t = 1.25 over the period 1.25 to 2. This was achieved in 8 iterations.

In the case of the method of Jamshedi [12], it is necessary to solve 84 coupled differential equations.

In order to apply the algorithm to the power system problem, it is necessary to discretise the system. The smallest integration interval used in solving the continuous time problem previously was 0.05 seconds (Jamshedi) and this will be used as the discrete time step. Thus the state equations for the discrete system could be written as :

$$x_1(k+1) = x_1(k) + 0.05\, x_2(k)$$

$$x_2(k+1) = (1 - 0.05\, c_1)\, x_2(k) - 0.05\, c_2 \sin x_1(k)\, x_3(k) - 0.025\, c_3 \sin 2x_1(k) + \frac{0.05}{M}\, x_5(k)$$

$$x_3(k+1) = (1 - 0.05\, c_4)\, x_3(k) + 0.05\, x_6(k) + 0.05\, c_5 \cos x_1(k)$$

$$x_4(k+1) = (1 - 0.05\, K_3)\, x_4(k) + 0.05\, K_2(k) + 0.05\, K_1\, u_1(k)$$

$$x_5(k+1) = (1 - 0.05\, K_5)\, x_5(k) + 0.05\, K_4\, x_4(k)$$

$$x_6(k+1) = (1 - 0.05\, K_7)\, x_6(k) + 0.05\, K_6\, u_2(k)$$

Next, it is necessary to reformulate the problem by putting the dynamic constraints in the form of equations 8.4.6 and discretising the cost function. Thus the problem is to minimise

$$J = \frac{0.05}{2} \sum_{k=0}^{k_f-1} \left[Q_{11}\left[x_1(k)-x_{f_1}\right]^2 + Q_{33}\left[x_3(k)-x_{f_3}\right]^2 + R_{11}\left[u_1(k)-u_{f_1}\right]^2 + R_{22}u_2(k)^2 \right] \tag{8.4.19}$$

where

$$Q_{11} = 4, \quad Q_{33} = 4, \quad R_{11} = R_{22} = 1$$

subject to

$$\begin{aligned}
x_1(k+1) &= x_1(k) + 0.05\, x_2^*(k) \\
x_2(k+1) &= (1-0.05\, c_1)\, x_2(k) - 0.05\, c_2 \sin x_1^*(k)\, x_3^*(k) - 0.025\, c_3 \sin 2x_1^*(k) + \frac{\Delta T}{M} x_5^*(k) \\
x_3(k+1) &= (1-0.05\, c_4)\, x_3(k) + 0.05\, x_6^*(k) + 0.05\, c_5 \cos x_1^*(k) \\
x_4(k+1) &= (1-0.05\, K_3)\, x_4(k) + 0.05\, K_2\, x_2^*(k) + 0.05\, K_1\, u_1(k) \\
x_5(k+1) &= (1-0.05\, K_5)\, x_5(k) + 0.05\, K_4\, x_4^*(k) \\
x_6(k+1) &= (1-0.05\, K_7)\, x_6(k) + 0.05\, K_6\, u_2(k)
\end{aligned} \tag{8.4.20}$$

$$\begin{aligned}
x_1^*(k) &= x_1(k) \\
x_2^*(k) &= x_2(k) \\
x_3^*(k) &= x_3(k) \\
x_4^*(k) &= x_4(k)
\end{aligned} \tag{8.4.21}$$

$$x_5^*(k) = x_5(k)$$
$$x_6^*(k) = x_6(k)$$

with $x(0)^T = \begin{bmatrix} 0.7105 & 0.0 & 4.2 & 0.8 & 0.8 & 0.5 \end{bmatrix}$

Writing the Hamiltonian and using the necessary conditions it is easy to see that

$$u_1(k) = -\frac{1}{R_{11}}\left[-R_{11}\,u_{f1} + 0.05\,K_1\;\lambda_4(k+1)\right]$$
$$u_2(k) = -\frac{1}{R_{22}}\left[-R_{22}\,u_{f2} + 0.05\,K_6\;\lambda_6(k+1)\right] \qquad 8.4.22$$

where $\underline{\lambda}$ is the costate vector

$$\lambda_1(k) = Q_{11}(x_1(k) - x_{f1}) + \lambda_1(k+1) + \beta_1(k)$$
$$\lambda_2(k) = (1-0.05\,c_1)\;\lambda_2(k+1) + \beta_2(k) \qquad 8.4.23$$
$$\lambda_3(k) = Q_{33}(x_3(k) - x_{f3}) + \lambda_3(k+1)\;(1-0.05\,c_4) + \beta_3(k)$$
$$\lambda_4(k) = (1-0.05\,K_3)\;\lambda_4(k+1) + \beta_4(k)$$
$$\lambda_5(k) = (1-0.05\,K_5)\;\lambda_5(k+1) + \beta_5(k)$$
$$\lambda_6(k) = (1-0.05\,K_7)\;\lambda_6(k+1) + \beta_6(k)$$

with $\lambda_i(k_f) = 0 \qquad i = 1, 2, \ldots 6.$

$$\beta_1(k) = -0.05\,c_2\;\lambda_2(k+1)\;x_3^*(k)\;\cos x_1^*(k) - 0.05\,c_3\,\lambda_2(k+1)$$
$$\cos 2x_1^*(k) - 0.05\,c_5\;\lambda_3(k+1)\;\sin x_1^*(k)$$
$$\beta_2(k) = 0.05\,\lambda_1(k+1) - 0.05\,K_2\,\lambda_4(k+1)$$
$$\beta_3(k) = -0.05\,c_2\,\lambda_2(k+1)\;\sin x_1^*(k) \qquad 8.4.24$$
$$\beta_4(k) = 0.05\,K_4\,\lambda_5(k+1)$$
$$\beta_5(k) = \frac{0.05}{M}\,\lambda_2(k+1)$$
$$\beta_6(k) = 0.05\;\lambda_3(k+1)$$

$$x_1^*(k) = x_1(k)$$
$$x_2^*(k) = x_2(k)$$
$$x_3^*(k) = x_3(k) \qquad 8.4.25$$
$$x_4^*(k) = x_4(k)$$

$$x_5^*(k) = x_5(k)$$
$$x_6^*(k) = x_6(k)$$

where $\beta_i(k)$, i = 1, ... 6 are the Lagrange multipliers corresponding to $(x_i(k) - x_i^*(k) = 0)$.

Then the algorithm as applied to this problem involves :

(a) guessing the vector sequences $\underline{\lambda}^L$, $\underline{x}^{*L}$ for iteration index L = 1.

(b) Substituting $\lambda_4(k)$, $\lambda_6(k)$ from (a) $(1 \leq k \leq k_f)$ into equations 8.4.22 to obtain $u_1^L(k), u_2^L(k)$ and these latter as well as x^{*L} from (a) into the R.H.S. of equation 8.4.20 to obtain $\underline{x}^L(k)$. Note that to do this the given initial state $\underline{x}(0)$ is also used.

(c) Substituting $\underline{\lambda}^L$, $\underline{x}^{*L}$ in the R.H.S. of equation 8.4.24 to obtain $\underline{\beta}^L$.

(d) For the next iteration $\underline{x}^{*L+1} = \underline{x}^L$ (equation 8.4.25) and $\underline{\lambda}^{L+1}$ is obtained by substituting the values of $\underline{x}^L, \underline{\beta}^L$ into the R.H.S. of equation 8.4.23 and using the fact that $\underline{\lambda}(k_f) = 0$.

(e) If $\underline{\lambda}^{L+1}(k) \simeq \underline{\lambda}^L(k)$ and $\underline{x}^{*L+1} \simeq \underline{x}^{*L}$ $(1 \leq k \leq k_f)$ then $\underline{u}^L$ is the optimum control. Else go to step (b).

Remark

The present approach involves substituting values of given variables into the right hand side of two equations for the controls, six for $\underline{x}$ and six for $\underline{\beta}$, i.e. a total of 14 independent single equations for level one and six equations for $\underline{\lambda}$ for level two (since $\underline{x}^{*L+1} = \underline{x}^L$). Thus the computational task per iteration is an infinitesimally small fraction of the corresponding task in any of the previous methods.

Simulation results

For $k_f = 40$ (i.e. $t_f = 2$ seconds) $\Delta T = 0.05$ as in the work of Mukhopadhyay et al. [11] as well as in Jameshedi [12] the problem was solved from the poor initial guess $\underline{\lambda} = \underline{x}^* = \underline{0}$ to an accuracy of 10^{-4} in 46 second level iterations which took 0.64 seconds to execute on the IBM 370/165 computer.

Since the computational burden increases rapidly with increasing optimisation intervals in previous methods, the optimisation interval was increased 4 times to $k_f = 160$. For this problem, convergence from the initial guess $\underline{\lambda} = \underline{x}^* = \underline{0}$ took place in 52 second level iterations which took 2.54 seconds to execute on the IBM 370/165 computer. Figs. 8.8 to 8.13 show the resulting optimum state trajectories and Figs. 8.14, 8.15 the optimum controls. Thus the computation time only increases proportionately to the increase in the optimisation interval.

Although the algorithm described previously is of much interest for discrete time systems, it is difficult to use it for continuous time systems by discretisation since accuracy requires that the discretisation be fine whereas this obviously increases the computational burden. Let us now consider a continuous time version of this method which uses a three level structure.

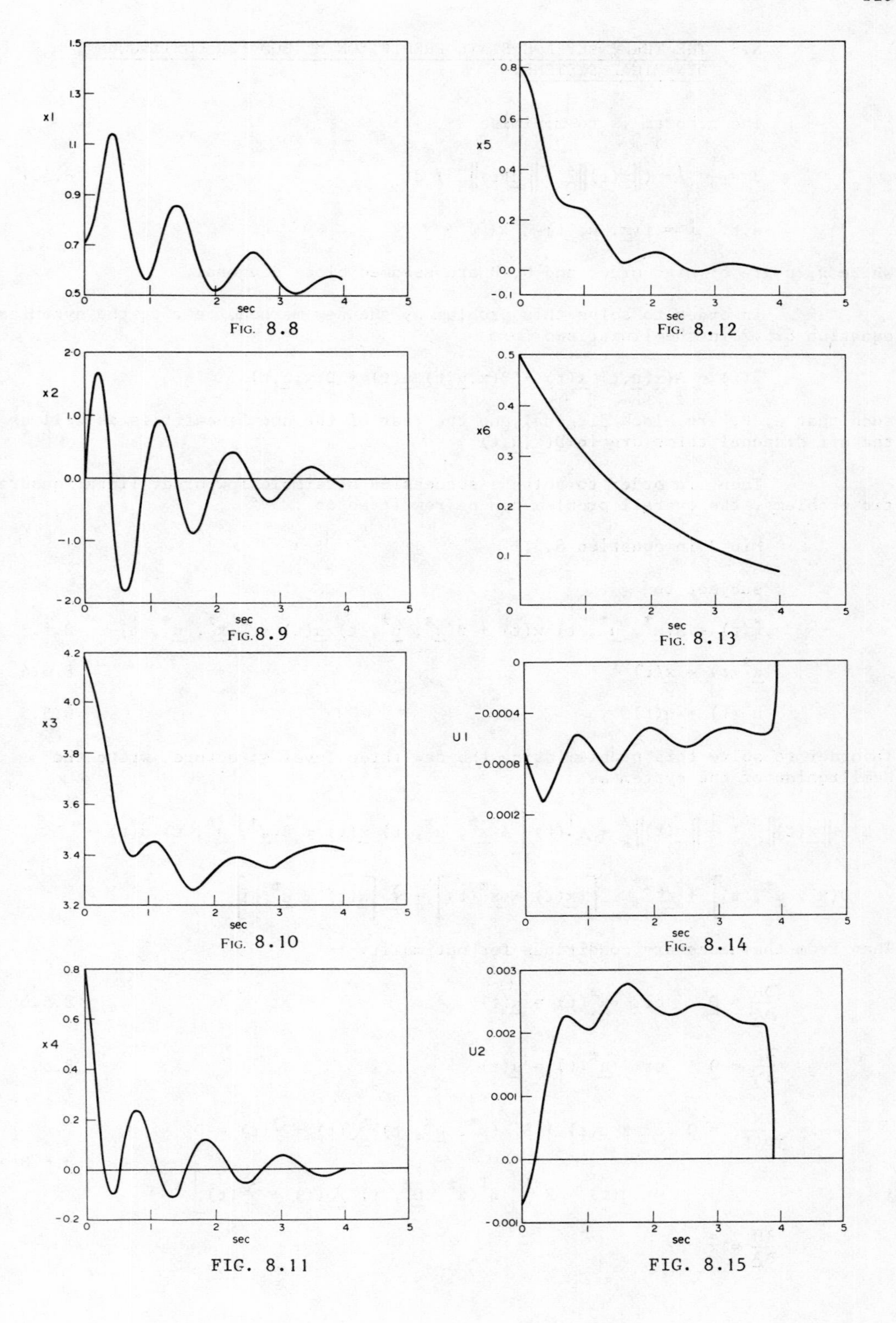

FIG. 8.8

FIG. 8.9

FIG. 8.10

FIG. 8.11

FIG. 8.12

FIG. 8.13

FIG. 8.14

FIG. 8.15

8.5 THE THREE LEVEL COSTATE PREDICTION METHOD FOR CONTINUOUS DYNAMICAL SYSTEMS.

The problem is to minimise

$$J = \frac{1}{2} \int_0^T (\|\underline{x}(t)\|_Q^2 + \|\underline{u}(t)\|_R^2) \, dt \qquad 8.5.1$$

$$\text{s.t.} \quad \dot{\underline{x}} = f(\underline{x}, \underline{u}, t) \; ; \; \underline{x}(0) = \underline{x}_o$$

where $\underline{x}$, $\underline{u}$ are of high order and Q, R are assumed block diagonal.

In order to solve this problem by the new method, rewrite the dynamical equation 8.5.2 in the linearised form

$$\dot{\underline{x}}(t) = A(\underline{x},\underline{u},t)\, \underline{x}(t) + B(\underline{x},\underline{u},t)\, \underline{u}(t) + D(\underline{x},\underline{u},t)$$

such that A, B, are block diagonal and the rest of the non linearities as well as the off diagonal terms are in $\underline{D}(\underline{x},\underline{u},t)$

Then, in order to solve a succession of simple low order linear quadratic problems, the overall problem can be rewritten as :

Min J in equation 8.5.1

subject to

$$\dot{\underline{x}}(t) = A(\underline{x}^*, \underline{u}^*, t)\, \underline{x}(t) + B(\underline{x}^*, \underline{u}^*, t)\, \underline{u}(t) + \underline{D}(\underline{x}^*, \underline{u}^*, t) \qquad 8.5.3$$

$$\underline{x}^*(t) = \underline{x}(t) \qquad 8.5.4$$

$$\underline{u}^*(t) = \underline{u}(t) \qquad 8.5.5$$

In order to solve this problem using the new three level structure, write the Hamiltonian of the system as

$$H = \frac{1}{2}\|\underline{x}(t)\|_Q^2 + \frac{1}{2}\|\underline{u}(t)\|_R^2 + \underline{\lambda}^T(t)\left[A(\underline{x}^*, \underline{u}^*, t)\, \underline{x}(t) + B(\underline{x}^*, \underline{u}^*, t)\, \underline{u}(t) + \underline{D}(\underline{x}^*, \underline{u}^*, t)\right] + \underline{\beta}^T\left[\underline{x}(t) - \underline{x}^*(t)\right] + \underline{\gamma}^T\left[\underline{u}(t) - \underline{u}^*(t)\right]$$

Then from the necessary conditions for optimality :

$$\frac{\partial H}{\partial \underline{\beta}} = \underline{0} \quad \text{or} \quad \underline{x}^*(t) = \underline{x}(t) \qquad 8.5.6$$

$$\frac{\partial H}{\partial \underline{\gamma}} = \underline{0} \quad \text{or} \quad \underline{u}^*(t) = \underline{u}(t) \qquad 8.5.7$$

$$\frac{\partial H}{\partial \underline{u}(t)} = \underline{0} \quad \text{or } R\,\underline{u}(t) + B^T(\underline{x}^*, \underline{u}^*, t)\,\underline{\lambda}(t) + \underline{\gamma}(t) = \underline{0}$$

$$\text{or } \underline{u}(t) = R^{-1}\left[B^T(\underline{x}^*, \underline{u}^*, t)\,\underline{\lambda}(t) + \underline{\gamma}(t)\right] \qquad 8.5.8$$

$$\frac{\partial H}{\partial \underline{\lambda}} = \dot{\underline{x}} \quad \text{or}$$

$$\dot{\underline{x}}(t) = A(\underline{x}^*, \underline{u}^*, t)\ \underline{x}(t) - B(\underline{x}^*, \underline{u}^*, t)\ R^{-1}\left[B^T(\underline{x}^*, \underline{u}^*, t)\underline{\lambda}(t) + \underline{\nu}(t)\right] + \underline{D}(\underline{x}^*, \underline{u}^*, t)\ ;\ \underline{x}(0) = \underline{x}_o \qquad 8.5.9$$

$$\frac{\partial H}{\partial \underline{x}(t)} = -\dot{\underline{\lambda}}(t) \quad \text{or}$$

$$\dot{\underline{\lambda}}(t) = -Q\underline{x}(t) - A^T(\underline{x}, \underline{u}, t)\ \underline{\lambda}(t) - \underline{\beta}(t) \qquad 8.5.10$$

$$\underline{\lambda}(T) = \underline{0}$$

$$\frac{\partial H}{\partial \underline{x}^*} = \underline{0} \quad \text{or}$$

$$\underline{\beta}(t) = \left[\frac{\partial \mathbb{A}(\underline{x}^*, \underline{u}^*, \underline{x}, t)}{\partial \underline{x}^*} + \frac{\partial \mathbb{B}(\underline{x}^*, \underline{u}^*, \underline{u}, t)}{\partial \underline{x}^*} + \frac{\partial \underline{D}^T(\underline{x}^*, \underline{u}^*, t)}{\partial \underline{x}^*}\right]\underline{\lambda}(t) \qquad 8.5.11$$

where

$$\mathbb{A}(\underline{x}^*, \underline{u}^*, \underline{x}, t) = A(\underline{x}^*, \underline{u}^*, t)\ \underline{x}\ (t)$$

$$\mathbb{B}(\underline{x}^*, \underline{u}^*, \underline{u}, t) = B(\underline{x}^*, \underline{u}^*, t)\ \underline{u}\ (t)$$

$$\frac{\partial H}{\partial \underline{u}^*(t)} = \underline{0} \quad \text{or}$$

$$\underline{\nu}(t) = \left[\frac{\partial \mathbb{A}(\underline{x}^*, \underline{u}^*, \underline{x}, t)}{\partial \underline{u}^*} + \frac{\partial \mathbb{B}(\underline{x}^*, \underline{x}^*, \underline{u}, t)}{\partial \underline{u}^*} + \frac{\partial \underline{D}^T(\underline{x}^*, \underline{u}^*, t)}{\partial \underline{u}^*}\right]\underline{\lambda}(t) \qquad 8.5.12$$

Now let

$$\underline{\beta}(t) = E(\underline{x}^*, \underline{u}^*, \underline{x}, \underline{u}, t)\ \underline{\lambda}(t) \qquad 8.5.13$$

$$\underline{\nu}(t) = L(\underline{x}^*, \underline{u}^*, \underline{x}, \underline{u}, t)\ \underline{\lambda}(t) \qquad 8.5.14$$

Using equation 8.5.13 in equtation 8.5.10, we have

$$\dot{\underline{\lambda}}(t) = -Q\ \underline{x}(t) - A^T(\underline{x}^*, \underline{u}^*, t)\underline{\lambda}(t) - E(\underline{x}^*, \underline{u}^*, \underline{x}, \underline{u}, t)\underline{\lambda}(t)$$

or

$$\dot{\underline{\lambda}}(t) = -Q\ \underline{x}(t) - F(\underline{x}^*, \underline{u}^*, \underline{x}, \underline{u}, t)\ \underline{\lambda}(t) \qquad 8.5.15$$

with $\underline{\lambda}(T) = \underline{0}$

where

$$F(\underline{x}^*, \underline{u}^*, \underline{u}, t) = A^T(\underline{x}^*, \underline{u}^*, t) + E(\underline{x}^*, \underline{u}^*, \underline{x}, \underline{u}, t)$$

Essentially in this method, equations 8.5.6. - 8.5.15 are iteratively satisfied within a three level structure. The various steps of the algorithm are :

Step 1 : Set the iteration index of the third level, i.e. k = 1 and provide a guess of the initial trajectories of $\underline{x}^*$, $\underline{u}^*$ to the lower levels

Step 2 : At level 2, guess the initial trajectories $\underline{\nu}(t)$, $\underline{\lambda}(t)$ and send these to level 1 and set the iteration index of level 2, i.e. L to be 1.

Step 3 : At level 1, integrate equation 8.5.9 forward in time with $\underline{x}(0) = \underline{x}_o$ and calculate $\underline{u}(t)$ and send to level 2.

Step 4 : At level 2, integrate equation 8.5.10 backwards from $\underline{\lambda}(T) = \underline{0}$ and label the $\underline{\lambda}$ trajectory obtained, by the index L+1, i.e. $\underline{\lambda}(t)^{L+1}$

Also calculate $\underline{\nu}(t)$ and label this new trajectory $\underline{\nu}(t)^{L+1}$

If $\sqrt{\int_0^T \| \underline{\lambda}^{L+1}(t) - \underline{\lambda}^L(t) \| dt} \leqslant \epsilon_\lambda$ is not satisfied,

where ϵ_λ is a small prechosen positive number, go to step 3. Else go to step 5.

Step 5 : At level 3, use $\underline{x}$, $\underline{u}$ obtained from the first one to produce a new $\underline{x}^*$, $\underline{u}^*$, by using equations 8.5.6 - 8.5.7 and label these new trajectories $\underline{x}^*(t)^{k+1}$, $\underline{u}^*(t)^{k+1}$

If $\sqrt{\int_0^T \| \underline{x}^{*k+1}(t) - \underline{x}^{*k}(t) \|^2} \leqslant \epsilon_x$

and

$$\sqrt{\int_0^T \| \underline{u}^{*k+1}(t) - \underline{u}^{*k}(t) \|^2} \leqslant \epsilon_u$$

where ϵ_x, ϵ_u are also small positive prechosen constants, stop and record $\underline{x}$, $\underline{u}$ as the optimal state and control trajectories. Otherwise send $\underline{x}^*$, $\underline{u}^*$ to level 2 and go to step 2.

Remark : This three level algorithm would appear to be very efficient since at each of the levels of the hierarchy, merely vector equations are manipulated as opposed to the matrix equations in previous methods.

Having described the method, it is illustrated on the practical problem of the non-linear synchronous machine.

Optimal control of synchronous machine excitation

The problem as usual is to manipulate the voltage applied to a synchronous machine in order to bring it back to its steady state operation. The model for the system is given by Mukhopadhyay as (see the example of section 8.2)

$$\dot{y}_1 = y_2$$

$$\dot{y}_2 = B_1 - A_1 y_2 - A_2 y_3 \sin y_1 - \frac{B_2}{2} \sin 2 y_1$$

$$\dot{y}_3 = u_1 - C_1 y_3 + C_2 \cos y_1$$

where y_1 is the rotor angle (radians)

y_2 is the speed deviation ($\frac{d}{dt} y_1$) (rad/s)

y_3 is the field flux linkage

u_1 the control variable in the voltage applied to the machine

The parameter values are :

$$A_1 = 0.2703,\ A_2 = 12.012,\ B_1 = 39.1892,\ B_2 = -48.048,$$

$$C_1 = 0.3222,\ C_2 = 1.9$$

It is desired to minimise

$$J = \frac{1}{2} \int_0^2 ((y_1-0.7461)^2 + y_2^2 + 0.1(y_3-7.7438)^2 + 100(u_1-1.1)^2)\, dt$$

Results

The algorithm was programmed on a IBM 370/165 digital computer to solve the above problem. Convergence to the optimum took place in 1.67 seconds compared to 15 seconds using the global quasilinearisation solution on the same computer and 6.26 using the efficient hierarchical solution of section 8.2 on the same computer.

The trajectories in each case were identical. Thus the new method does appear to provide excellent convergence.

Table 1 gives the number of iterations for convergence at level 2 for each iteration of level 1.

The results can be interpreted as follows :

(a) since the jobs at each level are easier (given that we deal at each level with only a part of the vector to be predicted), and since convergence is rapid as seen in table 1, the computation time is small compared to previous methods.

(b) An important feature of the method is that unlike in previous methods where it was necessary to solve a two point boundary value problem at the first level, here it is merely necessary to integrate a linear differential equation with a given boundary condition.

This method can be compared in a certain sense to the relaxation algorithms [13].

The iteration n° for the 3rd level	The n° of iterations required for the 2nd level to converge
1	2
2	2
3	2
4	2
5	2
6	2
7	2
8	1

Table 1

Next we will consider a more practical approach to the optimisation of non-linear systems.

8.6 THE HIERARCHICAL MODEL FOLLOWING CONTROLLER

The problem is provide a control for a large non-linear plant described by

$$\dot{\underline{x}} = \underline{f}\,(\underline{x},\ t) + B\,\underline{u}_1\,(t) \qquad 8.6.1$$

where as usual $\underline{x}$ is an n dimensional state vector, $\underline{u}$ is an m dimensional control vector, $\underline{f}$ is a non-linear function in C^2. This model describes a system which is linear in the controls. This is an important class of systems.

Now, Control Engineers are most familiar with linear systems and are usually able to define their design objectives in terms of the locations of the poles and zeros, etc. In the present approach we will control our non-linear plant described by equation 8.6.1 by forcing it to track the output of a linear plant which has desirable poles and zeros and which thus incorporates realistic design objectives over the fixed horizon of interest 0 - T.

Let us assume that the "Engineering' design objectives for the plant could be incorporated in the state $\underline{x}_m$ of a linear model whose dynamics are defined by

$$\dot{\underline{x}}_m = A\,\underline{x}_m + B\,\underline{u}_2 \qquad 8.6.2$$

where A is the model matrix whose poles have desirable locations, $\underline{x}, \underline{u}_1$ and $\underline{x}_m\ \underline{u}_2$ assumed to be of the same dimensions respectively.

Let us define an error $\underline{e}$ between the actual state $\underline{x}$ of the non-linear plant and the desired state $\underline{x}_m$ that the plant should follow, i.e.

let $\underline{e} = \underline{x} - \underline{x}_m$ and $\underline{u} = \underline{u}_1 - \underline{u}_2$ 8.6.3

Then $\mathring{\underline{e}} = A\,\underline{e} + B\,\underline{u} + \underline{f}(\underline{x}, t) - A\,\underline{x}$

or $\mathring{\underline{e}} = A\,\underline{e} + B\,\underline{u} + \underline{\gamma}(\underline{x}, t)$ 8.6.4

where $\underline{\gamma}(\underline{x}, t) = \underline{f}(\underline{x}, t) - A\,\underline{x}$ 8.6.5

Now, we want to force our non-linear plant's state $\underline{x}$ to track the desired state $\underline{x}_m$ but at the same time we do not want to use excessive control effort. To do this, let us define a cost function of the form

$$J = \int_0^T \left(\frac{1}{2}\|\underline{e}\|_Q^2 + \frac{1}{2}\|\underline{u}\|_R^2\right) dt \qquad 8.6.6$$

where Q is a block diagonal positive semi-definite matrix and R is a block diagonal positive definite matrix. Let us assume that there are N blocks Q_i, R_i ($i = 1, 2, \ldots, N$) in Q and R. Then J in equation 8.6.6 can be rewritten as

$$J = \sum_{i=1}^{N} \int_0^T \frac{1}{2}\left(\|\underline{e}_i\|_{Q_i}^2 + \|\underline{u}_i\|_{R_i}^2\right) dt \qquad 8.6.7$$

and it is desired to minimise this J subject to the system dynamics 8.6.4. Now, it is possible to decompose 8.6.4 itself into the form

$$\mathring{\underline{e}}_i = A_i\,\underline{e}_i + B_i\,\underline{u}_i + C_i\,\underline{z}_i + \underline{\gamma}_i \qquad 8.6.8$$

where

$$\underline{z}_i = \sum_{j=1}^{N} L_{ij}\,\underline{e}_j \qquad 8.6.9$$

If $\underline{\gamma}_i$ was linear, it would have been possible to solve this problem by duality theory using the Goal Coordination method. However, with the non-linear equality constraint, this method may yield a solution of the dual problem which is different from that of the primal problem, i.e. a duality gap may arise. To avoid this, it will be necessary to use one of the methods described in previous sections which use the prediction principle. Here we will use the penalty function / prediction approach of section 8.2. To do this, define a vector $\underline{e}^*$ of dimension n such that the problem is modified to :

$$J^1 = \sum_{i=1}^{N} \frac{1}{2}\int_0^T \left(\|\underline{e}_i\|_{Q_i}^2 + \|\underline{u}_i\|_{R_i}^2 + \|\underline{e}_i - \underline{e}_i^*\|_{S_i}^2\right) dt \qquad 8.6.10$$

where S_i is positive definite matrix. J^1 in 8.6.10 is to be minimised subject to

$$\mathring{\underline{e}}_i = A_i\,\underline{e}_i + B_i\,\underline{u}_i + C_i\,\underline{z}_i^* + \gamma_i^* \qquad 8.6.11$$

where

$$\underline{z}_i^* = \sum_{j=1}^{N} L_{ij}\,\underline{e}_j^* \qquad 8.6.12$$

$$\underline{\gamma}_i = f_i(\underline{x}_i^*, t) - A_i\underline{x}_i^* = f_i(\underline{e}_i^* + \underline{x}_m, t) - A_i(\underline{e}_i^* + \underline{x}_m) \qquad 8.6.13$$

Note that if an $\underline{e}^*$ could be found such that $\underline{e} = \underline{e}^*$, then this modified problem becomes identical to the problem of minimising J in 8.6.7 subject to 8.6.8 and 8.6.9. To ensure this, write the Hamiltonian as (*)

$$H = \sum_{i=1}^{N} \frac{1}{2}\|\underline{e}_i\|^2_{Q_i} + \frac{1}{2}\|\underline{u}_i\|^2_{R_i} + \frac{1}{2}\|\underline{e}_i - \underline{e}_i^*\|^2_{S_i} + \underline{\lambda}_i^T\left[A_i\underline{e}_i + B_i\underline{u}_i + C_i\underline{z}_i^* + \underline{\gamma}_i^*\right] + \underline{p}_i^T(\underline{e}_i - \underline{e}_i^*) \qquad 8.6.14$$

where $\underline{\lambda}_i$ is the adjoint variable for the ith subsystem and $\underline{p}_i$ is a multiplier for the equality constraint

$$\underline{e}_i = \underline{e}_i^* \qquad 8.6.15$$

which must be satisfied at the optimum.

Now, given $\underline{e}^*$ and $\underline{p}$ the Hamiltonian in equation 8.6.14 is additively separable and this leads to a two level algorithm where on the first level a "linear quadratic" problem is solved independently for the N subsystems and at the second level $\underline{e}^*$ and $\underline{p}_i$ are improved using the prediction principle, i.e. from iteration k to k+1

$$\begin{bmatrix} \underline{e}^{*k+1} \\ \underline{p}^{k+1} \end{bmatrix} = \begin{bmatrix} \underline{e}^*(\underline{e}^k, \underline{u}^k, \underline{e}^{*k}, \underline{p}^k) \\ \underline{p}(\underline{e}^k, \underline{u}^k, \underline{e}^{*k}, \underline{p}^k) \end{bmatrix} \qquad 8.6.16$$

where the function on the R.H.S. is calculated by using the conditions

$$\frac{\partial H}{\partial \underline{e}} = \underline{0} \quad \text{and} \quad \frac{\partial H}{\partial \underline{p}} = \underline{0} \qquad 8.6.17$$

Remarks

We have seen in a previous Section that such an algorithm will converge uniformly to the optimum. The overall approach is attractive here because :

a) the non-linear problem has been converted into a linear-quadratic problem so that it is not necessary to solve a difficult two point boundary value problem.

b) The design is more practical since it is possible to include "Engineering" considerations in a natural way.

c) Experience shows that the two level algorithm is quite efficient.

This algorithm has been tested on a number of examples where good results have been obtained. We next describe the application to one of these examples which has been considered by other methods in previous sections, i.e. that of synchronous machine excitation.

(*) Since the problem is linear in $\underline{u}$, it is not necessary to add a penalty for $\underline{u}$.

Example : The application of the hierarchical model follower to synchronous machine excitation

The system model

A model for a single machine connected by a transmission line to an infinite bus bar was given as before by Mukhopadhayay et al. [2] to be

$$\dot{y}_1 = y_2$$
$$\dot{y}_2 = B_1 - A_1 y_2 - A_2 y_3 \sin y_1 - \frac{B_2}{2} \sin 2y_1 \qquad 8.6.18$$
$$\dot{y}_3 = u - C_1\, y_3 + c_2 \cos y_1$$

where we recall that y_1 is the rotor angle (radians)
y_2 is the speed deviation ($\frac{d}{dt} y_1$ in rad/s)
y_3 is the field flux linkage

and control u is the voltage applied to the field windings of the synchronous machine and is assumed to be available for optimal manipulation. Mukhopadhayay [2] gives desired values for this system to be

$$y_1^* = 0.7461 \text{ radians} \quad ; \quad y_2^* = 0 \text{ rad/s}$$
$$y_3^* = 7.7438 \quad ; \quad u^* = 1.1$$

where in the steady state $y_i \longrightarrow y_i^*$ $i = 1, 2, 3$ and $u \longrightarrow u^*$. Also

$A_1 = 0.27$, $A_2 = 12.01$, $B_1 = 39.18$, $B_2 = -48.04$, $C_1 = 0.32$, $C_2 = 1.9$

The Linear Model

It is desired that the system should come back to the given steady state values of y_i^* , i = 1, 2, 3, and u^* "rapidly". A suitable linear model which incorporates this objective and which has been constructed from steady state considerations could be written as

$$\dot{\underline{y}}_m = A_m \underline{y}_m + B u_m \qquad 8.6.19$$

where

$$A_m = \begin{bmatrix} 0 & 1.0 & 0 \\ -64.535 & -0.2703 & -8.1535 \\ -1.28968 & 0.0 & -0.3222 \end{bmatrix} \qquad B = \begin{bmatrix} 0 \\ 0 \\ 1 \end{bmatrix}$$

It is desired that the model should follow the plant.

Application to the problem

The algorithm was applied to solve the synchronous machine excitation problem using the non-linear machine model given by equations 8.6.18 and the linear model given by equation 8.6.19. The error dynamics equation could be written as :

$$\mathring{\underline{e}} = A\,\underline{e} + B\,\underline{u} + \underline{\gamma} \qquad 8.6.20$$

where
$$\underline{\gamma} = \underline{f}(\underline{y}, t) - A\,\underline{y}$$

with $A = A_m$ given by equation 8.6.19, $B^T = (0 \quad 0 \quad 1)$ and $\underline{f}$ being the R.H.S. of equation 8.6.18.

The error dynamic equation 8.6.20 can be split into two subsystems given by

$$\frac{d}{dt}\begin{bmatrix} e_1 \\ e_2 \end{bmatrix} = \begin{bmatrix} 0 & 1 \\ -64.535 & -0.2703 \end{bmatrix}\begin{bmatrix} e_1 \\ e_2 \end{bmatrix} + \begin{bmatrix} \gamma_1^* \\ \gamma_2^* \end{bmatrix} + \begin{bmatrix} 0 \\ -8.1535 \end{bmatrix} e_3^* \qquad 8.6.21$$

and
$$\frac{d}{dt} e_3 = -0.3222\,e_3 + u + \gamma_3^* - 1.28968\,e_1^* \qquad 8.6.22$$

where
$$\gamma_1^* = 0\ ;\ \gamma_2^* = B_1 - A_1(y_{m_2} - e_2^*) - A_2(y_{m_3} - e_3^*)\sin(y_{m_1} - e_2^*) - \frac{B_2}{2}\sin 2(y_{m_2} - e_1^*) - \left[-64.535(y_{m_1} - e_1^*) - 0.2703(y_{m_2} - e_2^*) - 8.1535(y_{m_3} - e_3^*)\right]$$

and $\gamma_3^* = 1.1 - C_1(y_{m_3} - e_3^*) + C_2\cos(y_{m_1} - e_1^*) - \left[-1.28968\,(y_{m_1} - e_1^*) - 0.3222(y_{m_3} - e_3^*)\right]$

The cost function was chosen to be

$$J = \sum_{i=1}^{2} J_i = \sum_{i=1}^{2} \frac{1}{2}\int_0^{12} (\underline{e}_i^T Q_i \underline{e}_i + \underline{u}_i^T R_i \underline{u}_i + (\underline{e}_i - \underline{e}_i^*)^T S(\underline{e} - \underline{e}_i^*))\,dt$$

where
$$Q_1 = \begin{bmatrix} 1 & 0 \\ 0 & 1 \end{bmatrix} \quad Q_2 = 1, \quad R_2 = 1 \quad S_1 = \begin{bmatrix} 1 & 0 \\ 0 & 1 \end{bmatrix} \quad S_2 = 1$$

First level problems

For given e_i^*, p_i, $i = 1, 2, 3$, it is necessary to minimise J_1 and J_2 independently subject respectively to equations 8.6.21 and 8.6.22. This can be done by solving a Riccati type equation.

Second level

The second level algorithm here becomes from iteration j to j+1

$$\underline{e}^{*j+1} = \underline{e}^j$$

and

$$\underline{p}^{j+1} = \left[-S(\underline{e} - \underline{e}^*) - A^T \dot{\underline{\lambda}} + D^T \underline{\lambda} - \frac{\partial \underline{f}^T}{\partial \underline{e}^*}\underline{\lambda}\right]^j$$

where

$$D = \begin{bmatrix} 0 & 0 & -1.28968 \\ 0 & 0 & 0.0 \\ 0 & -8.1535 & 0 \end{bmatrix}$$

and $\dfrac{\partial \underline{f}^T}{\partial \underline{e}^*} =$

$$\begin{bmatrix} 0 & A_2(y_{m_3} - e_3^*)\cos(y_{m_1} - e_1^*) + p_2\cos 2(y_{m_1} - e_1^*) & C_2\sin(y_{m_1} - e_1^*) \\ -1 & A_1 & 0 \\ 0 & A_2\sin(y_{m_1} - e_1^*) & C_1 \end{bmatrix}$$

Simulation results

The above two level algorithm was programmed on a IBM 370/165 digital computer. The system was started from the initial conditions $y_1(0) = 0.5$, $y_2(0) = -0.4$, $y_3(0) = 8$. The model states were taken to be the same at time 0. Note that these are very far from the desired steady state.

Convergence took place in 54 iterations of the second level and at this point

$$\begin{bmatrix} \sqrt{\int_0^{12} \| \underline{e}^{*j+1} - \underline{e}^{*j}\|^2 \, dt} \\ \sqrt{\int_0^{12} \| \underline{p}^{j+1} - \underline{p}^{j}\|^2 \, dt} \end{bmatrix} < \begin{bmatrix} 10^{-5} \\ 10^{-5} \end{bmatrix}$$

Figs. 8.16 to 8.18 show the 3 states $\underline{y}$ and $\underline{y}_m$. Note that the model follows the plant quite closely. Fig. 8.19 shows the control and it is seen that it gets back to the steady state quite rapidly. Figs. 8.20 to 8.22 show the error function $\underline{e}$ on a very exaggerated scale. It is seen that the error eventually does become very small. The reason why the error e_3 is slightly larger than e_1 and e_2 is that the state y_3 has a large value for the steady state and larger deviations from it.

As a comparison for the calculation time, some results were available to the authors for an optimal solution of the problem using a similar quadratic cost function over the period $[0-12]$ where the problem was solved by quasilinearisation. There, the global solution required 59.72 seconds to compute compared to 26.01 seconds for the hierarchical solution. The reason for this large computational saving is that the present method merely requires a second order and a scaler Riccati equation to be solved at the first level compared to an iterative quasi-linearisation solution of a 3rd order problem for the overall system.

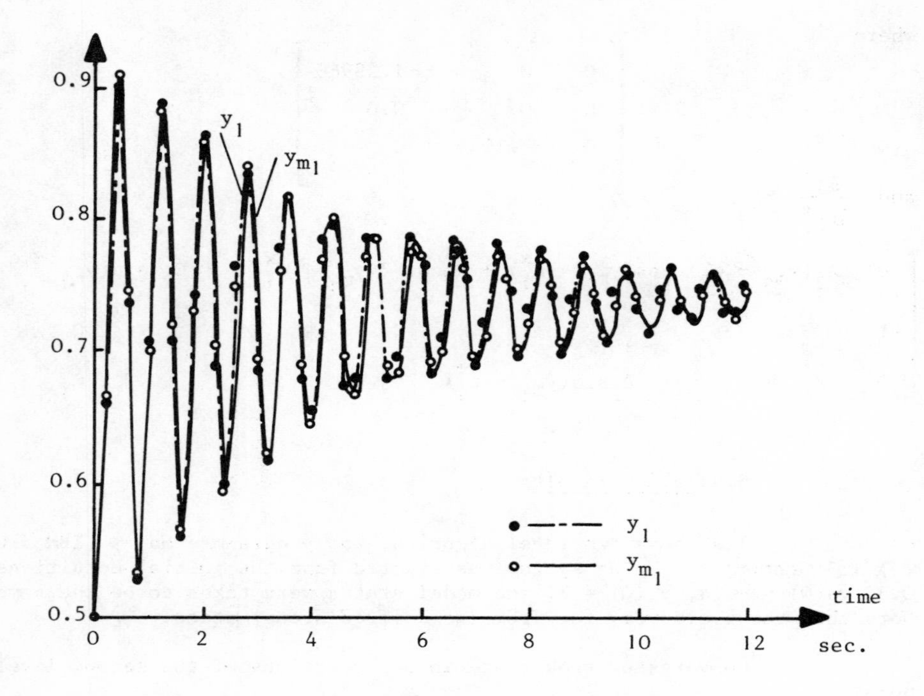

Fig. 8.16

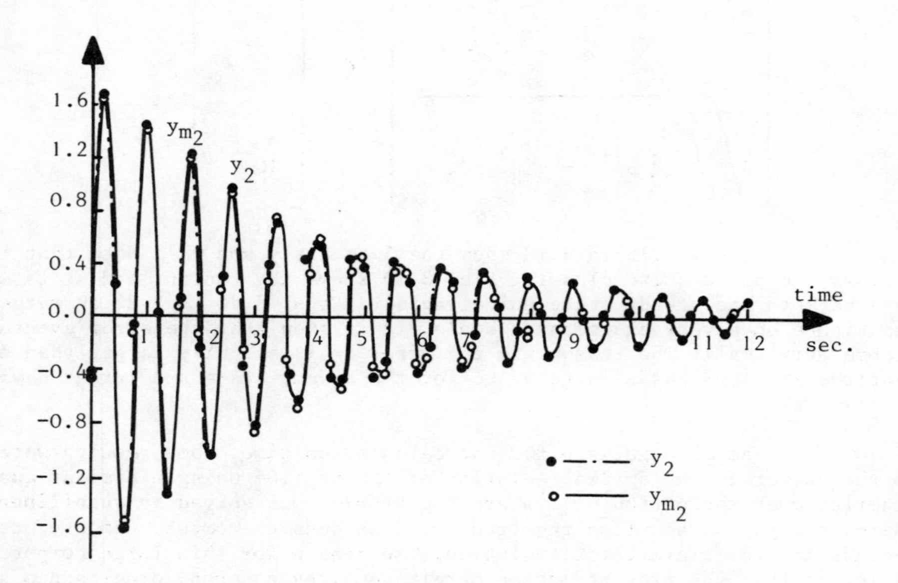

Fig. 8.17

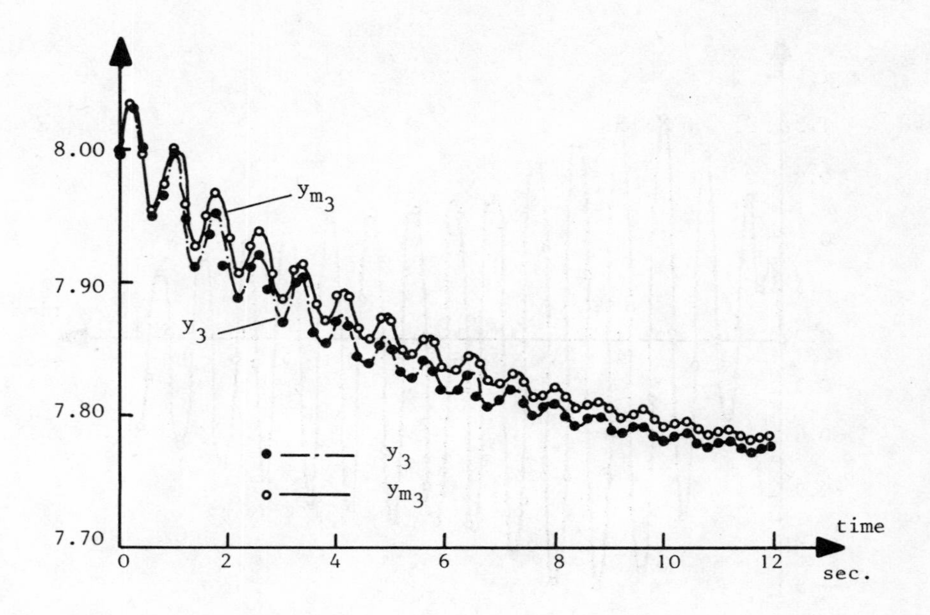

Fig. 8.18

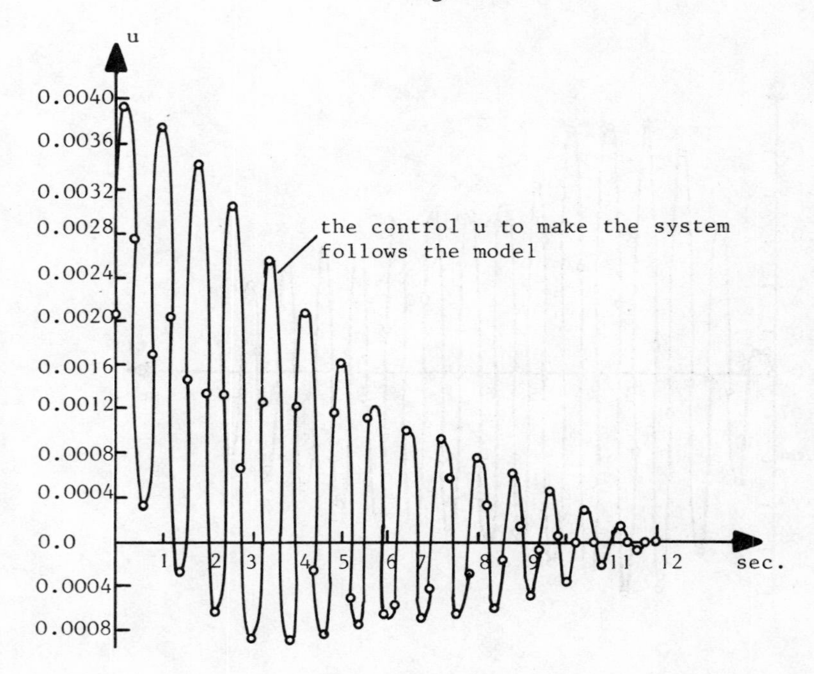

Fig. 8.19

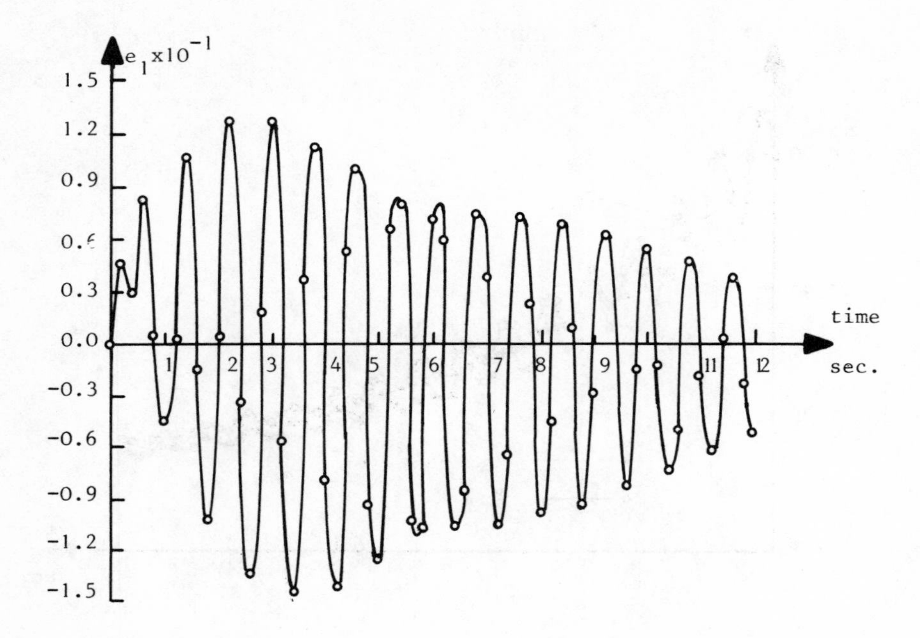

Fig. 8.20

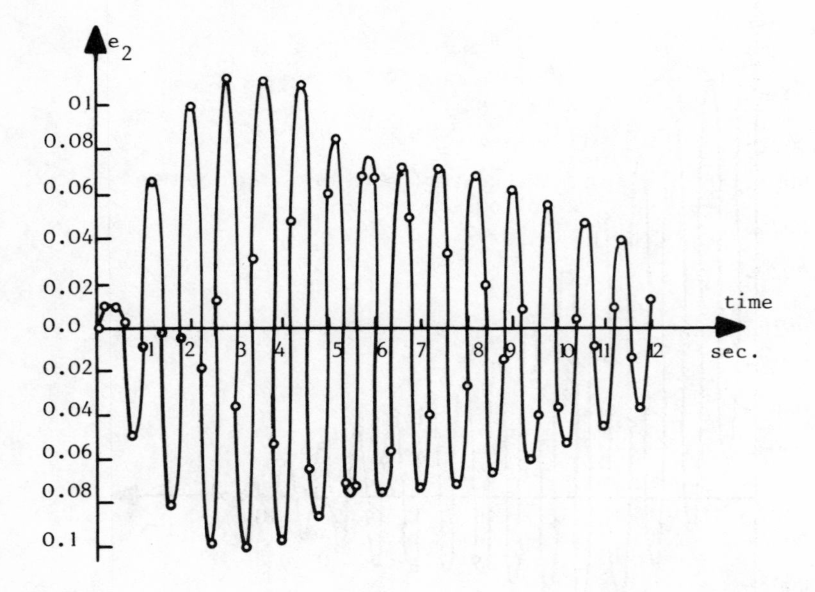

Fig. 8.21

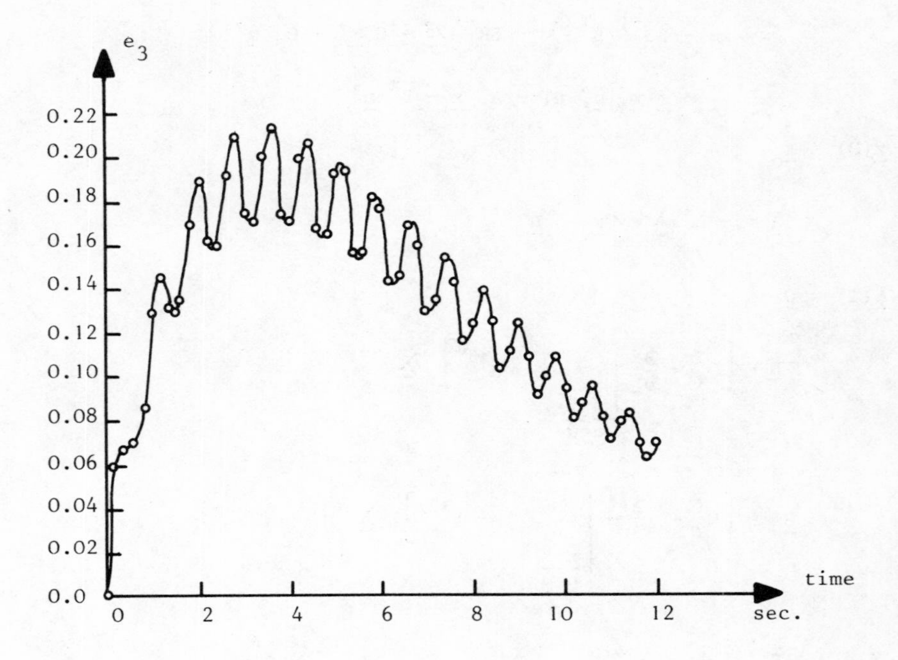

Fig. 8.22

Finally we consider the development of closed loop control for non-linear systems.

8.7 CLOSED LOOP CONTROL FOR NON-LINEAR SYSTEMS

The problem of computing optimal feedback control for large scale non-linear dynamical systems has received very little attention. This is not surprising since at the present time it is not possible, in general, to compute the feedback control law for even the decentralised case for non-linear systems, if it is desired that the feedback parameters be independent of the initial conditions. Thus the best that we could hope to achieve is to compute feedback controls which are relatively insensitive to initial state variations.

The basic approach that we will use relies on the fact that if for a given initial condition, the problem is solved using one of the methods described in section 8.2, then from the resulting state-costate trajectories it will be possible to obtain a control which is a function of the current state. If it appears that the system is too sensitive, it is possible to correct the feedback gain matrices on-line using a two level computational structure.

We begin our analysis by considering the computation of feedback control starting from the stationarity conditions given by equations 8.2.17 to 8.2.22 which we reproduce here, i.e.

$$\dot{\underline{x}} = A\underline{x} - BR^{-1} B^T \underline{\lambda} - BR^{-1} \underline{\beta} + C_1 \underline{x}^\circ + C_2 \underline{u}^\circ + \underline{f}(\underline{x}^\circ, \underline{u}^\circ, t) - A^* \underline{x}^\circ - B^* \underline{u}^\circ \quad 8.7.1$$

with $\underline{x}(0) = \underline{x}_o$

$$\dot{\underline{\lambda}} = - A^T \underline{\lambda} - Q\underline{x} - \underline{\pi} \quad 8.7.2$$

with $\underline{\lambda}(T) = \underline{0}$

$$\underline{\pi} = \left[C_1^T + \frac{\partial \underline{f}}{\partial \underline{x}} \Bigg|_{\substack{\underline{x}=\underline{\mathring{x}} \\ \underline{u}=\underline{\mathring{u}}}} - A^{*T} \right] \underline{\lambda} \quad 8.7.3$$

$$\underline{\beta} = \left[C_2^T + \frac{\partial \underline{f}^T}{\partial \underline{u}} \Bigg|_{\substack{\underline{x}=\underline{\mathring{x}} \\ \underline{u}=\underline{\mathring{u}}}} - B^{*T} \right] \underline{\lambda} \quad 8.7.4$$

$$\underline{x}^\circ = \underline{x} \quad 8.7.5$$

$$\underline{u}^\circ = \underline{u} = -R^{-1} \left[B^T \underline{\lambda} - R^{-1} \underline{\beta} \right] \quad 8.7.6$$

Here, at the lower level, it is necessary to solve the linear two point boundary value problem in $\underline{x}$, $\underline{\lambda}$, given by equations 8.7.1 and 8.7.2 for given $\underline{x}^\circ$, $\underline{u}^\circ$, $\underline{\pi}$, $\underline{\beta}$. Let us assume a solution of the form

$\underline{\lambda} = P \underline{x} + \underline{s}$ where P is a block diagonal matrix.

Then $\dot{\underline{\lambda}} = P \dot{\underline{x}} + \dot{P} \underline{x} + \dot{\underline{s}}$ substituting this into equation 8.7.1 and 8.7.2 we obtain after minor manipulations (it is left as an exercice for the reader to show that equation 8.7.7 is indeed valid) the feedback control :

$$\underline{u} = - R^{-1} \left[B^T P \underline{x} + B^T \underline{s} - \underline{\beta} \right] \quad 8.7.7$$

where

$$\dot{P} + A^T P + PA - P B R^{-1} B^T P + Q = 0 \quad 8.7.8$$

$P(T) = 0$ and

$$\dot{\underline{s}} + A^T \underline{s} - P B R^{-1} B^T \underline{s} + PD(\underline{x}^\circ, \underline{u}^\circ) + \underline{\pi} P B R^{-1} B^T \underline{\beta} + Q\underline{x} = \underline{0} \quad 8.7.9$$

with $\underline{s}(T) = \underline{0}$

where $D(\underline{x}^\circ, \underline{u}^\circ) = C_1 \underline{x}^\circ + C_2 \underline{u}^\circ + f(\underline{x}^\circ, \underline{u}^\circ, t) - A^* \underline{x}^\circ - B^* \underline{u}^\circ$

To compute the feedback control law given by equation 8.7.7, it is necessary to integrate the N blocks of the Riccati equation 8.7.8 independently, backwards in time from $P(T) = 0$ and equation 8.7.9 also backwards from $\underline{s}(T) = \underline{0}$

using the given values of the trajectories $\underline{x}^\circ$, $\underline{u}^\circ$, π, β supplied by the second level. When $\underline{x}^\circ$, $\underline{u}^\circ$, π, β converge to their optimal values, these are used to compute the optimal feedback control $\underline{u}$ from equation 8.7.7.

Note that this control is dependent on the initial state.

Next we consider an example to illustrate the approach.

Example : Control of two coupled synchronous machines

The problem is to control the excitation voltages of two coupled synchronous machines optimally. A model for the system is given by the 6 non-linear differential equations (see section 8.1)

$$\dot{y}_1 = y_2 \; ; \; \dot{y}_2 = B_1 - A_1 y_2 - A_2 y_3 \sin y_1 - \frac{B_2}{2} \sin 2y_1$$

$$\dot{y}_3 = M_1 - C_1 y_3 + C_2 \cos y_1$$

$$\dot{y}_4 = y_5$$

$$\dot{y}_5 = B_4 - A_4 y_5 - A_5 y_6 \sin y_4 - \frac{B_5}{2} \sin 2y_4$$

$$\dot{y}_6 = M_2 - C_4 y_6 - C_5 \cos y_4$$

where $y_1 - y_6$ are the 6 state variables and M_1, M_2 are the controls. The parameters are

$$A_1 = A_4 = 0.2703 \; ; \; A_2 = 12.012 \; ; \; A_5 = 14.4144$$

$$B_1 = B_4 = 39.1892 \; ; \; B_2 = -48.048 \; ; \; B_5 = -57.6576$$

$$C_1 = C_4 = 0.3222 \; ; \; C_2 = 1.9 \; ; \; C_5 = 2.28$$

and the system comprises two subsystems coupled by $y_1 = y_4$.

It is desired to minimise J where

$$J = \int_0^T \frac{1}{2} \left[\| \underline{y} - \underline{y}_p \|_Q^2 + \| \underline{M} - \underline{M}_c \|_R^2 \right] dt$$

where

$$\underline{y}_c^T = \begin{bmatrix} 0.7461 & 0 & 7.7438 & 0.7461 & 0 & 7.7438 \end{bmatrix}$$

$$\underline{M}_c^T = \begin{bmatrix} 1.1 & 1.1 \end{bmatrix}$$

$R = I_2$ the second order identity matrix and Q is a diagonal matrix with the diagonal given by

$$Q_D = \begin{bmatrix} 20 & 20 & 2 & 20 & 20 & 2 \end{bmatrix}$$

Simulation results

Using the initial conditions

$$y^T(0) = \begin{bmatrix} 0.7347 & -0.2151 & 7.7443 & 0.7347 & -0.2151 & 6.9483 \end{bmatrix}$$

the problem was solved and the optimum cost J was found to be J = 1. 1658.

The time varying optimal control law for this initial condition was of the form

$$M_1 = H_{11}\, y_1 + H_{12}\, y_2 + H_{13}\, y_3 + q_1$$

$$M_2 = H_{21}\, y_1 + H_{22}\, y_2 + H_{23}\, y_3 + q_2$$

Next the initial conditions were changed to

$$y(0)^T = \begin{bmatrix} 1 & 0.1 & 10 & 1 & 0.1 & 8.69 \end{bmatrix}$$

Note that these new initial conditions are far from the old ones. Here the optimal control gave a cost of J = 78.5048.

Using the previously calculated gains etc, the control was recalculated and this gave a cost of J = 80.1304, i.e. loss of optimality of 2 %.

Next we examine how we could modify the feedback parameters on-line in order to reduce sensitivity to initial state perturbations.

Improvement of the feedback control

From equation 8.7.7 it is clear that the optimal closed loop control is of the form

$$\underline{u} = H\,\underline{x} + \underline{q} \qquad 8.7.10$$

where H is independent of initial conditions whilst $\underline{q}$ is not. One possible way of making the system less sensitive to initial state variations is to modify $\underline{q}$ on-line as the "initial" states change as a result of unknown perturbations. This can be done within a two level structure as follows :

Let the original problem be

$$\min_{\underline{u}} \int_{t_o}^{t_f} f(\underline{x}, \underline{u}, t)\, dt \qquad 8.7.11$$

subject to

$$\dot{\underline{x}} = g(\underline{x}, \underline{u}, t) \;;\; \underline{x}\,(t_o) = \underline{x}_o \qquad 8.7.12$$

and assume that the optimal closed loop control of the form of equation 8.7.10 has been determined using the method described previously. Suppose now, that, at instant τ, the state changes to $\underline{x}(\tau)$, as a result of some perturbation and we wish to determine the new optimal $\underline{u}$ quickly on-line. In order to do this, consider equations 8.7.11 and 8.7.12 as

$$\underset{\underline{q}}{\text{Min}} \int_{\tau}^{t_f} f(\underline{x}, H\underline{x} + \underline{q}, t)\, dt$$

s.t. $\quad \mathring{\underline{x}} = \underline{g}(\underline{x}, H\underline{x} + \underline{q}, t) \; ; \; \underline{x}(t_o) = \underline{x}(\tau)$

In order to perform this minimisation to obtain a new $\underline{q}$, write the Hamiltonian

$$H^* = f(\underline{x}, H\underline{x} + \underline{q}, t) + \underline{\Psi}^T \left[\underline{g}(\underline{x}, H\underline{x} + \underline{q}, t) \right]$$

and minimise it w.r.t. $\underline{q}$. From the necessary conditions for optimality

$$-\dot{\underline{\Psi}} = \frac{\partial H^*}{\partial \underline{x}}, \quad \frac{\partial H^*}{\partial \underline{q}} = \underline{0} \text{ and } \mathring{\underline{x}} = \underline{g}(\underline{x}, H\underline{x} + \underline{q}, t),\; \underline{x}(t_o) = \underline{x}(\tau),$$

we wish basically to solve this two point boundary value problem iteratively to obtain a new $\underline{q}$. This can in fact be done using any of the standard hierarchical methods described previously using $\underline{q}$ instead of $\underline{u}$. The idea is that since in practice initial conditions do not change substantially, the precalculated value of $\underline{q}$ could serve as an excellent initial guess and thus convergence would take place rapidly.

8.8 CONCLUSIONS

In this chapter, we have seen how we can go from the methods for low order non-linear problems described in the previous chapter to the case where decomposition-coordination yields solution of high order problems. These decomposition-coordination methods are certainly very attractive from a computational point of view. However, as for all non-linear problem solution methods, ultimately the convergence conditions are rather restrictive and we have to rely on our experience and judgement to apply these methods to any particular problem. We must emphasise that the non-linear optimisation problem is a difficult one even for low order problems and a great deal of research is currently going on in this area. The methods described here are, in the opinion of the authors, the best available at the moment although this will change in due course as results of the current research efforts bear fruit.

REFERENCES FOR CHAPTER 8

[1] GEOFFRION, A.M. "Duality in non-linear programming".
SIAM Review, 13, 1, 1-37, 1971.

[2] MUKHOPADHYAY, B.K. and MALIK, O.P. "Optimal control of synchronous machine excitation by quasi-linearisation"
Proc. I.E.E. vol. 119, 1, 91-98, 1972.

[3] SMITH, N. and SAGE, A.P. "An introduction to hierarchical systems theory".
Computers and Electrical Engineering, 1, 55-71, 1973.

[4] HASSAN, M. and SINGH, M.G. "The optimisation of non-linear systems using a new two level method".
Automatica, July 1976.

[5] SINGH, M.G. "Dynamical hierarchical control".
North Holland Publishing Co., 1977.

[6] SINGH, M.G. and HASSAN, M. "A two level prediction algorithm for non-linear systems".
Automatica, Jan. 1977.

[7] SINGH, M.G. and HASSAN, M. "Hierarchical optimisation of non-linear dynamical systems with non separable cost functions".
LAAS report, Toulouse, France, 1976.

[8] MAHMOUD, M., VOGT, W and MICKLE, M. "Multi-level optimisation and control using generalised gradients".
Int. J. Control, 1977.

[9] HASSAN, M. and SINGH, M.G. "A two level costate prediction algorithm for non-linear systems".
Automatica, Nov. 1977.

[10] IYER, S.N. and CORY, B.J. "Optimisation of turbo-generator transient performance by differential dynamic programming".
I.E.E.E. Trans. Power Appar. Syst. 90.2149-2157, 1971.

[11] MUKHOPADHYAY, B.K. and MALIK, O.P. "Optimal control of non-linear power systems by an imbedding method".
I.J.C., 17, 1041-1058 (1973).

[12] JAMSHEDI, M. "Optimal control of non-linear power systems by an imbedding method".
Automatica 11, 633-636, 1975.

[13] LHOTE, F., MIELLOU, J.C., COMTE, P, HENRIOUD, J.M., LARY, B. SPITERI, P. "Algorithmes de relaxation-décentralisation en contrôle optimal".
Rapport de fin de contrat C.N.R.S., n° 1L, 9901, 1975 (ATP Commande).

PROBLEMS FOR CHAPTER 8

1. Write the jobs of the two levels for the continuous time version of the power systems example of section 8.4 example 3 using the Goal coordination method. Write a program to solve this problem. Is there a duality gap here ?

2. Repeat problem 1 using this time the new prediction method of Hassan and Singh.

3. Repeat problem 1 using the three level costate prediction method for continuous dynamical systems (section 8.5).

4. Develop a closed loop control structure for a single synchronous machine.

5. Write a program to solve the example of section 8.3

PART 3

Stochastic Problems

CHAPTER 9

Introduction to Probability Theory and Stochastic Processes

In this chapter, we will begin by introducing some fundamental probabilistic notions and go on to apply them to problems of state estimation and stochastic control in the next chapter.

9.1 INTRODUCTION TO PROBABILITY THEORY

We will begin by introducing two basic notions. The first is of Ω, the set of possible outcomes of an experiment. We will call any particular member of this set ω. As an example if we consider the throwing of a dice, then $\Omega = \{1, 2, 3, 4, 5, 6\}$.

The second fundamental concept is that of an event. For example, "obtain the number 3" or "obtain a pair" are events. With all events E we associate a subset A_E of Ω and since the correspondence is one to one, we will not distinguish between E and A_E.

Thus we can describe an experiment by the set Ω and a class of parts of Ω that we call Q. This class Q should contain the "certain" event and should be stable for the set operations "AND", "OR", etc. Thus we can describe a random experiment by a measurable space (Ω, Q).

Let us next define the concept of "probability". This is a function defined on the class of events, i.e. $P[\cdot]$ is a mapping of Q on $[0, 1]$.

The probability function $P[\cdot]$ which defines the probability of the event A satisfies the following three axioms

$$P[A] \geqslant 0 \quad \text{for all } A$$

$$P[\Omega] = 1 \quad \text{for the certain event}$$

and $P[A \cup B] = P[A] + P[B]$ if A and B are mutually exclusive. For all events $\{A_n\}$ of Q which are such that $A_i \cap A_j = \emptyset$, $i \neq j$

$$P\left(\bigcup^{\infty} A_i\right) = \sum_{i=1}^{\infty} P(A_i)$$

where $\emptyset$ is the null set.

The triple (Ω, Q, P) is called the probability space.

Let us next consider some formulae which arise from these axioms.

a) $P[\emptyset] = 0$

Because Ω and $\emptyset$ are mutually exclusive and $\Omega \cup \emptyset = \Omega$ therefore

$$P[\Omega \cup \emptyset] = P[\Omega] = P[\Omega] + P[\emptyset]$$

b) $P[A^c] = 1 - P[A]$

Since $A^c \cap A = \emptyset$ and $A^c \cup A = \Omega$

$$P[\Omega] = P[A^c \cup A] = P(A^c) + P[A]$$

c) $P[A \cap B^c] = P[A] - P[A \cap B]$

Indeed $(A \cap B)$ and $(A \cap B^c)$ are mutually exclusive and their union is A. Hence

$$P[A] = P[A \cap B] + P[A \cap B^c]$$ which proves the assertion.

d) $P[A \cup B] = P[A] + P(B) - P[A \cap B]$

To see this, let us write $A \cup B$ as a union of the two mutually exclusive events A and $A^c \cap B$

$$P[A \cup B] = P[A] + P[A^c \cap B]$$

and $$P[A^c \cap B] = P[B] - P[A \cap B]$$

if we use the previous result c). Hence the result

$$P[A \cup B] = P[A] + P[B] - P[A \cap B]$$

Example

As a simple example to illustrate these formulae suppose $P[A] = P[B] = \frac{1}{3}$ and $P[A \cap B] = \frac{1}{6}$. Then what is the probability of having neither A nor B ?

Here $P[A \cup B^c] = 1 - P[A \cup B] = 1 - P[A] - P[B] + P[A \cap B] = 1 - \frac{2}{3} + \frac{1}{6} = \frac{1}{2}$.

Conditional probability

Given two events A and B, the conditional probability of B given A denoted by $P[B \mid A]$ is the probability of B occurring knowing that A has already occurred. We define

$$P[B \mid A] = \frac{P[A \cap B]}{P[A]}$$

provided that $P[A] > 0$

Example

Suppose $P[A] = \frac{1}{4}$ $\qquad P[B|A] = \frac{1}{2}$ $\qquad P[A|B] = \frac{1}{4}$

Are A and B mutually exclusive ?

The answer is no, because

$$P[A \cap B] = \frac{1}{8}$$

Next let us consider the notion of independence.

Two events A and B in the same probability space are independent if, and only if

$$P\left[A \cap B\right] = P\left[A\right] . P\left[B\right]$$

Note that in this case

$$P\left[B \mid A\right] = P\left[B\right]$$

which agrees with our intuitive idea of independence and conditional probability in that since B and A are independent, we don't need to know A to arrive at the probability $P\left[B \mid A\right]$.

It is important to distinguish between the notions of independence and of mutually exclusive events. It is left as an excercise for the reader to verify on the previous example that A and B are independent.

Next let us consider Random variables.

Random variables

A real valued function X(.) defined on Ω is called a real random variable if for all real numbers x, the inequality $X(\omega) \leqslant x$ defines a set whose probability is defined, i.e.

$X : \Omega \longrightarrow R$ is a random variable if and only if

$$X^{-1}(-\infty, x) = \left\{\omega \mid X(\omega) \leqslant x\right\} \in Q \quad \forall \; x \in R.$$

As an example of a random variable, suppose we toss a coin 3 times to obtain

ω_1 = H H H $\qquad$ ω_2 = H H T $\qquad$ ω_3 = H T H $\qquad$ ω_4 = H T T

ω_5 = T H H $\qquad$ ω_6 = T H T $\qquad$ ω_7 = T T H $\qquad$ ω_8 = T T T

where H represents heads and T tails.

Each of these events has a probability $\frac{1}{8}$

$$\Omega = \left\{\omega_1, \omega_2, \ldots, \omega_8\right\}$$

One could be interested for example only in the number of heads that occur (and not the whole sequence). This is a variable which can take on only the values 0, 1, 2, 3 and this is an example of a random variable.

9.1.1. Description of a random variable

Although, by definition a random variable is a function on a probability space we are very rarely interested in the form of the functional relationship X(.) but rather in the probability that a certain value of the random variable X(.) occurs in a given set.

One way of describing random variables is in terms of their distribution functions. We define the distribution function of a random variable by

$$F(x) = P\left[\left\{\omega \mid X(\omega) \leqslant x\right\}\right]$$

Example

As an example, consider the tossing of a coin. Here

$$\Omega = [T, H] \qquad P[H] = p \qquad P(T) = q \qquad p+q = 1$$

We define the random variable X(.) by

$$X(H) = 1 \qquad\qquad X(T) = 0$$

and we determine the distribution function as follows :

if $x \geqslant 1$ we have

$$\{X(\omega) \leqslant x\} = \{X(\omega) = 0\} \cup \{X(\omega) = 1\}$$

Therefore $F(x) = P[X(\omega) \leqslant x] = 1$

if $0 \leqslant x \leqslant 1$ $\qquad \{X(\omega) \leqslant x\} = \{x \mid \omega\} = 0$

Hence $F(x) = q$

if $x < 0$ $\qquad F(x) = 0$

$$\text{Hence } F(x) = \begin{cases} 1 & x \geqslant 1 \\ q & 0 \leqslant x < 1 \\ 0 & x < 1 \end{cases}$$

The distribution function possesses the following properties :

$$\lim_{x \to \infty} F(x) = 1$$

$$F(-\infty) = 0$$

F(x) is a monotonic non decreasing function.

Another way of describing random variables is in terms of their density functions.

Probability density functions

For the case where F(x) is continuous, we can write

$$F(x) = \int_{-\infty}^{x} p(u)\, du$$

and we can define the probability density as

$$p(x) = \frac{dF(x)}{dx}$$

Next let us define mathematical expectation.

Mathematical expectation

We can define a random variable by associating with all parts of $A \in Q$ a random variable with steps I_A such that

$$I_A(\omega) = 1 \qquad \text{if } \omega \in A$$

$$I_A(\omega) = 0 \qquad \text{if } \omega \notin A$$

We can thus write the random variable in the form

$$X = \sum_i x_i I_{A_i}$$

where A_i is a partition of Ω.

We will define the mathematical expectation by associating with all random variables X defined on (Ω , Q) the real number

$$\sum_i x_i P(A_i)$$

For a continuous random variable X, the mathematical expectation will be defined by

$$E\left[X\right] = \int_{-\infty}^{\infty} x\, p(x)\, dx$$

In the same way, a function g(X) of the random variable X will have the mathematical expectation

$$E\left[g(X)\right] = \int_{-\infty}^{\infty} g(x)\, p(x)\, dx$$

Let us next consider some properties of the mathematical expectation operator.

Properties of the mathematical expectation operator

1. For a constant a, $E\left[a\right] = a$

2. If $g_1(X)$ and $g_2(X)$ are two functions of the random variable X and a and b are two constants, then

$$E\left[a\, g_1(X) + b\, g_2(X)\right] = a\, E\left[g_1(X)\right] + b\, E\left[g_2(X)\right]$$

Next, let us consider moments of random variables.

9.1.2 Moment generating functions and characteristic functions

The nth order moment of the random variable X is defined as

$$E\left[X^n\right] = \int_{-\infty}^{\infty} x^n\, p(x)\, dx$$

We define also a second order moment centered around the mean value (the mathematical expectation) by

$$E\left[(x - E\left[X\right])^2\right] = \text{var}\,(X)$$

and this quantity is called the variance of X.

Now, the evaluation of moments often involves complex integration operations. The moment generating function enables us to get around these problems so that we can generate all necessary moments by performing one integration and then differentiations.

The moment generating function is defined for all real numbers t by

$$\varphi(t) = E\left[e^{tX}\right]$$

In the continuous case :

$$\varphi(t) = \int_{-\infty}^{\infty} e^{tx} \, p(x) \, dx$$

In this case all the moments can be expressed in the form of successive derivatives of $\varphi(t)$ at t = 0, for example

$$\frac{d\varphi}{dt} = E\left[\frac{\partial}{dt} e^{tX}\right] = E\left[X e^{tX}\right]$$

$$\frac{d^2\varphi}{dt^2} = E\left[X^2 e^{tX}\right]$$

These derivates give us $E[X]$, $E[X^2]$, etc. if we evaluate them at t = 0.

The <u>characteristic function</u> is defined by

$$\phi(u) = E\left[e^{juX}\right]$$

In this case one can also obtain all the moments by differentiating the characteristic function, i.e.

$$E\left[X^n\right] = \left[\frac{1}{j^n}\right]\left[\frac{d^n}{du^n} \phi(u)\right]_{u=0}$$

Note that according to the definition, the chatacteristic function is the Fourier Transform of the probability density function, i.e.

$$\phi(u) = \int_{-\infty}^{\infty} e^{jux} \, p(x) \, dx$$

Thus if the characteristic function is absolutely integrable, the inverse Fourier Transform will give the density function :

$$p(x) = \frac{1}{2\pi} \int_{-\infty}^{\infty} e^{-jux} \phi(u) \, du$$

Next we see how these ideas can be extended to the multivariable case.

<u>Multidimensional random variables</u>

The random variables $X_1, X_2, \ldots X_n$ have a joint distribution if it is defined on the same probability space. They could be characterised by their <u>joint distribution function</u> :

$$F(x_1, x_2, \ldots x_n) = P\left[X_1(\omega) \leqslant x_1, X_2(\omega) \leqslant x_2 \ldots X_n(\omega) \leqslant x_n\right]$$

where

$$\left\{X_1(\omega) \leqslant x_1, \ldots X_n(\omega) \leqslant x_n\right\} \triangleq \left\{X_1(\omega) \leqslant x_1\right\} \cap \left\{X_2(\omega) \leqslant x_2\right\} \cap \cdots$$
$$\cdots \cap \left\{X_n(\omega) \leqslant x_n\right\}$$

i.e. it is the probability of simultaneously having

$$X_1 \leqslant x_1 \quad \text{and} \quad X_2 \leqslant x_2 \ldots \quad X_n \leqslant x_n$$

The random variables could also be characterized by their joint probability density, i.e.

$$F(x_1, x_2, \ldots x_n) = \int_{-\infty}^{x_1} \cdots \int_{-\infty}^{x_n} p(\xi_1, \xi_2, \ldots \xi_n)\, d\xi_1 \ldots d\xi_n$$

where, if the appropriate derivatives exist, we can define the joint density function by :

$$p(x_1, x_2, \ldots x_n) = \frac{\partial^n F}{\partial x_1 \ldots \partial x_n}(x_1, x_2, \ldots x_n)$$

Example

As a simple example, consider the case of two variables. Let X and Y be two random variables having the distribution functions F(x, y). This function has the following properties :

$$F(+\infty, +\infty) = 1$$

$$F(-\infty, y) = F(x, -\infty) = F(-\infty, -\infty) = 0$$

$$F(x_2, y) - F(x_1, y) = P\left[x_1 < X(\omega) \leqslant x_2, \quad y(\omega) \leqslant y\right]$$

Next we introduce the concept of a marginal distribution.

The basic idea here is that one is often interested by only some of the random variables. Suppose we are only interested in $x_1, x_2, \ldots, x_m$ where $m < n$.

The distribution function is therefore

$$F_{x_1 \ldots x_m}(x_1, \ldots x_m) = F_{x_1 \ldots x_n}(x_1, \ldots x_m, +\infty, \ldots +\infty)$$

This distribution is called the marginal distribution.

To define the marginal density function, let us differentiate this expression to obtain

$$p(x_1, \ldots x_m) = \frac{\partial^m}{\partial x_1, \ldots \partial x_m} F(x_1, \ldots x_m, +\infty, \ldots +\infty)$$

$$\text{or } F(x_1, \ldots x_m, +\infty, \ldots +\infty) = \int_{-\infty}^{x_1} \cdots \int_{-\infty}^{x_m} \cdots \underbrace{\int_{-\infty}^{\infty} \cdots \int_{-\infty}^{\infty}}_{m+1 \text{ to } n} p(\xi_1, \ldots \xi_n)\, d\xi_1 \ldots d\xi_n$$

so that

$$p(x_1,\ldots x_m) = \int_{-\infty}^{\infty} \cdots \int_{-\infty}^{\infty} p(x_1,\ldots x_m, \xi_{m+1},\ldots \xi_n)\, d\xi_{m+1} \ldots d\xi_n$$

is the marginal probability density.

Example

As an example, consider the case of 2 random variables which are defined as follows : we take out a ball two times one after the other from an urn containing 5 balls two of which are white and 3 are black. The suffix of the random variables is the try number when we pull a ball, i.e. it is one the first time and two the second time. x_k, k = 1 or 2, takes on the value 1 or 0 depending on whether the kth time we pulled a ball gave us a white one or not. We assume that after the first try, the ball taken out is put back into the urn. How do we calculate the marginal probabilities ?

Table 9.1 gives the probabilities of occurrence of the various possibilities. In this table the marginal probabilities are obtained by adding a row or a column. For $p(x_1)$ for example, it is necessary to take the probability density at the point x_1 for all possible values of x_2.

x_2 \ x_1	0	1	$p(x_2)$
0	B B $\frac{3}{5}$ x 315 = 9125	W B $\frac{2}{5}$ x 315 = 6125	$\frac{15}{25} = \frac{3}{5}$
1	$\frac{3}{5} \times \frac{2}{5} = \frac{6}{25}$	$\frac{2}{5} \times \frac{2}{5} = \frac{4}{25}$	$\frac{10}{25} = \frac{2}{5}$
$p(x_1)$	$\frac{3}{5}$	$\frac{2}{5}$	

Table 9.1

9.1.3 Mathematical expectation, covariance, correlation, independence

We will consider for simplicity the case of two random variables X_1 and X_2. Then we define

$$E\left[X_1\right] = \int_{-\infty}^{\infty}\int_{-\infty}^{\infty} x_1 p(x_1, x_2)\, dx_1\, dx_2$$

and

$$E\left[X_2\right] = \int_{-\infty}^{\infty}\int_{-\infty}^{\infty} x_2 p(x_1, x_2)\, dx_1\, dx_2$$

For the second order moments we have

$$E\left[X_1^2\right] = \int_{-\infty}^{\infty}\int_{-\infty}^{\infty} x_1^2\ p(x_1, x_2)\, dx_1\, dx_2$$

$$E\left[X_2^2\right] = \int_{-\infty}^{\infty}\int_{-\infty}^{\infty} x_2^2\ p(x_1, x_2)\, dx_1\, dx_2$$

$$E\left[X_1, X_2\right] = \int_{-\infty}^{\infty}\int_{-\infty}^{\infty} x_1 x_2\ p(x_1, x_2)\, dx_1\, dx_2$$

Finally, we define a centered second order moment called the covariance i.e.

$$\mathrm{Cov}\ (X_1, X_2) = E\left[(X_1 - E\left[X_1\right])\ (X_2 - E\left[X_2\right])\right]$$

Next let us consider the concept of correlation by defining the coefficient of correlation between two random variables X_1, X_2 by

$$\rho(X_1, X_2) = \frac{\mathrm{Cov}\ (X_1, X_2)}{\sqrt{\mathrm{Var}(X_1)}\ \sqrt{\mathrm{Var}(X_2)}}$$

provided that X_1, X_2 have variances which are finite and are strictly positive.

Having defined the coefficient of correlation we discuss the important concept of un-correlated independent random variables.

We have already seen that the condition for two events A and B to be independent is

$$P\left[A \cap B\right] = P\left[A\right]\ .\ P\left[B\right]$$

The following conditions are equivalent to this one, i.e.

$$F_{X_1,X_2}(x_1, x_2) = F_{X_1}(x_1)\ .\ F_{X_2}(x_2)$$

$$p_{X_1,X_2}(x_1, x_2) = p_{X_1}(x_1)\ .\ p_{X_2}(x_2)$$

$$E\left[f(X_1)\ g(X_2)\right] = E\left[f(X_1)\right] .\ E\left[g(X_2)\right]$$

for all functions f(.) and g(.).

In the same way for the case of n random variables we will have

$$p_{X_1, X_2, \ldots X_n}(x_1, x_2, \ldots x_n) = p_{X_1}(x_1) \cdot p_{X_2}(x_2) \ldots p_{X_n}(x_n)$$

Two random variables are said to be un-correlated if their second order moments are finite and if

$$\text{Cov}(X_1, X_2) = 0$$

which from the definition of the coefficient of correlation implies that

$$\rho(X_1, X_2) = 0$$

Next we will see that two independent random variables are also un-correlated.

$$\begin{aligned}
\text{Cov}(X_1, X_2) &= \int_{-\infty}^{\infty}\int_{-\infty}^{\infty} x_1 x_2 \, p(x_1, x_2) \, dx_1 \, dx_2 - E[X_1] \cdot E[X_2] \\
&= \int_{-\infty}^{\infty}\int_{-\infty}^{\infty} x_1 x_2 \, p(x_1) \, p(x_2) dx_1 \, dx_2 - E[X_1] \cdot E[X_2] \\
&= \int_{-\infty}^{\infty} x_1 \, p(x_1) \, dx_1 \int_{-\infty}^{\infty} x_2 \, p(x) \, dx_2 - E[X_1] \cdot E[X_2] \\
&= 0
\end{aligned}$$

However, in general, the converse is not true, i.e. two un-correlated random variables need not necessarily be independent. The absence of correlation implies that the condition

$$E[f(X_1) \, g(X_2)] = E[f(X_1)] \cdot E[g(X_2)]$$

is satisfied for $f(X) = X$

whilst independence requires that this condition be satisfied for all functions f and g.

Next let us define the characteristic function for n random variables.

For the n random variables $X_1, X_2, \ldots X_n$, the characteristic function is defined as

$$\phi(u_1, u_2, \ldots u_n) = E\left[e^{j \sum_{k=1}^{n} u_k X_k} \right]$$

If $X_1, X_2, \ldots X_n$ are independent, we have

$$\phi(u_1, u_2, \ldots u_n) = \phi_{X_1}(u_1) \, \phi_{X_2}(u_2) \ldots \phi_{X_n}(u_n)$$

Next let us define conditional probabilities.

9.1.4 Conditional probability density functions and conditional expectations

We define conditional probability density functions by analogy with the previous definition of conditional probability of occurrence of an event.

The conditional probability density of a random variable Y given that the random variable X took the value x (where X and Y are defined on the same probability space) is

$$p(y \mid x) = \frac{p(x, y)}{p(x)} = \frac{p(x, y)}{\int_{-\infty}^{\infty} p(x,y)dy} \qquad p(x) > 0$$

In the same way we can define the conditional expectation of Y given that X has taken the value x as

$$E\left[Y \mid X = x\right] = E\left[Y \mid X\right] = \int_{-\infty}^{\infty} y\, p(y \mid x)\, dy$$

Next we give two important properties of conditional expectations

a) $E\left[Y\right] = E\left[E\left[Y|X\right]\right]$

To see this, let us write

$$E\left[Y\right] = \int_{-\infty}^{\infty}\int_{-\infty}^{\infty} y\, p(x, y)\, dx\, dy$$

$$= \int_{-\infty}^{\infty}\int_{-\infty}^{\infty} y\, p(y|x)\, p(x)\, dx\, dy$$

$$= \int_{-\infty}^{\infty} p(x)\, dx \int_{-\infty}^{\infty} y\, p(y \mid x)\, dx$$

$$= \int_{-\infty}^{\infty} p(x)\, E\left[Y \mid X\right] dx$$

Remark : We see thus that the conditional expectation is not a deterministic value but it is a random variable.

b) If the random variables X and Y are independent, then

$$p(y \mid x) = p(y) \quad \text{so that}$$

$$E\left[Y \mid X\right] = E\left[Y\right]$$

Before we go any further, it would be useful to use vector notation for multidimensional random variables. In future, we will write random vectors as

$$\underline{X} = \begin{bmatrix} X_1 \\ X_2 \\ \vdots \\ X_n \end{bmatrix}$$

The probability density functions are scalers which will be written as

$$p(\underline{x}) = p(x_1, x_2, \ldots x_n)$$

The mathematical expectation of the vector $\underline{X}$ will be written as

$$E\left[\underline{X}\right] = \begin{bmatrix} E\left[X_1\right] \\ E\left[X_2\right] \\ \vdots \\ E\left[X_n\right] \end{bmatrix}$$

and the n x n matrix which has as its elements the covariances will have the general form

$$P = E\left[\left[\underline{X} - E(\underline{X})\right]\left[\underline{X} - E(\underline{X})\right]^T\right]$$

with

$$P = \begin{bmatrix} Var(X_1) & Cov(X_1, X_2) & \cdots & Cov(X_1, X_n) \\ Cov(X_1, X_2) & Var(X_2) & \cdots & \vdots \\ \vdots & & & \vdots \\ Cov(X_1, X_n) & & \cdots & Var(X_n) \end{bmatrix}$$

Having clarified our notation, we next go on to consider Gaussian random vectors since most of our analysis for state and parameter estimation will assume that the probability distributions are Gaussian.

9.2 GAUSSIAN RANDOM VECTORS

Definition

An n dimensional random vector $\underline{X}$ with mean $\underline{m} = E[\underline{X}]$ and covariance matrix $P = E\left[(\underline{X} - E(\underline{X}))(\underline{X} - E(\underline{X}))^T\right]$ is distributed in accordance with a Gaussian law if its characteristic function is

$$\phi(\underline{u}) = e^{(j\underline{u}^T\underline{m} - \frac{1}{2}\underline{u}^T P \underline{u})} \qquad 9.2.1$$

On inverting 9.2.1 by using a Fourier transform we obtain the Gaussian probability density as

$$p(\underline{x}) = \frac{1}{(2\pi)^{n/2}(\det(P))^{1/2}} e^{-\frac{1}{2}\left[(\underline{x}-\underline{m})^T P^{-1}(\underline{x}-\underline{m})\right]} \qquad 9.2.2$$

which does not exist if P is singular. For this reason it is preferable to define the Gaussian distribution in terms of its characteristic function.

From these definitions we see why Gaussian distributions are so attractive. The Gaussian probability density function can be described uniquely in terms of the two quantities : mean and covariance so that we do not have to worry about higher order moments. This is fortunate since many physical phenomena can be described in terms of Gaussian distributions.

Let us next see how our description is effected when the random variables $X_1, X_2, \ldots X_n$ which comprise our vector $\underline{X}$ are un-correlated. In this case, our matrices P, P^{-1}, will be diagonal and we could write

$$\phi(\underline{u}) = e^{j\sum_i u_i m_i - \frac{1}{2}\sum_i u_i^2 \sigma_i^2}$$

$$= \phi(u_1) \cdot \phi(u_2) \cdots \phi(u_n) \qquad 9.2.3$$

and in the same way

$$p(\underline{x}) = p(x_1) \cdot p(x_2) \cdot \quad \cdots \quad p(x_n) \qquad 9.2.4$$

Thus we see that for Gaussian distributions, un-correlated random variables are also independent.

Since in our work on estimation and control in the next chapter we will consider only linear systems with Gaussian random disturbances, it would be useful to develop formulae for performing linear transformations on Gaussian random vectors

9.2.1 Linear transformation of Gaussian random variables

Let us consider a vector $\underline{Z}$ which is related to a Gaussian vector $\underline{X}$ by the linear transformation

$$\underline{Z} = A\,\underline{X} + \underline{b} \qquad 9.2.5$$

where $\underline{Z}$ and $\underline{b}$ are p vectors, and A is a p x n matrix. A and $\underline{b}$ are deterministic.

The characteristic function of $\underline{Z}$ is by definition

$$\phi_Z(\underline{v}) = E\left[e^{j\underline{v}^T\underline{Z}}\right] = E\left[e^{j\underline{v}^T[A\underline{X}+\underline{b}]}\right] = e^{j\underline{v}^T\underline{b}}\;E\left[e^{j\underline{v}^T A\underline{X}}\right] \qquad 9.2.6$$

We recall that the characteristic function of $\underline{X}$ is

$$\phi_X(\underline{u}) = E\left[e^{j\underline{u}^T\underline{X}}\right] = e^{(j\underline{u}^T\underline{m} - \frac{1}{2}\underline{u}^T P\underline{u})} \qquad 9.2.7$$

(since $\underline{X}$ is Gaussian)

On comparing 9.2.6 and 9.2.7, we see that

$$\phi_Z(\underline{v}) = e^{j\underline{v}^T}\;\phi_X(A^T\underline{v}) = e^{j\underline{v}^T\underline{b}}\;e^{j\underline{v}^T A\underline{m} - \frac{1}{2}\underline{v}^T APA^T\underline{v}}$$

or

$$\phi_Z(\underline{v}) = e^{j\underline{v}^T(A\underline{m}+\underline{b}) - \frac{1}{2}\underline{v}^T APA^T\underline{v}} \qquad 9.2.8$$

$\underline{Z}$ is therefore a Gaussian vector with mean $E\left[\underline{Z}\right] = A\,\underline{m} + \underline{b}$ and with the covariance matrix

$$A\,P\,A^T = E\left[(\underline{Z} - E(\underline{Z}))(\underline{Z} - E(\underline{Z}))^T\right]$$

We see thus that Gaussian random vectors retain their Gaussian character under linear transformation.

Next we will consider how we can make two random vectors independent by linear transformation.

Consider a Gaussian vector $\underline{J}$ which is partitioned into two vectors $\underline{X}$ and $\underline{Y}$, i.e.

$$\underline{J} = \begin{bmatrix} \underline{X} \\ \cdots \\ \underline{Y} \end{bmatrix} \qquad 9.2.9$$

The covariance matrix of $\underline{J}$ is also partitioned in the following way

$$\begin{bmatrix} P_{xx} & \vdots & P_{xy} \\ \cdots & \cdots & \cdots \\ P_{yx} & \vdots & P_{yy} \end{bmatrix}$$

where

$$P_{xx} = E\left[(\underline{X} - E(\underline{X}))\ (\underline{X} - E(\underline{X}))^T\right]$$

$$P_{xy} = E\left[(\underline{X} - E(\underline{X}))\ (\underline{Y} - E(\underline{Y}))^T\right] \qquad = P_{yx}^{\ T} \qquad 9.2.10$$

$$P_{yy} = E\left[(\underline{Y} - E(\underline{Y}))\ (\underline{Y} - E(\underline{Y}))^T\right]$$

We leave as an exercise for the reader to show by applying the previous results that $\underline{X}$ and $\underline{Y}$ are Gaussian although we could not say whether $\underline{J}$ was Gaussian if it had been formed by two Gaussian vectors $\underline{X}$, $\underline{Y}$.

Let us form a new vector $\underline{W} = \begin{bmatrix} \underline{W}_1 \\ \cdots \\ \underline{W}_2 \end{bmatrix}$ by doing a linear transformation on $\underline{J}$ such that $\underline{W}_1$ and $\underline{W}_2$ are of the same dimension as $\underline{X}$ and $\underline{Y}$ respectively i.e.

$$\underline{W}_1 = \underline{X} + M\,\underline{Y}$$
$$\underline{W}_2 = \underline{Y} \qquad 9.2.11$$

We would like to find the transformation M such that $\underline{W}_1$ and $\underline{W}_2$ are independent.

Now the condition for independence is

$$E\left[(\underline{W}_1 - E(\underline{W}_1))\ (\underline{W}_2 - E(\underline{W}_2))^T\right] = 0 \qquad 9.2.12$$

which implies that

$$\phi_W(\underline{u}) = \phi_{W_1}(\underline{u}_1)\ \phi_{W_2}(\underline{u}_2)$$

and

$$p(\underline{W}) = p(\underline{W}_1)\ .\ p(\underline{W}_2) \qquad 9.2.13$$

We will use condition 9.2.12 to determine M

$$E\left[(\underline{X} - E(\underline{X}) + M(\underline{Y} - E(\underline{Y}))\ (\underline{Y} - E(\underline{Y}))^T\right] = P_{xy} + M\,P_{yy} = 0$$

or

$$M = -\,P_{xy}\,P_{yy}^{-1} \qquad 9.2.14$$

We have thus obtained by this transformation two independent Gaussian random vectors

$$\underline{W}_1 = \underline{X} - P_{xy}\,P_{yy}^{-1}\,\underline{Y}$$
$$\underline{W}_2 = \underline{Y} \qquad 9.2.15$$

The variance of $\underline{W}_1$ is given by

$$E\ (\underline{X} - E[\underline{X}] - P_{xy}\,P_{yy}^{-1}(\underline{Y} - E[\underline{Y}]))\ (\underline{X} - E[\underline{X}] - P_{xy}P_{yy}^{-1}(\underline{Y} - E[\underline{Y}])^T$$

$$= P_{xx} - P_{xy}\,P_{yy}^{-1}\,P_{yx} - P_{xy}\,P_{yy}^{-1}\,P_{yx} + P_{xy}\,P_{yy}^{-1}\,P_{yy}\,P_{yy}^{-1}\,P_{yx}$$

$$= P_{xx} - P_{xy} P_{yy}^{-1} P_{yx} \tag{9.2.16}$$

so that the new covariance matrix is

$$\begin{bmatrix} P_{xx} - P_{xy} P_{yy}^{-1} P_{yx} & \vdots & 0 \\ \cdots & \cdot & \cdots \\ 0 & \vdots & P_{yy} \end{bmatrix} \tag{9.2.17}$$

Next, let us calculate the conditional expectation.

9.2.2 Calculation of the conditional expectation

From the above analysis we note that it is possible to write the vector $\underline{X}$ as a sum of two independent terms

$$\underline{X} = P_{xy} P_{yy}^{-1} \underline{Y} + \underline{W}_1 \tag{9.2.18}$$

To calculate the conditional expectation, we write

$$E[\underline{X} | \underline{Y}] = P_{xy} P_{yy}^{-1} E[\underline{Y} | \underline{Y}] + E[\underline{W}_1 | \underline{Y}]$$

But since $\underline{W}_1$ is independent of $\underline{Y}$, we have

$$E[\underline{X} | \underline{Y}] = P_{xy} P_{yy}^{-1} \underline{Y} + E[\underline{W}_1] \tag{9.2.19}$$

But from 9.2.15, we can write $E[\underline{W}_1]$ as

$$E[\underline{W}_1] = E[\underline{X}] - P_{xy} P_{yy}^{-1} E[\underline{Y}] \tag{9.2.20}$$

Substituting for $E[\underline{W}_1]$ from 9.2.20 in 9.2.19, we have

$$E[\underline{X} | \underline{Y}] = E[\underline{X}] + P_{xy} P_{yy}^{-1} (\underline{Y} - E[\underline{Y}]) \tag{9.2.21}$$

The <u>conditional variance</u> is defined as

$$P_{\underline{X}|\underline{Y}} = E[(\underline{X} - E[\underline{X}|\underline{Y}])(\underline{X} - E[\underline{X}|\underline{Y}])^T | \underline{Y}] \tag{9.2.22}$$

This expression can be simplified if we use 9.2.18 and 9.2.21, i.e.

$$\underline{X} - E[\underline{X}|\underline{Y}] = P_{xy} P_{yy}^{-1} \underline{Y} + \underline{W}_1 - P_{xy} P_{yy}^{-1} \underline{Y} - E[\underline{W}_1]$$

or

$$\underline{X} - E[\underline{X}|\underline{Y}] = \underline{W}_1 - E[\underline{W}_1] \tag{9.2.23}$$

Then using 9.2.16, we have

$$P_{\underline{X}|\underline{Y}} = P_{xx} - P_{xy} P_{yy}^{-1} P_{yx}$$

Remarks

a) From equation 9.2.21 we see that the conditional methematical expectation of the Gaussian vector $\underline{X}$ knowing the Gaussian vector $\underline{Y}$ is a random Gaussian vector which is a linear combination of the elements of $\underline{Y}$.

b) The vector $\underline{X} - E[\underline{X}|\underline{Y}]$ is independent of $\underline{Y}$ and of all linear combinations of elements of $\underline{Y}$.

Next we will see how the previously developed elements of probability theory extend to the dynamical system models that are of interest in this book. We will thus consider next stochastic processes.

9.3 STOCHASTIC PROCESSES

We begin by defining a stochastic process.

A stochastic process (scaler or vectorial) $\{ X(t,w),\ t \in T,\ w \in \Omega \}$ is a family of random variables (or random vectors) indexed by the time parameter $t \in T$.

For a fixed t, X(t, .) is a random variable or a random vector.

For a fixed w, X(., w) is a function of time.

If Ω is discrete, we say that the process has a discrete state space or has discrete states.

The set T could be discrete or continuous. If the set T is discrete, the stochastic process will have discrete parameters and in the opposite case it will have continuous parameters.

Having defined a stochastic process, we will next see that it is possible to describe it by a simple extension of the concepts used for random variables.

9.3.1 Description of a stochastic process

a) Distribution function

The distribution function depends here in general on t. Thus

$$F(x_t) = P\left[X(t, w) \leqslant x_t \right]$$

is the first order distribution function.

If we next consider two instants of time t_1, t_2, the joint distribution will depend on t_1 and t_2 and it could be written as

$$F(x_{t_1}, x_{t_2}) = P\left[X(t_1, w) \leqslant x_{t_1} \cap X(t_2, w) \leqslant X_{t_2} \right]$$

which is the second order distribution function.

In general, we see that in order to characterise a stochastic process, it is necessary to define a joint distribution function for all finite sets of random variables : $X(t_o, w)$, $X(t_1, w)$ ... $X(t_N, w)$.

$$F(x_{t_o}, x_{t_1}, \dots x_{t_N}) = P\left[X(t_o, w) \leqslant x_{t_o} \cap X(t_1, w) \leqslant x_{t_1} \quad \dots \quad \cap X(t_N, w) \leqslant x_{t_N}\right] \qquad 9.3.1$$

As for the case of random variables, we can also describe them in terms of the probability density.

Thus we can define the first order probability density as

$$p(x_t) = \frac{\partial F(x_t)}{\partial x_t} \qquad 9.3.2$$

and similarly the second order probability density as

$$p(x_{t_1}, x_{t_2}) = \frac{\partial^2 F(x_{t_1}, x_{t_2})}{\partial x_{t_1} \partial x_{t_2}} \qquad 9.3.3$$

and for all finite sets of the parameters $\left\{t_o, t_1, \dots t_N\right\}$ the joint probability density will be denoted by

$$p(x_{t_o}, x_{t_1}, x_{t_2}, \dots x_{t_N})$$

b) Mean and Moments

The mathematical expectation of X(t, w) will now be a function of time defined by

$$E\left[X(t, w)\right] = \int_X x_t \, p(x_t) \, dx_t = m(t) \qquad 9.3.4$$

The autocorrelation function is defined by

$$R(t_1, t_2) = E\left[X(t_1, w) \, . \, X(t_2, w)\right] = \iint_{X\,X} x_{t_1} x_{t_2} \, p(x_{t_1}, x_{t_2}) \, dx_{t_1} \, dx_{t_2}$$

This is a function with two arguments : t_1 and t_2.

The covariance (or autocovariance) function will be

$$P(t_1, t_2) = E\left[(X(t_1, w) - m(t_1)(X(t_2, w) - m(t_2))\right]$$

In the case where $\underline{X}(t, w)$ is a vector process, $\underline{m}(t)$ will be a vector and the autocorrelation matrix will be

$$E\left[\underline{X}(t_1, w) \, . \, \underline{X}^T(t_2, w)\right]$$

and the covariance matrix will be

$$E\left[\underline{X}(t_1, w) - \underline{m}(t_1))(\underline{X}(t_2, w) - \underline{m}(t_2))^T\right]$$

9.3.2 Gaussian stochastic processes

A scaler stochastic process X(t, w) is Gaussian if for some set $t_1, t_2, \dots t_N$, the vector formed by the N random variables $X(t_1, w)$, $X(t_2, w)$, ... $X(t_N, w)$ is Gaussian. On defining :

$$\underline{m} = \begin{bmatrix} E[X(t_1, w)] \\ E[X(t_2, w)] \\ \vdots \\ E[X(t_N, w)] \end{bmatrix}; \quad P = \begin{bmatrix} Var(X(t_1, w) & \cdots & cov(X(t_1, w) X(t_N, w) \\ \vdots & Var(X(t_2, w) & \vdots \\ & \ddots & \\ Cov(X(t_1, w) X(t_N, w)) & \cdots & Var(X(t_N, w) \end{bmatrix}$$

We should find that the characteristic function is of the form

$$\phi(u_1, u_2, \dots u_N) = \phi(\underline{u}) = e^{j\underline{u}^T\underline{m} - \frac{1}{2}\underline{u}^T P \underline{u}}$$

As we mentioned earlier in the context of random variables, this definition is less restrictive than the one where we use the probability density of $X(t_1, w) \dots X(t_N, w)$. Indeed, for the case where $X(t_1, w) = X(t_2, w)$ for $t_1 \neq t_2$, our matrix P will be singular and in that case the probability density function will be undefined.

In the same way a vector stochastic process will be Gaussian if for some set $t_1, t_2, \dots t_N$, the vector formed by the set of the N random vectors each of dimension n, $\underline{X}(t_1, w), \underline{X}(t_2, w), \dots \underline{X}(t_N, w)$ is a Gaussian vector.

9.3.3. Stationarity

Consider a stochastic process $\{X(t, w), t \in T\}$ where T is a linear parametric set (which means that if t_1 and $t_2 \in T$, $t_1+t_2 \in T$). The process will be said to be strictly stationary if it has the same probability law as the process

$$\{X(t + \tau, w), \quad t \in T\} \qquad \text{i.e. we have}$$

$$p(x_{t_1}, x_{t_2}, \dots x_{t_N}) = p(x_{t_1+\tau}, x_{t_2+\tau}, \dots x_{t_N+\tau})$$

for all finite sets $t_1, t_2, \dots t_N \in T$ and for each $\tau \in T$.

If we consider two first order densities, we have

$$p(x_t) = p(x_{t+\tau})$$

which implies that the first order density is in this case independent of t which means that the mean of the process is a constant

$$E[X(t_1, w)] = m(t) = \text{constant}$$

For the second order density we have

$$p(x_{t_1}, x_{t_2}) = p(x_{t_1+\tau}, x_{t_2+\tau})$$

from which we conclude that this density should be a function of $t_2 - t_1$ so that the autocorrelation function depends only on the difference $t_2 - t_1$.

Wide sense or covariance stationarity

A second order process $E[X^2(t, w)] < \infty$ is said to be wide sense (or covariance) stationary if its mathematical expectation is constant and if its autocorrelation function depends only on $t_2 - t_1$.

A strict sense stationary process is obviously also a wide sense stationary process although the inverse is not necessarily true.

9.3.4 Markov processes

Consider a set of ordered parameters $t_o < t_1 < t_2 \ldots < t_N$. A stochastic process $X(t, w)$ will be called a Markov process if we could write

$$P\left[X(t_N, w) \leqslant x_{t_N} \mid X(t_{N-1}, w) = x_{t_{N-1}}, \ldots X(t_o, w) = x_{t_o}\right]$$

$$= P\left[X(t_N, w) \leqslant x_{t_N} \mid X(t_{N-1}, w) = x_{t_{N-1}}\right]$$

i.e. the entire past history of the process is contained in the last state.

We would have in the same way for the conditional probability density function

$$p(x_{t_N} \mid x_{t_{N-1}}, \ldots x_{t_o}) = p(x_{t_N} \mid x_{t_{N-1}})$$

Next let us develop an expression for the joint probability density function for a Markov process.

Using Bayes theorem, we have

$$p(x_{t_N}, x_{t_{N-1}}, \ldots x_{t_o}) = p(x_{t_N} \mid x_{t_{N-1}}, \ldots x_{t_o}).\ p(x_{t_{N-1}}, \ldots x_{t_o})$$

If the process is Markovian

$$p(x_{t_N}, x_{t_{N-1}}, \ldots x_{t_o}) = p(x_{t_N} \mid x_{t_{N-1}}).\ p(x_{t_{N-1}}, \ldots x_{t_o})$$

Doing the same operation on $p(x_{t_{N-1}}, \ldots x_{t_o})$ etc, we obtain finally

$$p(x_{t_N}, x_{t_{N-1}}, \ldots x_{t_o}) = p(x_{t_N} \mid x_{t_{N-1}}).\ p(x_{t_{N-1}} \mid x_{t_{N-2}}), \ldots p(x_{t_1} \mid x_{t_o})\ p(x_{t_o})$$

We see that we can describe completely a Markov process in terms of its probability density of transition $p(x_{t_k} \mid x_{t_{k-1}})$ and the distribution of the initial state.

Example

As an example, consider a discrete time process described by

$$X(t_{k+1}) = X(t_k) + u(k)$$

where $u(k), u(k-1), \ldots u(o)$ is a sequence of independent random variables. We assume, in addition, that the initial state $X(t_o)$ is a random variable which is independent of the sequence $u(k), u(k-1), \ldots$

We would like to determine for this process the conditional probability

$$p(x_{t_{k+1}} \mid x_{t_k}, x_{t_{k-1}}, \dots x_{t_o})$$

From the dynamical equation we see that if we know the state $X(t_k)$, the only random quantity is $u(k)$. It is necessary however to see that we have a Markov process here, i.e. knowledge of previous states $X(t_{k-1})$, ... $X(t_o)$ does not change in any way the probability of this random variable $u(k)$.

Indeed,

$$X(t_{k-1}) = u(k-2) + u(k-3) + \dots u(o) + X(t_o)$$

is independent of $u(k)$ by our previous assumption and this is true also for $X(t_{k-2})$, ... $X(t_o)$. We could therefore write

$$P(x_{t_{k+1}} \mid x_{t_k}, x_{t_{k-1}}, \dots x_{t_o}) = P(x_{t_{k+1}} \mid x_{t_k})$$

and the process is Markovian.

9.3.5 White noise

Let us consider first of all the spectral density of a stationary process.

Consider a stationnary process $X(t, w)$ of second order. Its autocorrelation function is given by

$$R(\tau) = E\left[X(t+\tau, w)\, X(t, w)\right]$$

We could define its spectral density by taking the Fourier transform, i.e.

$$\phi(w) = \int_{-\infty}^{\infty} R(\tau)\, e^{-jw\tau}\, d\tau$$

or inversely

$$R(\tau) = \frac{1}{2\pi} \int_{-\infty}^{\infty} \phi(w)\, e^{jw\tau}\, dw$$

In the same way for a discrete parameter process having an autocorrelation function

$$R(n) = E\left[X(t_{k+n}, w)\, X(t_k, w)\right]$$

the spectral density will be defined as

$$\phi(w) = \sum_{n=-\infty}^{\infty} R(n)\, e^{-jwn}$$

or inversely

$$R(n) = \frac{1}{2\pi} \int_{-\pi}^{\pi} \phi(w)\, e^{jnw}\, dw$$

Discrete white noise

By analogy with white light which contains all frequencies in its spectrum, we will define white noise as a process which has a constant spectral density.

In the case of discrete white noise, we have

$$\Phi(w) = \text{constant} = C$$

from which we have

$$R(n) = C\left(\frac{\sin n\pi}{n\pi}\right)$$

and thus

$$R(n) = \begin{cases} C \text{ if } n = 0 \\ 0 \text{ if } n = \pm 1, \pm 2, \ldots \end{cases}$$

The numbers $X(t_k, w)$, $X(t_{k-1}, w)$ ... $X(t_o, w)$ constitute a sequence of uncorrelated random variables.

We define a discrete Gaussian white noise, $u(t_k) = u_k$ as a Gaussian process whose autocorrelation function is given by

$$E\left[u_k u_j\right] = C\,\delta_{kj}$$

where δ_{kj} is the Kronecker Delta.

In this case, the random variables will be <u>independent</u>.

In the vector case, we define in the same way a Gaussian white noise sequence as a sequence of independent random vectors such that

$$E\left[\underline{u}_k\,\underline{u}_j^T\right] = Q\,\delta_{kj}$$

<u>Continuous white noise</u>

Here $\Phi(w)$ = constant.

Since the Fourier transform of a constant is the Dirac Delta Function, we obtain

$$E\left[u(t)\,u(\tau)\right] = q\,\delta(t - \tau)$$

We note that a process defined in this way is physically unrealisable since its power $R(o)$ is infinite

$$R(o) = \frac{1}{2\pi}\int_{-\infty}^{\infty} C\,d\,w$$

We define a Gaussian white noise as a Gaussian process $u(t)$ with

$$E\left[u(t)\,u(\tau)\right] = q\,\delta(t - \tau)$$

and in the vector case

$$E\left[\underline{u}(t)\;\underline{u}^T(\tau)\right] = Q\,\delta(t - \tau)$$

<u>Example</u>

Let us reconsider the previous example except that we assume that the process $u(t_k)$ is a discrete Gaussian white noise sequence.

$$x(t_{k+1}) = x(t_k) + u(t_k)$$

Thus let us make the following assumptions

a) $u(t_k)$ is a discrete Gaussian white noise sequence. Therefore $\{u(t_k), u(t_{k-1}), \ldots u(t_o)\}$ is a sequence of independent random variables

b) $X(t_o)$ is a Gaussian random variable which is uncorrelated with $\{u(t_k), u(t_{k-1}), \ldots u(t_o),\}$ i.e.

$$E\left[X(t_o)\, u(t_k)\right] = 0 \quad \forall\ k$$

This ensures that $X(t_o)$ and the noise are independent of each other.

$X(t_k)$ is therefore, as in the previous example, a Markov process. In addition, since here we have linear transformations of Gaussian random variables, $X(t_k)$ is a Gaussian process. This is thus an example of a Gauss-Markov process.

This leads us naturally to consider system models and in particular Gauss-Markov processes.

9.4 DYNAMICAL SYSTEMS AND GAUSS-MARKOV PROCESSES

In this section, we will study the behaviour of linear dynamical systems subject to Gaussian white noise disturbances. We begin by considering discrete models.

Discrete Time Models

We will consider linear models described in state space form , i.e.

$$\underline{x}_{k+1} = \phi_k\, \underline{x}_k + \underline{u}_k + \Gamma_k\, \underline{v}_k \qquad 9.4.1$$

where $\underline{u}_k$ is a vector of known deterministic inputs and $\underline{v}_k$ is a vector of Gaussian white noise.

With a view to simplifying the expressions, we will study the evolution of the system around a known deterministic trajectory so that we can equally consider the system equation

$$\underline{x}_{k+1} = \phi_k\, \underline{x}_k + \Gamma_k\, \underline{v}_k \qquad 9.4.2$$

Fig. 9.1 Gives a block diagram of the process.

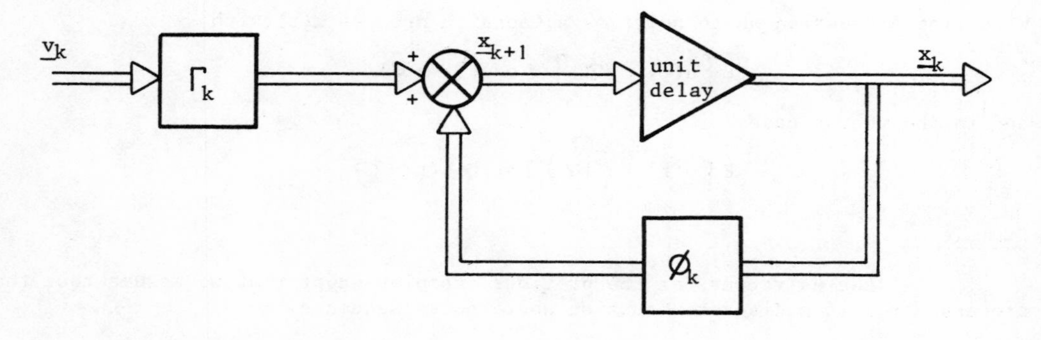

Fig. 9.1

We begin our study of discrete Gauss-Markov system models by making certain assumptions. These are :

a) $\underline{v}_k$ is a Gaussian white noise vector such that $E[\underline{v}_k] = 0 \quad E[\underline{v}_k \underline{v}_j^T] = Q\,\delta_{kj}$

b) The initial state $\underline{x}_o$ is a Gaussian random vector with $E[\underline{x}_o] = \underline{m}_o$;
$E[(\underline{x}_o - m_o)(\underline{x}_o - m_o)^T] = P_o$

c) The noise sequence $\underline{v}_k$ is independent of $\underline{x}_o$, i.e. $E[\underline{x}_o \underline{v}_k^T] = 0 \quad \forall\, k$

Remark

These assumptions are justified in many practical cases :

a - The perturbations are assumed to be Gaussian as a consequence of the central limit theorem if we consider that the disturbances arise from a large number of microscopic independent random actions.

b - The white noise assumption enables us to represent disturbances having "short" correlation times compared to the system.

c - Finally, we assume that the uncertainty in the case of the initial state is not related to the disturbances which act on the system.

With the above assumptions, the state of the system is a Gauss-Markov process. Indeed, if we study the distribution of $\underline{x}_{k+1}$ given $\underline{x}_k$, we could write

$$\underline{x}_1 = \Phi_o \underline{x}_o + \Gamma_o \underline{v}_o$$

$$\underline{x}_2 = \Phi_1 \underline{x}_1 + \Gamma_1 \underline{v}_1 = (\Phi_1 \Phi_o)\underline{x}_o + \Phi_1 \Gamma_o \underline{v}_o + \Gamma_1 \underline{v}_1$$

$$\vdots$$

$$\underline{x}_k = \Phi(k, o)\, \underline{x}_o + \sum_{\ell=0}^{k-1} \Phi(k, \ell+1)\, \Gamma_\ell\, v_\ell \qquad 9.4.3$$

where $\Phi(k, o)$ is the transition matrix from t_o to t_k,

$$\Phi(k, o) = \Phi_{k-1} \Phi_{k-2} \cdots \Phi_o \text{ with } \Phi(k, k) = I$$

$$\Phi(k, \ell+1) = \Phi_{k-1} \Phi_{k-2} \cdots \Phi_{\ell+1}$$

The previous relation shows that $\underline{x}_{k-1}$ depends on the first k-1 values of $\underline{v}$ and on $\underline{x}_o$. $\underline{x}_{k-1}$ is therefore independent of $\underline{v}_k$ and we can show in the same way that $\underline{x}_{k-2}, \ldots \underline{x}_o$ are independent of $\underline{v}_k$. We could thus write

$$P(\underline{x}_{k+1} \mid \underline{x}_{k-1}, \ldots \underline{x}_o) = P(\underline{x}_{k+1} \mid \underline{x}_k)$$

The process $\underline{x}_k$ is thus Markovian. In addition it is obtained by performing linear transformations on Gaussian random vectors. Thus $\underline{x}_k$ is Gaussian.

Remark

Note that a non-linear dynamical system will give with the above assumptions a Markovian process but the Gaussian nature of the process is not conserved by the non-linear transformations.

Calculation of the mean

We obtain

$$E\left[\underline{x}_{k+1}\right] = \Phi_k \, E\left[\underline{x}_k\right] \qquad 9.4.4$$

or

$$\underline{m}_{k+1} = \Phi_k \, \underline{m}_k \qquad 9.4.5$$

This is a recursive relationship which enables us to calculate the mean at each instant given the initial condition

$$E\left[\underline{x}_o\right] = \underline{m}_o \qquad 9.4.6$$

Calculation of the covariance matrix

At the t_{k+1} th instant, this matrix is defined as

$$P_{k+1} = E\left[(\underline{x}_{k+1} - \underline{m}_{k+1})(\underline{x}_{k+1} - \underline{m}_{k+1})^T\right] \qquad 9.4.7$$

we remark that

$$\underline{x}_{k+1} - \underline{m}_{k+1} = \Phi_k(\underline{x}_k - \underline{m}_k) + \Gamma_k \underline{v}_k \qquad 9.4.8$$

Thus

$$P_{k+1} = E\left[(\Phi_k(\underline{x}_k - \underline{m}_k) + \Gamma_k \underline{v}_k)((\underline{x}_k - \underline{m}_k)^T \Phi_k^T + \underline{v}_k^T \Gamma_k^T)\right] \qquad 9.4.9$$

We have already seen that $\underline{x}_k$ and $\underline{v}_k$ are independent so that

$$E\left[\underline{x}_k \underline{v}_k^T\right] = 0 = E\left[\underline{v}_k \underline{x}_k^T\right]$$

In addition,

$$E\left[\underline{m}_k \underline{v}_k^T\right] = \underline{m}_k \; E\left[\underline{v}_k^T\right] = \underline{0} \qquad \text{and}$$

also $E\left[\underline{v}_k \underline{m}_k^T\right] = 0$ since the noise is assumed to have zero mean.

This leaves therefore

$$P_{k+1} = \Phi_k P_k \Phi_k^T + \Gamma_k Q \Gamma_k^T \qquad 9.4.10$$

This is a recursive relation which enables us to calculate the covariance matrix of the process at the instant t_{k+1} given the covariance matrix at the instant t_k.

Given that the process is Gaussian, the two quantities $\underline{m}_{k+1}$ and P_{k+1} are sufficient to determine the probability law for the state $\underline{x}_{k+1}$.

We can equally determine a general expression for P_{k+1} from equation 9.4.3. This is given by

$$P_{k+1} = \Phi(k+1,o)P_o\Phi^T(k+1,o) + \sum_{\ell=0}^{k} \Phi(k+1,\ell+1) \quad \Gamma_\ell Q \Gamma_\ell^T \Phi^T(k+1,\ell+1) \qquad 9.4.11$$

Next we see how we can compute the covariance matrix between two instants t_k and t_{k+n}. We would like to compute

$$C(k+n,\ k) = E\left[(\underline{x}_{k+n} - \underline{m}_{k+n})(\underline{x}_k - \underline{m}_k)^T\right] \qquad n > 0 \qquad 9.4.12$$

To do this, we will express $\underline{x}_{k+n}$ as a function of $\underline{x}_k$, i.e.

$$\underline{x}_{k+n} = \Phi(k+n,\ k)\ \underline{x}_k + \sum_{\ell=k}^{k+n-1} \Phi(k+n,\ \ell+1)\ \Gamma_\ell\ \underline{v}_\ell \qquad 9.4.13$$

where

$$\Phi(k+n,\ k) = \Phi_{k+n-1} \cdot \Phi_{k+n-2} \cdot\ \ldots\ \Phi_k \qquad 9.4.14$$

On noting that $\underline{x}_k$ depends only on $\underline{x}_o$ and $\underline{v}_o$, $\underline{v}_1$, ..., $\underline{v}_{k-1}$, we obtain the final result

$$C(k+n,\ k) = \Phi(k+n,\ k)\ P_k \qquad n=1,\ 2,\ \ldots \qquad 9.4.15$$

Using a similar reasoning we can show that for the matrix

$$C(k,\ k+n) = E\left[\underline{x}_k - \underline{m}_k)\ (\underline{x}_{k+n} - \underline{m}_{k+n})^T\right]$$

the relation

$$C(k,\ k+n) = P_k\ \Phi^T(k+n,\ k) \qquad n=1,\ 2,\ \ldots \qquad 9.4.16$$

holds.

Consider next an example.

Example

Let us consider a scaler process described by the difference equation

$$x_{k+1} = a\ x_k + v_k$$

v_k is a Gaussian white noise with zero mean ($E\left[v_k\right] = 0$) and covariance q ($E(v_k\ v_j) = q\ \delta_{kj}$). The initial state x_o is zero mean Gaussian (i.e. $E\left[x_o\right] = 0$) with covariance P_o $\left[E(x_o^2) = P_o\right]$ and it is uncorrelated with v (i.e. $E\left[x_o v_k\right] = 0$ $\forall\ k$.

The variance of the process at an instant t_{k+1} is therefore given by the relation

$$P_{k+1} = a^2\ P_k + q$$

We could equally use the Global formula which gives

$$P_{k+1} = a^{2k+2}\ P_o + q(1+a^2 + \ldots + a^{2k})$$

or

$$P_{k+1} = a^{2k+2}\ P_o + q\ \frac{1-a^{2k+2}}{1-a^2}$$

It is interesting to see what happens when $k \rightarrow \infty$. If $|a| < 1$, $P_\infty = \frac{q}{1-a^2}$.

On the other hand, using the assumption $E[x_{k+1}] = 0$, we have therefore a stationary Gaussian process. It should be noted though that even if $E[x_o] = m_o \neq 0$, we would still obtain a zero mean process when $k \to \infty$, provided that $|a| < 1$.

Another thing to note is that if we want to produce a stationary process from the very beginning, it would be necessary to choose, as the initial state x_o, a random variable such that $E[x_o] = 0$, $E[x_o^2] = P_o = \frac{q}{1-a^2}$

Stationary Processes

The previous relationships enable us to compute the mean and the covariance matrix for the general case when Φ_k, Γ_k are functions of time. In the case where Φ_k, Γ_k are constant, it is interesting to find out under what additional conditions, we obtain a stationary process.

For the case of constant parameters, the mean and covariance equations can be written as

$$\underline{m}_k = \Phi^k \underline{m}_o \qquad 9.4.17$$

where Φ is a matrix with constant elements

and

$$P_k = \Phi^k P_o (\Phi^T)^k + \sum_{\ell=0}^{k-1} \Phi^\ell \, Q \, \Gamma^T (\Gamma^T)^\ell \qquad 9.4.18$$

We see that we need to find a matrix Φ such that the initial conditions $\underline{m}_o$, P_o don't have any influence when $k \to \infty$ and that the limit P_k exists. This will be so if Φ has eigenvalues whose modulus is strictly less than unity.

This result is similar to a classical result on the stability of linear systems. In fact the equation

$$P_\infty = \Phi P_\infty \Phi^T + \Gamma Q \Gamma \qquad 9.4.19$$

is used in the study of stability of the system

$$\underline{x}_{k+1} = \Phi \underline{x}_k \qquad 9.4.20$$

and we can show that the system is asymptotically stable if and only if for any positive definite matrix $\Gamma Q \Gamma^T$ there exists a positive definite matrix P which is the unique solution of equation 9.4.19.

In practice, we determine the parameters of the stationary regime by solving equation 9.4.19 which is a linear system of $\frac{n(n+1)}{2}$ equations where n is the dimension of the state vector of the system under study.

Density of Probability of transition

We can alternatively define a Gauss-Markov process

$$\underline{x}_{k+1} = \Phi_k \underline{x}_k + \Gamma_k \underline{v}_k \qquad 9.4.21$$

in terms of the probability distribution of the initial state $p(\underline{x}_o)$ and the transition probability density $p(\underline{x}_{k+1} | \underline{x}_k)$.

Since $\underline{x}_{k+1}$ and $\underline{x}_k$ are two Gaussian vectors (which are jointly Gaussian) it is sufficient to calculate the conditional mathematical expectation $E[\underline{x}_{k+1} | \underline{x}_k]$ and the conditional covariance matrix.

For the conditional expectation we have

$$E\left[\underline{x}_{k+1} \mid \underline{x}_k\right] = \Phi_k \underline{x}_k + \Gamma_k E\left[\underline{v}_k \mid \underline{x}_k\right] \quad 9.4.22$$

But $\underline{x}_k$ and $\underline{v}_k$ are independent so that $E\left[\underline{v}_k \mid \underline{x}_k\right] = E\left[\underline{v}_k\right] = \underline{0}$

so that

$$E\left[\underline{x}_{k+1} \mid \underline{x}_k\right] = \Phi_k \underline{x}_k \quad 9.4.23$$

For the conditional covariance matrix

$$E\left[(\underline{x}_{k+1} - E\left[\underline{x}_{k+1} \mid \underline{x}_k\right])(\underline{x}_{k+1} - E\left[\underline{x}_{k+1} \mid \underline{x}_k\right]) \mid \underline{x}_k\right]$$

$$= E\left[\Gamma_k \underline{v}_k \underline{v}_k^T \Gamma_k \mid \underline{x}_k\right]$$

$$= E\left[\Gamma_k \underline{v}_k \underline{v}_k^T \Gamma_k^T\right] = \Gamma_k Q \Gamma_k \quad 9.4.24$$

For the case where $\Gamma_k Q \Gamma_k$ is non singular, we will have

$$P(\underline{x}_{k+1} \mid \underline{x}_k) = \frac{1}{(2\pi)^{n/2}\left[\det[(\Gamma_k Q \Gamma_k^T)]\right]^{1/2}} e^{-\frac{1}{2}(\underline{x}_{k+1} - \Phi_k \underline{x}_k)^T (\Gamma_k Q \Gamma_k)^{-1} (\underline{x}_{k+1} - \Phi_k \underline{x}_k)} \quad 9.4.25$$

Next we will consider continuous dynamical systems.

9.5 CONTINUOUS DYNAMICAL SYSTEMS

Modelling

We will consider systems described by state space equations and subjected to Gaussian white noise disturbances.

Thus the system is represented by

$$\frac{d\underline{x}}{dt} = F(t)\underline{x} + G(t)\underline{u}(t) \quad 9.5.1$$

where $\underline{u}(t)$ is a Gaussian white noise such that

$$E\left[\underline{u}(t)\right] = \underline{0}, \; E\left[\underline{u}(t)\, \underline{u}^T(\tau)\right] = Q\,\delta(t-\tau)$$

This formulation leads to difficulties when we define integrals of white noise processes. To avoid these we will define a Weiner process with independent increments

$$d\underline{x} = F(t)\, \underline{x}\, dt + G(t)\, d\,\underline{v}(t) \quad 9.5.2$$

with

$$E\left[d\underline{v}\right] = \underline{0} \qquad E\left[d\underline{v}\, d\underline{v}^T\right] = Q\, dt$$

The process $\underline{v}(t)$ is assumed to be Gaussian. We assume in addition that the initial state $\underline{x}(t_o)$ is Gaussian with $E\left[\underline{x}(t_o)\right] = \underline{m}_o$ and

$$E\left[(\underline{x}(t_o) - \underline{m}_o)(\underline{x}(t_o) - \underline{m}_o)^T\right] = P_o$$

is also assumed to be independent of $\underline{v}(t)$.

The solution of equation 9.5.2 is

$$\underline{x}(t) = \phi(t, t_o)\,\underline{x}(t_o) + \int_{t_o}^{t} \phi(t, \tau)\,G(\tau)d\underline{v}(\tau) \qquad 9.5.3$$

It should be noted that the above integral can not be treated as an ordinary integral since the Weiner process has unbounded variations.

Stochastic integrals

We will next study integrals of the form $\int f(t)\,dv(t)$ where $dv(t)$ is an infinitesimal increment of Brownian motion (or Weiner process) $v(t)$. These are stochastic integrals which were initially defined for deterministic functions $f(t)$ and whose definition was extended to cover random functions $F(t,\omega)$. Thus we could define a stochastic integral as the limit of a sum -the deterministic convergence being replaced by a quadratic mean convergence-. We can show that if $f(t,\omega)$ are independent of future sample functions (realisations) of the Brownian motion and $\int E\left|f(t,\omega)\right|^2 dt < \infty$, then

$$E\left[\int_a^b f(t,\omega)\,dv(t)\right] = 0 \qquad 9.5.4$$

and

$$E\left[\int_a^b f_1(t,\omega)dv(t) \int_a^b f_2(t,\omega)dv(t)\right] = v^2 \int_a^b E\left[f_1(t,\omega)\,f_2(t,\omega)\right]dt \qquad 9.5.5$$

where $E\left[dv\right] = 0$ and $E\left[(dv)^2\right] = v^2\,dt$

This definition of stochastic integrals can be extended to vector processes and this clarifies the expression

$$\underline{x}(t) = \phi(t, t_o)\,\underline{x}(t_o) + \int_{t_o}^{t} \phi(t,\tau)\,G(\tau)d\underline{v}(\tau)$$

where $\phi(t, t_o)$ is the transition matrix defined by

$$\frac{d\,\phi(t, t_o)}{dt} = F(t)\,\phi(t, t_o)$$

Calculation of the mean

Taking the mathematical expectation of equation 9.5.3 we obtain

$$\underline{m}(t) = E\left[\underline{x}(t)\right] = \phi(t, t_o)\,E\left[\underline{x}(t_o)\right] + E\left[\int_{t_o}^{t}\phi(t,\tau)\,G(\tau)d\underline{v}(\tau)\right] \qquad 9.5.6$$

Using the properties of stochastic integrals, this reduces to

$$\underline{m}(t) = \phi(t, t_o)\,\underline{m}_o \qquad 9.5.7$$

On differentiating this expression with respect to time we obtain the vector differential equation

$$\frac{d\underline{m}(t)}{dt} = F(t)\,\underline{m}(t) \qquad 9.5.8$$

Covariance matrix

We need to calculate

$$P(t) = E\left[(\underline{x}(t) - \underline{m}(t))(\underline{x}(t) - \underline{m}(t))^T\right]$$

But

$$\underline{x}(t) - \underline{m}(t) = \phi(t, t_o)(\underline{x}(t_o) - \underline{m}_o) + \int_{t_o}^{t} \phi(t,\tau)\, G(\tau)\, d\underline{v}(\tau)$$

so that

$$P(t) = \phi(t, t_o)\, P(t_o)\, \phi^T(t, t_o) + E\left[\left(\int_{t_o}^{t} \phi(t,\tau)\, G(\tau)\, d\underline{v}(\tau)\right)\left(\int_{t_o}^{t} \phi(t,\tau)\, G(\tau)\, d\underline{v}(\tau)\right)^T\right] \qquad 9.5.9$$

But $\underline{x}(t_o)$ is independent of $d\underline{v}(t)$ for $t \geqslant t_o$ and $E\left[\int_{t_o}^{t} \phi(t,\tau) G(\tau)\, d\underline{v}(\tau)\right] = 0$.

The second term in 9.5.9 can be calculated using the properties of stochastic integrals so we obtain after some minor manipulations

$$P(t) = \phi(t, t_o)\, P(t_o)\, \phi^T(t, t_o) + \int_{t_o}^{t} \phi(t,\tau)\, G(\tau)\, Q\, G^T(\tau)\, \phi^T(t,\tau)\, d\tau \qquad 9.5.10$$

On differentiating 9.5.10 with respect to t we can obtain a matrix differential equation which describes the evolution of P(t), i.e.

$$\frac{dP}{dt} = F(t)\phi(t, t_o)\, P(t_o)\phi^T(t, t_o) + \phi(t, t_o)\, P(t_o)\phi^T(t, t_o) F^T(t) + G(t)\, Q\, G^T(t) + F(t)\int_{t_o}^{t} \phi(t,\tau)\, G(\tau)\, Q\, G^T(\tau)\, \phi^T(t,\tau)\, d\tau + \left(\int_{t_o}^{t} \phi(t,\tau)\, G(\tau)\, Q\, G^T(\tau)\phi^T(t,\tau)\, d\tau\right) F^T(t)\, d\tau$$

from which, after minor manipulations we obtain

$$\frac{dP(t)}{dt} = F(t)\, P(t) + P(t) F^T(t) + G(t)\, Q\, G^T(t) \qquad 9.5.11$$

Using a similar reasoning to the one used before we can also calculate the covariance matrix taken at two instants s and t

$$C(s, t) = E\left[(\underline{x}(s) - \underline{m}(s))(\underline{x}(t) - \underline{m}(t))^T\right] \qquad 9.5.12$$

For $s \geqslant t$, we have

$$C(s,\ t) = \phi(s,\ t)\ P(t) + E\left[\left(\int_t^s \phi(s,\tau)\ G(\tau)d\underline{v}(\tau)(\underline{x}(t) - \underline{m}(t))^T\right]\right.$$

Since $\underline{x}(t)$ is independent of $d\underline{v}(t)$ for $\tau \geqslant t$, we have the final results

$$C(s,\ t) = \phi(s,\ t)\ P(t) \qquad (s \geqslant t) \qquad 9.5.13$$

and for $t \geqslant 0$

$$C(s,\ t) = P(s)\ \phi^T(t,\ s) \qquad 9.5.14$$

Stationary Processes

We will assume to begin with that the matrices F(t) and G(t) are independent of time. The mean and covariance matrix are then given by

$$\underline{m}(t) = e^{F(t-t_o)}\ \underline{m}(t_o) \qquad 9.5.15$$

and

$$P(t) = e^{F(t-t_o)}\ P_o\ e^{F^T(t-t_o)} + \int_{t_o}^{t} e^{F(t-\tau)} GQG^T\ e^{F^T(t-\tau)}\ d\tau \qquad 9.5.16$$

When $t \longrightarrow \infty$, the mean and the covariance matrix will become independent of the initial conditions and they will converge to a steady state provided the matrix F is stable, i.e. if it has eigenvalues which have negative real parts.

The steady state value of the covariance matrix is given by the expression

$$F\ P_\infty + P_\infty\ F^T + G\ Q\ G^T = 0 \qquad 9.5.17$$

which gives a unique positive definite solution provided the system $\mathring{\underline{x}} = F\underline{x}$ is stable.

For a stationary process, the covariance matrix between two instants $t + \tau$ and t is

$$C(\tau) = E\left[(\underline{x}(t+\tau) - \underline{m}(t+\tau))\ (\underline{x}(t) - \underline{m}(t))^T\right] = e^{F\tau}\ P_\infty$$

or

$$C(\tau) = e^{F\tau}\ P_\infty \qquad \text{for } \tau > 0 \qquad 9.5.18$$

If we consider next the case $\tau < 0$

$$C(\tau) = E\left[(\underline{x}(t+\tau) - \underline{m}(t+\tau))\ (\underline{x}(t) - \underline{m}(t))^T\right] = P_\infty\ e^{-F^T\tau} \qquad 9.5.19$$

Spectral density

In the case of a stationary process, we could calculate its spectral density by taking the Bilateral Laplace Transform of its covariance matrix.

Indeed, for a stable system $\underline{m} = \underline{0}$

and

$$C(\tau) = E\left[\ \underline{x}(t+\tau)\ \underline{x}^T(t)\right]$$

Hence

$$S(p) = \int_{-\infty}^{\infty} e^{-\tau p}\, C(\tau)\, d\tau \qquad 9.5.20$$

We can decompose this into

$$S(p) = \int_{-\infty}^{0} e^{-\tau p}\, P_\infty\, e^{-F^T\tau}\, d\tau + \int_{0}^{\infty} e^{-\tau p}\, e^{F\tau}\, P_\infty\, d\tau$$

On putting $\tau' = -\tau$, the first term becomes

$$\int_{0}^{\infty} e^{\tau' p}\, P_\infty\, e^{F^T\tau'}\, d\tau'$$

and we have to evaluate two ordinary Laplace transforms if we put

$$R(p) = \int_{0}^{\infty} e^{-\tau p} e^{F\tau}\, P_\infty\, d\tau = \int_{0}^{\infty} e^{-\tau p}\, C(\tau)\, d\tau$$

Thus we have

$$S(p) = R(p) + R^T(-p)$$

The calculation of $R(p)$ can be done if we note that $C(\tau) = e^{F\tau} P_\infty$ obeys the differential equation

$$\frac{d\, C(\tau)}{d\tau} = F\, C(\tau)$$

so that $p\, R(p) - C(0) = F\, R(p)$

or

$$R(p) = (p\, I - F)^{-1}\, P_\infty$$

Thus

$$S(p) = (p\, I - F)^{-1}\, P_\infty + P_\infty\, (-pI - F^T)^{-1}$$

On premultiplying by $(pI - F)$ and post multiplying by $(-pI - F^T)$ we have

$$(pI-F)\, S(p)(-pI - F^T) = P_\infty(-pI - F^T) + (pI - F)\, P_\infty$$

$$= P_\infty\, F^T - F\, P_\infty = G\, Q\, G^T$$

from the relation defining P_∞ .

Thus

$$S(p) = (pI - F)^{-1}\, G\, Q\, G^T\, (-pI - F^T)^{-1} \qquad 9.5.21$$

This expression shows that the stationary process has a rational spectral density in p.

Inversely, for the problem of representing a process with a rational spectral density by a stationary Gauss-Markov process, it will be necessary to do a spectral factorisation in the form

$$S(p) = H(p)\, H^T(-p) \qquad 9.5.22$$

Finally in this chapter we consider how to discretise a continuous equation.

Discretisation of a continuous equation

Let us consider once again the stochastic differential equation

$$d\underline{x} = F(t)\ \underline{x}\ dt + G(t)\ d\underline{v}(t) \qquad 9.5.23$$

For discrete instants t_k, t_{k+1}, we have

$$\underline{x}(t_{k+1}) = \phi(t_{k+1}, t_k)\ \underline{x}(t_k) + \int_{t_k}^{t_{k+1}} \phi(t_{k+1},\tau)G(\tau)d\underline{v}(\tau) \qquad 9.5.24$$

Let us put

$$\underline{\omega}_k = \int_{t_k}^{t_{k+1}} \phi(t_{k+1},\tau)\ G(\tau)\ d\underline{v}(\tau) \qquad 9.5.25$$

On applying the results on stochastic integrals, we have

$$E\left[\underline{\omega}_k\right] = \underline{0}$$

$$E\left[\underline{\omega}_k \omega_k^T\right] = \int_{t_k}^{t_{k+1}} \phi(t_{k+1},\tau)\ G(\tau)\ Q\ G^T(\tau)\phi^T(t_{k+1},\tau)d\tau \qquad 9.5.26$$

Example

As an example, consider a scaler continuous system

$$dx = -\ ax\ dt + dv$$

$$\text{with} \quad E\left[dv\right] = 0\ ; \quad E\left[dv^2\right] = q\ dt$$

Discretisation with a time step Δt leads to the model

$$x_{k+1} = e^{-a\Delta t}\ x_k + \omega_k$$

with

$$E\left[\omega_k\right] = 0$$

$$E\left[\omega_k^{\ 2}\right] = \frac{q}{2a}\ (1 - e^{2a\Delta t})$$

9.6 CONCLUSIONS

In this chapter, we have introduced some basic probabilistic concepts in the context of estimation and identification. We have considered Gaussian random vectors and how to manipulate them, stochastic processes and Gauss-Markov processes in discrete and continuous time. The results that we have derived in this chapter will be invaluable in the development in the next two chapters.

REFERENCES FOR CHAPTER 9

Most of the material described here is well known. We have drawn heavily on the course notes in French of Professor G. Alengrin when writing this chapter and we are grateful to him for allowing us to do so.

Other useful references are :

[1] MEDITCH, J.S. (1969) "Stochastic linear estimation and control". Mc Graw Hill.

[2] PAPOULIS, A. (1965) "Probability random variables and stochastic processes". Mc Graw Hill.

PROBLEMS FOR CHAPTER 9

1. Show that

$$Cov(X_1, X_2) = E[X_1, X_2] - E[X_1] . E[X_2]$$

2. Develop an expression for the variance of the sum of two random variables X_1 and X_2 and show that

$$Var(X_1+X_2) = Var(X_1) + Var(X_2) + 2 Cov(X_1, X_2)$$

3. Consider a Gaussian random vector $\begin{bmatrix} x_1 \\ x_2 \\ x_3 \end{bmatrix}$ having a covariance matrix

$$P = \begin{bmatrix} 1 & 2 & 3 \\ 2 & 1 & 0 \\ 3 & 0 & 2 \end{bmatrix}$$

Form a new vector $\begin{bmatrix} \omega_1 \\ \underline{\omega}_2 \end{bmatrix}$ such that $\omega_1 = x_1 + [M_1 \;\; M_2] \begin{bmatrix} x_2 \\ x_3 \end{bmatrix}$

$$\underline{\omega}_2 = \begin{bmatrix} x_2 \\ x_3 \end{bmatrix}$$

Find the matrix $M = [M_1 \;\; M_2]$ such that ω_1 is independent of $\underline{\omega}_2$ and find the covariance matrix for

$$\begin{bmatrix} \omega_1 \\ \underline{\omega}_2 \end{bmatrix}$$

4. Consider the system

$$\overset{\circ}{x} = -\frac{1}{t+1} x$$

where x(0) is a zero mean Gaussian random variable with a positive variance $\sigma_o^2 > 0$. Show that the process is Gauss-Markov and that its probability density function is

$$f(x, t) = \frac{t+1}{\sqrt{2\pi}\,\sigma_o} e^{-(t+1)^2 x^2 | 2\sigma^2}$$

5. Consider a linear system described by

$$\begin{bmatrix} x_1(k+1) \\ x_2(k+1) \\ x_3(k+1) \end{bmatrix} = \begin{bmatrix} 1 & 0 & 2 \\ 2 & 1 & 0 \\ 3 & 2 & 1 \end{bmatrix} \begin{bmatrix} x_1(k) \\ x_2(k) \\ x_3(k) \end{bmatrix} + \begin{bmatrix} v_1(k) \\ v_2(k) \\ v_3(k) \end{bmatrix}$$

where $E[\underline{x}_o] = \begin{bmatrix} 1 \\ 1 \\ 1 \end{bmatrix}$; $E[\underline{v}] = \underline{0}$; $E[\underline{x}\,\underline{v}^T] = 0$

$$P_{xx}(0) = P(0) = \begin{bmatrix} 1 & 0 & 0 \\ 0 & 2 & 0 \\ 0 & 0 & 3 \end{bmatrix} ; \quad Q = \begin{bmatrix} 1 & 0 & 0 \\ 0 & 1 & 0 \\ 0 & 0 & 1 \end{bmatrix}$$

Calculate (i) $E\begin{bmatrix} x_1(4) \\ x_2(4) \\ x_3(4) \end{bmatrix}$

(ii) $P(4)$

CHAPTER 10

State and Parameter Estimation

In this chapter, we will describe some basic notions of state and parameter estimation. We will use the results of the previous chapter in our development. We will begin by giving some elements of estimation theory and then we will go on to consider the problem of parameter estimation. We will describe some of the methods in this area like least squares, recursive least squares, instrumental variable, etc. Next, we will consider the problem of state estimation and develop the Kalman filter both for the discrete time case as well as for continuous time. Finally, we will describe the celebrated separation theorem and discuss, very briefly, stochastic optimal control for linear systems.

10.1 ELEMENTS OF ESTIMATION THEORY

In systems analysis, a fundamental problem is to provide values for the unknown states or parameters of a system given noisy measurements which are some functions of these states and parameters.

If we consider a certain number of measurements (or sampled values) $z_1, z_2, \ldots z_n$ which depend on a parameter θ, we could define a function $T_N(z_1, z_2, \ldots z_N)$ which will be called the estimate of $\underline{\theta}$. Since the measurements z_i are in general random, the estimate $T_N(.)$ will also be a random variable.

Since all functions of z_i could be the estimates, the problem would be to find an estimate (which is a function of z_i) such that it is optimal with respect to some criterion. In addition, it is necessary to ensure that the estimates possess certain convergence properties with respect to the real value of the parameters.

We will call θ the real value of the parameter and $\hat{\theta}_N$ its estimate which is determined using the N measurements $z_1, z_2, \ldots z_N$. We will treat in the same way a vector of parameters $\underline{\theta}$ and its estimate $\hat{\underline{\theta}}_N$.

10.1.1 Principal properties of the estimate

We will begin by defining the notion of bias of an estimate. We will consider two cases : first where we know the real value of the parameter and second where we only know its mathematical expectation.

In the first case we will define an estimate which is conditionally unbiased by the relation

$$E\left[\, \hat{\underline{\theta}}_N \mid \underline{\theta}\, \right] = \underline{\theta} \qquad 10.1.1$$

which could also be written as

$$\int \hat{\underline{\theta}}_N \, p(\underline{Z}_N \mid \underline{\theta}) \, d\underline{Z}_N = \underline{\theta} \qquad 10.1.2$$

where

$$\underline{Z}_N = \left[z_1, z_2, \dots z_N \right]$$

In the second case we will define an unconditionally unbiased estimate as

$$E\left[\underline{\theta}_N \right] = E\left[\underline{\theta} \right]$$

or

$$E\left[\underline{\theta}_N \right] = \iint \underline{\hat{\theta}}_N \, \underline{p}(\underline{Z}_N, \underline{\theta}) \, d\underline{Z}_N \, d\underline{\theta} \qquad 10.1.3$$

Since $\underline{\hat{\theta}}_N$ is a function of $\underline{Z}_N$, we could also write :

$$E\left[\underline{\theta}_N \right] = \int \underline{\hat{\theta}}_N \, p(\underline{Z}_N) \, d\,\underline{Z}_N \qquad 10.1.4$$

The difference $E\left[\underline{\hat{\theta}}_N\right] - \underline{\theta}$ or $E\left[\underline{\hat{\theta}}_N\right] - E\left[\underline{\theta}\right]$

is the bias of the consistent estimate.

$\underline{\hat{\theta}}_N$ will be a consistent estimate of $\underline{\theta}$ if it converges to $\underline{\theta}$ in probability when N becomes large.

$$P\left[\| \underline{\hat{\theta}}_N - \underline{\theta} \|^2 < e \right] > 1 - \delta \quad \text{for } N > N_0 \qquad 10.1.5$$

Efficient Estimates

In $\underline{\hat{\theta}}_N$ is an unbiased estimate of $\underline{\theta}$ and if there is no other unbiased estimate which has a lower variance than that of $\underline{\theta}$, then $\underline{\hat{\theta}}_N$ is called an efficient estimate or an unbiased minimum variance estimate. The condition for an unbiased minimum variance estimate can be written as

$$E\left[(\underline{\hat{\theta}}_N - \underline{\theta})^T (\underline{\hat{\theta}}_N - \underline{\theta}) \right] \leqslant E\left[(\underline{\hat{\theta}}^* - \underline{\theta})^T (\underline{\hat{\theta}}^* - \underline{\theta}) \right] \qquad 10.1.6$$

where $\underline{\hat{\theta}}^*$ is any other unbiased estimate of $\underline{\theta}$.

10.1.2 Principal methods of obtaining estimates

a) Maximum likelihood estimate

We can define the conditional probability density function $p(\underline{Z}_N | \underline{\theta})$ as the likelihood function. This is a function of $\underline{\theta}$ whose maximum indicates the most likely value of the set $\underline{Z}_N = \left\{ z_1, z_2, \dots z_N \right\}$ that we obtain using the parameters $\underline{\theta}$.

In many cases we will maximize the Logrithm of this function. If the Logrithm has a continuous first derivative then a necessary condition for a maximum likelihood estimate can be obtained by differentiating

$$\text{Log } p(\underline{Z}_N | \underline{\theta}) \quad \text{w.r.t. } \underline{\theta}$$

so that we have

$$\left[\frac{\partial}{\partial \underline{\theta}} \text{ Log } p(\underline{Z}_N \mid \underline{\theta})\right]_{\underline{\theta} = \hat{\underline{\theta}}_{mL}} = \underline{0} \qquad 10.1.7$$

We note that this method can also be used in the case where $\underline{\theta}$ is not a random vector.

b) Bayesian Methods

Such methods consider the inverse problem and study the a posteriori probability density function $p(\underline{\theta} \mid \underline{Z}_N)$. We will study two general types of estimates.

- A posteriori maximum likelihood estimates

The a posteriori maximum likelihood estimate is the value of $\underline{\theta}$ which maximises the distribution $p(\underline{\theta} \mid \underline{Z}_N)$.

It is obtained, as in the previous method, and subject to the same conditions of validity by :

$$\left[\frac{\partial \text{Log } p(\underline{\theta} \mid \underline{Z}_N)}{\partial \underline{\theta}}\right]_{\underline{\theta} = \hat{\underline{\theta}}_{map}} = \underline{0} \qquad 10.1.8$$

We can develop the relationship which exists between these two estimates since

$$p(\underline{\theta} \mid \underline{Z}_N) = \frac{p(\underline{\theta}, \underline{Z}_N)}{p(\underline{Z}_N)} = \frac{p(\underline{Z}_N \mid \underline{\theta})\, p(\underline{\theta})}{p(\underline{Z}_N)} \qquad 10.1.9$$

We note that when the a priori distribution is uniform, both the estimates are identical.

- The conditional mean estimate

Here the criterion is different from that for the previous cases. We basically minimise the mean square error, i.e. for the scalar case we minimise

$$E\left[(\theta - \hat{\theta})^2\right] \qquad 10.1.10$$

In the case where $\underline{\theta}$ is a vector, we minimise

$$E\left[(\underline{\theta} - \hat{\underline{\theta}})^T Q (\underline{\theta} - \hat{\underline{\theta}})\right] \qquad 10.1.11$$

where Q is a non negative definite symmetrical weighting matrix.

We note that we can rewrite 10.1.11 as :

$$E\left[(\underline{\theta} - \hat{\underline{\theta}})^T Q(\underline{\theta} - \hat{\underline{\theta}})\right] = E\left[E\left[(\underline{\theta} - \hat{\underline{\theta}})^T Q(\underline{\theta} - \hat{\underline{\theta}}) \mid \underline{Z}_N\right]\right] \qquad 10.1.12$$

and we would like to satisfy the inequality

$$E\left[(\underline{\theta}-\hat{\underline{\theta}})^T Q(\underline{\theta}-\hat{\underline{\theta}})\,|\,\underline{Z}_N\right] \leqslant E\left[(\underline{\theta}-\underline{F})^T Q(\underline{\theta}-\underline{F})\,|\,\underline{Z}_N\right] \qquad 10.1.13$$

where F is any other function defined on the observations.

On writing $\underline{\theta}-\underline{F}=\underline{\theta}-\hat{\underline{\theta}}+\hat{\underline{\theta}}-\underline{F}$ and on expanding the right hand side of 10.1.13, we have

$$E\left[(\underline{\theta}-\hat{\underline{\theta}})^T Q(\underline{\theta}-\hat{\underline{\theta}})\,|\,\underline{Z}_N\right] \leqslant E\left[(\underline{\theta}-\hat{\underline{\theta}})^T Q(\underline{\theta}-\hat{\underline{\theta}})\,|\,\underline{Z}_N\right] \qquad 10.1.14$$
$$+(\hat{\underline{\theta}}-\underline{F})^T Q\, E\left[(\underline{\theta}-\hat{\underline{\theta}})\,|\,\underline{Z}_N\right] + E\left[(\underline{\theta}-\hat{\underline{\theta}})^T\,|\,Z_N\right] Q(\hat{\underline{\theta}}-\underline{F}) + (\hat{\underline{\theta}}-\underline{F})^T Q(\hat{\underline{\theta}}-\underline{F})$$

This inequality will be satisfied for any $Q \geqslant 0$ if

$$E\left[(\underline{\theta}-\hat{\underline{\theta}})\,|\,\underline{Z}_N\right]=\underline{0} \qquad 10.1.15$$

so that we find the conditional mean estimate as

$$\hat{\underline{\theta}}_{cm}=E\left[\underline{\theta}\;|\;\underline{Z}_N\right] \qquad 10.1.16$$

It should be noted that we developed the expression for the conditional mean estimates using a quadratic cost function of the form

$$C(\underline{\theta}-\hat{\underline{\theta}})=(\underline{\theta}-\hat{\underline{\theta}})^T Q(\underline{\theta}-\hat{\underline{\theta}})$$

In fact, it is possible to show that the conditional mean estimate is optimal for a much wider class of cost functions. Indeed, for all convex cost functions which are symmetrical with respect to $\underline{\theta}-\hat{\underline{\theta}}=\underline{0}$ and for all probability density function $p(\underline{\theta}\,|\,\underline{Z}_N)$, which are symmetrical with respect to the conditional mean and such that the product $C(\underline{\theta}-\hat{\underline{\theta}})\,p(\underline{\theta}\,|\,\underline{Z}_N)$ goes to zero as $N \longrightarrow \infty$, $\underline{\theta}_{cm}$ is an optimal estimate.

Next, we use these basic ideas to develop algorithms for parameter estimation.

10.2 PARAMETER ESTIMATION IN LINEAR STATIC SYSTEMS

Let us consider a static model described by the relationship

$$\underline{Z}=H\,\underline{\theta}+\underline{e} \qquad 10.2.1$$

where $\underline{Z}$ represents the vector of observations, $\underline{\theta}$ the vector of parameters and $\underline{e}$ a random vector which enables us to take into account the errors of observation. H is called the observation matrix. We will consider the case which is encountered rather frequently in practice where the dimension of $\underline{Z}$ is very high compared to that of $\underline{\theta}$. H is a transformation matrix with known elements.

We will successively examine the different hypotheses which lead to the principal methods of estimation.

Least squares estimation

We could develop this method either from purely deterministic considerations, i.e. minimisation of the sum of the squares of the errors : $\mathrm{Min}(\underline{e}^T\underline{e})$, or as we do here, i.e. by applying the maximum likelihood approach. For this purpose we make the following assumptions :

Assumptions

$\underline{e}$ is a zero mean Gaussian random vector whose elements have the same variance and which are uncorrelated with each other, i.e.

$$E\left[\underline{e}\right] = \underline{0} \qquad E\left[\underline{e}\,\underline{e}^T\right] = \sigma^2 I \qquad 10.2.2$$

$\underline{\theta}$ is either an unknown deterministic vector or a random vector and in the latter case it is assumed to be independent of $\underline{e}$.

The likelihood function will then be calculated on noting that

$$E\left[\underline{Z} \mid \underline{\theta}\right] = H\underline{\theta} \qquad 10.2.3$$

$$E\left[(\underline{Z} - H\underline{\theta})(\underline{Z} - H\underline{\theta})^T \mid \underline{\theta}\right] \qquad 10.2.4$$

$$= E\left[\underline{e}\,\underline{e}^T \mid \underline{\theta}\right] = E\left[\underline{e}\,\underline{e}^T\right] = \sigma^2 I$$

We obtain thus

$$p(\underline{Z} \mid \underline{\theta}) = K\, e^{-\frac{1}{2}\left[(\underline{Z}-H\underline{\theta})^T(\sigma^2 I)(\underline{Z}-H\underline{\theta})\right]} \qquad 10.2.5$$

and we need to find the value of $\underline{\theta}$ which maximizes this probability density function, i.e.

$$\frac{\partial\, \mathrm{Log}(p(\underline{Z} \mid \underline{\theta}))}{\partial \underline{\theta}} = 0 \Rightarrow \frac{\partial}{\partial \underline{\theta}}\left[(\underline{Z} - H\underline{\theta})^T \frac{1}{\sigma^2}(\underline{Z}-H\underline{\theta})\right] = \underline{0} \qquad 10.2.6$$

from which we obtain the least squares estimate as

$$\underline{\hat{\theta}} = (H^T H)^{-1} H^T \underline{Z} \qquad 10.2.7$$

Remark

From the above assumptions, the least squares estimate is the same as the maximum likelihood estimate. We can obtain the same expression for the least squares estimate as 10.2.7 if we consider the elements of $\underline{e}$ to be independent random variables having the same distribution, zero mean and with moments which exist up to 4th order and which are finite.

Bias

The least squares estimate is an unbiased estimate if $\underline{e}$ is a zero mean vector. Indeed, for the case where $\underline{\theta}$ is a deterministic vector, we have

$$E\left[\underline{\hat{\theta}}\right] = \underline{\theta} + (H^T H)^{-1} H^T E\left[\underline{e}\right] = \underline{\theta} \qquad 10.2.8$$

for $E\left[\underline{e}\right] = \underline{0}$

We also examine the case where the elements of the observation matrix are random variables. This case arises in the study of dynamical systems. In this case, from 10.2.8, $E[\underline{\hat{\theta}}] = \underline{\theta}$ if $E[\underline{e}] = \underline{0}$ and H is independent of $\underline{e}$.

Variance of the estimation error

The variance of the estimation error will be given by

$$E\left[(\underline{\theta} - \underline{\hat{\theta}})(\underline{\theta} - \underline{\hat{\theta}})^T\right] = E\left[(H^TH)^{-1} H^T \underline{e}\, \underline{e}^T H(H^TH)^{-1}\right] = \sigma^2 (H^T H)^{-1} \qquad 10.2.9$$

for the case where H is deterministic or where H and θ are independent.

Estimation of the variance of the noise

If σ^2 is not known, it would be desirable to estimate it. We will call this estimate $\hat{\sigma}^2$. To estimate $\hat{\sigma}^2$, we will study the statistics of the residual term $\underline{Z} - H\underline{\hat{\theta}}$.

We can write this as

$$\underline{Z} - H\underline{\hat{\theta}} = H\underline{\theta} + \underline{e} - H(H^T H)^{-1} H^T(H\underline{\theta} + \underline{e}) = (I - H(H^TH)^{-1} H^T)\, \underline{e} \qquad 10.2.10$$

where I is an identity matrix.

Since

$$E\left[\underline{e}^T \underline{e}\right] = n\sigma^2 \qquad 10.2.11$$

(where n is the dimension of $\underline{Z}$) we will determine the quantity

$$E\left[(\underline{Z} - H\underline{\hat{\theta}})^T(\underline{Z} - H\underline{\hat{\theta}})\right] = E\left[\underline{e}^T(I - H(H^T H)^{-1} H^T)^2\, \underline{e}\right] \qquad 10.2.12$$

where we have used the fact that

$$I - H(H^T H)^{-1} H^T$$

is a symmetrical matrix. We can also verify that this matrix is idempotent, i.e.

$$(I - H(H^T H)^{-1} H^T)\,(I - H(H^T H)^{-1} H^T) = I - H(H^T H)^{-1} H^T \qquad 10.2.13$$

from which

$$E\left[(\underline{Z} - H\underline{\hat{\theta}})^T (\underline{Z} - H\underline{\hat{\theta}})\right] = E\left[\underline{e}^T(I - H(H^T H)^{-1} H^T)\,\underline{e}\right] \qquad 10.2.14$$

Since the quantity inside the expectation operator $E\left[\,.\,\right]$ is a scalar, we can replace it by its trace, i.e. by

$$E\left[\text{trace}\,(\underline{e}^T(I - H(H^T H)^{-1} H^T)\,\underline{e})\right]$$

so that by using the properties of the trace of a product we have

$$E\left[\text{trace}\,(I - H(H^T H)^{-1} H^T)\,\underline{e}\,\underline{e}^T\right] = \text{trace}\left[(I - H(H^T+1)^{-1} H^T \sigma^2\right]$$

so that

$$E\left[(\underline{Z} - H\hat{\underline{\theta}})^T (\underline{Z} - H\hat{\underline{\theta}})\right] = \sigma^2 \left[n\text{-trace}\left[H(H^T H)^{-1} H^T\right]\right]$$

$$= \sigma^2 \left[n\text{-trace}\left[(H^T H)(H^T H)^{-1}\right]\right] = \sigma^2 (n - p) \qquad 10.2.15$$

where p is the number of parameters to be estimated. Thus the estimate $\hat{\sigma}^2$ will be given by

$$\hat{\sigma}^2 = \frac{1}{n-p} (\underline{Z} - H\hat{\underline{\theta}})^T (\underline{Z} - H\hat{\underline{\theta}}) \qquad 10.2.16$$

such that

$$E\left[\hat{\sigma}\right] = \sigma^2 \qquad 10.2.17$$

10.3 APPLICATION OF THE LEAST SQUARES METHOD TO PARAMETER ESTIMATION FOR A DYNAMICAL MODEL

We will consider a single input-single output model described by its transfer function in Z, i.e.

$$\frac{y(Z)}{U(Z)} = \frac{b_o + b_1 Z^{-1} + \ldots + b_n Z^{-n}}{1 + a_1 Z^{-1} + \ldots + a_n Z^{-n}} \qquad 10.3.1$$

This can be rewritten as

$$y(k) = -a_1 y(k-1) - \ldots - a_n y(k-n) + \ldots + b_n u(k-n) + e(k) \qquad 10.3.2$$

Here, e(k) is a random variable which takes into account the uncertainty or noise in the model. Subsequently we will study in some depth how we introduce noise into models.

We assume that $\{e(k)\}$ is a sequence of independent zero mean random variables having the same distribution. If we study the evolution of the system for N samples (where $N \gg n$), we can put the problem in a form which is analogous to that given by equation 10.2.1.

$$\underbrace{\begin{bmatrix} y(k+1) \\ \vdots \\ y(k+N) \end{bmatrix}}_{\underline{Y}_N} = \underbrace{\begin{bmatrix} -y(k) & \ldots & -y(k-n+1) & u(k+1) & \ldots & u(k-n+1) \\ \vdots & & \vdots & \vdots & & \vdots \\ -y(k+N-1) & & & u(k+N) & & \end{bmatrix}}_{H} \underbrace{\begin{bmatrix} a_1 \\ \vdots \\ a_n \\ b_o \\ \vdots \\ b_n \end{bmatrix}}_{\underline{\theta}} + \underbrace{\begin{bmatrix} e(k+1) \\ \vdots \\ e(k+N) \end{bmatrix}}_{\underline{e}} \qquad 10.3.3$$

It is thus quite straight forward to write an expression for the least squares estimate for $\underline{\theta}$ when its elements are the parameters of the transfer function of the system, i.e.

$$\underline{\hat{\theta}}_N = (H^T H)^{-1} H^T \underline{Y}_N \tag{10.3.4}$$

The $H^T H$ which is a (2n+1)(2n+1) symmetrical matrix can be written as

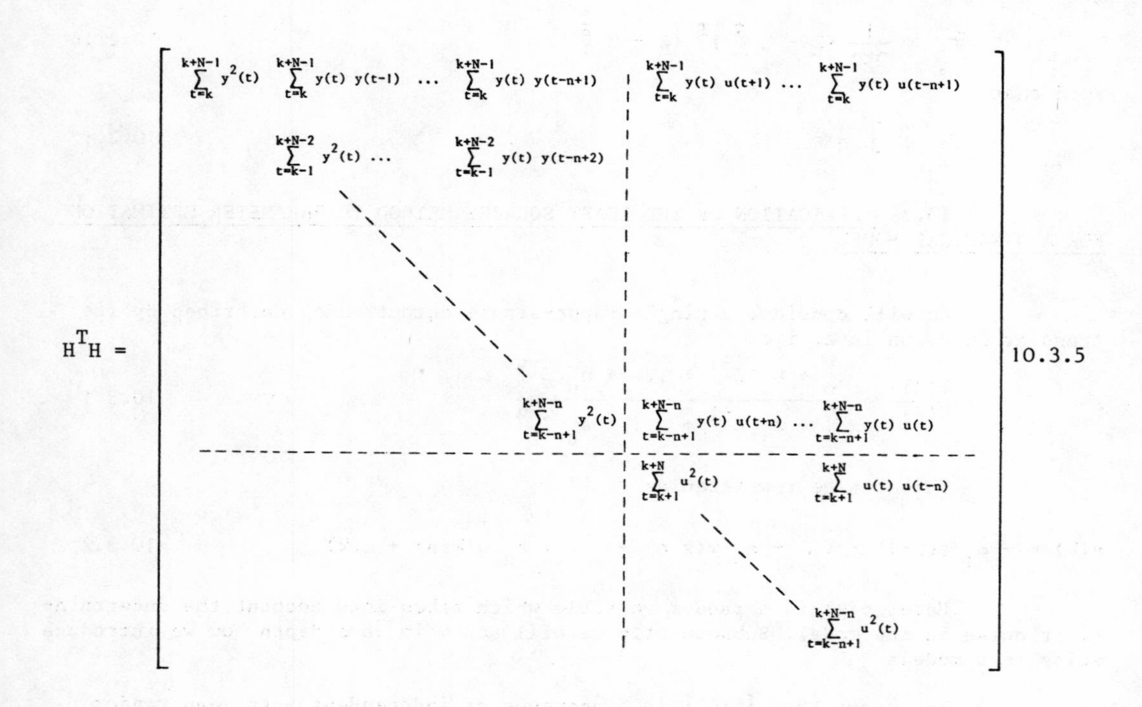

$$H^T H = \left[\begin{array}{cccc|ccc} \sum_{t=k}^{k+N-1} y^2(t) & \sum_{t=k}^{k+N-1} y(t)\,y(t-1) & \dots & \sum_{t=k}^{k+N-1} y(t)\,y(t-n+1) & \sum_{t=k}^{k+N-1} y(t)\,u(t+1) & \dots & \sum_{t=k}^{k+N-1} y(t)\,u(t-n+1) \\ & \sum_{t=k-1}^{k+N-2} y^2(t) & \dots & \sum_{t=k-1}^{k+N-2} y(t)\,y(t-n+2) & & & \\ & & \ddots & & & & \\ & & & \sum_{t=k-n+1}^{k+N-n} y^2(t) & \sum_{t=k-n+1}^{k+N-n} y(t)\,u(t+n) & \dots & \sum_{t=k-n+1}^{k+N-n} y(t)\,u(t) \\ \hline & & & & \sum_{t=k+1}^{k+N} u^2(t) & & \sum_{t=k+1}^{k+N} u(t)\,u(t-n) \\ & & & & & \ddots & \\ & & & & & & \sum_{t=k-n+1}^{k+N-n} u^2(t) \end{array}\right] \tag{10.3.5}$$

Similarly, the vector $H^T \underline{Y}_N$ could be written as

$$\begin{bmatrix} \sum_{t=k+1}^{k+N} y(t)\,y(t-1) \\ \sum_{t=k+1}^{k+N} y(t)\,y(t-2) \\ \vdots \\ \sum_{t=k+1}^{k+N} y(t)\,y(t-n) \\ \sum_{t=k+1}^{k+N} y(t)\,u(t) \\ \vdots \\ \sum_{t=k+1}^{k+N} y(t)\,u(t-n) \end{bmatrix} \tag{10.3.6}$$

From a practical point of view we see that an algorithm for computing the least squares estimates requires the calculation of sums of products of the type

$$y(t).y(t-\tau),\ y(t).u(t-\tau),\ u(t)\ u(t-\tau)$$

This is conceptually similar to correlation methods.

The properties of this estimate were studied by Astrom [1] who showed that under the above assumptions concerning the noise e(k) and hypotheses concerning u(k), the estimate $\underline{\hat{\theta}}_N$ is asymptotically unbiased.

Recursive least squares

From an algorithmic point of view, the above method is somewhat unsatisfactory since data often comes in a sequential form and the above method requires recalculation with all previous data every time an additional piece of data becomes available. It would be highly desirable to calculate the parameters recursively as new data becomes available.

Let us consider once again the vector equation

$$\underline{Y}_N = H\,\underline{\theta} + \underline{e} \qquad 10.3.7$$

and let us determine the estimate $\underline{\hat{\theta}}_{N+1}$ after N+1 observations. From equation 10.3.3, we have

$$\begin{bmatrix} \underline{Y}_N \\ --- \\ y(k+n+1) \end{bmatrix} = \begin{bmatrix} H \\ --- \\ \underline{h}^T_{N+1} \end{bmatrix} \underline{\theta} + \begin{bmatrix} \underline{e} \\ ---- \\ e(k+N+1) \end{bmatrix} \qquad 10.3.8$$

In equation 10.3.8, the new observation is of the form

$$y(k+N+1) = (-y(k+N) - y(k+N-1) \ldots - u(k+N+1) \ldots) \begin{bmatrix} a_1 \\ \vdots \\ a_n \\ b_o \\ \vdots \\ b_n \end{bmatrix} + e(k+N+1) \qquad 10.3.9$$

Then the estimate $\underline{\hat{\theta}}_{N+1}$ can be written as

$$\underline{\hat{\theta}}_{N+1} = (H^T H + \underline{h}_{N+1} h^T_{N+1})^{-1} (H^T \underline{Y}_N + h_{N+1}\, y(k+N+1)) \qquad 10.3.10$$

In order to expand the right hand side of this equation, we will use the following matrix inversion lemma : consider the non-singular square matrices

A and C. If the square matrix $A + BCB^T$ and $C^{-1} + B^T A^{-1} B$ are also non-singular we have the identity

$$(A+BCB^T)^{-1} = A^{-1} - A^{-1}B(C^{-1} + B^T A^{-1} B)^{-1} B^T A^{-1} \qquad 10.3.11$$

In the present case we obtain

$$(H^TH + \underline{h}_{N+1} \underline{h}_{N+1}^T)^{-1} = (H^T H)^{-1} - (H^T H)^{-1} \underline{h}_{N+1}(1 + \underline{h}_{N+1}^T (H^T H)^{-1} \underline{h}_{N+1})^{-1} \underline{h}_{N+1}^T (H^T H)^{-1} \qquad 10.3.12$$

We note that $1 + \underline{h}_{N+1}^T (H^T H)^{-1} \underline{h}_{N+1}$ is a scalar.

Then

$$\hat{\underline{\theta}}_{N+1} = \hat{\underline{\theta}} + (H^T H)^{-1} \underline{h}_{N+1} y(k+N+1) - (H^T H)^{-1} \underline{h}_{N+1}(1 + \underline{h}_{N+1}^T (H^T H)^{-1} \underline{h}_{N+1})^{-1} \underline{h}_{N+1}^T (H^T H)^{-1} (H^T \underline{y}_N + \underline{h}_{n+1} y(k+N-1) \qquad 10.3.13$$

On regrouping the second and last term on the R.H.S., we have

$$(H^T H)^{-1} \underline{h}_{N+1} (H^T H)^{-1} \underline{h}_{N+1})^{-1} (H \underline{h}_{N+1}^T (H^T H)^{-1} \underline{h}_{N+1} - \underline{h}_{N+1}^T (H^T H)^{-1} \underline{h}_{N+1}) y(k+N+1) \qquad 10.3.14$$

$$= (H^T H)^{-1} \underline{h}_{N+1} (1+\underline{h}_{N+1} (H^T H)^{-1} \underline{h}_{N+1})^{-1} y(k+N+1)$$

from which we obtain the expression for the estimate after N+1 measurements, i.e.

$$\hat{\underline{\theta}}_{N+1} = \hat{\underline{\theta}}_N + (H^TH)^{-1} \underline{h}_{N+1}(1+\underline{h}_{N+1}^T (H^TH)^{-1} \underline{h}_{N+1})^{-1} (y(k+N+1)-\underline{h}_{N+1}^T \hat{\underline{\theta}}_N) \qquad 10.3.15$$

Let us put

$$K_N = (H^T H)^{-1} \underline{h}_{N+1} (1 + \underline{h}_{N+1}^T (H^T H)^{-1} \underline{h}_{N+i})^{-1}$$

which leads to the final expression :

$$\hat{\underline{\theta}}_{N+1} = \hat{\underline{\theta}}_N + K_N(y(K+N+1) - \underline{h}_{N+1}^T \hat{\underline{\theta}}_N) \qquad 10.3.16$$

This expression shows that we calculate the value of the estimate after (N+1) measurements using the previous value of the estimate and a corrective term which is proportional to the difference between the <u>predicted</u> value, i.e. $\underline{h}_{N+1}^T \hat{\underline{\theta}}_N$ and the value of <u>measured output</u> y(k+N+1). The factor K_N could be considered as a gain and we will also determine a relation which will enable us to calculate it for every new observation.

By analogy with the expression for the error covariance matrix for the

Gaussian case (10.2.9), let us write :

$$P_N = \alpha\,(H^T H)^{-1} \qquad 10.3.17$$

so that

$$K_N = P_N\, \underline{h}_{N+1} (\alpha + \underline{h}_{N+1}^T P_N\, \underline{h}_{N+1})^{-1}$$

At the next sampling point we need to calculate

$$P_{N+1} = \alpha\,(H^T H + \underline{h}_{N+1}\, \underline{h}_{N+1}^T)^{-1} \qquad 10.3.18$$

on applying once again the matrix inversion lemma we obtain :

$$P_{N+1} = P_N - P_N\, \underline{h}_{N+1} (\alpha + \underline{h}_{N+1}^T P_N\, \underline{h}_{N+1})^{-1}\, \underline{h}_{N+1}^T P_N \qquad 10.3.19$$

we note that the term $\alpha + \underline{h}_{N+1}^T P_N\, \underline{h}_{N+1}$ is a scalar so that the calculation of 10.3.16 and 10.3.19 does not involve any matrix inversions.

In addition we need to store only the elements of the previous estimate vector $\hat{\underline{\theta}}_N$ and the elements of the P_N matrix (which is a symmetrical matrix).

It should be noted that we assumed in our development of the least squares method that the random variables $\underline{e}(k)$ are independent. We will next study different models for linear dynamical systems and we will examine what consequence this has on the nature of the noise $\underline{e}(k)$.

10.4 INPUT-OUTPUT RELATIONSHIP FOR A NOISY DYNAMICAL SYSTEM

We will consider the state space model

$$\begin{aligned} \underline{x}_{k+1} &= \Phi\, \underline{x}_k + \Gamma\, \underline{v}_k \\ \underline{y}_k &= H\, \underline{x}_k \end{aligned} \qquad 10.4.1$$

where we assume, in the interest of simplicity, that the deterministic input $\underline{u}_k$ is zero. $\underline{v}_k$ is a white noise vector. We can obtain a relation comprising output values at different instants on writing

$$\begin{aligned} \underline{y}_{k+1} &= H\, \Phi\, \underline{x}_k + H\, \Gamma\, \underline{v}_k \\ \underline{y}_{k+2} &= H\, \Phi^2\, \underline{x}_k + H\, \Gamma\, \underline{v}_k + H\, \Gamma\, \underline{v}_{k+1} \\ \underline{y}_{k+3} &= H\, \Phi^3\, \underline{x}_k + H \Phi \Gamma^2\, \underline{v}_k + H \Phi \Gamma\, \underline{v}_{k+1} + H\, \Gamma\, \underline{v}_{k+1} \end{aligned} \qquad 10.4.2$$

In order to clarify our ideas, let us assume that our system is of third order. Then, by the Cayley-Hamilton theorem we have

$$\Phi^3 = - \lambda_3 \Phi^2 - \lambda_2 \Phi - \lambda_1 I \qquad 10.4.3$$

We obtain thus

$$y_{k+3} = -\lambda_3 y_{k+2} - \lambda_2 y_{k+1} - \lambda_1 y_k + (\lambda_3 H \Phi \Gamma + \lambda_2 H$$
$$+ H \Phi^2 \Gamma)\underline{v}_k + (\lambda_3 H\Gamma + H\Phi\Gamma)\ \underline{v}_{k+1} + H\Gamma\ \underline{v}_{k+2} \qquad 10.4.4$$

which could be rewritten as

$$y_{k+3} = -\lambda_3 y_{k+2} - \lambda_2 y_{k+1} - \lambda_1 y_k + \epsilon_{k+3} \qquad 10.4.5$$

In 10.4.5, ϵ_{k+3} is a <u>correlated</u> noise term. We see thus that the input-output relationship of a system whose dynamics are subjected to white noise, contains a correlated noise term so that if we were to apply the least squares method here, we would obtain a biased estimate.

<u>General input-output model</u>

From the above results, we see that we can put the input-output relationship for a single input-single output system into the form

$$A(Z^{-1})y(k) = B(Z^{-1})u(k) + \lambda C(Z^{-1})\ e(k) \qquad 10.4.6$$

where e(k) is a sequence of independent zero mean Gaussian random variables having a variance of unity. (The coefficient λ enables us to represent the case where e(k) has a variance different from one). $A(Z^{-1})$, $B(Z^{-1})$, $C(Z^{-1})$ are polynomials given by

$$A(Z^{-1}) = 1 + a_1 Z^{-1} + \ \dots + a_n Z^{-n} \qquad 10.4.7$$

$$B(Z^{-1}) = b_o + b_1 Z^{-1} + \dots + b_n Z^{-n} \qquad 10.4.8$$

$$C(Z^{-1}) = 1 + C_1 Z^{-1} + \ \dots + C_n Z^{-n} \qquad 10.4.9$$

We will assume that

a) the system described by equation 10.4.6 is stable
b) there are no common factors amongst the three polynomials $A(Z^{-1})$,$B(Z^{-1})$,$C(Z^{-1})$, i.e.

we assume that the system is completely controllable and completely observable.

This system can be represented by Fig. 10.1.

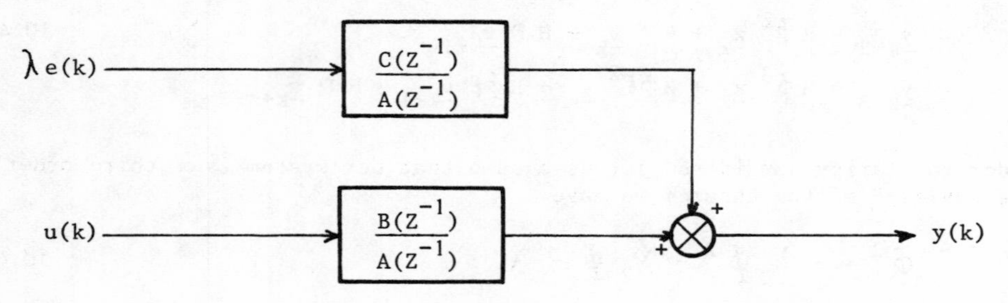

Fig. 10.1

10.5 THE GENERALISED LEAST SQUARES METHOD

This method will be applied to the general model of Fig. 10.1. The above model is unsuitable for the application of the least squares method since here we have a correlated noise term. Nevertheless, it is interesting to retain the "structure" of the least squares method since the method is easy to implement numerically.

In the development of the generalised least squares method we begin by writing the expression

$$A(Z^{-1})\ y(k) = B(Z^{-1})\ u(k) + C(Z^{-1})\ e(k) \qquad 10.5.1$$

in the form

$$A(Z^{-1})\ \left(\frac{1}{C(Z^{-1})}\ y(k)\right) = B(Z^{-1})\ \left(\frac{1}{C(Z^{-1})}\ u(k)\right) + e(k) \qquad 10.5.2$$

which can be represented by the block diagram shown in Fig. 10.2.

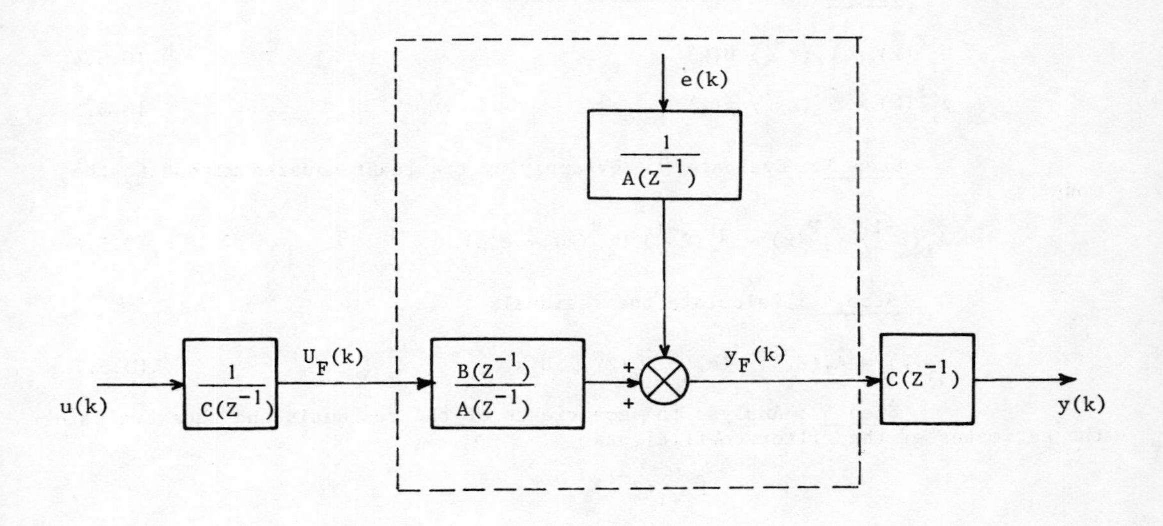

Fig. 10.2

We see from Fig. 10.2 that we can apply the standard least squares method to the block diagram inside the dotted lines since e(k) is a white noise sequence. Thus if we know $C(Z^{-1})$, we could determine a filter $\frac{1}{C(Z^{-1})}$ and we could operate on the quantities $U_F(k)$ and $y_F(k)$.

We note that knowledge of the polynomial $C(Z^{-1})$ involves knowing the covariance R of the correlated noise which acts on the system.

In practice we won't know $C(Z^{-1})$ and the generalised least squares method will enable us to determine it by an iterative technique. The different steps in this method are summed up below :

(a) Take an a priori value R_i for the covariance matrix R (if we take $R_o = \sigma^2 I$, we have from the very first iteration a least squares estimate).

(b) Calculate the estimate $\underline{\hat{\theta}}_i$ of the a_j and b_j parameters using the Markov estimate for the value R_i.

(c) Calculate the residuals $\underline{\epsilon}_i = \underline{Y} - H\,\underline{\hat{\theta}}_i$ and deduce the new value R_{i+1} for the covariance of the noise.

(d) Go back to step (b) with R_{i+1}.

The practical implementation requires :

Step 1 : Take an a priori "filter"

$$\hat{F}_i(Z^{-1}) = 1 + \hat{f}_{i1}\, Z^{-1} + \;\dots + \hat{f}_{in}\, Z^{-n} \tag{10.5.3}$$

Step 2 : Calculate the filtered values

$$U_i^F(k) = \hat{F}_i(Z^{-1})\, U(k) \tag{10.5.4}$$

$$y_i^F(k) = \hat{F}_i(Z^{-1})\, y(k) \tag{10.5.5}$$

Step 3 : Evaluate $\underline{\hat{\theta}}_i$ by applying the least squares method to the model :

$$\hat{A}_i(Z^{-1})\, y_i^F(k) = \hat{B}_i(Z^{-1})\, U_i^F(k) + e(k) \tag{10.5.6}$$

Step 4 : Calculate the residuals

$$e_i(k) = \hat{A}_i(Z^{-1})\, y(k) - \hat{B}_i(Z^{-1})\, U(k) \tag{10.5.7}$$

Step 5 : Analyse the covariance of the residuals and thus evaluate the estimates of the filter coefficients

$$\hat{F}_{i+1}(Z^{-1})$$

In fact, we see from the previous relations that

$$\epsilon_i\,(P) = \frac{1}{\hat{F}_i(Z^{-1})}\; e(k) \tag{10.5.8}$$

from which we obtain the relationships

$$\begin{aligned} \epsilon_i(k) &= -\hat{f}_{i1}\, \epsilon_i(k-1) \;\;\dots\;\; - \hat{f}_{in}\,\epsilon_i(k-n) + e(k) \\ \epsilon_i(k+1) &= -\hat{f}_{i1}\, \epsilon_i(k) \;\;\dots\;\; - \hat{f}_{in}\,\epsilon_i(k-n+1) + e(k+1) \\ &\;\;\vdots \end{aligned} \tag{10.5.9}$$

We can apply the least squares method here to determine the coefficients of the filter since e(k), e(k+1) ... are independent random variables. These estimated values give us the filter $\hat{F}_{i+1}(Z^{-1})$.

Step 6 : Re-start the procedure with $\hat{F}_{i+1}(Z^{-1})$, $\hat{F}_{i+2}(Z^{-1})$, ... until the estimates converge.

The convergence of this algorithm has not been proved so far although on the applications that have been tested, convergence did take place.

Note that we can also determine, as in the least squares method, the recursive equations for the estimation of $\underline{\theta}$ and of the coefficients of the filter.

10.6 THE INSTRUMENTAL VARIABLE METHOD

Let us reconsider the model

$$A(Z^{-1})\, y(k) = B(Z^{-1})\, U(k) + C(Z^{-1})\, e(k) \qquad 10.6.1$$

which can be put into the form

$$y(k) = - a_1 y(k-1) + \ldots - a_n y(k-n) + b_o u(k) + \ldots + b_n u(k-n) + \epsilon_k \qquad 10.6.2$$

where ϵ_k is a correlated noise.

If we have N equations of this type, we can summarise them as before in the form

$$\underline{Y}_N = H\, \underline{\theta} + \underline{\epsilon} \qquad 10.6.3$$

On premultiplying this equation by a matrix Φ which is of the same dimension as H, we will have

$$\Phi^T\, \underline{Y}_N = (\Phi^T H)\, \underline{\theta} + \Phi^T \underline{\epsilon} \qquad 10.6.4$$

If $\Phi^T H$ is non-singular, we could write :

$$\underline{\theta} = (\Phi^T H)^{-1}\, \Phi^T\, \underline{Y}_N - (\Phi^T H)^{-1}\, \Phi^T \underline{\epsilon} \qquad 10.6.5$$

We seek thus, a matrix Φ such that the estimate of $\underline{\theta}$ which is defined by

$$\hat{\underline{\theta}} = (\Phi^T H)^{-1}\, \Phi^T\, \underline{Y}_N \qquad 10.6.6$$

converges in some way towards the parameter vector $\underline{\theta}$.

We note that for the case where all the components of $\underline{\epsilon}$ are independent, we could apply the least squares method and in that case the instrumental matrix will be H itself.

We can show that if Φ satisfies the following conditions :

(1) $\frac{1}{N} \Phi^T \underline{\epsilon}$ tends (in probability) towards a vector with zero components.

(2) $\frac{1}{N} \Phi^T H$ tends in probability towards a non-singular matrix M.

Then the estimate $\hat{\underline{\theta}} = (\Phi^T H)^{-1} \Phi^T \underline{Y}_N$ converges in probability to $\underline{\theta}$.

Indeed,

$$\hat{\underline{\theta}} - \underline{\theta} = (\Phi^T H)^{-1} \Phi^T \underline{\epsilon} \qquad 10.6.7$$

will approach, in probability, $M^{-1} . 0$ i.e. towards a vector with zero components so that

$$\hat{\underline{\theta}} \xrightarrow{P} \underline{\theta} \qquad 10.6.8$$

Example

As an example let us consider how we would determine the instrumental matrice for a first order system. The system is described by

$$\begin{aligned} x_{k+1} &= a\, x_k + b\, u_k \\ y_k &= x_k + e_k \end{aligned} \qquad 10.6.9$$

which leads to the input-output relation

$$y_k = a\, y_{k-1} + b\, u_{k-1} + (e_k - a\, e_{k-1}) \qquad 10.6.10$$

If we choose H itself as the instrumental matrix, we will have

$$H = \begin{pmatrix} y_{k-1} & u_{k-1} \\ y_k & u_k \\ \vdots & \vdots \end{pmatrix} \qquad H^T \underline{\epsilon} = \begin{pmatrix} y_{k-1} & y_k & \cdots \\ u_{k-1} & u_k & \cdots \end{pmatrix} \begin{pmatrix} e_k - a\, e_{k-1} \\ e_{k+1} - a\, e_k \\ \vdots \end{pmatrix}$$

Let us consider the first row of the $H^T \underline{\epsilon}$ vector :

$$y_{k-1}(e_k - a\, e_{k-1}) + y_k(e_{k+1} - a\, e_k) + \ldots$$

If we expand the first term in this expression, we obtain :

$$y_{k-1}\,(e_k - a\, e_{k-1}) = (a\, y_{k-2} + b\, u_{k-2} + e_{k-1} - a\, e_{k-2})\,(e_k - a\, e_{k-1}) \qquad 10.6.11$$

Equation 10.6.11 shows that there are terms here which are correlated with each other. To eliminate this correlation, it is necessary to displace the observation y_{k-1} and we will thus choose as our instrumental matrix

$$\Phi = \begin{pmatrix} y_{k-2} & u_{k-1} \\ y_{k-1} & u_k \\ \vdots & \vdots \end{pmatrix} \qquad 10.6.12$$

In that case the first term of the vector $\Phi^T \underline{\epsilon}$ will be

$$y_{k-2} (e_k - a e_{k-1}) + \dots$$

and we have

$$y_{k-2}(e_k - a e_{k-1}) = (a y_{k-3} + b y_{k-3} + e_{k-2} - a e_{k-3}) (e_k - a e_{k-1})$$

We see thus that we have as our instrumental matrice the matrix of delayed observations. In general we will delay the observations in the Φ matrix by at least an order equal to the order of the system. In the case of an nth order system, the H and Φ matrices will thus be :

$$H = \left(- y_{k-1} \ - y_{k-2} \ \dots \ - y_{k-n} \quad u_k \dots u_{k-n} \right)$$

$$\Phi = \left(- y_{k-1-n} \ \dots \ - y_{k-2-n} \quad u_k \dots u_{k-n} \right)$$

It can be shown that Φ^T H converges in probability towards a non-singular matrix.

Next we consider state estimation and develop the celebrated Kalman filter.

10.7 THE KALMAN FILTER

In the development of the state estimation algorithm we will use some of the properties of Gaussian random variables that we studied in the previous chapter. We begin by summarising the properties that we will be using.

10.7.1 Some useful properties of Gaussian random vectors

Consider two random Gaussian vectors $\underline{x}$ and $\underline{y}$ such that $\begin{bmatrix} \underline{x} \\ \hline \underline{y} \end{bmatrix}$ is Gaussian. These vectors have the following properties :

Property 1

The conditional mathematical expectation of $\underline{x}$ given $\underline{y}$ is

$$E\left[\underline{x} \mid \underline{y}\right] = E\left[\underline{x}\right] + P_{xy} P_{yy}^{-1} \left[\underline{y} - E(\underline{y})\right] \qquad 10.7.1$$

In addition, $\underline{x} - E\left[\underline{x} \mid \underline{y}\right]$ is independent of $\underline{y}$ and has zero mean.

Property 2

Consider the Gaussian vector $\begin{bmatrix} \underline{x} \\ \underline{y}_1 \\ \underline{y}_2 \end{bmatrix}$ where $\underline{y}_1$ and $\underline{y}_2$ are independent vectors. The conditional mathematical expectation of $\underline{x}$ given $\underline{y}_1$ and $\underline{y}_2$ is :

$$E\left[\underline{x} \mid \underline{y}_1, \underline{y}_2\right] = E\left[\underline{x} \mid \underline{y}_1\right] + E\left[\underline{x} \mid \underline{y}_2\right] - E\left[\underline{x}\right] \qquad 10.7.2$$

Proof

To prove the validity of equation 10.7.2, let us form the vector $\underline{y} = \begin{bmatrix} \underline{y}_1 \\ \underline{y}_2 \end{bmatrix}$. From the hypothesis on the independence of $\underline{y}_1$ and $\underline{y}_2$ we have

$$P_{yy} = \begin{bmatrix} P_{y_1y_1} & 0 \\ 0 & P_{y_2y_2} \end{bmatrix} \qquad 10.7.3$$

In addition, the P_{xy} matrix could be partitioned as :

$$P_{xy} = \begin{bmatrix} P_{xy_1} & P_{xy_2} \end{bmatrix} \qquad 10.7.4$$

From property 1, we have :

$$E\left[\underline{x} \mid \underline{y}_1, \underline{y}_2\right] = E\left[\underline{x}\right] + \begin{bmatrix} P_{xy_1} & P_{xy_2} \end{bmatrix} \begin{bmatrix} P^{-1}_{y_1y_1} & 0 \\ 0 & P^{-1}_{y_2y_2} \end{bmatrix} \begin{bmatrix} \underline{y}_1 - E[\underline{y}_1] \\ \underline{y}_2 - E[\underline{y}_2] \end{bmatrix} \qquad 10.7.5.$$

Then on expanding the R.H.S. of 10.7.5, we have :

$$E\left[\underline{x} \mid \underline{y}_1, \underline{y}_2\right] = E\left[\underline{x}\right] + \underbrace{P_{xy_1} P^{-1}_{y_1y_1}\left[\underline{y}_1 - E\left[\underline{y}_1\right]\right]}_{E\left[\underline{x} \mid \underline{y}_1\right] - E\left[\underline{x}\right]} + \underbrace{P_{xy_2} P_{y_2y_2}\left[\underline{y}_2 - E[\underline{y}_2]\right]}_{E\left[\underline{x} \mid \underline{y}_2\right] - E\left[\underline{x}\right]} \qquad 10.7.6$$

Q.E.D.

Property 3

The conditional probability density of the vector $\underline{x}$, knowing the vector $\underline{y}$, is Gaussian with mean $E\left[\underline{x} \mid \underline{y}\right]$ and covariance $P_{xx} - P_{xy} P^{-1}_{yy} \overline{P}_{yx}$

These various properties which were essentially studied in chapter 9 will enable us to develop the principal equations of the Kalman Filter. We begin the development by considering a discrete model.

10.7.2 Discrete model

The problem is to estimate, at each instant, in an optimal way, the state of a dynamical system using noisy measurements of the output of the system.

The general model of the system is of the form :

$$\begin{aligned} \underline{x}_{k+1} &= \Phi_k \underline{x}_k + \Gamma_k \underline{v}_k \\ \underline{y}_k &= H_k \underline{x}_k + \underline{\omega}_k \end{aligned} \qquad 10.7.7$$

The assumptions concerning the model and the disturbances are summed up below.

Assumptions

(1) $\underline{v}_k$ and $\underline{\omega}_k$ are Gaussian white noise sequences such that

$$E[\underline{v}_k] = \underline{0}, \quad E[\underline{\omega}_k] = \underline{0}$$

$$E[\underline{v}_k \underline{v}_j^T] = Q\,\delta_{kj}, \quad E[\underline{\omega}_k \underline{\omega}_j^T] = R\,\delta_{kj}$$

(2) The disturbances $\underline{v}_k$ and $\underline{\omega}_k$ are uncorrelated with each other :

$$E[\underline{v}_k \underline{\omega}_j^T] = 0 \qquad \forall \quad k \text{ and } j$$

Note that this assumption is not strictly necessary. It is convenient though since the final expression for the filter is much simpler in this case.

(3) The initial state $\underline{x}_o$ is a Gaussian random vector with mean $E[\underline{x}_o] = \underline{\mu}_o$ and with covariance

$$E[(\underline{x}_o - \underline{\mu}_o)(\underline{x}_o - \underline{\mu}_o)^T] = P_o$$

(4) The initial state $\underline{x}_o$ and the noises $\underline{v}_k$ and $\underline{\omega}_k$ are uncorrelated

$$E[\underline{x}_o \underline{v}_k^T] = 0, \quad E[\underline{x}_o \underline{\omega}_k^T] = 0 \qquad \forall \quad k$$

(5) The elements of Φ_k, Γ_k, H_k are known.

With these hypotheses we can formulate the filter problem.

10.7.3 The optimal filter problem

The optimal filter problem is to determine the "best" estimate of $\underline{x}_k$ given the sequence of measurements

$$\{\underline{y}_o, \underline{y}_1, \cdots \underline{y}_k\} = Y_k \qquad 10.7.8$$

The "best" estimate is in the sense of minimisation of the following quadratic error criterion :

$$E[(\underline{x}_k - \hat{\underline{x}}_k)^T S(\underline{x}_k - \hat{\underline{x}}_k)] \qquad 10.7.9$$

where S is a symmetrical non-negative definite matrix and $\hat{\underline{x}}_k$ is a function of the observations $\{\underline{y}_o, \underline{y}_1, \cdots, \underline{y}_k\}$

We saw in the beginning of this chapter that the optimal estimate with respect to this criterion is the conditional mean estimate, i.e.

$$\hat{\underline{x}}_{k|k} = E[\underline{x}_k | \underline{y}_o, \underline{y}_1, \cdots \underline{y}_k] = E[\underline{x}_k | Y_k] \qquad 10.7.10$$

We note that given our assumptions, the conditional probability density $P(\underline{x}_k | Y_k)$ is Gaussian and the maximum a posteriori estimate is identical to the conditional mean estimate for this problem.

We also define the 1 step prediction which is the estimation of the state $\underline{x}_k$ at the instant t_k given the sequence of measurements up to t_{k-1}, i.e.

$$\hat{\underline{x}}_{k|k-1} = E[\underline{x}_k | \underline{y}_o, \underline{y}_1, \dots \underline{y}_{k-1}] = E[\underline{x}_k | y_{k-1}] \qquad 10.7.11$$

From this definition we see that we could also solve two other problems which are close to the filter problem. For example, the determination of $\hat{\underline{x}}_{k|i}$ with $i < k$ is the prediction or extrapolation problem. On the other hand, the determination of $\hat{\underline{x}}_{k|i}$ with $i > k$ is the smoothing problem.

Next we develop the filter equations.

10.8 DEVELOPMENT OF THE FILTER EQUATIONS

The approach to the development of the filter equations can be divided up into a number of distinct steps.

Step 1 : Transition of the state $\underline{x}_{k-1}$ to $\underline{x}_k$

We assume that we know the estimate $\hat{\underline{x}}_{k-1|k-1}$ and we want to determine the one step predictor $\hat{\underline{x}}_{k|k-1}$.

Let us consider the dynamical equation

$$\underline{x}_k = \Phi_{k-1} \underline{x}_{k-1} + \Gamma_{k-1} \underline{v}_{k-1} \qquad 10.8.1$$

On taking the conditional mean of the two sides of this equation, we obtain :

$$E[\underline{x}_k | \underline{y}_o, \underline{y}_1, \dots \underline{y}_{k-1}] = \Phi_{k-1} E[\underline{x}_{k-1} | \underline{y}_o, \dots \underline{y}_{k-1}] + \Gamma_{k-1}[\underline{v}_{k-1} | \underline{y}_o, \dots \underline{y}_{k-1}] \qquad 10.8.2$$

But since $\underline{v}_{k-1}$ is independent of $\underline{\omega}_k$, $\underline{v}_{k-2}$, ... $\underline{v}_o$ and $\underline{x}_o$, it is independent of $\underline{y}_o$, $\underline{y}_1$, ... $\underline{y}_{k-1}$ so that

$$E[\underline{v}_{k-1} | \underline{y}_o, \underline{y}_1, \dots \underline{y}_{k-1}] = E[\underline{v}_{k-1}] = 0$$

so that from 10.8.2.

$$\boxed{\hat{\underline{x}}_{k|k-1} = \Phi_{k-1} \hat{x}_{k-1|k-1}} \qquad 10.8.3$$

Next we determine the covariance matrix for the one step prediction error, i.e.

$$P_{k|k-1} = E[(\underline{x}_k - \hat{\underline{x}}_{k|k-1})(\underline{x}_k - \hat{\underline{x}}_{k|k-1})^T \mid \underline{y}_o, \underline{y}_1, \dots y_{k-1}] \qquad 10.8.4$$

From property 1, $(\underline{x}_k - \hat{\underline{x}}_{k|k-1})$ is independent of the sequence $\underline{y}_o$, $\underline{y}_1$, ... $\underline{y}_{k-1}$ so that covariance matrix in 10.8.4 can be rewritten as

$$P_{k|k-1} = E[(\underline{x}_k - \hat{\underline{x}}_{k|k-1})(\underline{x}_k - \hat{\underline{x}}_{k|k-1})^T] \qquad 10.8.5$$

The one step prediction error can be written as

$$\underline{x}_k - \hat{\underline{x}}_{k|k-1} = \Phi_{k-1}(\underline{x}_{k-1} - \hat{\underline{x}}_{k-1|k-1}) + \Gamma_{k-1} \underline{v}_{k-1} \qquad 10.8.6$$

so that we obtain

$$P_{k|k-1} = \Phi_{k-1} P_{k-1|k-1} \Phi^T_{k-1} + \Phi_{k-1} E[(\underline{x}_{k-1} - \hat{\underline{x}}_{k-1|k-1}) v^T_{k-1}|$$

$$\underline{y}_{k-1}]\Gamma^T_{k-1} + \Gamma_{k-1} E[\underline{v}_{k-1}(\underline{x}_{k-1} - \hat{\underline{x}}_{k-1|k-1})^T | Y_{k-1}]\Phi^T_{k-1} + \Gamma_{k-1} Q \Gamma^T_{k-1}$$

We note that $E[\underline{x}_{k-1}|\underline{v}^T_{k-1}] = 0$ since $\underline{x}_{k-1}$ is a function of $\underline{x}_o$ and of $\underline{v}_o, \underline{v}_1, \ldots \underline{v}_{k-2}$ but not of $\underline{v}_{k-1}$. In addition, $\underline{v}_{k-1}$ has zero mean. Also :

$$E[\hat{\underline{x}}_{k-1|k-1} \underline{v}^T_{k-1} | Y_{k-1}] = \hat{\underline{x}}_{k-1|k-1} E[\underline{v}^T_{k-1|k-1}] = \hat{x}_{k-1|k-1} E[\underline{v}^T_{k-1}] = 0$$

so that the covariance matrix could be determined from the expression

$$\boxed{P_{k|k-1} = \Phi_{k-1} P_{k-1|k-1} \Phi^T_{k-1} + \Gamma_{k-1} Q \Gamma^T_{k-1}} \qquad 10.8.7$$

Step 2 : One step prediction of the filtered estimate

Next we want to express the estimate of $\underline{x}_k$ given the measurements up to t_k, i.e.

$$Y_k = \{ \underline{y}_o, \underline{y}_1, \ldots \underline{y}_k \}$$

To do this, we will determine the conditional probability density

$$p(\underline{x}_k|\underline{y}_o, \ldots \underline{y}_k) = p(\underline{x}_k | Y_k)$$

Note that we can rewrite this density as

$$p(\underline{x}_k|Y_k) = p(\underline{x}_k|Y_{k-1}, \underline{y}_k)$$

where we separate out the last measurement from all the previous measurements Y_{k-1}. If we apply BAYES theorem to this expression, this leads to the relationship

$$p(\underline{x}_k|Y_k) = p(\underline{x}_k|Y_{k-1}, \underline{y}_k) = \frac{p(\underline{x}_k, Y_{k-1}, \underline{y}_k)}{p(Y_{k-1}, \underline{y}_k)} = p(\underline{y}_k|\underline{x}_{k-1}) \cdot \frac{p(\underline{x}_k, Y_{k-1})}{p(Y_{k-1}, \underline{y}_k)}$$

$$= p(\underline{y}_k|\underline{x}_k, Y_{k-1}) \frac{p(\underline{x}_k|Y_{k-1})}{p(\underline{y}_k|Y_{k-1})} \qquad 10.8.8$$

Again, if we consider the observation equation

$$\underline{y}_k = H_k \underline{x}_k + \underline{\omega}_k$$

we see that knowledge of $\underline{x}_k$ implies that the only random quantity left is $\underline{\omega}_k$ which is independent of $\{\underline{y}_o, \underline{y}_1, \ldots \underline{y}_{k-1}\}$ (since $\{\underline{\omega}_k\}$ is a sequence of independent random variables and $\underline{\omega}_k$ is independent of $\underline{x}_o$ and of $\underline{v}_o, \underline{v}_1, \ldots \underline{v}_k$). We can thus write

$$p(\underline{y}_k|\underline{x}_k, Y_{k-1}) = p(\underline{y}_k|\underline{x}_k)$$

On substituting in 10.8.8 we obtain

$$p(\underline{x}_k| Y_k) = \frac{p(\underline{y}_k|\underline{x}_k) \; p(\underline{x}_k|Y_{k-1})}{p(\underline{y}_k|Y_{k-1})} \qquad 10.8.9$$

In order to determine the maximum a posteriori estimate using the above expression, it will only be necessary to evaluate the probability densities of the numerator since the denominator is not an explicit function of $\underline{x}_k$.

To evaluate $p(\underline{y}_k | \underline{x}_k)$ we consider the observation equation. For given $\underline{x}_k$, $\underline{y}_k$ is a Gaussian random vector of mean

$$E\left[\underline{y}_k \mid \underline{x}_k\right] = H_k\left[\underline{x}_k\right]$$

since $E\left[\underline{\omega}_k \mid \underline{x}_k\right] = E\left[\underline{\omega}_k\right] = 0$

The covariance matrix is given by

$$E\left[\underline{y}_k - H_k\,\underline{x}_k)(\underline{y}_k - H_k\,\underline{x}_k)^T\right] = E\left[\underline{\omega}_k\,\underline{\omega}_k^T\right] = R$$

we could thus write the probability density $p(\underline{y}_k \mid \underline{x}_k)$ as

$$p(\underline{y}_k | \underline{x}_k) = K\,e^{-\frac{1}{2}(\underline{y}_k - H_k\,\underline{x}_k)^T R^{-1}(\underline{y}_k - H_k\,\underline{x}_k)} \qquad 10.8.10$$

where K is an appropriate normalising constant.

To evaluate the a priori probability density $p(\underline{x}_k | \underline{y}_{k-1})$, we note that from property 3, this probability density is Gaussian of mean $\hat{\underline{x}}_{k|k-1}$ and covariance $P_{k|k-1}$ so that we have

$$p(\underline{x}_k | \underline{Y}_{k-1}) = K'\,e^{-\frac{1}{2}(\underline{x}_k - \hat{\underline{x}}_{k|k-1})^T P_{k|k-1}^{-1}(\underline{x}_k - \hat{\underline{x}}_{k|k-1})}$$

so that the a posteriori probability density $p(\underline{x}_k | Y_k)$ could be written as

$$p(\underline{x}_k | Y_k) = K''\,e^{-\frac{1}{2}\left[(\underline{y}_k - H_k\,\underline{x}_k)^T R^{-1}(\underline{y}_k - H_k\,\underline{x}_k) + (\underline{x}_k - \hat{\underline{x}}_{k|k-1})^T P_{k|k-1}^{-1}(\underline{x}_k - \hat{\underline{x}}_{k|k-1})\right]} \qquad 10.8.11$$

Here K" also takes into account the denominator $p(\underline{y}_k | Y_{k-1})$ in 10.8.9.

In order to develop the maximum a posteriori estimate (which is the same as the conditional mean estimate for this case), we can differentiate the logarithm of 10.8.11 with respect to $\underline{x}_k$ and set this to zero to obtain the estimate $\hat{\underline{x}}_{k|k}$. We thus obtain

$$H_k^T R^{-1}(\underline{y}_k - H_k\,\underline{x}_k) - P_{k|k-1}^{-1}(\underline{x}_k - \hat{\underline{x}}_{k|k-1}) = 0$$

For $\underline{x}_k = \hat{\underline{x}}_{k|k}$

$$(H_k^T R^{-1} H_k + P_{k|k-1}^{-1})\,\hat{\underline{x}}_{k|k} = H_k^T R^{-1}\,\underline{y}_k + P_{k|k-1}^{-1}\,\hat{\underline{x}}_{k|k-1}$$

which could be rewritten as

$$\boxed{\hat{\underline{x}}_{k|k} = \hat{\underline{x}}_{k|k-1} + (H_k^T R^{-1} H_k + P_{k|k-1}^{-1})^{-1} H_k^T R^{-1}(\underline{y}_k - H_k\,\hat{\underline{x}}_{k|k-1})} \qquad 10.8.12$$

This expression enables us to obtain the new value of the state estimate given a new observation.

Finally, let us calculate the variance of the estimation error. Note that

$$\underline{\hat{x}}_{k|k} = (H_k^T R^{-1} H_k + P_{k|k-1}^{-1})^{-1} (H_k^T R^{-1} \underline{y}_k + P_{k|k-1}^{-1} \underline{\hat{x}}_{k|k-1})$$

so that

$$\underline{x}_k - \underline{\hat{x}}_{k|k} = \underline{x}_k - (H_k^T R^{-1} H_k + P_{k|k-1}^{-1})^{-1} (H_k^T R^{-1} H_k \underline{x}_k + H_k^T R^{-1} \underline{\omega}_k +$$

$$P_{k|k-1}^{-1} \underline{\hat{x}}_{k|k-1} + P_{k|k-1}^{-1} \underline{x}_k - P_{k|k-1}^{-1} \underline{x}_k)$$

which on simplification becomes

$$\underline{x}_k - \underline{\hat{x}}_{k|k} = - (H_k^T R^{-1} H_k + P_{k|k-1}^{-1})^{-1} (H_k^T R^{-1} \underline{\omega}_k - P_{k|k-1}^{-1} (\underline{x}_k - \underline{\hat{x}}_{k|k-1})$$

Using this expression and noting that $\underline{\omega}_k$ and $\underline{x}_k$ are independent and that $\underline{\omega}_k$ is a zero mean noise and also that $\underline{\omega}_k$ and $\underline{\hat{x}}_{k|k-1}$ are independent, we obtain

$$\boxed{P_{k|k} = (H_k R^{-1} H_k + P_{k|k-1}^{-1})^{-1}} \qquad 10.8.13$$

The expressions 10.8.12 and 10.8.13 involve the inversion of an n x n matrix where n is the order of the state vector. However, since the observation vector is usually of lower order, it is possible to convert these matrix inversions to lower order ones using the matrix inversion Lemma that we have already used in the recursive least squares method.

Thus we can rewrite for the expression in equation 10.8.13

$$(H_k R^{-1} H_k + P_{k|k-1}^{-1}) = P_{k|k-1} - P_{k|k-1} H_k^T (H_k P_{k|k-1} H_k^T + R)^{-1} H_k P_{k|k-1}$$

and for the expression 10.8.12 we need to evaluate

$$(H_k R^{-1} H_k + P_{k|k-1}^{-1})^{-1} H_k^T R^{-1} = P_{k|k-1} H_k^T (H_k P_{k|k-1} H_k^T + R)^{-1}$$

Using these, we obtain the two new relations

$$\boxed{\underline{\hat{x}}_{k|k} = \underline{\hat{x}}_{k|k-1} + P_{k|k-1} H_k^T (H_k P_{k|k-1} H_k^T + R)^{-1} (\underline{y}_k - H_k \underline{\hat{x}}_{k|k-1})} \qquad 10.8.14$$

and

$$\boxed{P_{k|k} = P_{k|k-1} - P_{k|k-1} H_k^T (H_k P_{k|k-1} H_k^T + R)^{-1} H_k P_{k|k-1}} \qquad 10.8.15$$

The equations 10.8.3, 10.8.7, 10.8.14, 10.8.15 constitute the equations of the optimal linear minimum variance filter developed by Kalman and Bucy.

The equation 10.8.14 enables us to calculate the estimate at the instant t_k, given the measures $\underline{y}_o$, $\underline{y}_1$, ... $\underline{y}_k$ using the one step prediction and the difference between the actual output $\underline{y}_k$ and the predicted output $H_k \hat{\underline{x}}_{k|k-1}$. This difference is weighted by the term

$$K_k = P_{k|k-1} H_k^T (H_k P_{k|k-1} H_k^T + R)^{-1} \qquad 10.8.16$$

The matrix (or vector) K_k plays the role of a "gain" and is usually called the filter gain.

The difference $\underline{y}_k - H_k \hat{\underline{x}}_{k|k-1}$ is called the innovations process.

The equations 10.8.14 and 10.8.15 are evaluated in the following order.

Gain : $K_k = P_{k|k-1} H_k^T (H_k P_{k|k-1} H_k^T + R)^{-1}$

Calculation of the estimate : $\hat{\underline{x}}_{k|k} = \hat{\underline{x}}_{k|k-1} + K_k (\underline{y}_k - H_k \hat{\underline{x}}_{k|k-1})$

Calculation of the covariance : $P_{k|k} = (I - K_k H_k) P_{k|k-1}$

These equations can be represented as shown in Fig. 10.3

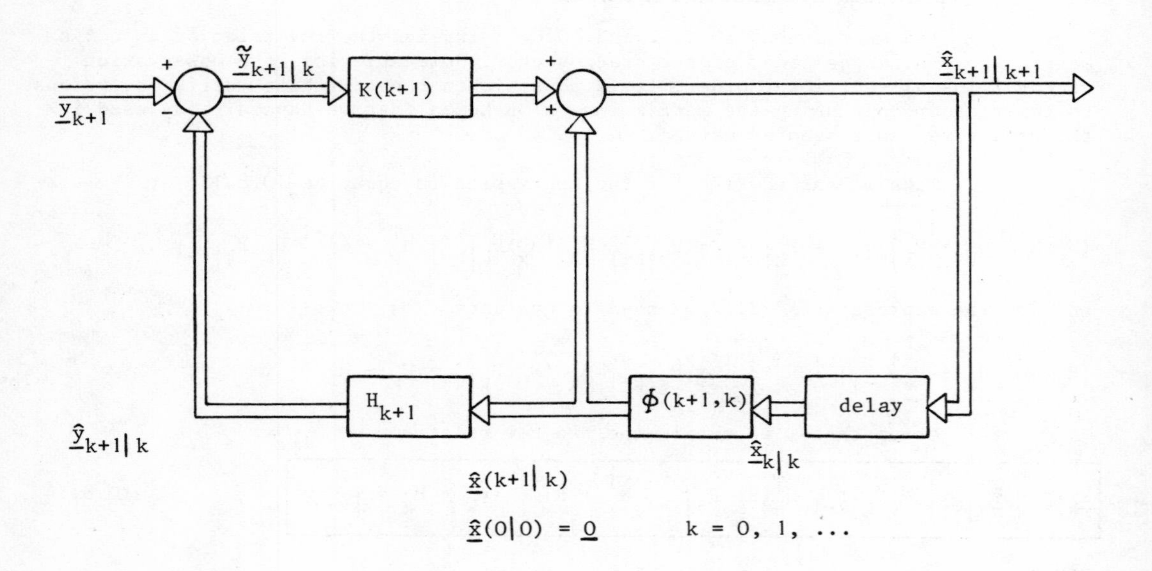

Fig. 10.3 : Block diagram of the Kalman Filter

We note that the filter has the same structure as the process. Equation 10.8.15 which enables us to compute the error covariance is a matrix equation of the Riccati type. We note that this equation as well as equation 10.8.7 are independent of the observations so that these equations as well as the filter gain can be pre-calculated.

Next we consider some illustrative examples.

Example 1

We consider a constant (x) of which we take n successive measurements (z_k) each one of which has a random zero mean error with known covariance $R_k = R$. The initial estimate of x is given to be x_o where x_o is a Gaussian random variable of covariance P_o. We want to show by using the Kalman filter method that

$$\hat{x}_{k|k} = \frac{R\, x_o + P_o \sum_{i=1}^{k} z_i}{R + K\, P_o}$$

and

$$P_{k|k} = \frac{R\, P_o}{R + k\, P_o}$$

To solve this problem, we note that our a priori information is :

State variable : x

State equation : $\mathring{x} = 0$ or $x = a$ (constant)

$$\phi = 1, \quad \Gamma = 0, \quad Q = 0$$

Measurement equation : $z_k = x + v_k$

Thus $H = 1$

Known initial conditions : initial state : x_o

variance of the initial state P_o

For this case, the Kalman filter equations reduce to :

$$K_k = P_{k|k-1}\left[P_{k|k-1} + R\right]^{-1} = \frac{P_{k|k-1}}{P_{k|k-1}+R}$$

$$P_{k|k-1} = P_{k-1|k-1}$$

$$P_{k|k} = P_{k|k-1} - K_k\, P_{k|k-1}$$

Thus

$$K_k = \frac{P_{k-1}}{P_{k|k-1}+R} \quad \text{and} \quad P_{k|k} = \frac{R\, P_{k-1|k-1}}{P_{k-1|k-1}+R}$$

This enables us to put the estimate in the form :

$$\hat{x}_{k|k} = \hat{x}_{k-1|k-1} + \frac{P_{k-1|k-1}}{P_{k-1|k-1}+R}\left[z_k - \hat{x}_{k-1|k-1} \right]$$

or

$$\hat{x}_{k|k} = \frac{P_{k-1|k-1}\, z_k + R\, \hat{x}_{k-1|k-1}}{P_{k-1|k-1} + R}$$

This expression shows that the new estimate $\hat{x}_{k|k}$ is a linear combination of

(a) the old estimate weighted by the variance of the new measurement and

(b) the new measurement weighted by the variance of the old estimate.

This is a consequence of the trade off between the confidence we have in the old estimates and those in the new measurements.

If we use the above relation in conjunction with the initial conditions $\hat{x}_o$ and P_o, we obtain the desired results

$$\hat{x}_{k|k} = \frac{Rx_o + P_o \sum_{i=1}^{k} z_i}{R + k\, P_o}$$

$$P_{k|k} = \frac{R\, P_o}{R+k\, P_o}$$

These results show that :

(a) as we use new measurements, the variance $P_{k|k}$ of the estimation errors decreases

(b) in the limit when $k \to \infty$, all traces of the initial conditions disappear and we have

$$\lim_{k \to \infty} \hat{x}_{k|k} = \frac{\sum_{1}^{k} z_i}{k}$$

$$\lim_{k \to \infty} P_{k|k} = \lim_{k \to \infty} \frac{R}{R} = 0$$

i.e. the estimate of x reduces to the arithmetic mean of the measurements.

Example 2

We consider the frictionless motion of a particle (x) in a straight line. A priori we know neither its position nor its velocity. We only know that (x) obeys the differential equation $\ddot{x} = 0$ and that the initial conditions $(x_o, \dot{x}_o)$ could be considered to be zero mean Gaussian random variables of variance $E\left[x_o^2\right] = 1(m^2)$, $E\left[\dot{x}_o^2\right] = 3(m/s)^2$.

We assume in addition that x_o and $\dot{x}_o$ are independent, i.e. $E\left[x_o\, \dot{x}_o\right] = 0$. The only measurements that are made are of the position x and these give the following results :

at

$t = 0 \qquad z_o = 0.08$

$t = 1 \qquad z_1 = 2.05$

$t = 2 \qquad z_2 = 3.8$

$t = 3 \qquad z_3 = 6.01$

These measurements are a zero mean random variable of variance $R = E\left[v_k^{\,2}\right] = 0.1(m)^2$. We want to estimate the position and velocity at time t = 3 and the corresponding covariance $P_{3|3}$.

The application of the Kalman filter to this problem will involve the following conceptual steps :

a) formulation of the equations
b) calculation of the K_k, $P_{k|k-1}$, $P_{k|k}$ matrices
c) calculation of $\hat{x}_{k|k}$ and $\hat{\dot{x}}_{k|k}$

a) Formulation of the equations

We could write the state equations as

$x_1 = x \qquad$ (position)

$x_2 = \dot{x} \qquad$ (velocity)

From the differential equation $\ddot{x} = 0$, we have

$$x = \dot{x}\, t + x(0)$$

so that for any t_1, t_2 we have

$$x(t_2) = x(t_1) + \dot{x}(t_2 - t_1)$$

$$\dot{x}(t_2) = \dot{x}(t_1)$$

We can rewrite these equations in the matrix form

$$\begin{bmatrix} x_1(t_2) \\ x_2(t_2) \end{bmatrix} = \begin{bmatrix} 1 & t_2-t_1 \\ 0 & 1 \end{bmatrix} \begin{bmatrix} x_1(t_1) \\ x_2(t_1) \end{bmatrix}$$

The transition matrix from t_1 to t_2 is thus

$$\Phi(t_2, t_1) = \begin{bmatrix} 1 & t_2 - t_1 \\ 0 & 1 \end{bmatrix}$$

If we consider the successive instants to be 0, 1 second, 2 seconds, 3 seconds, then the transition matrix becomes the constant matrix

$$\Phi = \begin{bmatrix} 1 & 1 \\ 0 & 1 \end{bmatrix}$$

The measurement equation here is

$$z = \begin{bmatrix} 1 & 0 \end{bmatrix} \begin{bmatrix} x_1 \\ x_2 \end{bmatrix} + v$$

where $R = E\left[v^2\right] = 0.1\ m^2$

The measurement matrix H is thus

$$H = \begin{bmatrix} 1 & 0 \end{bmatrix}$$

We note finally that the system under consideration is not subject to noise so that

$$Q = \begin{bmatrix} 0 & 0 \\ 0 & 0 \end{bmatrix}$$

Initial conditions

Let us call (-1) the index of the instant before the first measurement and (0,1,2,3) the instants (0 s, 1 s, 2 s, 3 s). Then

$$P_{0|-1} = \begin{bmatrix} 1 & 0 \\ 0 & 3 \end{bmatrix}$$

$$\begin{bmatrix} x_1 \\ x_2 \end{bmatrix}_{0|-1} = \begin{bmatrix} 0 \\ 0 \end{bmatrix}$$

We assume that $\hat{\underline{x}}_{0|-1} = E\left[\underline{x}_o\right]$

b) Calculation of $K_{k|k-1}$, $P_{k|k}$

At t = 0 $\quad K_o = P_{0|-1}\, H^T \left[H\, P_{0|-1}\, H^T + R\right]^{-1}$

$$= \begin{bmatrix} 1 & 0 \\ 0 & 3 \end{bmatrix}\begin{bmatrix} 1 \\ 0 \end{bmatrix}\left[(1 \quad 0)\begin{pmatrix} 1 & 0 \\ 0 & 3 \end{pmatrix}\begin{pmatrix} 1 \\ 0 \end{pmatrix} + 0.1\right]^{-1} = \begin{pmatrix} 0.9091 \\ 0 \end{pmatrix}$$

$$P_{0|0} = P_{0|-1} - K_0\, H\, P_{0|-1}$$

$$= \begin{pmatrix} 1 & 0 \\ 0 & 3 \end{pmatrix}\begin{pmatrix} 0.9091 \\ 0 \end{pmatrix}\begin{pmatrix} 1 & 0 \end{pmatrix}\begin{pmatrix} 1 & 0 \\ 0 & 3 \end{pmatrix} = \begin{pmatrix} 0.091 & 0 \\ 0 & 3 \end{pmatrix}$$

We see that the first measurement reduces the variance of the estimate of x from 1 m^2 to 0.091 m^2. However, there is no change in the variance of $\mathring{x}$. This is reasonable since a single measurement of x does not allow us to estimate $\mathring{x}$ in the absence of correlation between x and $\mathring{x}$.

Going on with our calculations we obtain :

$$P_{1|0} = \begin{pmatrix} 3.091 & 3 \\ 3 & 3 \end{pmatrix}$$

$$K_1 = \begin{pmatrix} 0.969 \\ 0.94 \end{pmatrix}$$

$$P_{1|1} = \begin{pmatrix} 0.096 & 0.095 \\ 0.095 & 0.180 \end{pmatrix}$$

At t = 2

$$P_{2|1} = \begin{pmatrix} 0.466 & 0.275 \\ 0.275 & 0.180 \end{pmatrix}$$

$$K_2 = \begin{pmatrix} 0.823 \\ 0.486 \end{pmatrix}$$

$$P_{2|2} = \begin{pmatrix} 0.083 & 0.049 \\ 0.049 & 0.0465 \end{pmatrix}$$

At t = 3

$$P_{3|2} = \begin{pmatrix} 0.0228 & 0.096 \\ 0.096 & 0.047 \end{pmatrix}$$

$$K_3 = \begin{pmatrix} 0.695 \\ 0.2824 \end{pmatrix}$$

$$P_{3|3} = \begin{pmatrix} 0.07 & 0.029 \\ 0.029 & 0.02 \end{pmatrix}$$

c) Successive estimates of x and $\mathring{x}$

$$\begin{pmatrix} \hat{x}_1 \\ \hat{x}_2 \end{pmatrix}_{0|0} = \begin{pmatrix} \hat{x}_1 \\ \hat{x}_2 \end{pmatrix}_{0|-1} + K_o \left[Z_o - H \begin{pmatrix} \hat{x}_1 \\ \hat{x}_2 \end{pmatrix}_{0|1} \right]$$

$$= \begin{pmatrix} 0 \\ 0 \end{pmatrix} + \begin{pmatrix} 0.9091 \\ 0 \end{pmatrix} \left(0.08 - (1, 0) \begin{pmatrix} 0 \\ 0 \end{pmatrix} \right) = \begin{pmatrix} 0.0727 \\ 0 \end{pmatrix}$$

We see that the first estimate of the position (0.0727) is much closer to the first measurement (0.08 m) than to the initial estimate (0 m) since the latter was rather poorly estimated a priori (variance of 3 m^2) compared to the measurements (0.1 m^2).

The first estimate of the velocity remains zero since a measurement of the position gives no new information on the velocity.

Continuing our estimation we have

$$\begin{pmatrix} \hat{x}_1 \\ \hat{x}_2 \end{pmatrix}_{1|1} = \begin{pmatrix} 1.988 \\ 1.859 \end{pmatrix}$$

$$\begin{pmatrix} \hat{x}_1 \\ \hat{x}_2 \end{pmatrix}_{2|2} = \begin{pmatrix} 3.808 \\ 1.836 \end{pmatrix}$$

$$\begin{pmatrix} \hat{x}_1 \\ \hat{x}_2 \end{pmatrix}_{3|3} = \begin{pmatrix} 5.9 \\ 1.94 \end{pmatrix}$$

Next we will consider continuous time estimation.

10.9 CONTINUOUS TIME ESTIMATION

The model in this case is of the form :

$$\dot{\underline{x}}(t) = F(t)\,\underline{x}(t) + G(t)\,\underline{\omega}(t) \qquad 10.9.1$$

$$\underline{z}(t) = H(t)\,\underline{x}(t) + \underline{v}(t) \qquad \forall\, t \geq t_o \qquad 10.9.2$$

where $\underline{x}$ is n dimensional state vector, $\underline{\omega}$ a p dimensional zero mean Gaussian white noise vector of known covariance. $\underline{z}$ is an m dimensional observation vector and $\underline{v}$ is an m dimensional zero mean Gaussian random white noise vector of known covariance, F(t), G(t), H(t) are respectively continuous n x n, n x p and m x n matrices. The initial time t_o is fixed. The covariances of $\underline{\omega}$ and $\underline{v}$ are Q and R respectively, i.e.

$$E\left[\underline{\omega}(t)^T\,\underline{\omega}(\tau)\right] = Q(t)\;\delta(t-\tau)$$

and

$$E\left[\underline{v}(t)\,\underline{v}^T(\tau)\right] = R(t)\;\delta(t-\tau) \qquad 10.9.3$$

$$\forall\; t,\tau \geq t_o$$

where δ is the Dirac Delta function. The white noise processes are assumed to be independent of each other, i.e.

$$E\left[\underline{\omega}(t)\,v(\underline{\tau})^T\right] = 0 \qquad 10.9.4$$

Again, as in the case of discrete time processes, this is not a restrictive assumption since we can develop the filter equations without this assumption, although the expressions in that case are somewhat more complicated.

We will denote by $\underline{\hat{x}}(t_1 \mid t)$ the estimate of $\underline{x}$ at some time $t_1 \geq t_o$ which is based on measurements $\underline{z}(\tau)$ over the interval $t_o \leq \tau \leq t$ and the error of estimation by $\underline{\tilde{x}}(t_1 \mid t)$ where

$$\underline{\tilde{x}}(t_1 \mid t) = \underline{x}(t_1) - \underline{\hat{x}}(t_1 \mid t) \qquad 10.9.5$$

and the estimation problem is to determine the optimal estimate of $\underline{x}(t_1)$ given the system described by equation 10.9.1 and observations described by equation 10.9.2.

In order to solve this estimation problem, we will reformulate the problem as a discrete time problem and then go from the discrete to the continuous case using some form of a limiting procedure.

The equivalent discrete equation to equation 10.9.1 is the zero mean Gauss-Markov sequence

$$\underline{x}(t+\Delta t) = \Phi(t+\Delta t,\, t)\,\underline{x}(t) + \Gamma(t+\Delta t,\, t)\,\underline{\omega}(t) \qquad 10.9.6$$

where

$$\Phi(t+\Delta t,\, t) = I + F(t)\Delta t + O(\Delta t^2) \qquad 10.9.7$$

$$\Gamma(t+\Delta t, t) = G(t)\,\Delta t + O(\Delta t^2) \qquad 10.9.8$$

where $\left\{ \underline{\omega}(t),\ t = t_o + j\,\Delta t \quad j = 0, 1, \ldots \right\}$

is a zero mean Gaussian white noise sequence whose covariance is given by

$$E\left[\underline{\omega}(t)\,\underline{\omega}^T(\tau)\right] = \frac{Q(t)}{\Delta t}\,\delta_{jk}$$

where t is the discrete time index which we defined above whilst τ is the discrete time index $\{\tau = t_o + k\,\Delta t,\ k = 0, 1, \ldots\}$, $Q(t)$ is a positive semi-definite matrix $\forall\ t \geq t_o$ and δ_{jk} is the Kronecker Delta. The initial state $\underline{x}(t_o)$ is a zero mean Gaussian random n vector which is independent of $\{\underline{\omega}(t),\ t = t_o + j\,\Delta_t,\ j = 0, 1, \ldots\}$ and has the covariance $E\left[\underline{x}(t_o)\,\underline{x}^T(t_o)\right] = P(t_o)$ where $P(t_o)$ is an n x n positive semi-definite matrix.

Next we can write the discrete form of the measurement equation as

$$\underline{z}(t+\Delta t) = H(t+\Delta t)\,\underline{x}(t+\Delta t) + \underline{v}(t+\Delta t) \qquad 10.9.9$$

The stochastic process $\left\{\underline{v}(t+\Delta t),\ t = t_o + j\,\Delta t,\ j = 0, 1, \ldots\right\}$ is a zero mean Gaussian white sequence which is independent of $\underline{x}(t_o)$ and which has a covariance given by

$$E\left[\underline{v}(t+\Delta t)\,\underline{v}^T(\tau+\Delta t)\right] = \frac{R(t+\Delta t)}{\Delta t}\,\delta_{jk} \qquad 10.9.10$$

where $R(t+\Delta t)$ is a positive definite matrix $\forall\ t \geq t_o$.

We further assume that $\underline{v}$ and $\underline{\omega}$ are independent.

Thus our estimation problem is to determine $\hat{\underline{x}}(t_1 \mid t)$ of $\underline{x}(t_1)$, where $t_1 = t_o + i\,\Delta t,\ i = 0, 1, \ldots,$ given the system 10.9.6 and measurements

$$\left\{\underline{z}(\tau),\quad \tau = t_o + \Delta t,\ t_o + 2\,\Delta t,\ \ldots\ t\right\}$$

defined by 10.9.9.

Having discretised our problem, we next develop the continuous time Kalman Filter equations.

The continuous time filtering equations

From the previous section, the optimal filter for the system of equations 10.9.6 and 10.9.9 is given by

$$\hat{\underline{x}}(t+\Delta t\ t+\Delta t) = \Phi(t+\Delta t, t)\,\hat{\underline{x}}(t \mid t) + K(t+\Delta t)\left[\underline{z}(t+\Delta t) - H(t+\Delta t)\,\Phi(t+\Delta t, t)\,\hat{\underline{x}}(t \mid t)\right]$$

where $\hat{\underline{x}}(t_o \mid t_o) = \underline{0}$

Using equation 10.9.7 :

$$\Phi(t+\Delta t, t)\,\hat{\underline{x}}(t \mid t) = \left[I + F(t)\,\Delta t + O(\Delta t^2)\right]\hat{\underline{x}}(t \mid t)$$

or

$$\hat{\underline{x}}(t+\Delta t \mid t+\Delta t) - \hat{\underline{x}}(t \mid t) = F(t)\,\hat{\underline{x}}(t \mid t)\,\Delta t + K(t+\Delta t)$$
$$\left[\underline{z}(t+\Delta t) - H(t+\Delta t)\,\Phi(t+\Delta t)\,\hat{\underline{x}}(t \mid t)\right] + O(\Delta t^2) \qquad 10.9.11$$

Dividing through by Δt and taking the limit as $\Delta t \longrightarrow 0$, we obtain

$$\overset{\circ}{\hat{\underline{x}}} = K(t)\,\hat{\underline{x}} + \lim_{\Delta t \to 0} K(t+\Delta t)\left[\underline{z}(t+\Delta t) - H(t+\Delta t)\frac{\Phi(t+\ t,t)}{\Delta t}\hat{\underline{x}}(t \mid t)\right] \qquad 10.9.12$$

where $\hat{\underline{x}} = \hat{\underline{x}}(t \mid t)$ and $t \geqslant t_o$.

We can evaluate the limit on the right hand side of equation 10.9.12 as the product of the limits

$$\lim_{\Delta t \to 0} \frac{K(t+\Delta t)}{\Delta t} \quad \text{and} \quad \lim_{\Delta t \to 0}\left[\underline{z}(t+\Delta t) - H(t+\Delta t)\,\Phi(t+\Delta t, t)\,\hat{\underline{x}}\right]$$

provided that these limits exist.

Let us first consider the second limit. Using the expansion of Φ from 10.9.7, we see that

$$\lim_{\Delta t \to 0}\left[\underline{z}(t+\Delta t) - H(t+\Delta t)\,\Phi(t+\Delta t, t)\,\hat{\underline{x}}\right] = \underline{z}(t) - H(t)\,\hat{\underline{x}}(t \mid t) \qquad 10.9.13$$

For the other limit, we note that from the discrete filter equations we have

$$\begin{aligned} K(t+\Delta t) &= P(t+\Delta t \mid t)\,H^T(t+\Delta t)\left[H(t+\Delta t)\,P(t+\Delta t \mid t)\,H^T(t+\Delta t) + \frac{R(t+\Delta t)}{\Delta t}\right]^{-1} \\ &= P(t+\Delta t \mid t)\left[H^T(t+\Delta t)\,P(t+\Delta t \mid t)\,H^T(t+\Delta t)\,\Delta t + R(t+\Delta t)\right]^{-1}\Delta t \end{aligned} \qquad 10.9.14$$

and

$$P(t+\Delta t \mid t) = \Phi(t+\Delta t, t)\,P(t \mid t)\,\Phi^T(t+\Delta t, t) + \Gamma(t+\Delta t, t)\,\frac{Q(t)}{\Delta t}\,\Gamma^T(t+\Delta t, t) \qquad 10.9.15$$

Substituting into 10.9.15 from 10.9.7, 10.9.8, expanding the result and regrouping terms we obtain the relation

$$\begin{aligned} P(t+\Delta t, t) &= \left[I + F(t)\Delta t + O(\Delta t^2)\right] P(t \mid t) \left[I + F(t)\Delta t + O(\Delta t^2)\right]^T \\ &\quad + \left[G(t)\Delta t + O(\Delta t^2)\right] \frac{Q(t)}{\Delta t}\left[G(t)\Delta t + O(\Delta t^2)\right]^T \\ &= P(t \mid t) + F(t)\,P(t \mid t) + P(t \mid t)\,F^T(t) + G(t)\,Q(t)\,G^T(t)\Delta t + O(\Delta t^2) \end{aligned} \qquad 10.9.16$$

Thus for any $t \geqslant t_o$ and $\Delta t \geqslant 0$,

$$\lim_{\Delta t \to 0} P(t+\Delta t \mid t) = P(t \mid t) \qquad 10.9.17$$

As a consequence of this result and the assumptions that H(t) and R(t) are continuous, it follows that

$$\begin{aligned} \lim_{\Delta t \to 0} \frac{K(t+\Delta t)}{\Delta t} &= P(t \mid t)\,H^T(t) \lim_{\Delta t \to 0}\left[H(t+\Delta t)\,P(t+\Delta t \mid t)\,H^T(t+\Delta t)\,\Delta t + R(t+\Delta t)\right]^{-1} \\ &= P(t \mid t)\,H^T(t)\,R^{-1}(t) \end{aligned}$$

Hence we obtain the continuous filter equation

$$\boxed{\dot{\hat{\underline{x}}} = F(t)\,\hat{\underline{x}} + K(t)\left[\,\underline{z}(t) - H(t)\,\hat{\underline{x}}\,\right]} \qquad 10.9.18$$

for $t \geqslant t_o$ with $\hat{\underline{x}}(t_o \mid t_o) = 0$ where

$$\boxed{K(t) = P(t \mid t)\, H^T(t)\, R^{-1}(t)} \qquad 10.9.19$$

To obtain the equation for the evolution of the error covariance, let us consider the filter error equation

$$\tilde{\underline{x}}(t+\Delta t \mid t) = \left[I - K(t+\Delta t)\, H(t+\Delta t)\right] \Phi(t+\Delta t, t)\, \tilde{\underline{x}}(t \mid t) + \left[I - K(t+\Delta t)\, H(t+\Delta t)\right]$$
$$\Gamma(t+\Delta t, t)\, \underline{\omega}(t) - K(t+\Delta t)\, \underline{v}(t+\Delta t) \qquad 10.9.17$$

On substituting for Φ, Γ, dividing through by zero and taking the limit $\Delta t \rightarrow 0$, we obtain

$$\mathring{\tilde{\underline{x}}} = \left[F(t) - K\,H(t)\right] \tilde{\underline{x}} + G(t)\, \underline{\omega}(t) - K(t)\, \underline{v}(t) \qquad 10.9.20$$

Since $\{\tilde{x}(t+\Delta t \mid t + \Delta t)\ t = t_o + j\,\Delta t,\ j = 0, 1, \ldots\}$ is a zero mean Gauss-Markov sequence, x is also a zero mean Gauss-Markov process. After manipulations of this, we obtain the differential equation for $P(t \mid t)$ as

$$\dot{P}(t \mid t) = F(t)\, P(t \mid t) + P(t \mid t)\, F^T(t) + G(t)\, Q(t)\, G^T(t)$$

$$- \lim_{\Delta t \rightarrow 0} \underline{K(t+\Delta t)\, H(t+\Delta t)\, P(t \mid t)}$$

or $\underline{\dot{P} = F(t)\, P + PF^T(t) - PH^T(t)\, R^{-1}(t)\, H(t)\, P + G(t)\, Q(t)\, G^T(t)}$ 10.9.21

where $P = P(t \mid t)$ and $P(t_o \mid t_o) = P(t_o)$.

Thus 10.9.18 and 10.9.21 are the continuous time equations respectively for the evolution of the filtered estimate $\hat{\underline{x}}(t \mid t)$ and of the error covariance of this estimate.

In Fig. 10.4 we have a block diagram of the Kalman filter for continuous time systems.

Here again we see that the filter has the same structure as the system model. The filter could be considered to be a model of the system dynamics $\mathring{\underline{x}} = F(t)\, \underline{x}$ which is forced by the feedback correction signal $K(t)\, \tilde{\underline{z}}(t \mid t)$. Equation 10.9.21 is a matrix Riccati equation and it is analogous to the matrix Riccati equation of optimal deterministic control. In fact a duality relationship exists between the two. $P(t \mid t)$ is a symmetric matrix since it is a covariance matrix so in order to evaluate it, it is necessary only to solve $\frac{n(n+1)}{2}$ instead of n^2 non-linear differential equations.

Next let us consider an example of the use of this filter.

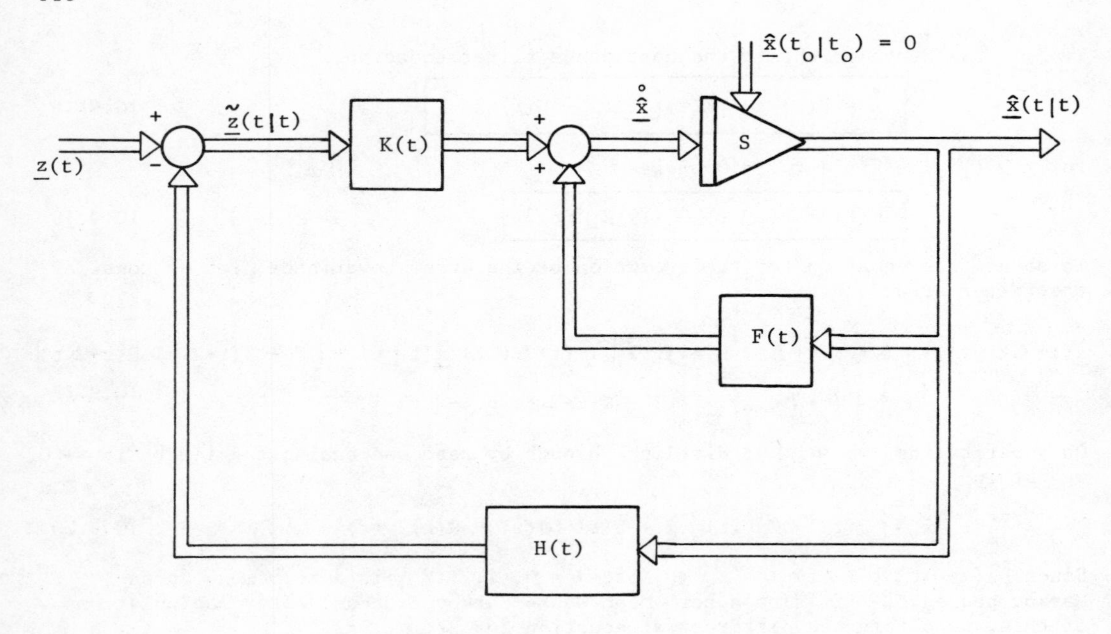

Fig. 10.4

Example

Consider the scalar process

$$\dot{x} = a\,x + \eta$$
$$z = c\,x + \xi$$

where η, ξ are uncorrelated zero mean Gaussian random white noise vectors of covariances q and r respectively. x(0) is a zero mean Gaussian random vector with covariance p(0).

Here, the optimal filtered estimate is given by

$$\dot{\hat{x}} = a\,\hat{x} + K\left[z - c\,\hat{x}\right]$$

where $K = \frac{p\,c}{r}$
and p is the solution of the Riccati equation

$$\dot{p} = 2\,a\,p - \frac{p^2\,c^2}{r} + q$$

In the steady state, $\mathring{p} \longrightarrow 0$ so that we have

$$p^2 - \frac{2\,a\,r}{c^2}\,p - \frac{r}{c^2}\,q = 0$$

or
$$p = \frac{a\,r}{c^2} + \frac{2}{c^2}\sqrt{(a^2\,r^2 + q\,r)}$$

and $K = \frac{p\,c}{r} = \frac{a}{c} \pm \frac{2}{c}\sqrt{a^2 + \frac{q}{r}}$

Example 2

Consider the system

$$\frac{d}{dt}\begin{bmatrix} x_1 \\ x_2 \end{bmatrix} = \begin{bmatrix} -0.8 & 0.6 \\ 0.6 & -0.8 \end{bmatrix}\begin{bmatrix} x_1 \\ x_2 \end{bmatrix} + \begin{bmatrix} \eta_1 \\ \eta_2 \end{bmatrix}$$

$$\begin{pmatrix} z_1 \\ z_2 \end{pmatrix} = \begin{pmatrix} 1 & 0 \\ 0 & 1 \end{pmatrix} \begin{pmatrix} x_1 \\ x_2 \end{pmatrix} + \begin{pmatrix} \xi_1 \\ \xi_2 \end{pmatrix} \quad ; \quad Q = R = I$$

$$\hat{x}_1(0) = \hat{x}_2(0) = 0, \qquad x_1(0) = x_2(0) = 1$$

$$P_{11}(0) = P_{22}(0) = 1$$

Solving this problem, Fig. 10.5 shows the evolution of the estimates $\hat{x}_1$ and $\hat{x}_2$ for a particular sample function of the noise.

Finally, in this chapter, we consider the problem of optimal stochastic control.

10.10 OPTIMAL STOCHASTIC CONTROL

In our filter development, we had ignored control inputs. Let us now consider what happens if these are included. We will now consider a stochastic control problem where the cost function is the expected value of a quadratic form in states and controls, and the linear dynamical system constraints contain random terms as in the present chapter and also control inputs. We will see that for the case where the noise terms are Gaussian white noise vectors, it is possible to separate out the calculation of the controller gain from that of the filter gain as shown in Fig. 10.6. Thus the filter could be calculated using the methods developed in this chapter whilst the controls could be calculated using the deterministic methods outlined in chapter 5.

In this section we will only explain the nature of the main results without going into any proofs. We will consider both the discrete time and continuous time cases.

10.10.1 The discrete time stochastic controller

The system model in this case also includes a control input, i.e. our system model is

$$\underline{x}(k+1) = \Phi(k+1,k)\, \underline{x}(k) + \Gamma(k+1,k)\, \underline{\omega}(k) + \Psi(k+1,k)\, \underline{u}(k) \qquad 10.10.1$$

and

$$\underline{z}(k+1) = H(k+1)\, \underline{x}(k+1) + \underline{v}(k+1) \qquad 10.10.2$$

where $\underline{u}(k)$ is the r dimensional control vector, Ψ is a n x r control transition matrix and the other terms are as before.

The control problem is to minimise the following average cost function

$$J_N = E \sum_{i=1}^{N} \left[\| \underline{x}(i) \|^2_{A(i)} + \| \underline{u}(i-1) \|^2_{B(i-1)} \right] \qquad 10.10.3$$

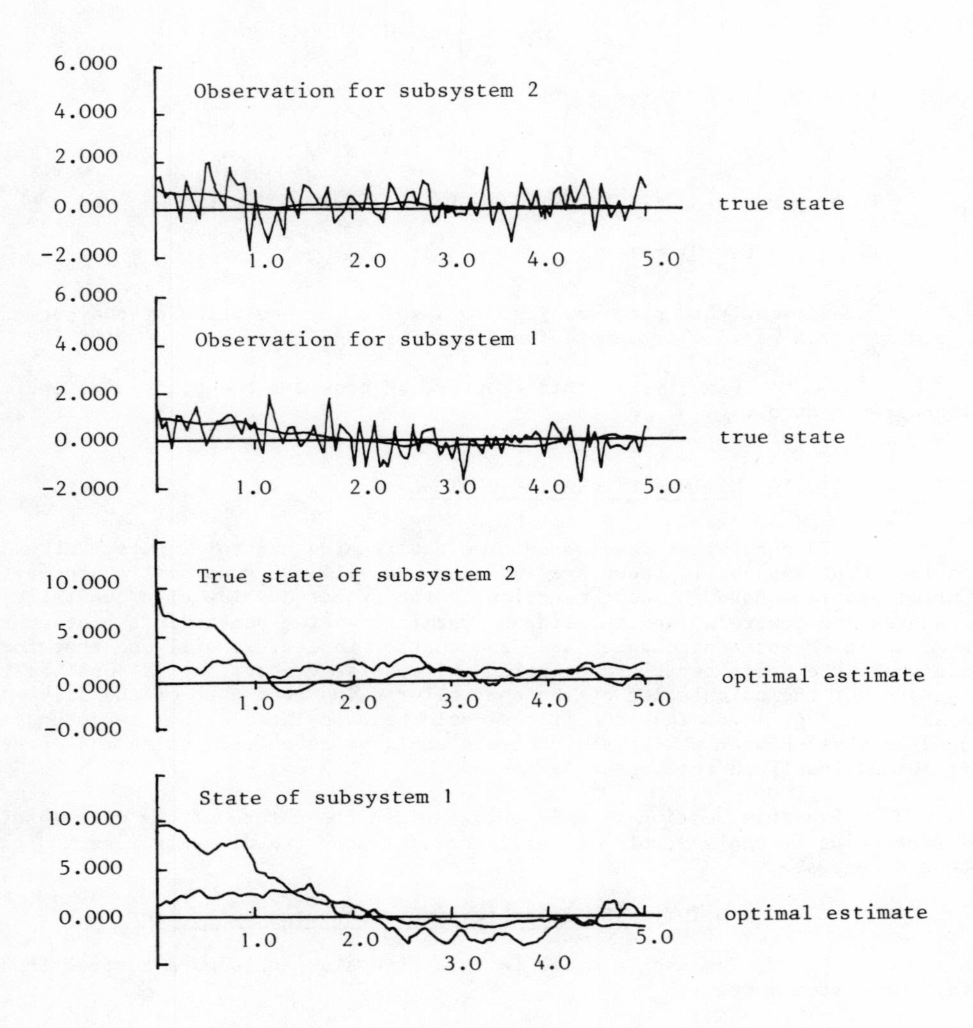

Fig. 10.5 : Simulation results for example 2

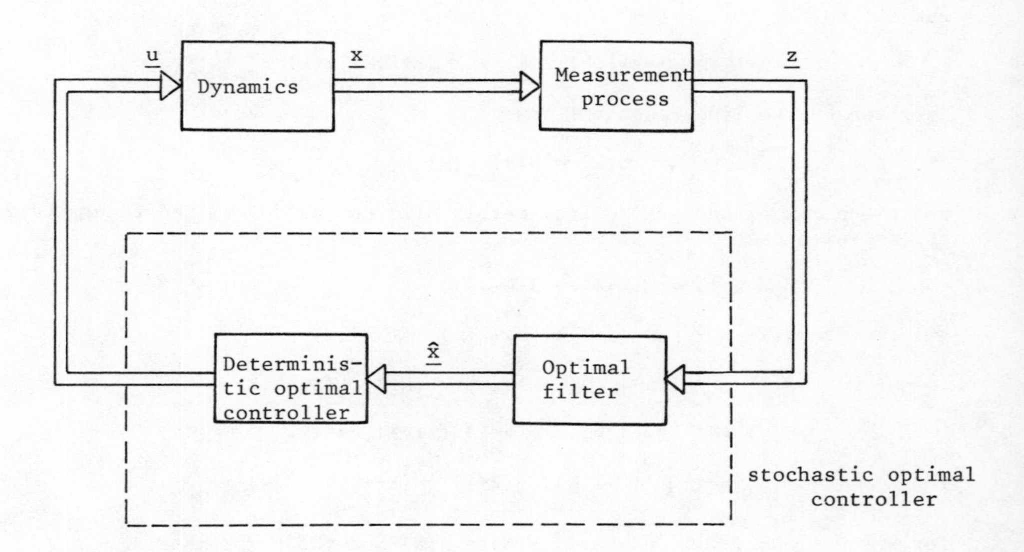

Fig. 10.6

subject to the constraints 10.10.1 and 10.10.2 Here A(i) and B(i-1) are symmetric positive semi-definite matrices which are n x n and r x r respectively.

The expected value taken in the cost function in 10.10.3 implies that we are assessing performance over an ensemble of systems. The cost function is chosen such that only those states which can be controlled and only those controls which can affect them appear in J_N. Thus $\underline{u}(0)$ is present in the cost function since its choice affects $\underline{x}(1)$, $\underline{x}(2)$, ... , $\underline{x}(N)$ while neither $\underline{x}(0)$ is present (since it can't be affected by any control choice) nor is $\underline{u}(N)$ (since $\underline{u}(N-1)$ affects $\underline{x}(N)$ at which point the control problem is terminated).

It must be emphasised that the solution that we are seeking of this problem is a physically realisable one, i.e. we are seeking a solution which is such that it only uses the available information (and not for example future measurements). For any given k = 0, 1, ..., N-1, the available information on the system state consists of the measurement sequence $\underline{z}(1), \underline{z}(2), \ldots \underline{z}(k)$ and the mean value $\underline{x}(0)$ of the initial state $\underline{x}(0)$. Thus we are seeking a control law of the form

$$\underline{u}(k) = \underline{f}\left[\underline{z}(1), \underline{z}(2), \ldots \underline{z}(k), \underline{x}(0)\right]$$

We saw in chapter 5 the solution for a deterministic version of this problem and we recapitulate it in the present notation as :

The solution of the problem of minimising

$$J_N = \left\{ \sum_{i=1}^{N} \left[\| \underline{x}(i) \|^2_{A(i)} + \| \underline{u}(i-1) \|^2_{B(i-1)} \right] \right\} \qquad 10.10.4$$

subject to

$$\underline{x}(k+1) = \Phi(k+1,k)\ \underline{x}(k) + \Psi(k+1,k)\ \underline{u}(k) \qquad 10.10.5$$

is given by the linear control law

$$\underline{u}(k) = S(k)\ \underline{x}(k) \qquad 10.10.6$$

where the r x n feedback control matrix S(k) can be determined recursively from the following set of relations :

$$W(k+1) = M(k+1) + A(k+1) \qquad 10.10.7$$

$$S(k) = -\left[\Psi^T(k+1,k)\ W(k+1)\ \Psi(k+1,k) + B(k)\right]^{-1} \Psi^T(k+1,k)\ W(k+1)\ \Phi(k+1,k) \qquad 10.10.8$$

$$M(k) = \Phi^T(k+1,k)\ W(k+1)\ \Phi(k+1,k) + \Phi^T(k+1,k)\ W(k+1)\ \Psi(k+1,k)\ S(k) \qquad 10.10.9$$

for k = N-1, N-2, ..., 0 where W(N) = A(N) and the r x r matrix

$$\left[\Psi^T(k+1,k)\ W(k+1)\ \Psi(k+1,k) + B(k)\right]$$

must be positive definite $\forall$ k.

If we now consider the original stochastic problem, it is possible to show using dynamic programming that the optimal control is given by

$$\boxed{\underline{u}(k) = S(k)\ \hat{\underline{x}}(k \mid k)} \qquad 10.10.10$$

for k = 0, 1, ..., N-1

where S(k) is determined using the algorithm in equations 10.10.7, 10.10.8 and 10.10.9 whilst $\hat{x}(k \mid k)$ is calculated using the filter equations, i.e.

$$\hat{\underline{x}}(k \mid k) = \hat{\underline{x}}(k \mid k-1) + K(k)\left[\underline{z}(k) - H(k)\ \hat{\underline{x}}(k \mid k-1)\right] \qquad 10.10.11$$

where

$$\hat{\underline{x}}(k \mid k-1) = \Phi(k,k-1)\ \hat{\underline{x}}(k-1 \mid k-1) + \Psi(k, k-1)\ \underline{u}(k-1) \qquad 10.10.12$$

Note that $\hat{\underline{x}}(N \mid N)$ need not be determined since the last control acts at k = N-1. Also $\hat{\underline{x}}(0 \mid 0) = \underline{x}(0)$ implies that $\underline{u}(0) = S(0)\ \hat{\underline{x}}(0 \mid 0)$ is zero if $\underline{x}(0)$ has zero mean.

Fig. 10.7 shows the structure of this controller.

To summarise, the optimal controller for the stochastic linear regulator comprises the optimal linear filter cascaded with the optimal feedback gain matrix of the deterministic linear regulator. The filter gains and the controller gains are computed separately.

This result is the celebrated separation theorem of linear stochastic control and was proved independently by Joseph and Tou[2] and by Gunkel and Franklin[3].

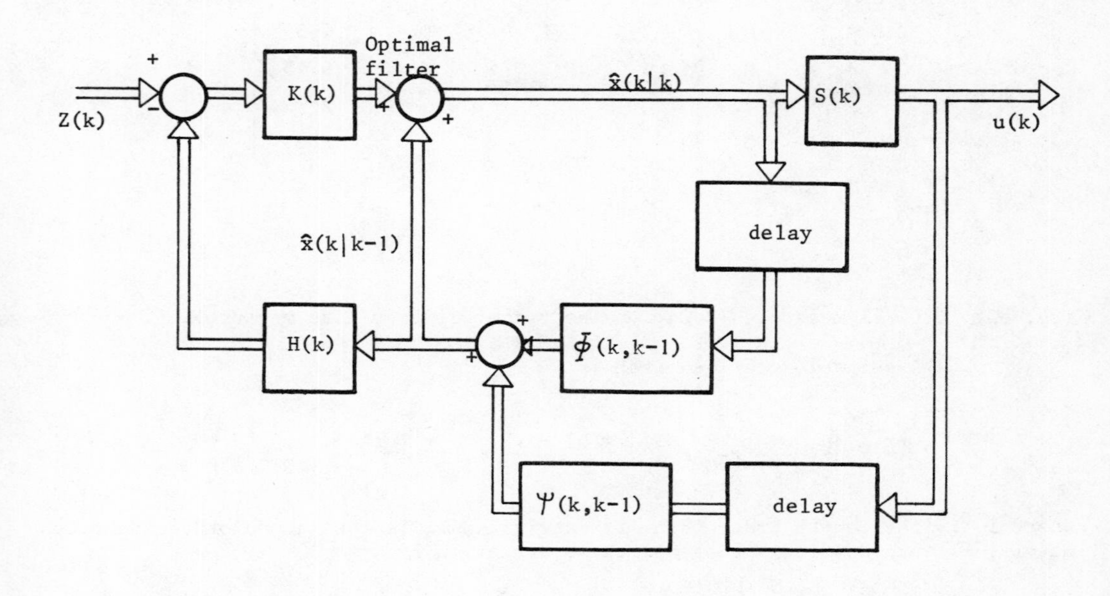

Fig. 10.7

Next, as an example, we consider a two reach river problem.

Example : Stochastic control for a river problem

The dynamic behaviour of a two reach river system is given by

$$\underline{x}(k+1) = A\,\underline{x}(k) + B\,\underline{u}(k) + \underline{C} + \underline{\xi}(k)$$

$$\underline{y}(k+1) = D\,\underline{x}\,(k+1) + \underline{\eta}\,(k+1)$$

where $\underline{\xi}$, $\underline{\eta}$ are uncorrelated zero mean Gaussian random vectors, x_1, x_3 are the concentrations of B.O.D. (mg/l) in reaches 1 and 2 of the stream and x_2, x_4 are the concentrations of D.O. in reaches 1 and 2 of the stream. The control is given by

$$\underline{u}(k) = \begin{bmatrix} \pi_1(k) \\ \pi_2(k) \end{bmatrix} \qquad k = 0, 1, \ldots, k-1$$

where π_1, π_2 are the maximum fractions of B.O.D. removed from the effluent in reaches 1 and 2.

For river Cam near Cambridge, A, B, $\underline{C}$, D are given by

$$A = \left[\begin{array}{cc|cc} 0.18 & 0 & 0 & 0 \\ -0.25 & 0.27 & 0 & 0 \\ \hline 0.55 & 0 & 0.18 & 0 \\ 0 & 0.55 & -0.25 & 0.27 \end{array}\right]$$

$$B = \begin{bmatrix} 2.0 & 0 \\ 0 & 0 \\ 0 & 2.0 \\ 0 & 0 \end{bmatrix} \qquad \underline{C} = \begin{bmatrix} 4.5 \\ 6.15 \\ 2 \\ 2.65 \end{bmatrix}$$

$$D = \begin{bmatrix} 0 & 1 & 0 & 0 \\ 0 & 0 & 0 & 1 \end{bmatrix}$$

i.e. only D.O. is measured (B.O.D. being rather difficult to measure).

A suitable cost function is

$$J = E\left\{ \frac{1}{2} \left\| \underline{x}(K) \right\|^2_{I_4} + \sum_{k=0}^{K-1} \frac{1}{2}\left(\left\| \underline{x}(k) - \underline{x}^d \right\|^2_{I_4} + \left\| \underline{u}(k) \right\|^2_{100I_2} \right) \right\}$$

where I_4 is the fourth order identity matrix and I_2 is the second order identity matrix, $\underline{x}^d$ are desired values and these are given by

$$\underline{x}^d = \begin{bmatrix} 5 \\ 7 \\ 5 \\ 7 \end{bmatrix}$$

This problem can be physically interpreted as : it is desired on average to maintain B.O.D. around 5 mg/l, D.O. around 7 mg/l whilst minimising the treatment at the sewage works.

Simulation results

For k = 159, the system was simulated on the IBM 370/165 digital computer ; the controller gave a constant gain matrix, i.e.

$$\underline{u} = G\,\underline{x} + \underline{d}$$

where

$$G = \begin{bmatrix} 0.0074 & -0.0011 & 0.0006 & -0.0001 \\ -0.0126 & -0.0015 & 0.0042 & -0.0004 \end{bmatrix}$$

$$\underline{d} = \begin{bmatrix} 0.05449 \\ 0.00668 \end{bmatrix}$$

Using this gain in the filter hierarchy, the filter was simulated. Fig. 10.8 to 10.11 show the states and estimates of the four states while Fig. 10.12 and 10.13 show the control. Note that the estimation of x_1, x_3 is poorer than that of x_2, x_4 since there is no measurement of B.O.D.

Next, let us consider the stochastic controller for continuous dynamical systems.

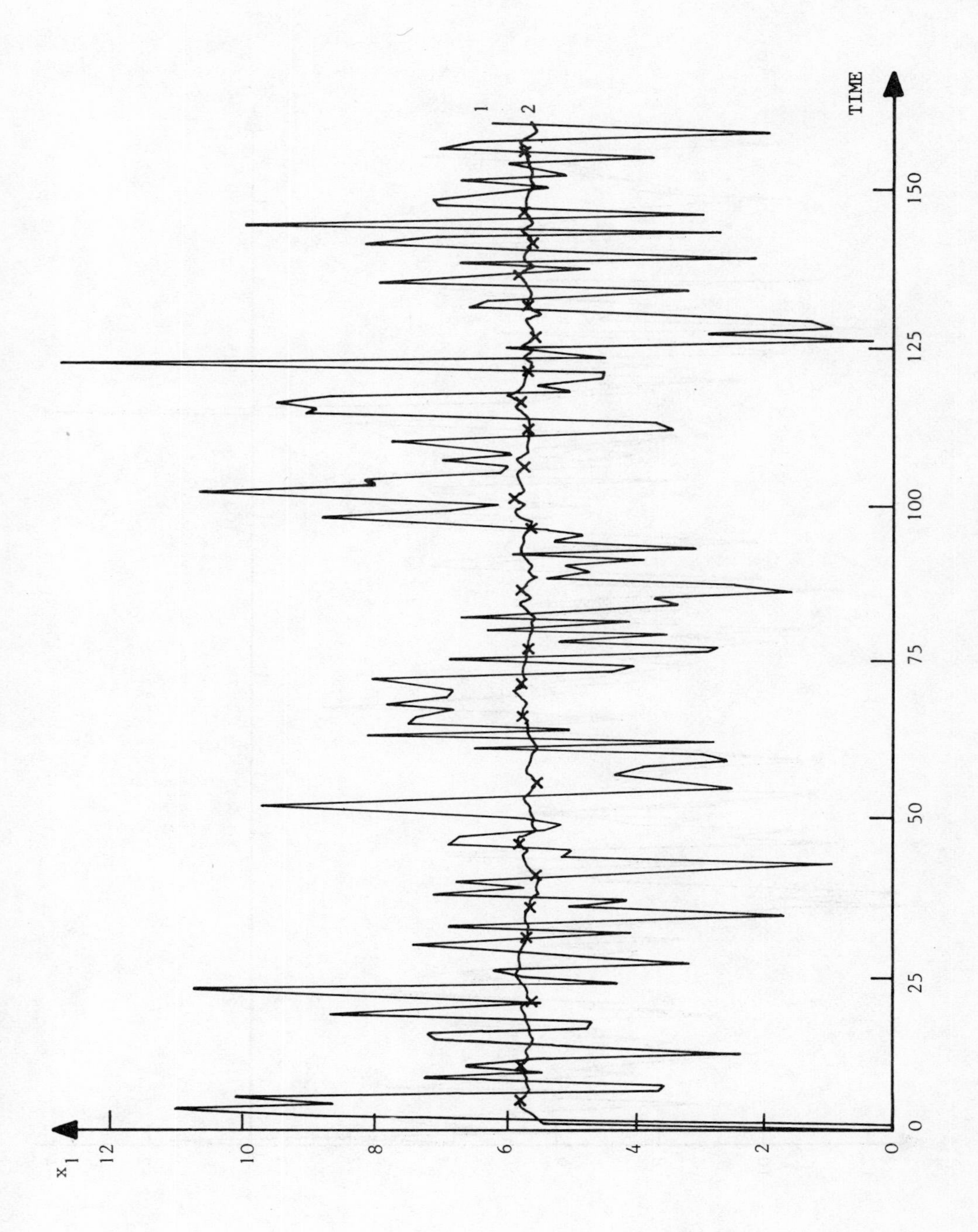

Fig. 10.8

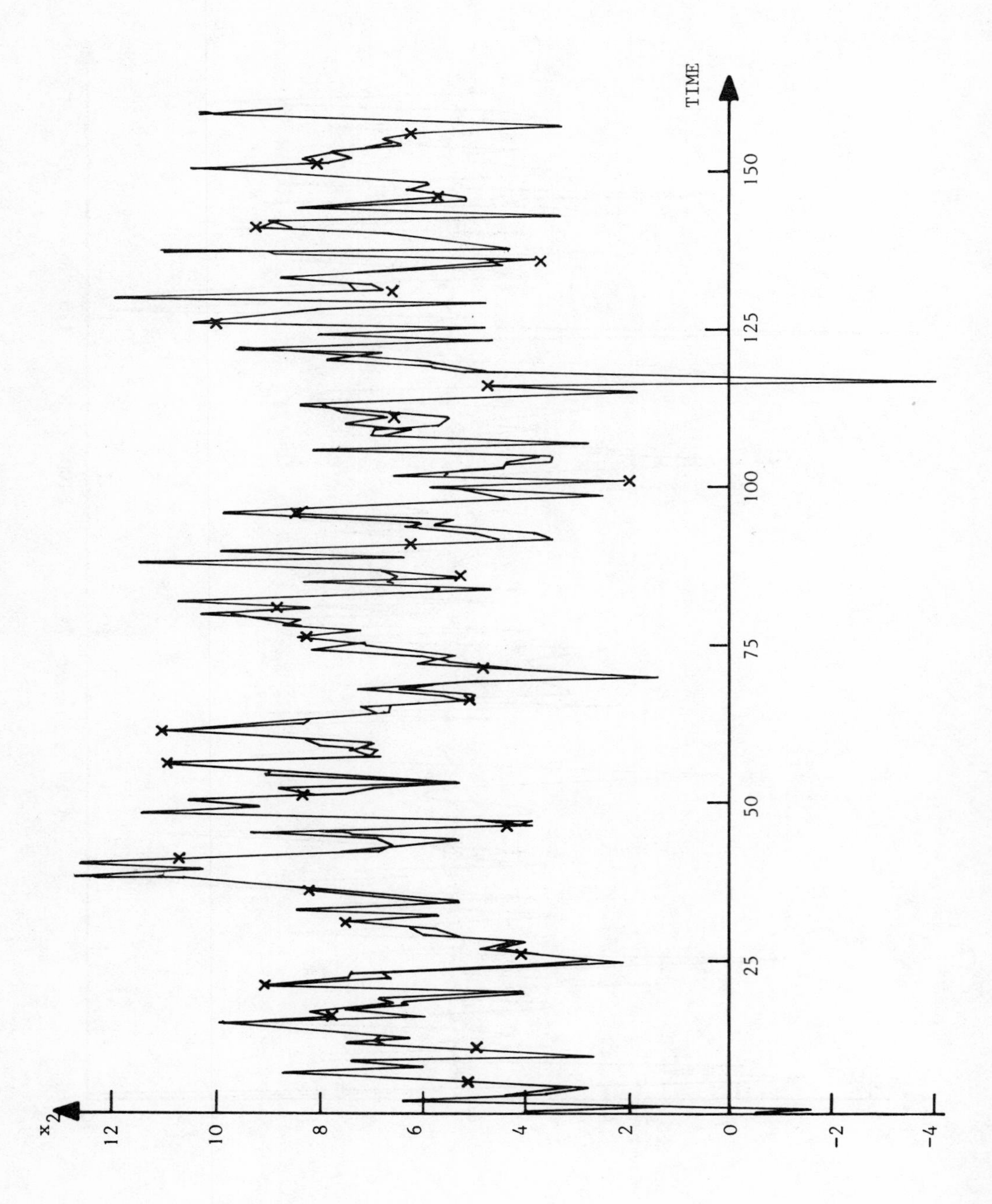

Fig. 10.9

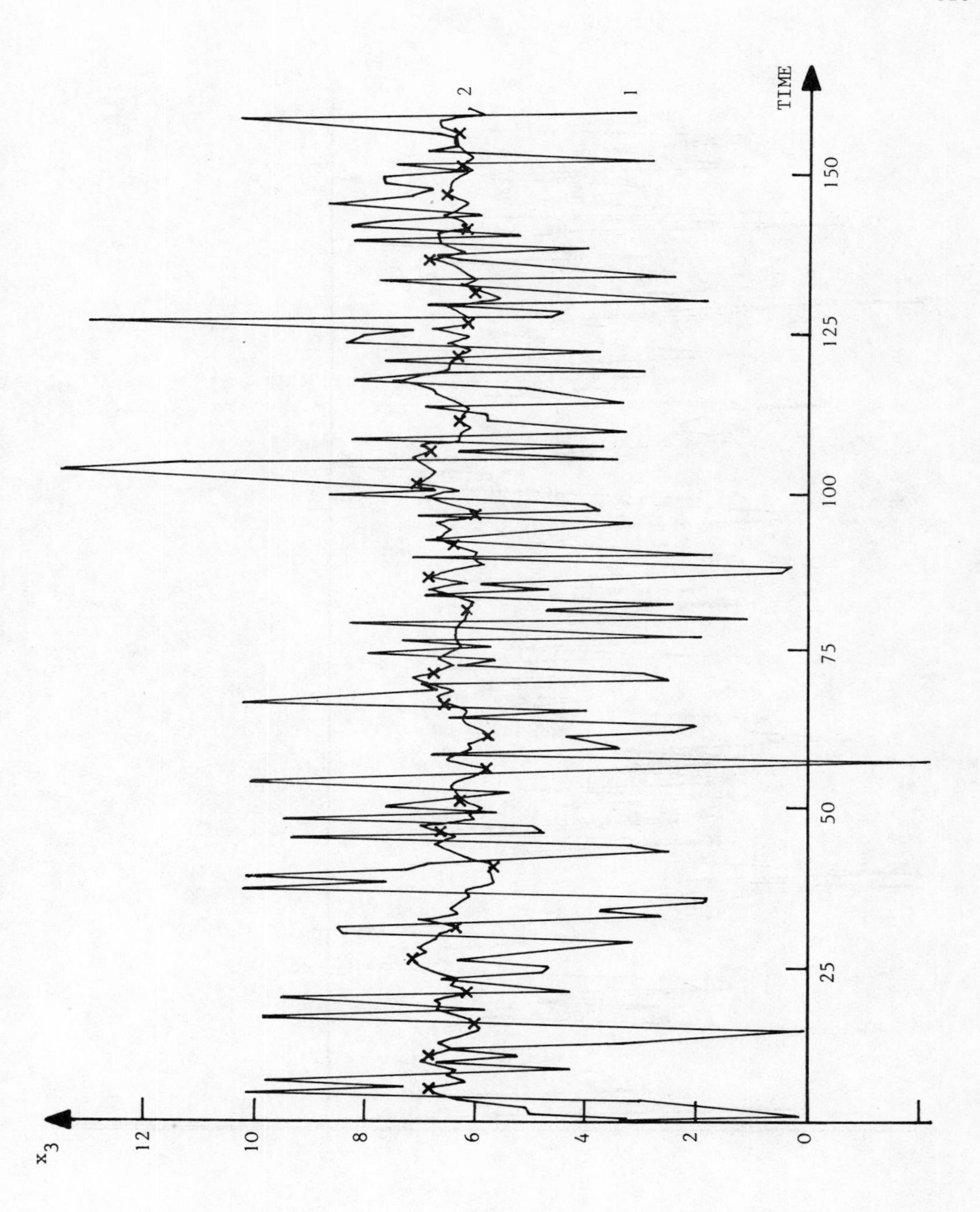

Fig. 10.10

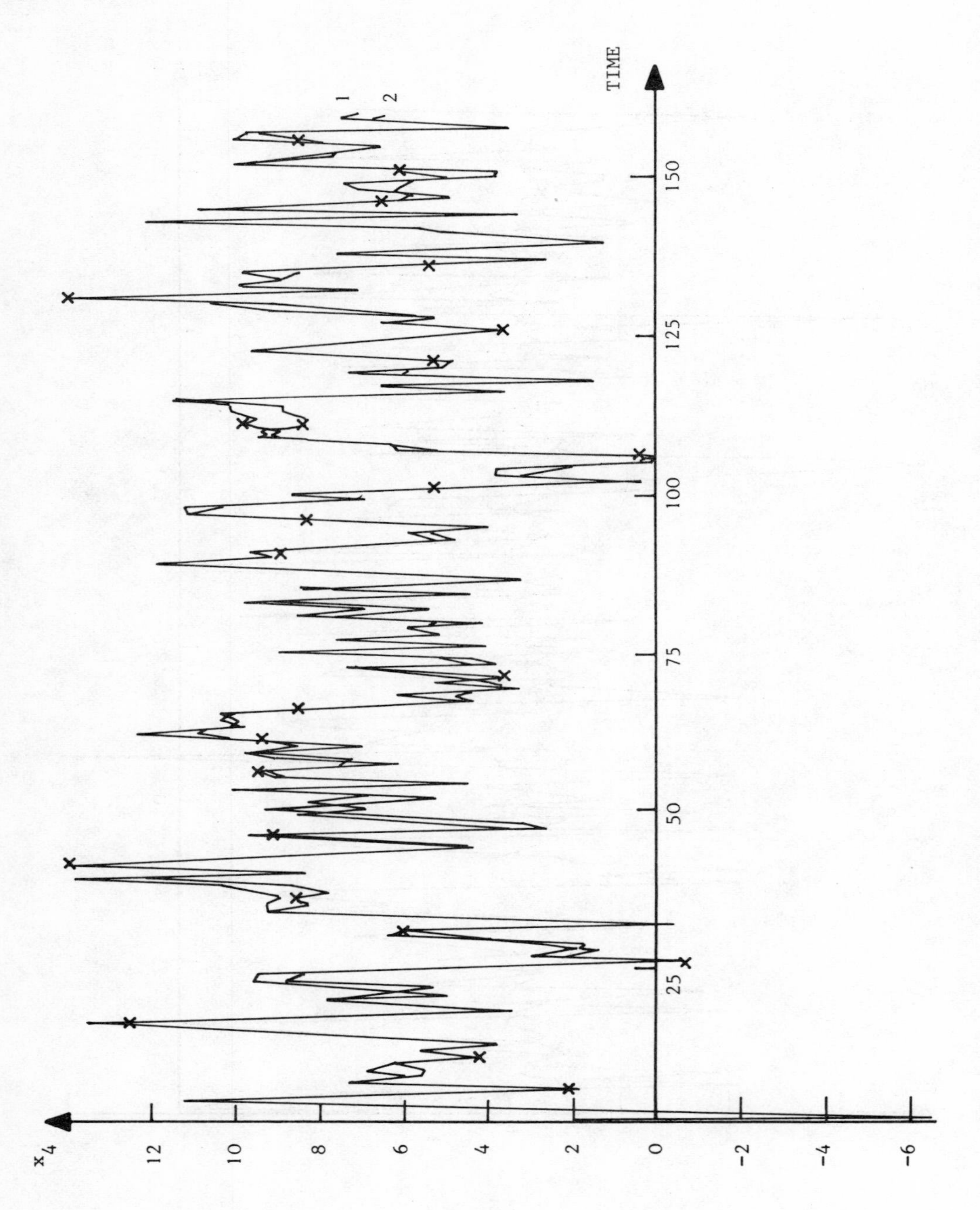

Fig. 10.11

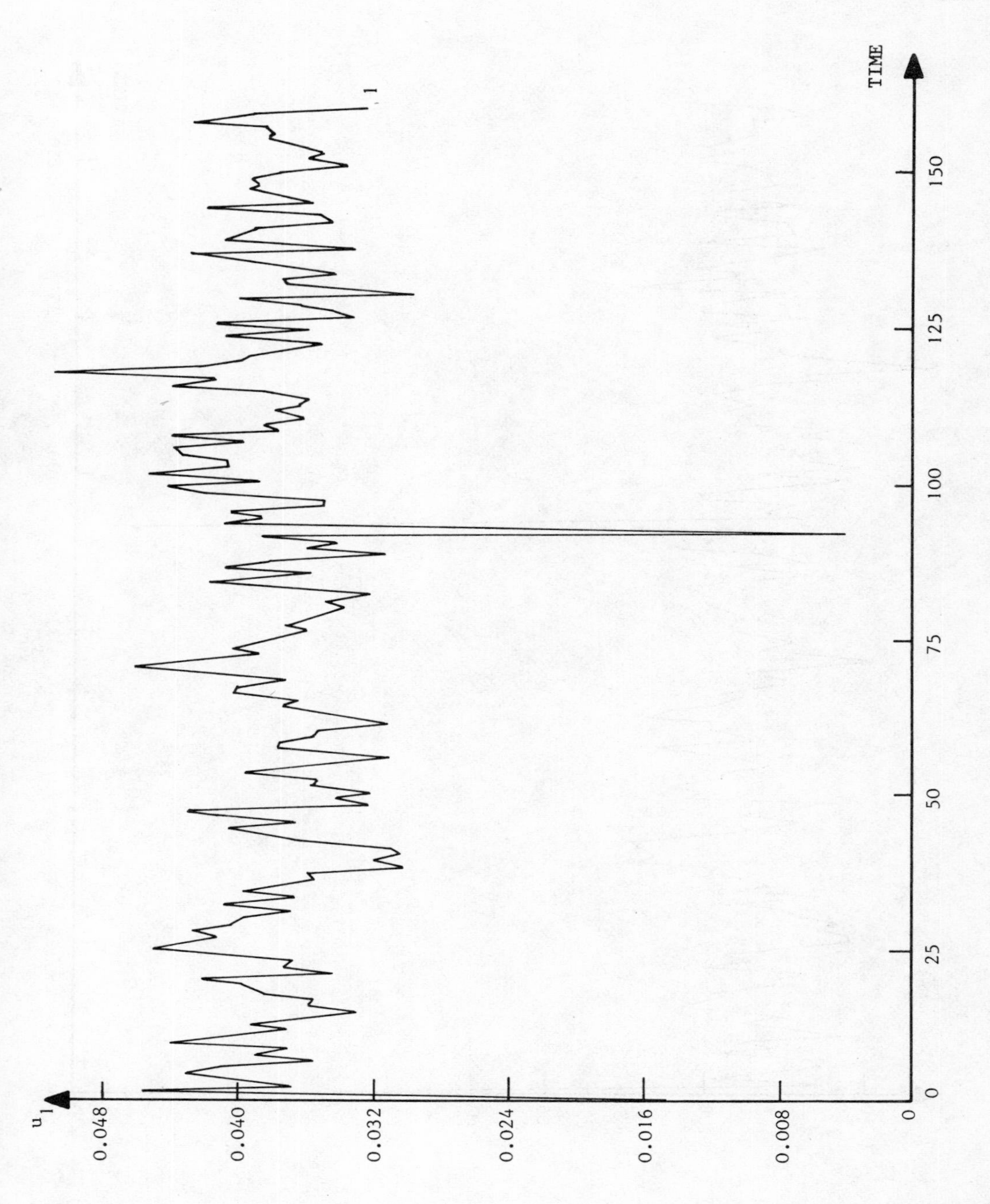

Fig. 10.12

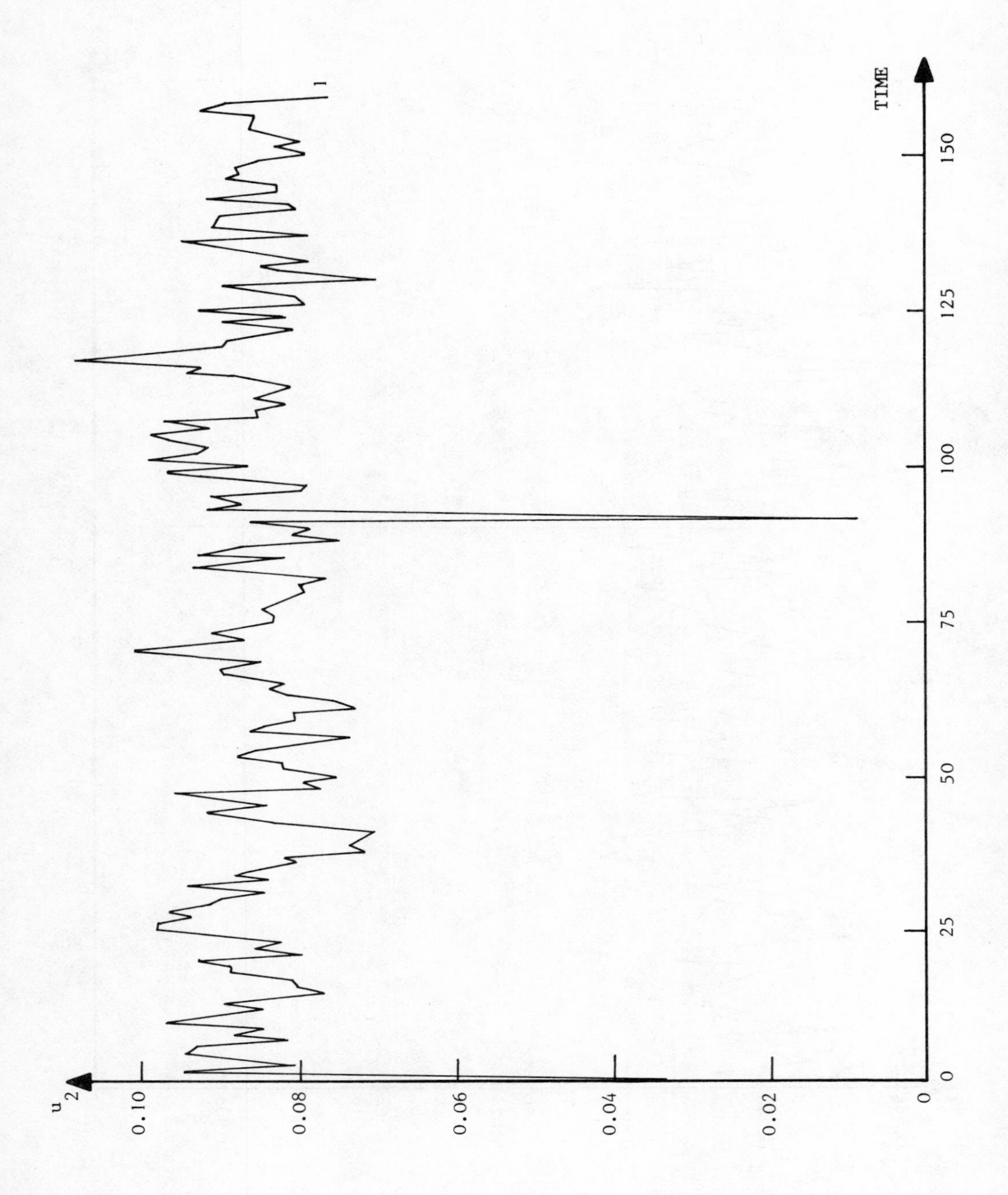

Fig. 10.13

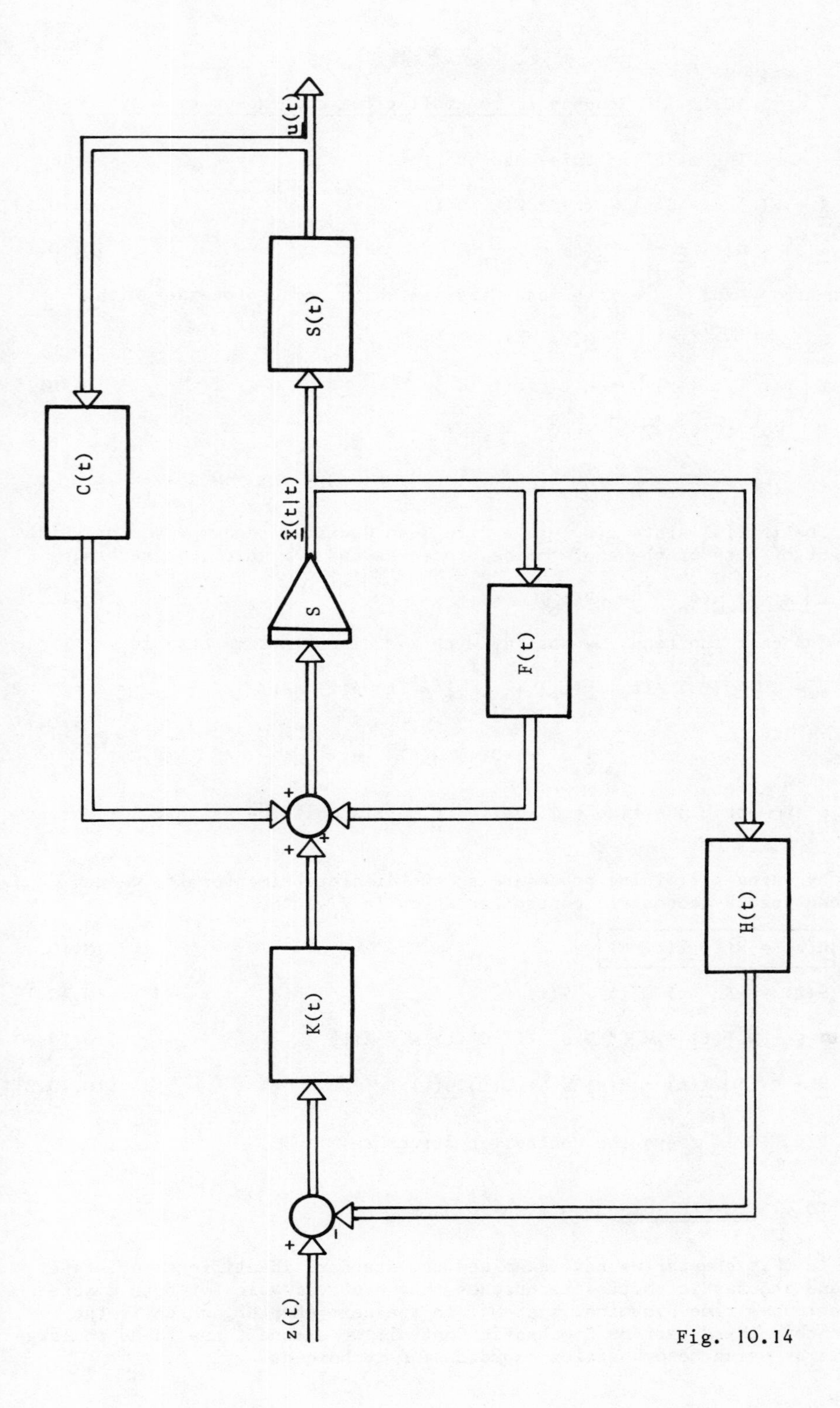

Fig. 10.14

10.10.2 Stochastic controller for continuous time systems

The model in this case is

$$\dot{\underline{x}} = F(t)\ \underline{x} + G(t)\ \underline{\omega}(t) + C(t)\ \underline{u}(t) \qquad 10.10.13$$

$$\underline{z}(t) = H(t)\ \underline{x}(t) + \underline{v}(t) \qquad 10.10.14$$

for $t \geq t_o$ where $\underline{\omega}$ and $\underline{v}$ are zero mean Gaussian white noise processes with

$$E[\underline{\omega}(t)\ \underline{\omega}^T(\tau)] = Q(t)\ \delta(t-\tau)$$

$$E[\underline{v}(t)\ \underline{v}^T(\tau)] = R(t)\ \delta(t-\tau) \qquad 10.10.15$$

and

$$E[\underline{\omega}(t)\ \underline{v}^T(\tau)] = 0$$

$\forall\ t, \tau > t_o$

The initial state $\underline{x}(t_o)$ is a zero mean Gaussian random n vector which is independent of both of the above noise processes and for which, as before,

$$E[\underline{x}(t_o)\ \underline{x}(t_o)^T] = P(t_o) \qquad 10.10.16$$

The cost function, by analogy with the discrete time case is

$$J = E\Big[\underline{x}^T(t_f)\ A(t_f)\ \underline{x}(t_f) + \int_{t_o}^{t_f} [\underline{x}^T(t)\ A(t)\ \underline{x}(t) + \underline{u}^T(t)\ B(t)\ \underline{u}(t)]\ dt\Big] \qquad 10.10.17$$

A is an n x n symmetric positive semi-definite matrix whilst B is an r x r positive definite matrix.

By using a limiting procedure to the discrete time result, we can obtain the continuous stochastic controller which is given by

$$\boxed{\underline{u}(t) = S(t)\ \hat{\underline{x}}(t \mid t)} \qquad 10.10.18$$

$$S(t) = -B^{-1}(t)\ C^T(t)\ W(t) \qquad 10.10.19$$

$$\dot{W} = -F^T(t)\ W - W\ F(t) + W\ C(t)\ B^{-1}(t)\ C^T(t)\ W - A(t) \qquad 10.10.20$$

$$\dot{\hat{\underline{x}}} = F(t)\ \hat{\underline{x}} + K(t)[\underline{z}(t) - H(t)\ \hat{\underline{x}}] + C(t)\ \underline{u}(t) \qquad 10.10.21$$

Fig. 10.14 shows the controller structure.

10.11 CONCLUSIONS

In this chapter we have examined the standard identification, state estimation and stochastic control techniques that are available for both discrete time and continuous time dynamical systems. In the next chapter, which is the final one in this third part on Stochastic Control, we extend these ideas to large scale systems by using decomposition-coordination techniques.

REFERENCES FOR CHAPTER 10

The material described in this chapter is all quite well known. We have extensively used for this chapter the course notes in French of Professor G. ALENGRIN, Maître de Conférences at the University of Nice. We would recommend as further reading

[1] MEDITCH, J., 1969, "Stochastic optimal linear estimation and control". Mc Graw Hill Book Company.

[2] JAZWINSKI, A.H., 1970, "Stochastic processes and filtering theory". Academic Press.

[3] SAGE, A.P., 1968, "Optimum systems control". Prentice Hall.

SPECIFIC REFERENCES

[1] ÅSTROM, K.J. : "Lecture notes on the identification problem, the least squares method". Report n° 6806, Lund Institute of Technology, 1968.

[2] JOSEPH, P. and TOU, J. : "On linear control theory". Trans. AIEE pt II, vol. 80, p. 193, 1961.

[3] GUNCKEL, T., and FRANKLIN, G. : "A general solution for linear sampled data control". J. Basic Eng. Vol. 85, p. 197, 1963.

PROBLEMS FOR CHAPTER 10

1. Consider the system

$$y(k) = -a_1 y(k-1) - a_2 y(k-2) + b\,u(k) + e(k)$$

where $\{e(k)\}$ is a sequence of independent random variables having the same distribution and having a zero mean. Find the values of a_1, a_2, b using the least squares method given the following sequence of values for y(k) and u(k)

$y(0) = 5$ $y(1) = 0$ $y(2) = 5$ $y(3) = 0$ $y(4) = 10$

$y(5) = 10$ $u(5) = 0$ $u(4) = 0$ $u(3) = 6$ $u(2) = 0$

2. Calculate the variance of the noise in excercise 1 and thus the estimation error.

3. How do the parameters in question 1 change if we receive the new measurements

$y(6) = 0$ $u(6) = 0$ $y(7) = 0$ $u(7) = 0$

Calculate the new parameters using the recursive least squares method.

4. Consider the system described by

$$x(k+1) = a\,x(k) + b\,u(k)$$

$$y(k) = x(k) + e(k)$$

Find the values of a, b, using the instrumental variable method given the following values for y(k), u(k).

$y(1) = 2$ $y(2) = 3$ $y(3) = 0$ $y(4) = 0$

$u(3) = 2$ $u(2) = 1$

5. Compute $P_{3|3}$ and $\begin{bmatrix} \hat{x}_1 \\ \hat{x}_2 \end{bmatrix}_{3|3}$ using the Kalman filter in example 2 in section 10.8

if the measured values are :

$t = 0$ $z_0 = 0.1$

$t = 1$ $z_1 = 2.0$

$t = 2$ $z_2 = 3.7$

$t = 3$ $z_3 = 6.0$

6. Consider the system

$$\dot{x} = -2.5\,x + \xi + u$$

$$y = 3x + \eta$$

where ξ and η are zero mean Gaussian white noise sequences with covariances

$$E\left[\xi(t)\ \ \xi(\tau)\right] = 1\ \delta(t-\tau)$$

$$E\left[\eta(t)\ \ \eta(\tau)\right] = 2\ \delta(t-\tau)$$

Also $x(0) = 1$, $P(x_o) = 1$.

Compute the estimate of x given the following sequence of measurements :

$$y(0) = 1,\ y(.1) = .9,\ y(.2) = .8,\ y(.3) = .7,\ y(.4) = .6,\ y(.5) = .5,$$

and compute also the optimal stochastic control u.

CHAPTER 11

Estimation Theory and Stochastic Control for Large Scale Systems

Although the theory of estimation, as outlined in chapter 10, has found major applications in the aero-space field, there have been rather fewer applications for the case of industrial systems, socio-economic problems, etc. The main reason for this is that the relevant models in the case of industrial and other systems are usually of much higher dimension so that practical utilisation of estimation techniques is hindered by computational problems. In this chapter, we consider the problem of state and parameter estimation using decomposition and a multi-level structure. We will begin by considering parameter estimation and then go on to study state estimation and subsequently the problem of stochastic control.

We begin our analysis by considering the maximum a posteriori approach to parameter estimation as developed by Arafeh and Sage [1].

Next, we consider the explicit solution of the large scale state estimation problem. We will begin by describing the approach of Pearson [2] and see that this approach is somewhat impractical. We will then consider the more practical, albeit suboptimal approach of Shah [3]. Next, we will consider the optimal decentralised filter calculation structure of Hassan et al. [4] and the dual approach using the control algorithms. Finally, we will consider stochastic optimal control for large scale linear-quadratic-Gaussian problems.

11.1 THE MAXIMUM A POSTERIORI APPROACH [1]

In this approach, we consider the following n^{th} order discrete time non-linear system

$$\underline{x}(k+1) = \underline{\phi}\left[\underline{x}(k), k\right] + \underline{\omega}(k) \tag{11.1.1}$$

$$\underline{z}(k) = \underline{h}\left[\underline{x}(k), k\right] + \underline{v}(k) \tag{11.1.2}$$

where $\underline{x}$ is the n dimensional vector of the states and parameters to be estimated, $\underline{\phi}\left[\underline{x}(k), k\right]$ is the n dimensional vector valued non-linear function which describes the structure of the system and includes any deterministic inputs, $\underline{\omega}(k)$ is the n dimensional system noise vector and this is assumed to be a Gaussian white noise sequence with

$$E\left[\underline{\omega}(k)\right] = \underline{q}(k) \;;\; E\left[\underline{\omega}(k)\,\underline{\omega}(\ell)^T\right] = Q(k)\,\delta(k-\ell)$$

where δ is the Kronecker Delta, Q is a positive definite block diagonal matrix, $\underline{z}(k)$ is the m dimensional observation or measurement vector. $\underline{h}\left[\underline{x}(k), k\right]$ is the m dimensional vector valued non-linear observation function and $\underline{v}(k)$ is the m dimensional observation noise vector.

The measurement noise $\underline{v}(k)$ is also a Gaussian white noise sequence with

$$E\left[\underline{v}(k)\right] = \underline{r}(k) \;;\; E\left[\underline{v}(k)\,\underline{v}(\ell)^T\right] = R(k)\,\delta(k-\ell)$$

where R(k) is a positive definite block diagonal matrix.

The initial state is also assumed to be a Gaussian random vector with

$$E\left[\underline{x}(k_o)\right] = \underline{x}(k) \text{ and } E\left[\underline{x}(k_o)\,\underline{x}(k_o)^T\right] = P(k_o)$$

where $P(k_o)$ is also assumed to be block diagonal.

Now, for this problem, the maximum a posteriori estimate can be obtained by minimising

$$\underset{\text{w.r.t.}\,\underline{\omega}(k)}{J} = \frac{1}{2}\left\|\underline{x}(k_o) - \overline{\underline{x}}(k_o)\right\|^2_{P^{-1}(k_o)} + \frac{1}{2}\sum_{k=k_o}^{k_f-1}\left\{\left\|\underline{z}(k+1) - \underline{v}(k+1) - \underline{h}[\underline{x}(k+1), k+1]\right\|^2_{R^{-1}(k+1)} + \left\|\underline{\omega}(k) - \underline{q}(k)\right\|^2_{Q^{-1}(k)}\right\} \quad 11.1.3$$

subject to the dynamical constraints given by equation 11.1.1 with k_o, k_f fixed.

To perform this minimisation, let us write the Hamiltonian of the problem as

$$H = \frac{1}{2}\left\|\underline{z}(k+1) - \underline{v}(k+1) - \underline{h}[\underline{x}(k+1), k+1]\right\|^2_{R^{-1}(k+1)} + \frac{1}{2}\left\|\underline{\omega}(k) - \underline{q}(k)\right\|^2_{Q^{-1}(k)} + \underline{\lambda}^T(k+1)\left\{\underline{\phi}[\underline{x}(k), k] + \underline{\omega}(k)\right\} \quad 11.1.4$$

Here, $\underline{\lambda}$ is the costate vector.

Using the discrete minimum principle outlined in chapter 5, we can write the two point boundary value problem as :

$$\underline{x}(k+1) = \underline{\phi}[\underline{x}(k), k] + \underline{q}(k) - Q(k)[\underline{\lambda}(k+1) - \Phi(k)\, H^T(k+1)\, R^{-1}(k+1)\, \underline{\nu}(k+1)] \quad 11.1.5$$

$$\underline{\lambda}(k) = \underline{\phi}^T(k)\, H^T(k+1)\, R^{-1}(k+1)\, \underline{\nu}(k+1) + \Phi^T(k)\, \underline{\lambda}(k+1) \quad 11.1.6$$

where

$$\underline{\nu}(k+1) = \underline{z}(k+1) - \underline{v}(k+1) - \underline{h}[\underline{x}(k+1, k+1)] \quad 11.1.7$$

$$\Phi(k) = \frac{\partial \underline{\phi}^T[\underline{x}(k), k]}{\partial \underline{x}(k)} \quad 11.1.8$$

$$H(k+1) = \frac{\partial \underline{h}^T[\underline{x}(k+1), k+1]}{\partial \underline{x}(k+1)} \quad 11.1.9$$

and from the transversality conditions

$$\underline{\lambda}(k_o) = P^{-1}(k_o)[\underline{x}(k_o) - \overline{\underline{x}}(k_o)] \quad 11.1.10$$

$$\underline{\lambda}(k_f) = \underline{0} \quad 11.1.11$$

By solving this discrete time two point boundary value problem, we can obtain the fixed interval smoothing solution. Is is possible to obtain the filtering solution using invariant imbedding. We will do this to a decomposed version of our original problem. To perform the decomposition, let us introduce interconnection variables as

$$\underline{\pi}_i(k) = \underline{g}_i(\underline{x}_j, k) \qquad j = 1, 2, \ldots N \tag{11.1.12}$$

We can thus decompose the cost function as

$$J^o = \sum_{i=1}^{N}\left\{\frac{1}{2}\left\|\underline{x}_i(k_o) - \bar{\underline{x}}_i(k_o)\right\|^2_{P_i^{-1}(k_o)} + \sum_{k=k_o}^{k_f-1}\frac{1}{2}\left\|\underline{z}_i(k+1) - \underline{v}_i(k+1) - \underline{h}_i(\underline{x}_i, \underline{\pi}_i, k+1)\right\|^2_{R_i^{-1}(k+1)} + \frac{1}{2}\left\|\underline{\omega}_i(k) - \underline{q}_i(k)\right\|^2_{Q_i^{-1}(k)}\right\} \tag{11.1.13}$$

This can be rewritten as

$$J = \sum_{i=1}^{N} J_i$$

where

$$J_i = \frac{1}{2}\left\|\underline{x}_i(k_o) - \bar{\underline{x}}_i(k_o)\right\|^2_{P_i^{-1}(k_o)} + \sum_{k=k_o}^{k_f-1}\left\{\frac{1}{2}\left\|\underline{z}_i(k+1) - \underline{v}_i(k+1) - \underline{h}_i(\underline{x}_i, \underline{\pi}_i, k+1)\right\|^2_{R^{-1}(k+1)} + \frac{1}{2}\left\|\underline{\omega}_i(k) - \underline{q}_i(k)\right\|^2_{Q_i^{-1}(k)}\right\} \tag{11.1.14}$$

The system model can also be decomposed, i.e.

$$\underline{x}_i(k+1) = \underline{\phi}_i\left[\underline{x}_i, \underline{\pi}_i, k\right] + \underline{\omega}_i(k)$$
$$\underline{z}_i(k) = \underline{h}_i\left[\underline{x}_i, \underline{\pi}_i, k\right] + \underline{v}_i(k) \tag{11.1.15}$$

for i = 1, 2, ..., N.

To solve the decomposed problem using say the interaction prediction principle, the ith subsystem's Hamiltonian can be written as

$$H_i = \frac{1}{2}\left\|\underline{z}_i(k+1) - \underline{v}_i(k+1) - \underline{h}_i(\underline{x}_i, \underline{\pi}_i, k+1)\right\|^2_{R_i^{-1}(k+1)} + \frac{1}{2}\left\|\underline{\omega}_i(k) - \underline{q}_i(k)\right\|^2_{Q_i^{-1}(k)} + \underline{\beta}_j(k)^T\left[\underline{\pi}_j(k) - \underline{g}_j(\underline{x}_i, k)\right] + \underline{\lambda}_i^T(k+1)\left[\underline{\phi}_i(\underline{x}_i, \underline{\pi}_i, k) + \underline{\omega}_i(k)\right] \tag{11.1.16}$$

$$j = 1, 2, \ldots N$$

where $\underline{\beta}_j$ is a Lagrange multiplier vector.

From the Hamiltonian, the ith susbystem's two point boundary value problem can be written as

$$\underline{x}_i(k+1) = \underline{\phi}_i\left[\underline{x}_i, \underline{\pi}_i, k\right] + \underline{q}_i(k) - Q_i(k)\Phi_i^{-T}(k)\left[\underline{\lambda}_i(k) - \underline{B}_i(k)\right]$$
$$\underline{\lambda}_i(k+1) = \Phi_i^T\left[\underline{\lambda}_i(k) - \underline{B}_i(k)\right] + H_i^T(k+1)\, R_i^{-1}(k+1)\, \underline{\nu}_i(k+1) \tag{11.1.17}$$

where

$$\Phi_i^T(k) = \frac{\partial \underline{\phi}_i}{\partial \underline{x}_i(k)}\left[\underline{x}_i, \underline{\pi}_i\ k\right]$$

$$H_i(k+1) = \frac{\partial \underline{h}_i(\underline{x}_i, \underline{\pi}_i, k+1)}{\partial \underline{x}_i(k+1)}$$

$$\underline{\nu}_i(k+1) = \underline{z}(k+1) - \underline{v}_i(k+1) - \underline{h}_i(\underline{x}_i, \underline{\pi}_i, k+1)$$

$$\underline{B}_i(k) = \frac{\partial}{\partial \underline{x}_i(k)} \left[\underline{\beta}_j(k)^T \left[\underline{\pi}_j(k) - \underline{g}_j(\underline{x}_i, k) \right] \right]$$

From the transversality conditions, the boundary conditions are given by

$$\underline{\lambda}_i(k_o) = \underline{P}_i^{-1}(k_o) \left[\underline{x}_i(k_o) - \overline{\underline{x}}_i(k_o) \right] \qquad 11.1.18$$

$$\underline{\lambda}_i(k_f) = \underline{0}$$

Arafeh and Sage [1] have developed the following sequential estimation algorithm for the ith subsystem by using the principle of invariant imbedding :

$$\hat{\underline{x}}_i(k+1) = \underline{\phi}_i(\hat{\underline{x}}_i, \underline{\pi}_i, k) + \underline{q}_i(k) + K_i(k+1) \hat{\underline{\nu}}_i(k+1) + N_i(k+1) \underline{B}_i(k)$$

$$K_i(k+1) = P_i(k+1) H_i^T(k+1) R_i^{-1}(k+1) \qquad 11.1.19$$

$$P_i(k+1) = P_i(k+1/k) + D_i(k) \left[I - \frac{\partial M_i(k+1)}{\partial \underline{x}_i(k+1/k)} P_i(k+1/k) + C_i(k+1) \right]^{-1}$$

$$P_i(k+1/k) = \Phi_i(k) P_i(k) \Phi_i^T(k) + Q_i(k)$$

where

$$N_i(k+1) = -\left\{ P_i(k+1) \left[I - \frac{\partial M_i(k+1)}{\partial x_i(k+1/k)} Q_i(k) \right] - Q_i(k) \right\} \Phi_i^{-T}(k)$$

$$C_i(k+1) = \frac{\partial}{\partial \underline{x}_i(k)} \left\{ \Phi_i^{-T}(k) B_i(k) - \frac{\partial M_i(k+1)}{\partial \underline{x}_i(k+1/k)} Q_i(k) \Phi_i^{-T}(k) \underline{B}_i(k) \right\} P_i(k) \Phi^T(k)$$

$$D_i(k) = \frac{\partial}{\partial \underline{x}_i(k)} \left\{ Q_i(k) \Phi_i^{-T}(k) \underline{B}_i(k) \right\} P_i(k) \Phi_i^T(k)$$

$$M_i(k+1) = H_i^T(k+1) R_i^{-1}(k+1) \hat{\underline{\nu}}_i(k+1)$$

Arafeh and Sage [1] make no attempt to optimally coordinate the subsystem filters. Rather they suggest a strategy of "coordination for improvement". The idea is to use sequential coordination for improvement with the passing of time since for many systems initial results are not as important as obtaining acceptable results near the final time.

For this purpose it is possible to use a sequential predictor-corrector scheme for the improvement of the coordination. In this method, first of all the coordination variables $\underline{\pi}_i$ and $\underline{\beta}_i$ can be predicted by linear extrapolation, i.e.

$$\underline{\beta}_i(k+1/k) = \underline{\beta}_i(k) + \underline{\beta}_i^*(k)$$

$$\underline{\pi}_i(k+1/k) = \underline{\pi}_i(k) + \underline{\pi}_i^*(k)$$

where $\underline{\beta}_i^*(k) = \underline{\beta}_i(k) - \underline{\beta}_i(k-1)$

$$\underline{\pi}_i^*(k) = \underline{\pi}_i(k) - \underline{\pi}_i(k-1)$$

The predicted value of $\underline{\beta}_i$ is corrected by a gradient procedure, i.e.

$$\underline{\beta}_i(k+1) = \underline{\beta}_i(k+1/k) + \alpha \left[\underline{\pi}_i(k) - \underline{g}_i \left[\hat{\underline{x}}_j(k), k \right] \right]$$

where α is the step length and $\hat{\underline{x}}_j$ is obtained from solving equation 11.1.19 at the first level. $\underline{\pi}_i$ is corrected by averaging the predicted and estimated value of $\underline{\pi}_i$ according to

$$\underline{\pi}_i(k+1) = \frac{1}{2}\left[\underline{\pi}_i(k+1/k) + \underline{g}_i(\underline{x}_j(k),k)\right]$$

Arafeh and Sage [1] have successfully applied this approach to the estimation problem of an interconnected power system. Next we consider Pearson's approach.

11.2 THE OPTIMAL FILTER OF PEARSON [2]

Pearson was almost certainly the first to suggest a two level structure for state estimation for linear interconnected dynamical systems. Although, as we will see in the sequel, the approach is impractical, the basic idea is certainly an interesting one since it continues on directly from the idea of hierarchical optimisation for linear-quadratic problems as outlined in chapter 6.

The original formulation of Pearson is in continuous time. However, here we outline a discrete time version of this, as extended by Singh [5] since the latter avoids certain difficulties of representation.

Thus, let us consider the system

$$\underline{x}(k+1) = \phi(k+1,k)\ \underline{x}(k) + \Gamma(k+1,k)\ \underline{m}(k) + C(k+1,k)\ \underline{z}(k) \qquad 11.2.1$$

where $\underline{x}$ is the state vector, $\underline{z}$ is the interconnection vector which is linked to the output vector $\underline{y}$ by the matrix L, i.e.

$$\underline{z}(k) = L\ \underline{y}(k) \qquad 11.2.2$$

$\underline{z}(k)$ is assumed to be completely observable through the vector $\underline{w}(k)$ which comprises $\underline{z}(k)$ plus white noise, i.e.

$$\underline{w}(k) = \underline{z}(k) + \underline{\xi}(k) \qquad 11.2.3$$

where $\underline{\xi}(k)$ is a zero mean Gaussian white noise vector of known covariance, i.e.

$$E\left[\underline{\xi}(j)\ \underline{\xi}^T(k)\right] = S\,\delta_{jk} \qquad 11.2.4$$

$\underline{m}(k)$ in equation 11.2.1 is also a zero mean Gaussian white noise vector of known covariance, i.e.

$$E\left[\underline{m}(j)\ \underline{m}^T(k)\right] = Q\,\delta_{jk} \qquad 11.2.5$$

The output equation for the system is of the form

$$\underline{y}(k) = M\ \underline{x}(k) + N\ \underline{m}(k) \qquad 11.2.6$$

We note that the output vector $\underline{y}(k)$ in Pearson's formulation is not used as an observation vector but rather, it serves the purpose of generating the interconnections $\underline{z}(k)$ in the model. For the observation of the state, we have another equation, i.e.

$$\underline{v}(k) = H(k)\ \underline{x}(k) + \underline{\eta}(k) \qquad 11.2.7$$

where H is a block diagonal matrix and $\underline{\eta}(k)$ is a zero mean Gaussian white noise vector of known covariance

$$E\left[\underline{\eta}(j)\ \underline{\eta}^T(k)\right] = R\,\delta_{jk} \qquad 11.2.8$$

In this formulation, we assume that all the system matrices are block diagonal except L. This is a reasonable assumption if our system comprises interconnected dynamical subsystems.

The main idea used by Pearson is to convert the problem of estimation of $\underline{x}$ to one of minimisation of an appropriate functional and this latter minimisation problem could be solved in a hierarchical way using one of the methods outlined in Chapter 6. As in the previous section, the appropriate functional to be minimised is

$$J = \frac{1}{2}\left\|\underline{x}(o) - \bar{\underline{x}}_o\right\|^2_{P_o^{-1}} + \frac{1}{2}\sum_{k=o}^{n}\left\{\left\|\underline{m}(k)\right\|^2_{Q^{-1}} + \left\|\underline{v}(k) - H(k)\,\underline{x}(k)\right\|^2_{R^{-1}} + \left\|\underline{z}(k) - \underline{w}(k)\right\|^2_{S^{-1}}\right\} \qquad 11.2.9$$

subject to the dynamical constraints

$$\underline{x}(k+1) = \phi(k+1,k)\,\underline{x}(k) + \Gamma(k+1,k)\,\underline{m}(k) + C(k+1,k)\,\underline{z}(k) \qquad 11.2.10$$

$$\underline{z}(k) = L\,M\,\underline{x}(k) + LN\,\underline{m}(k) \qquad 11.2.11$$

Now, in this formulation, the interconnections between the subsystems only appear through $\underline{z}$ (the matrices ϕ, Γ, C, Q, R, S, H being block-diagonal). Thus it is possible to decompose this problem by the Goal coordination method of chapter 6 and this involves, in this case, the maximisation w.r.t. $\underline{\pi}$ of the dual function $\phi(\underline{\pi})$ where

$$\phi(\underline{\pi}) = \min_{\underline{x},\underline{m}}\left[L(\underline{x}, \underline{m}, \underline{\pi}) \text{ s.t. } 11.2.10\right]$$

where $L(\underline{x}, \underline{m}, \underline{\pi})$ is given by

$$L(\underline{x}, \underline{m}, \underline{\pi}) = \frac{1}{2}\left\|\underline{x}(o) - \bar{\underline{x}}_o\right\|^2_{P_o^{-1}} + \sum_{k=0}^{n}\left\{\frac{1}{2}\left\|\underline{m}(k)\right\|^2_{Q^{-1}} + \frac{1}{2}\left\|\underline{v}(k) - H(k)\,\underline{x}(k)\right\|^2_{R^{-1}} + \frac{1}{2}\left\|\underline{z}(k) - \underline{w}(k)\right\|^2_{S^{-1}} + \underline{\pi}^T(k)\left[L\,H\underline{x}(k) + LN\,\underline{m}(k) - \underline{z}(k)\right]\right\} \qquad 11.2.12$$

For a given $\underline{\pi}(k)$, we could decompose 11.2.12 since L is separable and we could solve this decomposed problem by independant minimisation for each subproblem at level 1 and improve the $\underline{\pi}(k)$ trajectory at each iteration on the second level using a gradient technique. This two level solution gives us the optimal estimate $\hat{\underline{x}}$ of $\underline{x}$. However, in order to solve this problem in practice, we need to know all the observations $\underline{v}(k)$ and $\underline{w}(k)$, k = 0 to n, and these are not available a priori. Thus although such an approach may be useful in the case of parameter estimation, it is not of much significance as far as state estimation is concerned.

Next we consider a suboptimal approach which is quite attractive for state estimation, i.e. the approach of Shah.

II.3 THE SUBOPTIMAL FILTER OF SHAH [3]

A possible approach to state estimation for linear interconnected dynamical systems is the Supplemented Partitionning Approach (S.P.A.) of Shah. To simplify the analysis in this section, we will consider a system which is decomposed into two subsystems. The dynamical equations for the two subsystems could be written as :

$$\underline{x}_1(k+1) = \phi_{11}(k+1,k)\ \underline{x}_1(k) + \phi_{12}(k+1,k)\ \underline{x}_2(k) + \underline{w}_1(k)$$

$$\underline{y}_1(k+1) = H_{11}(k+1)\ \underline{x}_1(k) + H_{12}(k+1)\ \underline{x}_2(k+1) + \underline{v}_1(k+1)$$

$$\underline{x}_2(k+1) = \phi_{21}(k+1,k)\ \underline{x}_1(k) + \phi_{22}(k+1,k)\ \underline{x}_2(k) + \underline{w}_2(k) \qquad 11.3.1$$

$$\underline{y}_2(k+1) = H_{21}(k+1)\ \underline{x}_1(k+1) + H_{22}\ (k+1)\ \underline{x}_2(k+1) + \underline{v}_2(k+1)$$

Here, $\underline{x}_1$ and $\underline{x}_2$ are respectively of dimension n_1 and n_2, $\phi_{11}, \phi_{12}, \phi_{21}, \phi_{22}$ are transition matrices of dimension $n_1 x n_1$, $n_1 x n_2$, $n_2 x n_1$ and $n_2 x n_2$ respectively. w_1, w_2 are zero mean Gaussian white noise vectors of known covariance, i.e.

$$E\left[\underline{w}_i(j)\ \underline{w}_i^T(k)\right] = Q_{ii}\ \delta_{jk} \qquad i = 1, 2, \ldots \qquad 11.3.2$$

$\underline{y}_1$, $\underline{y}_2$ are respectively m_1 and m_2 dimensional output vectors. $\underline{v}_1$ and $\underline{v}_2$ are zero mean Gaussian white observation noise vectors of known covariance, i.e.

$$E\left[\underline{v}_i(j)\ \underline{v}_i^T(k)\right] = R_{ii}\ \delta_{jk} \qquad i = 1, 2, \ldots \qquad 11.3.3$$

H_{11}, H_{12}, H_{21}, H_{22} are observation matrices of dimensions $m_1 x n_1$, $m_1 x_2$, $m_2 x n_1$ and $m_2 x n_2$ respectively. We assume that $\underline{w}_i(k)$, $\underline{v}_i(k)$ are uncorrelated and that $\underline{w}_i(k)$, $\underline{w}_j(k)$ and $\underline{v}_i(k)$ are independent where $i = 1, 2$, $j = 1, 2$, $i \neq j$, i.e.

$$E\left[\underline{w}_i(j)\ \underline{v}_i^T(k)\right] = 0 \qquad j = 0 \text{ to } N,\ k = 1 \text{ to } N$$

$$E\left[\underline{w}_i(j)\ \underline{w}_\ell^T(k)\right] = 0 \qquad k, j = 0 \text{ to } N \qquad 11.3.4$$

$$E\left[\underline{v}_i(j)\ \underline{v}_\ell^T(k)\right] = 0 \qquad j, k = 0 \text{ to } N$$

The initial states of the system, i.e. $\underline{x}_i(k_o)$, $i = 1, 2$, are zero mean Gaussian random variables of covariance

$$P_{x_i x_i}(k_o) = E\left[\underline{x}_i(k_o)\ \underline{x}_i^T(k_o)\right]$$

where $P_{x_i x_i}$ is a positive semi-definite matrix.

We assume that the model given by equations 11.3.1 has the following properties :

a - the stochastic processes $\left\{\underline{x}(k)\ ;\ k = 0, 1, \ldots\right\}$ and $\left\{\underline{y}(i)\ ;\ i = 1, 2, \ldots\right\}$ are zero mean Gaussian.

b - $E\left[\underline{x}(j)\ \underline{w}^T\ (k)\right] = 0$ for $k \geqslant j$; $j = 0, 1, \ldots$

c - $E\left[\underline{y}(j)\ \underline{w}^T\ (k)\right] = 0$ for $k \geqslant j$; $j = 1, 2, \ldots$

d - $E\left[\underline{x}(j)\ \underline{v}^T\ (k)\right] = 0$ for $j = 0, 1, \ldots$ and $k = 1, 2, \ldots$

e - $E\left[\underline{y}(j)\ \underline{v}^T\ (k)\right] = 0$ for $k > j$; $k, j = 1, 2, \ldots$

Let us next define the estimation error to be of the form

$$\tilde{\underline{x}}_i\ (k|k) = \underline{x}_i(k) - \hat{\underline{x}}_i(k|k) \qquad 11.3.5$$

where $\hat{\underline{x}}_i(k|k)$ is the estimate of $\underline{x}_i(k)$ given $\underline{y}(1)$, $\underline{y}(2)$, ... $\underline{y}(k)$.

Next let us modify the equations of the first subsystem as :

$$\underline{x}_1(k+1) = \phi_{11}(k+1,k)\ \underline{x}_1(k) + \phi_{12}(k+1,k)\ \hat{\underline{x}}_2(k|k) + \underline{w}_1^*(k)$$

$$z_{11}(k+1) = H_{11}(k+1)\ \underline{x}_1(k+1) + \underline{v}_1^*(k+1) \qquad 11.3.6$$

where

$$\underline{w}_1^*(k) = \underline{w}_1(k) + \phi_{12}(k+1,k)\ \tilde{\underline{x}}_2(k|k)$$

$$\underline{v}_1^*(k+1) = \underline{v}_1(k) + H_{12}(k+1)\ \tilde{\underline{x}}_2(k+1|k) \qquad 11.3.7$$

and $\underline{z}_{11}$ (k+1) will be considered as the observation process for the first subsystem at k+1.

If $\hat{\underline{x}}_2(k|k)$ and $\hat{\underline{x}}_2(k+1|k)$ are known and $\underline{w}_1^*(k)$, $\underline{v}_1^*(k+1)$ are treated as white noise sequences, then we could use the standard Kalman filter to solve this new problem. The covariances of these white noise terms $\underline{w}_1^*(k)$, $\underline{v}_1^*(k+1)$ are taken by Shah [3] to be :

$$Q_{11}^*(k) = E\left[\underline{w}_1^*(k)\ \underline{w}_1^{*T}(k)\right]$$

$$= Q_{11}(k) + \phi_{12}(k+1,k)\ P_{22}(k|k)\ \phi_{12}^T(k+1,k) \qquad 11.3.8$$

and

$$R_{11}^*(k+1) = E\left[\underline{v}_1^*(k+1)\ \underline{v}_1^{*T}(k+1)\right]$$

$$= R_{11}(k+1) + H_{12}(k+1)\ P_{22}(k+1|k) H_{12}^T(k+1) \qquad 11.3.9$$

Using these covariances, our filtered estimates are given by

$$\hat{\underline{x}}_1(k+1|k) = \phi_{11}(k+1,k)\ \hat{\underline{x}}_1(k|k) + \phi_{12}(k+1,k)\ \hat{\underline{x}}_2(k|k) \qquad 11.3.10$$

$$\hat{\underline{x}}_1(k+1|k+1)_1 = \hat{x}_1(k+1|k) + K_{11}(k+1)_1\left[\underline{z}_{11}(k+1) - H_{11}(k+1)\ \hat{\underline{x}}_1(k+1|k)\right] \qquad 11.3.11$$

In 11.3.11, K_{11} is the filter gain matrix and this can be obtained using the equation :

$$K_{11}(k+1)_1 = P_{11}(k+1|k)\ H_{11}^T(k+1)\left[H_{11}(k+1)\ P_{11}(k+1|k)\ H_{11}^T(k+1) + R_{11}^*(k+1)\right]^{-1} \qquad 11.3.12$$

where

$$P_{11}(k+1|k) = \phi_{11}(k+1,k)\ P_{11}^*(k|k)\ \phi_{11}^T(k+1,k) + Q_{11}^*(k) \qquad 11.2.13$$

where

$$P_{11}^*(k+1|k+1)_1 = \left[I - K_{11}(k+1)\ H_{11}(k+1)\right]\ P_{11}(k+1|k)$$

$$\left[I - K_{11}(k+1)_1\ H_{11}(k+1)\right]^T + K_{11}(k+1)_1\ R_{11}^*(k+1)\ K_{11}^T(k+1)_1 \qquad 11.3.14$$

Up to now, we have used the set of measurements $\underline{y}_1(k+1)$, i.e. the observations $z_{11}(k+1)$ to estimate the error covariance matrix $P_{11}^*(k+1|k+1)$ and the a posteriori state estimate $\hat{\underline{x}}_1\ (k+1|k+1)_1$. The same quantities need to be calculated also for the second subsystem, i.e. $P_{22}^*(k+1|k+1)$ and $\hat{\underline{x}}_2(k+1|k+1)_1$ using the measurements $\underline{y}_2(k+1)$ or $\underline{z}_{22}(k+1)$. Let us assume that this has been done. The next step is to treat the set $\underline{y}_2(k+1)$ for the first subsystem and $\underline{y}_1(k+1)$ for the second.

Let us form the new observation process

$$\underline{z}^{*}_{12}(k+1) = \underline{y}_2(k+1) - H_{22}(k+1)\ \hat{\underline{x}}_2(k+1|k+1)$$
$$= H_{21}(k+1)\ \underline{x}_1(k+1) + \underline{v}_2(k+1) + H_{22}(k+1)\ \tilde{\underline{x}}_2(k+1|k+1)_1 \qquad 11.3.15$$

We have thus

$$\hat{\underline{x}}_1(k+1\ |\ k+1)_2 = \hat{\underline{x}}_1(k+1\ |\ k+1)_1 + K_{11}(k+1)_2\ [\underline{z}^{*}_{12}(k+1)$$
$$- H_{21}(k+1)\ \hat{\underline{x}}_1(k+1\ |\ k+1)_1] \qquad 11.3.16$$

where

$$K_{11}(k+1)_2 = P^{*}_{11}(k+1\ |\ k+1)_1\ H^{T}_{21}(k+1)\ [\ H_{21}(k+1)$$
$$P^{*}_{11}(k+1\ |\ k+1)_1 H^{T}_{21}(k+1) + R_{22}(k+1) + H_{22}(k+1)\ P^{*}_{22}\ (k+1|\ k+1)_1\ H^{T}_{22}(k+1)]^{-1} \qquad 11.3.17$$

and

$$P^{*}_{11}(k+1\ |\ k+1) = [\ I - K_{11}(k+1)_2\ H_{21}(k+1)\ ]\ P^{*}_{11}(k+1\ |\ k+1)_1$$
$$[\ I - K_{11}(k+1)_2\ H_{21}(k+1\ ]^{T} + K_{11}(k+1)_2\ [\ R_{22}(k+1) + \qquad 11.3.18$$
$$H_{22}(k+1)\ P^{*}_{22}(k+1\ |\ k+1)_1\ H^{T}_{22}(k+1)\ ]\quad K^{T}_{11}(k+1)_2$$

We could thus predict the a priori error covariance matrix over the period $k+1 \longrightarrow k+2$ by using $P^{*}_{11}(k+1|\ k+1)$ as

$$P^{*}_{11}(k+1\ |\ k+1) = \emptyset_{11}(k+2,k+1)\ P^{*}_{11}(k+1|k+1)_1\emptyset^{T}_{11}(k+2,k+1)$$
$$+ Q_{11}(k+1) + \emptyset_{21}(k+2,k+1)\ P^{*}_{22}(k+1|\ k+1)_1\emptyset^{T}_{12}(k+2,k+1 \qquad 11.3.19$$

We note that this filter is suboptimal since we did not take into account the off-diagonal terms in the covariance matrix. However, it is computationally very attractive to implement since it requires a substantially smaller number of elementary multiplication operations compared to the global Kalman filter and it also provides savings in computer storage.

Next we consider a continuous time version of this approach.

11.4 THE CONTINUOUS TIME S.P.A. FILTER [6]

For the continuous time case for two subsystems, the dynamical equations could be written as :

$$\dot{\underline{x}}_1(t) = A_{11}(t)\ \underline{x}_1(t) + A_{12}(t)\ \underline{x}_2(t) + \underline{\eta}_1(t)$$
$$\underline{y}_1(t) = C_{11}(t)\ \underline{x}_1(t) + \underline{\xi}_1(t)$$
$$\dot{\underline{x}}_2(t) = A_{21}(t)\ \underline{x}_1(t) + A_{22}(t)\ \underline{x}_2(t) + \underline{\eta}_2(t) \qquad 11.4.1$$
$$\underline{y}_2(t) = C_{22}(t)\ \underline{x}_2(t) + \underline{\xi}_2(t)$$

where the symbols mean: $\underline{x}_1$ is the l_1 state vector of subsystem I ; $\underline{x}_2$ is the l_2 state vector of subsystem II ; A_{11}, A_{12}, A_{21}, A_{22} are, respectively, $l_1 x l_1$, $l_1 x l_2$, $l_2 x l_1$, $l_2 x l_2$ matrices. $\underline{\eta}_1$, $\underline{\eta}_2$ are white noise vectors of dimensions l_1

l_2, respectively. $\underline{y}_1$, $\underline{y}_2$ are the output vectors of the two subsystems and are $p_1 x 1$, and $p_2 x 1$, respectively. $\underline{\xi}_1$ and $\underline{\xi}_2$ are $p_1 x 1$ and $p_2 x_1$ vectors of measurement noise for the two subsystems. C_{11}, C_{12} are measurement matrices of the two subsystems and are of dimension $p_1 x l_1$ and $p_2 x l_2$, respectively.

The stochastic processes $\underline{\eta}_i(t)$, $t \geq t_o$, $i = 1, 2$ and $\underline{\xi}_i(t)$, $t \geq t_o$, $i = 1, 2$, are zero mean Gaussian white noise of known covariance, i.e.

$$E\left[\underline{\eta}_i(t)\, \underline{\eta}_i^T(\tau)\right] = Q_i(t)\,\delta(t-\tau) \;;\quad i = 1, 2$$

and 11.4.2

$$E\left[\underline{\xi}_i(t)\, \underline{\xi}_i^T(\tau)\right] = R_i(t)\,\delta(t-\tau) \;;\quad i = 1, 2$$

for all $t, \tau \geq t_o$. The $l_i x l_i$ $(i = 1, 2)$ matrices Q_i are positive semidefinite for $t, \tau > t_o$ while the $(p_i x p_i)$ $(i = 1, 2)$ matrices R_i are continuous and positive definite for $t \geq t_o$. It is assumed that the stochastic processes $\underline{\eta}_i$, $\underline{\xi}_i$ are independent of each other, i.e.

$$E\left[\underline{\eta}_i(t)\, \underline{\xi}_i(\tau)^T\right] = 0 \qquad \text{for all } \tau,\ t \geq t_o.$$

The initial states $\underline{x}_i(t_o)$, $i = 1, 2$, are assumed to be zero mean Gaussian random (l_i) vectors. They are independent of $\underline{\eta}_i(t)$, $t \geq t_o$, and of $\eta_i(t)_i$, $t \geq t_o$ and their $(l_i x l_i)$ covariance matrices: $E\left[\underline{x}_i(t_o)\underline{x}_i^T(t_o)\right] = P(t_o)$ are positive semidefinite. It is also assumed that the overall system noise covariance matrix is block diagonal, i.e.

$$E\left[\underline{\eta}_i(t)\, \underline{\eta}_j^T(\tau)\right] = 0 \qquad \begin{matrix} i = 1, 2 \;;\; j = 1, 2 \\ i \neq j \end{matrix}$$

This is a realistic assumption for practical systems. The estimation error is defined as :

$$\underline{\tilde{x}}_i(t_1/t) = \underline{x}_i(t_1) - \underline{\hat{x}}_i(t_1/t) \qquad i = 1, 2 \qquad 11.4.3$$

where $\underline{\hat{x}}_i(t_1/t)$ is the estimate at t_1 given all previous observations. Consider subsystem II. Since $\hat{x}_1(t_1) = \hat{x}_1(t_1|t) + \tilde{x}_1(t_1|t)$ by equation 11.4.3 above, the subsystem equations could be written as :

$$\underline{\dot{x}}_2(t) = A_{21}(t)\, \underline{\hat{x}}_1(t|t) + A_{21}(t)\, \underline{\tilde{x}}_1(t|t) + A_{22}(t)\, \underline{x}_2(t) + \underline{\eta}_2(t)$$

$$\underline{y}_2(t) = C_{22}(t)\, \underline{x}_2(t) + \underline{\xi}_2(t)$$

or

$$\underline{\dot{x}}_2(t) = A_{21}(t)\, \underline{\hat{x}}_1(t|t) + A_{22}(t)\, \underline{x}_2(t) + \underline{\eta}_2^*(t) \qquad 11.4.4$$

where

$$\underline{\eta}_2^*(t) = A_{21}(t)\, \underline{\tilde{x}}_1(t|t) + \underline{\eta}_2(t) \qquad 11.4.5$$

Similarly subsystem I equations could be written as :

$$\underline{\dot{x}}_1(t) = A_{11}(t)\, \underline{x}_1(t) + A_{12}(t)\, \underline{\hat{x}}_2(t|t) + A_{12}(t)\, \underline{\tilde{x}}_2(t|t) + \underline{\eta}_1(t)$$

or

$$\underline{\dot{x}}_1(t) = A_{11}(t)\, \underline{x}_1(t) + A_{12}(t)\, \underline{\hat{x}}_2(t|t) + \underline{\eta}_1^*(t) \qquad 11.4.6$$

where

$$\underline{\eta}_1^*(t) = A_{12}(t)\, \underline{\tilde{x}}_2(t|t) + \underline{\eta}_1(t) \qquad 11.4.7$$

Now consider the simple filter structure for this system consisting of two subsystems :

$$\dot{\hat{\underline{x}}}_1(t|t) = A_{11}(t)\ \hat{\underline{x}}_1(t|t) + K_1\left[\underline{y}_1(t) - C_{11}(t)\ \hat{\underline{x}}_1(t|t)\right] + A_{12}(t)\ \hat{\underline{x}}_2(t|t) \quad 11.4.8$$

$$\dot{\hat{\underline{x}}}_2(t|t) = A_{22}(t)\ \hat{\underline{x}}_2(t|t) + K_2\left[\underline{y}_2(t) - C_{22}(t)\ \hat{\underline{x}}_2(t|t)\right] + A_{21}\ \hat{\underline{x}}_1(t|t) \quad 11.4.9$$

where

$$K_1 = P_{11}C_{11}^T R_1^{-1} \quad 11.4.10$$

$$K_2 = P_{22}C_{22}^T R_2^{-1} \quad 11.4.11$$

and P_{11}, P_{22} are the solutions of the subsystem matrix Riccati equations :

$$\dot{P}_{11} = A_{11}P_{11} + P_{11}A_{11}^T - P_{11}C_{11}^T R_1^{-1} C_{11}P_{11} + Q_1\ ;\ P_{11}(0) = P_{11_o} \quad 11.4.12$$

$$\dot{P}_{22} = A_{22}P_{22} + P_{22}A_{22}^T - P_{22}C_{22}^T R_2^{-1} C_{22}P_{22} + Q_2\ ;\ P_{22}(0) = P_{22_o} \quad 11.4.13$$

Figure 11.1 shows the structure of the filter.

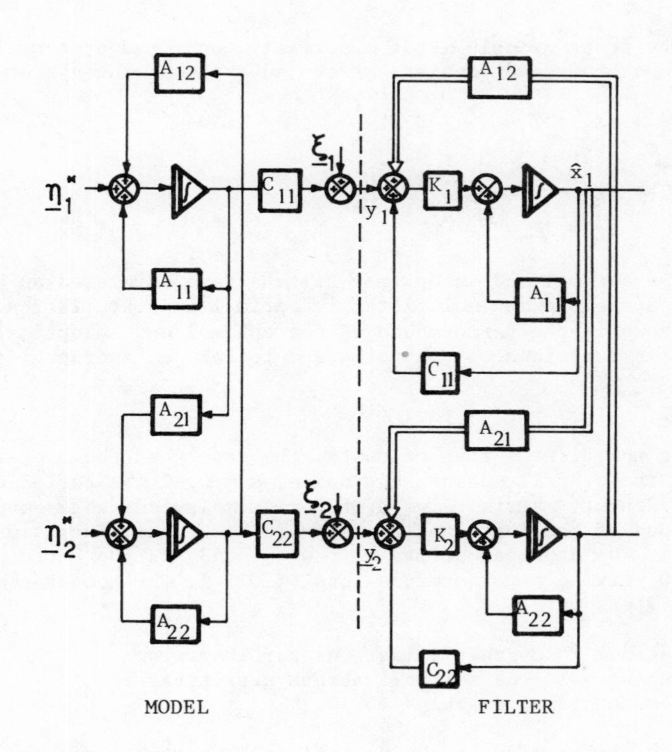

Fig. 11.1

The principal attraction of the above S.P.A. filter is its computational simplicity. Shah [3] derived expressions for the computational requirements of the optimal and the suboptimal filter. However, his formulae were for a discrete time system whereas here a continuous time system is being considered. Still it is easy to show that the computational requirements, in terms of elementary multiplication operations, for the two continuous time filters, are for each iteration :

Optimal

$$20l^3 + 4p^3 + 5l^2p + 5lp^2 + 4lm^2 + 8l^2 + 12lp + 4lm + m^2$$

Suboptimal

$$\sum_{i=1}^{2} (20l_i^3 + 4p_i^3 + 5_ip_i^2 + 5l_i^2p_i + 5l_im_i^2 + 5m_i^2 + 8l_i^2 + 4l_im_i + 12l_im_i^2) + 10l_i^2l_2 + 10l_il_2^2$$

since $l \gg l_i$, $m \gg m_i$, $p \gg p_i$, the computational savings can be substantial.

To compare the performance of the optimal and suboptimal filters, the square of the estimation error was taken as the criterion function, i.e. for a particular system, at each iteration, the optimal and suboptimal estimates were compared with the states obtained from a simulation of the overall system.

Example 1

The first example is of a serially connected system. The overall system is of second order and consists of two subsystems. The system equations are :

$$\dot{x}_1 = -x_1 + \eta_1$$
$$y_1 = x_1 + \xi_1$$
$$\dot{x}_2 = -x_1 - x_2 + \eta_2$$
$$y_2 = x_2 + \xi_2$$

The optimal and suboptimal filters were simulated on an ICL 4130 digital computer. Since the optimal filter is optimal in a statistical sense, it is necessary to compare the performance of the optimal and suboptimal filters for a number of sample functions of the noise and to take an average.

Results

In the first series of tests, the case when the overall system noise was Q = I and the overall measurement noise, R = 0.5I was tested where I is the second-order identity matrix. Five different simulations were done for different sample functions of the noise and the average was taken. Each simulation was for 100 iterations. The average optimal cost was 14.43 and the corresponding suboptimal cost was 14.50, giving a performance loss of 0.5 %. The computational requirements for the two filters were :

Optimal = 404 multiplications per iteration
Suboptimal = 88 multiplications per iteration
Computational saving = 460 %

The performance of the optimal and suboptimal filters for any given sample function of the noise was very similar.

Example 2

This is a continuous time version of one of Shah's examples. The system equations were :

$$\dot{x}_1 = -0.8\ x_1 + 0.6\ x_2 + \eta_1\ ; \qquad y_1 = x_1 + \zeta_1$$

$$\dot{x}_2 = 0.6\ x_1 - 0.8\ x_2 + \eta_2\ ; \qquad y_2 = x_2 + \zeta_2$$

For this example, the computational requirements for the two filters are the same as for example 1, i.e.

Optimal = 404 multiplications per iteration
Suboptimal = 88 multiplications per iteration
Saving = 460 %

The filters were simulated for a variety of system and measurement noise levels. In each case, five different simulations were performed for different sample functions of the noise. The initial conditions on the states and estimates were :

$$x_1(0) = x_2(0) = 1\ ; \qquad \hat{x}_1(0) = \hat{x}_2(0) = 0$$

$$P_{11}(0) = P_{22}(0) = 1$$

Table 11.1 gives the optimal and suboptimal performances.

Q	R	Optimal average cost for 100 iterations	Suboptimal average cost for 100 iterations	Performance loss
I	0.1I	4.47	4.77	6.9 %
I	0.2I	7.59	8.27	9 %
I	0.5I	12.55	13.90	10 %
I	I	15.48	18.51	20 %
I	2I	25.36	31.88	26 %

Table 11.1

So far we have considered suboptimal approaches to state and parameter estimation. Next, let us consider the decentralised optimal filter calculation structure of Hassan et al. [4].

11.5 THE DECENTRALISED CALCULATION STRUCTURE FOR THE OPTIMAL KALMAN FILTER

In this section, we will develop an optimal filtering algorithm which uses a two level hierarchical structure. This filter is applicable for systems comprising N subsystems. We will see that the calculation time required for this filter is smaller than that for the global Kalman filter. In addition, since the calculations are done on low order subsystem blocks, these calculations will be more accurate and this will lead to a more numerically stable filter.

Let us begin by considering the basis of the decentralised filter calculation structure.

The most appealing property of the global Kalman filter from a practical point of view is its recursive nature. Essentially, this recursive property of the filter arises from the fact that if an estimate exists based on measurements up to that instant, then when receiving another set of measurements, one could subtract out from these measurements that part which could be anticipated from the results of the first measurements, i.e. the updating is based on that part of the new data which is orthogonal to the old data. In the new filter for systems comprising lower order interconnected subsystems this orthogonalisation is performed subsystem by subsystem, i.e. the optimal estimate of the state of subsystem one is obtained by successively orthogonalising the error based on a new measurement for subsystems 1, 2, 3, ... N w.r.t. the Hilbert space formed by all measurements of all the subsystems up to that instant. Much computational saving results using this successive orthogonalisation procedure since at each stage only low order subspaces are manipulated.

The actual orthogonalisation procedure that is performed in the Kalman filter is based on the following theorem (cf. Luenberger [7]).

Theorem 1

Let β be a member of space H of random variables which is a closed subspace of L_2 and let $\hat{\beta}$ denote its orthogonal projection on a closed subspace Y_1 of H (thus $\hat{\beta}_1$ is the best estimate of β in Y_1). Let $\underline{y}_2$ be an m vector of random variables generating a subspace Y_2 of H and let $\hat{\underline{y}}_2$ devote the m-dimensional vector of the projections of the components of $\underline{y}_2$ on to Y_1 (thus $\hat{\underline{y}}_2$ is the vector of best estimates of $\underline{y}_2$ in Y_1). Let $\tilde{\underline{y}}_2 = \underline{y}_2 - \hat{\underline{y}}_2$.

Then the projection of β on to the subspace $Y_1 \oplus Y_2$, denoted $\hat{\beta}$ is

$$\hat{\beta} = \hat{\beta}_1 + E\,(\beta\,\tilde{\underline{y}}_2^{\,T})\left[E\,(\tilde{\underline{y}}_2\,\tilde{\underline{y}}_2^{\,T})\right]^{-1}\tilde{y}_2$$

where E is the expected value.

For Proof c.f. Luenberger [7]. The above equation can be interpreted as :

$\hat{\beta}$ is $\hat{\beta}_1$ plus the best estimate of
β in the subspace $\tilde{Y}_2$ generated by $\tilde{\underline{y}}_2$

Consider next the system comprising N interconnected linear dynamical subsystems defined by

$$\underline{x}_i(k+1) = \phi_{ii}\,\underline{x}_i\,(k) + \sum_{j=1;i\neq j}^{N} \phi_{ij}\,\underline{x}_j\,(k) + \underline{w}_i\,(k) \qquad 11.5.1$$

$$i = 1, 2, \ldots N$$

with the outputs given by

$$\underline{y}_i(k+1) = H_i\,\underline{x}_i\,(k+1) + \underline{v}_i\,(k+1) \qquad 11.5.2$$

$$i = 1, \ldots N$$

where $\underline{v}_i$, $\underline{w}_i$ are uncorrelated zero mean gaussian white noise sequences with covariances Q_i, R_i respectively. Consider the Hilbert space Y formed by the measurements of the overall systems. At the instant k+1, this space is denoted by Y(k+1). The optimal minimum variance estimate $\hat{\underline{x}}(k+1|k+1)$ is given by

$$\hat{\underline{x}}(k+1 \mid k+1) = E\left\{\underline{x}(k+1) \mid Y(k+1)\right\} = E\left\{\underline{x}(k+1) \mid Y(k)\right\} + E\left\{\underline{x}(k+1) \mid \tilde{\underline{y}}(k+1 \mid k)\right\} \qquad 11.5.3$$

This equation states algebraically the geometrical result of theorem 1. The idea of the new filter is to decompose the second term, i.e. $E\{\underline{x}(k+1|\tilde{\underline{y}}(k+1|k)\}$ such that the optimal estimate $\hat{\underline{x}}(k+1 \mid k+1)$ is given using the two terms by considering the estimate as the orthogonal projection of $\underline{x}_i(k+1)$ taken on the Hilbert space generated by

$$Y(k) \oplus \tilde{y}_1(k+1 \mid k) \oplus \tilde{y}_2^1(k+1 \mid k+1) \oplus \tilde{y}_3^2(k+1 \mid k+1) \oplus \ldots \oplus \tilde{y}_N^{N-1}(k+1 \mid k+1)$$

where $\tilde{y}_i^{i-1}(k+1 \mid k+1)$ is the subspace generated by the subspace of measurements $y_i(k+1)$ and the projection of it on the subspaces generated by $Y(k) + Y_1(k+1) + Y_2(k+1) + \ldots + Y_{i-1}(k+1)$ which leads to theorem 2.

Theorem 2

The optimal estimate $\hat{\underline{x}}_i(k+1|k+1)$ of the ith subsystem is given by the projection of $\underline{x}_i(k+1)$ on the space generated by all measurements upto k (Y(k)) and the projection of $\underline{x}_i(k+1)$ on the subspace generated by $\tilde{Y}_1(k+1 \mid k) \oplus \tilde{y}_2^1(k+1|k+1) \oplus \ldots \oplus \tilde{y}_N^{N-1}(k+1 \mid k+1)$

Proof : rewrite equation 11.5.3 as

$$\hat{\underline{x}}_i(k+1|k+1) = E\left\{\underline{x}_i(k+1 \mid Y(k), \underline{y}_1(k+1), \underline{y}_2(k+1), \ldots \underline{y}_i(k+1), \underline{y}_{i+1}(k+1), \ldots \underline{y}_N(k+1)\right\} = E\left\{\underline{x}_i(k+1 \mid Y(k), \underline{y}_1(k+1), \underline{y}_2(k+1) + \ldots + \underline{y}_i(k+1), \underline{y}_{i+1}(k+1), \ldots \underline{y}_{N-1}(k+1)\right\} + E\left\{\underline{x}_i(k+1) \mid \tilde{\underline{y}}_N^{N-1}(k+1 \mid k+1)\right\}$$

where $\tilde{\underline{y}}_N^{N-1}(k+1 \mid k+1) = \underline{y}_N(k+1) - E\left\{\underline{y}_N(k+1)|Y(k), \underline{y}_1(k+1), \ldots \underline{y}_{N-1}(k+1)\right\}$

or $$\hat{\underline{x}}_i(k+1 \mid k+1) = E(\underline{x}_i(k+1 \mid Y(k) + E\left\{x_i(k+1) \mid \tilde{\underline{y}}_1(k+1 \mid k)\right\} + \sum_{r=2}^{N} E\left\{\underline{x}_i(k+1) \mid \tilde{\underline{y}}_r^{r-1}(k+1 \mid k+1)\right\}$$

which proves the assertion.

Using the idea of successive orthogonalisation of the spaces defined above, the algebraic structure of the new decentralised filter is next described.

11.6 THE ALGEBRAIC STRUCTURE OF THE NEW FILTER

In order to develop the filter equations for the overall system comprising N interconnected subsystems, write the equations for the overall system as :

$$\underline{x}(k+1) = \emptyset(k+1, k)\, \underline{x}(k) + \underline{w}(k)$$
$$\underline{y}(k+1) = H(k+1)\, \underline{x}(k+1) + \underline{v}(k+1)$$

where the subsystem structure is seen more clearly by decomposing those equations as

$$\underline{x}_i(k+1) = \sum_{j=1}^{N} \phi_{ij}(k+1, k)\, \underline{x}_j(k) + \underline{w}_i(k)$$

$$\underline{y}_i(k+1) = H_i(k+1)\, \underline{x}_i(k+1) + \underline{v}_i(k+1)$$

Then the optimal state prediction for the ith subsystem is given by

$$\hat{\underline{x}}_i(k+1 \mid k) = \sum_{j=1}^{N} \phi_{ij}(k+1, k)\, \hat{\underline{x}}_j(k \mid k) \qquad 11.6.1$$

Now, by definition of the prediction errors, $\tilde{\underline{x}}_i(k+1 \mid k) = \underline{x}_i(k+1) - \hat{\underline{x}}_i(k+1 \mid k)$

A recursive expression for the covariance of the prediction error can be written as :

$$P_{ii}(k+1 \mid k) = \sum_{j=1}^{N}\sum_{r=1}^{N} \phi_{ij}(k+1, k)\, P_{jr}\, \phi^T_{ir}(k+1,k) + Q_i(k)$$

$$= \sum_{j=1}^{N} \phi_{ij}(k+1,k) \left\{ \sum_{r=1}^{N} P_{jr}(k \mid k)\phi^T_{ir}(k+1,k) \right\} + Q_i(k) \qquad 11.6.2$$

Also,

$$P_{ij}(k+1 \mid k) = \sum_{r=1}^{N}\sum_{\ell=1}^{N} \phi_{ir}(k+1,k)\, P_{r\ell}(k \mid k)\, \phi^T_{j\ell}(k+1,k)$$

$$= \sum_{r=1}^{N} \phi_{ir}(k+1,k) \left\{ \sum_{\ell=1}^{N} P_{r\ell}(k|k)\, \phi^T_{j\ell}(k+1,k) \right\} \qquad 11.6.3$$

Now, using the proof of theorem 2, Hassan [8] has shown that

$$\hat{\underline{x}}_i(k+1|k+1) = \hat{\underline{x}}_i(k+1|k+1)_\ell + \sum_{r=\ell+1}^{N} P_{x_i \tilde{y}_r^{r-1}}(k+1|k+1)\, P^{-1}_{\tilde{y}^{r-1}\tilde{y}^{r-1}}(k+1|k+1)\, \tilde{\underline{y}}_r^{r-1}(k+1|k+1) \qquad 11.6.4$$

where $\hat{\underline{x}}_i(k+1|k+1)_\ell = \hat{\underline{x}}_i(k+1|k+1)_{\ell-1} + K^{\ell-1}_{i\ell}(k+1)\, \tilde{\underline{y}}^{\ell-1}_\ell(k+1|k+1)$ 11.6.5

and

$$P_{ii}(k+1|k+1)_\ell = P_{ii}(k+1|k+1)_{\ell-1} - K^{\ell-1}_{i\ell}(k+1)\, P_{\tilde{y}^{\ell-1}_\ell \tilde{x}_{i_{\ell-1}}}(k+1|k+1) \qquad 11.6.6$$

where $K^{\ell-1}_{i_\ell}(k+1) = P^{\ell-1}_{\tilde{x}_i \tilde{y}^{\ell-1}_\ell}(k+1|k+1)\, P^{-1}_{\tilde{y}^{\ell-1}_\ell \tilde{y}^{\ell-1}_\ell}(k+1|k+1)$ 11.6.7

$$\tilde{\underline{y}}^{\ell-1}_\ell(k+1|k+1) = \tilde{\underline{y}}^{\ell-2}_\ell(k+1|k+1) - K^{\ell-2}_{\ell_{\ell-1}}(k+1)\, \tilde{\underline{y}}^{\ell-2}_{\ell-1}(k+1|k+1) \qquad 11.6.8$$

$$K^{\ell-2}_{\ell_{\ell-1}} = P_{\tilde{y}^{\ell-2}_\ell \tilde{y}^{\ell-2}_\ell}(k+1|k+1)\, P^{-1}_{\tilde{y}^{\ell-2}_{\ell-1} \tilde{y}^{\ell-2}_{\ell-1}}(k+1|k+1) \qquad 11.6.9$$

$$P_{\tilde{y}_\ell^{\ell-1}\tilde{y}_\ell^{\ell-1}}(k+1|k+1) = P_{\tilde{y}_\ell^{\ell-2}\tilde{y}_\ell^{\ell-2}}(k+1|k+1) - K_{\ell_{\ell-1}}^{\ell-2}(k+1)\, P_{\tilde{y}_{\ell-1}^{\ell-2}\tilde{y}_{\ell-1}^{\ell-2}}(k+1|k+1) \qquad 11.6.10$$

$$P_{\tilde{y}_\ell^{\ell-1}\tilde{y}_\ell^{\ell-1}}(k+1|k+1) = H_\ell P_{\tilde{x}_\ell^{\ell-1}\tilde{x}_\ell^{\ell-1}}(k+1|k+1)\, H_\ell^T + R_\ell(k+1)$$

$$\qquad 11.6.11$$

$$P_{\tilde{x}_i^{\ell-1}\tilde{y}_\ell^{\ell-1}}(k+1|k+1) = P_{\tilde{x}_i^{\ell-1}\tilde{x}_\ell^{\ell-1}}(k+1|k+1)\, H_\ell^T\,(k+1)$$

$$P_{ij}(k+1|k+1)_\ell = P_{ij}(k+1|k+1)_{\ell-1} - K_{i\ell}^{\ell-1}(k+1)\, P_{\tilde{y}_\ell^{\ell-1}\tilde{x}_j^{\ell-1}}(k+1|k+1) \qquad 11.6.12$$

Thus equations 11.6.1 to 11.6.3 and 11.6.5 to 11.6.12 give the algebraic equations of the filter.

The mechanization of the algorithm for one step of the filter is

1. From equations 11.6.1 to 11.6.3 we calculate the prediction estimate as well as its covariance matrix.

2. Put $l = 1$ (note that $\underline{\hat{x}}_i(k+1|k+1)_o = \underline{\hat{x}}_i(k+1|k)$ and $P_{ij}(k+1|k+1)_o = P_{ij}(k+1|k)$, $i = 1, \ldots N$; $j = 1, \ldots N$.
From equations 11.6.5 to 11.6.15 calculate the filtered estimate $\underline{\hat{x}}_i(k+1|k+1)_\ell$ and the corresponding covariance matrix.

3. If $l = N$ the resulting estimate is the optimal Kalman estimate and the covariance matrix is the minimum error covariance matrix, otherwise go to step 2.

Note that although this algorithm and the global Kalman filter are algebraically equivalent the numerical properties of the decomposed filter are significantly better.

We next consider the computational requirements of this new filter as compared to those of the global Kalman filter.

The computational requirements of the new filter as compared to those of the global Kalman filter can be divided into two categories, i.e. storage requirements and computational time requirements. The storage requirements of the new filter are roughly similar to those of the global Kalman filter although if the processing is done on a multi-processor then the new filter's storage can be conveniently distributed between the computers. The computational time advantage of the new filter requires further elaboration.

Now, a good measure of the computation time requirements of the global Kalman filter and the new decentralised filter is given by the number of elementary multiplication operations involved. Consider first the number of elementary multiplications required for the global Kalman filter.

Number of multiplications required for the global Kalman filter

Assume that $\underline{x}$ is of dimension n, $\underline{y}$ is of dimension m, then the number of multiplications required under the assumption that H is block diagonal and each subsystem has the same number of states and outputs is :

$$1.5\, n^2 + 1.5\, n^3 + nm\left(\frac{1}{N} + \frac{2m+1}{2N} + m + 1 + \frac{n+1}{2}\right) + \frac{m^2(3m+1)}{2}$$

where N is the number of subsystem.

Number of multiplications required for the new filter

Assume that all subsystem have equal number of state variables $\frac{n}{N}$ and equal number of measurements $\frac{m}{N}$ where N is the number of subsystems.

Then number of multiplication required is :

$$1.5\, n^2 + 1.5\, n^3 + N\left\{ \frac{mn}{2} + \frac{mn\; 2m+n)}{2N^3} + \frac{m^2(\frac{3m}{N} + 1)}{2N^2} + N\left[\frac{n^2 m}{N^3} + \frac{nm^2}{N^3} + \frac{nm}{N^2} + \frac{nm(n+N)}{2N^3}\right] + \frac{N(N-1)}{2} \cdot \frac{n^2 m}{N^3}\right\}$$

It is easy to show that for high order systems, the new filter will give substantial savings in computation time.

Example 1 : State estimation for a system comprising 11 coupled synchronous machines

The multimachine power system under consideration consists of 11 coupled machines. The model of an n-machine system consists of a set of non-linear equations which can be written for the ith machine as (cf. for example Darwish and Fantin [9] for further details on modelling of such systems) :

$$M_i \ddot{\delta}_i = P_i - \sum_{\substack{j=1 \\ j\neq i}}^{n} b_{ij} \sin \delta_{ij} \qquad i = 1, \ldots, n$$

where M is the inertia, P is the power injected, b_{ij} is the interconnection variable and δ is the angle.

For the system to be completely controllable and completely observable, the nth machine is taken as reference. Then, by subtracting the nth equation from the equation of each machine, a model can be constructed for the n machine system.

For small perturbations, the non-linear model can be linearised about the equilibrium point so that the linear equations which result can be written in discrete form

$$\underline{x}(k+1) = A\, \underline{x}(k) + \underline{\xi}$$

For the 11 machine system, A is the 20 x 20 matrix given in table 11.2 and $\underline{\xi}$ is a zero mean Gaussian white noise vector. For the ith machine, the observation equation is given by

$$y_i = \begin{bmatrix} 0 & 1 \end{bmatrix} \begin{bmatrix} x_{1i} \\ x_{2i} \end{bmatrix} + \begin{bmatrix} u_i \end{bmatrix}$$

where y_i is the speed of the ith machine and u_i is also a zero mean Gaussian white noise vector sequence.

The covariances of $\underline{\xi}$ and $\underline{u}$ are given by Q_i and R_i respectively and these were taken to be $R_i = 1$ and Q_i diagonal with the diagonal elements given by 5, P_o the initial covariance was also taken to be diagonal with the diagonal given by 25. The initial estimate was taken to be zero whilst the initial states were all

	1	2	3	4	5	6	7	8	9	10	11	12	13	14	15	16	17	18	19	20
1	0	1	0	0	0	0	0	0	0	0	0	0	0	0	0	0	0	0	0	0
2	-93.931	0	3.74383	0	5.37437	0	2.62699	0	4.05603	0	-1.64764	0	-2.7157	0	-6.33619	0	-14.37576	0	-12.59014	0
3	0	0	0	1	0	0	0	0	0	0	0	0	0	0	0	0	0	0	0	0
4	15.9499	0	-140.367	0	9.60086	0	14.026	0	6.684	0	2.67515	0	1.51862	0	-1.99922	0	-7.02907	0	-9.53736	0
5	0	0	0	0	0	1	0	0	0	0	0	0	0	0	0	0	0	0	0	0
6	10.6933	0	10.258	0	-116.061	0	6.21809	0	4.3133	0	0.41594	0	-1.06214	0	-4.6919	0	-11.74903	0	-11.52972	0
7	0	0	0	0	0	0	0	1	0	0	0	0	0	0	0	0	0	0	0	0
8	8.51556	0	18.07436	0	7.65584	0	-13.972	0	4.075	0	1.7033	0	1.16265	0	-3.055	0	-7.66032	0	-10.25109	0
9	0	0	0	0	0	0	0	0	0	1	0	0	0	0	0	0	0	0	0	0
10	11.2953	0	8.60465	0	5.98992	0	4.3135	0	-124.466	0	2.5982	0	-0.0644	0	-1.3065	0	-9.4617	0	-9.0164	0
11	0	0	0	0	0	0	0	0	0	0	0	1	0	0	0	0	0	0	0	0
12	1.96937	0	1.96213	0	1.68471	0	1.2125	0	1.6449	0	-99.694	0	1.9443	0	-0.933	0	-9.1423	0	-9.939	0
13	0	0	0	0	0	0	0	0	0	0	0	0	0	1	0	0	0	0	0	0
14	7.0993	0	10.2119	0	5.69536	0	7.5948	0	5.5155	0	11.7763	0	-166.755	0	5.63471	0	7.2399	0	-4.4116	0
15	0	0	0	0	0	0	0	0	0	0	0	0	0	0	0	1	0	0	0	0
16	0.19444	0	-0.31102	0	0.05862	0	-0.397	0	0.4133	0	1.255	0	-0.689	0	-89.399	0	-6.811	0	-8.578	0
17	0	0	0	0	0	0	0	0	0	0	0	0	0	0	0	0	0	1	0	0
18	5.084	0	7.478	0	3.9759	0	5.6047	0	4.0148	0	6.0777	0	7.286	0	8.009	0	-141.137	0	-0.465	0
19	0	0	0	0	0	0	0	0	0	0	0	0	0	0	0	0	0	0	0	1
20	1.582	0	1.400	0	0.9652	0	0.8061	0	1.8019	0	1.577	0	0.5298	0	1.739	0	-1.011	0	-110.087	0

TABLE 11.2

taken to be 10.0.

Simulation results

The global and the new decentralised filter were simulated on an IBM 370/165 digital computer over a time horizon of 80 discrete points.

Figures 11.1 to 11.3 show the first 3 states and the corresponding estimates using the global Kalman filter and figures 11.4 to 11.6 the corresponding states and estimates using the new decentralised filter. Note that the global Kalman filter is numerically unstable whilst the hierarchical solution is stable. Essentially, numerical errors build up to make the global 20th order filter unstable whilst in the case of the hierarchical solution, since only 2nd order subsystem are used at each stage, these numerical inaccuracies are avoided so that the resulting filter remains stable.

Example 2

As a second example, we consider the problem of state estimation for the 22nd order 3 reach river pollution control model.

Let us assume that the state of the system is corrupted by a zero mean Gaussian white noise vector $\underline{w}(k)$ and that the outputs are corrupted by the zero mean Gaussian white noise vector $\underline{v}(k)$. The initial state vector $\underline{x}(k_o)$ is also assumed to be a zero mean Gaussian white noise vector. Since in river pollution control, the only variable that can be easily measured is the D.O. The output vector is the vector of measurements of D.O. in the first, second and third reaches.

This 22nd order system could thus be described by the state equations

$$\underline{x}(k+1) = A\,\underline{x}(k) + \underline{D} + \underline{w}(k)$$

where A is given in table 11.3 and

$$\underline{D}^T = \begin{bmatrix} 0.13375 & 0 & 0.2725 & 0.10475 & 0 & 0 & 0 & 0 \\ 0.0475 & 0 & 0 & 0 & 0.039775 & 0 & 0 & 0 \\ 0 & 0.0475 & 0 & 0 & 0 & 0 & 0 & 0 \end{bmatrix}$$

The output model is of the form

$$\underline{u}(k+1) = H\,\underline{x}(k+1) + \underline{v}(k+1)$$

where H =

$$\begin{bmatrix} 0 & 1.0 & 0 & 0 & 0 & 0 & 0 & 0 & 0 & 0 & 0 & 0 & 0 & 0 & 0 & 0 & 0 & 0 & 0 & 0 & 0 & 0 \\ 0 & 0 & 0 & 0 & 0 & 0 & 0 & 1.0 & 0 & 0 & 0 & 0 & 0 & 0 & 0 & 0 & 0 & 0 & 0 & 0 & 0 & 0 \\ 0 & 0 & 0 & 0 & 0 & 0 & 0 & 0 & 0 & 0 & 0 & 0 & 0 & 0 & 0 & 0 & 0 & 1.0 & 0 & 0 & 0 & 0 \end{bmatrix}$$

We choose the following numerical values for our problem :

final time : 2 days
number of time points : 81
$Q = 5\,I_{22}$
$R = I_3$
$P_o = 25\,I_{22}$

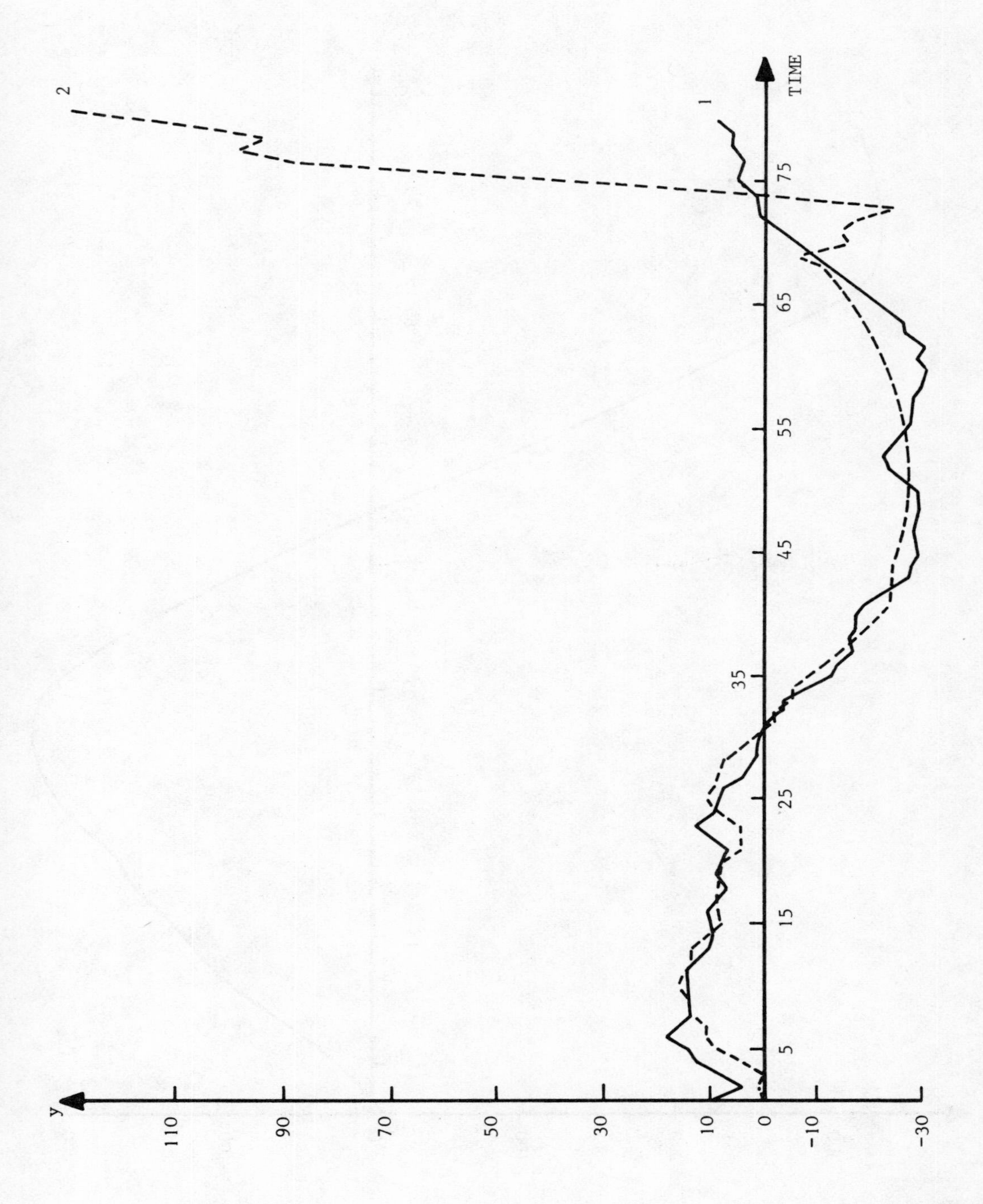

Figure 11.1

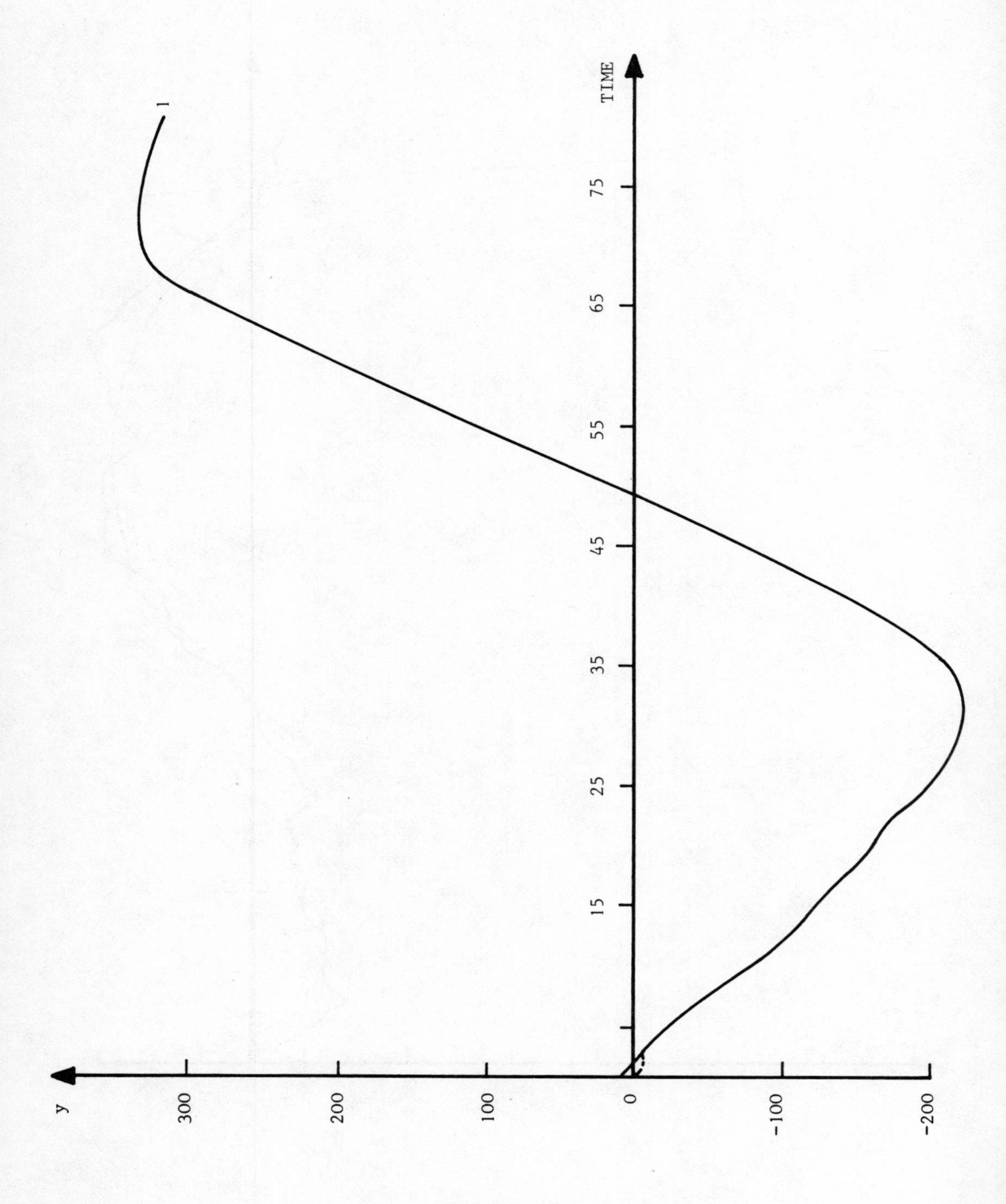

Figure 11.2

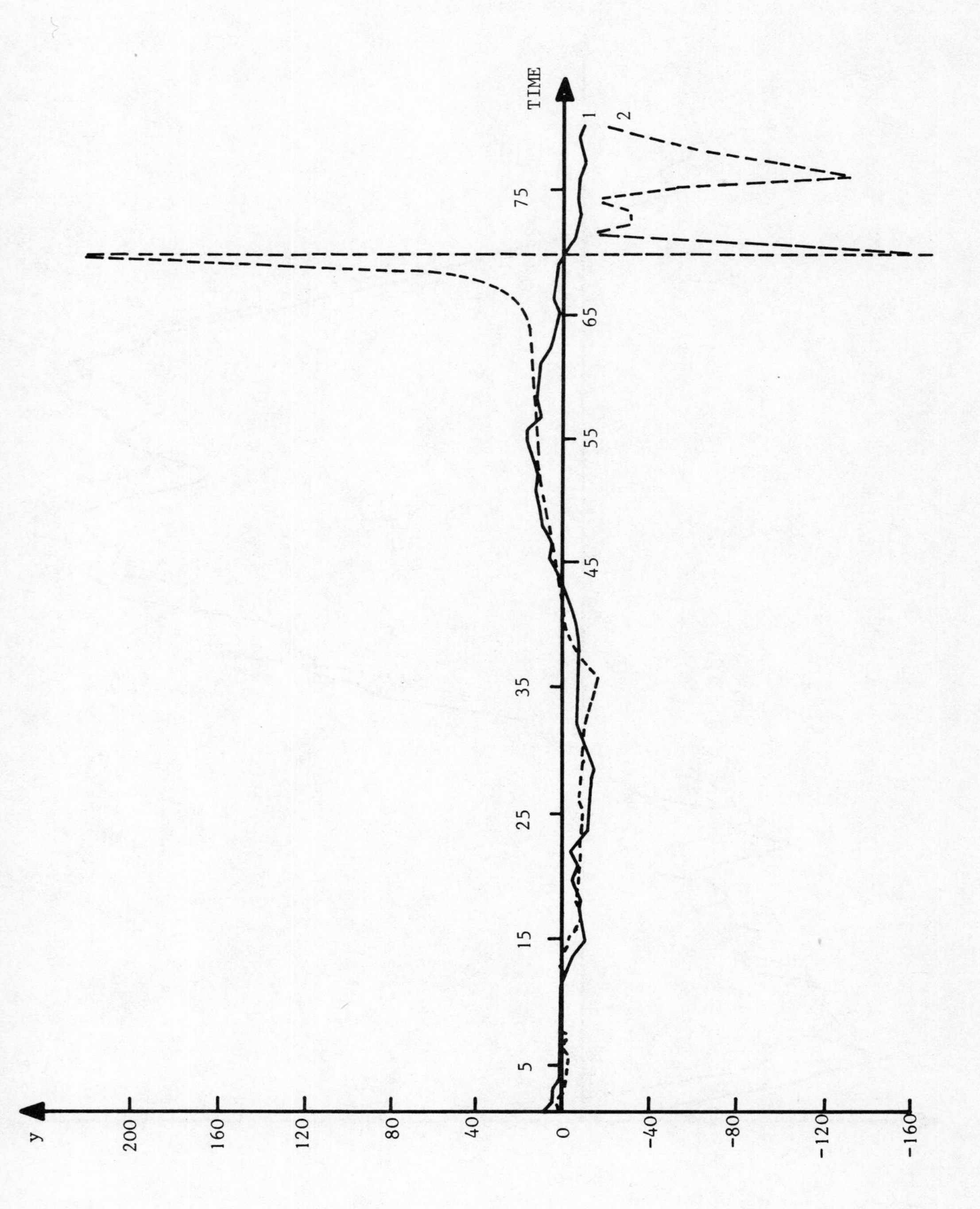

Figure 11.3

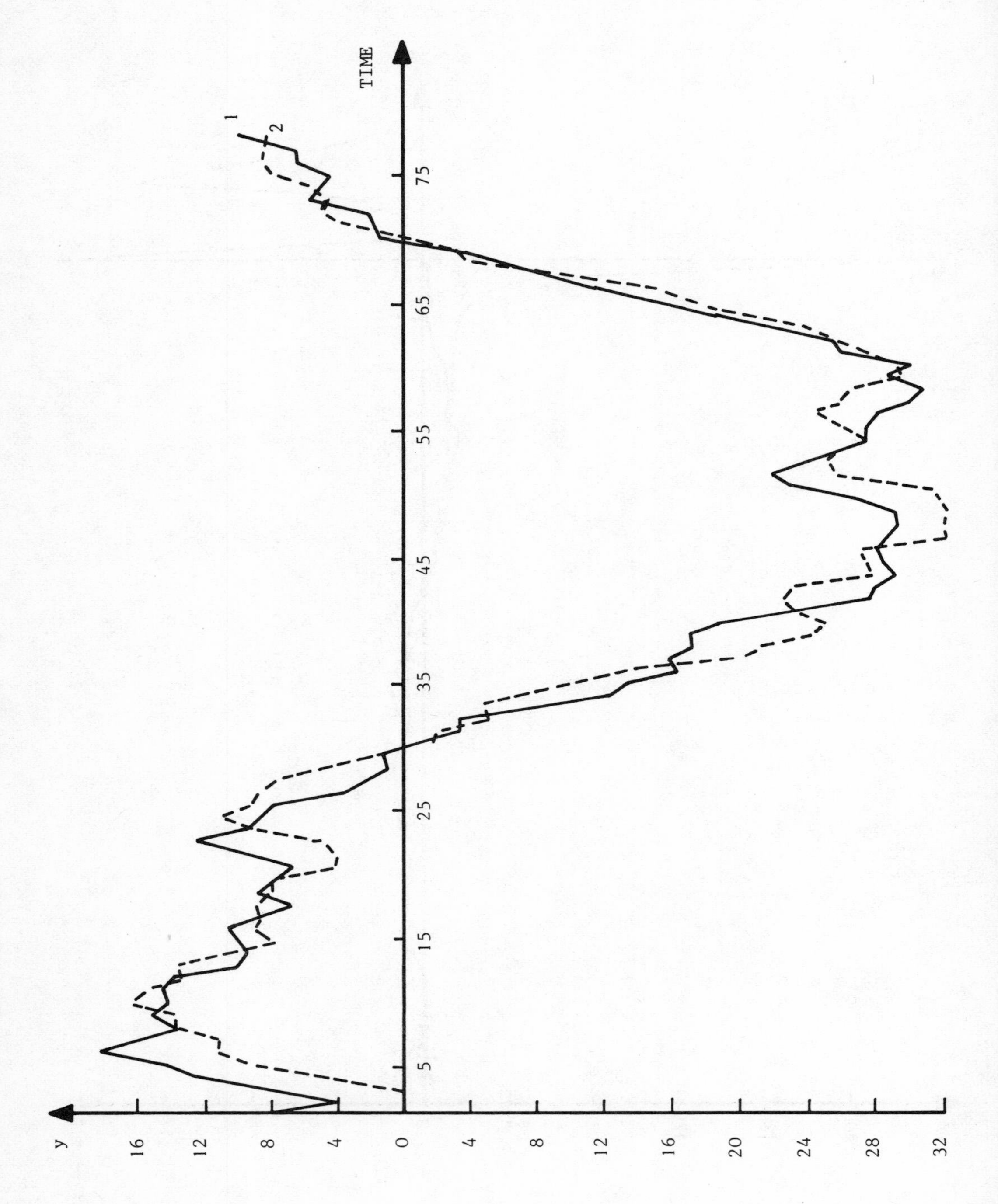

Figure 11.4

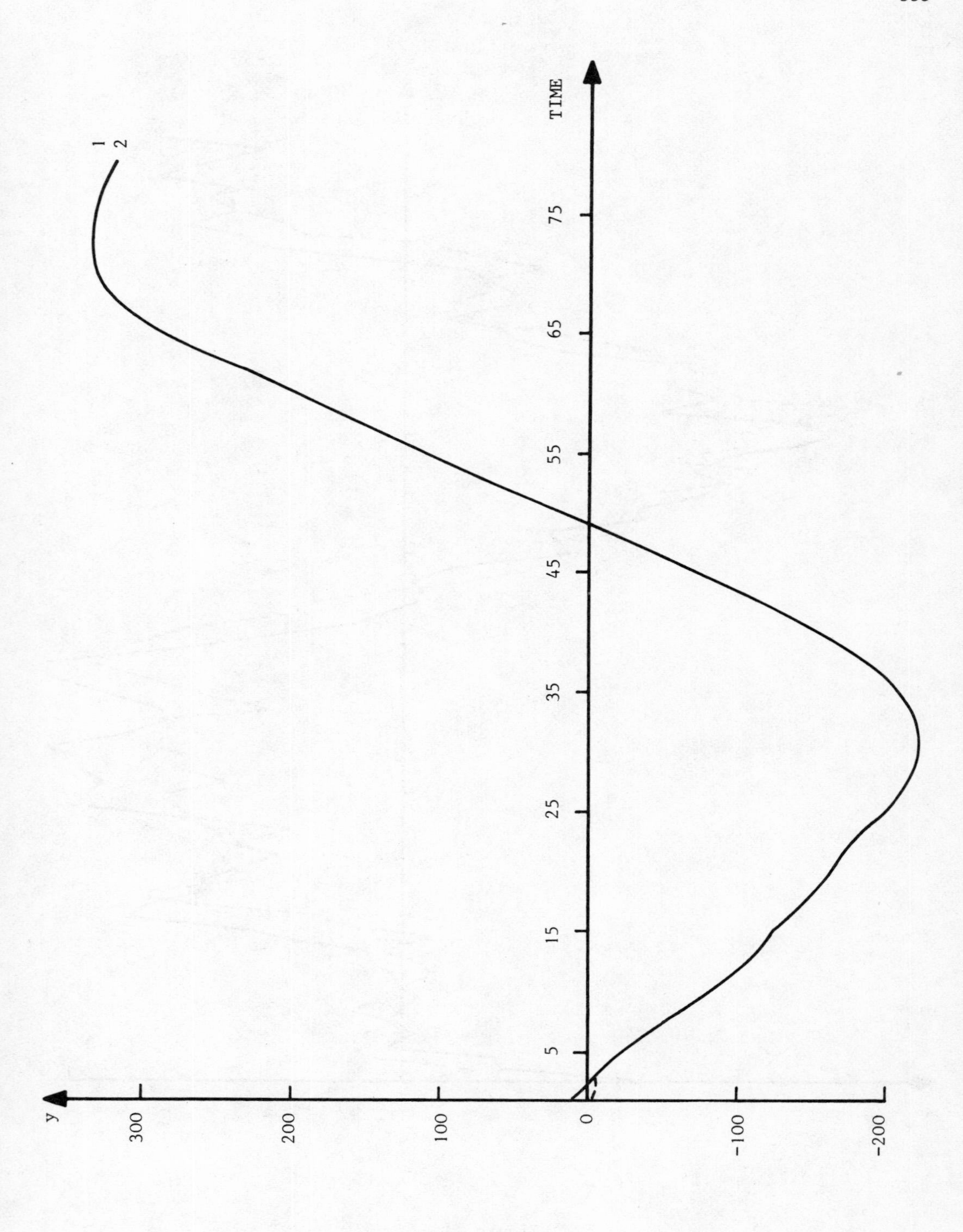

Figure 11.5

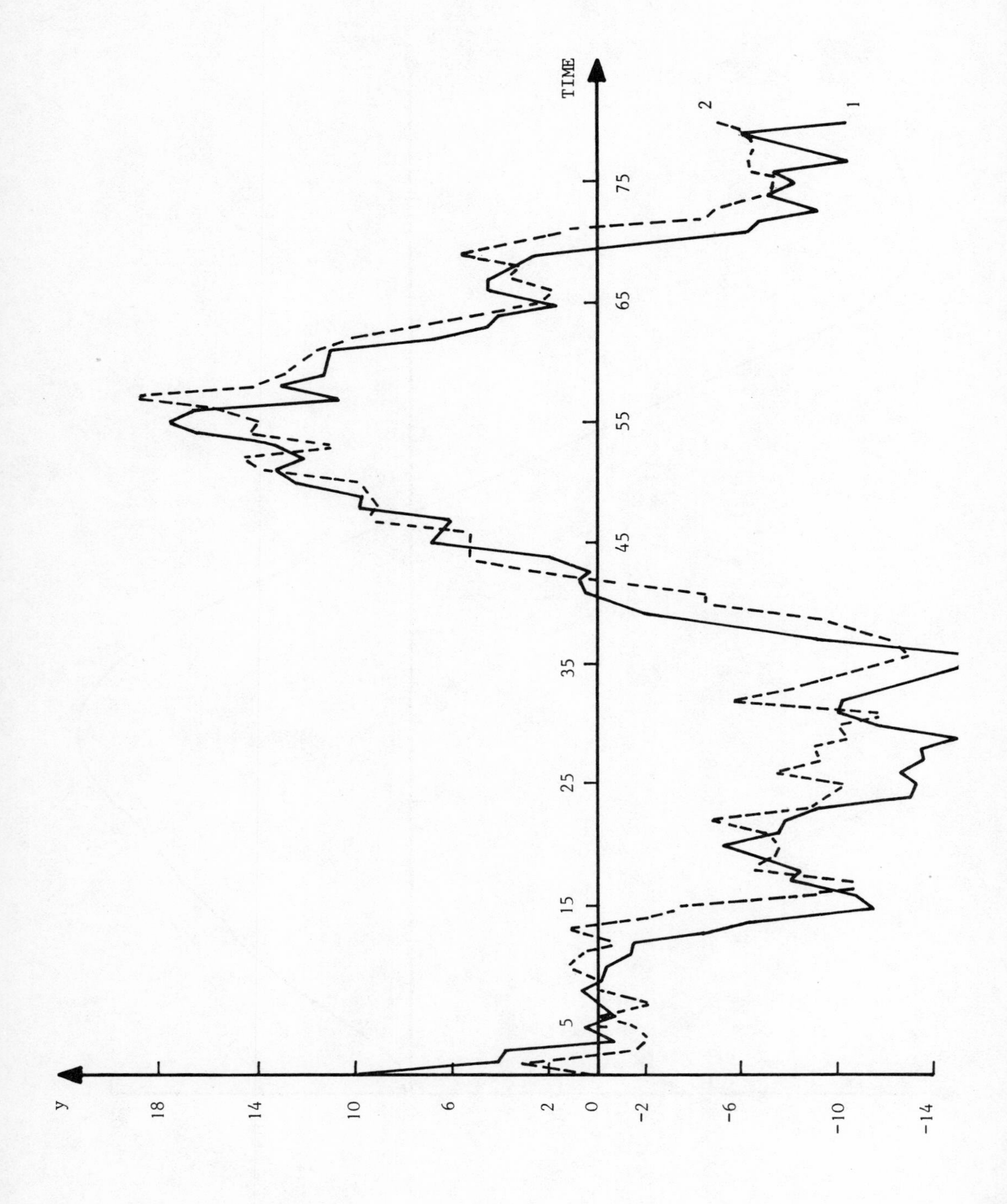

Figure 11.6

0.967	0	0	0	0	0	0	0	0	0	0	0	0	0	0	0	0	0	0	0	0	0
-0.008	0.97	0	0	0	0	0	0	0	0	0	0	0	0	0	0	0	0	0	0	0	0
0.00337	0.0	0.967	0.0157	0	0.0055	0	0	0	0	0	0	0	0	0	0	0	0	0	0	0	0
0	0	0	1.0	0.025	0	0	0	0	0	0	0	0	0	0	0	0	0	0	0	0	0
0.2	0	0	-0.2	0.9	0	0	0	0	0	0	0	0	0	0	0	0	0	0	0	0	0
0	0	0	0	0	1.0	0.025	0	0	0	0	0	0	0	0	0	0	0	0	0	0	0
0.05	0	0	0	0	-0.05	0.95	0.0157	0	0.00337	0	0	0	0	0	0	0	0	0	0	0	0
0	0.00337	-0.008	0	0	0	0	0.97	0.0157	0	0.00337	0	0	0	0	0	0	0	0	0	0	0
0	0	0	0	0	0	0	0	1.0	0.025	0	0	0	0	0	0	0	0	0	0	0	0
0	0.2	0	0	0	0	0	0	-0.2	0.9	0	0	0	0	0	0	0	0	0	0	0	0
0	0	0	0	0	0	0	0	0	0	1.0	0.025	0	0	0	0	0	0	0	0	0	0
0	0.05	0	0	0	0	0	0	0	0	0.05	0.95	0	0	0	0	0	0	0	0	0	0
0	0	0.00337	0	0	0	0	0	0	0	0	0	0.967	0.0157	0	0.00337	0	0	0	0	0	0
0	0	0	0	0	0	0	0	0	0	0	0	0	1.0	0.025	0	0	0	0	0	0	0
0	0	0.2	0	0	0	0	0	0	0	0	0	0	-0.2	0.9	0	0	0	0	0	0	0
0	0	0	0	0	0	0	0	0	0	0	0	0	0	0	1.0	0.025	0	0	0	0	0
0	0	0	0	0	0	0	0	0	0	0	0	0	0	0	-0.05	0.95	0	0	0	0	0
0	0	0	0	0	0	0	0.00337	0	0	0	0	-0.008	0	0	0	0	0.97	0.0157	0	0.00337	0
0	0	0	0	0	0	0	0	0	0	0	0	0	0	0	0	0	0	1.0	0.025	0	0
0	0	0	0	0	0	0	0.2	0	0	0	0	0	0	0	0	0	0	-0.2	0.9	0	0
0	0	0	0	0	0	0	0	0	0	0	0	0	0	0	0	0	0	0	0	1.0	0.025
0	0	0	0	0	0	0	0	0	0	0	0	0	0	0	0	0	0	0	0	-0.05	0.95

Table 11.3

$$\underline{x}(k_o) = \underline{10} \quad ; \quad \underline{\hat{x}}(k_o \mid k_o) = \underline{0}$$

where Q, R, P_o are respectively the covariance matrices for $\underline{w}(k)$, $\underline{v}(k)$ and $\underline{x}(k_o)$, and I_i is the ith order identity matrix.

Results

This problem was solved once using the global Kalman filter and once using the above hierarchical filter on an IBM 370/165 digital computer. In the hierarchical solution, the system was decomposed into three subsystems. Fig. 11.7 shows the observation of D.O. for the first reach whilst fig. 11.8 shows the real value and estimated value of B.O.D. for this reach, and fig. 11.9 shows the corresponding values for D.O. using the two approaches. We see that here both the filters are numerically stable. However the hierarchical filter only required 15.21 sec. compared to 17.98 sec. for the global filter.

We have so far considered the state and parameter estimation problems for large scale linear interconnected dynamical systems. Finally, we consider the stochastic control problem.

11.7 THE STOCHASTIC CONTROL PROBLEM USING THE NEW FILTER

The problem of interest is to minimise the cost function

$$J = E\left\{\frac{1}{2}\sum_{k=k_o}^{k_f-1} \|\underline{x}(k+1)\|^2_{Q_1} + \|\underline{u}(k)\|^2_{R_1}\right\} \qquad 11.7.1$$

subject to the constraints

$$\underline{x}(k+1) = \emptyset\underline{x}(k) + \Psi\underline{u}(k) + C\underline{z}(k) + \underline{n}(k) \qquad 11.7.2$$

where $\quad \underline{z}(k) = L\,M\underline{x}(k) + L\,N\underline{u}(k) \qquad 11.7.3$

and $\quad \underline{y}(k) = H\,\underline{x}(k) + \underline{v}(k) \qquad 11.7.4$

Here, $\underline{n}$, $\underline{v}$, are uncorrelated zero mean Gaussian random white noise vectors of known covariance. Q_1, R_1, $\emptyset$, Ψ, C, M, N, H are all block diagonal matrices with blocks corresponding to a distinct subsystem structure whilst L is a full matrix. Thus essentially we have a system comprising N linear interconnected dynamical systems whose outputs are corrupted by noise as shown by equation 11.7.4 and whose interaction inputs $\underline{z}$ are formed by a linear combination of the states and controls of all the other subsystems.

Having briefly described the problem we are now in a position to examine how we can incorporate this filter hierarchy into the deterministic control hierarchy in order to compute optimal stochastic control.

The hierarchical controller for optimal stochastic control

The stochastic optimal controller can be developed using the following two theorems which arise from the fact that since our information pattern is classical, the separation theorem of chapter 10 works.

Theorem 3

For large scale linear interconnected dynamical systems with quadratic cost functions of the type

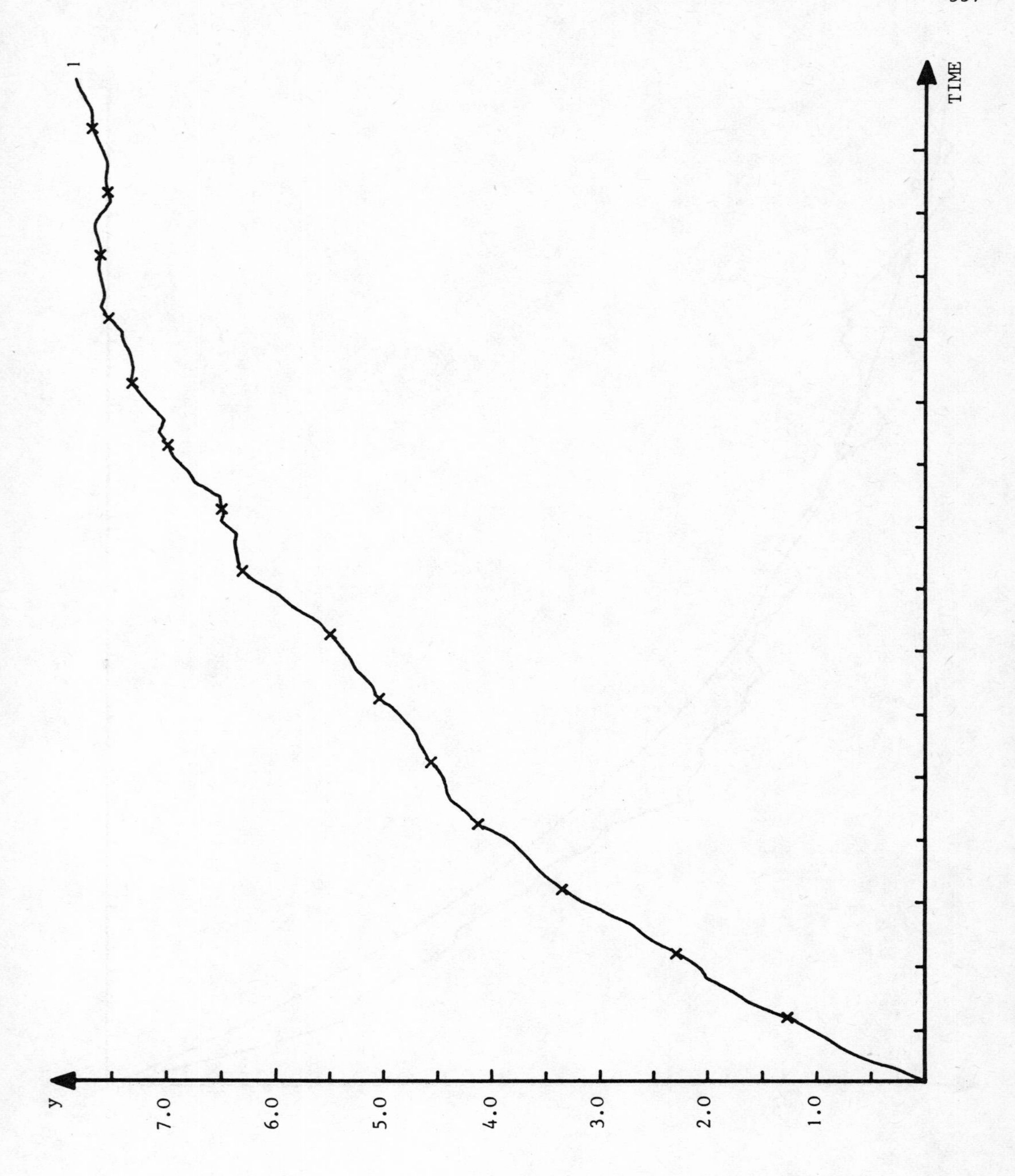

Figure 11.7

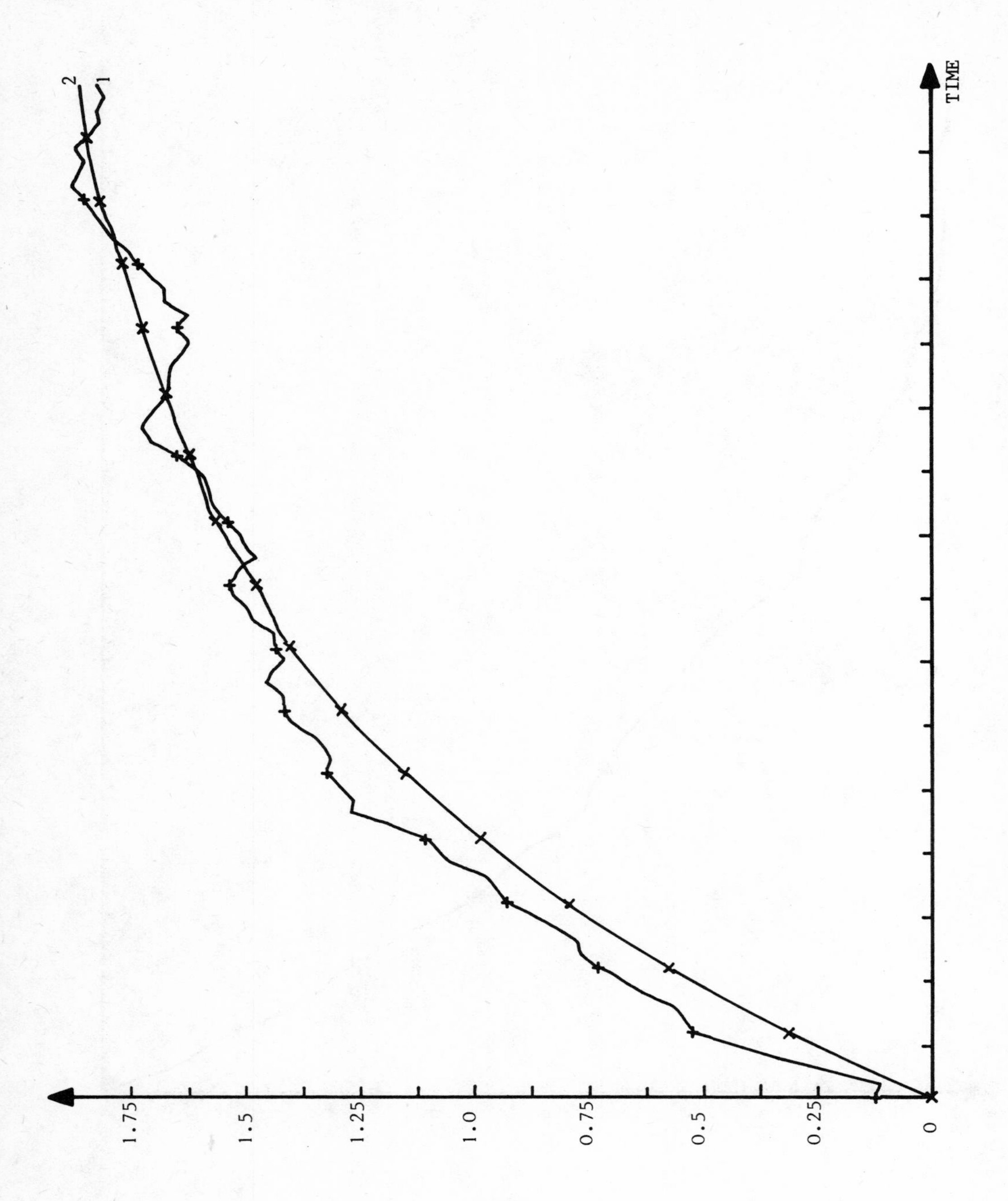

Figure 11.8

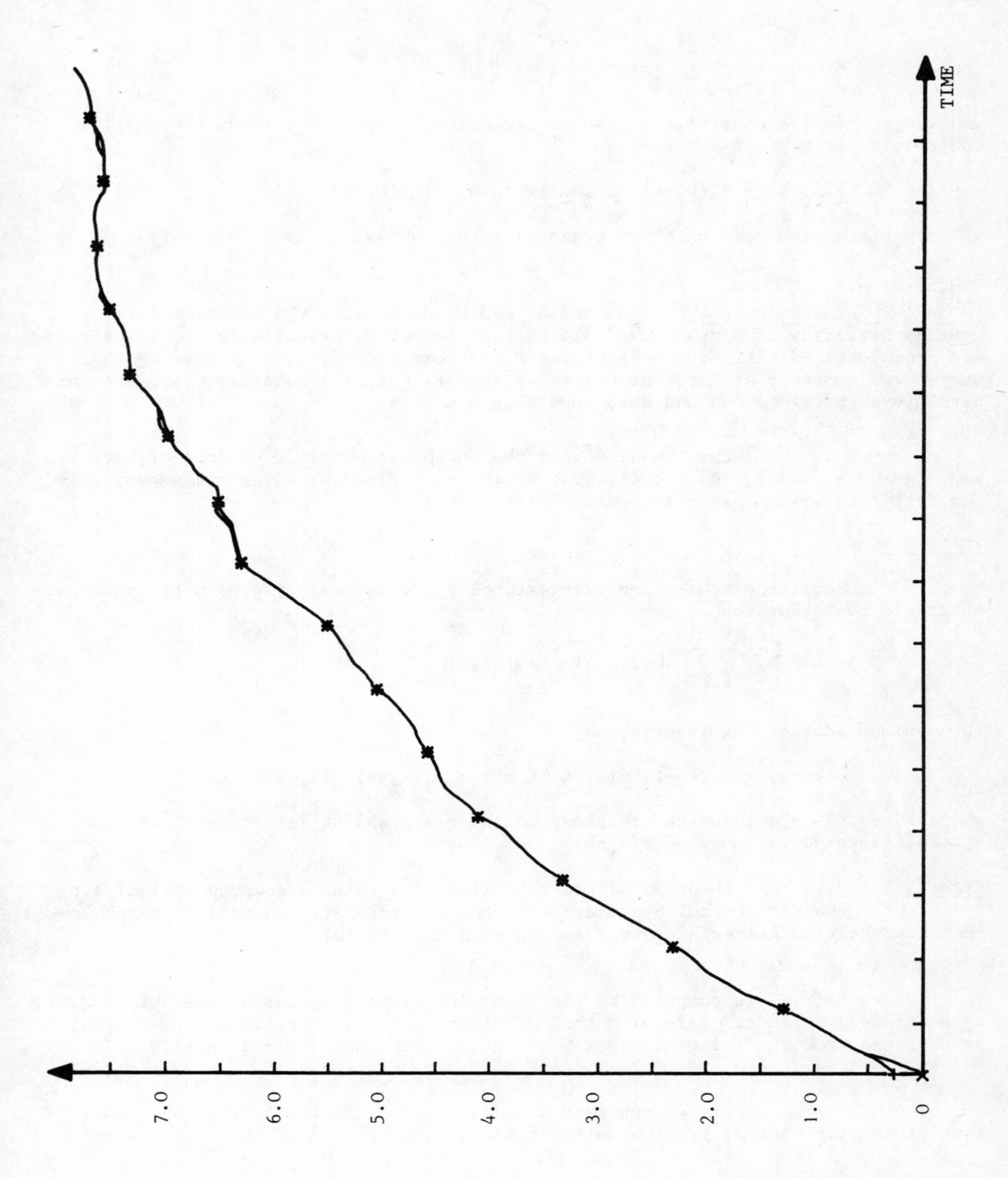

Figure 11.9

$$\text{Min } J = \frac{1}{2} \sum_{k=k_o}^{k_f-1} \|\underline{x}(k+1)\|^2_{Q_1} + \|\underline{u}(k)\|^2_{R_1} \qquad 11.7.5$$

and subject to the constraints given by equations 11.7.2 to 11.7.4, the optimum control law is given by

$$\underline{u}(k_f-k) = S_1(k_f-k)\ \underline{x}(k_f-k) + S_2(k_f-k)\ \underline{x}(k_f-k) \qquad 11.7.6$$

where S_i is a time varying block diagonal matrix whilst S_2 is a full matrix.

Remark

The calculation of S_1 and S_2 can be done using the deterministic control hierarchy of Singh [10], outlined in chapter 6. Essentially, S_1 is calculated from local Riccati equations whilst S_2 is computed off-line by storing the states and controls at certains points of the trajectories obtained using a standard hierarchical method and then inverting a matrix.

Next, we know(*) that the optimal stochastic controller uses the same "gain" S_1 and S_2 and the filtered estimate obtained by using the decentralised filter equations given in section 11.6.

Theorem 4

For large scale linear interconnected dynamical systems with an average quadratic cost function of the type

$$J = E\left\{\frac{1}{2} \sum_{k=k_o}^{k_f-1} \|\underline{x}(k+1)\|^2_{Q_1} + \|\underline{u}(k)\|^2_{R_1}\right\}$$

the optimum control law is given by

$$\underline{u}(k_f-k) = S_1(k_f-k)\ \hat{\underline{x}}(k_f-k\,/\,k_f-k) + S_2(k_f-k)\ \hat{\underline{x}}(k_f-k/k_f-k)$$

where S_1 and S_2 are identical to those in theorem 3 whilst $\hat{\underline{x}}(k_f-k/k_f-k)$ is the optimal filtered estimate of $\underline{x}(k_f-k)$.

Remarks : With this theorem, we see that it is possible to compute optimal stochastic control by superposing the deterministic control hierarchy on the decentralised filter hierarchy as shown in Fig. 11.10.

Here S_1 is computed by the local deterministic controllers whilst S_2 is determined using the 2 level method of Singh [10]. The estimate $\hat{\underline{x}}(k_f-k/k_f-k)$ is obtained from the filter hierarchy using the equations defined in 11.6.

Next we consider a hierarchical solution to the continuous time stochastic control problem using Kalman's duality concepts [11].

(*) by applying the separation principle

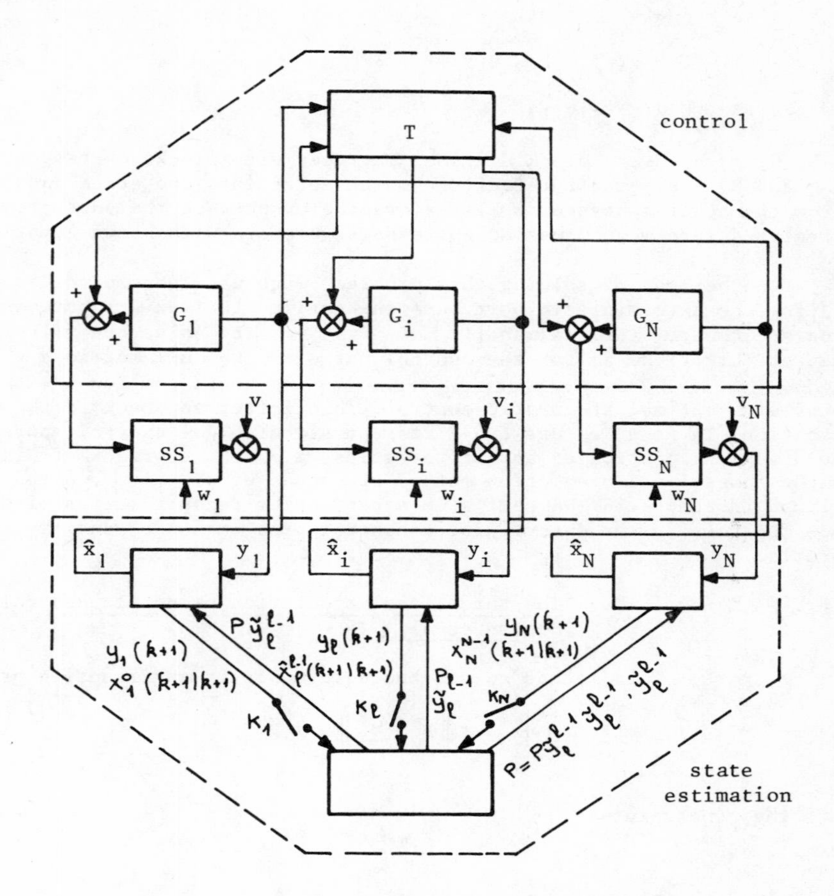

Fig. 11.10 : Stochastic control structure

11.8 THE DUALITY APPROACH TO OPTIMAL STOCHASTIC CONTROL OF L.G.G. PROBLEMS

11.8.1 General considerations

We recall that the basic problem is to minimise

$$J=\frac{1}{2}\, E\left\{\sum_{i=1}^{N} \int_{0}^{t_f} \left(\|\underline{x}_i(t)\|^2_{Q_i} + \|\underline{u}_i(t)\|^2_{R_i} \right) dt \right\} \tag{11.8.1}$$

where Q_i are at least positive semidefinite matrices whilst R_i are positive definite matrices, subject to the dynamical constraints :

$$\dot{\underline{x}}(t) = A\,\underline{x}(t) + B\,\underline{u}(t) + C\,\underline{z}(t) + \underline{D}(t) \qquad 11.8.2$$

$$\underline{z}(t) = L\,M\,\underline{x}(t) + L\,N\,\underline{u}(t) \qquad 11.8.3$$

$$\underline{y}(t) = H\,\underline{x}(t) + \underline{v}(t) \qquad 11.8.4$$

where A, B, C, M, N, H are block diagonal matrices with blocks corresponding to those in Q and R, $\underline{z}$ are the interactions which enter into the state equation 11.8.2 from the other subsystem whilst $\underline{y}$ are the outputs of the subsystems. $\underline{w}$, $\underline{v}$ are uncorrelated zero mean Gaussian white noise vectors with known covariances.

Another way of solving this problem is to use the controller of Singh et al.[13] for the deterministic part, use the duality that exists between linear estimation and control (cf. Kalman [11] or Sage [12]) to solve the filter problem using the same algorithm as for the control but with the dual matrices. Then, by analogy with the separation theorem, we could combine the two structures to provide optimal stochastic control. This latter approach is the one used in this section. In fact, we use two different hierarchical control approaches for calculating the controller and filter gains, i.e. the approach of Singh et al. [13] and the three level prediction method of Hassan et al. [14], both of which were outlined in chapter 6. We next give a very brief recapitulation of these two approaches for solving the deterministic control problem and by duality, the filter problem.

11.8.2 The three level prediction principle controller of Hassan et al. [14]

This method solves the deterministic optimisation problem

$$J = \frac{1}{2}\int_0^{t_f} \left(\|\underline{x} - \underline{x}^*\|_Q^2 + \|\underline{u} - \underline{u}^*\|_R^2 \right) dt \qquad 11.8.5$$

subject to the constraints

$$\dot{\underline{x}} = A\underline{x} + B\underline{u} + C\underline{z} + \underline{D} \qquad 11.8.6$$

$$\underline{y} = M\underline{x} + N\underline{u} \qquad 11.8.7$$

$$\underline{z} = L\underline{y} \qquad 11.8.8$$

where A, B, C, Q, M, N, are block diagonal matrices and L is a full matrix. Here, $\underline{y}$ are the system outputs and $\underline{z}$ is an interaction input vector which allows outputs of various subsystems to be fed back as inputs into other subsystems.

In order to solve this problem, write the Hamiltonian of the overall system as

$$H = \frac{1}{2}\|\underline{x} - \underline{x}^*\|_Q^2 + \frac{1}{2}\|\underline{u} - \underline{u}^*\|_R^2 + \underline{\lambda}^T(A\underline{x}+B\underline{u}+C\underline{z}+\underline{D}) + \underline{\pi}^T(LM\underline{x} + LN\underline{u} - \underline{z}) \qquad 11.8.9$$

Then, from the necessary conditions for optimality :

$$\frac{\partial H}{\partial \underline{u}} = 0 \quad \text{or } \underline{u} = -R^{-1}\left[B^T \underline{\lambda} + N^T L^T \underline{\pi}\right] + \underline{u}^* \qquad 11.8.10$$

$$\frac{\partial H}{\partial \underline{\lambda}} = \dot{\underline{x}} = A\underline{x} + B\underline{u} + C\underline{z} + \underline{D} \qquad 11.8.11$$

$$\frac{\partial H}{\partial \underline{x}} = -\dot{\underline{\lambda}} = A^T \underline{\lambda} + Q(\underline{x}-\underline{x}^*) + M^T L^T \underline{\pi} \qquad 11.8.12$$

$$\frac{\partial H}{\partial \pi} = \underline{0} \text{ or } \underline{Z} = LM\,\underline{x} + LN\,\underline{u} \qquad 11.8.13$$

$$\frac{\partial H}{\partial \underline{z}} = \underline{0} \text{ or } \underline{\pi} = C^T \underline{\lambda} \qquad 11.8.14$$

Now any algorithm must eventually satisfy these necessary conditions for optimality. The three level costate prediction algorithm does this within a three level structure as follows.

Step 1

Set iteration index k for the 3rd level as k = 1 and provide an initial guess for $\underline{\pi}^k$ and $\underline{z}^k$ (eg. $\underline{\pi}^1(t) = \underline{0}$, $\underline{z}^1(t) = \underline{0}$).

Step 2

Set iteration index ℓ for the second level as $\ell = 1$ and provide an initial guess for $\underline{\lambda}^{k\ell}$.

Step 3

Calculate $u^{k\ell}$ from equation 11.8.10 and integrate equation 11.8.11 forward using this $\underline{u}^{k\ell}$ and $\underline{x}(0) = \underline{x}_o$ to obtain $\underline{x}^{k\ell}$.

Step 4

At level 2, integrate equation 11.8.12 backwards with $\underline{\lambda}(t_f) = \underline{0}$ to obtain $\underline{\lambda}^{k(\ell+1)}$.

Step 5

Calculate the error

$$E_\lambda = \sqrt{\int_o^{t_f} \| \underline{\lambda}^{k(\ell+1)} - \underline{\lambda}^{k\ell} \| \, dt}$$

If $E_\lambda > \varepsilon_\lambda$ where ε_λ is a small prechosen constant, then put $\ell = \ell_1$ and go to step 3. Else

Step 6

Calculate $\underline{z}^{k+1}$ and $\underline{\pi}^{k+1}$ from equations 11.8.13 and 11.8.14 respectively

Step 7

Calculate the errors

$$E_z = \sqrt{\int_0^{t_f} \| \underline{z}^{k+1} - \underline{z}^k \| dt}$$

$$E_\pi = \sqrt{\int_0^{t_f} \| \underline{\pi}^{k+1} - \underline{\pi}^k \| dt}$$

If $E_z > \varepsilon_z$ or $E_\pi > \varepsilon_\pi$, where ε_z, ε_π, are also small prechosen constants, then put $k \overset{z}{=} k+1$ and go to step 2 setting

$$\underline{\lambda}^k = \underline{\lambda}^{(k+1)}.$$

If $E_z \leqslant \varepsilon_2$, $E_\pi \leqslant \varepsilon_\pi$, stop and record the control and state trajectories as the optimum ones.

Next we recapitulate the two level prediction method of chapter 5 .

The two level prediction method [13]

In this method, we resolve equations 11.8.10 to 11.8.12 at the lower level by providing $\underline{\pi}$, $\underline{z}$ and improve $\underline{\pi}$, $\underline{z}$ at level 2 using equations 11.8.13 and 11.8.14. At the optimum, we can use the first n value of $\underline{x}$ and $\underline{u}$ (for an nth order system) to compute the feedback gain matrix. Details of the feedback gain matrix calculation are given by Singh et al. [13] and also in chapter 6.

Having recapitulated the control methods, we consider the filter calculation.

The filter calculation using duality

The basis of this calculation is the observation of Kalman that the dual of the optimal estimation problem is the optimal regulator problem. Thus for the estimation problem of the system

$$\dot{\underline{x}} = F\underline{x} + G\underline{w}$$

$$\underline{y} = H\underline{x} + \underline{v}$$

with the noise inputs $\underline{w}$, $\underline{v}$ being uncorrelated zero mean gaussian vectors with covariances Q and R, and covariance of initial state as P_o then an alternative way of solving this problem is to define the dynamical system

$$\dot{\hat{\underline{x}}} = F\,\hat{\underline{x}} + G\,\hat{\underline{w}}$$

$$\underline{y} = H\,\hat{\underline{x}} + \hat{\underline{v}}$$

by replacing

$$F(t) = A^T(t^*)$$
$$G(t) = C^T(t^*)$$
$$H(t) = B^T(t^*)$$

where $t = -t^*$.

The dual dynamical system is then defined as

$$\frac{d\underline{x}^*}{dt} = A(t^*)\,\underline{x}^*(t^*) + B(t^*)\,\underline{u}^*(t^*)$$

$$\underline{y}^*(t^*) = C(t^*)\ \underline{x}^*(t^*)$$

which has the transition matrix

$$\phi^*(t^*, t_o^*) = \phi^T(t_o, t)$$

The optimal regulator problem is to minimise subject to the above dynamical constraints

$$J = \frac{1}{2}\|\underline{x}(t_o^*)\|_P^2 + \frac{1}{2}\int_{t^*}^{t_o^*}\left\{ \|\underline{x}^*(\tau^*)\|^2_{Q(\tau^*)} + \|\underline{u}^*(\tau^*)\|^2_{R(\tau^*)} \right\} d\tau$$

with the solution $\underline{x}^*$ of this regulator giving the optimal filtered estimate $\hat{\underline{x}}$.

Remarks

1. We can clearly solve the above regulator problem, which is the equivalent dual of the estimation problem that we are interested in, by either of the two hierarchical methods described previously.

2. By analogy with the separation theorem, it is possible to compute the filter gain independently from the controller gain, using one of the two methods described previously and to superpose the two to achieve optimal stochastic control.

Next we describe the river problem and apply these algorithms to it in order to provide optimal stochastic control.

11.9 APPLICATION TO THE 52nd ORDER RIVER POLLUTION CONTROL PROBLEM

Let us consider the 6 reach 52nd order model described previously, except that now we also have noise terms :

$$\dot{z}_1 = -1.32\ z_1 + 0.1\ u_1 + 0.9\ z_o + \eta_1$$

$$\dot{q}_1 = -0.32\ z_1 - 1.2\ q_1 + 0.9\ q_o + 1.9 + \eta_2$$

$$\dot{z}_2 = 0.9\sum_{i=1}^{3} a_i\ z_1(t-\tau_i) - 1.32\ z_2 + 0.1\ u_2 + \eta_3$$

$$\dot{q}_2 = 0.9\sum_{i=1}^{3} a_i\ q_i(t-\tau_i) - 0.32\ z_2 - 1.2\ q_2 + 1.9 + \eta_4$$

$$\dot{z}_3 = 0.9\sum_{i=1}^{3} a_i\ z_2(t-\tau_i) - 1.32\ z_3 + 0.1\ u_3 + \eta_5$$

$$\dot{q}_3 = 0.9\sum_{i=1}^{3} a_i\ q_2(t-\tau_i) - 0.32\ z_3 + 1.2\ q_3 + 1.9 + \eta_6$$

$$\dot{z}_4 = 0.9\sum_{i=1}^{3} a_i\ z_3(t-\tau_i) - 1.32\ z_4 + 0.1\ u_4 + \eta_7$$

$$\dot{q}_4 = 0.9 \sum_{i=1}^{3} a_i\, q_3(t-\tau_i) - 0.32\, z_4 + 1.2\, q_4 + 1.9 + \eta_8$$

$$\dot{z}_5 = 0.9 \sum_{i=1}^{3} a_i\, z_4(t-\tau_i) - 1.32\, z_4 + 0.1\, u_5 + \eta_9$$

$$\dot{q}_5 = 0.9 \sum_{i=1}^{3} a_i\, q_4(t-\tau_i) - 0.32\, z_5 + 1.2\, q_5 + 1.9 + \eta_{10}$$

$$\dot{z}_6 = 0.9 \sum_{i=1}^{3} a_i\, z_5(t-\tau_i) - 1.32\, z_6 + 0.1\, u_6 + \eta_{11}$$

$$\dot{q}_6 = 0.9 \sum_{i=1}^{3} a_i\, q_5(t-\tau_i) - 0.32\, z_6 + 1.2\, q_5 + 1.9 + \eta_{12}$$

with the outputs given by

$$\begin{bmatrix} y_1 \\ \cdot \\ \cdot \\ \cdot \\ y_6 \end{bmatrix} = \begin{bmatrix} 1 & 0 & 0 & \cdots\cdots\cdots & 0 \\ 0 & 0 & 0 & \cdots\cdots\cdots & 0 \\ 0 & 0 & 1 & 0 \cdots\cdots & 0 \\ 0 & 0 & 0 & 0 \cdots\cdots & 0 \\ 0 & 0 & 0 & 0\ 1\ 0 \cdots & 0 \\ 0 & 0 & 0 & 0\ 0\ 0 \cdots & 0 \end{bmatrix} \begin{bmatrix} q_1 \\ z_1 \\ \cdot \\ \cdot \\ q_6 \\ z_6 \end{bmatrix} + \begin{bmatrix} v_1 \\ v_2 \\ \cdot \\ \cdot \\ \cdot \\ v_6 \end{bmatrix}$$

These equations describe the dynamics of B.O.D. (z_i) and D.O. (q_i) in each of the 6 reaches with the interconnection between the reaches being provided by the distributed delays. These distributed delays account for the dispersion of pollutents and enables the overall description to be realistic. The control variable is the B.O.D. content of the effluent which enters each reach and this is inversely related to the level of treatment in the treatment plant prior to discharge of the effluent in the river.

$\eta_1, \ldots, \eta_{12}$ and $v_1, v_2, \ldots v_6$ are uncorrelated zero mean gaussian white noise vectors of known covariance. The river pollution control problem is essentially how to control the effluent discharges in order to maintain desired levels of B.O.D. and D.O. in the river despite the stochastic effects.

As before, the pure delays can be approximated by a 2nd order Taylor series expansion, i.e. let

$$z_7(t) = z_1(t-\tau_i)$$

then $\quad z_4(s) = z_1(s)\, e^{-s\tau i}$

where s is the Laplace variable.

$$\text{Now } z_1(s)\, e^{-s\tau i} = z_1(s) \left[1 + s\tau_i + \frac{s^2 \tau_i^{\,2}}{2} \right]^{-1}$$

Taking only the first 3 terms $z_1(t) = z_7(t) + \tau_i\, \dot{z}_7(t) + \frac{\tau_i^{\,2}}{2} z_7(t)$

introducing 2 additional state variables z_7 and z_8 where

$$\mathring{z}_7(t) = z_8(t)$$

it is possible to model the delayed variable $z_1(t-\tau_i)$.

Note that since $\tau_1 = 0$, there are 2 delays in each of the equations. Then, using the above approximation a system of 52 variables is obtained with subsystem one having 2 variables and subsystem 2 to 6 having 10 variables each.

The control problem is to bring the system back to the steady state desired values :

(concentrations) (in mg/1)

$z_1^* = 2.10$ $q_1^* = 8.5$

$z_2^* = 2.47$ $q_2^* = 7.3$

$z_3^* = 2.47$ $q_3^* = 6.4$

$z_4^* = 2.56$ $q_4^* = 5.7$

$z_5^* = 2.47$ $q_5^* = 5.2$

$z_6^* = 2.94$ $q_6^* = 4.7$

whilst the control deviate as

$$u_1 = \Delta u_1 + 28.90$$

$$u_2 = \Delta u_2 + 12.90$$

$$u_3 = \Delta u_3 + 10.40$$

$$u_4 = \Delta u_4 + 11.60$$

$$u_5 = \Delta u_5 + 9.50$$

$$u_6 = \Delta u_6 + 16.50$$

where Δu_i is the deviation around the nominal steady state value. The cost was chosen as

$$J = \left\{ E \sum_{i=1}^{6} \int_0^{12} \left[(z_i - z_i^t)^2 Q_i + (q_i + q_i^t)^2 s_i + \Delta u_i^2 \right] dt \right\}$$

$$Q_i = 2 \qquad s_i = 1$$

The two hierarchical optimisation algorithms were used to solve the control and the filtering problems independently. The computational requirements of the two methods were the following :

For the controller :

a) The two level prediction method required 186 K of storage and 80 seconds of computation on the IBM 370/165 digital computer.

b) The three level costate prediction method of Hassan et al. [14] required 136 K storage and 35 seconds of computation on the IBM 370/165 digital computer in comparison.

c) The global solution using a single level method (i.e. integration of a 52 x 52 Riccati equation) required 482 K of storage and 150 seconds of computation on the IBM 370/165 digital computer.

The computation time and storage requirements for the filter calculation were identical to the above requirements for the controller for all the three methods. The optimal state trajectories for the system when optimal stochastic control is applied are shown in Figs. 11.11 to 11.16 and the controls are shown in Fig. 11.17.

These results show that (a) it is possible to generate optimal stochastic control using one of the above two hierarchical structures and (b) that its computationally extremely attractive to do so for large scale systems like our river system.

11.10 CONCLUSIONS

In this chapter we have treated problems of state and parameter estimation of large scale systems by decomposition techniques. Some preliminary results have been presented. We think this is a fruitful area for further research and we expect many more interesting results to emerge in the years ahead.

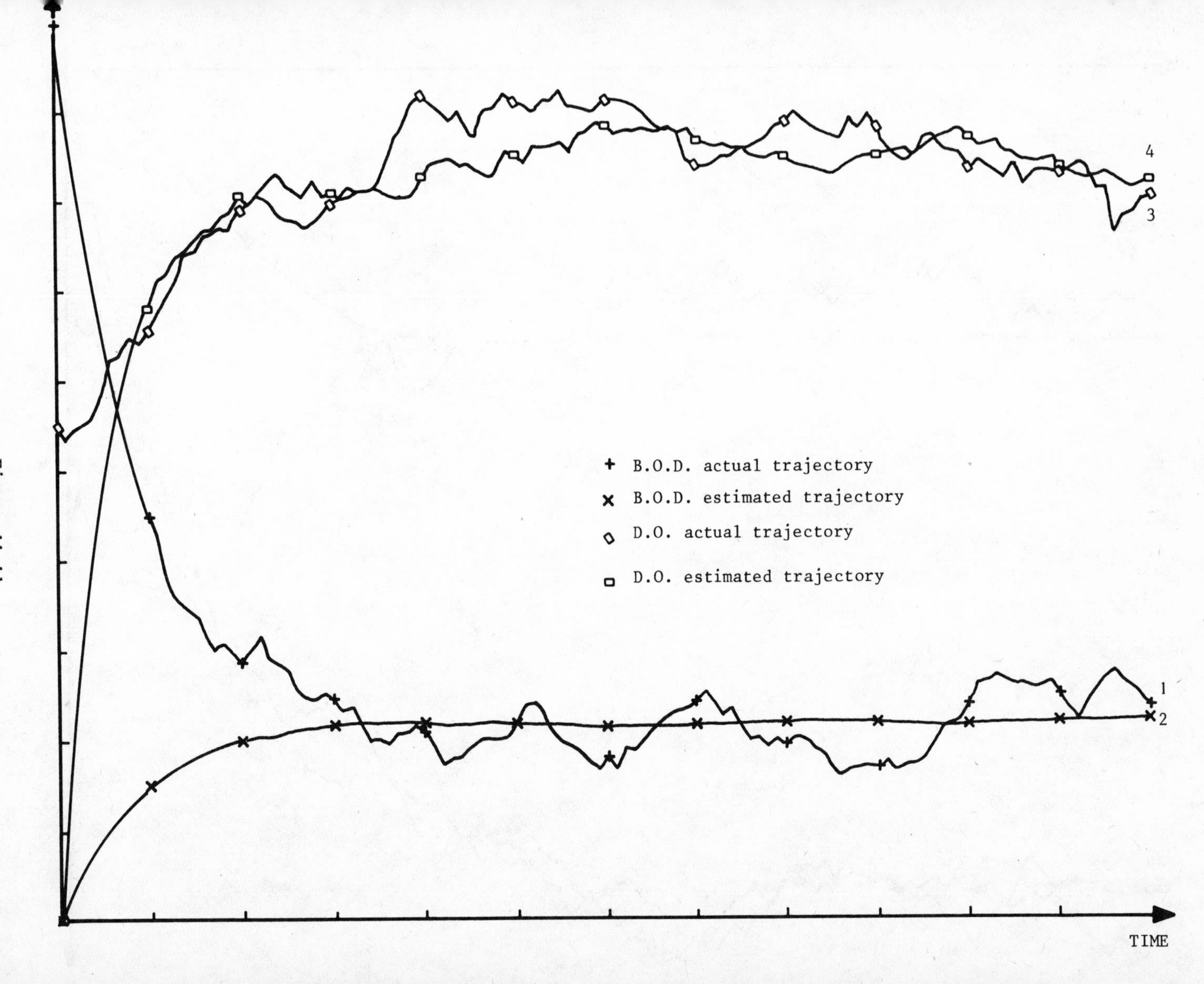
+ B.O.D. actual trajectory
× B.O.D. estimated trajectory
◇ D.O. actual trajectory
□ D.O. estimated trajectory
TIME
1
2
3
4

Figure 11.11

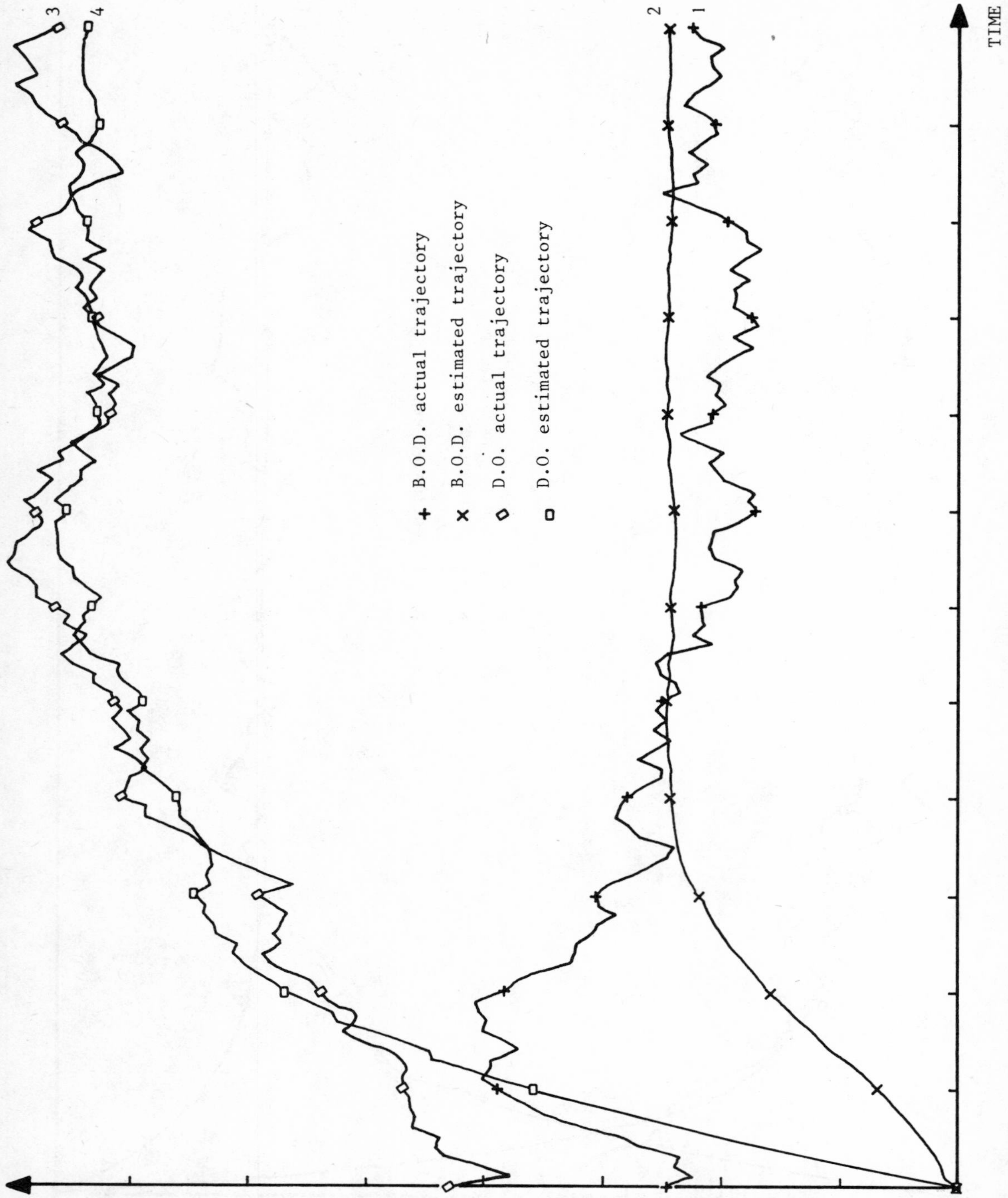

+ B.O.D. actual trajectory
× B.O.D. estimated trajectory
◇ D.O. actual trajectory
□ D.O. estimated trajectory

1 2 3 4

TIME

Figure 11.12

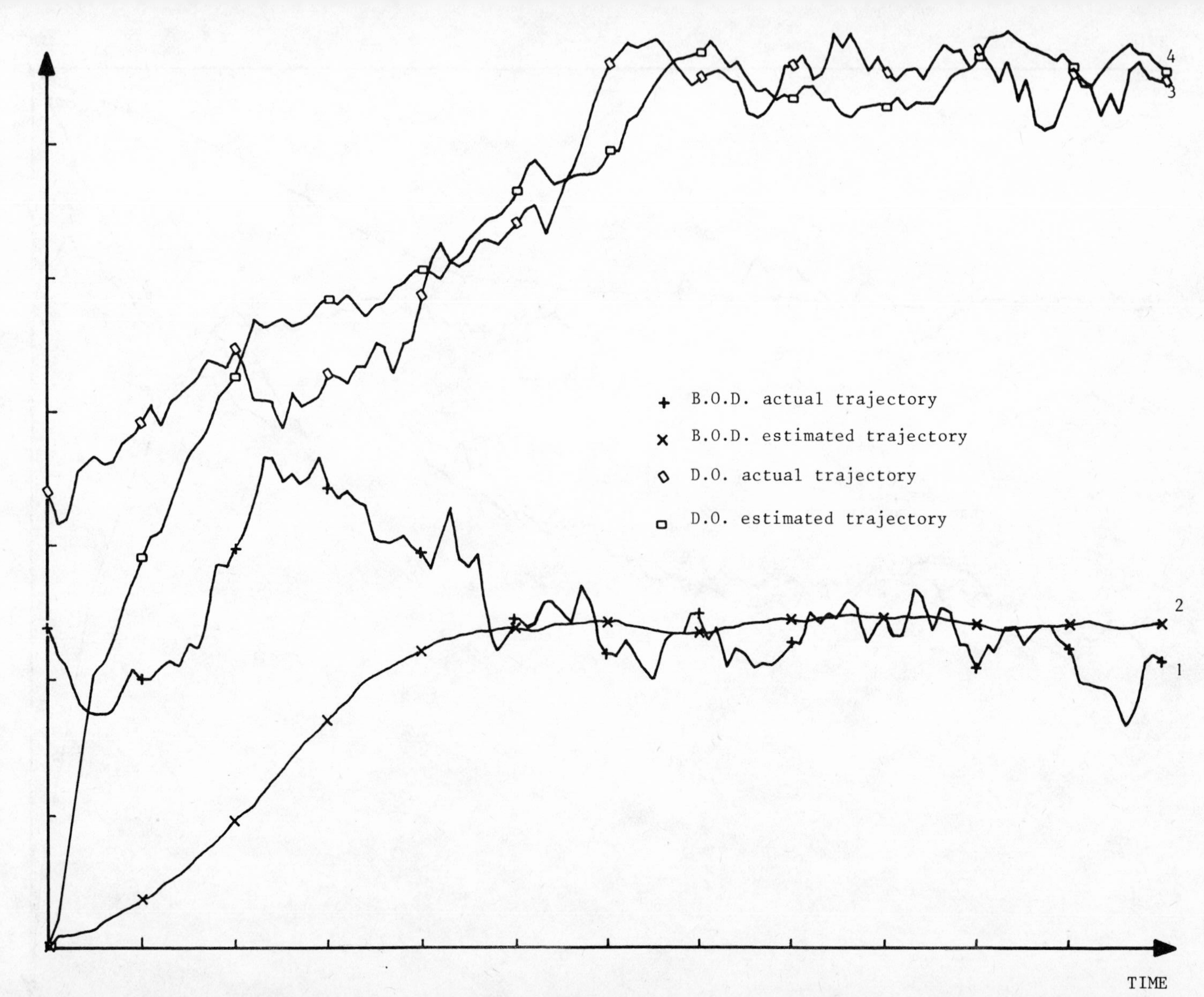

4
3
B.O.D. actual trajectory
B.O.D. estimated trajectory
D.O. actual trajectory
D.O. estimated trajectory
2
1
TIME

Figure 11.13

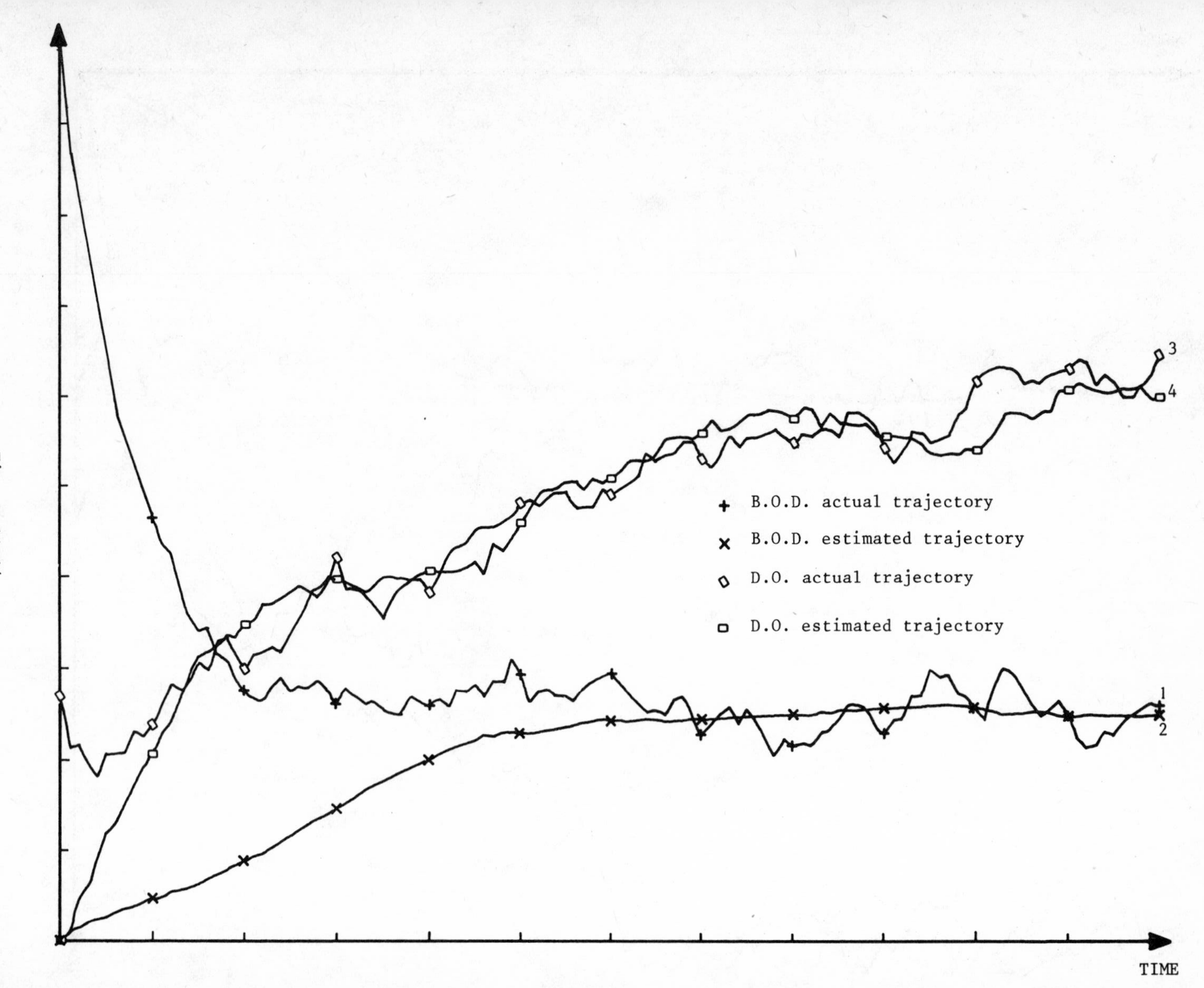
B.O.D. actual trajectory
B.O.D. estimated trajectory
D.O. actual trajectory
D.O. estimated trajectory
3
4
1
2
TIME

Figure 11.14

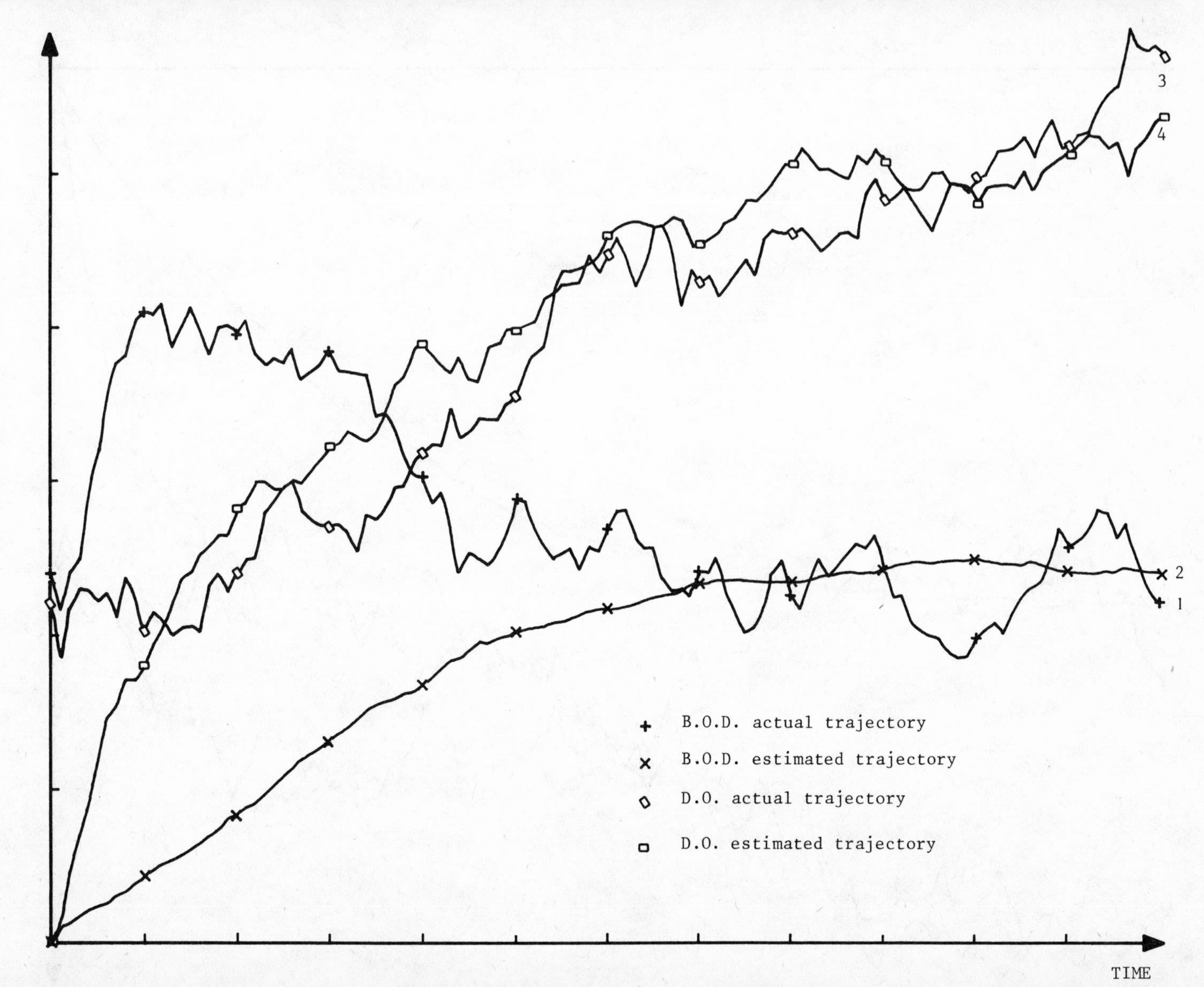
B.O.D. actual trajectory
B.O.D. estimated trajectory
D.O. actual trajectory
D.O. estimated trajectory
TIME
1
2
3
4

Figure 11.15

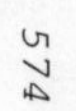

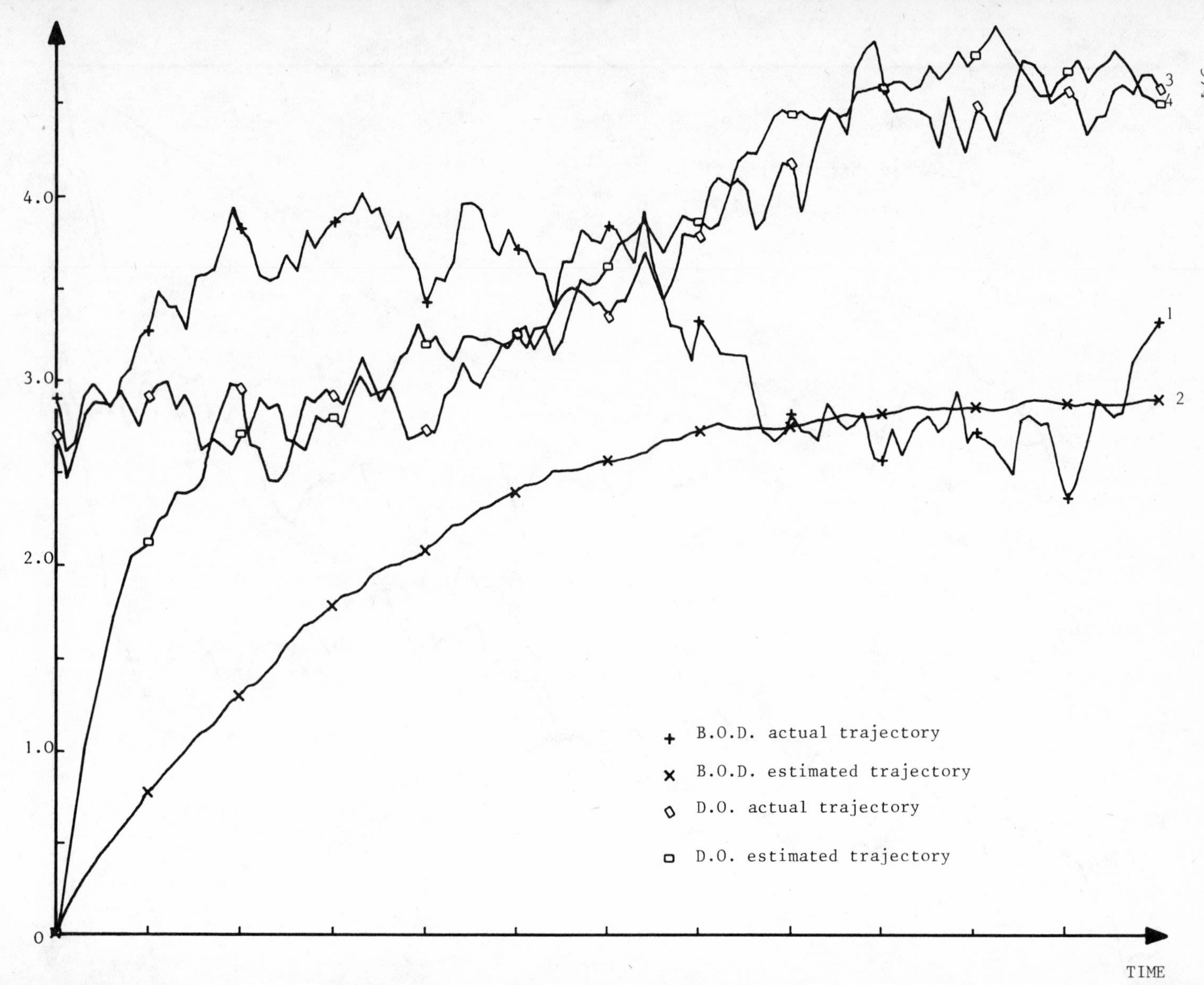

4.0
3.0
2.0
1.0
0
TIME
1
2
3
4
B.O.D. actual trajectory
B.O.D. estimated trajectory
D.O. actual trajectory
D.O. estimated trajectory

Figure 11.16

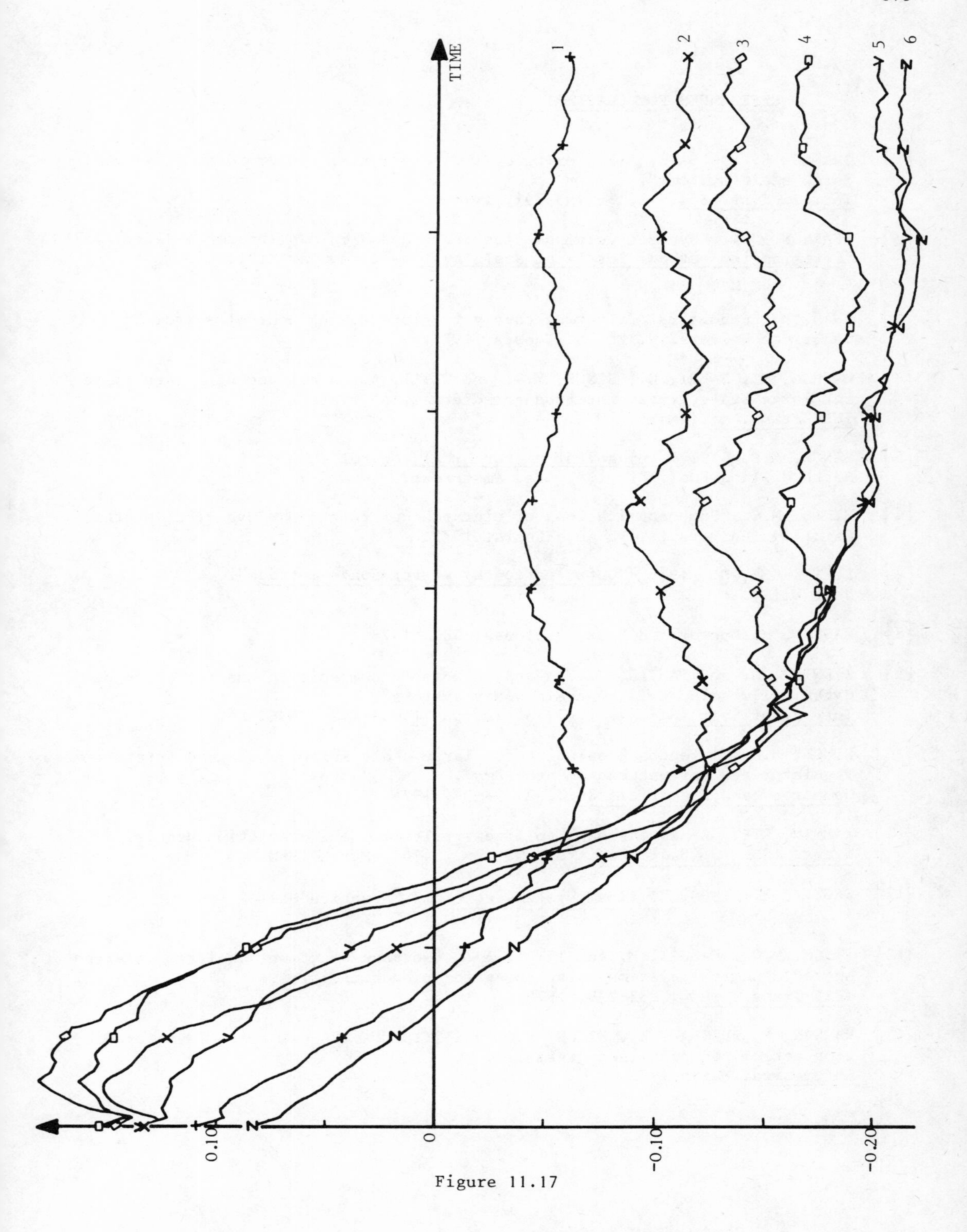

Figure 11.17

REFERENCES FOR CHAPTER 11

[1] ARAFEH, S. and SAGE, A.P. "Multilevel discrete time system identification in large scale systems". Int. J. Syst. Sci., 5, 8, 753-791, 1974.

[2] PEARSON, J.D. "Dynamic decomposition techniques" in D.A. Wismer (editor), 1971, "Optimisation methods for large scale systems", Mc Graw Hill.

[3] SHAH, M. "Suboptimal filtering theory for interacting control systems". Cambridge University, Ph. D. Thesis, 1971.

[4] HASSAN, M., SALUT, G., SINGH, M.G. and TITLI, A. "A new decentralised filter for large scale linear interconnected dynamical systems". IEEE Trans. Ac., April 1978.

[5] SINGH, M.G., 1977, "Dynamical hierarchical control". North Holland Publishing Co, Ltd, Amsterdam.

[6] SINGH, M.G. "Some applications of hierarchical control for dynamical systems". Cambridge University, Ph. D. Thesis, 1973.

[7] LUENBERGER, D., 1969, "Optimisation by vector space methods". John Wiley.

[8] HASSAN, M. Doctorat d'Etat, Toulouse, June 1978.

[9] DARWISH, M. and FANTIN, J. "An approach for decomposition and reduction of dynamical models of large scale power systems". Int. J. of Syst. Sci., vol. 7, n° 10, pp. 1101-1112, 1976.

[10] SINGH, M.G. "A feedback solution for large scale infinite stage discrete time regulator and servomechanism problems". Computers and Electrical Eng., 3, 93-99, 1978.

[11] KALMAN, R.E. "A new approach to linear filtering and prediction problems". Trans. ASME. J. Basic Eng. Vol. 82, pp. 34-45, March 1960.

[12] SAGE, A.P., 1968, "Optimum systems control". Prentice Hall

[13] SINGH, M.G., HASSAN, M. and TITLI, A."A feedback solution for large interconnected dynamical systems using the prediction principle". IEEE Trans. SMC 6, 233-239, 1976.

[14] HASSAN, M. HURTEAU, R., SINGH, M.G. and TITLI, A. "A new three level prediction method for non-linear systems". Automatica, March 1978.

PROBLEMS FOR CHAPTER 11

1. Use the S.P.A. filter to calculate the estimate $\begin{bmatrix} \hat{x}_1(3|3) \\ \hat{x}_2(3|3) \end{bmatrix}$ for the system :

$$x_1(k+1) = 2\, x_1(k) + 3\, x_2(k) + \xi_1(k)$$

$$x_2(k+1) = 3\, x_1(k) - 2\, x_2(k) + \xi_2(k)$$

$$y_1(k) = 4\, x_1(k) + \eta_1(k)$$

$$y_2(k) = x_2(k) + \eta_2(k)$$

where ξ and η are zero mean independent Gaussian random vectors with covariance matrices respectively :

$$\begin{bmatrix} 1 & 0 \\ 0 & 1 \end{bmatrix}, \quad \begin{bmatrix} 0.5 & 0 \\ 0 & 0.5 \end{bmatrix}$$

The initial state $\underline{x}(\underline{0})$ is also a zero mean Gaussian random vector having the covariance matrix :

$$\begin{bmatrix} 1.5 & 0 \\ 0 & 1.5 \end{bmatrix}$$

The following observations are available for the process :

$$\begin{bmatrix} y_1(0) \\ y_2(0) \end{bmatrix} = \begin{bmatrix} 0 \\ 0 \end{bmatrix} \quad ; \quad \begin{bmatrix} y_1(1) \\ y_2(1) \end{bmatrix} = \begin{bmatrix} 2 \\ 2 \end{bmatrix}$$

$$\begin{bmatrix} y_1(2) \\ y_2(2) \end{bmatrix} = \begin{bmatrix} 3 \\ 3 \end{bmatrix} \quad ; \quad \begin{bmatrix} y_1(3) \\ y_2(3) \end{bmatrix} = \begin{bmatrix} 5 \\ 5 \end{bmatrix}$$

2. Use Hassan's method to solve problem 1. Compare the computations required for this method with those required for the global Kalman filter solution.

3. We consider the frictionless displacement of a point (x) on a line. A priori, we know neither its position nor its velocity. We only know that x obeys the equation

$$\ddot{x} = 0$$

and that the initial conditions could be considered to be zero mean Gaussian random variables with variances

$$E\left[x_o^{\,2}\right] = 1\,(m)^2, \quad E\left[\dot{x}_o^{\,2}\right] = 3\,(m/s)^2$$

x_o and $\dot{x}_o$ are assumed to be independent. The only measurements that are made are those of the displacement (x) and they lead to the following results :

at t = 0, we have $z_o = 0.08$
at t = 1, we have $z_1 = 2.05$
at t = 2, we have $z_2 = 3.8$
at t = 3, we have $z_3 = 6.01$

These measurements are a zero mean Gaussian random variable having the variance

$$R = E\left[V_k^2\right] = 0.1\ (m)^2$$

Use the discrete Filter of Shah to estimate $\begin{bmatrix} x(3|3) \\ \dot{x}(3|3) \end{bmatrix}$

4. Use the continuous time filter of Shah to solve problem 3.

5. Use the optimal filter of Hassan to solve problem 3.

6. Use the duality approach to solve problem 3.

PART 4

Robust Decentralised Control

CHAPTER 12

Robust Decentralised Control

12.1 INTRODUCTION

A major difficulty in the implementation of feedback controls for large scale systems arises from the fact that the susbystems are often rather widely distributed in space. Feedback control requires exchange of state information but providing this information exchange capability is often the most significant cost in the control implementation. In this final chapter, we examine ways of eliminating this state information exchange on-line for the case of linear interconnected dynamical systems. We do this by separating the off-line design phase from the on-line implementation phase. In the design phase we compute the "best" block diagonal gain matrix which takes into account the fact that each subsystem interacts with the other subsystems. In the implementation phase, each independent block of the global gain matrix provides local decentralised control for a subsystem and no state information is transferred between subsystem controllers on-line.

We develop three different approaches to tackling the decentralised control problem [1, 2, 3, 4]. The first approach is the "brute force" one where by constraining the global gain matrix to be diagonal we obtain a non-linear two point boundary value problem which we solve using a three level hierarchical structure [1, 4]. Unfortunately, the resulting gains are initial state dependent. To provide a decentralised controller whose gains are independent of initial conditions, we next consider a different algorithm where we provide a low order model of the interactions for each subsystem [2, 4]. The resulting controller aside from having gains independent of initial conditions, is also able to reject constant unknown disturbances. This is achieved by modifying the original cost function. To avoid such cost function modification, we present a third method which achieves decentralised control through a model following technique [3].

It should be emphasised that all the controllers developed in this chapter are suboptimal although simulation examples will show that in fact the degree of suboptimality is quite insignificant. We are now in a position to examine our first method.

12.2 A HIERARCHICAL STRUCTURE FOR COMPUTING DECENTRALISED CONTROL [1,4]

The control problem for a linear interconnected dynamical system can be written as

$$\text{Min } J = \frac{1}{2}\int_0^T \left(\|\underline{x}\|^2_{Q^*} + \|\underline{u}\|^2_R\right) dt \qquad 12.2.1$$

subject to

$$\dot{\underline{x}} = A\underline{x} + B\underline{u} + C\underline{z} \qquad \underline{x}(0) = \underline{x}_0 \qquad 12.2.2$$

$$\underline{z} = L\underline{x} \qquad 12.2.3$$

where $\|\cdot\|^2_S = .^T S.$, Q^*, R, A, B and C are block diagonal matrices with N blocks corresponding to the subsystems and L is a full matrix. $\underline{x}$ is the n dimensional state vector of the overall system and $\underline{u}$ is the n dimensional overall control vector. Note that if $\underline{u}$ is of lower dimension than n, we can add additional pseudo controls.

Now, for this problem, Singh et al. [5] have shown that the optimal control is of the form

$$\underline{u}(t) = - G_b \underline{x}(t) - T\underline{x}(t) \qquad 12.2.4$$

where G_b is the block diagonal matrix obtained by solving a Riccati equation independently for each subsystem and T is a full matrix obtained by a hierarchical technique.

The minimal value of the criterion is

$$J_{opt} = \frac{1}{2}\int_0^T (\|\underline{x}\|^2_{Q^*} + \|G_b\underline{x} + T\underline{x}\|^2_R)dt = \frac{1}{2}\int_0^T (\|\underline{x}\|^2_{Q^*+w})dt \qquad 12.2.5$$

$$\text{where } w = G^T_b R G_b + G^T_b RT + T^T R G_b + T^T RT \qquad 12.2.6$$

Now since T is a full matrix, optimal control requires feedback of all the states. Often, this is too expensive to implement, i.e. whereas it is possible to provide local feedback around each subsystem, it is impractical to provide feedback of states between the subsystems. One way of avoiding this state information transfer is to constrain the matrix T to be a diagonal matrix T_d. Thus we need to find the T_d which minimises

$$J = \frac{1}{2}\int_0^T (\|\underline{x}\|^2_{Q^*} + \|\underline{x}\|^2_W) dt \qquad 12.2.7$$

$$\text{where } W = G^T_b R G_b + G^T_b R T_d + T^T_d R G_b + T^T_d R T_d \qquad 12.2.8$$

subject to

$$\dot{\underline{x}} = (A - B G_b)\underline{x} + C\underline{z} - BT_d \underline{x} \qquad 12.2.9$$

$$\underline{z} = L \underline{x} \qquad 12.2.10$$

where B is an n x n matrix.

Let us rewrite equation 12.2.8 as

$$W = (F^T + G^{O\,T}_b) R(F + G^O_b) \qquad 12.2.11$$

where $F = G^d_b + T_d$ and $G_b = G^d_b + G^O_b$

where G^d_b is the diagonal part of G_b and G^O_b is the off-diagonal part of G_b. Let us next define $\mathcal{A}$ as the matrix comprising the diagonal parts of $(A - B G_b)$, $\mathcal{A}^\circ$ as the matrix comprising the off-diagonal parts of $(A - B G_d)$. Q, B_d are respectively the diagonal parts of Q^* and B. Q^O, B^O are the off-diagonal parts of Q^* and B respectively.

Then we can rewrite equation 12.2.7 as

$$\text{Min } J = \frac{1}{2}\int_0^T \left\{ \|\underline{x}\|^2_Q + \underline{x}^T F^T R F \underline{w} + g(\underline{x}, F, G^O_b) \right\} dt \qquad 12.2.12$$

where

$$g(\underline{x}, F, G_b^0) = \|\underline{x}\|^2_{Q^0} + \underline{x}^T F^T R G_b^0 \underline{x} + \underline{x}^T G_b^{0T} R F \underline{x} + \underline{x}^T G_b^{0T} R G_b^0 \underline{x} \quad 12.2.13$$

Also the dynamical equation (12.2.9) can be rewritten as

$$\dot{\underline{x}} = \mathcal{A} \underline{x} - B_d T_d \underline{x} + \underline{y}(\underline{x}, \underline{z}, T_d) \quad 12.2.14$$

where

$$\underline{y}(\underline{x}, \underline{z}, T_d) = \mathcal{A}^0 \underline{x} + C\underline{z} - B^0 T_d \underline{x} \quad 12.2.15$$

12.2.1 The three level calculation structure [1, 4]

In order to calculate the decentralised gain matrices we envisage a 3 level computational structure where we "fix" certain variables at the 2nd and third levels and this enables us to decompose at the 1st level into a number of independent low order subproblems.

Let us add additional equality constraints

$$\underline{x}^* = \underline{x} \quad 12.2.16$$

and

$$T_d^* = T_d \quad 12.2.17$$

into our original problem so that it becomes

$$\text{Min } J = \frac{1}{2} \int_0^T \left\{ \|\underline{x}\|^2_Q + \underline{x}^{*T} F R F \underline{x}^*, + g(\underline{x}^*, F, G_b^0) \right\} dt \quad 12.2.18$$

where

$$g(\underline{x}^*, F^*, G_b^0) = \|\underline{x}^*\|^2_{Q^0} + \underline{x}^{*T} F^T R G_b^0 \underline{x}^* + \underline{x}^{*T} G_b^{0T} R F \underline{x}^* + \underline{x}^* G_b^{0T} R G_b^0 \underline{x}^* \quad 12.2.19$$

$$\dot{\underline{x}} = \mathcal{A} \underline{x} - B_d T_d \underline{x}^* + \underline{y}(\underline{x}^*, \underline{z}, T^*_d) \quad 12.2.20$$

where

$$\underline{y}(\underline{x}^*, \underline{z}, T^*_d) = \mathcal{A}^0 \underline{x}^* + C\underline{z} - B^0 T^*_d \underline{x}^* \quad 12.2.21$$

$$\underline{z} = L\underline{x} \quad 12.2.22$$

$$T^*_d = T_d \quad 12.2.23$$

$$\underline{x}^* = \underline{x} \quad 12.2.24$$

In order to solve this problem, let us write the Hamiltonian for this modified problem as

$$H = \frac{1}{2} \|\underline{x}\|^2_Q + \frac{1}{2} \underline{x}^{*T} F^T R F \underline{x}^* + \frac{1}{2} g(\underline{x}^*, F, G_b^0) + \underline{\lambda}^T [\mathcal{A}\underline{x} - B_d T_d \underline{x}^* + \underline{y}(\underline{x}^*, \underline{z}, T_d^*)]$$

$$+ \underline{\pi}^T [L\underline{x} - \underline{z}] + \underline{\beta}^T [\underline{x} - \underline{x}^*] + \sum_{i=1}^n \nu_i (T_{d_i} - T^*_{d_i})$$

where $\underline{\pi}$, $\underline{\beta}$ and ν_i $(i = 1, 2, \ldots, n)$ are Lagrange multipliers and $\underline{\lambda}$ is the costate vector.

The necessary conditions for optimality yield

$$\underline{z} = L\underline{x} \qquad 12.2.25$$

$$\underline{\pi} = C^T \underline{\lambda} \text{ or } \underline{k} = L^T\underline{\pi} \qquad 12.2.26$$

$$\underline{x}^* = \underline{x} \qquad 12.2.27$$

$$T^*_{d_i} = T_{d_i} \qquad 12.2.28$$

$$\underline{\beta} = \left[F^T RF + Q^0 + F^T RG^0_b + G^{OT}_b RF + G^{OT}_b RG^0_b\right] \underline{x}^* + \left[\mathcal{A}^{OT} - T^*_d B^{OT} - T^T_d B^T_d\right] \underline{\lambda} \qquad 12.2.29$$

$$\underline{\nu} = \text{diag}\left\{ (R G^0_b \underline{x} - B^{OT} \underline{\lambda}) \underline{x}^{*T} \right\} \qquad 12.2.30$$

The basis of the 3 level algorithm is to fix $\underline{z}$ and $L^T \underline{\pi} = \underline{k}$ at level 3, $\underline{x}^*, \underline{\beta}, \underline{\nu}$ and T_{d_i} (i = 1 to n) at level 2 and this enables us to decompose the Hamiltonian such that at level 1 we treat only one variable at a time. Thus we have

$$\frac{\partial H}{\partial T_{d_i}} = 0 \text{ or } x^{*2}_i (G^d_{b_i} + T_{d_i})R_i - B_{d_i} \lambda_i x^*_i + \nu_i = 0$$

or

$$T_{d_i} = G^d_{b_i} - \frac{1}{R_i x_i^2} (-B_{d_i} \lambda_i x^*_i + \nu_i) \qquad 12.2.31$$

$$\frac{\partial H}{\partial \lambda_i} = \dot{x}_i \text{ or}$$

$$\dot{x}_i = \mathcal{A}_i x_i - B_{d_i} \left\{ -G^d_{b_i} - \frac{1}{x^{*2}_i} \left[\nu_i - B_{d_i} \lambda_i x^*_i\right]\right\} x_i + y_i(\underline{x}^*, \underline{z}, T^*_d) \qquad 12.2.32$$

$$\frac{\partial H}{\partial x_i} = - \dot{\lambda}_i \text{ or } \dot{\lambda}_i = -Q_i x_i - \mathcal{A}_i \lambda_i - k_i - \beta_i \qquad 12.2.33$$

with $\lambda_i(T) = 0$

Also we have

$$\dot{x}_i = \mathcal{A}_i x_i + B_{d_i} G^d_{b_i} x^*_i + \frac{B_{d_i}}{R_i x^*_i} \left[\nu_i - B_{d_i} \lambda_i x^*_i\right] + y_i(\underline{x}^*, \underline{z}, T^*_d) \qquad 12.2.34$$

with $x_i(0) = x_i(0)$

To solve the two point boundary value problem 12.2.33, 12.2.34, let us write

$$\lambda_i = P_i x_i + \eta_i \qquad 12.2.35$$

Then by simple manipulations we obtain

$$\dot{P}_i = - 2\mathcal{A}_i P_i + \frac{B^2_{d_i}}{R_i} P^2_i - Q_i \qquad 12.2.36$$

with $P_i(T) = 0$

$$\dot{\eta}_i = (-\mathcal{A}_i + \frac{P_i B_{d_i}^2}{R_i})\eta_i - P_i(B_{d_i} G_{b_i}^d x_i^* + \frac{B_{d_i}\nu_i}{R_i x_i^*} + y_i(\underline{x}^*, \underline{z}, T_d^*) - k_i - \beta_i \qquad 12.2.37$$

Thus our three level algorithm is:

Step 1 : We start with an initial guess for the trajectories $\underline{z}^{\ell}$ and $\underline{k}^{\ell}$ at level 3 and set the iteration index $\ell = 1$.

Step 2 : At level 2 we guess initially the trajectories $\underline{x}^{*\alpha}$, $T_{d_i}(i = 1 \text{ to } n)$, $\underline{\beta}^{\alpha}$ and $\underline{\nu}^{\alpha}$, and we set the iteration index $\alpha = 1$.

Step 3 : Using the trajectories $\underline{x}^{*\alpha}$, $T_d^{*\alpha}$, $\underline{\beta}^{\alpha}$ and $\underline{\nu}^{\alpha}$, we calculate P_i and η_i using equations 12.2.36 and 12.2.37 for each variable and $\underline{\lambda}$ from the relation $\underline{\lambda} = P\underline{x} + \underline{\eta}$. Finally, we calculate T_d using equation 12.2.31.

Step 4 : Using equations 12.2.27 to 12.2.30, the values of $\underline{x}$ and $\underline{\lambda}$ calculated at the first level and those of $\underline{z}^{\ell}$ and $\underline{k}^{\ell}$ supplied by the 3rd level we calculate at level 2 the new trajectories $\underline{x}^{*\alpha+1}$, $T_{d_i}^{*\alpha+1}$, $\underline{\beta}^{\alpha+1}$ and $\underline{\nu}^{\alpha+1}$. If the integral of the norm of the difference between $\underline{x}^{*\alpha+1}$, $\underline{x}^*$ as well as respectively between $(T_d^{*\alpha+1}, T_d^{*\alpha})$, $(\underline{\beta}^{\alpha+1}, \underline{\beta}^{\alpha})$, $(\underline{\nu}^{\alpha+1}, \underline{\nu}^{\alpha})$, is not sufficiently small, we go back to step 3. Otherwise we go to the 3rd level and calculate the new trajectories for $\underline{k}^{\ell+1}$ and $\underline{z}^{\ell+1}$ using equations 12.2.25 and 12.2.26. Here again we test if the integral of the norm of the difference respectively between $\underline{z}^{\ell+1}$, $\underline{z}^{\ell}$ is small. If so, we record T_d as the decentralised gain matrix. Otherwise we go to step 2 and use $\underline{z}^{\ell+1}$, $\underline{k}^{\ell+1}$ as the new predictions.

Remarks

1. This is a prediction type of algorithm and its convergence can be proved using a technique similar to that used by Hassan and Singh [6] for non - linear systems.

2. The algorithm is very simple since at the lowest level mere substitution into single variable formulae is required.

3. At the optimum, we obtain T_d and thus the decentralised gain matrix.

4. It should be noted that the entire calculation is done off-line and only the decentralised gains are used on-line to compute and implement the optimal decentralised control.

5. Although the decentralised gains are initial state dependent, computational experience indicates that the control is relatively insensitive to small variations in initial conditions. This arises from the fact that on examining equation 12.2.14 for T_{d_i}, we find that T_{d_i} depends on the ratio of the states and costates. Intuitively, since the dynamics dictate the relationships between the various states and costates and these do not change significantly for small variations in initial conditions, the optimal T_{d_i} will only be changed slightly if the initial conditions change so that not changing T_{d_i} would still give us a good suboptimal control as we see in an example latter on.

Next, let us put a bound on the suboptimality of the controller.

12.2.2 Suboptimality bounds

Since our control is suboptimal, it would be useful to have an expression which bounds the possible performance degradation.

For the optimal system we have

$$\dot{\hat{\underline{x}}} = (A - B\,G_b)\,\hat{\underline{x}} + c\,L\hat{\underline{x}} - B\,T\,\hat{\underline{x}}$$

or

$$\hat{\underline{x}}\,(t) = \phi(t,\,0)\,\underline{x}(0)$$

For the suboptimal case let $T_d = T - T_d''$, then $\underline{x}_{sub}(t)$ can be written as

$$\underline{x}_{sub}(t) = \phi\,(t,\,0)\underline{x}(0) + \int_0^T \phi\,(t,\tau)\;B\;T_d''\;\underline{x}_{sub}(\tau)\;d\tau$$

$$= \hat{\underline{x}}(t) + \left[\int_0^T \phi\,(t,\tau)\;BT_d''\;\phi^T(t,\tau)\;d\tau\right]\underline{x}(0)$$

$$= \hat{\underline{x}}(t) + \tilde{\underline{x}}(t)$$

$$= \left[I + \lambda\,(t,\,0)\right]\hat{\underline{x}}(t) = \Gamma\;(t,\,0)\;\hat{\underline{x}}(t)$$

Here $\dot{\phi}^T(t,\tau) = (A - BG_b + CL - BT_d)\,\phi^T(t,\tau)$

$$\lambda(t,\,0) = \left[\int_0^T \phi(t,\tau)\;BT_d''\phi^T(t,\tau)\;d\tau\right]\phi\;(0,\,t)$$

Thus we have :

$$J_{sub} = \frac{1}{2}\int_0^T\left\{\|\underline{x}_{sub}\|_Q^2 + \underline{x}_{sub}^T(G_b + T - T_d'')^T\,R(G_b + T - T_d'')\underline{x}_{sub}\right\}dt$$

$$= \frac{1}{2}\int_0^T\Big\{\|\underline{x}_{sub}\|_Q^2 + \underline{x}_{sub}^T(G_b + T)^T\,R(G_b + T)\,\underline{x}_{sub}$$

$$+ \underline{x}_{sub}^T T_d''^T\,RT_d''\,\underline{x}_{sub} - 2\underline{x}_{sub}^T T_d''^T R(G_b+T)\underline{x}_{sub}\Big\}\,dt$$

$$J_{sub} = \frac{1}{2}\int_0^T\left(\|\underline{x}_{sub}\|_{Q+w}^2 + \underline{x}_{sub}^T\;T_d''^T\;RT_d''\;\underline{x}_{sub} - 2\underline{x}_{sub}^T T_d''^T R(G_b+T)\underline{x}_{sub}\right)dt$$

$$= \frac{1}{2}\int_0^T\Big(\|\hat{\underline{x}}\|_{Q+w}^2 + \|\tilde{\underline{x}}\|_{Q+w}^2 + 2\,\tilde{\underline{x}}^T(Q+w)\,\hat{\underline{x}}$$

$$+ \|\underline{x}_{sub}\|_{T_d''^T RT_d''}^2 - 2\,\|\underline{x}_{sub}\|_{T_d''^T R(G_b+T)}^2\Big)\,dt$$

$$J_{sub} = \frac{1}{2}\int_0^T \|\hat{\underline{x}}\|_{Q+w}^2\;dt + \frac{1}{2}\int_0^{t_f}\Big\{\|\hat{\underline{x}}\|_{(\Gamma-I)^T(Q+w)(\Gamma-I)}^2$$

$$+ 2\,\|\hat{\underline{x}}\|_{(\Gamma-I)^T(Q+w)}^2 + \|\hat{\underline{x}}\|_{\Gamma^T T_d''^T RT_d''\Gamma}^2 - 2\,\|\hat{\underline{x}}\|_{\Gamma^T T_d''^T R(G_b+T)\Gamma}^2\Big\}\;dt$$

$$= J_{opt} + \frac{1}{2}\int_0^T \|\hat{x}\|^2_{F(t)}$$

where $F(t) = (\Gamma(t, 0) - I)^T (Q+w) (\Gamma - I) + [(\Gamma(t, 0) - I]^T (Q + w)$

$$+\Gamma^T(t, 0)\, T''^T_d RT''_d \Gamma(t, 0) - 2\Gamma^T(t, 0)\, T''^T_d R(G_b+T)\Gamma(t, 0)$$

Now since $\int_0^T \|\hat{x}\|^2_F\, dt \leqslant \mu_M(F) \int_0^T \|\hat{x}\|^2\, dt$

and $\int_0^T \|\hat{x}\|^2\, dt \leqslant \frac{1}{\mu_m(Q+w)} \int_0^T \|\hat{x}\|^2_{Q+w}\, dt$

$$J_{sub} \leqslant (1+\mathcal{E})\, J_{opt}$$

$$0 \leqslant \mathcal{E} \leqslant \frac{\mu_M(F)}{\mu_m(Q+w)}$$

where $\mu_M(F)$ = the max norm of F

$\mu_m(Q+w)$ = minimum eigenvalue of Q + w

Next we apply the algorithm to a river pollution control model to provide decentralised control.

Example : River pollution control

The problem of control of pollution in a river system has previously been treated by Singh et al. [7] and extensively in this book. The problem, as we have seen, is essentially to control optimally the treated sewage discharges from sewage stations on a river system in order to reach a compromise between the cost of treatment and the cost incurred by having too polluted a river. This problem can be treated as a "linear quadratic" problem. If the problem is solved as is done by Singh et al. [7] then the resulting optimal control requires feedback of all the states. The implementation of the feedback between the subsystems is very costly in this case because the subsystems, which are "reaches" of the river are often many kilometers distant from each other and the cost of telemetry links is the main cost of implementation of the control. If we implement the decentralised controller developed here, we would clearly have substantial savings and the controller would moreover be more robust. In the treatment in this section we will use the two reach no delay model.

The original problem here is to minimise

$$J = \int_0^8 \left\{ 2(x_1 - 4.06)^2 + (x_2-8)^2 + 2(x_3-5.94)^2 + (x_4-6)^2 + u_1^2 + u_2^2 \right\} dt$$

subject to

$$\frac{d}{dt}\begin{bmatrix} x_1 \\ x_2 \\ x_3 \\ x_4 \end{bmatrix} = \begin{bmatrix} -1.32 & 0 & 0 & 0 \\ -0.32 & -1.2 & 0 & 0 \\ 0.9 & 0 & -1.32 & 0 \\ 0 & 0.9 & -0.3 & -1.2 \end{bmatrix}\begin{bmatrix} x_1 \\ x_2 \\ x_3 \\ x_4 \end{bmatrix} + \begin{bmatrix} 0.1 & 0 \\ 0 & 0 \\ 0 & 0.1 \\ 0 & 0 \end{bmatrix}\begin{bmatrix} u_1 \\ u_2 \end{bmatrix} + \begin{bmatrix} 5.35 \\ 10.9 \\ 4.19 \\ 1.9 \end{bmatrix}$$

For this case, the optimal and the decentralised control gains were calculated. The diagonal gain matrix T_d was calculated for the initial conditions $[10 \quad 7 \quad 5 \quad 7]$ and the convergence to 10^{-4} was achieved in 4 iterations of the 3rd level which took 18.65 seconds to execute on the IBM 370/165 computer. To test the sensitivity of the decentralised controller to variations in initial conditions, the initial conditions were perturbed by 50 % to $[5 \quad 3.5 \quad 2.5 \quad 3.5]$. For this case, the optimal centralised control gave a cost of 6.79 whilst the decentralised controller $(G_b + T_d)$ calculated using the first set of initial conditions gave a cost of 7.98 showing that the decentralised control is indeed relatively insensitive to small variations in the initial conditions. For comparison, the completely decentralised control obtained by ignoring all interactions gave a cost of 8.45.

Figs. 12.1 to 12.4 give the trajectories of the optimal and suboptimal states. As we can see, these are very close to each other thus showing that the suboptimal controller is a good one.

Next we see how we can extend the controller design such that the system has a prespecified degree of stability.

12.2.3 Decentralised controller with a prespecified degree of stability. Problem formulation.

Consider an interconnected dynamical system comprising N subsystems described by :

$$\dot{\underline{x}}_i = A_i \underline{x}_i + B_i \underline{u}_i + \sum_{j=1}^{N} e_{ij} A_{ij} \underline{x}_j + \underline{d}_i \qquad i = 1, 2, \ldots, N \qquad 12.2.38$$

where e_{ij} are the elements of the interconnection matrix E which is continuous in time with $0 \leqslant e_{ij}(t) \leqslant 1$ and we would like to design a controller for this system by minimising a quadratic performance index and ensuring at the same time that the controlled system exibits a pre-specified degree of stability (in the sense of Anderson and Moore [8], chapter IV), i.e.

$$\text{Min } J = \sum_{i=1}^{N} \frac{1}{2} \int_0^T e^{2\alpha t} \left\{ \| \underline{x}_i - \underline{x}_i^d \|^2_{Q_i^*} + \| \underline{u}_i \|^2_{R_i} \right\} dt \qquad 12.2.39$$

Here, T is at least 3-4 time constants of the system so that it reaches the steady state. In order to convert this problem into the standard "linear quadratic" form, let us define

$$\hat{\underline{x}}(t) = e^{\alpha t} \underline{x}(t) \; ; \; \hat{\underline{u}}(t) = e^{\alpha t} \underline{u}(t) \qquad 12.2.40$$

Then we can rewrite our problem as

$$\text{Min } J = \sum_{i=1}^{N} \frac{1}{2} \int_0^T \| \hat{\underline{x}}_i - \hat{\underline{x}}_i^d \|^2_{Q_i^*} + \| \underline{u}_i \|^2_{R_i} \;) \, dt \qquad 12.2.41$$

subject to

$$\dot{\hat{\underline{x}}}_i = \hat{A}_i \hat{\underline{x}}_i + B_i \hat{\underline{u}}_i + \sum_{j=1}^{N} e_{ij} A_{ij} \hat{\underline{x}}_j + \hat{\underline{d}}_i(t) \qquad 12.2.42$$

where $\hat{A}_i = (A_i + \alpha_i I)$ where I is the identity matrix,

$$\hat{\underline{d}}_i(t) = \underline{d}_i e^{\alpha t}$$

For this system, we can write the optimal control $\hat{\underline{u}}_i(t)$ as :

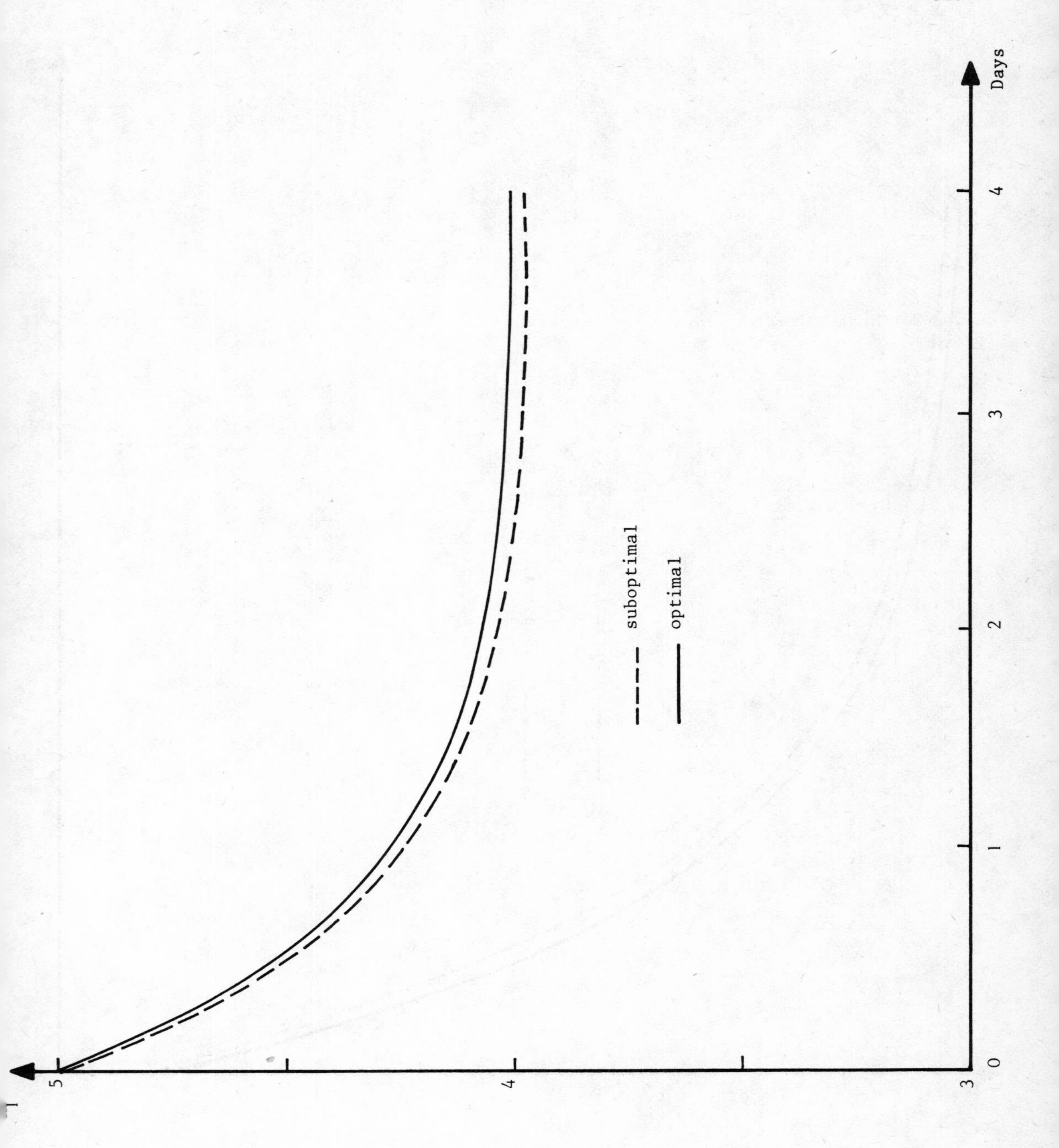

Fig. 12.1

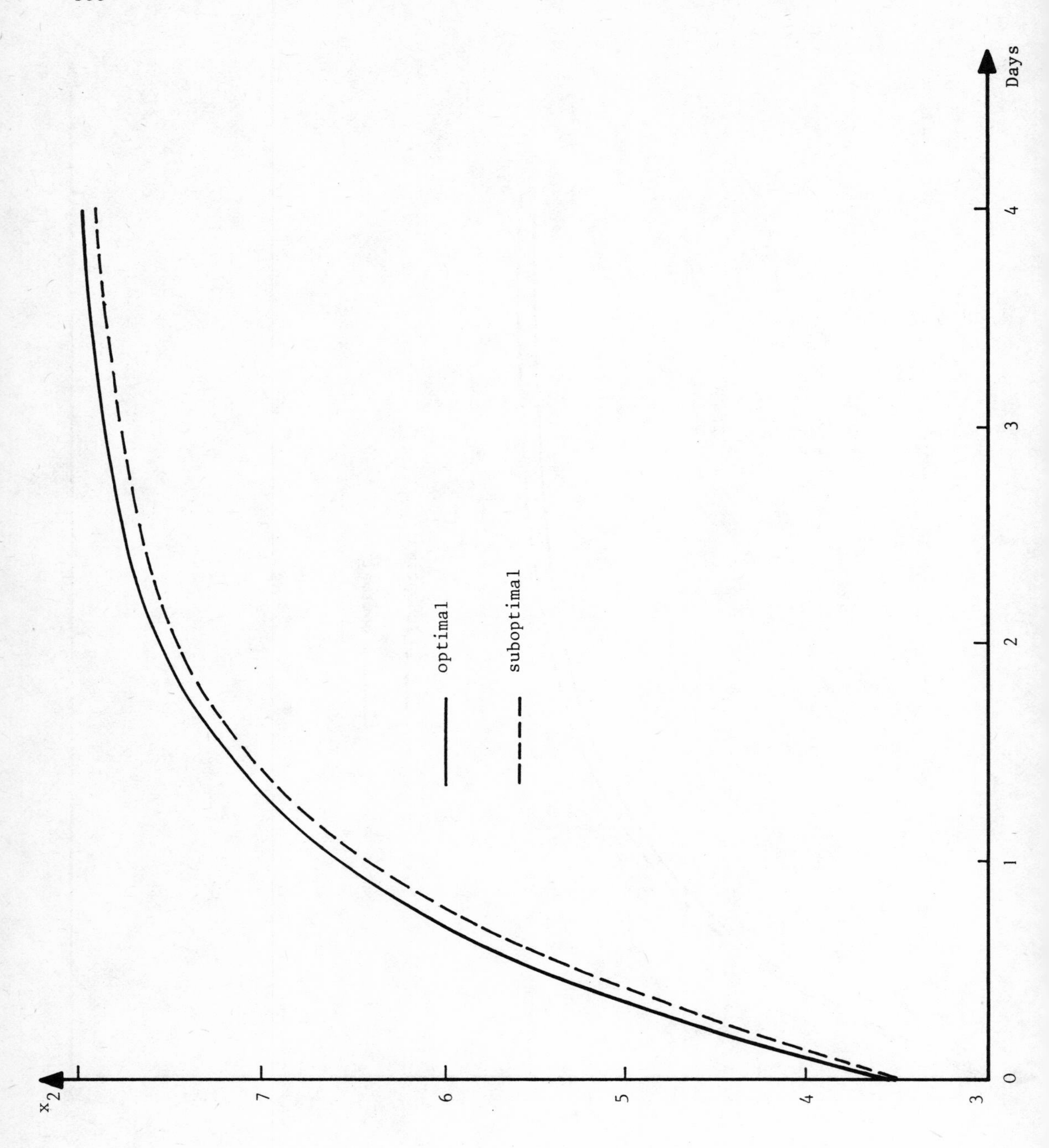

Fig. 12.2

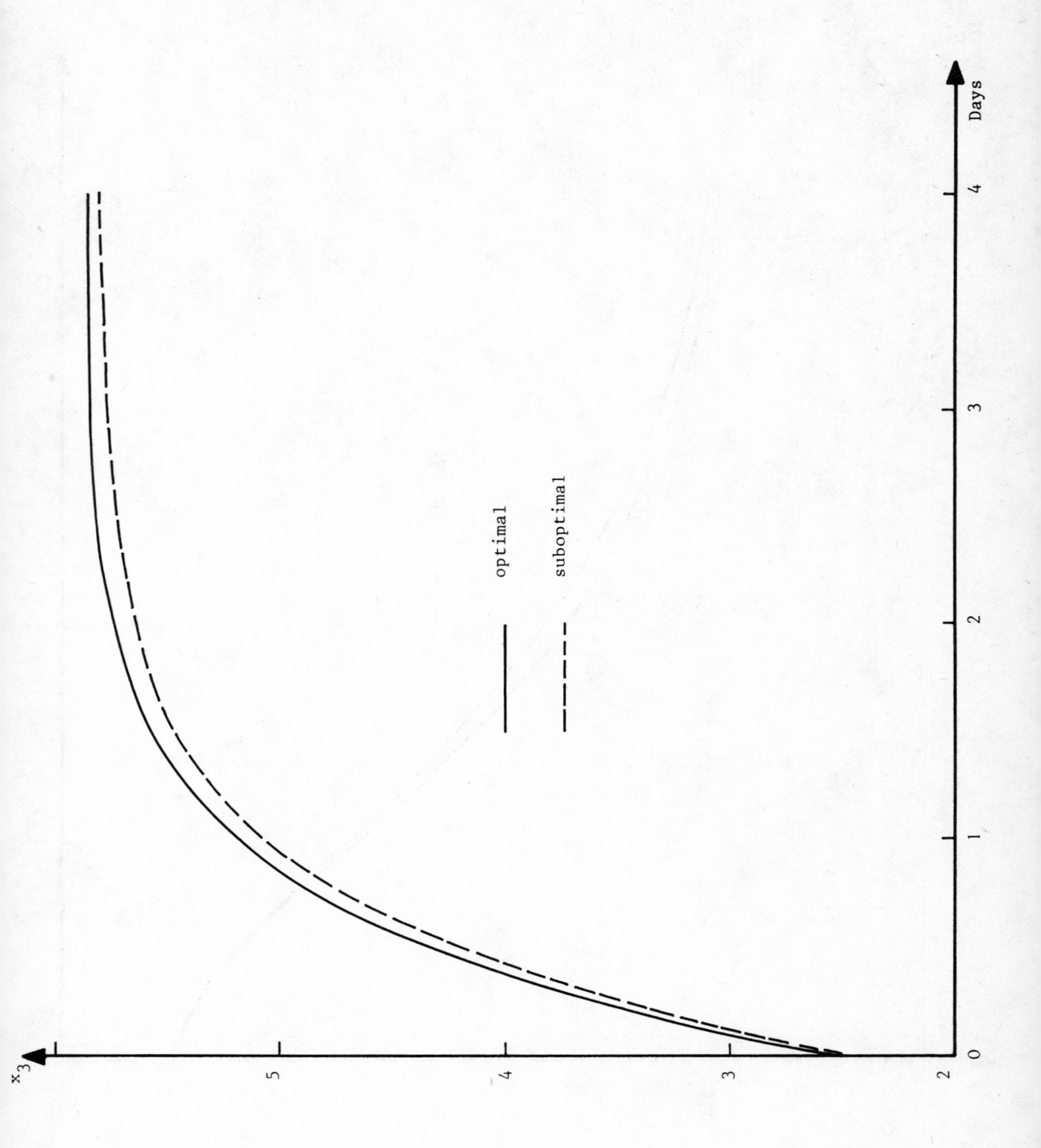

Fig. 12.3

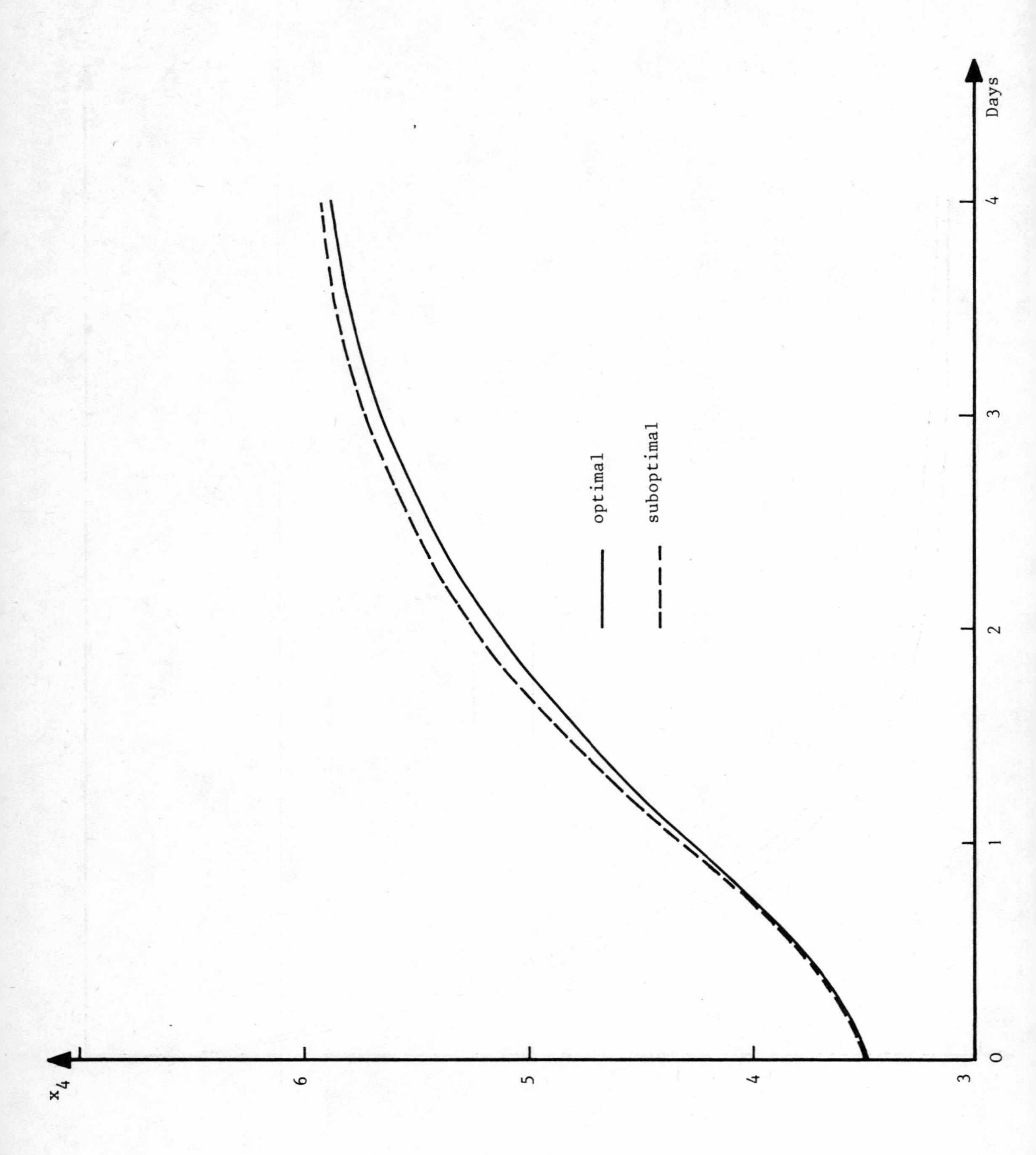

Fig. 12.4

$$\underline{\hat{u}}_i(t) = - G_{bi}\, \underline{\hat{x}}_i - \sum_{j=1}^{N} \left[e_{ij}\, T_{ij}\, \underline{\hat{x}}_j \right] - \underline{\hat{\xi}}_i \qquad 12.2.43$$

where $G_{bi} = - R_i^{-1} B_i^T P_i$ where P_i is the solution of the local (i.e. decomposed) Riccati equation for the ith subsystem.

Define $\mathcal{T} = \left[e_{ij}\ T_{ij} \right]$

then in the global form :

$$\underline{\hat{u}}(t) = - G_b\, \underline{\hat{x}} - \mathcal{T}\underline{\hat{x}} - \underline{\hat{\xi}} \qquad 12.2.44$$

and the minimal value of the cost function can be written as

$$J_{opt} = \frac{1}{2} \int_0^T \Big[\| \underline{\hat{x}}_o - \underline{\hat{x}}^d \|^2_{Q^*} + \| \underline{\hat{x}}_o \|^2_{W^*} + 2\, \underline{\hat{x}}_o^T\, G_b^T\, \underline{\hat{\xi}}$$

$$+ 2\, \underline{\hat{x}}_o^T\, \mathcal{T}^T R\, \underline{\hat{\xi}} + \underline{\hat{\xi}}^T R\, \underline{\hat{\xi}} \,)\Big]\, dt \qquad 12.2.45$$

where $\underline{\hat{x}}_o$ is the optimal state resulting from the application of the globally optimal control given by equation 12.2.44.

Let us write

$$W^* = G_b^T R G_b + G_b^T R\, \mathcal{T} + \mathcal{T}^T R G_b + \mathcal{T}^T R\, \mathcal{T} \qquad 12.2.46$$

In order to compute a completely decentralised control let us replace in equation 12.2.46 the matrix $\mathcal{T}$ by T_d which is constrained to be a diagonal matrix. We would therefore like to choose the matrix T_d in order to minimise the cost function

$$J = \frac{1}{2} \int_0^T \| \underline{\hat{x}} - \underline{\hat{x}}^d \|^2_{Q^*} + \| \underline{\hat{x}} \|^2_W + 2\underline{\hat{x}}^T\, G_b\, R\, \underline{\hat{\xi}}$$

$$+ 2\, \underline{\hat{x}}\, T_d^T\, R\, \underline{\hat{\xi}} + \underline{\hat{\xi}}^T R\, \underline{\hat{\xi}}\,)\, dt \qquad 12.2.47$$

subject to $\dot{\underline{\hat{x}}} = (\hat{A} - BG_b)\, \underline{\hat{x}} + C\, \underline{\hat{z}} + BT_d\, \underline{\hat{x}} + \underline{\hat{D}}$

$\underline{\hat{z}} = L\, \underline{\hat{x}}$

where $W = G_b^T R G_b + G_b^T R T_d + T_d^T R G_b + T_d^T R T_d$

B is a matrix of dimension n x n (if in practice B is of lower dimension than n x n we can introduce additional fictitious controls such that the corresponding elements in B which multiply these controls are zero)

$$\underline{\hat{D}}(t) = \underline{\hat{d}}(t) - B\underline{\hat{\xi}}$$

let us rewrite W as

$$W = (G_b^d + T_d + G_b^o)^T R (G_b^d + T_d + G_b^o) = (F + G_b^o)^T R (F + G_b^o) \qquad 12.2.48$$

where G_b^d is the matrix consisting of the diagonal elements of G_b, G_b^o is the matrix consisting of the off-diagonal elements of G_b. Let $\mathcal{A}$ be the matrix consisting of the diagonal part of $(\hat{A}-BG_d)$, $\mathcal{A}^o$ the matrix consisting of the off-diagonal elements of $(\hat{A}-BG_b)$, Q, B_d be the matrices consisting respectively of the diagonal elements of Q^* and B and Q^o, B^o the matrices consisting of the off-diagonal elements of Q^* and B respectively. Then we can write our optimisation problem as

$$\text{Min } J = \frac{1}{2}\int_0^T \Big\{ \| \underline{\hat{x}} - \underline{\hat{x}}^d \|_Q^2 + \underline{\hat{x}}^T F^T RF \underline{\hat{x}} + 2 \underline{\hat{x}}^T G_b^T R\underline{\hat{\xi}} + 2\underline{\hat{x}}^T T_d^T R\underline{\hat{\xi}} + \underline{\hat{\xi}}^T R\underline{\hat{\xi}} + g(\underline{\hat{x}}, F, G_b^o) \Big\} dt \qquad 12.2.49$$

subject to $\underline{\dot{\hat{x}}} = \mathcal{A}\,\underline{\hat{x}} - B_d T_d \underline{\hat{x}} + \underline{y}(\underline{\hat{x}}, \underline{\hat{z}}, T_d)$

$\underline{\hat{z}} = L\underline{\hat{x}}$

where $g(\underline{\hat{x}}, F, G_b^o) = \| \underline{\hat{x}} - \underline{\hat{x}}^d \|_{Q^o}^2 + \underline{\hat{x}}^T F^T R G_b^o \underline{\hat{x}} + \underline{\hat{x}}^T G_b^{o\,T} R F \underline{\hat{x}} + \underline{\hat{x}}^T G_b^{oT} R G_b^o \underline{\hat{x}}$

$\underline{y}(\underline{\hat{x}}, \underline{\hat{z}}, T_d) = \mathcal{A}^o \underline{\hat{x}} + C \underline{\hat{z}} - B^o T_d \underline{\hat{x}} + \underline{\hat{D}}$

12.2.4 The hierarchical computational structure to compute decentralised control

We will use an extension of the algorithm in section 12.2.1 to solve this problem. Thus let us introduce the additional linear constraints

$$T_d^* = T_d$$
$$\underline{\hat{x}}^* = \underline{\hat{x}}$$

Let us substitute these values into our problem so that it becomes

$$\text{Min } J = \frac{1}{2}\int_0^T (\| \underline{\hat{x}} - \underline{\hat{x}}^d \|_Q^2 + \underline{\hat{x}}^{*T} F^T R\hat{F} \underline{x}^* + 2 \underline{\hat{x}}^T G_b^T R \underline{\xi} + 2 \underline{\hat{x}}^{*T} T_d^T R\underline{\hat{\xi}} + \underline{\hat{\xi}}^T R\underline{\hat{\xi}}^T + g(\underline{x}^*, F^*, G_b^o)) \, dt$$

$$12.2.50$$

subject to $\underline{\dot{\hat{x}}} = \mathcal{A}\,\underline{\hat{x}} - B_d T_d \underline{x}^* + \underline{y}(\underline{x}^*, \underline{z}, T_d)$

$\underline{\hat{z}} = L \underline{\hat{x}}$

$T_d^* = T_d$

$\underline{\hat{x}}^* = \underline{\hat{x}}$

where

$$g(\underline{x}^*, F^*, G_b^o) = \| \underline{\hat{x}}^* - \underline{\hat{x}}^d \|_{Q^o}^2 + \underline{\hat{x}}^{*T} F^T R G_b^o \underline{\hat{x}}^* + \underline{\hat{x}}^{*T} G_b^{oT} RF \underline{\hat{x}}^* + \underline{\hat{x}}^{*T} G_b^{oT} RG_b^o \underline{\hat{x}}^* \qquad 12.2.51$$

$$\underline{y}(\underline{\hat{x}}^*, \underline{\hat{z}}, T_d^*) = \mathcal{A}^o \underline{\hat{x}}^* + C\underline{\hat{z}} - B^o T_d^* \underline{x}^* + \underline{\hat{D}}(t)$$

In order to solve this modified problem, let us write the overall Hamiltonian as

$$H = \frac{1}{2}\| \hat{\underline{x}} - \hat{\underline{x}}^d \|^2_Q + \frac{1}{2}\ \underline{\hat{x}}^{*}F^T RF\ \underline{\hat{x}}^{*} + \underline{\hat{x}}^T\ G_b^T\ R\ \hat{\underline{\xi}} + \underline{\hat{x}}^{*T}T_d^T\ R\ \hat{\underline{\xi}} + \frac{1}{2}\hat{\underline{\xi}}^T R\ \hat{\underline{\xi}} + \frac{1}{2}\ g(\underline{\hat{x}}^{*},\ F\ , G_b^o)$$

$$+ \underline{\lambda}^T[\mathcal{A}\ \underline{\hat{x}} - B_d\ T_d\ \underline{\hat{x}}^{*} + \underline{y}(\underline{x}^{*},\ \underline{z},\ T_d^{*})] + \underline{\pi}^T[L\underline{\hat{x}} - \underline{\hat{z}}] + \underline{\beta}^T(\underline{\hat{x}} - \underline{\hat{x}}^{*}) \qquad 12.2.52$$

$$+ \sum_{i=1}^{n} \gamma_i\ (T_{d_i} - T_{d_i}^{*})$$

where π, β, γ_i are Lagrange multipliers. Note that because we had constrained T_d to be a diagonal matrix, we can attach the scalar Lagrange multipliers γ_i to each constraint $T_{di} = T_{di}^{*}$ which means that we do not have to improve matrix functions at the second level.

The necessary conditions for optimality for this problem can be written as :

$$\left.\begin{aligned} \frac{\partial H}{\partial \pi} &= \underline{0} \qquad \text{or} \qquad \underline{\hat{z}} = L\ \underline{\hat{x}} \\ \frac{\partial H}{\partial \underline{\hat{z}}} &= \underline{0} \qquad \text{or} \qquad \underline{\pi} = C^T \underline{\lambda} \end{aligned}\right\} \qquad 12.2.53$$

$$\frac{\partial H}{\partial \underline{\beta}} = \underline{0} \qquad \text{or} \qquad \underline{\hat{x}}^{*} = \underline{\hat{x}}$$

$$\frac{\partial H}{\partial \gamma_i} = \underline{0} \qquad \text{or} \qquad T_{d_i}^{*} = T_{d_i}$$

$$\frac{\partial H}{\partial \underline{x}^{*}} = \underline{0} \qquad \text{or}$$

$$\underline{\beta} = F^T\ RF\ \underline{\hat{x}}^{*} + T_d^T\ R\ \hat{\underline{\xi}} + Q^o(\underline{\hat{x}}^{*} - \underline{\hat{x}}^d) + F^T\ \ RG_b^o\ \underline{\hat{x}}^{*} + G_b^{oT}RF\ \underline{x}^{*} + G_b^{oT}RG_b^o\ \underline{x}^{*}$$

$$- T_d^T\ B_d^T\ \underline{\lambda} + (\mathcal{A}^{oT} - T_d^{*T}\ B_o^T\ )\ \underline{\lambda}$$

$$\text{or}\ \ \underline{\beta}(t) = [F^T RF + Q^o + F^T RG_b^o + G_b^o\ RF + G_b^{oT}RG_b^o]\underline{\hat{x}}^{*} - Q^o\ \underline{\hat{x}}^d + T_d^T\ R\ \hat{\underline{\xi}} \qquad 12.2.54$$

$$+ (\mathcal{A}^{oT} - T_d^{*T}\ B^{oT} - T_d^T\ B_d^T)\ \underline{\lambda}$$

$$\frac{\partial H}{\partial T^{*}{}_d} = 0$$

$$\text{or}\qquad \gamma = \text{diag}\ [\frac{1}{2}\ R\ G_b^o\ \underline{\hat{x}}^{*}\ \underline{\hat{x}}^{*T} + \frac{1}{2}\ R\ G_b^o\ \underline{\hat{x}}^{*}\ \underline{\hat{x}}^{*T} - B^{oT}\ \underline{\lambda}\,\underline{\hat{x}}^{*T}\ ]$$

$$\text{or}\qquad \gamma = \text{diag}\ [(R\ G_b^o\ \underline{\hat{x}}^{*} - B^{oT}\ \underline{\lambda})\ \underline{\hat{x}}^{*T}]$$

Now, if we assume that $\underline{\hat{x}}^{*}$, T_d^{*}, $\underline{\beta}$, $\underline{\gamma}$, $\underline{\hat{z}}$ and $\underline{k} = L^T\underline{\pi}$ are given, then we can decompose the Hamiltonian into subsystems where each subsystem has only one variable.

Let $R\hat{\xi} = \underline{v}$, $G_b^T R \hat{\xi} = \underline{r}$. Then $\frac{\partial H}{\partial T_{d_i}} = 0$ gives

$$T_{d_i} = -G_{b_i}^d - \frac{1}{R_i x_i^2}\left[\hat{x}_i^* v_i - B_{d_i}\lambda_i \hat{x}_i^* + \nu_i\right]$$

$$\frac{\partial H}{\partial \lambda_i} = \mathring{\hat{x}}_i$$

or $$\dot{\hat{x}}_i = \mathcal{A}_i \hat{x}_i - B_{d_i}\left\{-G_{b_i}^d - \frac{1}{R_i \hat{x}_i^{*2}}\left[\hat{x}_i^* v_i + \nu_i - B_{d_i}\lambda_i \hat{x}_i^*\right]\right\}\hat{x}_i^* + y_i(\underline{\hat{x}}, \underline{\hat{z}}, T_d^*) \qquad 12.2.55$$

$$\frac{\partial H}{\partial \hat{x}_i} = -\dot{\lambda}_i$$

or $$-\dot{\lambda}_i = Q(\hat{x}_i - \hat{x}_i^d) + r_i + \mathcal{A}_i \lambda_i + k_i + \beta_i$$

or $$\dot{\lambda}_i = -Q_i(\hat{x}_i - \hat{x}_i^d) - r_i - \mathcal{A}_i \lambda_i - k_i - \beta_i \qquad 12.2.56$$

and $$\dot{\hat{x}}_i = \mathcal{A}_i \hat{x}_i + B_{d_i} G_{b_i}^d \hat{x}_i^* + \frac{B_{d_i}}{R_i} v_i + \frac{B_{d_i}}{R_i \hat{x}_i^*}\nu_i - \frac{B^2_{d_i}}{R_i}\lambda_i + y_i(\underline{\hat{x}}^*, \underline{\hat{z}}, T_d^*) \qquad 12.2.57$$

Now to solve the two point boundary value problem given by equations 12.2.56 and 12.2.57, with known terminal conditions, let us put

$$\lambda_i = P_i x_i + \eta_i$$

Then it is easy to see that P_i and η_i are given by

$$\dot{P}_i = -2\mathcal{A} P_i + \frac{B^2_{d_i}}{R_i} P_i^2 - Q_i \qquad 12.2.58$$

$$\dot{\eta}_i = \left(-\mathcal{A}_i + \frac{P_i B^2_{d_i}}{R_i}\right)\eta_i - P_i(B_{d_i} G_{b_i}^d \hat{x}_i^*) + B_{d_i}\frac{\nu_i}{R_i} + \frac{B_{d_i}\nu_i}{R_i \hat{x}_i^*} + y_i(\underline{\hat{x}}^*, \underline{\hat{z}}, T_d^*) + Q_i \hat{x}_i^d - r_i - k_i - \beta_i \qquad 12.2.59$$

Thus we envisage a three level hierarchical algorithm for solving this problem.

The algorithm is given by :

Step 1 : Guess the initial trajectories $\underline{\hat{z}}^\ell$ and $\underline{k}^\ell$ at level 3 for the iteration index $\ell = 1$.

Step 2 : Guess the initial trajectories $\underline{x}^{*\alpha}$, $T_d^{*\alpha}$, $\underline{\beta}^\alpha$, $\underline{\nu}^\alpha$ at the second level and set the iteration index $\alpha = 1$.

Step 3 : Using $\underline{x}^{*\alpha}$, $T_d^{*\alpha}$, $\underline{\beta}^\alpha$, $\underline{\nu}^\alpha$ obtained from step 2, calculate P_i, η_i, from equations 12.2.58 and 12.2.59 and thus $\underline{\hat{x}}$ from equation 12.2.57 and $\underline{\lambda}$ from equation 12.2.56. Calculate also T_d.

Step 4 : Substitute the $\underline{x}$, $\underline{\lambda}$, obtained at level one into the R.H.S. of equation 12.2.54 to obtain $\hat{\underline{x}}^{*\alpha+1}$, $T_{d_i}^{*\alpha+1}$, $\beta^{\alpha+1}$ and $\gamma^{\alpha+1}$. If the norm of the difference between $\underline{x}^{*\alpha+1}$ and $\underline{x}^{*\alpha}$ as well as between $(T_{d_i}^{*\alpha+1}, T_{d_i}^{*\alpha})$, $(\beta^{\alpha+1}, \beta^{\alpha})$, $(\gamma^{\alpha+1}, \gamma^{\alpha})$ is not sufficiently small, go to step 2. Otherwise go to level 3 to calculate $\underline{\pi}^{\ell+1}$, $\hat{\underline{z}}^{\ell+1}$ from equations 12.2.53. If the norm of the difference between $(\underline{\pi}^{\ell+1}, \underline{\pi}^{\ell})$ and between $(\hat{\underline{z}}^{\ell+1}, \underline{z}^{\ell})$ is sufficiently small, record T_d as the decentralised gain matrix, otherwise go to step 2 using $\underline{\pi}^{\ell+1}$, $\hat{\underline{z}}^{\ell+1}$ as the new guess.

Remarks

1. This is a prediction type algorithm whose convergence can be proved as for the previous method using a technique similar to the one used by Hassan and Singh [6].

2. The algorithm is very simple since at the lowest level mere substitution into single variable formulae is all that is required.

3. At the optimum, we obtain T_d and thus the decentralised gain matrix.

4. The entire calculation is done off line and only the decentralised gains are implemented on line to produce the decentralised control.

5. Although the decentralised gains are initial state dependent, computational experience indicates that the control is relatively insensitive to small variations.

6. In practice, we will use a set of "average" initial conditions to compute the decentralised gains off-line and then implement the gains on-line to control the system even when the initial conditions change.

7. The controller has a prespecified degree of stability α and this stability aspect will be studied in the next section.

Note that when the interconnection matrix e_{ij} is considered, the closed loop system will be

$$\dot{\underline{x}}_i = A_i \underline{x}_i - B_i G_{b_i} \underline{x}_i - B_i e_{ii} T_{d_i} \underline{x}_i + \sum_{j=1}^{N} e_{ij} A_{ij} \underline{x}_j + \underline{d}_i - B_i \underline{\xi}_i \qquad 12.2.60$$

where T_{d_i} is the diagonal part of T_d corresponding to subsystem i.

12.2.5 Stability of the decentralised control system

The stability properties of the controlled system can be studied via the following theorem.

Theorem

Part 1 : The system represented by equation 12.2.60 will be exponentially stable with degree α if the following condition is satisfied

$$\sum_{i=1}^{N} \sum_{j=1}^{N} e_{ij} f_{ij} + \sum_{i=1}^{N} \sum_{j=1}^{N} b_{ij} \leqslant \frac{1}{2} \frac{\min_i \lambda_m(W_i)}{\max_i \lambda_M(P_i)} + \frac{\min_i a_i \sum_{i=1}^{N} e_{ii}}{\max_i \left\{ \lambda_M(P_i) \right\}} \qquad 12.2.61$$

where W_i, P_i, $\lambda_M(W_i)$, $\lambda_M(P_i)$, f_{ij}, a_i, b_{ij} are defined in the proof below.

Part 2 : The system represented by equation 12.2.60 will be more stable than the corresponding system where the decentralised gain matrix G_{b_i} is used only.

Proof

Consider first the minimisation of the criterion 12.2.41 subject to equation 12.2.42 under the assumption that the structural connection between the subsystem is ideal and as a result we obtain the closed loop system

$$\dot{\hat{\underline{x}}} = (\hat{A} - B\, G_b - B\,\tau)\,\hat{\underline{x}} - B\,\hat{\underline{\xi}}(t) + \underline{d}(t)$$

or

$$\dot{\hat{\underline{x}}} = (\hat{A} - B\, G_b - B\,\tau)\,\hat{\underline{x}} + \hat{\underline{D}}(t)$$

If the system is stable and if the final time T is sufficiently large such that $\dot{\hat{x}} \rightarrow 0$ and $\hat{x} \rightarrow \hat{x}^d$ when $t \rightarrow T$, then at $t = T$ we have

$$\hat{\underline{D}}(T) = -(\hat{A} - B\, G_b - B\tau)\,\hat{\underline{x}}^d(T)$$

or

$$\hat{\underline{D}}(t) = -(\hat{A} - B\, G_b - B\tau)\,\hat{\underline{x}}^d(t)$$

Now, we will study the stability of the system when only the local feedback gain G_{b_i} is applied and the interactions are ignored and where the gain T_d is used as well so that the interactions are taken into account.

Case 1 : Stability of the system controlled using the gain G_{b_i} only

To study this stability, consider the Lyapunov function

$$V = \sum_{i=1}^{N} \hat{\underline{x}}_i^T P_i \hat{\underline{x}}_i \qquad 12.2.62$$

where P_i is the solution of the decomposed Riccati equation from which we obtained G_{b_i}. Since P_i is positive definite, let us find the condition under which $\dot{V}$ is at least negative semidefinite. Let us assume that the final time T is sufficiently large and the interconnection matrix E has elements $\bar{e}_{ij}$ which are zero or one. Then

$$\frac{dV}{dt} = \sum_{i=1}^{N} \left\{ \left[\hat{\underline{x}}_i^T (\hat{A}_i^T - P_i B_i R_i^{-1} B_i^T) + \sum_{j=1}^{N} \bar{e}_{ij}\, \hat{\underline{x}}_j^T A_{ij}^T + \hat{\underline{D}}_i^T \right] P_i \hat{\underline{x}}_i + \right.$$

$$\left. \hat{\underline{x}}_i^T P_i \left[(\hat{A}_i - B_i R_i^{-1} B_i^T P_i)\,\hat{\underline{x}}_i + \sum_{j=1}^{N} \bar{e}_{ij} A_{ij} \hat{\underline{x}}_j + \hat{\underline{D}}_i \right] \right\}$$

$$= \sum_{i=1}^{N} \hat{\underline{x}}_i^T (- Q_i - P_i B_i R_i^{-1} B_i^T P_i) \hat{\underline{x}}_i + \sum_{i=1}^{N} 2\hat{\underline{x}}_i^T P_i \sum_{j=1}^{N} \bar{e}_{ij} A_{ij} \hat{\underline{x}}_j + 2 \sum_{i=1}^{N} \hat{\underline{x}}_i^T P_i \underline{D}_i$$

12.2.63

$$= -\sum_{i=1}^{N} \| \hat{\underline{x}}_i \|^2_{W_i^*} + 2 \sum_{i=1}^{N} \hat{\underline{x}}_i^T P_i \sum_{j=1}^{N} \bar{e}_{ij} A_{ij} \hat{\underline{x}}_j - 2 \sum_{i=1}^{N} \hat{\underline{x}}_i^T P_i \sum_{j=1}^{N} H_{ij} \hat{\underline{x}}_j^d (t)$$

where

$$H = (- \hat{A} + B G_b + B\tau) \quad ; \quad W_i^* = Q_i + P_i B_i R_i^{-1} B_i^T P_i$$

Now since $\sum_{i=1}^{N} \hat{\underline{x}}_i^T P_i \sum_{j=1}^{N} \bar{e}_{ij} A_{ij} \hat{\underline{x}}_j \leqslant \sum_{i=1}^{N} \lambda_M(P_i) \sum_{j=1}^{N} \bar{e}_{ij} f_{ij} \| \hat{\underline{x}}_j \|^2$

12.2.64

$$\leqslant \max_i \lambda_M (P_i) \sum_{i=1}^{N} \sum_{i=j}^{N} \bar{e}_{ij} f_{ij} \| \hat{\underline{x}}_j \|^2$$

where $\lambda_M (P_i)$ = max eigen-value of P_i

$$f_{ij} = \lambda_M^{1/2} (A_{ij}^T A_{ij})$$

12.2.65

$$\sum_{i=1}^{N} \| \hat{\underline{x}}_i \|^2_{W_i^*} \geqslant \min_i \lambda_m(W_i^*) \sum_{i=1}^{N} \| \hat{\underline{x}}_i \|^2$$

where $\lambda_M (W_i^*)$ = min eigen-value of W_i^*

$$\sum_{i=1}^{N} \hat{\underline{x}}_i^T P_i \sum_{j=1}^{N} H_{ij} \underline{x}_j^d (t) \leqslant \max \lambda_M (P_i) \sum_{i=1}^{N} \sum_{j=1}^{N} b_{ij} \| \hat{\underline{x}}_j \|^2$$

where

$$b_{ij} = \lambda_M^{1/2} (H_{ij}^T H_{ij})$$

Then for $\frac{dV}{dt} \leqslant 0$, i.e. for the system to be asymptotically stable, we have

$$2 \max_i \lambda_M (P_i) \sum_{i=1}^{N} \sum_{j=1}^{N} \bar{e}_{ij} f_{ij} + 2 \max_i \lambda_M (P_i) \sum_{i=1}^{N} \sum_{j=1}^{N} b_{ij} \leqslant \min_i \lambda_M (W_i)$$

12.2.66

$$\text{or} \quad \sum_{i=1}^{N} \sum_{j=1}^{N} \bar{e}_{ij} f_{ij} + \sum_{i=1}^{N} \bar{e}_{ii} a_i \leqslant \frac{1}{2} \frac{\min_i \lambda_m(W_i^*)}{\max_i \lambda_M(P_i)}$$

Since $|e_{ij}| \leqslant |\bar{e}_{ij}|$ element by element, the above equation holds for any interconnection matrix and the system will be asymptotically stable with degree α.

<u>Case 2 : Stability of the system when the gain T_d is applied as well</u>

Considering the same Lyapunov function as above, we have :

$$\frac{dV}{dt} = - \sum_{i=1}^{N} \| \hat{\underline{x}}_i \|^2_{W_i^*} + 2 \sum_{i=1}^{N} \hat{\underline{x}}_i^T P_i \sum_{j=1}^{N} \bar{e}_{ij} A_{ij} \hat{\underline{x}}_j + 2 \sum_{i=1}^{N} \hat{x}_i^T P_i \qquad 12.2.67$$

$$\sum_{j=1}^{N} H_{ij} \hat{\underline{x}}_j^d (t) - 2 \sum_{i=1}^{N} \hat{\underline{x}}_i^T P_i B_i e_{ii} T_{di} \hat{\underline{x}}_i$$

and since $\sum_{i=1}^{N} \| \underline{x}_i \|^2 \bar{e}_{ii} \geqslant \min_i a_i \sum_{i=1}^{N} \| \underline{x}_i \|^2 \bar{e}_{ii}$ 12.2.68

where $\Gamma_i = P_i B_i T_{d_i}$ and $a_i = \lambda_M^{1/2} (\Gamma_i^T \Gamma_i)$

then for dV/dt to be negative semidefinite :

$$\sum_{i=1}^{N} \sum_{j=1}^{N} \bar{e}_{ij} f_{ij} + \sum_{i=1}^{N} \sum_{j=1}^{N} b_{ij} \leqslant \frac{1}{2} \frac{\min_i \lambda_m (W_i^*)}{\max_i \lambda_M (P_i)} + \frac{\min_i a_i \sum_{i=1}^{N} \bar{e}_{ii}}{\max_i \lambda_M (P_{ii})} \qquad 12.2.69$$

Here again, since $(\bar{e}_{ij}) \leqslant (e_{ij})$ element by element, the above expression holds for any interconnection matrix E. In addition, if it is satisfied, the original system will be exponentially stable with degree α . This proves the first part of the theorem.

Comparing the expressions in equations 12.2.66 and 12.2.69, we see that expression 12.2.69 is less restrictive. Also, if 12.2.66 is satisfied, then the L.H.S. of 12.2.69 will be much less than the R.H.S. of the inequality 12.2.69 and hence the system will be more stable. This proves the second part of the theorem.

Example

In order to illustrate this approach and its stability property, we shall consider the following simple problem :

$$\min J = \frac{1}{2} \int_0^{10} \| \underline{x} \|_Q^2 + \| \underline{u} \|_R^2 \, dt$$

s.t. $\qquad \dot{\underline{x}} = A\underline{x} + B\underline{u}$

where $A = \begin{bmatrix} 0 & 1 & 1 \\ -2 & -3 & -0.5 \\ 1 & 2 & 3 \end{bmatrix}$; $B = \begin{bmatrix} 1 & 0 \\ 1 & 0 \\ 0 & 1 \end{bmatrix}$; $Q = I_3$, $R = I_2$, $\alpha = 0.0$

where I_i is the identity matrix of order i.

This problem has been solved for three different cases :

i - by using the globally optimal closed loop gain matrix
ii - by decomposing the problem into two subproblems, the first one with

$A_1 = \begin{bmatrix} 0 & 1 \\ -2 & -3 \end{bmatrix}$; $B_1 = \begin{bmatrix} 1 \\ 1 \end{bmatrix}$ and the second with $A_2 = [3]$; $B_2 = [1]$

and solving these two problems as if they were completely decoupled to obtain the block diagonal gain matrix G_b. The matrix, G_b, was used in the feedback loop and the suboptimal closed loop trajectories were obtained.

iii - by decomposing the problem into two subproblems and calculating the diagonal gain matrix T_d which has been used in addition to the matrix G_b in the feedback loop. The suboptimal trajectories for this case have been also calculated.

Figures 12.5 to 12.7 show the optimal and suboptimal responses for the above three cases. From these, one can conclude the following :

1. The system is globally stable in all the three cases.

2. Although for the second case the system is also stable, one can see from the trajectories that the system in this case has the tendency to oscillate which illustrates part 2 of theorem.

3. The suboptimal trajectories with the closed loop gain matrix $G_b + T_d$ are very close to the optimal ones.

We note that the present approach has two basic drawbacks, i.e. (a) the gains are initial state dependent and (b) we have to operate on the overall system as opposed to the subsystems.

In the approach developed in the next section, we provide a simple reduced order model for the interactions and using this interaction model, we compute a decentralised controller whose parameters are independent of the initial states and which can also accomodate unknown constant external disturbances. In addition, by a simple modification, the controller can be designed to have a prespecified degree of stability and be connectively asymptotically stable. This controller thus avoids most of the drawbacks of the ones studied up to now.

12.3 THE DECENTRALISED CONTROLLER DESIGN

12.3.1 The general approach

Let us consider the optimisation problem of large scale systems comprising N linear interconnected dynamical subsystems :

$$\text{Min } J = \frac{1}{2} \int_0^\infty (\|\underline{x}'\|_Q^2 \; \|\underline{u}'\|_R^2) \, dt \qquad 12.3.1$$

subject to

$$\dot{\underline{x}}' = A\underline{x}' + B\underline{u}' + C\underline{z}' \quad \text{(dynamical constraints)} \qquad 12.3.2$$

$$\underline{z}' = L\,\underline{x}' \quad \text{(interconnection constraints)} \qquad 12.3.3$$

where $Q \geqslant 0$, $R > 0$; A, B, C, are also block diagonal with the N blocks in A, B, C corresponding to those in Q, R. L is a complete matrix (the matrices B and C having ranks r and m where r, m, are the dimensions of $\underline{u}$, $\underline{z}$ respectively).

The above optimisation problem represents the problem of computing the N optimal control vectors $\underline{u}' = \begin{bmatrix} u'^1 \\ \vdots \\ u'^N \end{bmatrix}$

In order to make the resulting controller more robust, let us add the additional term $\Psi(\underline{u}'_2, \dot{\underline{u}}'_2)$ in the cost function. This has the effect of making the controller less sensitive to constant unknown disturbances [9]. Thus, we

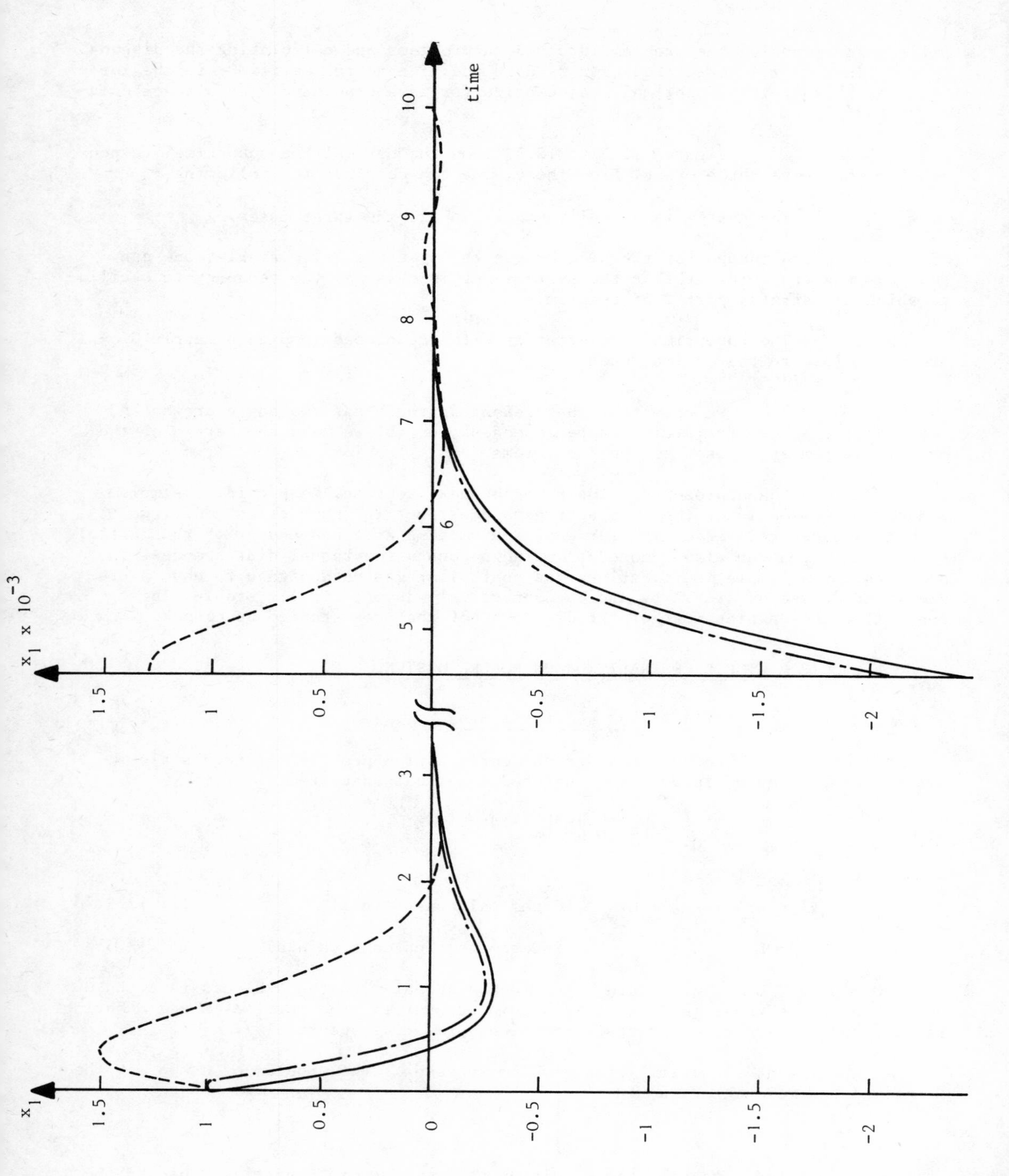

Fig. 12.5

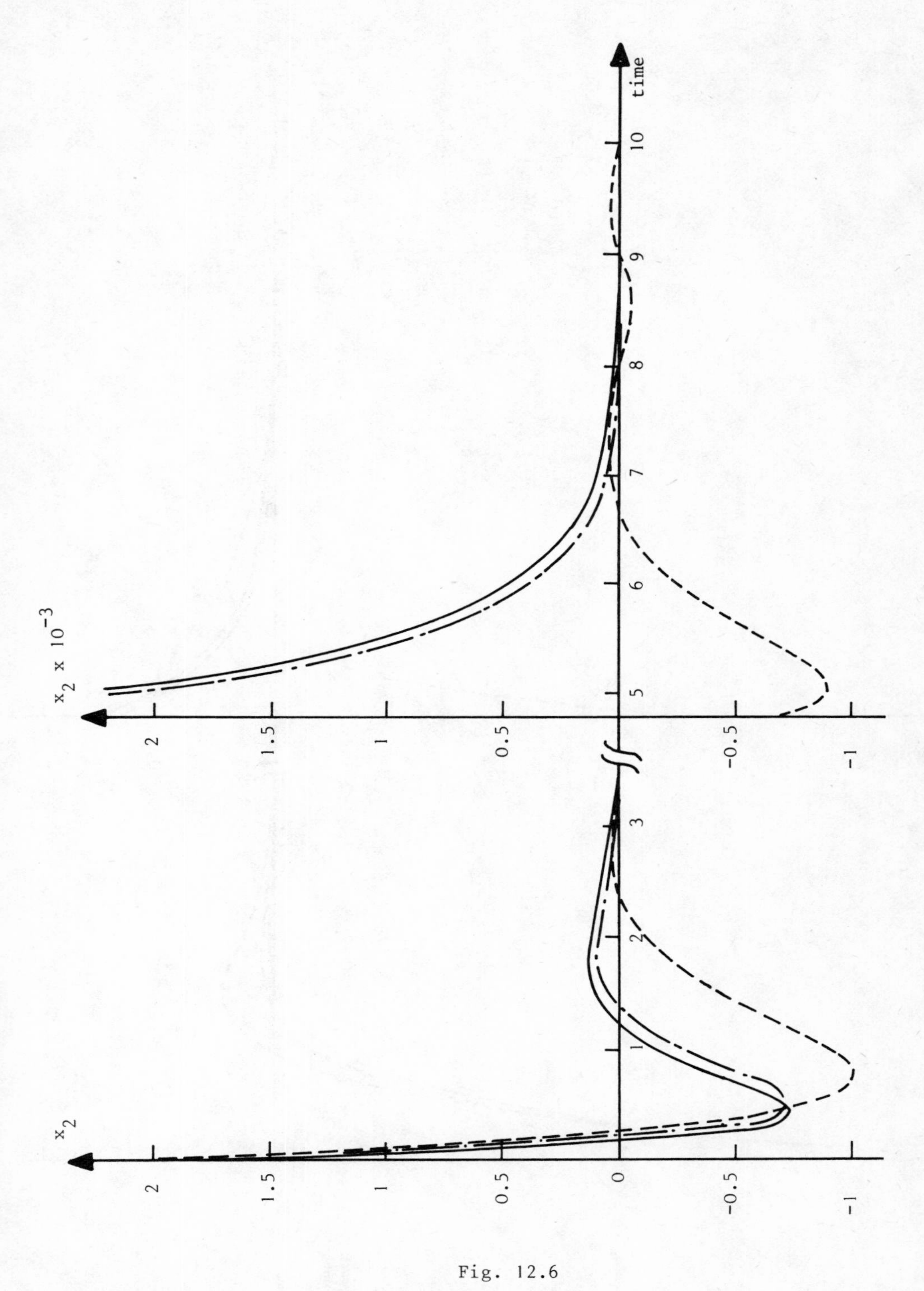

Fig. 12.6

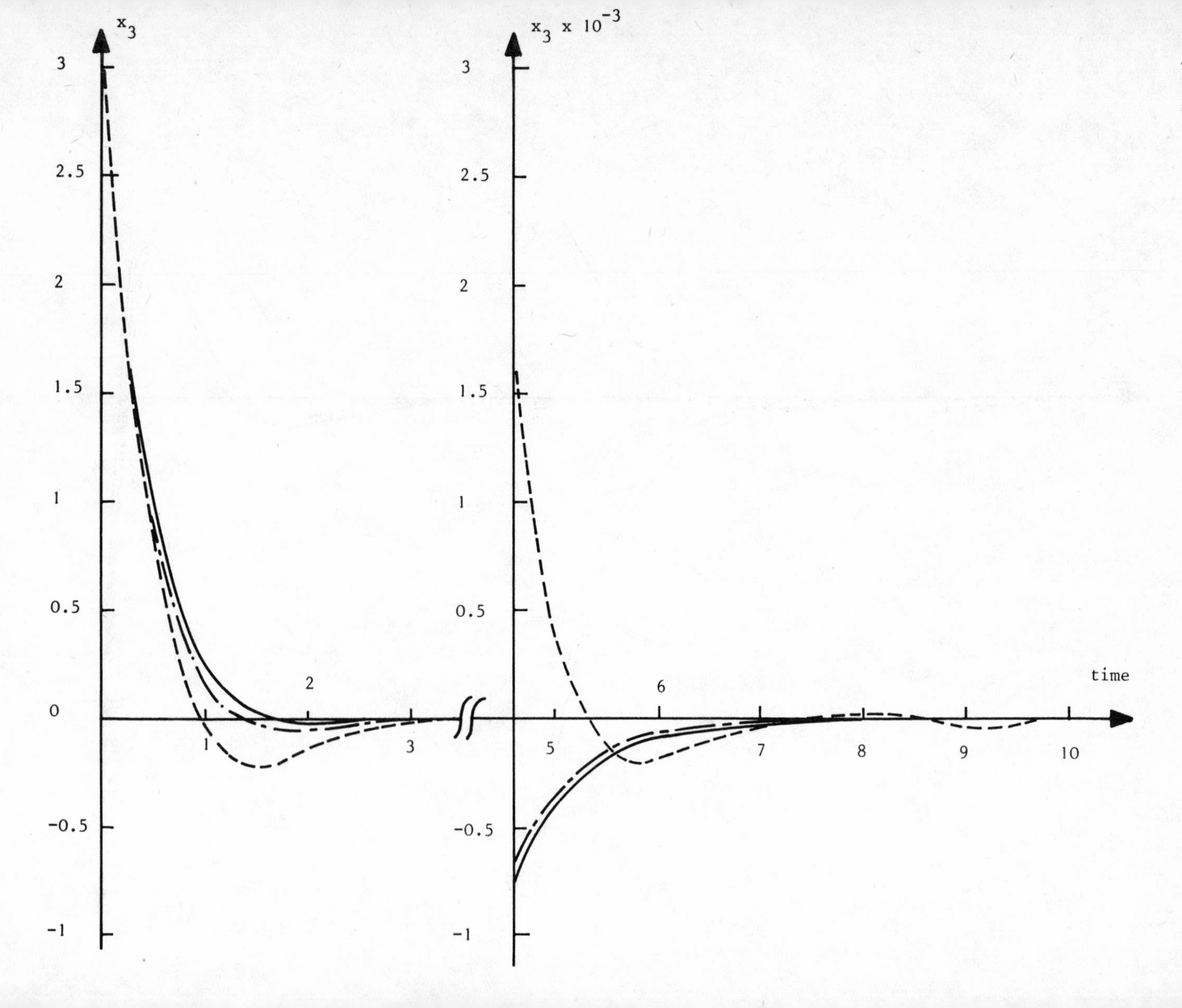

Fig. 12.7

rewrite the control as

$$\underline{u}' = \underline{u}'_1 + \underline{u}'_2 \qquad 12.3.4$$

and the control problem as

$$\text{Min } J' = \int_0^\infty (\| \underline{x}' \|_Q^2 + \| \underline{u}'_1 \|_R^2 + \Psi(\underline{u}'_2, \dot{\underline{u}}'_2))\, dt \qquad 12.3.5$$

where Ψ is an appropriate function (to be chosen latter) of $\underline{u}_2$ and $\dot{\underline{u}}_2$ subject to

$$\dot{\underline{x}}' = A\underline{x}' + B\underline{u}'_1 + C\underline{z}' + B\underline{u}'_2 \qquad 12.3.6$$

$$\underline{z}' = L\underline{x}' \qquad 12.3.7$$

Let us choose the vector $\underline{\gamma}_i$ which minimises the Euclidean norm of the vector $B_i\ \underline{u}'_{2i} - C_i \underline{\gamma}_i$, thus we have :

$$\underline{\gamma}_i = (C_i^T C_i)^{-1} C_i^T B_i \underline{u}'_{2i}$$

Substituting for $B_i \underline{u}'_{2i}$ in the dynamical constraint equation 12.3.6 by $C_i \underline{\gamma}_i$ from the above relation and replacing $\mathring{\underline{x}}'_i$, $\underline{u}'_{1i}$, $\underline{u}'_{2i}$, $\underline{z}'_i$ and $\mathring{\underline{u}}'_2$ by $\underline{x}_i$, $\underline{u}_{1i}$, $\underline{u}_{2i}$, $\underline{z}_i$ and $\mathring{\underline{u}}_{2i}$ the problem can be written in the following form :

$$\text{Min } J = \sum_{i=1}^{N} \frac{1}{2} \int_0^\infty \| \underline{x}_i \|_{Q_i}^2 + \| \underline{u}_{1i} \|_{R_i}^2 + \Psi_i(\underline{u}_{2i}, \mathring{\underline{u}}_{2i})\, dt \qquad 12.3.8a$$

$$\text{subject to } \dot{\underline{x}}_i = A_i \underline{x}_i + B_i \underline{u}_i + C_i \left[(C_i^T C_i)^{-1} C_i^T B_i \underline{u}_{2i} + \underline{z}_i \right] \qquad 12.3.8b$$

Now let us model the interactions that any particular subsystem "sees" as

$$\dot{\underline{z}} = A_z \underline{z} \qquad 12.3.9$$

Then in equation 12.3.8b let us put

$$\underline{y} = (C^T C)^{-1} C^T B \underline{u}_2 + \underline{z} \qquad 12.3.10$$

$$= \mathbb{B} \underline{u}_2 + \underline{z}$$

where $(C^T C)^{-1} C^T B = \mathbb{B}$

Differentiating equation 12.3.10, we have

$$\dot{\underline{y}} = \mathbb{B} \dot{\underline{u}}_2 + \dot{\underline{z}}$$

$$= \mathbb{B} \dot{\underline{u}}_2 + A_z \underline{z}$$

or

$$\dot{\underline{y}} = \mathbb{B} \dot{\underline{u}}_2 + A_z \underline{y} - A_z \mathbb{B} \underline{u}_2$$

Let us define $\underline{v} = \mathbb{B} \dot{\underline{u}}_2 - A_z \mathbb{B} \underline{u}_2$

Then we have :

$$\dot{\underline{x}} = A\underline{x} + B\underline{u}_1 + C\underline{y}$$

$$\dot{\underline{y}} = A_z \underline{y} + \underline{v}$$

Let $\underline{\tilde{x}} = \begin{pmatrix} \underline{x} \\ \underline{y} \end{pmatrix}$

Then we can write the following dynamical equation in $\underline{\tilde{x}}$, i.e.

$$\dot{\underline{\tilde{x}}} = \begin{bmatrix} A & C \\ 0 & A_z \end{bmatrix} \underline{\tilde{x}} + \begin{bmatrix} B & 0 \\ 0 & I \end{bmatrix} \begin{bmatrix} \underline{u}_1 \\ \underline{v} \end{bmatrix} \quad 12.3.11$$

where I is the identity matrix.

We can rewrite equation 12.3.11 as :

$$\dot{\underline{\tilde{x}}} = \tilde{A}\,\underline{\tilde{x}} + \tilde{B}\,\underline{\tilde{U}} \quad 12.3.12$$

where $\tilde{A} = \begin{bmatrix} A & C \\ 0 & A_z \end{bmatrix}$; $\tilde{B} = \begin{bmatrix} B & 0 \\ 0 & I \end{bmatrix}$; $\underline{U} = \begin{bmatrix} \underline{u}_1 \\ \underline{v} \end{bmatrix}$

An appropriate choice of Ψ in the function to be minimised in equation 12.3.5 is

$$\Psi = \underline{v}^T R\,\underline{v}$$

so that the cost function to be minimised subject to the dynamical constraints 12.3.12 is

$$\text{Min } J = \int_0^\infty \left(\|\underline{\tilde{x}}\|^2_{Q^o} + \begin{bmatrix} \underline{u}_1^T & \underline{v}^T \end{bmatrix} \begin{bmatrix} R_1 & 0 \\ 0 & R_2 \end{bmatrix} \begin{bmatrix} \underline{u}_1 \\ \underline{v} \end{bmatrix} \right) dt \quad 12.3.13$$

where $\tilde{Q}^o = \left[\begin{array}{c|c} Q & 0 \\ \hline 0 & 0 \end{array}\right]$

On solving this problem, we obtain

$$\underline{\tilde{U}} = -\tilde{R}^{-1}\tilde{B}^T\,\tilde{P}\,\underline{\tilde{x}}$$

or, for the ith subsystem :

$$\underline{\tilde{u}}_i = -\tilde{G}_{1i}\,\underline{x}_i - \tilde{G}_{2i}\,\underline{y}_i$$

Thus

$$\begin{bmatrix} \underline{u}_{1i} \\ \underline{v}_i \end{bmatrix} = \begin{bmatrix} -\tilde{G}_{1i_1} & \underline{x}_i \\ \hline -\tilde{G}_{1i_2} & \underline{x}_i \end{bmatrix} - \begin{bmatrix} \tilde{G}_{2i_1} & \underline{y}_i \\ \hline \tilde{G}_{2i_1} & \underline{y}_i \end{bmatrix} \quad 12.3.14$$

or

$$\underline{u}_{1i} = -\tilde{G}_{1i_1}\,\underline{x}_i - \tilde{G}_{2i_1}\,\underline{y}_i \quad 12.3.15$$

$$\underline{v}_i = -\tilde{G}_{1i_2}\,\underline{x}_i - \tilde{G}_{2i_2}\,\underline{y}_i \quad 12.3.16$$

Now from the definition of $\underline{v}$,

$$\mathbb{B}_i\,\underline{\mathring{u}}_{2i} = \underline{v}_i + A_{2i}\,\mathbb{B}_i\,\underline{u}_{2i} \quad 12.3.17$$

As before, let us chose the two vectors α_i and β_i which minimise the Euclidean norm of the two vector equations

$$\mathbb{B}_i \underline{\alpha}_i - A_{2i} \mathbb{B}_i \underline{u}_{2i} \quad \text{and} \quad \mathbb{B}_i \underline{\beta}_i - \underline{v}_i$$

Then using this in the equation of $\dot{\underline{u}}_2$ after replacing $\underline{u}_2$ by $\underline{u}''_2$ we have :

$$\dot{\underline{u}}''_{2i} = (\mathbb{B}_i^T \mathbb{B}_i)^{-1} \mathbb{B}_i^T A_{z_i} \mathbb{B}_i \underline{u}''_{2i} + (\mathbb{B}_i^T \mathbb{B}_i)^{-1} \mathbb{B}_i^T (-\tilde{G}_{1i_2} \underline{x}_i - \tilde{G}_{2i_2} \underline{y}_i)$$

$$= A_{u_i} \underline{u}''_{2i} - H_i \underline{x}_i - F_i \underline{y}_i$$

where $A_{u_i} = (\mathbb{B}_i^T \mathbb{B}_i)^{-1} \mathbb{B}_i^T A_{z_i} \mathbb{B}_i$

$$H_i = (\mathbb{B}_i^T \mathbb{B}_i)^{-1} \mathbb{B}_i^T \tilde{G}_{1i_2}$$

$$F_i = (\mathbb{B}_i^T \mathbb{B}_i)^{-1} \mathbb{B}_i^T \tilde{G}_{2i_2}$$

Thus, we can rewrite the equations for computing the controls as

$$\dot{\underline{x}}_i = (A - B_i \tilde{G}_{1i_1}) \underline{x}_i - B_i \tilde{G}_{2i_1} \underline{y}_i + \tilde{C}_i \underline{z}_i + B_i \underline{u}''_{2i} \qquad 12.3.18$$

and

$$\dot{\underline{y}}_i = (A_{z_i} - \tilde{G}_{2i_2}) \underline{y}_i - \tilde{G}_{1i_2} \underline{x}_i \qquad 12.3.19$$

$$\dot{\underline{u}}''_{2_i} = A_{u_i} \underline{u}''_{2_i} - H_i \underline{x}_i - F_i \underline{y}_i$$

By these equations, it is possible to compute the controls using the analogue structure given in Fig. 12.8.

Remarks

1. The computation of the control is extremely simple since the analogue computation (and the pre-calculation of the blocks in the analogue diagrams) is done subsystem by subsystem so that it is not necessary to solve the global problem off-line within a hierarchical structure as in the previous method.

2. The controls are independent of initial conditions.

3. The controller can accomodate unknown non-zero constant disturbances and still reach the desired steady state.

4. The controller is completely decentralised.

We illustrate the new controller on the three reach 22nd order river pollution control problem which is a good example for the application of decentralised control because of the long distance between the control centers.

Application of the controller to the river problem (three reach distributed delay model).

As we have seen, the model is the following :

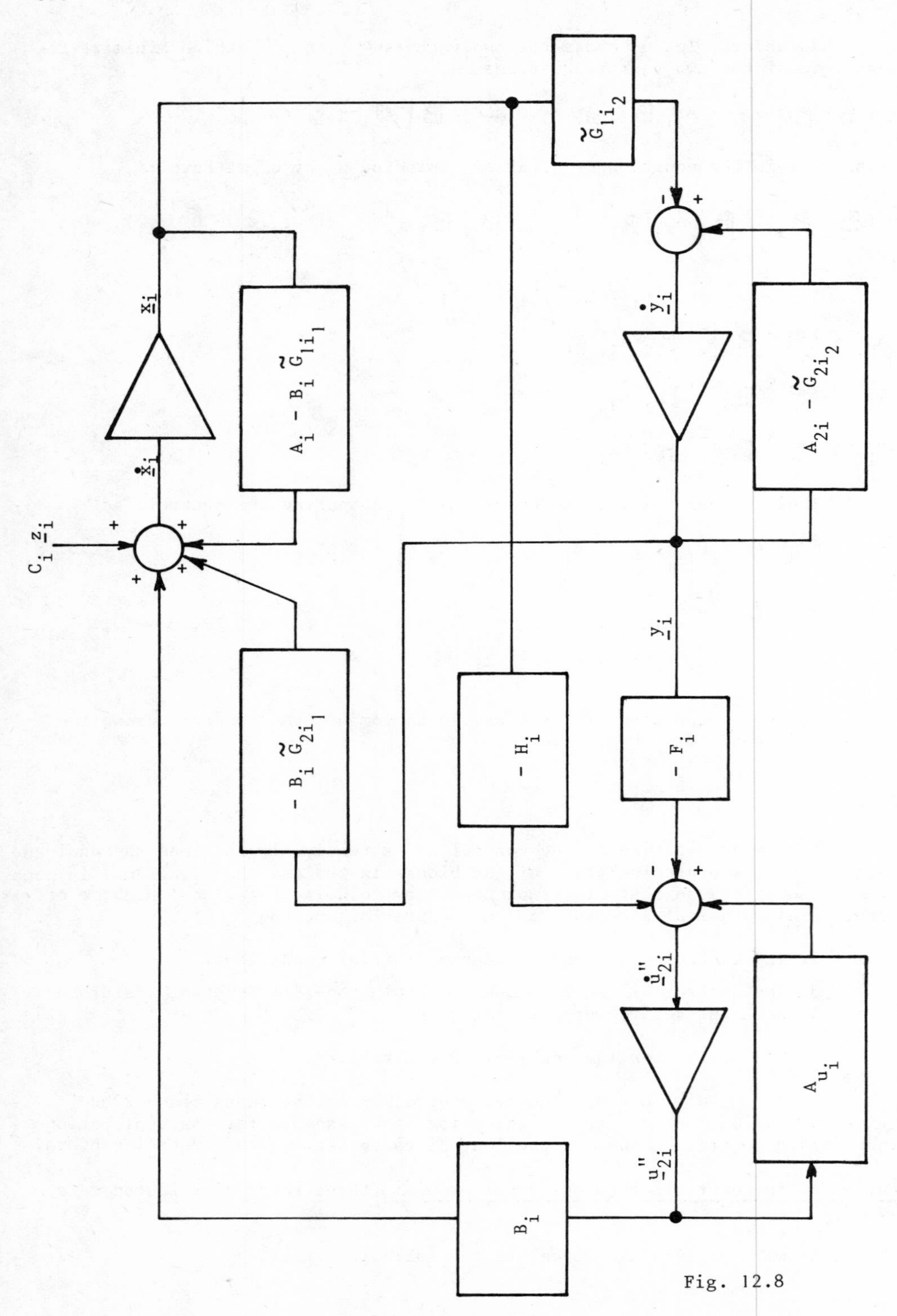

Fig. 12.8

$$\dot{z}_1 = -\ 1.32\ z_1 + 0.1\ m_1 + 0.9\ z_0 + 5.35 + 0.1\ u_1$$

$$\dot{q}_1 = -\ 0.32\ z_1 - 1.29\ q_1 + 0.9\ q_0 + 1.9$$

$$\dot{z}_2 = 0.9 \sum_{j=1}^{s} a_j\ q_1\ (t-\tau_j) - 1.32\ z_2 + 0.1\ m_2 + 4 + 0.1\ u_2$$

$$\dot{q}_2 = 0.9 \sum_{j=1}^{s} a_j\ q_1\ (t-\tau_j) - 0.32\ z_2 - 1.29\ q_2 + 1.9 \qquad 12.3.20$$

$$\dot{z}_3 = 0.9 \sum_{j=1}^{s} a_j\ z_2\ (t-\tau_j) - 1.32\ z_3 + 0.1\ m_3 + 4 + 0.1\ u_3$$

$$\dot{q}_3 = 0.9 \sum_{j=1}^{s} a_j\ q_2\ (t-\tau_j) - 0.32\ z_3 - 1.29\ q_3 + 1.9$$

where q_1, q_2, q_3 are the concentrations of D.O. in the three reaches, z_1, z_2, z_3 are the concentrations of B.O.D. in the three reaches and m_1, m_2, m_3 are the controls. Suitable values for s, τ, a, for the river Cam near Cambridge are :

$$s = 3, \tau_1 = 0, \tau_2 = 1/2 \text{ day}, \tau_3 = 1 \text{ day},\ z_0 = 0,\ q_0 = 10,\ a_1 = 0.15$$

$$a_2 = 0.7,\ a_3 = 0.15$$

The system as described by equations 12.3.20 is nominally of infinite dimension in the state space. It is possible to obtain a good finite dimensional approximation by expanding the delayed terms in a Taylor series and taking the first two terms, as we saw before. This makes the overall system of order 22.

For the above system, a cost function with weights

$$R_{1_i} = 1,\ i = 1, 2, 3\ ;\ R_{2_i} = 1000,\ i = 2, 3\ ;\ Q_i = \overset{\longleftrightarrow\, 6}{\begin{bmatrix} 2.0\ldots0\ldots\ldots\ldots0 \\ 0\ldots10\ldots0\ldots0 \\ 0 \qquad \ddots \quad 0 \end{bmatrix}}\begin{matrix}\updownarrow \\ 6 \\ \end{matrix}_{20\times10}$$

was chosen.

For the model reduction, A_{z_i} was chosen to be

$$A_{z_i} = \begin{bmatrix} 1.30232 & -0.00053 \\ -0.32 & -1.2 \end{bmatrix} \qquad i = 1, 2, \ldots$$

The initial conditions used were

$$\underline{x}(0) = \begin{bmatrix} 10.0 & 7.0 & 5.94 & 10.0 & 10.0 & 10.0 & 10.0 \\ 6.0 & 7.0 & 7.0 & 7.0 & 7.0 & 5.2370 & 5.94 \\ 5.94 & 5.94 & 5.94 & 4.69 \end{bmatrix}$$

$$\underline{y}(0) = \begin{bmatrix} 10.0 & 7.0 & 5.94 & 6.0 \end{bmatrix}$$

$$\underline{u}_2(0) = \begin{bmatrix} 0 & 0 \end{bmatrix}$$

Figs. 12.9 to 12.14 show the optimal and decentralised suboptimal trajectories for B.O.D. and D.O. for the three reaches. We see that the optimal and

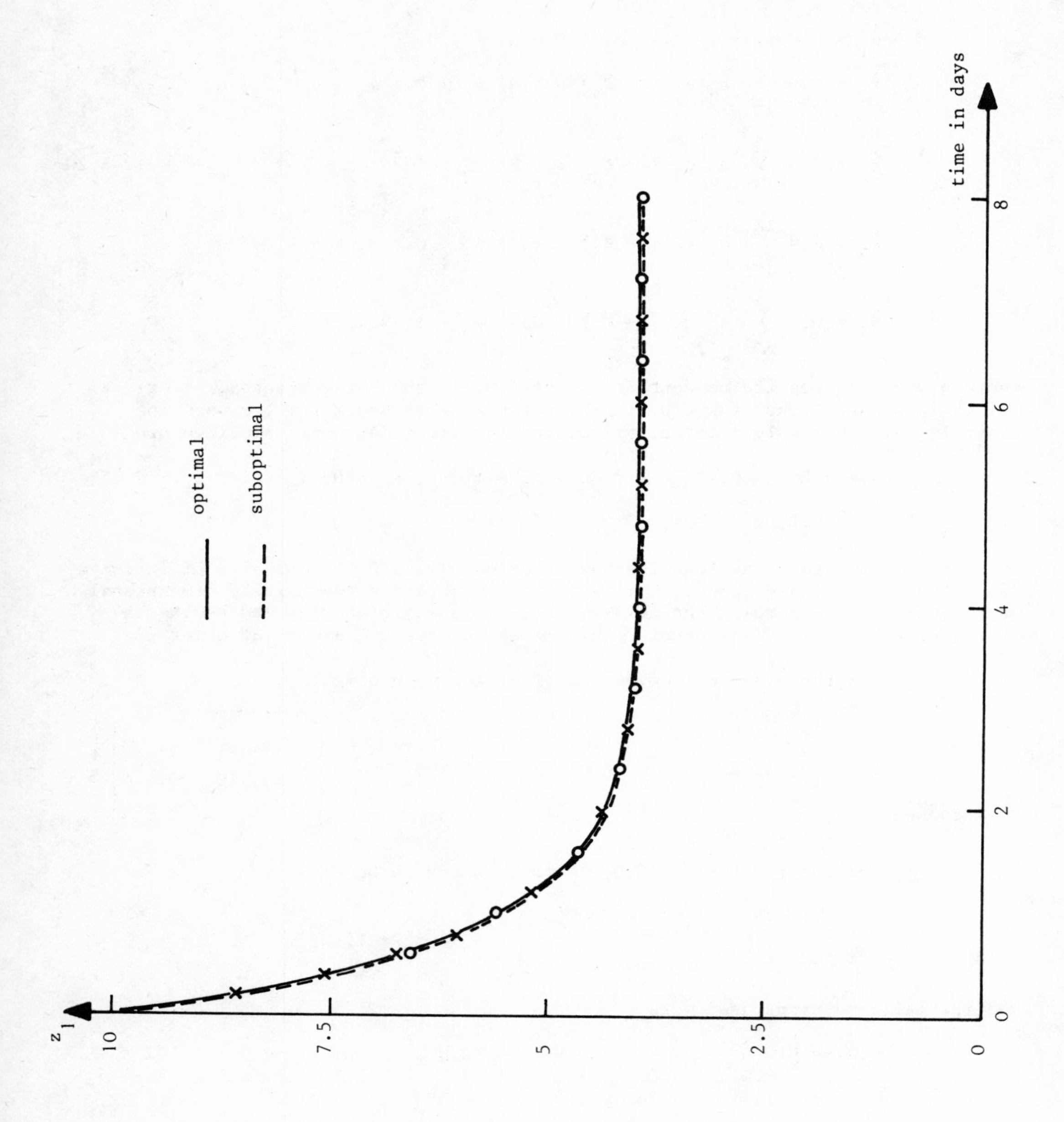

Fig. 12.9

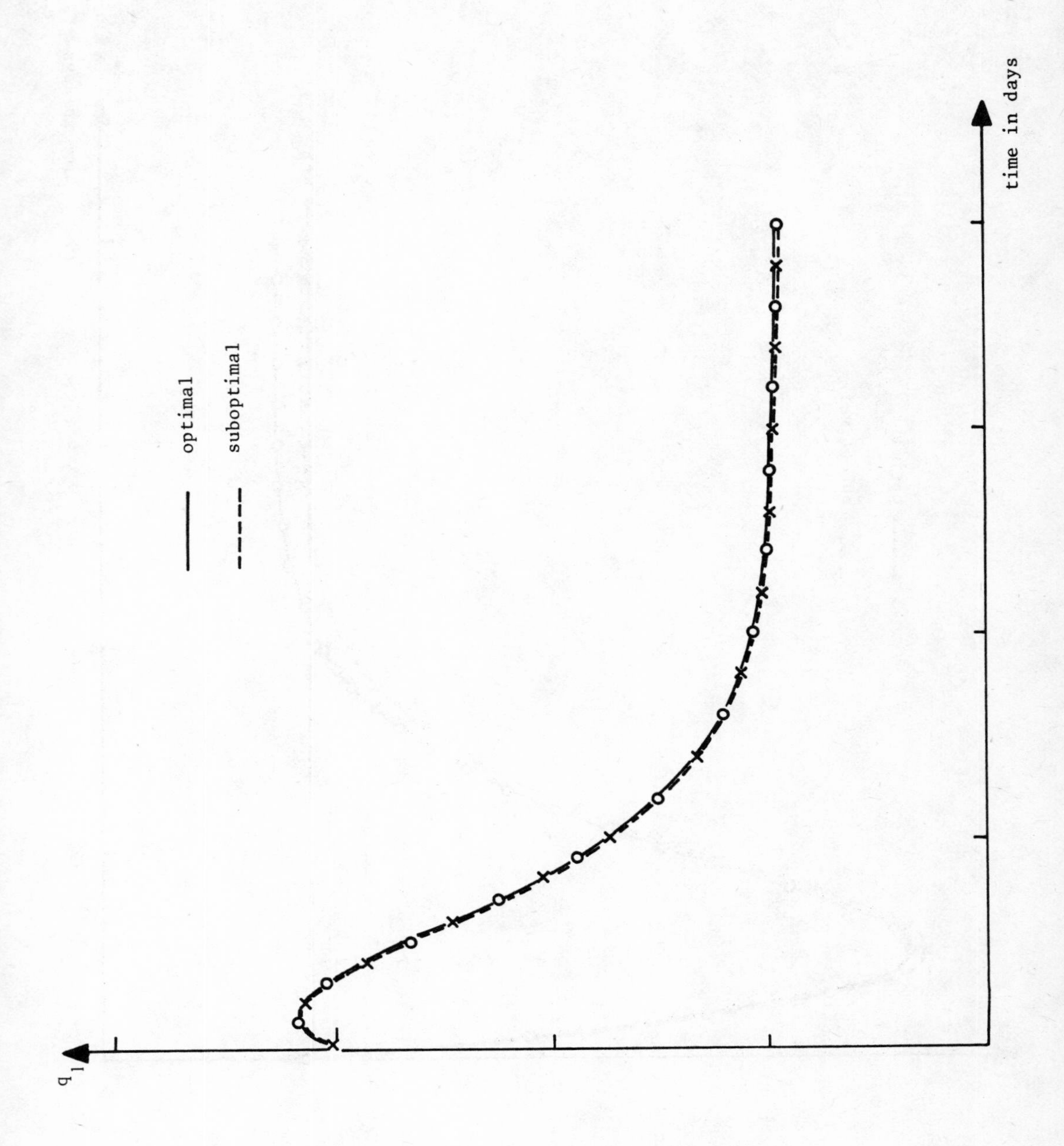

Fig. 12.10

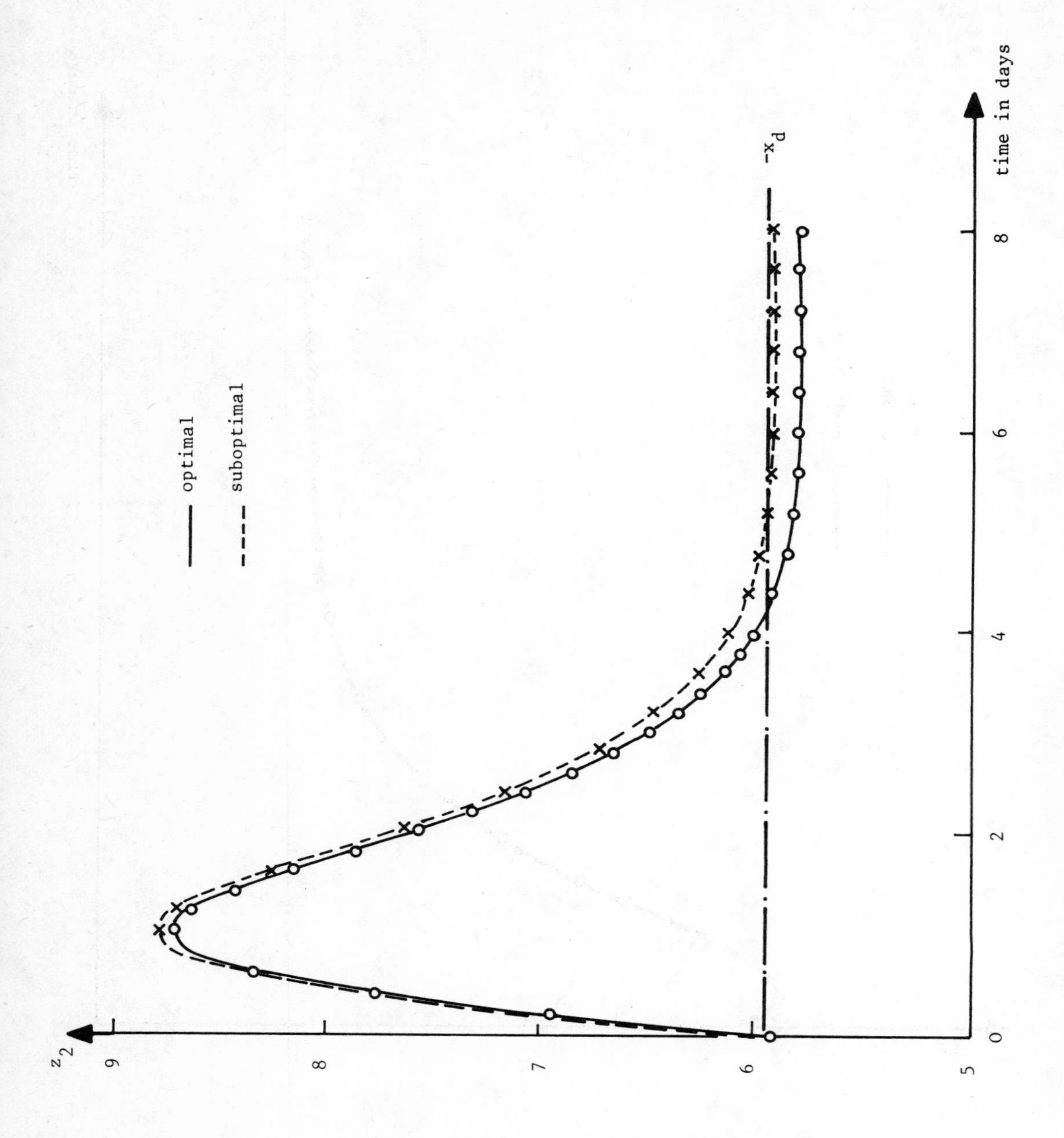

Fig. 12.11

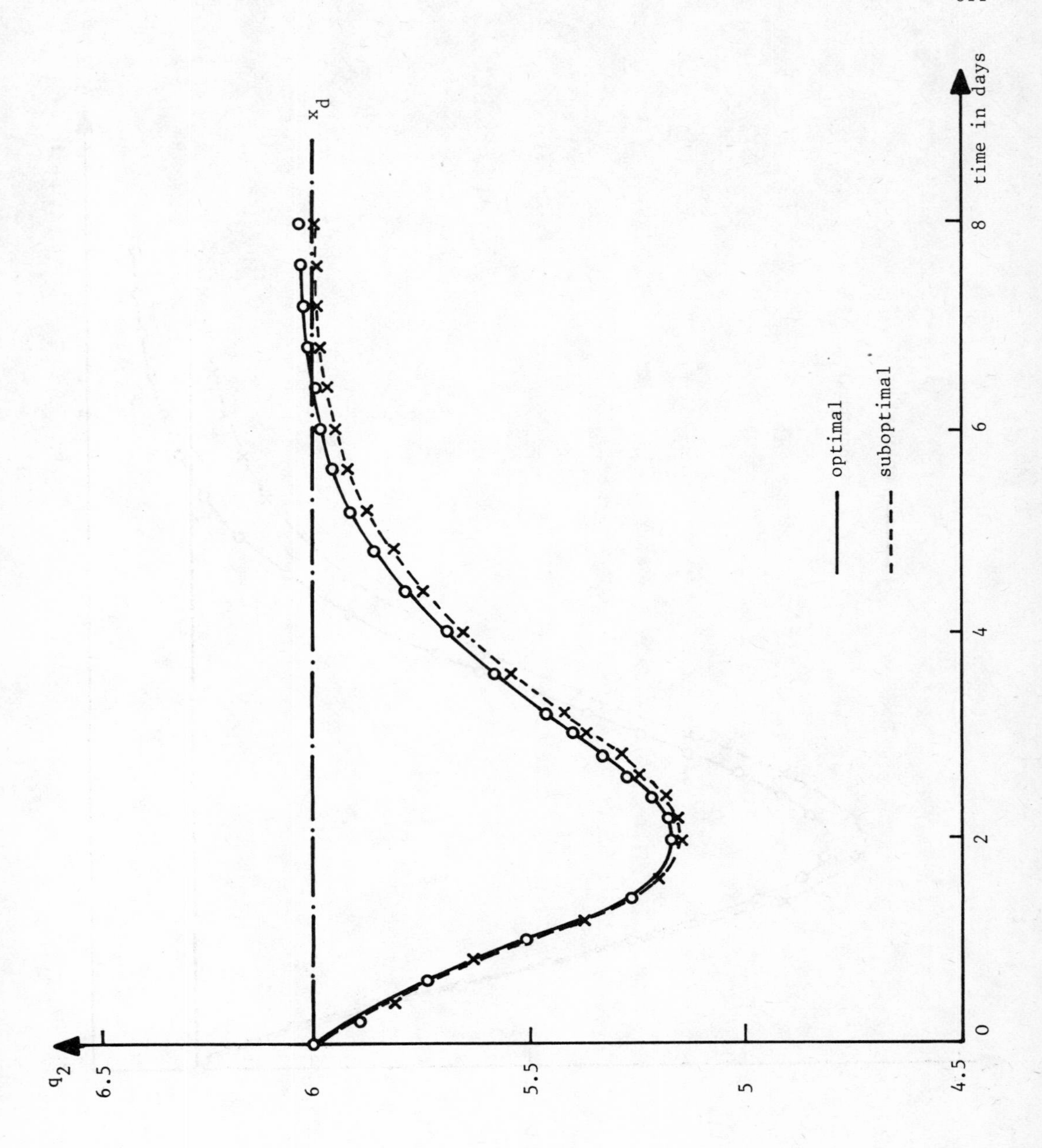

Fig. 12.12

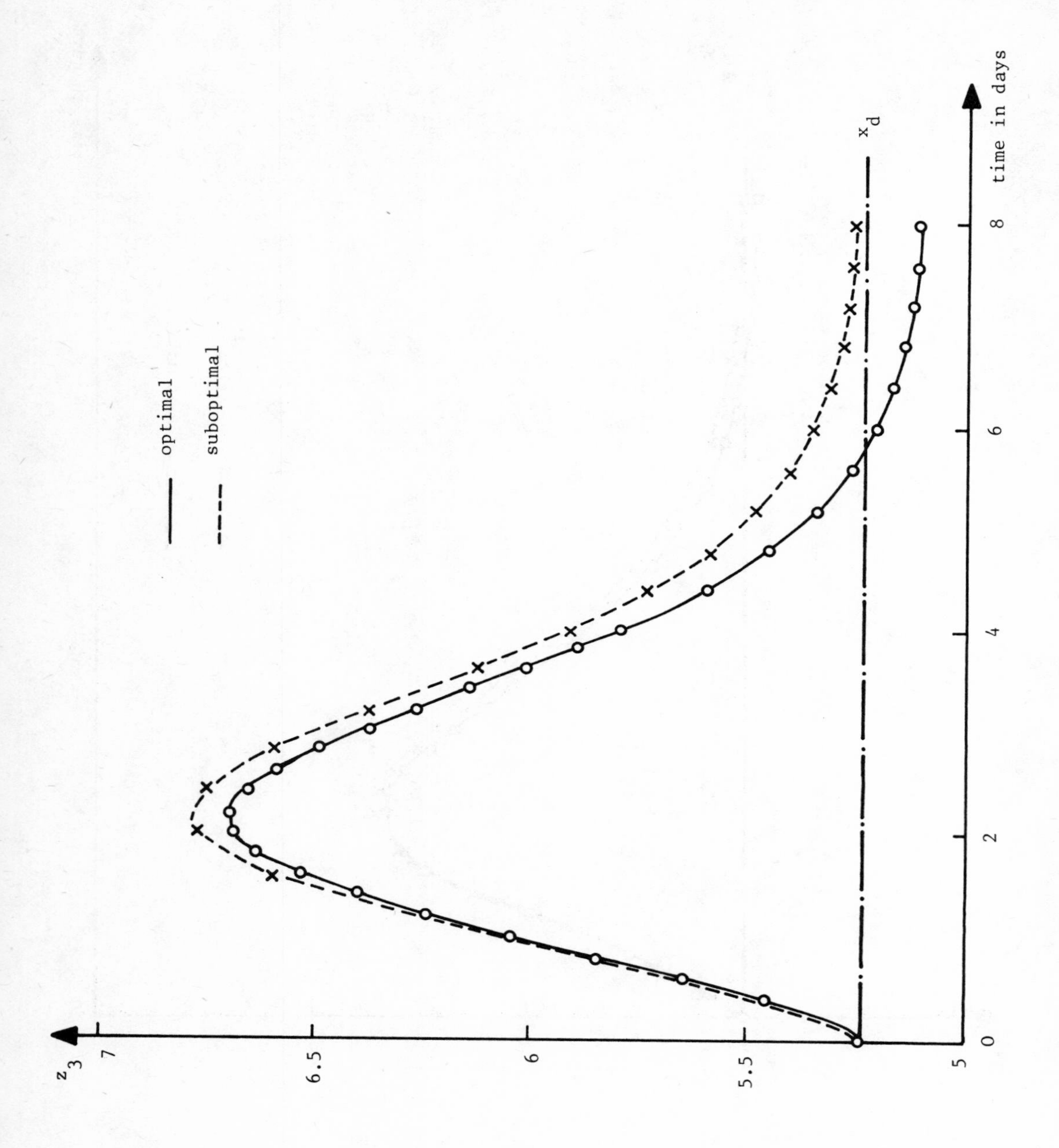

Fig. 12.13

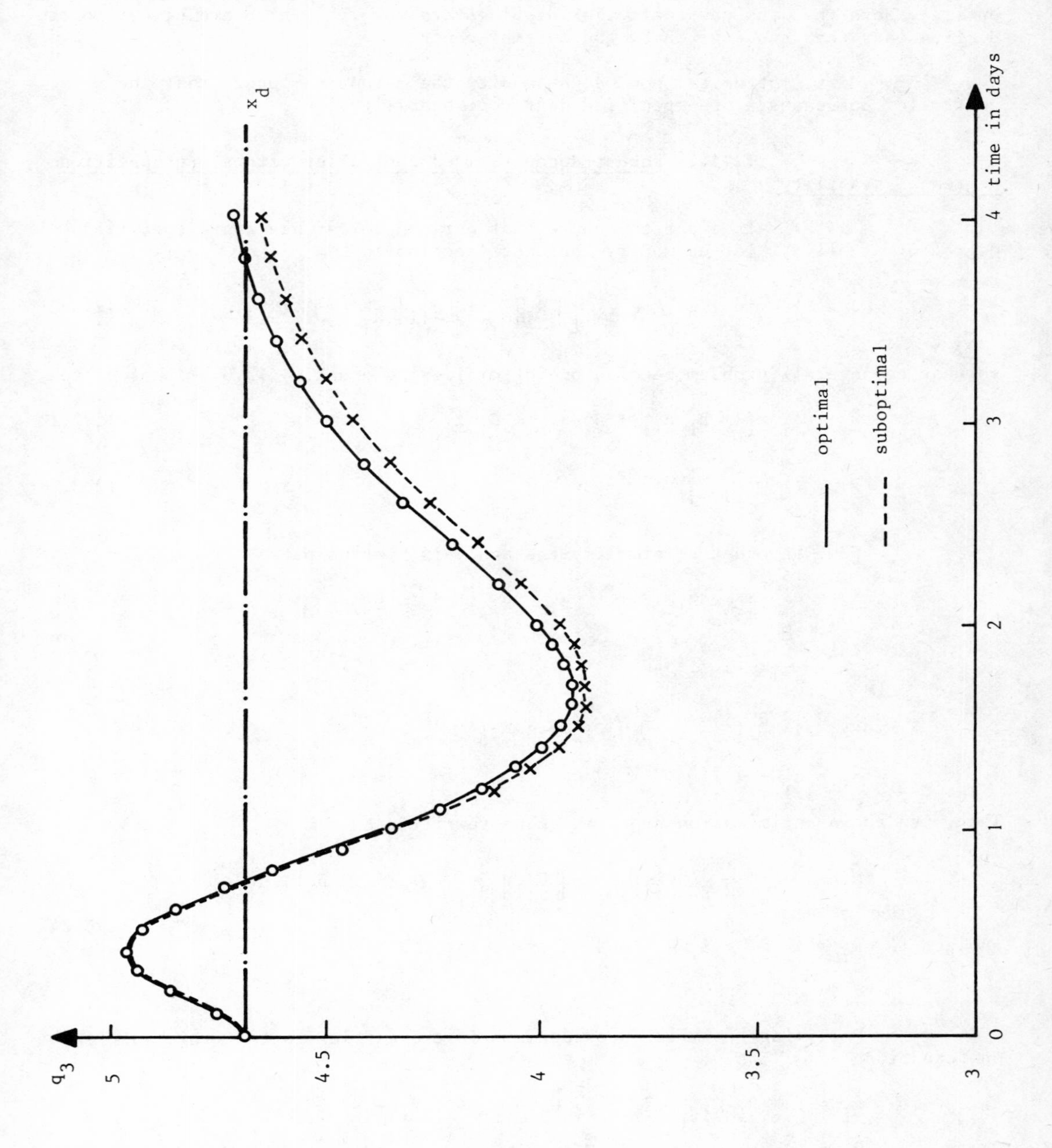

Fig. 12.14

suboptimal trajectories are very close showing that the control is indeed near optimal. Note that the new controller provides trajectories which are closer to the desired ones than those given by the optimal control.

Next, let us see how we can modify the method to ensure that the controller possesses a pre-specified degree of stability.

12.3.2 Robust decentralised controller with a pre-specified degree of stability

In order to ensure that our controller has a pre-specified degree of stability, let us modify the cost function to

$$J = \sum_{i=1}^{N} \min 1/2 \int_0^{\infty} e^{2\alpha t}\left[\|\underline{x}_i\|^2_{Q_i} + \|\underline{u}_{1i}\|^2_{R_{1i}} + \Psi_i(\underline{u}_{2i},\mathring{\underline{u}}_{2i})\right] dt \qquad 12.3.21$$

so that the overall problem becomes one of minimising equation 12.3.21 subject to

$$\dot{\underline{x}}_i = A_i\,\underline{x}_i + B_i\,\underline{u}_{1i} + B_i\,\underline{u}_{2i} + C_i\,\underline{z}_i \qquad 12.3.22$$

$$\underline{z}_i = \sum_{j=1}^{N} L_{ij}\,\underline{x}_j \qquad 12.3.23$$

Now, in order to simplify the analysis, let us put

$$\hat{\underline{x}}_i = e^{\alpha t}\,\underline{x}_i,\quad \hat{\underline{u}}_{1i} = e^{\alpha t}\,\underline{u}_{1i},\quad \hat{\underline{u}}_{2i} = e^{\alpha t}\,\underline{u}_{2i}$$

$$\hat{\underline{z}}_i = \sum_{j=1}^{N} L_{ij}\,\hat{\underline{x}}_j$$

$$\theta_i(\hat{\underline{u}}_{2i},\ \mathring{\hat{\underline{u}}}_{2i}) = e^{2\alpha t}\,\Psi_i(\underline{u}_{2i},\ \mathring{\underline{u}}_{2i})$$

$$\hat{A}_i = A_i + \alpha I$$

Then, the above optimisation problem can be rewritten as

$$\text{Min } \hat{J} = \sum_{i=1}^{N} \frac{1}{2} \int_0^{\infty} \|\hat{\underline{x}}_i\|^2_{Q_i} + \|\hat{\underline{u}}_{1i}\|^2_{R_{1i}} + \theta_i(\hat{\underline{u}}_{2i},\ \hat{\underline{u}}_{2i})\, dt$$

$$\text{subject to}\quad \dot{\hat{\underline{x}}}_i = A_i\,\hat{\underline{x}}_i + B_i\,\hat{\underline{u}}_{1i} + B_i\,\hat{\underline{u}}_{2i} + C_i\,\hat{\underline{z}}_i \qquad 12.3.24$$

$$\hat{\underline{z}} = L\,\hat{\underline{x}}$$

Following a similar procedure as in the earlier part of this section, one can find

$$\dot{\hat{\underline{x}}}_i = (\hat{A}_i - B_i\,\tilde{G}_{1i_1})\,\hat{\underline{x}}_i - B_i\,\tilde{G}_{2i_1}\hat{\underline{y}}_i + C_i\,\hat{\underline{z}}_i + B_i\,\hat{\underline{u}}''_{2i}$$

$$\dot{\hat{\underline{y}}}_i = (\hat{A}_{z_i} - \tilde{G}_{2i_2})\,\hat{\underline{y}}_i - \tilde{G}_{1i_2}\,\hat{\underline{x}}_i \qquad 12.3.25$$

$$\dot{\hat{\underline{u}}}''_{2i} = \hat{A}_{u_i}\,\hat{\underline{u}}''_{2i} - H_i\,\hat{\underline{x}}_i - F_i\,\hat{\underline{y}}_i$$

or

$$\dot{\underline{x}} = (A_i - B_i \tilde{G}_{1i_1}) \underline{x}_i - B_i \tilde{G}_{2i_1} \underline{y}_i + C_i \underline{z}_i + B_i \underline{u}_{2i}$$

$$\dot{\underline{y}}_i = (A_{z_i} - \tilde{G}_{2i_2}) \underline{y}_i - \tilde{G}_{1i_2} \underline{x}_i \qquad 12.3.26$$

$$\dot{\underline{u}}_{2i} = A_{u_i} \underline{u}_{2i} - H_i \underline{x}_i - F_i \underline{y}_i$$

where $A_{u_i} = \hat{A}_{u_i} - \alpha I$

Thus we have designed a robust controller with a prespecified degree of stability α. But as we can see, this controller is not able to adapt itself to any structural perturbations. To do this in an ideal manner, it is necessary to define an interconnection variable with each element in the C_i matrix and another variable with each element of the matrix A_{z_i}. However, in this case it is necessary to solve the problem completely to calculate the new gain matrices every time we have a perturbation in the system. To avoid this difficulty, let us define an interconnection variable e_{ij} with each element of the vector $\underline{z}_i$ where $0 \leqslant e_{ij} \leqslant 1$; $i = 1, \ldots, N$ and $j = 1, \ldots, q_i$. (We note that this definition of the interconnection variables is different from that of Siljak [10] since here we define an interconnection variable with each off diagonal element different from zero in the global A matrix).

Therefore, let us define $\underline{\mathfrak{z}}_i$ to be :

$$\underline{\mathfrak{z}}_i = \begin{bmatrix} e_{i1} & & & \\ & e_{i2} & & 0 \\ & & \ddots & \\ 0 & & & e_{iq_i} \end{bmatrix} \underline{z}_i \quad \text{or} \quad \underline{\mathfrak{z}}_i = \begin{bmatrix} e_{i_1} & Z_{i1} \\ \vdots & \\ e_{iq_i} & Z_{iq_i} \end{bmatrix}$$

Also we shall define a scalar quantity $\mathcal{E}_i$ with each subsystem which represents the sum of the actual interconnection variables divided by the ideal values, i.e. :

$$\mathcal{E}_i = \frac{\sum_{j=1}^{q_i} e_{ij}}{\sum_{j=1}^{q_i} \bar{e}_{ij}} = \frac{\sum_{j=1}^{q_i} e_{ij}}{q_i}$$

where $\bar{e}_{ij}$ is the ideal interconnection variable and is equal to one. Then we shall modify our state equation to take into consideration the effect of any structural perturbations as follows :

$$\dot{\underline{x}}_i = (A_i - B_i \tilde{G}_{i2_i}) \underline{x}_i - \mathcal{E}_i B_i \tilde{G}_{2i_1} \underline{y}_i + C_i \underline{z}_i + \mathcal{E}_i B_i \underline{u}''_{2i} \qquad 12.3.27$$

or

$$\dot{\hat{\underline{x}}}_i = (\hat{A}_i - B_i \tilde{G}_{i1_i}) \underline{x}_i - \mathcal{E}_i B_i \tilde{G}_{i2_i} \hat{\underline{y}}_i + C_i \hat{\underline{z}}_i + \mathcal{E}_i B_i \hat{\underline{u}}''_{2i} \qquad 12.3.28$$

The analogue implementation of this system is given in Figure 12.15.

Next let us study the stability of the system on implementing the decentralised controllers.

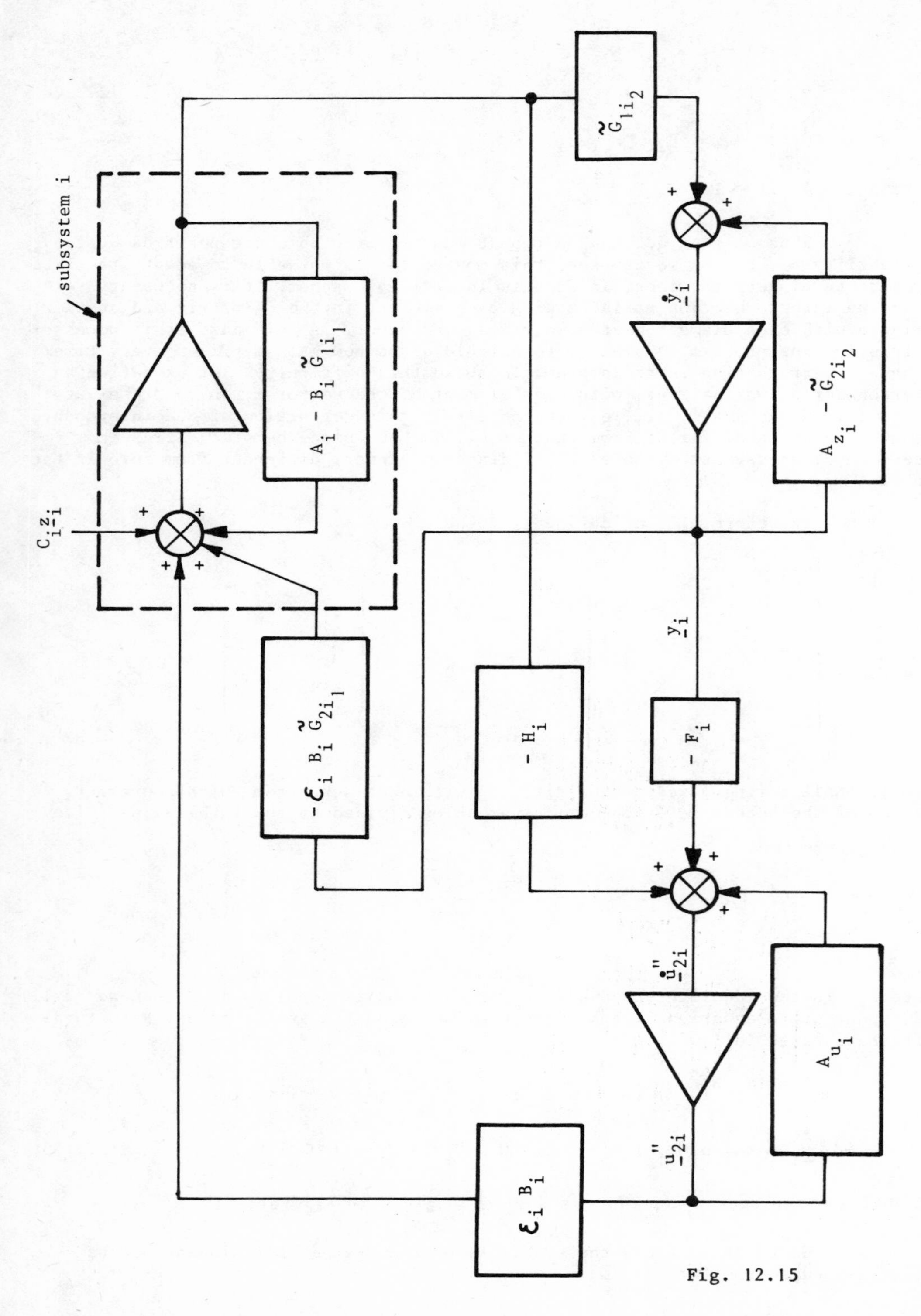
subsystem i
$C_i z_i$
$A_i - B_i \tilde{G}_{1i_1}$
$-\mathcal{E}_i B_i \tilde{G}_{2i_1}$
$\tilde{G}_{1i_2}$
$\dot{y}_i$
$A_{z_i} - \tilde{G}_{2i_2}$
y_i
$-H_i$
$-F_i$
$\dot{u}''_{2i}$
u''_{2i}
A_{u_i}
$\mathcal{E}_i B_i$

Fig. 12.15

12.3.3 Stability of the system on implementing the decentralised controllers

In this section, we shall derive a sufficient condition which guarantees the asymptotic stability of the system. Consider the following theorem :

Theorem

The global system with the closed loop decentralised controllers will be asymptotically stable with degree α if the following sufficient condition is satisfied :

$$2\sum_{i=1}^{N} \gamma_i + 2\max_i \lambda_M(P_i) \sum_{i=1}^{N}\sum_{j=1}^{N} \beta_{ij} \leq \lambda_m(w_i) \qquad 12.3.29$$

where $P_i = \begin{bmatrix} \tilde{P}_i & | & 0 \\ --- & | & --- \\ 0 & | & K_i \end{bmatrix}$

$\tilde{P}_i$ is the solution of the Riccati equation for the modified optimisation problem,
K_i is a positive definite matrix to be defined later
$\lambda_M(P_i)$ = max eigenvalue of P_i
$\lambda_m(w_i)$ = min eigenvalue of w_i

$$\gamma_i = \| \hat{\mathcal{A}}_{i2}^T P_i \| \qquad \beta_{ij} = \| \mathcal{C}_i \Xi_{ij} \|$$

$$w_i = \begin{bmatrix} \tilde{Q}_i + \tilde{P}_i \tilde{B}_i \tilde{R}_i^{-1} \tilde{B}_i^T \tilde{P}_i & | & 0 \\ ------------- & | & --- \\ & | & S_i \end{bmatrix}$$

S_i is a positive definite matrix
$\hat{\mathcal{A}}_{i2}$, Ξ_{ij} and $\mathcal{C}_i$ will be defined later.

Proof

To prove this theorem we shall begin with the system described by equations 12.3.25, 12.3.28, and then we shall prove that this system is asymptotically stable. Consequently the original system will be asymptotically stable with degree α . To do this, let us group the equations 12.3.25, 12.3.28 into one vector equation. Thus we have :

$$\dot{\hat{\underline{x}}}_i = \hat{\mathcal{A}}_i \hat{\underline{x}}_i + \mathcal{C}_i \hat{\underline{z}}_i \qquad 12.3.20$$

where $\hat{\underline{x}}_i = \begin{bmatrix} \hat{\underline{x}}_i \\ \underline{y}_i \\ \underline{u}''_{2i} \end{bmatrix}$, $\hat{\mathcal{A}}_i = \begin{bmatrix} \hat{A}_i - B_i \tilde{G}_{1i_1} & -\mathcal{E}_i B_i \tilde{G}_{2i_1} & \mathcal{E}_i B_i \\ -\tilde{G}_{1i_2} & A_{z_i} - \tilde{G}_{2i_2} & 0 \\ -\mathcal{H}_i & -F_i & \hat{A}_{u_i} \end{bmatrix}$

$$\mathcal{C}_i = \begin{bmatrix} C_i \\ 0 \\ \\ 0 \end{bmatrix}$$

$$\mathfrak{z}_i = \begin{bmatrix} e_{i1} & & \\ & \ddots & \\ & & e_{iq_i} \end{bmatrix} \sum_{j=1}^{N} L_{ij}\underline{\hat{x}}_j = \sum_{j=1}^{N} E_{ij}\underline{\hat{x}}_j = \sum_{j=1}^{N} \Xi_{ij}\, \underline{\hat{X}}_j$$

where

$$E_{ij} = \begin{bmatrix} e_{i1} & & 0 \\ & \ddots & \\ 0 & & e_{iq_i} \end{bmatrix} ; \Xi_{ij} = \begin{bmatrix} E_{ij} & 0 & 0 \end{bmatrix}$$

Thus we have :

$$\underline{\mathring{\hat{X}}}_i = \hat{\mathcal{A}}_i\, \underline{\hat{X}}_i + \mathcal{C}_i \sum_{j=1}^{N} \Xi_{ij}\, \underline{\hat{X}}_j \qquad 12.3.31$$

Let us consider the following Lyapunov function :

$$V = \sum_{i=1}^{N} \underline{\hat{X}}_i^T\, P_i\, \underline{\hat{X}}_i \qquad 12.3.32$$

$$\text{where } P_i = \left[\begin{array}{c|c} \tilde{P}_i & 0 \\ \hline 0 & K_i \end{array}\right]$$

K_i is a positive definite matrix to be defined.
Since P_i is positive definite, then V is positive definite.

Therefore for the system to be asymptotically stable, $\dot{V}$ must be negative semidefinite. Thus we have :

$$\mathring{V} = \sum_{i=1}^{N} \underline{\mathring{\hat{X}}}_i^T\, P_i\, \underline{\hat{X}}_i + \hat{X}_i P_i \mathring{\hat{X}}_i \leqslant 0 \qquad 12.3.33$$

Substituting from 12.3.31 in 12.3.33, we obtain :

$$\sum_{i=1}^{N} \Big\{ \underline{\hat{X}}_i^T \hat{\mathcal{A}}_i^{\,T} P_i \underline{\hat{X}}_i + \sum_{j=1}^{N} \underline{\hat{X}}_j^T \Xi_{ij}^T\, \mathcal{C}_i^T P_i \underline{\hat{X}}_i + \underline{\hat{X}}_i^T P_i \hat{\mathcal{A}}_i \underline{\hat{X}}_i$$
$$+ \underline{\hat{X}}_i^T P_i\, \mathcal{C}_i \sum_{j=1}^{N} \Xi_{ij}\, \underline{\hat{X}}_j \qquad 12.3.34$$

Now let :

$$\mathcal{E}_i = 1 - \xi_i \qquad \text{where } 0 \leqslant \xi_i \leqslant 1$$

and $\hat{\mathcal{A}}_i = \hat{\mathcal{A}}_{i1} + \hat{\mathcal{A}}_{i2} - \hat{\mathcal{A}}_{i3}$

where

$$\hat{\mathcal{A}}_{i1} = \begin{bmatrix} \hat{A}_i & C_i & 0 \\ 0 & \hat{A}_{z_i} & 0 \\ 0 & 0 & \hat{A}_{u_i} \end{bmatrix}, \quad \hat{\mathcal{A}}_{i2} = \begin{bmatrix} 0 & -C_i + \xi_i B_i \tilde{G}_{2i_1} & \epsilon_i B_i \\ 0 & 0 & 0 \\ -\mathcal{H}_i & -F_i & 0 \end{bmatrix}$$

$$\hat{\mathcal{A}}_{i3} = \begin{bmatrix} B_i \tilde{G}_{1i_1} & B_i \tilde{G}_{2i_1} & 0 \\ \tilde{G}_{1i_2} & \tilde{G}_{2i_2} & 0 \\ 0 & 0 & 0 \end{bmatrix} = \left[\begin{array}{c|c} \tilde{B}_i & 0 \\ \hline 0 & 0 \end{array}\right] \left[\begin{array}{c|c} \tilde{G}_i & 0 \\ \hline 0 & 0 \end{array}\right] = \bar{B}_i \bar{G}_i$$

Then equation 12.3.34 can be rewritten in the following form :

$$\sum_{i=1}^{N} \hat{\underline{X}}_i^T \hat{\mathcal{A}}_{i1} P_i \hat{\underline{X}}_i + \hat{\underline{X}}_i^T P_i \hat{\mathcal{A}}_{i1} \hat{\underline{X}}_i + \hat{\underline{X}}_i^T \hat{\mathcal{A}}_{i2} P_i \hat{\underline{X}}_i + \hat{\underline{X}}_i^T P_i \hat{\mathcal{A}}_{i2} \hat{\underline{X}}_i$$

$$+ \sum_{j=1}^{N} \hat{\underline{X}}_j^T \Xi_{ij}^T \mathcal{C}_i^T P_i \hat{\underline{X}}_i^T + \hat{\underline{X}}_i^T P_i \mathcal{C}_i \sum_{j=1}^{N} \Xi_{ij} \hat{\underline{X}}_j \qquad 12.3.35$$

$$\leqslant \left\{ \hat{\underline{X}}_i^T \left[G_i^{*T} B_i^{*T} P_i + P_i B_i G_i \right] \hat{\underline{X}}_i \right\}$$

Now, if A_{u_i} is stable one can calculate the positive definite matrix of Lyapunov K_i, i.e.

$$A_{u_i}^T K_i + K_i A_{u_i} = - S_i$$

where S_i is a negative definite matrix.

Then, using the definition of $\hat{\mathcal{A}}_{i1}$ equation 12.3.35 can be rewritten as follows :

$$\sum_{i=1}^{N} 2\hat{\underline{X}}_i^T \hat{\mathcal{A}}_{i2}^T P_i \hat{\underline{X}}_i + 2\hat{\underline{X}}_i^T P_i e_i \sum_{j=1}^{N} \Xi_{ij} \hat{\underline{X}}_j \Bigg\} \leqslant \sum_{i=1}^{N} \left\{ \hat{\underline{X}}_i^T w_i \hat{\underline{X}}_i \right\} \qquad 12.3.36$$

where

$$w_i = \left[\begin{array}{c|c} \tilde{Q}_i + \tilde{P}_i \tilde{B}_i \tilde{R}_i^{-1} \tilde{B}_i^T \tilde{P}_i & 0 \\ \hline 0 & S_i \end{array}\right]$$

Now since :

$$\sum_{i=1}^{N} \hat{\underline{X}}_i^T P_i \mathcal{C}_i \sum_{j=1}^{N} \Xi_{ij} \hat{\underline{X}}_j^T \leqslant \sum_{i=1}^{N} \lambda_M(P_i) \sum_{j=1}^{N} \beta_{ij} \| \hat{\underline{X}}_j \|^2$$

$$\leqslant \max_i \lambda_M(P_i) \sum_{i=1}^{N} \sum_{j=1}^{N} \beta_{ij} \| \hat{\underline{X}}_j \|^2$$

where $\lambda_M(P_i)$ = max eigenvalue of P_i

$$\beta_{ij} = \lambda_M^{1/2} \left[\Xi_{ij}^T \mathcal{C}_i^T \mathcal{C}_i \Xi_{ij} \right]$$

Also let :

$$\lambda_m(w_i) = \text{min eigenvalue of } w_i$$

$$\gamma_i = \lambda_M^{1/2} \left[P_i \hat{\mathcal{A}}_{i2}^T \hat{\mathcal{A}}_{i2} P_i \right]$$

Then we have :

$$\sum_{i=1}^{N} 2\gamma_i + 2 \max_i \lambda_M(P_i) \sum_{i=1}^{N} \sum_{j=1}^{N} \beta_{ij} \leqslant \lambda_m(w_i) \qquad 12.3.37$$

Thus, if 12.3.37 is satisfied, the system described by equation 12.3.31 will be asymptotically stable and hence the original system will be asymptotically stable with degree α and the theorem is proved.

Next, we consider our final approach to computing near optimal decentralised control.

12.4 ROBUST DECENTRALISED CONTROLLER USING A MODEL FOLLOWER [3][4]

In the development of the previous method we note that :

a - we added an additional term in the cost function $\Psi(u_{2i}, \dot{u}_{2i})$ so that our original cost function was modified.

b - we used approximations during the design of the controller algorithm. These were : we replaced $B\, u_{2i}$, $\hat{A}_{zi}\, \mathbb{B}_i\, \hat{u}_{2i}$ by $(C_i^T C_i)\, C_i^T B_i\, \hat{u}_{2i}$ and $(\mathbb{B}_i^T \mathbb{B}_i)^{-1} \mathbb{B}_i^T A_{zi}\, \mathbb{B}_i\, u_{2i}$ which minimised the Euclidean norms of $B_i \hat{u}_{2i} - C_i \gamma_i$ and $\mathbb{B}_i \alpha_i - \hat{A}_{z_i} \mathbb{B}_i\, \hat{u}_{2i}$.

Now, although both these assumptions were physically reasonable ones, we see in this final part of the chapter how we can avoid using these assumptions.

The basic idea of the new method is to construct a relatively crude dynamical model for the interactions and use the output of this interaction model for each subsystem as the interaction input. Thus the subsystem control will be a function of the subsystem states and of the interconnections.

However, since our interaction model is a crude one, we will formulate another optimisation problem which will enable us to modify the interaction model parameters based on system measurements. In this way, we will converge to a good interaction model and thus a good decentralised feedback control.

12.4.1 Problem formulation

Let us consider the following optimisation problem :

$$\text{Min } J = \sum_{i=1}^{N} \frac{1}{2} \int_0^{\infty} \left\{ \|\underline{x}_i\|_{Q_i}^2 + \|\underline{u}_i\|_{R_i}^2 \right\} dt \qquad 12.4.1$$

subject to

$$\dot{\underline{x}}_i = A_i\, \underline{x}_i + B_i\, \underline{u}_i + C_i\, \underline{z}_i \qquad 12.4.2$$

$$\underline{z}_i = \sum_{j=1}^{N} L_{ij}\, \underline{x}_j \qquad 12.4.3$$

where $\underline{x}_i$ is the n_i dimensional state vector of the ith subsystem, $\underline{u}_i$ is the m_i dimensional control vector and $\underline{z}_i$ is the q_i dimensional interconnection vector.

Let us suppose initially that $\underline{z}_i$ can be represented by the following dynamical model

$$\dot{\underline{z}}_i = A_{z_i} \underline{z}_i \quad 12.4.4$$

Then we can modify our optimisation problem to the form

$$\min J = \frac{1}{2} \sum_{i=1}^{N} \int_0^\infty (\| \underline{y}_i \|^2_{\tilde{Q}_i} + \| \underline{u}_i \|^2_{R_i}) \, dt \quad 12.4.5$$

subject to

$$\dot{\underline{y}}_i = \tilde{A}_i \underline{y}_i + \tilde{B}_i \underline{u}_i \quad 12.4.6$$

$$\text{where } \underline{y}_i = \begin{bmatrix} \underline{x}_i \\ \hline \underline{z}_i \end{bmatrix} ; \quad \tilde{A}_i = \left[\begin{array}{c|c} A_i & C_i \\ \hline 0 & A_{z_i} \end{array}\right] ; \quad \tilde{B}_i = \begin{bmatrix} B_i \\ \hline 0 \end{bmatrix} ; \quad \tilde{Q}_i = \left[\begin{array}{c|c} Q_i & 0 \\ \hline 0 & 0 \end{array}\right]$$

The solution of this new problem is given by :

$$\underline{u}_i^* = - R_i^{-1} \tilde{B}_i P_i \underline{y}_i$$

where P_i is the solution of a matrix Riccati equation. We can rewrite the optimal control as :

$$\underline{u}_i^* = - G_{i1} \underline{x}_i - G_{i2} \underline{z}_i \quad 12.4.7$$

From the above equation we see that the gain matrix is a function of the approximate system model and the optimal control $\underline{u}_i^*$ is a function of the subsystem state $\underline{x}_i$ and the interconnection input vector $\underline{z}_i$.

Now although we can obtain $\underline{z}_i$ from our interconnection model, this model is a relatively crude one. It is therefore necessary to improve the model on-line.

To improve the modelled interaction inputs, let us construct a subsystem model which is forced by the interconnections, i.e.

$$\dot{\hat{\underline{x}}}_i = A_i \hat{\underline{x}}_i + B_i \underline{u}_i + C_i \hat{\underline{z}}_i \quad 12.4.8$$

where $\hat{\underline{x}}_i$ is the state vector of the subsystem model. Then if we substitute for $\underline{u}_i$ in equation 12.4.2 and 12.4.8 after having replaced $\underline{z}_i$ by $\hat{\underline{z}}_i$ in equation 12.4.7 where $\hat{\underline{z}}_i$ is the out-put of the interconnection model, we have :

$$\dot{\underline{x}}_i' = (A_i - B_i G_{i1}) \underline{x}_i' + C_i \underline{z}_i - B_i G_{i2} \hat{\underline{z}}_i \quad 12.4.9$$

or

$$\dot{\underline{x}}_i' = \hat{A}_i \underline{x}_i' + C_i \underline{z}_i - B_i G_{i2} \underline{z}_i \quad 12.4.10$$

$$\text{where } \hat{A}_i = A_i - B_i G_{i1} \quad 12.4.11$$

$\underline{x}_i'$ is the suboptimal state of subsystem i resulting from using $\hat{\underline{z}}_i$ instead of $\underline{z}_i$ and

$$\dot{\hat{\underline{x}}}_i' = \hat{A}_i \hat{\underline{x}}_i' + C_i \hat{\underline{z}}_i - B_i G_{i2} \hat{z}_i \quad 12.4.12$$

Let $\underline{\tilde{x}}_i = \underline{x}'_i - \underline{\hat{x}}'_i$ 12.4.13

$\underline{\tilde{z}}_i = \underline{z}_i - \underline{\hat{z}}_i$ 12.4.14

Thus, subtracting equation 12.4.12 from equation 12.4.10, we obtain

$$\dot{\underline{\tilde{x}}}_i = \hat{A}_i \underline{\tilde{x}}_i + C_i \underline{\tilde{z}}_i \qquad 12.4.15$$

and our aim is to minimise the error vectors $\underline{\tilde{x}}_i$ and $\underline{\tilde{z}}_i$. For this problem, we can construct another optimisation problem which is :

$$\min \tilde{J}_i = \frac{1}{2} \int_0^\infty (\| \underline{\tilde{x}}_i \|^2_{H_i} + \| \underline{\tilde{z}}_i \|^2_{S_i}) \, dt \qquad 12.4.16$$

subject to the constraint 12.4.15 where H_i and S_i are positive semidefinite and positive definite matrices respectively.

The solution of this problem is

$$\underline{\tilde{z}}_i^{*} = - S_i^{-1} C_i^T K_i \underline{\tilde{x}}_i \qquad 12.4.17$$

where K_i is a solution of an appropriate Riccati type equation.

Using equation 12.4.17 and 12.4.14, we obtain :

$$\underline{z}_i^{*} = \underline{\hat{z}}_i - S_i^{-1} C_i^T K_i \underline{\tilde{x}}_i \qquad 12.4.18$$

and we will use this to generate the control. Thus, the dynamical equation for the subsystem will be :

$$\overset{\circ}{\underline{x}}_i = \hat{A}_i \underline{x}_i + C_i \underline{z}_i - B_i G_{i2} \left[\underline{\hat{z}}_i - S_i^{-1} C_i^T K_i (\underline{x}_i - \underline{\hat{x}}_i) \right] \qquad 12.4.19$$

$$\overset{\circ}{\underline{\hat{x}}}_i = \hat{A}_i \underline{x}_i + (C_i - B_i G_{i2}) \underline{\hat{z}}_i \qquad 12.4.20$$

$$\overset{\circ}{\underline{\hat{z}}}_i = A_{z_i} \underline{\hat{z}}_i \qquad 12.4.21$$

and we could represent these equations by Fig. 12.16.

To enable the controller to adapt to structural perturbations, we define an interconnection variable with the elements of the $\underline{z}_i$ vector, i.e. e_{ij} where

$$0 \leqslant e_{ij} \leqslant 1, \qquad i = 1, 2, \ldots, N, \; j = 1, 2, \ldots, q_i$$

thus we let :

$$\mathfrak{z}_i = \begin{bmatrix} e_{ij} & Z_{i1} \\ & \\ e_{iq_i} & Z_{iq_i} \end{bmatrix} \qquad \hat{\mathfrak{z}}_i = \begin{bmatrix} e_{i1} & \hat{Z}_{i1} \\ & \\ e_{iq_i} & \hat{Z}_{iq_i} \end{bmatrix}$$

or $\mathfrak{z}_i = \sum_{j=1}^{N} E_{ij} \underline{x}_j$ 12.4.22

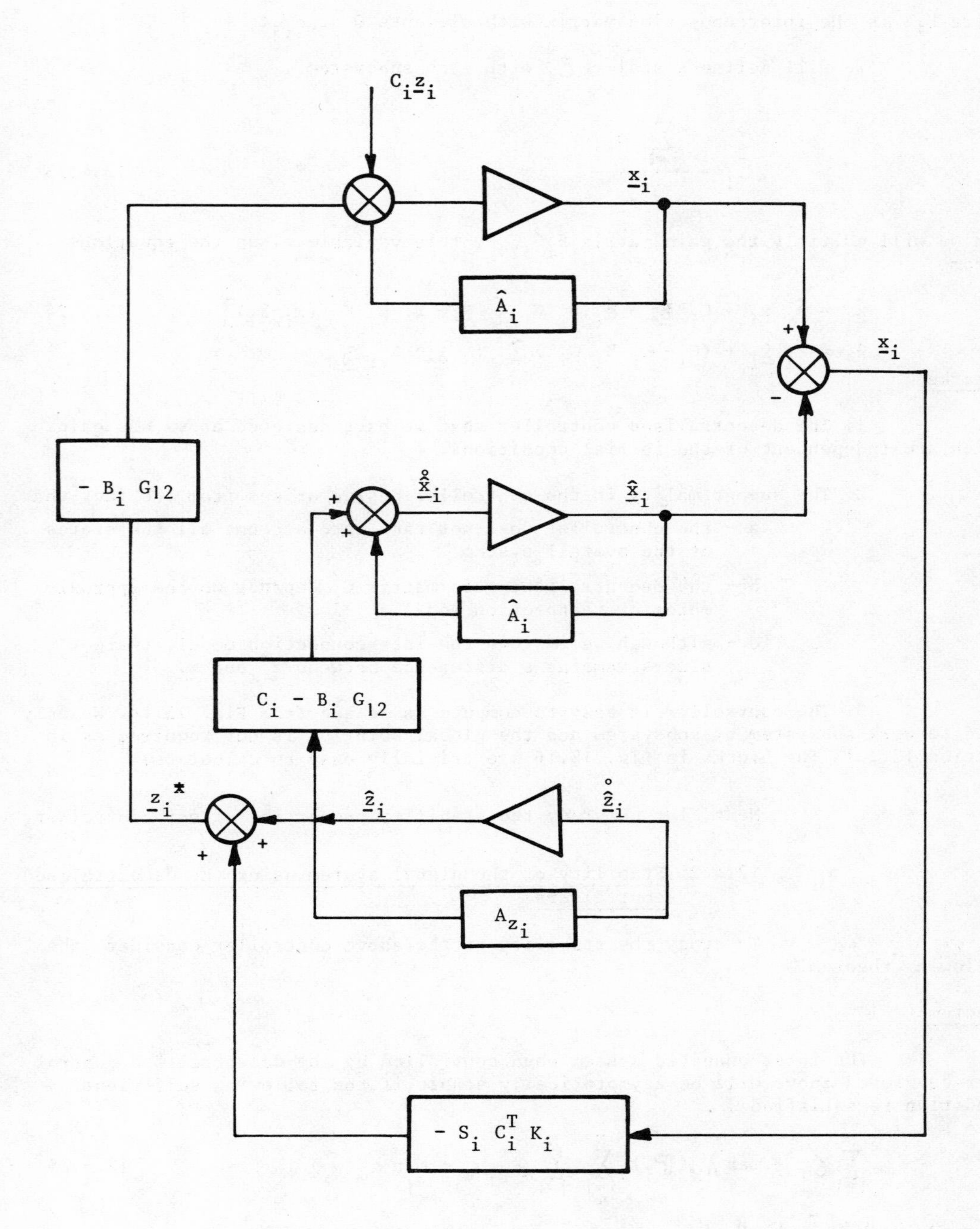

Fig. 12.16

where E_{ij} is the interconnection matrix with elements 0 or e_{ij}.

We will define a scalar $\mathcal{E}_i$ with each subsystem.

$$\mathcal{E}_i = \frac{\sum_{j=1}^{q_i} e_{ij}}{q_i} \qquad 12.4.23$$

and we will multiply the gain matrix $B_i G_{i2}$ by this variable. Thus the equations become

$$\dot{\underline{x}}_i = \hat{A}_i \underline{x}_i + C_i \underline{z}_i - \mathcal{E}_i B_i G_{i2} [\underline{z}_i - S_i C_i^T K_i (\underline{x}_i - \hat{\underline{x}}_i)] \qquad 12.4.24$$
$$\dot{\hat{\underline{x}}}_i = \hat{A}_i \hat{\underline{x}}_i + (C_i - \epsilon_i B_i G_{i2}) \hat{\underline{z}}_i \; ; \; \hat{\underline{z}}_i = A_{zi} \underline{z}_i$$

Remarks

1. The decentralised controller that we have designed above has gains which are independent of the initial conditions.

2. The suboptimality in the controller design arises from the fact that

   a - the controller does not take into account all the states of the overall system

   b - the decentralised gain matrix G_i depends on the approximate interconnection model

   c - although we improve the interconnection model, there always remains a difference between $\underline{z}^*$ and $\underline{z}$.

3. The controller is easy to compute as we see from Fig. 12.16. We only need to work subsystem by subsystem and the global solution is not required as in section 12.2.1. The blocks in Fig. 12.16 are trivially easy to calculate.

Next, let us study the stability properties of our controller.

12.4.2 Stability of the global system using the decentralised controllers

To study the stability of the above controller consider the following theorem.

Theorem

The interconnected system when controlled by the decentralised controllers developed above will be asymptotically stable if the following sufficient condition is satisfied :

$$\sum_{i=1}^{N} \gamma_i + \max_i \lambda_M(\mathbb{P}_i) \sum_{i=1}^{N} \sum_{j=1}^{N} \beta_{ij} \leqslant \frac{1}{2} \min_i \lambda_m(w_i) \qquad 12.4.25$$

where $\gamma_i = \| \mathcal{A}_{i3}^T \mathbb{P}_i \|$

$\beta_{ij} = \| \mathcal{C}_i \, \Xi_{ij} \|$

$$\mathbb{P}_i = \left[\begin{array}{c|c} P_i & 0 \\ \hline 0 & K_i \end{array}\right], \quad w_i = \left[\begin{array}{c|c} \tilde{Q}_i & 0 \\ \hline 0 & H_i \end{array}\right] + \left[\begin{array}{c|c} P_i \tilde{B}_i R_i^{-1} \tilde{B}_i^T P_i & 0 \\ \hline 0 & K_i C_i S_i^{-1} C_i^T K_i \end{array}\right]$$

$\lambda_M(\mathbb{P}_i)$ = maximum eigenvalue of the matrix P

$\lambda m(w_i)$ = minimum eigenvalue of the matrix w_i

<u>Proof</u>

From equation 12.4.24, 12.4.20 and 12.4.21, we have :

$$\dot{\underline{x}}_i = \mathcal{A}_i \underline{X}_i + \mathcal{C}_i \underline{z}_i \qquad 12.4.26$$

where

$$\mathcal{A}_i = \left[\begin{array}{c|c|c} A_i - B_i G_{i1} + \mathcal{E}_i B_i G_{i2} S_i^{-1} C_i^T K_i & -\mathcal{E}_i B_i G_{i2} & -\mathcal{E}_i B_i G_{i2} S_i^{-1} C_i^T K_i \\ 0 & A_{z_i} & 0 \\ 0 & C_i - B_i G_{i2} & A_i - B_i G_{i1} \end{array}\right]$$

$$\mathcal{C}_i = \begin{bmatrix} C_i \\ 0 \\ 0 \end{bmatrix}; \quad \underline{X}_i = \begin{bmatrix} \underline{x}_i \\ \hat{\underline{z}}_i \\ \hat{\underline{x}}_i \end{bmatrix}$$

$$\bar{z}_i = \sum_{j=1}^{N} E_{ij} \underline{x}_j = \sum_{j=1}^{N} \Xi_{ij} \underline{X}_j \qquad 12.4.27$$

where $\Xi_{ij} = [E_{ij} \quad 0 \quad 0]$

Therefore $\dot{\underline{X}}_i = \mathcal{A}_i \underline{X}_i + \mathcal{C}_i \sum_{j=i}^{N} \Xi_{ij} \underline{X}_j \qquad 12.4.28$

Let us now consider the following Lyapunov function :

$$v = \sum_{i=1}^{N} \underline{X}_i^T \mathbb{P}_i \underline{X}_i \qquad 12.4.29$$

where $\mathbb{P}_i = \left[\begin{array}{c|c} P_i & 0 \\ \hline 0 & K_i \end{array}\right]$

Where P_i, K_i are the solutions of the Riccati equations for the optimisation problem 12.4.5 subject to 12.4.6 and 12.4.16 subject to 12.4.15 respectively.

Since v is positive definite, we will seek the condition for which $\mathring{v}$ will be negative semidefinite. Thus we have :

$$\dot{v} = \sum_{i=1}^{N} \dot{\underline{X}}_i^T \mathbb{P}_i \underline{X}_i + \underline{X}_i^T \mathbb{P}_i \mathring{\underline{X}}_i \leqslant 0$$

On using the augmented equation 12.4.28 of the dynamical system, we obtain :

$$\dot{v} = \sum_{i=1}^{N} \Big\{ \underline{x}_i^T \mathcal{A}_i^T \mathbb{P}_i \underline{x}_i + \sum_{j=1}^{N} \underline{x}_j^T \Xi_i^T \mathcal{C}_i^T \mathbb{P}_i \underline{x}_i + \underline{x}_i^T \mathbb{P}_i \mathcal{A}_i \underline{x}_i + \underline{x}_i^T \mathbb{P}_i \mathcal{C}_i \sum_{j=1}^{N} \Xi_{ij} \underline{x}_j \Big\} \tag{12.4.30}$$

Let us put $\mathcal{E}_i = 1 - \xi_i$ and rewrite $\mathcal{A}_i$ as $\mathcal{A}_i = \mathcal{A}_{i1} - \mathcal{A}_{i2} + \mathcal{A}_{i3}$ where

$$\mathcal{A}_{i1} = \left[\begin{array}{c|c|c} A_i & C_i & 0 \\ \hline 0 & A_{z_i} & 0 \\ \hline 0 & 0 & A_i - B_i G_{i1} \end{array}\right]$$

$$\mathcal{A}_{i2} = \left[\begin{array}{c|c|c} B_i G_{i1} & B_i G_{i2} & 0 \\ \hline 0 & 0 & 0 \\ \hline 0 & 0 & C_i S_i^{-1} C_i^T K_i \end{array}\right]$$

$$\mathcal{A}_{i3} = \left[\begin{array}{c|c|c} \epsilon_i B_i G_{i2} S_i^{-1} C_i^T K_i & \xi_i B_i G_{i2} - C_i & -\mathcal{E}_i B_i G_{i2} S_i^{-1} C_i^T K_i \\ \hline 0 & 0 & 0 \\ \hline 0 & C_i - B_i G_{i2} & C_i S_i^{-1} C_i^T K_i \end{array}\right]$$

Then the relation 12.4.30 becomes

$$\sum_{i=1}^{N} \Big\{ \underline{x}_i^T \mathcal{A}_i \mathbb{P}_i \underline{x}_i - \underline{x}_i^T \mathcal{A}_{i2}^T \mathbb{P}_i \underline{x}_i + \underline{x}_i^T \mathcal{A}_{i3}^T \mathbb{P}_i \underline{x}_i + \underline{x}_i^T \mathbb{P}_i \mathcal{A}_{i1} \underline{x}_i - \underline{x}_i^T \mathbb{P}_i \mathcal{A}_{i2} \underline{x}_i + \underline{x}_i^T \mathbb{P}_i \mathcal{A}_{i3} \underline{x}_i + 2\underline{x}_i^T \mathbb{P}_i \mathcal{C}_i \sum_{j=1}^{N} \Xi_{ij} \underline{x}_j \Big\} \leqslant 0 \tag{12.4.31}$$

On using the solution P_i, K_i of the Riccati equations, we have :

$$\sum_{i=1}^{N} \Big\{ \underline{x}_i^T Q_i \underline{x}_i - \underline{x}_i^T \mathbb{P}_i \mathcal{A}_{i2} \underline{x}_i + 2\underline{x}_i^T \mathcal{A}_{i3}^T \mathbb{P}_i \underline{x}_i + 2\underline{x}_i^T \mathbb{P}_i \mathcal{C}_i \sum_{j=1}^{N} \Xi_{ij} \underline{x}_j \Big\} \leqslant 0$$

Let $w_i = Q_i + \mathbb{P}_i \mathcal{A}_{i2} = \left[\begin{array}{c|c} \tilde{Q}_i & 0 \\ \hline 0 & H_i \end{array}\right] + \left[\begin{array}{c|c} P_i \tilde{B}_i R_i^{-1} \tilde{B}_i^T P_i & 0 \\ \hline 0 & K_i C_i S_i^{-1} C_i^T K_i \end{array}\right]$

Then

$$\sum_{i=1}^{N} 2\left\{ \underline{x}_i^T \mathcal{A}_{i3}^T \mathbb{P}_i \underline{x}_i + \underline{x}_i^T \mathbb{P}_i \mathcal{C}_i \sum_{j=1}^{N} \Xi_{ij} \underline{x}_j \right\} \leqslant \sum_{i=1}^{N} \underline{x}_i^T w_i \underline{x}_i \qquad 12.4.32$$

since

$$\sum_{i=1}^{N} \underline{x}_i^T \mathbb{P}_i \mathcal{C}_i \sum_{j=1}^{N} \Xi_{ij} \underline{x}_j \leqslant \sum_{i=1}^{N} \lambda_M(\mathbb{P}_i) \sum_{j=1}^{N} \beta_{ij} \| \underline{x}_j \|^2$$

$$\leqslant \max_i \lambda_M(\mathbb{P}_i) \sum_{i=1}^{N} \sum_{j=1}^{N} \beta_{ij} \| \underline{x}_j \|^2$$

where $\beta_{ij} = \| \mathcal{C}_i \Xi_{ij} \|$ and writing $\gamma_i = \| \mathcal{A}_{i3}^T \mathbb{P}_i \|$

we have the final result

$$\sum_{i=1}^{N} \gamma_i + \max_i \lambda_M(\mathbb{P}_i) \sum_{i=1}^{N} \sum_{j=1}^{N} \beta_{ij} \leqslant \frac{1}{2} \min_i \lambda_m(w_i) \qquad 12.4.33$$

which completes the proof.

Next, we illustrate the approach on the 22nd order river pollution control example.

River pollution control example

The three reach problem was solved on the IBM 370/165 digital computer with H and S taken as identity matrices. Using the computed decentralised gains, Figs. 12.17 to 12.21 show the optimal and suboptimal trajectories of z_i and q_i (i = 1, 2, 3) and Figs. 12.22 to 12.26 show the optimal and estimated trajectories for the interconnections. These figures show that :

a) the suboptimal solution obtained by this method is closer to the optimal one than by using the interaction modelling method of section 12.3.

b) with the decentralised controllers we have succeeded in providing near optimal estimates for the interconnection variables.

12.5 CONCLUSIONS

In this chapter, we have given an overview of some preliminary results that have been obtained in the area of robust decentralised control for large scale linear interconnected dynamical systems. The control is robust in the sense that no state information is passed between the subsystemswhen the control is applied although the fact that we are dealing with an interconnected dynamical system is taken into account in the design state. Thus the control is completely decentralised with local regulators controlling each subsystem.

We have given three methods for the synthesis of the decentralised gains. In the first approach we constrain the global gain matrix to be block

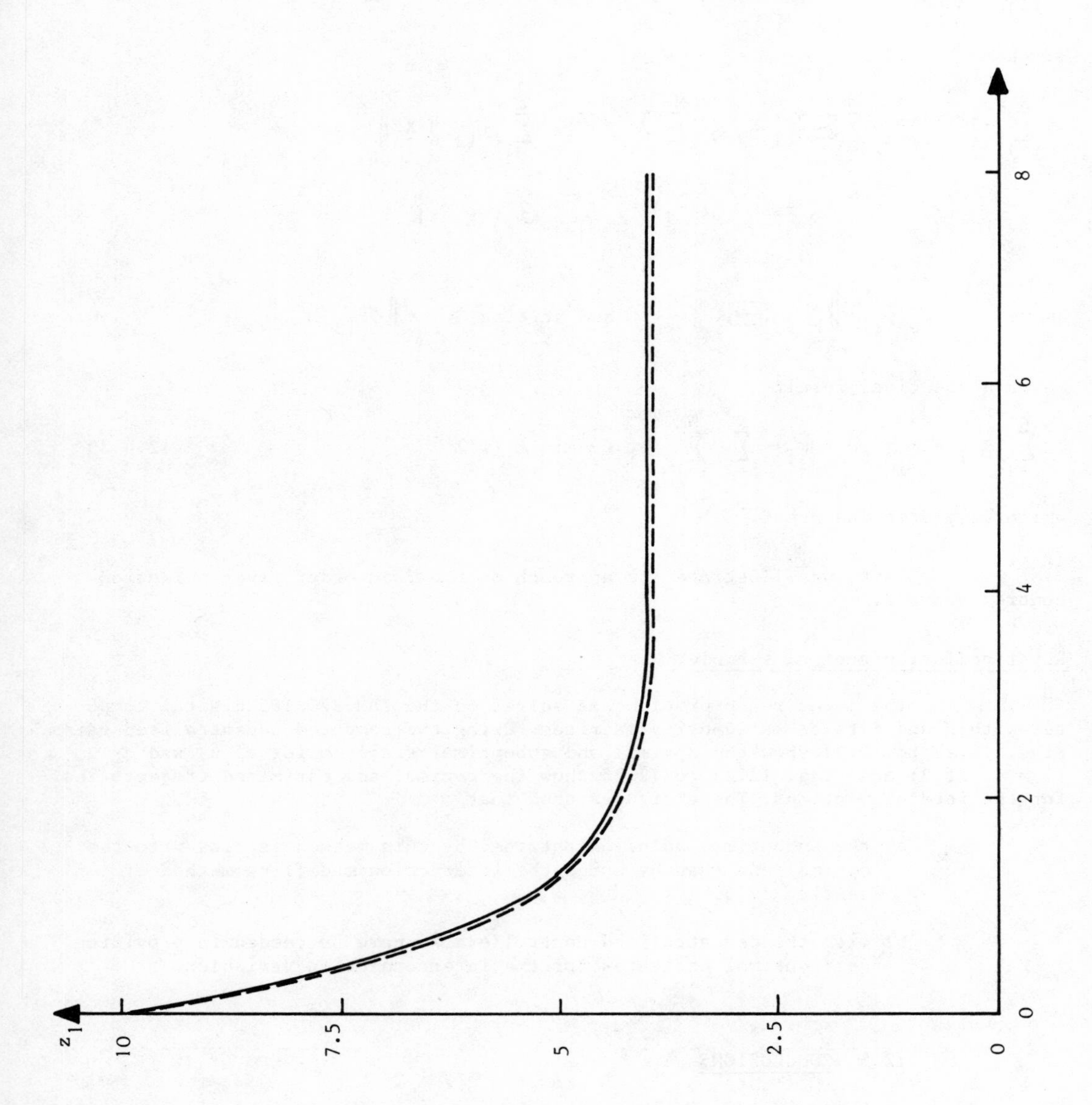

Fig. 12.17

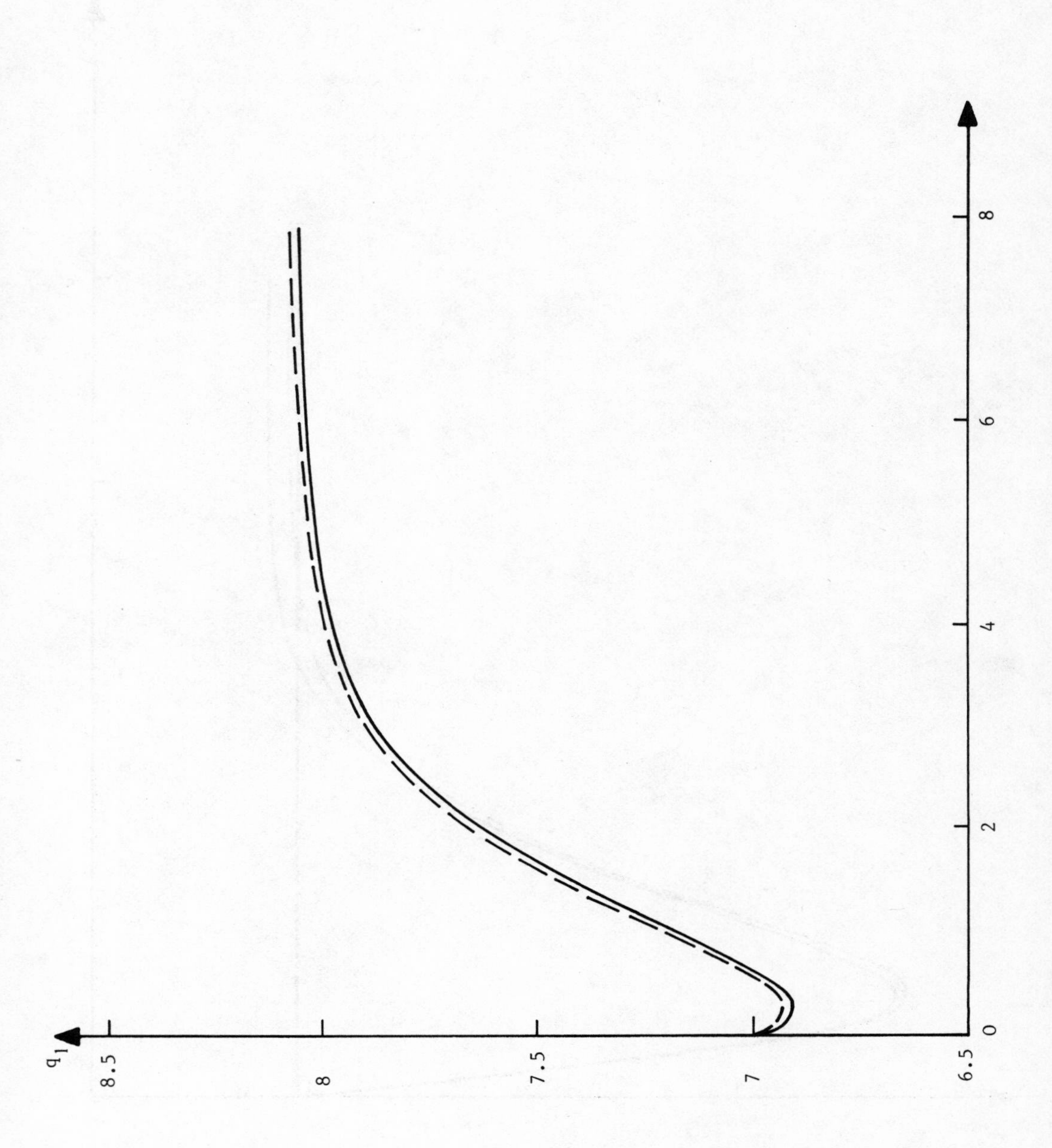

Fig. 12.18

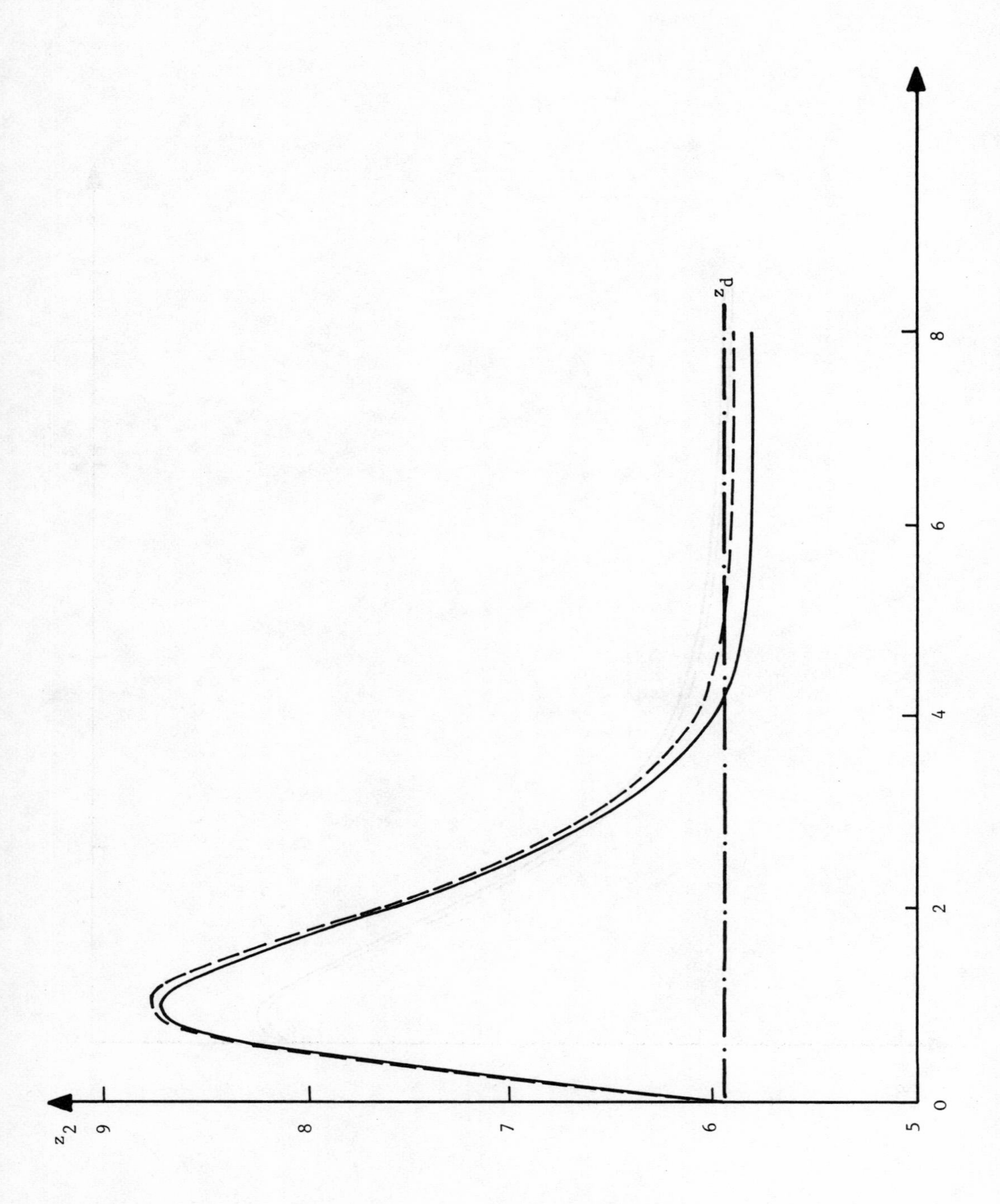

Fig. 12.19

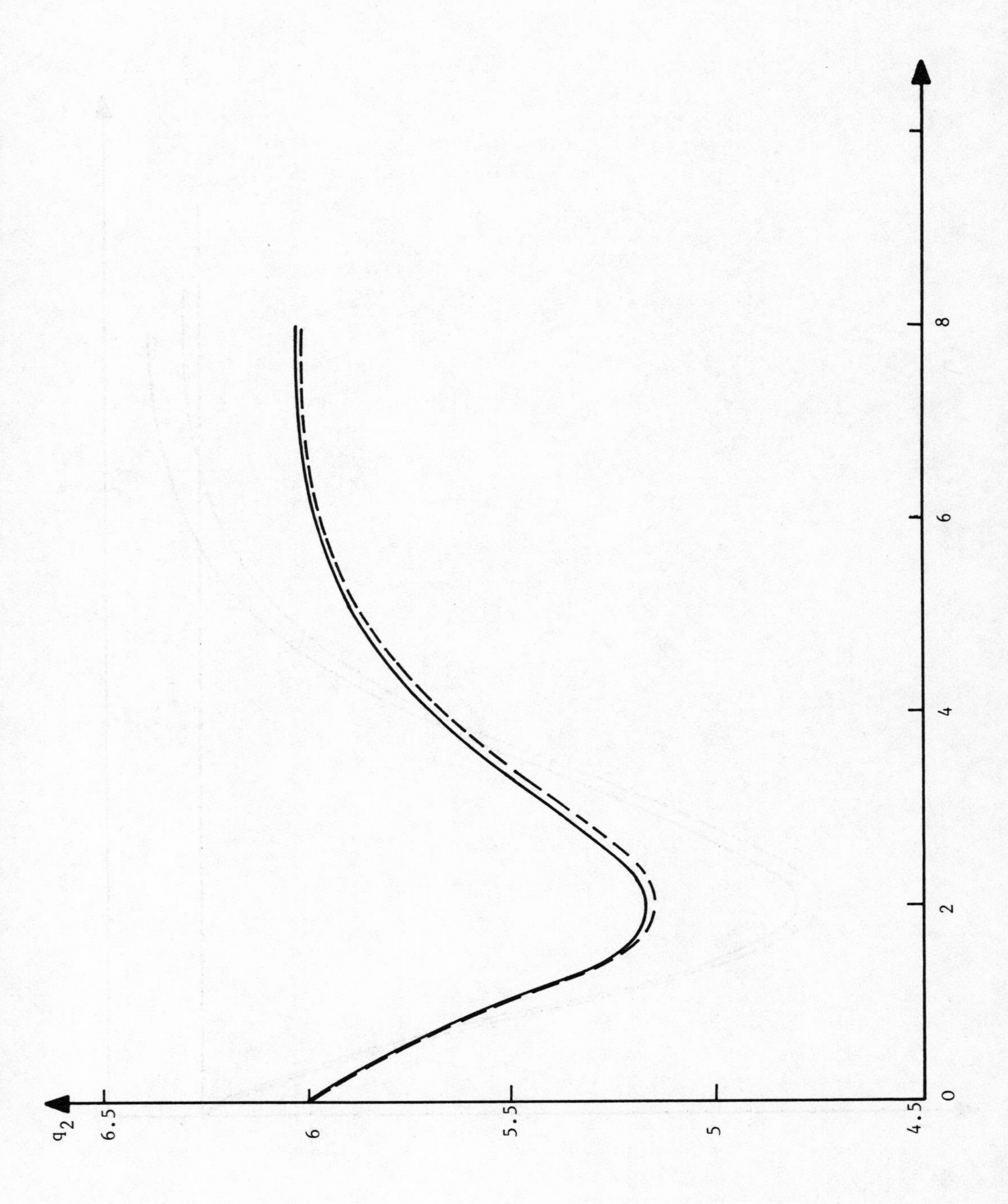

Fig. 12.20

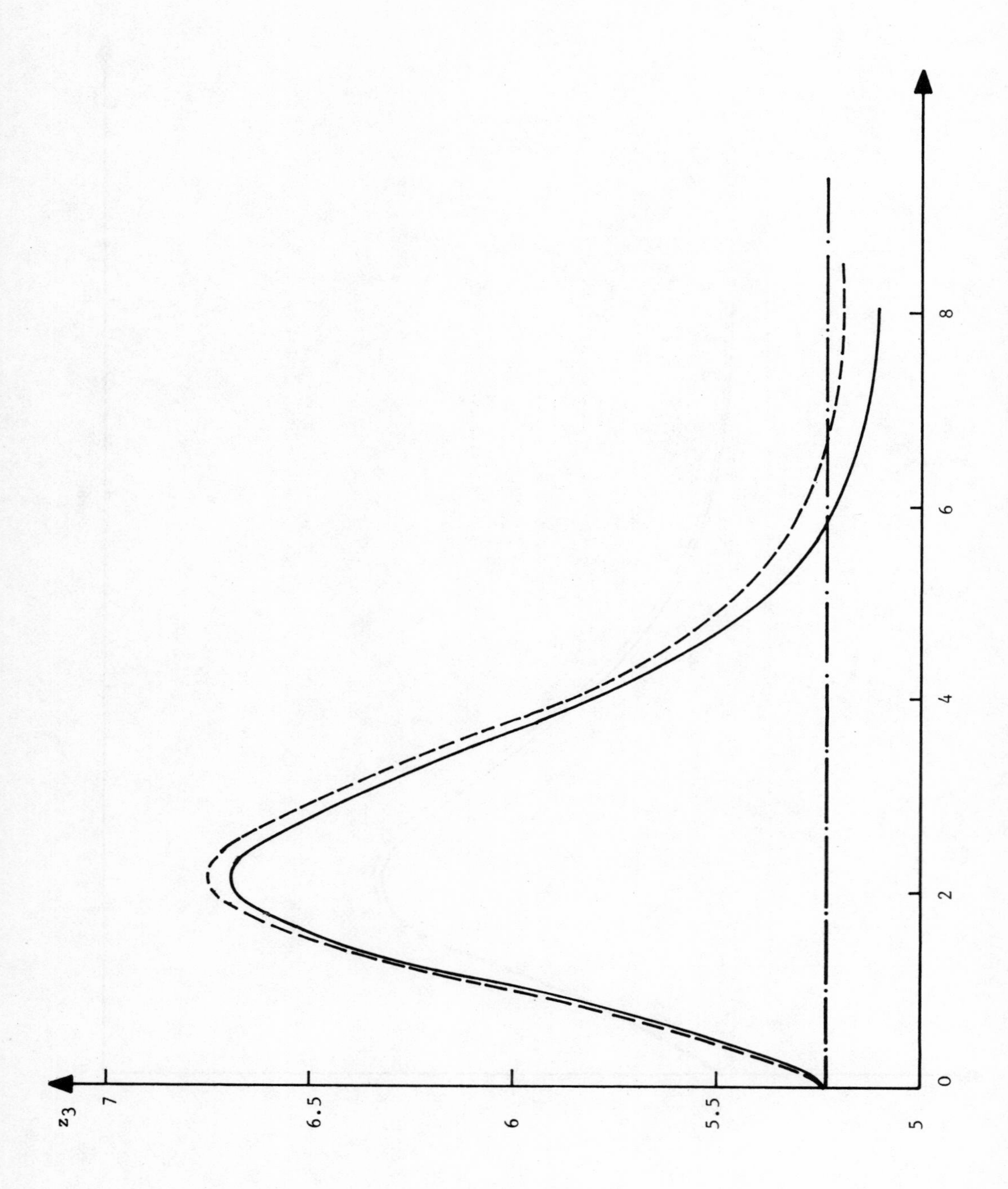

Fig. 12.21

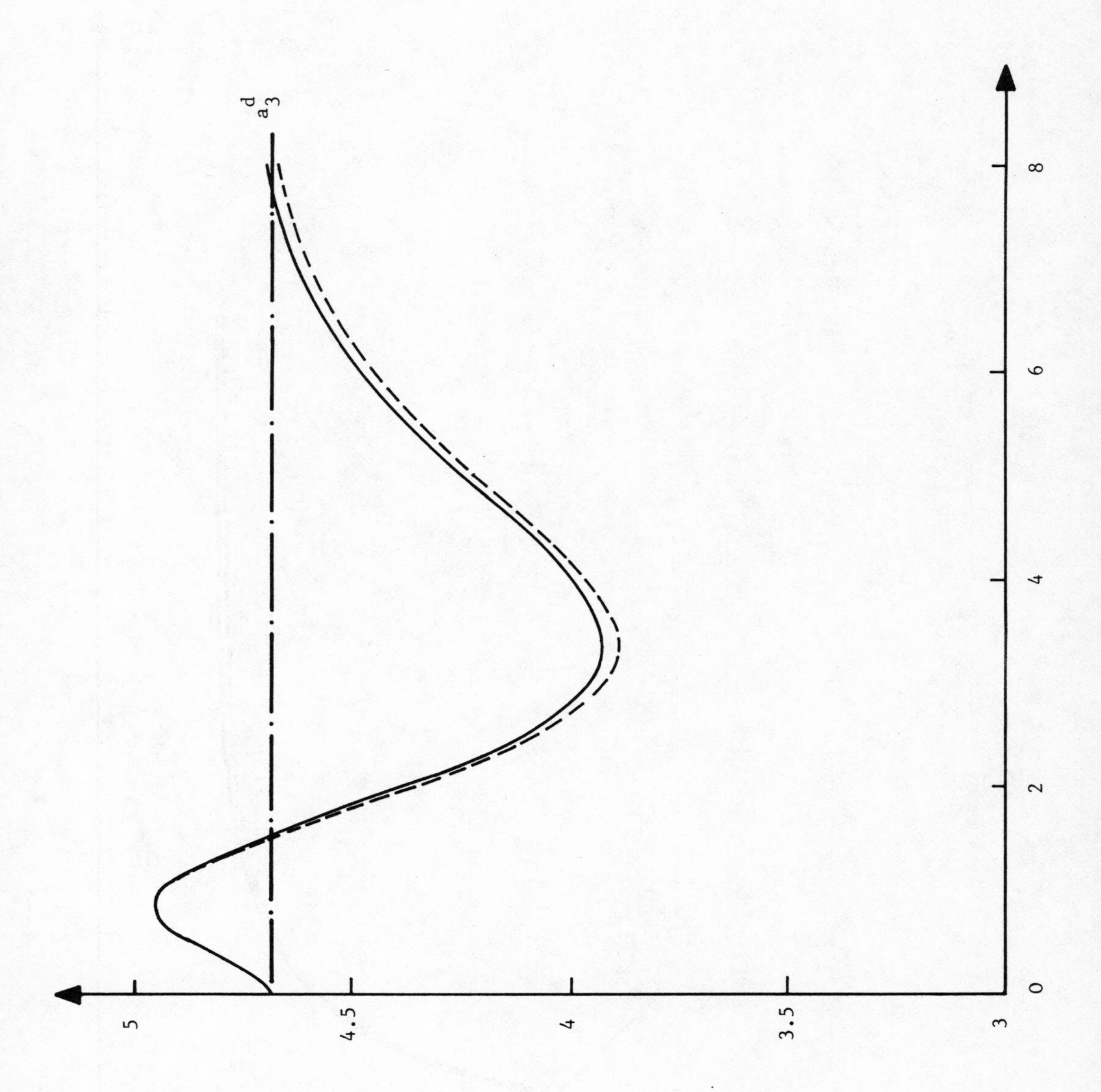

Fig. 12.22

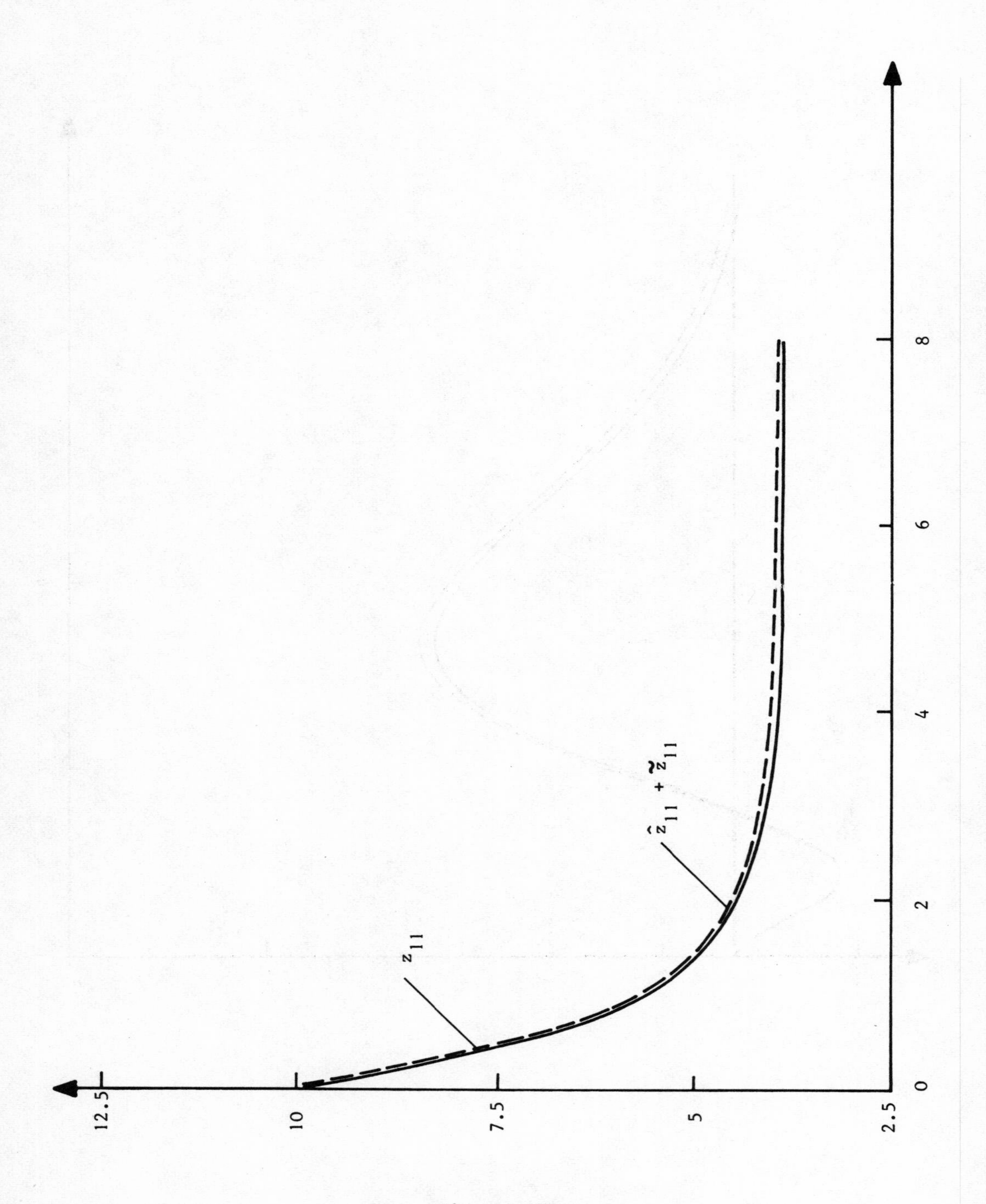

Fig. 12.23

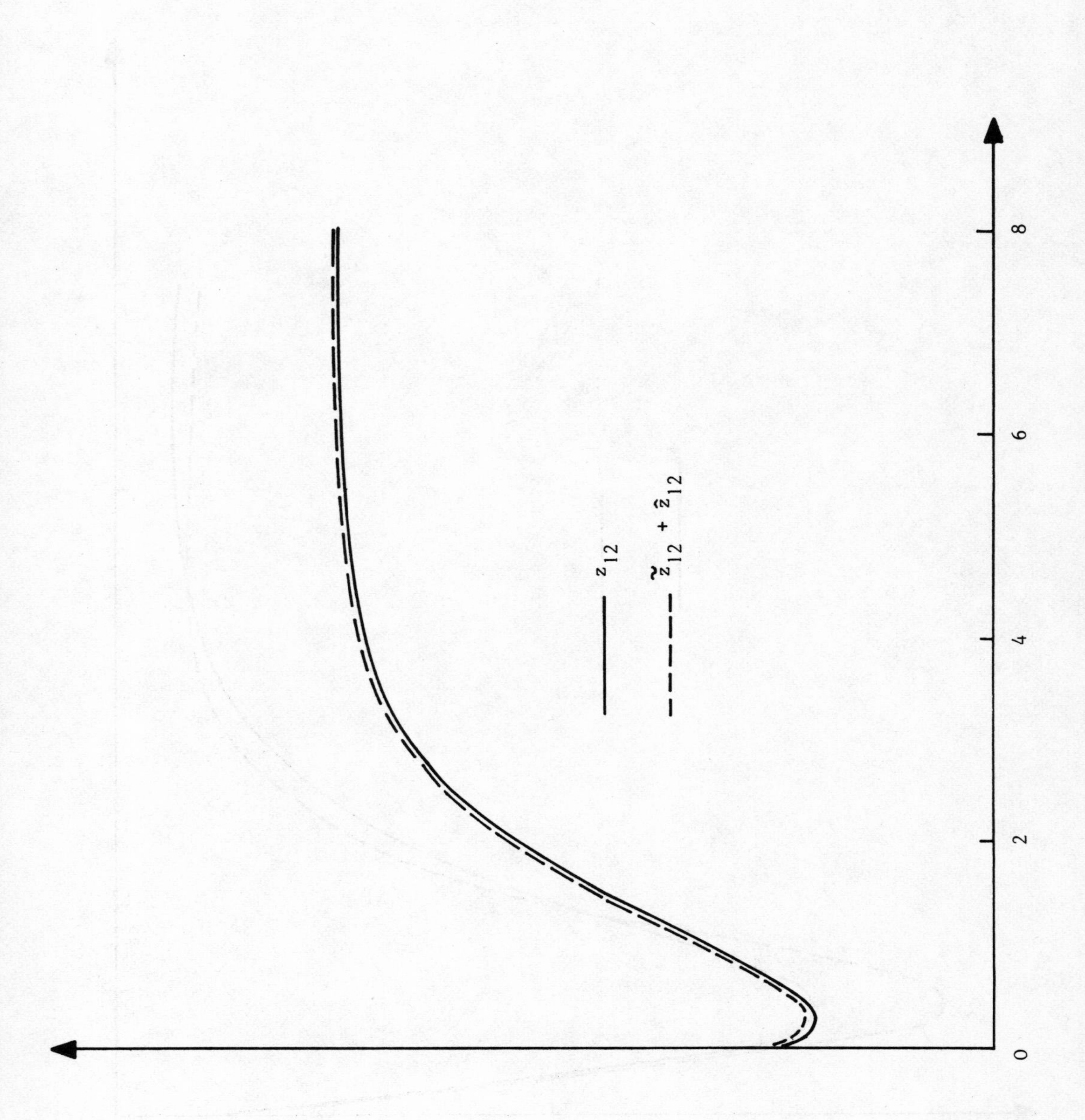

Fig. 12.24

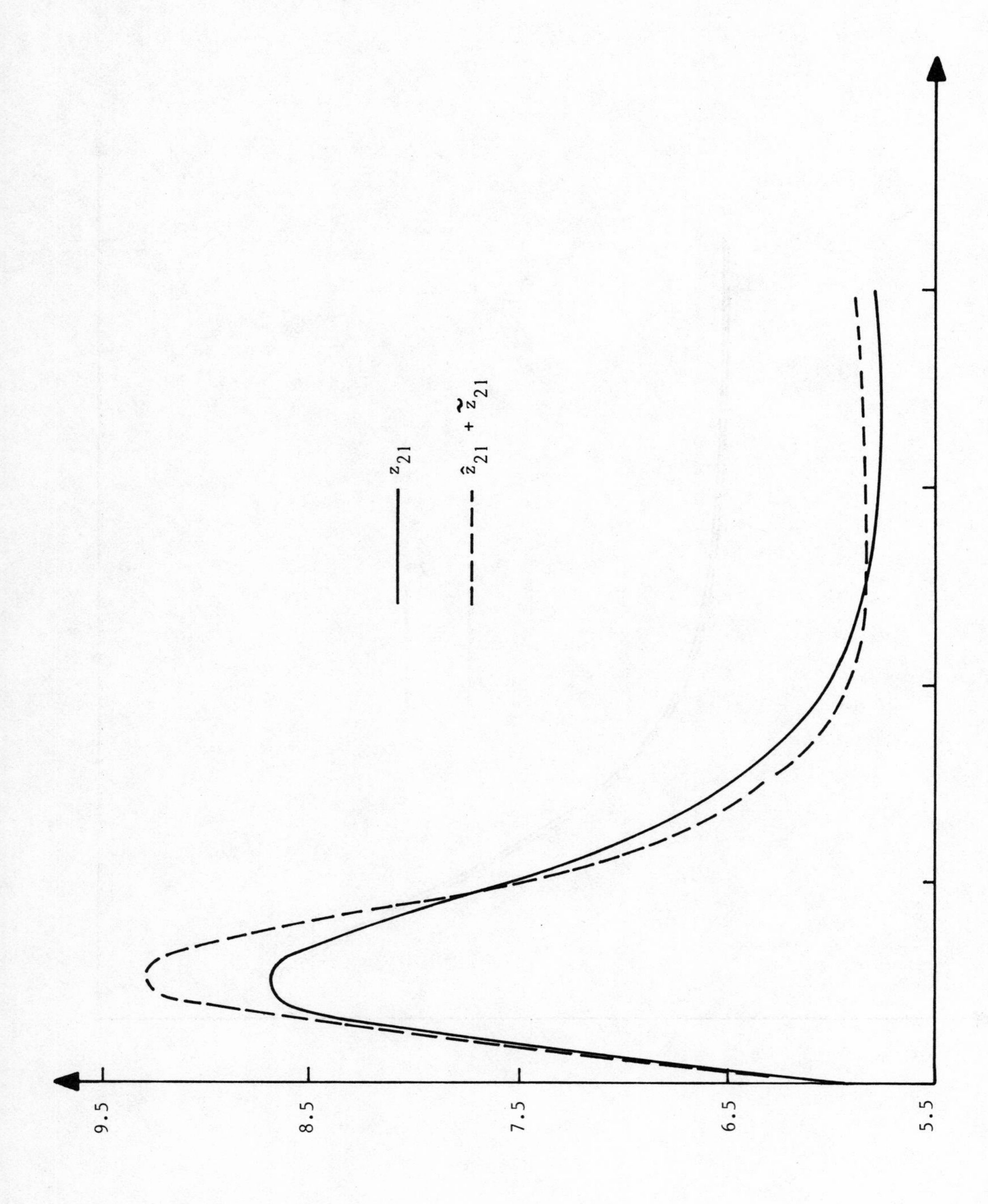

9.5
8.5
7.5
6.5
5.5
z_{21}
$\hat{z}_{21} + \tilde{z}_{21}$

Fig. 12.25

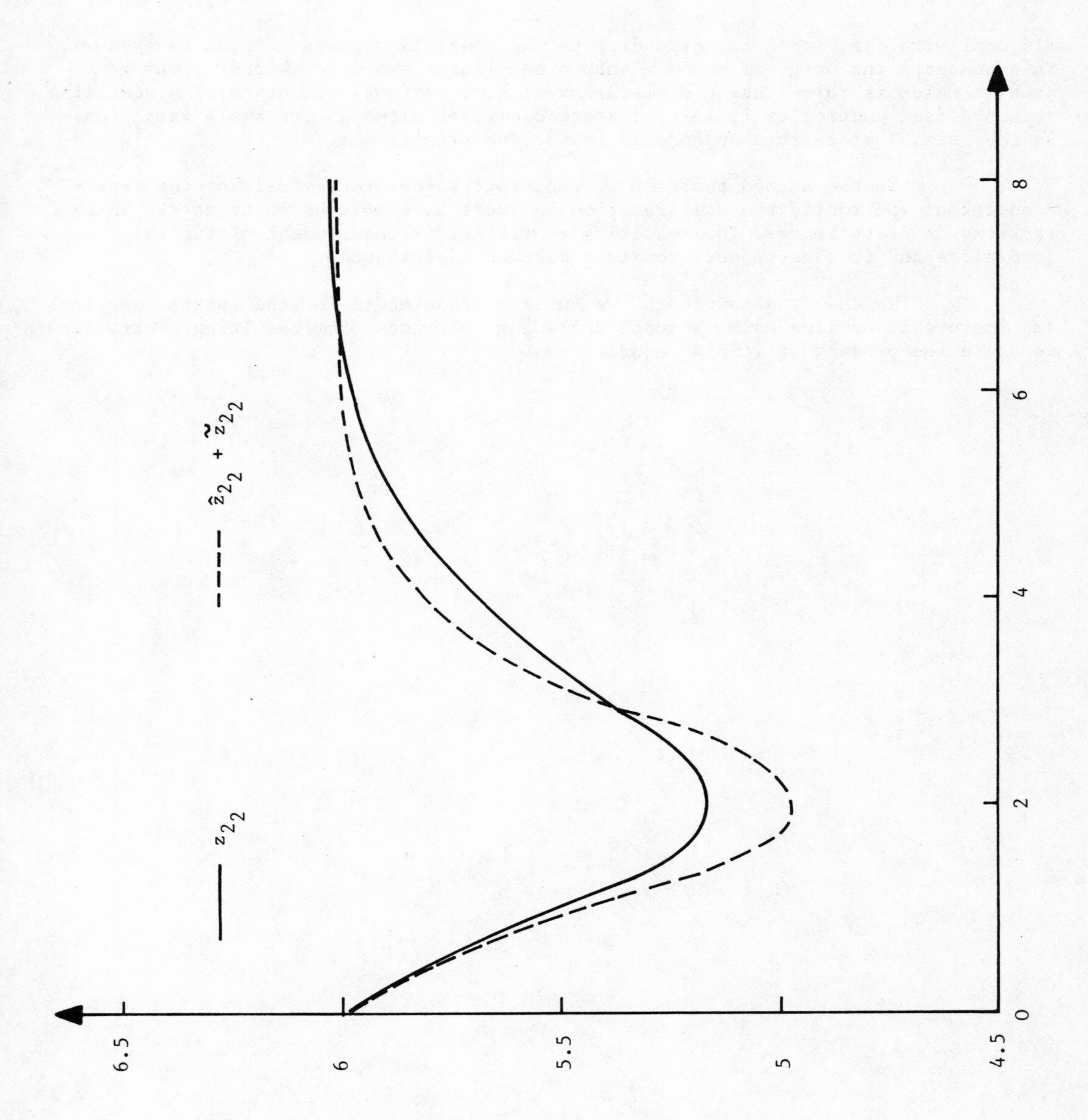

Fig. 12.26

diagonal with the blocks corresponding to the controller gains for the subsystems. This converts the original problem into a non-linear two point boundary value problem which is solved using a hierarchical computational structure. The resulting decentralised controller is initial state dependent although for small variations in the initial state this dependence is not too significant.

In the second approach we construct a low order model for the interconnections and modify the cost function in order to enable us to treat the interactions as disturbances. The resulting controller is independent of initial conditions and it also rejects constant unknown disturbances.

In the final approach, we build a crude model for the interconnections and improve it on-line using a model following technique. The resulting controller is again independent of initial conditions.

REFERENCES FOR CHAPTER 12

[1] HASSAN, M. and SINGH, M.G. "A hierarchical computational structure for near optimal decentralised control". IEEE Trans. SMC. July 1978.

[2] HASSAN, M. and SINGH, M.G. "A robust decentralised controller for linear interconnected dynamical systems". To appear in Proc. IEE. 1978.

[3] HASSAN, M. and SINGH, M.G. "A new robust model following controller". Submitted for publication

[4] HASSAN, M. Thèse d'Etat, Toulouse (Université Paul Sabatier), June 1978.

[5] SINGH, M.G., HASSAN, M. and TITLI, A. "A feedback solution for large scale interconnected dynamical systems using the prediction principle". IEEE Trans. SMC 6, 233-239, 1976.

[6] HASSAN, M. and SINGH, M.G."The optimisation of non-linear systems using a new two level method". Automatica 12, 359-363, 1976.

[7] SINGH, M.G. and HASSAN, M. "A closed loop hierarchical solution for the continuous time river pollution control problem". Automatica 12, 261-264, 1976.

[8] ANDERSON, B. and MOORE, J., 1971, "Linear optimal control". Prentice Hall.

[9] JOHNSON, C.D. "Further study of the linear regulator with disturbances : the case of vector disturbances satisfying a linear differential equation". IEEE Trans. AC 222-228, April 1970.

[10] SILJAK, D.D. "On stability of large scale systems under structural perturbations". IEEE Trans. SMC, July 1973, p. 415-417.

PROBLEMS FOR CHAPTER 12

1. Write down the equations for the two reach pure delay river pollution model of chapter 5 and test the approach of section 12.2 on this model.

2. How can one ensure that the design in excercise 1 has a fixed degree of stability α.

3. Design the decentralised controller for excercise 1 using the approach of section 3. Compare the results with those of excercise 1 when a 10 % perturbation is given in the initial conditions.

4. How is the design modified if we want to ensure that the system has a prespecified degree of stability α .

5. Design the robust decentralised model following controller of section 12.4 for the system of excercise 1. Compare the performance of this controller with that designed in excercise 1 and with the one designed in excercise 3.

Author Index

Subject Index